NEAR ZERO: NEW FRONTIERS OF PHYSICS

William Martin Fairbank

NEAR ZERO:
NEW FRONTIERS OF PHYSICS

Edited by

J. D. Fairbank

B. S. Deaver, Jr.

C. W. F. Everitt

P. F. Michelson

W. H. FREEMAN AND COMPANY
New York

Sources of Photographs

F. Alkemade for W. H. Freeman and Company—pp. ii, xxiv, xxv, 2, 66 (*bottom*), 372 (*top*), 570 (*bottom*), 840 (*top*), 882, 921 (*middle*), 922 (*top and bottom photos*).

F. Alkemade—pp. 66 (*top*), 496, 840 (*bottom*), 917 (*bottom right*), 918 (*middle and bottom photos*), 920 (*two bottom photos*), 923, 924, 925 (*top and middle photos*), 926.

Stanford University News and Publications Service—pp. 372 (*bottom*), 570 (*top and middle photos*), 921 (*top and bottom photos*), 925 (*bottom*).

Whitman College Archives—p. 917 (*bottom left*).

Duke University Archives—p. 919 (*bottom right*).

J. Mercado, Stanford—p. 248

M. Perryman—p. 922 (*middle*).

Library of Congress Cataloging-in-Publication Data

Near zero
   Includes index.
   1. Low temperatures—Congresses. 2. Superconductivity
—Congresses. 3. Superfluidity—Congresses.
4. Critical phenomena (Physics)—Congresses.
5. Astrophysics—Congresses. 6. Gravitation—Congresses.
I. Fairbank, J. D.
QC277.9.N43 1987 536′.56 87-8455
ISBN 0-7167-1831-6

Printed in the United States of America

1 2 3 4 5 6 7 8 9 0   KP   6 5 4 3 2 1 0 8 9 8

Contents

CHAPTER III
SUPERCONDUCTIVITY

CHAPTER IV
APPLICATIONS OF SUPERCONDUCTIVITY

CHAPTER V
FUNDAMENTAL PARTICLES

CHAPTER VI
GRAVITATION AND ASTROPHYSICS

CHAPTER VII
SURFACE SHIELDING

CHAPTER VIII
SOME THOUGHTS ON FUTURE FRONTIERS OF PHYSICS

Contributors

E. D. Adams, Department of Phyiscs, University of Florida, Gainesville, Florida

E. Amaldi, Dipartimento di Fisica, Università 'La Sapienza,' Rome, Italy

J. T. Anderson, Hewlett–Packard, Palo Alto, California

M. A. Bagshaw, Department of Therapeutic Radiology, Stanford University School of Medicine, Stanford, California

J. Bardeen, Department of Physics, University of Illinois at Urbana-Champaign, Illinois

M. Bassan, Dipartimento di Fisica, II Università di Roma, Rome, Italy

F. Bloch, Department of Physics, Stanford University, Stanford, California

S. P. Boughn, Department of Astronomy, Haverford College, Haverford, Pennsylvania

W. J. Bowers, Jr., Temperature and Pressure Division, National Bureau of Standards, Gaithersburg, Maryland

D. P. Boyd, Department of Radiology, University of California at San Francisco, and Imatron, Inc., South San Francisco, California

J. V. Breakwell, Department of Aeronautics and Astronautics, Stanford University, Stanford, California

M. J. Buckingham, Department of Physics, University of Western Australia, Perth, Western Australia

B. Cabrera, Department of Physics, Stanford University, Stanford, California

E. A. Cornell, Department of Physics, Massachusetts Institute of Technology, Cambridge, Massachusetts

E. P. Day, Gray Freshwater Biological Institute, University of Minnesota, Navarre, Minnesota

B. S. Deaver, Jr., Department of Physics, University of Virginia, Charlottesville, Virginia

D. B. DeBra, Department of Aeronautics and Astronautics, Stanford University, Stanford, California

C. W. F. Everitt, High Energy Physics Laboratory, Stanford University, Stanford, California

H. A. Fairbank, Department of Physics, Duke University, Durham, North Carolina

J. D. Fairbank, Menlo Park, California

W. M. Fairbank, Department of Physics, Stanford University, Stanford, California

W. M. Fairbank, Jr., Department of Physics, Colorado State University, Fort Collins, Colorado

S. B. Felch, Varian Research Center, Palo Alto, California

P. Fessenden, Department of Therapeutic Radiology, Stanford University School of Medicine, Stanford, California

C. R. Fisel, Department of Radiology, Massachusetts General Hospital, Boston, Massachusetts and Harvard University Medical School, Brookline, Massachusetts

S. J. Freedman, Physics Division, Argonne National Laboratory, Argonne, Illinois

R. P. Giffard, Hewlett–Packard, Palo Alto, California

A. M. Goldman, School of Physics and Astronomy, University of Minnesota, Minneapolis, Minnesota

J. M. Goodkind, Department of Physics, University of California, San Diego, La Jolla, California

W. Gordy, Department of Physics, Duke University, Durham, North Carolina

H. Griffiths, P.G.C., Inc., Black Mountain, North Carolina

W. O. Hamilton, Department of Physics and Astronomy, Louisiana State University, Baton Rouge, Louisiana

A. F. Hebard, AT & T Bell Laboratories, Murray Hill, New Jersey

G. B. Hess, Department of Physics, University of Virginia, Charlottesville, Virginia

R. Hofstadter, Department of Physics, Stanford University, Stanford, California

J. N. Hollenhorst, AT & T Bell Laboratories, Murray Hill, New Jersey

G. M. Keiser, High Energy Physics Laboratory, Stanford University, Stanford, California

D. S. Langley, Department of Physics, Whitman College, Walla Walla, Washington

G. S. LaRue, The Boeing Electric Company, Seattle, Washington

M. C. Leifer, Spectroscopy Imaging Systems, Fremont, California

P. D. Levine, Lockheed Missiles and Space Company, Inc., Sunnyvale, California

J. A. Lipa, Department of Physics, and High Energy Physics Laboratory, Stanford University, Stanford, California

W. A. Little, Department of Physics, Stanford University, Stanford, California

J. M. Lockhart, Department of Physics and Astronomy, San Francisco State University, and High Energy Physics Laboratory, Stanford University, Stanford, California

C. M. Lyneis, Lawrence Berkeley Laboratory, University of California, Berkeley, California

J. M. J. Madey, Stanford Photon Research Laboratory, Stanford University, Stanford, California

E. R. Mapoles, Inertial Confinement Fusion Program, Lawrence Livermore National Laboratory, Livermore, California

P. L. Marston, Department of Physics, Washington State University, Pullman, Washington

M. S. McAshan, High Energy Physics Laboratory, Stanford University, Stanford, California, and Supercollider Central Design Group, Lawrence Berkeley Laboratory, Berkeley, California

D. F. McQueeney, Department of Physics, Cornell University, Ithaca, New York

P. F. Michelson, Department of Physics, Stanford University, Stanford, California

J. H. Miller, Space Science Division, Ames Research Center, National Aeronautics and Space Administration, Moffett Field, California

B. E. Moskowitz, Department of Physics and Astronomy, University of Rochester, Rochester, New York

J. Napolitano, Physics Division, Argonne National Laboratory, Argonne, Illinois

H. J. Paik, Department of Physics and Astronomy, University of Maryland, College Park, Maryland

J. D. Phillips, Department of Physics, Stanford University, Stanford, California

J. S. Philo, Department of Molecular and Cell Biology, University of Connecticut, Storrs, Connecticut

P. B. Pipes, Department of Physics and Astronomy, Dartmouth College, Hanover, New Hampshire

D. A. Pistenmaa, Department of Radiation Oncology, Fairfax Hospital, Falls Church, Virginia

R. E. Rand, Department of Radiology, University of California at San Francisco and Imatron, Inc., South San Francisco, California

K. W. Rigby, Department of Physics, Stanford University, Stanford, California

R. Ruffini, International Center for Relativistic Astrophysics and Dipartimento di Fisica, Università 'La Sapienza,' Rome, Italy

H. A. Schwettman, Department of Physics, Stanford University, Stanford, California

T. I. Smith, High Energy Physics Laboratory, Stanford University, Stanford, California

S. R. Stein, Ball Aerospace Systems Division, Boulder, Colorado

M. Taber, High Energy Physics Laboratory, Stanford University, Stanford, California

R. C. Taber, High Energy Physics Laboratory, Stanford University, Stanford, California

K. S. Thorne, Theoretical Astrophysics, California Institute of Technology, Pasadena, California

W. P. Trower, Department of Physics, Virginia Polytechnic Institute and State University, Blacksburg, Virginia

V. S. Tuman, School of Liberal Arts and Sciences, California State University, Stanislaus, Turlock, California

J. P. Turneaure, High Energy Physics Laboratory, Stanford University, Stanford, California

C. T. Van Degrift, Temperature and Pressure Division, National Bureau of Standards, Gaithersburg, Maryland

R. A. Van Patten, High Energy Physics Laboratory, and Department of Aeronautics and Astronautics, Stanford University, Stanford, California

C. F. von Essen, Department of Radiation Oncology, Southwood Community Hospital, Norfolk, Massachusetts

R. V. Wagoner, Department of Physics, Stanford University, Stanford, California

J. D. Walecka, Department of Physics, Stanford University, Stanford, California, and Continuous Electron Beam Accelerator Facility, Southeastern Universities Research Association, Newport News, Virginia

G. K. Walters, Department of Physics, Rice University, Houston, Texas

C. A. Waters, Pearson Electronics, Inc., Palo Alto, California

J. P. Webb, Photographic Research Laboratories, Eastman Kodak Company, Rochester, New York

M. W. Werner, Space Science Division, Ames Research Center, National Aeronautics and Space Administration, Moffett Field, California

J. C. Wheatley, Department of Physics, University of California at Los Angeles, California and Los Alamos National Laboratory, Los Alamos, New Mexico

J. P. Wikswo, Jr., Department of Physics and Astronomy, Vanderbilt University, Nashville, Tennessee

E. G. Wilson, Customer Training, Intel Corporation, Santa Clara, California

F. C. Witteborn, Space Science Division, Ames Research Center, National Aeronautics and Space Administration, Moffett Field, California

P. W. Worden, Jr., High Energy Physics Laboratory, Stanford University, Stanford, California

C. N. Yang, Department of Physics, State University of New York, Stony Brook, New York

Preface

The last forty years have provided physicists with unusual opportunities to explore new regions of physics with ever-advancing experimental tools. This book details the work of a group of physicists associated with William M. Fairbank who have explored the regions of physics opened up by making one or more of the variables of physics very nearly zero.

The idea of the book was conceived in August 1981 during the 16th International Conference on Low Temperature Physics (LT-16), at which time about twenty people, mostly former students and associates of Fairbank, met to plan a scientific conference to honor him on the occasion of his sixty-fifth birthday. The idea had been discussed among various members of the group for a number of years, and it was their unanimous opinion that the best way to honor him was to bring together a group of his present and former students and colleagues for a period of intense discussion of the current state and future potential of the ideas and experiments that he has originated or extended in some major way.

Fairbank has contributed seminally to a remarkable number of topics in physics. His research has been characterized by pushing various physical parameters to their smallest extreme in search of new fundamental phenomena. He has been a leader in the search for new phenomena at temperatures approaching absolute zero, in the production and use of regions of space with nearly zero magnetic field or nearly zero electric field, and in the use of low temperature techniques for extending experimental sensitivity to nearly fundamental limits in order to study gravitation and elementary particle physics. This thread of "near zero" is woven throughout the topics on which he has worked, and he has spoken about it on several occasions, one of them an invited talk at LT-16; it inspired the title for the conference and this book.

The conference, "Near Zero: New Frontiers of Physics," was held March 23-25, 1982 at Stanford University with nearly 150 participants. This book is perhaps its most important component. It is an attempt to convey to a wider audience some of the insights, stimulation, and enthusiasm for which Fairbank is well known and which were much in evidence at the conference. The book is a study in creative collaboration, a testament to Fairbank's belief that physics is most exciting and successful when widely shared. As he is first to point out, many of his ideas have come to fruition through his interactions with large numbers of students and colleagues who share his enthusiasm for physics and who have worked together on tough but important experiments. This book itself is the product of a large group of people, most of whom share a common bond of association with Fairbank.

We note with sorrow and a profound sense of loss the deaths of three distinguished contributors to this volume, Felix Bloch, Walter Gordy and John Wheatley.

Some of the papers here are almost verbatim transcriptions of talks given at the conference. Others are expanded versions of the talks, which offer more complete reviews of given topics. Other papers have been created specifically for this book. Some of the work described has been done in Fairbank's low temperature research group, some has an element of its genesis there, some has been done in cooperation therewith, some has been done entirely independently.

A few words about the organization of the papers may be helpful to the reader. The first chapter provides a personal setting for the scientific theme by supplying biographical data about Fairbank and an analysis of his particular creative style.

The arrangement of subsequent papers is essentially chronological with respect to Fairbank's involvement in the field, dealing first with a series of events from 1941 through his early years at Duke University, then with the properties of ^{3}He and ^{4}He on which he concentrated at Duke, then with basic superconductivity to which he turned at Stanford, and finally with the ever-broadening applications of superconductivity and other low temperature techniques to fundamental problems in elementary particle physics and gravitation.

In each section there is an attempt to trace the ideas from their earliest days to the present and to extrapolate to the future. The pioneering work on nuclear magnetic resonance in ^{3}He, the discovery of the phase separation in ^{3}He–^{4}He solutions, and the subsequent implications of these ideas for current research with ^{3}He are examined. Extremely detailed measurements of the specific heat of ^{4}He at the lambda point that led to a whole area of physics studying the nature of phase transitions, particularly those with logarithmic singularities, are reported, and their implications are discussed.

Experiments on macroscopic quantum effects both in superfluid helium and in superconductors are discussed. Fluxoid quantization, the London moment, the production of nearly zero magnetic field regions, and the recent closely related experiments to measure very precisely the ratio h/m in superconductors to a precision sufficient to see relativistic corrections to the electron mass, are described.

Early work that Fairbank did on microwave properties of superconducting tin led subsequently to studies of high-Q superconducting cavities that in turn led to development of the superconducting accelerator and superstable oscillators. A subsequent outgrowth is the free electron laser.

The desire for ever more sensitive magnetic measurements led from the early thermally and mechanically modulated inductance detectors to exploitation of SQUID's for experiments in medical physics, fundamental particle physics, and gravitation, where they will be challenged at the fundamental limits of their sensitivity. The realization of the potential of a completely superconducting system consisting of persistent current magnet, magnetic shields, and magnetometer led to work on the superconducting susceptometer that is now being used for measuring magnetic properties of materials including biological ones.

Low temperature techniques are involved in a number of experiments designed to study fundamental particles. The search for magnetic monopoles using superconducting shields and magnetometers that produce a unique signature for a monopole, and the recording of a candidate event, have motivated much new research on monopoles, which are of paramount importance in grand unification theories. Searches for free fractional charges using a variety of techniques are described.

Also reviewed in the book are experiments to measure the electric dipole moment of the neutron through extremely sensitive magnetic measurements on the orientation of the spins in a very dilute solution of ^{3}He in ^{4}He. There are also experiments to study the free fall of a single electron and of a single positron to compare the gravitational force on particle and antiparticle. These experiments have led to the discovery of an unusual electrostatic shielding effect, apparently due to some kind of transition on the surface of oxidized copper. This is a highly speculative topic but one which could lead to a new area of surface physics experiments.

The experiments in gravitational physics bring together the most innovative experimental techniques from low temperature physics in an attempt to produce new fundamental tests of the general theory of relativity. Experiments to measure the orientation of the spin axis of a precision gyroscope in a zero-g satellite have made enormous progress from their inception more than two decades ago and show promise for success. The low temperature gravitational wave detector is another example of the role that low temperature techniques can play in the quest for the ultimate in measurement sensitivity as applied to an important fundamental problem.

The book concludes with Fairbank's own views of some new directions with potential for important discoveries in the future.

It is the hope of the editors that the collection in a single volume of these apparently diverse topics, linked by the common thread of "near zero," will provide insight and stimulation to many physicists in other fields. The potential of "near zero" techniques—and supersensitive precise measurement in general—for opening new frontiers of physics seems to be far more extensive than even the many varied papers in this volume would suggest.

Acknowledgments

This book is the work of a very large number of people, to all of whom we express our deep gratitude for their contributions. We note particularly the crucial role of the sponsors and organizers of the Near Zero conference, of which this book is a major component. Sponsors who provided major financial support for the conference were: Donald C. Freeman; William S. Goree, Inc.; IBM Corporation; Koch Process Systems, Inc.; Rockwell International; *Scientific American*/W. H. Freeman and Company; Alfred P. Sloan Foundation; Westinghouse Electric Corporation; and the Xerox Corporation, Palo Alto Research Center.

Members of the conference organizing committee were: E. Dwight Adams, Blas Cabrera, Bascom S. Deaver, Jr., C. W. Francis Everitt, Henry A. Fairbank, J. D. Fairbank, Allen M. Goldman, John M. Goodkind, William O. Hamilton, Arthur F. Hebard, James M. Lockhart, John M. J. Madey, Peter F. Michelson, Ho Jung Paik, Arthur L. Schawlow, H. Alan Schwettman, and Walter Trela. Arthur Schawlow was general chairman. John Madey headed the local committee and was assisted by many members of the Stanford Physics Department. Bascom Deaver, Jr., coordinated the effort outside Stanford, including the contacting of sponsors. Polly Crandall served as conference secretary.

In the preparation of this book, a number of individuals have played vital roles. The editors are indebted to Alan Schwettman for invaluable aid at a critical moment in the completion of the book; to Norva Shick for her dedication, wise judgment, and great care in the preparation of the camera-ready copy; to Jenifer Conan Tice for her willing cooperation and tireless attention to details throughout the course of the book; and to Mark Kent for programming the book's design in LaTeX.

The editors express their appreciation to John Lipa, Norman Bettini and Hans Wehrli for assistance with difficult drawings and brochures, and to Frans Alkemade for his valuable contributions to the photographic account of both the Near Zero experiments and the people associated with them. In the case of several old photographs, acknowledgment was not possible because the editors have no record of the sources. The editors extend special thanks to Jeremiah J. Lyons, Philip McCaffrey and the rest of the W. H. Freeman team for guidance and assistance at the various stages of the publishing process.

April 1988   — The Editors

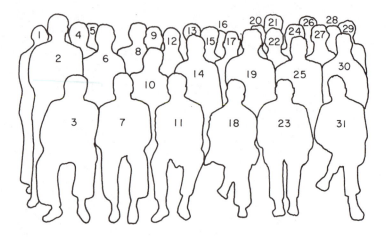

Near Zero Conference Contributors

1. D. B. DeBra
2. R. V. Wagoner
3. C. N. Yang
4. J. P. Turneaure
5. G. A. Westenskow
6. J. M. Goodkind
7. W. M. Fairbank
8. J. M. Lockhart
9. L. Halpern
10. S. B. Felch

11. H. A. Fairbank
12. E. G. Wilson
13. P. L. Marston
14. C. A. Waters
15. J. T. Anderson
16. J. R. Henderson
17. P. B. Pipes
18. F. Bloch
19. A. M. Goldman
20. M. C. Leifer
21. A. F. Hebard

22. E. D. Adams
23. J. Bardeen
24. H. A. Schwettman
25. J. P. Webb
26. G. Fisher
27. M. S. McAshan
28. J. N. Hollenhorst
29. J. D. Phillips
30. J. A. Lipa
31. B. S. Deaver, Jr.

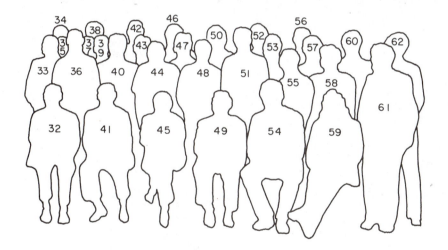

Stanford University March 23–25, 1982

32. W. Gordy
33. R. E. Rand
34. R. C. Taber
35. G. K. Walters
36. P. W. Worden, Jr.
37. E. R. Mapoles
38. B. Cabrera
39. F. C. Witteborn
40. L. B. Holdeman
41. M. J. Buckingham

42. J. M. J. Madey
43. W. O. Hamilton
44. G. S. LaRue
45. E. Amaldi
46. G. M. Keiser
47. E. P. Day
48. J. C. Wheatley
49. R. Hofstadter
50. P. D. Levine
51. J. P. Wikswo, Jr.
52. R. A. Van Patten

53. J. Napolitano
54. R. Ruffini
55. S. J. Freedman
56. C. M. Lyneis
57. T. I. Smith
58. H. J. Paik
59. C. W. F. Everitt
60. W. M. Fairbank, Jr.
61. M. Bassan
62. P. F. Michelson

NEAR ZERO: NEW FRONTIERS OF PHYSICS

CHAPTER I

Dimensions of a Physicist

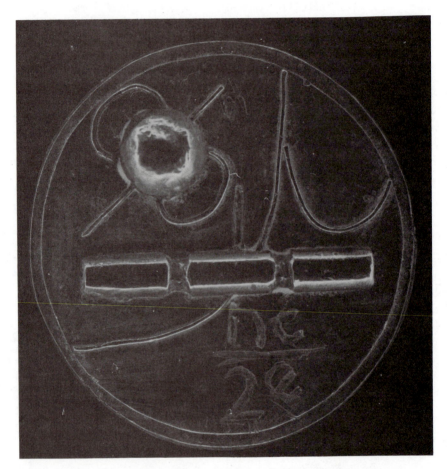

Tie tack fabricated for William Fairbank by Francis Everitt in 1968 combining representations of the London moment, the lambda point, and three sections of the superconducting accelerator, with the expression for the flux quantum including the factor of 1/2 revealed by the quantized flux experiment.

I.1

Introductory Remarks

The Editors and J. D. Walecka

Ever-advancing experimental and theoretical techniques are opening up new horizons in physics, with consequences which could affect our most basic ideas about the fundamental laws of physics. One such experimental technique involves reducing one or more of the variables of physics nearly to zero. This book details the work of a group of physicists associated with William M. Fairbank who have helped pioneer this frontier, and is an indication of the worldwide effort now pushing back the boundaries of this field.

As described in this book, making one or more of the variables of physics nearly zero enables one to probe regions of physics previously inaccessible. This adventure takes us to very low temperatures, to outer space, to unexpectedly low electric and magnetic fields, to single atoms, to incredibly small changes in length and temperature, to the smallest possible energy and charge, exploring such subjects as superconductivity and superfluidity, gravitation and general relativity, fundamental particles and astrophysics, superconducting accelerators and free electron lasers, cancer therapy and magnetic measurements in the human body, liquid and solid helium, phase transitions and organic superconductors.

Chapters II through VII of this book detail the adventures of this group of physicists exploring "near zero" frontiers. Chapter VIII presents some thoughts on future frontiers of physics.

Chapter I features an essay on creativity by C. W. F. Everitt, and two other papers offering historical background on William M. Fairbank's involvement in physics, written by J. D. Walecka, former chairman of the Stanford Physics Department, and by Walter Gordy, longtime colleague of Fairbank at the MIT Radiation Laboratory and Duke University.

The conference entitled "Near Zero: New Frontiers of Physics" held at Stanford University March 25–27, 1982 in honor of Fairbank's sixty-fifth birthday provided the springboard for this book. Walecka's speech opening the conference is excerpted below:

On behalf of Stanford University and the Department of Physics, I would like to welcome all of you to this conference entitled "Near Zero: New Frontiers of Physics." This is a scientific conference in honor of William M. Fairbank on the occasion of his sixty-fifth birthday. The program is a tribute to Bill's many contributions to the field and covers a wide-ranging variety of topics, all on the frontiers of physics. There are talks on fluxoid quantization, near zero fields both magnetic and electric, superconductivity, quantum fluids, the lambda transition in ^{4}He, a test of general relativity through the orbiting gyroscope, gravitational radiation detectors, and the search for the quark. The unifying theme of this conference as outlined by the organizers is that of "the new frontiers which W. M. Fairbank and his colleagues over the years have opened up by probing further and further 'near zero' in temperature, magnetic field, electric field, charge, noise, weak-signal detection, *etc.*" They also add, "It is hoped that each paper will help bring out the cross fertilization of fields and the collaboration and constructive interaction which have been hallmarks of W. M. Fairbank's research." It is gratifying, and a fine tribute to Bill, to note the very distinguished group of scientists, including five Nobel laureates, speaking at this conference.

I think it is appropriate for me to give a brief biographical sketch. Bill was born in Minneapolis, Minnesota, on February 24, 1917. He received his A.B. from Whitman College in 1939 and Ph.D. from Yale in 1948. From 1940 to 1942 he was a teaching fellow at the University of Washington, and from 1942 to 1945 he was a staff member of the Radiation Laboratory at the Massachusetts Institute of Technology. He was a Sheffield Fellow at Yale from 1945 to 1946, and an assistant professor at Amherst from 1947 to 1952. From Amherst he went to Duke where he was associate professor of physics from 1952 to 1958 and professor from 1958 to 1959. Bill came to Stanford in September, 1959, and in one sense it is appropriate that I give these introductory remarks because that is exactly the same time that I came to Stanford.

Bill has received many honors over the years. He was California Scientist of the Year in 1962. He was elected to the National Academy of Sciences in 1963. He received the Oliver E. Buckley Solid State Physics Prize in 1963, and the Research Corporation Award in 1965. He was elected to the American Academy of Arts and Sciences in 1967, and received the Wilbur Lucius Cross Medal from the Yale Graduate School in 1968. Bill received the Fritz London Award in 1968. He was elected to the American Philosophical Society in 1978, and served as chairman of the section on physics of the American Association for the Advancement of Science in 1979. He has received honorary Doctor of Science degrees from Whitman College, Amherst College and Duke University. Bill has served on the editorial council of the *Annals of Physics* and the editorial boards of the *Journal of Low Temperature Physics* and the *Review of Modern Physics*.

If I had to characterize Bill Fairbank's work, I would say that he always attacks fundamental problems, exciting problems, and hard problems, on the forefront of physics. He uses ingenious methods to study problems that most others would not even dare tackle. His work is characterized by enthusiasm, dedication, and long hard hours spent here in the physics department. Whenever I go elsewhere, the first few questions always include, "What are Bill Fairbank's latest results?" and "What is Bill Fairbank doing now?" His contributions to physics have brought great distinction to this physics department and to this university.

The citation for the Research Corporation Award in 1965 provides a good characterization of Bill Fairbank's work. This award was presented "For Outstanding Contributions to Science" and reads as follows:

> For his elegant and precise performance of several crucial experiments of fundamental importance in the field of very low temperature physics, and especially for the discovery of flux quantization which demonstrates a macroscopic quantum effect. His brilliantly conceived and executed investigations have increased greatly our understanding of the properties of superfluid liquids and superconductors and have opened the way into previously inaccessible regions of physical phenomena.

But perhaps Bill's greatest contribution, in my opinion, is his work with students and young people. He has turned out an exceptional collection of Ph.D. students from his group over the years at Stanford and Duke, many of whom are now leaders in the field. A complete list of students who earned Ph.D. degrees under W. M. Fairbank is presented here. Each of these students had his or her own project, which was probably a factor in attracting the students to Bill Fairbank's group; and that project involved an experiment which was on the frontier of physics. The result of any one of these experiments would make an important, if not profound, contribution to our understanding of the physical world. Bill's relationship with his students has always been characterized by encouragement, support and optimism. The fact that this conference was arranged by his former

students, postdocs and associates is a tribute to the esteem and regard they
have for Bill Fairbank.

Students who earned Ph.D. degrees under William M. Fairbank

1955 W. B. Ard, Jr.	1974 P. M. Selzer
(W. Gordy codirector)	1975 B. Cabrera
1956 A. J. Dessler	1975 H. J. Paik
1956 G. K. Walters	1975 S. R. Stein
1957 G. S. Blevins	(J. P. Turneaure codirector)
(W. Gordy codirector)	1976 S. P. Boughn
1959 W. D. McCormick	1976 D. E. Claridge
1960 E. D. Adams	1976 J. P. Wikswo, Jr.
1960 J. M. Goodkind	1976 E. G. Wilson
1960 C. F. Kellers	1977 R. S. Hanni
1960 J. N. Kidder	(J. M. J. Madey codirector)
1962 B. S. Deaver, Jr.	1977 J. M. Lockhart
1965 M. Bol	1977 P. L. Marston
1965 A. M. Goldman	1978 G. S. LaRue
1965 F. C. Witteborn	1978 J. S. Philo
1966 L. V. Knight	1979 D. A. G. Deacon
1967 H. D. Cohen	(J. M. J. Madey codirector)
1967 G. B. Hess	1979 M. A. Taber
1967 W. J. Trela	1980 C. A. Waters
1968 J. M. Pierce	1981 M. C. Leifer
1969 J. P. Webb	1982 E. R. Mapoles
1970 P. B. Pipes, Jr.	1982 G. A. Westenskow
1971 T. D. Bracken	1984 J. D. Phillips
1971 A. F. Hebard	1985 M. Bassan
1971 J. M. J. Madey	1986 C. R. Fisel
1971 D. K. Rose	1986 B. E. Moskowitz
1973 E. P. Day	1986 B. J. Neuhauser
1973 L. B. Holdeman	1986 K. W. Rigby
1973 K. L. Verosub	1987 J. R. Henderson

As I think back over the years that Bill Fairbank and I have shared at
Stanford, [an anecdote comes to mind which] occurred just a couple of years
ago. I was sitting in the stands at one of the big Stanford football games.
The stands were filled; there must have been 80,000 people in attendance.
It was announced that a special race had been arranged for the enjoyment
of the fans at halftime. About a half dozen sprinters were warming up on
the track. All eyes turned to them, and I picked up my field glasses to see
who was involved. I could not believe my eyes. There was Bill Fairbank
warming up for the 100 yard dash. He ran and did a creditable job. I think
his competition was Payton Jordan, who is one of the best in the world in
his age group. There are not many track athletes anywhere, at any age,
who have performed before a crowd of that size.

Again let me welcome you all to the conference. It is an interesting
program and a fine tribute to William M. Fairbank.

The Nature of the Man, William Martin Fairbank

Walter Gordy

The following is the text of the speech given by Walter Gordy after the banquet at the Near Zero conference on March 26, 1982:

I shall preface my remarks by mentioning the events that led to my good fortune in meeting the man to whom the Near Zero conference is dedicated, and then briefly recount some events occurring during our forty years of friendship. These events are selected to give glimpses into the nature of the man we honor, William Martin Fairbank.

The chain of events that led to my meeting Bill began on December 7, 1941. It was a bright Sunday morning. As my wife and I walked along a street in Pasadena a lone man approaching us shouted, "Pearl Harbor has been bombed!" Indeed, it *had* been bombed, as we heard from a thousand voices that afternoon when, out of curiosity, we visited the famous "Spit and Argue College" located on an endowed pier over the ocean's edge at Long Beach.

I had gone to Cal Tech earlier in the year to work with chemist Linus Pauling. Millikan, who measured the electron charge, was still in charge at the Institute. The only time I was recognized by him took place at the faculty reception in the fall when Pauling introduced me as a new National Research Fellow in physics. Upon hearing the word "physics," Millikan asked that the young man stand up again. Then he turned to

Pauling, "If the young man is a fellow in physics, what is he doing in the chemistry department?" Pauling's only reply was his characteristic broad grin.

On the Monday morning after that Sunday of December 7, I went back to the Gates and Crellin Laboratories to continue my experiments on electron diffraction. Soon I was interrupted by Pauling, who came in and told me that we must quickly wind down this work and that he hoped I would remain at the Institute to work on a new project he was starting under the NDRC. He could not tell me what kind of work I would be doing. He added that I could work in the physics department if I preferred, on a project being initiated there. I asked to have a few days to decide. I then went to talk with Willie Fowler, a young assistant professor, about the physics project, but he could tell me nothing except that physics would be glad to have me. Before I could reach a decision between these opportunities, I received a Class A offer from my draft board back home, and then a letter from the MIT Radiation Laboratory, offering a fourth opportunity. The MIT terms were satisfactory, but there was again no hint of the nature of the work except the name of the laboratory, which implied that it had something to do with radiation. I decided to accept this fourth opportunity, work unknown; and early in the New Year, 1942, my young wife and I left Pasadena for Boston. This more or less accidental choice proved to be a good one, especially since it led to the withdrawal of the Class A offer from the draft board and to a long and happy acquaintance with Bill and Jane Fairbank.

When I arrived at MIT, I found that a few hundred scientists from all over the country had preceded me and that more were arriving daily. An air of excitement and expectancy prevailed in the halls where microwave radar was beginning to be born. Some of those arriving were graduate students, like Bill and Jane, whose education was suddenly halted by the bombs which fell on Pearl Harbor. Some were postdoctoral fellows whose research projects were similarly halted. Some were young instructors or assistant professors, like Ed Purcell and Luis Alvarez, and some were famous professors like I. I. Rabi and Hans Bethe. The mixture was good.

During my first week there, as I was returning from an early lunch with Rufus Wright, a new acquaintance, the two of us paused under the rotunda of the great hall to greet an attractive young couple. Rufus introduced them as Jane and Bill Fairbank. I learned that they were both graduate students in physics at the University of Washington when the war began and that they were married the previous summer. Their excitement over the work of the Radiation Laboratory impressed me greatly. As they were approaching us I had noticed that they were walking rather fast, Bill slightly ahead of Jane. Their enthusiasm, their friendliness, their brisk pace—all evident at our first meeting—I came to realize were traits deeply ingrained in them.

As Rufus and I walked on down the hall to the Radiation Laboratory, I remarked that I was favorably impressed by that young couple. He agreed with me and advised me to get better acquainted with them if I wished to have a rapid introduction to the laboratory. He insisted that they already knew everybody and everything that was going on there. I followed his suggestion and found it good. Members of the staff told me that Jane was as smart as Bill, and perhaps even more persistent. Unfortunately, in her early months at the laboratory she was hit by a car and severely hurt as she walked out of MIT onto Massachusetts Avenue. For a long time she was too crippled to do experimental work—or to keep up with Bill's rapid gait—but she was exceedingly competent in analysis of data and in preparation of reports.

While Jane was more or less confined to desk work, I found that Bill and I were doing more and more experiments together. He imbued me with the significance of $kT\Delta\nu$, the thermal, or Johnson, noise that ultimately limited the sensitivity of our radar receivers. In this expression, T was the operating temperature (in Kelvin units) of the radar system, approximately the climatic temperature, which never approached absolute zero even in the Boston winters! As all of you know, the signal detectors and amplifiers raised the effective operating temperature, or noise level, well above the climatic, or room, temperature. Our objective, as I early learned from Bill, was to reduce the effective operating temperature to the actual operating temperature. This was the goal of other laboratory staff members who worked on radar receivers; but none, I think, pursued it with more vigor than did Bill Fairbank. I suspect that he is still in the business of reducing $kT\Delta\nu$.

In reducing the noise level one could, of course, do nothing about Boltzmann's constant k. Consequently, we took many hard looks at the effective bandwidth of the receiver $\Delta\nu$, as well as the effective noise temperature T_{eff}. In my pre-war research I had used bandwidths of much less than a cycle per second in infrared spectrometers, but I soon learned from Bill and others that in radar receivers we were compelled to use bandwidths of the order of thirty or more megacycles per second if we were to get the full information of pulsed radar signals through the receiver. We had many discussions about the best fitting of the receiver response curves to the transmitter pulse shape to achieve the lowest effective $\Delta\nu$.

One day we had a visitor from the theoretical section of the laboratory to advise us about the optimum combination of transmitter pulse shapes and receiver response curves. He claimed that he had worked out a combination that would allow us to see signals below the $kT\Delta\nu$ noise level. At that point Bill became noticeably excited. I looked up just in time to see the speaker take a peanut from his pocket, pitch it straight up almost to the ceiling and catch it in his mouth on the way down.

When riding home that evening on the clanging Massachusetts Avenue street car, I noticed that our visitor of the afternoon was sitting across from me in the car. I watched him closely for I found him an exceptionally interesting person. Soon I was rewarded. He took a peanut from his pocket, tossed it to the top of the moving car, and caught it in his mouth on its way down. The next morning when I asked Bill who the visitor was, he said, "Oh, that was Norbert Wiener."

Throughout the war the Fairbanks and I worked in the same division of the Radiation Laboratory, the one concerned with shipborne radar systems. Bill and I spent much of our time on joint projects testing radars over sea water. This work was done from trailers on the seashore rather than from ships at sea; the experimental systems with which we worked were not yet seaworthy! It was in these trailers down by the sea that I became acquainted with Bill's uncanny ability to make a complex, home-made, strung-around system operate with near perfection just at the critical time when a ship was passing. I give you an example. Once we were sent to the Boston harbor to take radar range data on a new battleship, the SS Massachusetts, as it departed from the harbor on a secret mission. For security reasons, I suppose, we were not told the specific time of departure. Suddenly a very strong, close-up signal appeared on our radar screen and moved rapidly out. Had the ship moved out one minute earlier we wouldn't have seen a thing. In this last minute Bill had made some critical adjustments. It is the nature of this man to come through at the last minute. I have never known him to waste time waiting for a train or a plane. He has an uncanny ability to arrive just in time. The one instance that I know when he did miss a train, he drove his car to the next stop and caught the train just in time. Jane knows, because she drove the car back home, from Mebane to Durham, North Carolina.

When testing radar by the sea, we once took our portable laboratory to the shores of Cape Cod in midwinter. The area was under military control at the time, and some of the military personnel had rooms in an old two-story beach house which in peacetime was occupied only in summer. Bill and I shared an unheated corner room on the second floor. That was when he introduced me to near zero temperatures (Fahrenheit). There were windows on both exposed walls. Bill gave me my choice of the two beds. Needless to say, I chose the one nearest an inside wall. He insisted on opening one window before going to bed. He said that he always slept with windows open, even in Walla Walla, Washington, during mid-winter. In the middle of the night I was awakened by a strong wind which was blowing sleet and snow completely across Bill's bed onto my face. I slipped quickly out of bed and quietly closed the window, hoping not to wake Bill, who seemed to be sleeping soundly. As soon as I was back under my covers I heard the faint squeak of the window being opened and again felt snow in my face. After giving Bill plenty of time to get soundly asleep, I again closed the window, only to have it re-opened as soon as I got under the covers. This process was repeated a number of times before I gave up, pulled the covers over my face and head, and went to sleep.

In the morning Bill graciously consented to close the window while we dressed. When we walked down the rickety stairs to go out, I was shivering inside my heavy top coat. A crusty, weathered native, whom I assumed to be a civilian employee of the Coast Guard, greeted us. "Have you heat?" he asked. I quickly replied, "Definitely not." "Well," he said, "you can heat at the Coast Guard restaurant. They have good heats down there." In my frozen state I failed to ask him whether the Coast Guard "heats" were hot.

Later in the war, when K-band (centimeter wave) radar reached a practical stage in its development and an urgent need arose for detection of such objects as submarine snorkels close to the water, Bill and I with two other staff members made extensive over-water comparisons of the three developed radar bands: S (10 cm), X (3 cm), and K (1 cm). There was more power available at the longer wavelengths, but the shorter waves were more effective close to the surface of the water because out-of-phase interference between their direct and sea-reflected components was less than that for the longer wavelengths.

In a single trailer we installed the three radar sets, S, X, and K, with measuring equipment for each. Our favorite location for this trailer laboratory was Fishers Island, a small island near the end of Long Island and across from the submarine base at New London, Connecticut. In peacetime the island was enjoyed by wealthy New Yorkers in their vacation homes. During the war it became a military base, with security clearance required for access to any part of it. On our first trip to the island, the MIT security officer arranged in advance for us to take our trailer laboratory across on the ferry from New London to Fishers Island. Nevertheless, the army sergeant in charge of security at the entrance to the ferry insisted on going through our trailer to see if we were smuggling anything that should not be on the island. He gazed around with a puzzled look and seemed to be almost ready to walk out without comment until his eyes fell upon a small black box having connecting knobs all over its surface. He pointed to the box and asked, "What is dat?" We told him that it was a thermistor bridge. The box actually contained three thermistor bridges, one for each wavelength, with separate sets of external connectors for each. The sergeant stood in silence, looking closely at the black box. Then he turned and said to us, "Dat don't look like no bridge to me," and promptly walked out of the trailer. When he refused to let us take the trailer on the ferry boat, we gave him the telephone number of the security officer at MIT and pleaded with him to call there for an explanation of our priorities. He took the number and went into a back room. Finally he came out and motioned to us to pull our trailer onto the boat, which had been held up for more than two hours past its scheduled departure time.

In his twenties Bill was an energetic, enthusiastic participant in sports. I think that he is, by nature, interested first in science and, by a close second, in sports. On our first evening on Fishers Island he insisted that we go in search of a bowling alley. He felt certain there must be one somewhere on the island because of the great number of military officers around. He was right; there was one. On that evening, and thereafter as long as we were on the island, Bill bowled vigorously until the alley was closed for the night.

I'll mention only one other example of his enjoyment of sporting activities. On one of our trips to the Cape shore to make over-water tests, we parked our laboratory trailer near a shallow marsh lake. When we came to the trailer early in the morning after a very cold night, we found that the lake was solidly frozen. Bill walked out on the ice and called back to me, "Walter, do you skate?" I answered that I was sorry to admit that where I grew up the lakes did not freeze over and that I had never learned to skate. "Well, then," he shouted, "You crank the generator, and I'll do the skating."

In a very real way, the MIT Radiation Laboratory was a security-guarded university with more than two thousand students and only one part-time professor who lectured once a week. Because of the exceptional competence of the professor and the high level of motivation among the students, this lecture course was probably the most effective ever given. The professor was one of the "greats" at Stanford, the late William W. Hansen, co-inventor of the klystron, a vital component of microwave radars in wartime, a power source in postwar microwave spectroscopy, and a driver of the pioneering linear electron accelerators.

When Hansen first came to lecture, the staff members of the laboratory attended in droves and frantically took notes on everything he said—until the powers-that-were decided that two staff members, Samuel Seeley and Sidney Millman, would be officially designated to take notes on the lectures. Their notes, after being carefully edited, were classified "Top Secret." Each staff member received his copy. Hansen's Notes rapidly became the most valuable source book in the Radiation Laboratory.

One of the very useful devices that came from Hansen's lectures probably influenced Bill in his choice of the subject of his Ph.D. thesis as well as his decision to go into low temperature physics. To test the performance of radar systems including those in our experimental trailer laboratory, it was necessary to measure, among other things, the power output of the source (usually a magnetron), the transmission line losses, and the superfluous noise generated in the receiver system. Hansen suggested that the ringing time of a simple high-Q echo cavity could be used to monitor the overall performance of the radars without measurement of their individual components: The higher the power, the longer the detectable ringing time. At a given transmitter power, the lower the overall receiver noise level (above $kT\Delta\nu$), the greater the detectable ringing time. The proposed cavity for X-band radar resembled an ordinary one-pound coffee can, but the ones we had made cost about three thousand dollars each—without the coffee. As I remember, they were reasonably well constructed and coated on the inside with gold. (Nevertheless, I could not help thinking that, before Pearl Harbor, I could have bought two fancy new automobiles for the price of one empty echo can.) The echo can worked so well that an uneducated G.I. from the backwoods of any state could test the performance of his radar system before, or after, going to sea.

The Q value of the echo cavity used in the Radiation Laboratory to monitor radar systems was a few hundred thousand, limited mainly by the surface losses of the interior walls. As Bill used these cavities, he must have been thinking about how much greater the Q could be, how much longer the echo signals would ring, if the cavity walls were made from a low temperature superconductor such as tin. This was to be the problem explored in his Ph.D. thesis research. He found that the Q value could be made several million. The corresponding resonance signal was so sharp that he began talking about locking a clock to the resonance to learn whether such a clock would keep better time than one controlled by a swinging pendulum and monitored by the stars. It probably would have; but before he could complete the experiment, even more accurate atomic and molecular clocks were invented. These superconducting tin cans were the beginning of Bill's explorations of the interesting phenomena which occur along the approaches to absolute zero.

Shortly after the end of the war Bill and Jane left the MIT Radiation Laboratory for Yale, Bill to complete the training for the Ph.D. degree and Jane to start a family. On their Christmas card to us in December 1946, Jane expressed her delight in her new role as the mother of Billy.

All of you know how well Bill has succeeded, but some of you may not know how well Jane has succeeded. You may take a look at Bill, Jr. (Billy), who has given a paper in this symposium, or ask Art Schawlow, who directed Bill, Jr.'s thesis research. Although Bill, Jr., is their only son to become a physicist, he is not necessarily the handsomest or smartest of the three boys to whom Jane taught Little League baseball at Duke. Bob, the second born, took a law degree at New York University and is now practicing law in Los Angeles. Richard, the youngest, last year took a degree from the Stanford Graduate School of Business and received the prize for finishing at the top of his class. He is now a consultant in Washington, D.C.

About the time Bill went to Yale to complete his graduate training, I became an associate professor at Duke and went there in January of 1946 to initiate a program in microwave spectroscopy. My more advanced status resulted purely from an age difference which, at this time, I would be glad to exchange with him.

From Duke I kept an eye on Bill's progress toward his Ph.D. at Yale. When it seemed that the time was drawing near for him to receive the degree, I persuaded Walter Nielsen, chairman of the physics department at Duke, to authorize me to telephone Bill and offer him an assistant professorship at Duke. At that time the authority for such action was vested in the chairman, with the approval of the academic dean. No staff vote or committee action was required. To secure an offer to a particular individual, one had only to convince the departmental chairman, who, in turn, had only to get the approval of the academic dean. This one-on-one action is much simpler than the full court press required to score a goal in the modern, democratic institutions we now have.

When I phoned Bill to tell him about the Duke offer, he informed me that the job at Duke was exactly the kind he most desired, that Professor Fritz London's presence made Duke an ideal place to start a low temperature laboratory, and that it would be a joy to resume associations with me. Then he said, "I want to accept, but I can't—because I have recently accepted an appointment at Amherst College, and I feel obligated to serve there for a few years." Knowing that it is the nature of Bill to live up to what he considers an honorable obligation, I didn't try to persuade him. I simply said, "I'll call you again in a few years." I did, and he accepted.

When Bill came to Duke in the summer of 1952, there was not, and had never been, a low temperature laboratory at Duke; he had to start from scratch to build one. He plunged into the work with the confidence and enthusiasm that I had seen in the 25 year old Bill whom I first met in the Radiation Laboratory. By the next spring he had liquid helium flowing at Duke, with rapid pumps to reduce its temperature to about 1 K.

Although he found no low temperature laboratory when he came to Duke, he did find one of the world's leading low temperature theorists, Fritz London. It is to Bill's credit that he recognized the tremendous advantage of Professor London's presence to the new low temperature laboratory he was building. London had many theoretical problems he wished Bill to put to experimental test. One of them in which he, as well as other leading theorists, had expressed considerable interest was the extent to which ^{3}He behaves as an ideal Fermi-Dirac gas. Bill proposed to find out by measurement of the strengths of the ^{3}He nuclear magnetic resonance signals as the temperature of the liquid ^{3}He is reduced. Particles of an ideal Fermi-Dirac gas at sufficiently low temperature are expected to have increasingly antiparallel alignment of their spins, which causes magnetic susceptibility to deviate from the classical $1/T$ Curie law and finally to become independent of temperature when all the spins become aligned.

During the first several months at Duke, Bill spent most of his time and funds in assembling the low temperature cryostat for attack on the ^{3}He problem. To speed up the experiment I offered my help and the use of a 12 inch Varian magnet (the third one made by Varian Associates) and electronic equipment from the microwave laboratory. Hans Dehmelt, first to detect nuclear quadrupole resonance, was a research associate in my laboratory at the time. He helped to construct an exceedingly sensitive nuclear magnetic resonance spectrometer. With this assistance and with a sample of ^{3}He more than 99% pure, which was supplied by the Oak Ridge National Laboratory, Bill was in the business of measuring the nuclear magnetic susceptibility of liquid ^{3}He by the beginning of the summer of 1953.

Accurate measurements of the ^{3}He nuclear magnetic susceptibility were made from 4.2 to 1.2 K, which was as low as could be obtained by simple pumping of the liquid helium surrounding the sample tube. These measurements showed no discernible deviation from the classical $1/T$ law. A plot of χ(nuclear) against $1/T$ appeared to be linear. The experiment was halted for a period while Bill constructed a magnetic cooling device that would permit measurements to lower temperatures. Cooling between 1.2 and 0.23 K was achieved by adiabatic demagnetization of 25 grams of potassium chrome alum. The measurements, when resumed below 1.2 K, revealed a definite departure from the $1/T$ Curie law expected from Boltzmann statistics. The curve fitted beautifully that for an ideal Fermi-Dirac gas having a degeneracy temperature of 0.45 K.

Professor London was excited and happy over the ^{3}He experiment. When measurements were in progress, he often stood in the door of the laboratory awaiting each new data point. It is a pity that he died just as the data for the critical range of 1.2 to 0.23 K were being obtained, the data which confirmed his expectations of the Fermi-Dirac statistical behavior of the ^{3}He liquid.

In the spring of 1954, the Duke Physics Department had the honor of a brief visit by Professor C. J. Gorter, director of the world-famous low temperature laboratory of Leiden. He asked Bill what he was planning to do in the new low temperature laboratory. Bill told him about the NMR experiments designed to test the statistical behavior of liquid ^{3}He at temperatures below 1 K. Professor Gorter, who evidently had not seen the first publication, in the *Physical Review* of October 1, 1953, commented that this was a very important experiment but that it was much too difficult to be undertaken at Duke. A few weeks after Gorter returned to Leiden, Bill sent him a prepublication copy of the paper describing the results mentioned for the critical range below 1 K. In response, he received a letter of congratulations from Gorter, with an apology for having grossly underrated the capability of the new laboratory at Duke.

The ^{3}He NMR experiment offers another example of Bill's charmed, last-minute "come-throughs," with which I had become familiar during our collaborations in the Radiation Laboratory. On the first trial run of the experiment, Hans Dehmelt had the NMR spectrometer working perfectly hours before the count down, but Bill's cryostat, associated pumps, vacuum gauges, temperature gauges, etc., were not ready until near zero time. The whole system, with a maze of glass connecting tubes, was mounted on a dolly which could be moved forward to put the ^{3}He sample tube between the poles of the magnet. Professor London was standing by the laboratory door, and several graduate students were in the hall outside waiting to see the first NMR signal at Duke. At time = 0, Bill, announcing that he was ready, began to roll the cryostat toward the magnet. The whole low temperature assembly began to sway back and forth. I turned my eyes away, to avoid seeing it crash into the magnet poles. When I looked back, the motion had ceased, and the ^{3}He cell was at the exact center between the magnet poles. Hans quickly turned up the magnetic field, and there before our eyes on the CRO scope was a strong ^{3}He signal. My mind flashed back to the strong radar signal of the battleship SS Massachusetts that Bill and I had seen years before.

Despite the fact that electrons are Fermi particles, Professor London spoke often of a Bose-Einstein condensation as being the probable cause of superconductivity. Although at the time there was no known mechanism for the coupling, he seemed to believe strongly that the electrons in a superconductor were coupled or coordinated in such a manner that they could be considered as an assembly of Bose particles in a Bose-Einstein condensation. All the electrons would fall into the same lowest-energy level in momentum space corresponding to a macroscopic quantum state of long range order. The removal of electrons from this lowest level would break up the coordination and destroy the long range order. It would require energy corresponding to that of the superconducting gap, i.e., the difference in energy between the electrons in the superconducting state and the thermal electrons of the normal state. The mechanism for this electronic coordination is now well understood from the work of Bardeen, Cooper, and Schrieffer, but it was not at the time of London's death. In the conclusion of his book, Superfluids, Volume II (1954), which he was proofreading on the evening before his death, London states:

> According to this point of view, both superfluidity in liquid helium and electric superconductivity would result from a macroscopic quantum mechanism producing a gradual condensation into a state of more or less rigid, long-range order of the momentum vector ... although the question of the mechanism producing this condensation of the electronic fluid is still unanswered—despite noticeable progress recently made in this field.

The critical temperature T_c for the onset of dc superconductivity for a number of superconductors, including the common metal tin, falls in the range of 4 to 1 K, easily covered in Bill's new low temperature laboratory at Duke. The energy gap between the superconducting quantum state and the normal state was thought at the time to be of the order of kT_c. This corresponds to a frequency of kT_c/h, which for tin is 77 GHz with $T_c = 3.72$ K. By 1954 we had in the Duke microwave laboratory instruments and techniques for measurements of frequencies up to and above 300 GHz. Therefore, Bill and I joined forces and, with a graduate student, Gilbert Blevins, set out to measure the quantum gap in tin. Figure 1 is a photograph of Bill and me with the essential components for these measurements.

We found that a resonant cavity coated with tin and tuned to 77 GHz does not become superconductive at the critical temperature, 3.72 K, but that there is evidence for partial conductivity, with some residual resistance, at temperatures much lower than T_c at frequencies of 77 GHz and

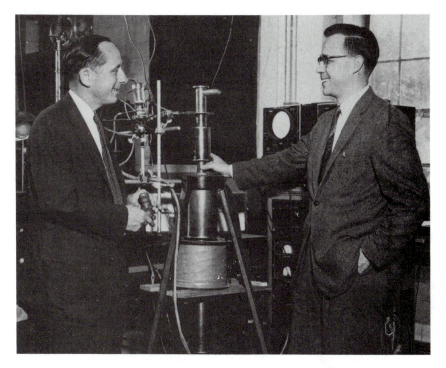

FIGURE 1. William M. Fairbank (right) and Walter Gordy with essential laboratory components for measurement of superconductivity at high millimeter wave frequencies (1955).

up to 150 GHz. The onset of this partial superconductivity occurred at lower temperatures as the frequency increased; and the residual resistance increased with frequency. The results were strongly suggestive of an energy gap that became more sharply defined as the temperature was lowered below T_c, but we could not sweep our microwave spectrometer across the evidently broad gap. We suffered from too much resolution and too little tuning range. Tinkham and Glover, approaching from the other side of the gap with a broadly tuned far-infrared instrument, beat us to the crossing. Bill graciously accepted defeat and started thinking of other ways to detect long-range order in superconductors.

Bill's spectacular success with Bascom Deaver, his first student at Stanford, in detecting macroscopic quantization of the magnetic field outside a superconductor is well known. Again, it is indeed a great pity that Professor London did not live to see this proof of the macroscopic quantization of the magnetic field that he had predicted years earlier. I doubt that anything in Bill's scientific career would have given him more pleasure than showing these data to Professor London.

A strong characteristic of Bill's nature is experimental boldness. He seems to enjoy most those experiments that offer the greatest challenge to his wits and endurance. Joe Reynolds, a classmate of his at Yale and later vice president for research at Louisiana State University, expressed this to me very effectively. Back in the sixties when government money was flowing freely and physics was booming, I met Joe at a physics meeting in the city where the money is passed out. Bill was there talking to other physicists in the halls, discussing his experiments and what he planned to do next. Joe remarked to me that Bill, unlike many physicists of that period, was not in the slightest secretive about what he was doing, or planning to do. According to Joe, secrecy for Bill makes no sense because the experiments he undertakes are much too difficult for anyone else.

I close these remarks with a word of caution to the physicists of the world. Simply because William Fairbank's laboratory is the only one—so far—to obtain definitive evidence for fractional electronic charges does not mean that they do not exist. Quite to the contrary.

I.3

The Creative Imagination
of an Experimental Physicist

C. W. F. Everitt

1. MEETING THE MAN

To have worked closely with a man for over twenty years is in one sense to
know him well. Whether such knowledge qualifies one to be an interpreter
of his scientific genius is another matter. One may be too close to the mountain; and the papers in this volume give some indication how mountainous
William Fairbank's influence and achievement have been. More subtle are
issues of scientific atmosphere and personal exchange. To what extent are
new ideas unique to an individual and to what extent are they part of the
wider scientific air? In what degree have a colleague's ideas and thought
processes so entered one's being that one no longer recognizes their prove-
nance or even their significance? Familiarity breeds unawareness. And even
if awareness can be recovered, assessment may be far off. Bill Fairbank is
a man of strong and unusual individuality; he is also unusually creative.
Creativity and character must be related; but can one hope to trace the
anatomy and physiology in their connections? Some years ago I wrote a paper on Maxwell's creativity. Then and later it seemed to me possible with
the perspective of a hundred years to reach some understanding of the links
between Maxwell's creative style and his education, scientific environment
and family background, but much harder to penetrate his imaginative process. The outer structure was clear; the inner human mystery remained.

Bill Fairbank once remarked to me that he had no idea where his ideas came from. They seemed to float in on him out of the air. We finally must share the same puzzlement.

Still, some understanding is worth seeking. My first reaction on being asked to provide a critical introduction to this volume was to cast my mind back to my earliest encounter with Bill in November 1961. At that time I was a young research associate at the University of Pennsylvania, not long out of England, working with Kenneth Atkins on third sound in liquid helium films. Bill, much in the news then because of his recent discovery with Bascom Deaver of quantized flux in superconductors, was visiting the physics department at Penn for a few days to give the annual Mary Amanda Wood lectures. Before Bill arrived, Atkins mentioned some of his achievements, which I as a neophyte in low temperature physics knew little about; then, in introducing Bill at the first of the three lectures Atkins characterized him by saying that whenever some theorist proposed an utterly impossible experiment the next thing you would hear a few months later was that Bill had done it with even higher precision than was asked. No doubt Atkins had in mind the Buckingham-Fairbank-Kellers work on the logarithmic discontinuity in the specific heat of liquid helium at the λ-point, where a challenge by Blatt, Butler and Schafroth to find a rounding of the specific heat curve within 10^{-3} K of the λ-point, and a bet by Feynman that there would be no rounding within 10^{-5} K, led Bill and his colleagues to do a 10^{-6} K experiment. Readers of this volume will learn that John Lipa and his colleagues have now extended the precision of the experiment to 10^{-8} K and are developing a space version that should give specific heat measurement to 0.1% within 10^{-10} K of the λ-point (paper II.9). The bearing of John's work on Wilson's renormalization group theory of critical point phenomena, and the historical ties of that theory to Bill's experiment, are reviewed by Michael Buckingham (paper II.8). Returning to 1961 and the University of Pennsylvania, I cannot forbear quoting another of Atkins' remarks, not made in public. This was to the effect that Bill finally had gone too far. He was talking about a new test of general relativity to be done in a satellite, with a cryogenic gyroscope having the preposterously low drift rate of 1/100 of an arc-second per year. Still, one must concede that the gestation period for that particular experiment has been more than a few months.

After such a build-up, Bill's lectures might have proved an anticlimax; in fact, as anyone would expect who has heard him speak, they were brilliantly successful. Bill's drive and breadth of vision, his ingenuity and almost boyish enthusiasm, reminded me of Blackett with whom I had worked in London; and, not by chance, I conceived the wish to join Bill at Stanford.

Two further characteristic Fairbank incidents marked those few days at Penn. The first was in the second lecture when Bill, who had seen our work that afternoon, at one point suddenly broke off in mid-sentence to urge anyone who had not done so to "go down to the basement of this building and see the beautiful experiment on third sound." At the time I was most struck by the generosity of this remark; in retrospect it seems even more telling of Bill's alertness. Too many physicists are so preoccupied with immediate concerns that they never look about them. Bill's curiosity is always active. The second incident came on the following day. Bill encountered me in a corridor, seized me by the arm, said there was something he had been wanting to talk to me about, marched me into the nearest empty office, and began discoursing at the blackboard on some utterly incomprehensible topic. It was flattering but I could not imagine what Bill thought I had to contribute. Only later did I come to understand, in Bascom Deaver's words, that there are times when Bill needs an impedance to think into.

Be that as it may, Bill's thoughts were manifestly worth impeding. Soon afterward I sent a letter in which, according to Bill's recollection, I stated that I wanted to do a far-out experiment, and, according to my own recollection, that I would not mind, indeed would prefer, attempting something challenging. Several months later, having heard nothing, with some trepidation I telephoned, was referred from Bill's office to his home, and had a conversation with one of the Fairbank boys that was rather more California casual than my English conservatism was used to. When finally Bill and I did connect, it was a relief to learn that he had been just on the point of sending me an offer. So it was that in October 1962 I found myself in the old physics building at Stanford, sharing an office with Alan Schwettman who had arrived from Rice University a month earlier, endeavoring to learn simultaneously four subjects—relativity, space technology, superconductivity and the design of inertial navigation instruments—whose only point in common seemed to be my total ignorance of all of them. Not to be afraid of one's own ignorance, however, is one of the great Fairbank themes, to which I shall return.

2. THE DEVELOPMENT OF A PHYSICIST: I. SEATTLE TO THE MIT RADIATION LABORATORY

Bill's scientific career divides conveniently into three phases. The first is the period from Roosevelt High School in Seattle, from which Bill graduated in 1934, through Whitman College, the University of Washington, and the years from 1942 to 1945 spent working on radar at the MIT Radiation Laboratory. The second, beginning in November 1945, just after World War II, and extending through August 1959, covers graduate school at Yale and the years as a professor at Amherst and Duke. The third, from 1959 to now, is the Stanford period. Soon Bill will be entering on a fourth

Evolution of W. M. Fairbank's scientific career and experiments that he and his associates have undertaken

February 24, 1917	Born, Minneapolis, Minnesota
1934	Graduated, Roosevelt High School, Seattle, Washington
1935	Entered Whitman College
1939	A.B., Chemistry, Whitman College
1940	Completed physics major, Whitman College
1940–42	Graduate school, physics, University of Washington
August 16, 1941	Married Jane Davenport, Seattle, Washington
1942–1945	Staff member, Radiation Laboratory, MIT
1945–1947	Graduate school and Sheffield Fellow, Yale University
1947	First published paper (on second sound, with C. T. Lane and H. A. Fairbank)
1948	Ph.D., Yale University; thesis, "The Surface Resistance of Superconducting Tin at 10,000 Megacycles."
1947–1952	Assistant professor, Amherst College
1947–1952	Construction of liquefier to separate ^{3}He from ^{4}He
1952–1959	Associate professor and professor, Duke University
1954	Observation of Fermi degeneracy in ^{3}He
1955–1956	^{4}He bubble chamber
1956	Phase separation in ^{3}He-^{4}He mixtures
1958	Logarithmic discontinuities in specific heat of helium at the lambda point
1959–Present	Professor, Stanford University
1961	First conceptual proposal, with Schiff, NASA gyroscope experiment
1961	Quantized flux in superconductors
1962	California Scientist of the Year for 1961
1962	Start of superconducting accelerator program
1963	Member, National Academy of Sciences
1963	Oliver E. Buckley solid state physics prize
1963	Observation of London moment
1964	Research Corporation award
1965	Richtmyer lecture: "Near Zero: A New Frontier of Physics"
1967	Published results of electron free fall experiment
1967	Member, American Academy of Arts and Sciences
1967	Quantized ^{4}He version of Newton's ice pail experiment
1968	Fritz London award
1969	Start of gravitational wave program
1969	Nuclear antiferromagnetism in solid ^{3}He
1970	Invention and demonstration of porous plug device for controlling liquid helium in dewars in space
1971	First data from fractional charge experiment
1972	Start of free electron laser project
1973	Studies of biological molecules using SQUID's
1974	Start of magnetocardiology program
1975	Large scale ultralow field magnetic shields
1975	First operation of low temperature gravitational wave antenna
1976	Quantized surface profile of rotating ^{4}He

Evolution of W. M. Fairbank's scientific career (*continued*)

1977	Surface shielding on copper oxide
1977	Start of research on lambda point experiment for space
1978	Member, American Philosophical Society
1979	Polarized ^{3}He in low magnetic fields
1982	"Near Zero: New Frontiers of Physics" conference, Stanford University
1984	Start of flight program of gyroscope experiment
1984	Design of self-shielded superconducting magnet for magnetic resonance imaging

phase of his career, "retirement," concerning which the only safe prediction is that Bill will bring to it the same flair for the unexpected that he has to each of the three earlier phases.

Bill has remarked that his high school physics teacher, Alfred E. Scheer, a gifted man who might have gone into research but for World War I, taught him two things, that physics was a rational subject and that the physics problems in the rather elementary text they used (by Charles Dull) were within Bill's capacity to solve. A teacher who elicits that response has done well.

Bill's father had graduated from Amherst College. Had times been ordinary, Bill, too, would probably have gone to Amherst, but the Depression dictated otherwise. So in 1935 Bill entered Whitman College, closer to home, to be followed there a year later by his younger brother Henry, with whom throughout their two lives he has had such a close personal and professional relationship. Knowing now, as Bill could not remotely have imagined in 1935, that twelve years later he would become an assistant professor at Amherst, we may be tempted to discern in all this an instance of what seems to be a recurring two-sided pattern in Bill's life, in which events that fail to happen in one period are fulfilled in another way later on, leaving Bill meanwhile with the opportunity of turning the apparent setback to advantage. Examples will appear in Bill's encounters with both Duke and Stanford Universities. Entering Whitman, Bill faced the normal undergraduate uncertainties about life and career, not lessened by the anxieties of the era. Physics receded into the background. One student whom Bill knew had it all figured out: even in a depression people need glasses; he would become an optometrist. For a while Bill, too, flirted with optometry.

Bill was earning his way through college. With Fairbank resourcefulness he took the job of busing tables in the women's dormitory. So occurred the most determining personal encounter of his life: he met Jane Davenport. Jane also was from Seattle, though from a different high school, and as a top student in her class had won a scholarship to Whitman. She and

Bill started going together in 1937 and were married in August 1941. Kip Thorne describes Bill the scientist as either lucky or clairvoyant. In sharing his life with someone as intelligent, strong minded and high-principled as Jane, Bill has indeed been lucky. Whether his clairvoyance extended to forseeing the success Jane and he would have in raising a family is a nice question.

Going together for Bill and Jane meant choosing the same major, chemistry. Jane planned to become a high school teacher in science and mathematics. Bill's path was less clear, but chemistry with some physics seemed a reasonable foundation for the one year's graduate training he would need to become an optometrist. Then toward the end of 1938 events took a new turn. Two professors, Carroll Zimmerman and Ivar Highberg, revitalized Whitman physics. Earlier, Whitman had had under Benjamin Brown, an Oersted medalist, one of the best undergraduate physics programs in the United States, but when he retired in 1931 it languished. Highberg, a Whitman graduate, returned to Whitman as a faculty member in 1936, half time in physics and half time in mathematics; Zimmerman came in 1937. One result of their teaching was to induce Henry to major in physics; another was to reignite Bill's enthusiasm for the subject.

In the Christmas vacation of 1938, Bill decided to find out more about optometry. Through talking with several optometrists and physicians in Seattle he became convinced that to have a satisfying career he would need to qualify as an ophthalmologist, that is, go to medical school. He aired his concerns with an old family friend, Will Thompson, head of the department of fisheries at the University of Washington. Thompson, who was preeminently successful both as a teacher and as a research worker, said (to quote Bill's own recollections): "Bill, don't pay any attention to what anyone says about what is a good, stable job. Do what you like. I like fish; my parents were absolutely horror-stricken. But because I like fish, I've done very well and now I'm one of the famous men in fish, and head of the fishery department here." Such advice to such a student could have only one result. Bill chose physics.

Who influenced whom next is not entirely clear. Bill and Jane completed their chemistry majors in 1939. Henry was already majoring in physics and would graduate in 1940. Jane, who had been planning an extra year at Whitman to earn her teaching credential, decided to combine that with a second major, in physics, not to mention the editing of the college newspaper for a semester. Bill also decided to stay the extra year for a physics major. Under Zimmerman all three began looking toward graduate school in physics. Apart from Harold Argo in 1939, who had gone on to George Washington University (Zimmerman's *alma mater*), they were among the very few Whitman students to do so since the days of Brown. So

Zimmerman found himself conducting a seminar on Leigh Page's *Introduction to Theoretical Physics* with a class of three: Jane, Henry and Bill.

Next came the choice of graduate school. Physics schools in those days were not the Ph.D. factories they now are; admitting a woman was unusual, while the idea of a woman teaching assistant (and neither Bill nor Jane could do without an assistantship) verged on the preposterous. Henry became one of the small entering class at Yale. Bill and Jane were granted two of the four teaching assistantships for that year at the University of Washington, one reason being that the university had a home economics major with a required physics course for which it was thought that a woman teaching assistant would be useful. One woman who preceded Jane as a teaching assistant in the University of Washington Physics Department was Georgeanne Robertson Caughlan, who became a professor at Montana State University.

Jane and Bill stayed at the University of Washington a year and a half, benefiting much from the teaching of Edwin Uehling; and Bill was just beginning to define a thesis topic, choosing with characteristic initiative a line of research no one at Washington was working on, a molecular physics technique developed by Bleakney at Princeton. Then in rapid succession came Pearl Harbor, a draft notice for Bill, and a mysterious offer transmitted through one of their professors, W. Cady, of defense work for Bill and Jane on the East Coast. A few weeks later they were at MIT.

Walter Gordy, in paper I.2, describes the impression this "attractive young couple" made at the Radiation Laboratory, an impression confirmed by Ernest Pollard in the glimpse he gives of them in his little book *Radiation*. Jane was, after Dorothy Montgomery, the second woman appointed to the Rad Lab scientific staff. She and Bill joined the seaborne radar division under Larry Marshall (later under Pollard), working with Emmett Hudspeth, Richard Emberson and others including Gordy. After initial responsibility for a demonstration apparatus, they became involved in overwater tests of 10 cm and 3 cm radar at Fishers Island and Cape Cod. Bill continued experimental work throughout the war. One of his contributions, which was patented though never widely used, was a new technique, codenamed RASCAL, for calibrating radar systems. After a severe car accident in 1943, Jane focused more and more on the editing of scientific reports about the overwater transmission of 10 cm, 3 cm and 1 cm radar, and the editing of instruction manuals for shipborne radar sets. Bill has remarked that the Rad Lab, besides supplying the knowledge of radar that made his doctoral research at Yale possible, gave him experimental self-confidence. It was indeed, for many others besides Bill and Jane, a first class educational experience. To be suddenly thrown together on equal terms with three thousand other physicists, many of them world-renowned, to do important work with solid funding and scant bureaucracy, must be thrilling. Too bad wars are needed to make such things happen.

3. THE DEVELOPMENT OF A PHYSICIST: II. YALE TO DUKE

In October 1945, on Pollard's recommendation, Bill was admitted to graduate school at Yale. Jane by then was starting her family, Bill, Jr., being born three months later. Meanwhile Henry, having completed his Yale Ph.D. in 1944, had spent the last fifteen months of the war at Los Alamos. Henry's doctorate had been earned in unusual circumstances. When he arrived at Yale in 1940, C. T. Lane, who had been on the faculty there since 1932 working on magnetism, was in the process of moving into low temperature physics and had for some years been building a large scale (1 liter/hr) helium liquefier based on the famous design by Peter Kapitza. Lane's was the first expansion engine helium liquefier built after the two made by Kapitza at Cambridge and Moscow; and under Lane, Yale was to become the first physics department in the U.S. with a significant low temperature group, though several chemists, notably W. F. Giauque at Berkeley, were doing important work. Henry decided to join Lane; but since Lane almost immediately went off to the Kidde Company to build oxygen liquefiers for the U.S. Navy, and the Kapitza liquefier was not yet working, Henry built an apparatus containing a hydrogen liquefier in which, with Lane's intermittent advice, he measured the resistivities of copper-tin alloys as a function of temperature, serving the while as an instructor on the Yale faculty. In September 1945, shortly before Bill's arrival, Henry returned to Yale, again as an instructor, and resumed work with Lane.

Bill at first was uncertain about a field of research. He considered doing nuclear physics with Pollard, but Pollard's interests were changing; nuclear physics, he advised Bill, would soon be ancient news; something less developed like low temperature physics was likely to be more rewarding. Later Pollard himself moved into biophysics. So Bill joined Lane and Henry in a field where there were thriving centers of research at Oxford, Cambridge, Leiden, Moscow and Toronto, but almost nothing in the U.S. With this Bill had his first encounter with the work of the London brothers, not Fritz, whose colleague he was to become at Duke, but Heinz.

Bill and I have summarized elsewhere[1] Heinz London's truly remarkable achievements, too often overlooked in the greater force of Fritz. Heinz' first

---

[1] See articles on Fritz and Heinz London in volume viii of *Dictionary of Scientific Biography* (Scribners, New York, 1973). Bill was responsible in discussions with the editor, Charles Gillispie, for the inclusion of the article on Heinz.

research interest under F. E. Simon at Breslau in 1932 had been to search for alternating current losses in superconductors. This work, though unsuccessful experimentally, led Heinz London to construct independently of Gorter and Casimir a two-fluid model of superconductivity, introducing the concept of a penetration depth and eventually producing with Fritz the famous 1935 paper containing the "London equations" of superconductivity. Throughout the 1930's, Heinz, now at Oxford, continued his experimental search, progressively pushing the measurements to higher frequencies; finally in 1939 he demonstrated significant ac dissipation at 1500 MHz (a radio wavelength of 20 cm) in tin samples cooled below the superconducting transition temperature.

Bill knew nothing about Heinz London's work but did know about radar. He also knew that while superconductors have vanishing electrical resistance at dc and in low frequency alternating fields, in the infrared they respond like normal metals. There must be a transition from normal to superconducting behavior in some intermediate range of frequencies. Applying his radar experience, Bill decided for his doctoral dissertation at Yale to search for ac losses in superconducting cavities at microwave frequencies. Unknown to him, Brian Pippard at Cambridge took up the same theme at the same time. Since Lane had no knowledge of radar, Bill's thesis topic, like the aborted one at the University of Washington, was very much of his own making. Alan Schwettman elucidates it admirably in his history of the superconducting accelerator (paper IV.2). The lion's share of discovery in microwave superconductivity was Pippard's. Through a combination of experimental and theoretical insight involving measurements on superconductors doped with varying amounts of impurity, he proved that there is a coherence length ℓ over which superconductivity acts, and that ℓ has an inverse relationship to the London penetration depth λ. In the Bardeen-Cooper-Schrieffer (BCS) theory of superconductivity, the Pippard length is identified with the pairing distance of the electrons. To Bill the publication of Pippard's work and the realization that Heinz London had come earlier must have been a slight shock. Nevertheless, his research had an importance of its own, bearing fruit in the later development by John Turneaure and others of high-Q cavities for the superconducting accelerator (paper IV.5).

Even at Yale, with thesis pressures and a family to support, Bill did not restrict himself to this one topic. Lane and Henry were engaged in two topics of their own: the detection of second sound in superfluid helium, and the study of ^{3}He-^{4}He mixtures. Bill involved himself in both. Second sound was an exciting subject, both from the novelty of the idea of a temperature wave and from the controversy that existed between Landau and Tisza over its precise character and temperature dependence. Lane and Henry and then Bill, not knowing that Peshkov in Moscow had already detected it, were eager to be first in the field. With advice from Lars Onsager not only about theory but about how second sound might be detected through its effect in generating first sound in the vapor above the liquid, they completed a beautiful experiment measuring the temperature dependence of the second sound velocity down to 1.3 K. Later, as Henry Fairbank describes in paper II.7, four groups, Maurer and Herlin, Pellam and Scott, Atkins and Osborne, and De Klerk, Hudson and Pellam, extended the measurements to lower temperatures and confirmed Landau's theory by showing that below 1 K the velocity of second sound rose asymptotically to $1/\sqrt{3}$ times the velocity of first sound in liquid ^{4}He. Henry himself, with his students, investigated the behavior of second sound in ^{3}He–^{4}He mixtures. They found that in mixtures with concentrations of order 1% the ^{3}He atoms replaced phonon excitations as the carrier of the second sound at the lower temperatures. Under these conditions the ^{4}He behaves like a vacuum, the ^{3}He behaves like an ideal gas, and the velocity of second sound goes down with temperature as $\sqrt{T}$.

For Bill, the most important outcome of the collaboration with Lane and Henry on second sound may have been in learning to appreciate the theoretical genius of Onsager. I shall return later to Bill's special knack for interacting with theorists. Onsager was important not only in the second sound experiment but in several others including the next subject to which Bill turned his attention, ^{3}He.

The start of Bill's work on ^{3}He coincided with his appointment to the Amherst faculty in September 1947, a year before the completion of his doctoral dissertation at Yale. The theoretical significance of ^{3}He as presumably a Fermi liquid (in contrast to the presumed Bose-Einstein characteristics of ^{4}He) was widely recognized; the difficulty was that pure ^{3}He was nowhere available. The Atomic Energy Commission could make it through the decay of tritium prepared in nuclear reactors, but this was not known to ordinary mortals. Well helium from Texas had a ^{3}He concentration of 1 part in 10^7, atmospheric helium a concentration of 1 part in 10^6; the strongest mixtures ordinarily available were those concentrated through thermal diffusion by Alfred Nier at the University of Minnesota, for which the ^{3}He abundance was about 1 part in 10^4. But there was one other avenue. When a temperature gradient is applied to bulk superfluid ^{4}He, it sets up a counterflow of the normal and superfluid components in the liquid. Lane and Henry Fairbank discovered that in ^{3}He–^{4}He mixtures the ^{3}He is transported with the normal fluid but not the superfluid. The ^{3}He concentration could therefore be increased by "heat flushing"; Lane and Henry reached concentration of 1 part in 10^3, and would no doubt have gone further had it not been for the miserably small amounts of liquid they had to work with. Starting from atmospheric helium one would have to process a thousand liters of liquid or 700,000 liters of gas at STP to get 1 cc of liquid ^{3}He.

With incredible boldness, Bill at Amherst decided to do just that. Already at Yale he had conceived a heat-flushing device for separating the ^{3}He from ^{4}He during the liquefaction process. At Amherst he, together with Theodore Soller and Bruce Benson, obtained a grant of $10,000 from the Research Corporation to build one of the new Collins helium liquefiers, partly in the physics department machine shop and partly from components purchased from the Arthur D. Little Company. Bill's own account of his strategy is appealing:

> I thought one could do either of two kinds of experiments at a small undergraduate college like Amherst: either experiments too trivial to be done at a university or experiments too hard to be done at a university. I opted for what I thought was one of the latter. I designed a heat flush separator to remove the one part in 10^7 of ^{3}He from the ^{4}He as fast as we could liquefy the ^{4}He. We could in principle obtain about 1 cc of ^{3}He gas for each large cylinder of natural well helium. This would, we thought, give us the only ^{3}He in the world with which to experiment. We succeeded in building the liquefier and separator and separating ^{3}He from ^{4}He by the end of my five years at Amherst. The reason I thought the experiment was too hard for a university was that one needed to dedicate the liquefier to the separation project. This I thought would be incompatible with [a program containing] several [students all doing] Ph.D. theses.

These remarks are singularly revealing of the praxis of science. The experiment was "too hard" for a university, not from innate technical difficulty but from the circumstantial difficulty of running a program with doctoral students. Bill was bold, but his boldness was founded on an astute grasp of the realities of the situation. The outcome was equally interesting. By superhuman energy Bill did, as he remarks, succeed in getting the liquefier and separator working by the end of his five years at Amherst. He should have cornered the market. The catch was that just then the Atomic Energy Commission decided to make small quantities of pure ^{3}He generally available. The vast effort of heat flush separation had gone for naught.

But not quite for naught. In building his own liquefier at Amherst, Bill had come to understand the operation of the Collins machine extremely well, and he found that it gave better results on using a much larger helium gas compressor than the one supplied with the Collins machines built by the Arthur D. Little Company. At Duke the physics department had raised money to buy a liquefier, so when Bill went there in the fall of 1952 he asked the Arthur D. Little Company to sell him a liquefier without the compressor so that he could buy his own compressor elsewhere. In Bill's words:

> They said a larger compressor would not do any good because the liquefier had been designed by Sam Collins for the size of compressor they were using. They wouldn't sell the liquefier without the compressor, but agreed to try two compressors running in parallel. I went

to Boston to watch the test, and much to the surprise of everyone, including Collins, the liquefier produced three times as much helium per hour in liquid form. This, then, became the standard for their liquefier. They sold it with two compressors.

This incident gave Bill a certain reputation, and helped other low temperature physicists to become more productive through having more helium to work with. It also stood Bill in good stead a dozen years later when he and Alan Schwettman were persuading Collins to undertake the bold project of designing a very large liquefier operating at superfluid temperatures for use in the superconducting accelerator.

Bill's transfer from Amherst to Duke was inspired by Walter Gordy. Gordy had tried to draw Bill to Duke in 1947, only to have Bill reply, with obvious regret, that while there was nothing he would have liked better than the chance to work with Gordy and Fritz London, he could not in honor break his existing agreement with Amherst. In fact, the Amherst years had increased Bill's sense of independence, and when he reached Duke he was raring to go. He proceeded at once with experiments on a small quantity (about 0.2 liquid cc) of ^{3}He condensed from gas obtained from the AEC. His idea from the beginning had been to look for evidence of Fermi degeneracy in ^{3}He by applying the technique of nuclear magnetic resonance, newly invented by Bloch and Purcell, to measure the magnetic susceptibility of the liquid as a function of temperature. At Amherst Robert Romer, then an undergraduate, had done an honors thesis with Bill on how to perform the NMR measurements. At Duke Bill had the advantage of Walter Gordy's presence and expertise in electron spin resonance techniques, as well as Gordy's generosity in making his Varian NMR magnet available for the ^{3}He work. Within four months Bill, together with William Ard, a student of Gordy's, and Hans Dehmelt of Germany, had set up the apparatus and begun NMR measurements. The work was continued by King Walters, who joined the group as a graduate student in 1953. With measurements down to 1.2 K showing nothing unusual, King and Bill cooled the sample further by adiabatic demagnetization, and at 0.45 K had the thrilling experience of observing the onset of Fermi degeneracy.

The result, says Bill, obtained in February 1954, "did wonders for my self-confidence," the more so because late in the previous year C. J. Gorter, the director of the Leiden Laboratory, on a visit to Duke, had shown great skepticism about their ability to make the measurement, let alone discover anything. Afterwards Gorter made generous amends. In April 1954, two months after the discovery, a distinguished physicist came to Bill at the Washington meeting of the American Physical Society and in Bill's account "said 'we are going to do a very exciting experiment at Los Alamos. We have got one of Purcell's students coming and we are going to do the nuclear resonance of ^{3}He' ... I just reached into my briefcase and said 'Would you like to see the data?'"

One man who would have liked to see the data was Fritz London, whose sudden death at the age of 54 came a few days before the discovery.

I shall come back later, in a discussion of the relationships between theory and experiment in physics, to this work on ^{3}He and the work on ^{3}He-^{4}He mixtures which followed it. Bill's career swiftly blossomed. Alex Dessler worked with him on shock waves of second sound in superfluid ^{4}He; Buckingham, Fairbank and Kellers did their amazing λ-point experiment; Martin Block and Bill cooperated in developing a liquid helium bubble chamber, later used by Block and others for many years in particle physics research; and again working together Bill and Michael Buckingham invented the persistatron, a switchable circuit element which they hoped would become the basis for a superconducting computer. Duke and Bill had done well by each other.

4. THE DEVELOPMENT OF A PHYSICIST: III. THE STANFORD YEARS

While at Yale, Bill had written to Stanford to the effect that low temperature physics was an exciting field, Stanford an important university, and himself a promising prospective faculty member. Months later, after he had gone to Amherst, Bill received an embarrassed reply from the then chairman of the Stanford Physics Department stating that the letter had become misplaced while going the rounds of the faculty but that anyway there was no opening. So much for that. Ten years later Bill, now at Duke, happened to meet Robert Hofstadter; Bill, without any thought of his own future, expatiated on the wonders of low temperatures, mentioning some of his own students as candidates for a faculty position should Stanford ever enter the field; Hofstadter was impressed. What Bill did not know was that the new physics department, transformed under Leonard Schiff, had already decided to embark on low temperature research, and a search was underway to fill assistant professor and full professor positions. William Little was appointed to the assistant professor position, and in 1958 Felix Bloch visited Duke and inquired of Fairbank's interest in a Stanford position. Bill has described his apprehension at the suggestion:

> I was very happy at Duke. I had a going laboratory. Even if my success fell off in the future, I had earned respect from my colleagues. My family was happy. The boys' school experiences were very good. We liked it at Duke. I was therefore apprehensive, knowing in my heart I would go to Stanford. It was a more challenging school. I liked the West Coast where I grew up. I would be entering an unknown situation and giving up a very good one at Duke.

Then there were the comments of a friend who said, "I hear you've got a chance to go to Stanford. You are 42 years old, and I know what happens to such people; I've made a study of it. You get an ultimate position that's very nice at a very good university and you don't do anything more for the rest of your life."

Thank God for blunt friends. If there is one thing Bill cannot resist it is a challenge, and the man who had built a ^{3}He-^{4}He separator at Amherst and rebutted Gorter's suggestion that experiments on Fermi degeneracy were too hard for American physicists, was not the man to lie prone before the threat of becoming a has-been at 42. Bill joined the Stanford faculty in September 1959. Within the next few months he had generated a flood of new ideas and programs. His own account of what happened is that breaking with Duke after he had established his reputation and starting over in a new place provided a golden opportunity to rethink his career. Manifestly that is true, but Stanford also contributed in ways that need to be understood for any right assessment of Bill Fairbank.

The work Bill has done, or inspired others to do, at Stanford falls within three categories. There have been experiments on superconductivity and liquid helium, which include the work with Bascom Deaver on quantized magnetic flux in superconductors, work on the Josephson effect, work on ^{3}He, and a range of different experiments on quantized vorticity in superfluid ^{4}He. All of these, though highly original, have been small scale experiments within the established tradition of low temperature physics. Then there have been the novel, but still small scale, investigations aimed not at low temperature phenomena themselves but at applying cryogenic techniques elsewhere, such as the experiments on the free fall of the electron, on the inverse square law of gravitation, on ultralow magnetic fields, on the ^{3}He gyroscope, on quarks, on blood flow and on biological molecules. Finally there have been "big physics" programs: the superconducting accelerator, the orbiting gyroscope experiment, the collaboration between Stanford and Louisiana State University on gravitational wave antennas, and various offshoots and extensions of these.

Specifics follow; meanwhile we must recognize that success in such experiments on whatever scale demands good students (whom Bill has found), while success in "big physics" demands both able individual colleagues and a certain corporate infrastructure, administrative and intellectual. Administrative infrastructures are widely, perhaps too widely, accepted in this bureaucratic age; not so common is the recognition that large ideas flourish only in the right intellectual soil. Yet the point is all-important. Bill Fairbank has often said that the first truth a graduate student must learn is that it is possible to do significant work, and that nine times out of ten the realization comes through being at a place where significant work is being done. A similar contagion spreads big physics. Having worked with

Martin Block on the liquid helium bubble chamber at Duke, Bill was ready for bigger things. Stanford supplied the setting.

It did so through the actions of two men, each only slightly known to Bill and one already ten years dead—Frederick Terman and W. W. Hansen. Terman, author of a classic text on radioscience and architect of Stanford's engineering fortunes, had formed the atmosphere of enterpreneurial thinking that was to extend beyond the lands of the Stanford campus in the phenomenon we now call Silicon Valley. When Bill came to Stanford the phenomenon, though not the name, was vigorously at work if only partly understood. That year one of Bill's colleagues in the physics department, Edward Ginzton, became chairman of Varian Associates. Silicon Valley has had its effect on Bill. More telling, however, has been the remote influence within the university of Bill Hansen, who, though he was only 40 when he died of a respiratory ailment in 1949, has had more effect on physics at Stanford than any other individual man.

Hansen, whose weekly lectures at the Rad Lab on microwave theory Bill and Jane attended regularly, invented the electron linear accelerator shortly after the end of the war. With Terman's support, he set up the Stanford Microwave Laboratory (ancestor of the present Ginzton and High Energy Physics Laboratories), and with a team of forty physicists and engineers started building the 160 foot long Mark III accelerator. He died before it was finished, but by 1951 it had begun its history of brilliant scientific success which culminated in Robert Hofstadter's Nobel-prize winning experiments on nuclear form factors, the first reliable measurements on the shapes of nuclei. The High Energy Physics Laboratory (HEPL), and the physics tradition connected with it, have been crucial to Bill.

Although the Mark III accelerator, extended to 300 feet, continued operating until 1971, plans for a much longer instrument were under way by 1956. Bill's arrival at Stanford came in the middle of the stormy discussions preceding the construction, from 1962 on, of the two mile long Stanford Linear Accelerator Center (SLAC). One consequence of the discussions was the severance of SLAC from the Stanford Physics Department and HEPL; another was to stimulate Bill to reexamine an idea he had had earlier, the possibility of operating a linear accelerator with resonant cavities formed from a superconductor rather than from copper at room temperature. The high Q's of superconducting cavities offered the hope of greatly increasing the duty cycle of the accelerator. With John Pierce as a graduate student and Perry Wilson of HEPL, Bill began research on accelerator cavities. Extended by Alan Schwettman and his associates from 1962 on, this work led to the construction of the superconducting accelerator within HEPL, and to the development of HEPL as, among other things, a center for large scale cryogenic research. The gyroscope experiment, started in the physics department, was transferred administratively to HEPL in 1968.

The cryogenic gravitational wave program, also developed within HEPL, was started in 1969.

The history of the gyroscope experiment, conceived jointly with Leonard Schiff of the physics department and Robert Cannon of the department of aeronautics and astronautics, is discussed in paper VI.3(A). It and other of Bill's activities, notably the blood flow and magnetic resonance imaging programs, both conceived jointly with members of the Stanford Medical School, illustrate a futher quality of his that has found a congenial home at Stanford: the ability to cooperate with members of other departments.

Some of the programs Bill has started since coming to Stanford have taken a long time, and occasionally one hears the suggestion that Bill is brilliant but never finishes anything. But the Fermi degeneracy and λ-point experiments at Duke and the quantized flux experiment at Stanford were finished with lightening speed. Truth lies elsewhere. Hard as the earlier experiments may have been, those like the quark, free fall and gyroscope experiments, in which cryogenic techniques are applied to other branches of physics, have involved challenges of a different order, sometimes including development of whole new technologies. My own interpretation of this phase of Bill's career is different. I like to think that when Bill at 42 pondered his career in response to that blunt friend, he chose, partly unconsciously, a number of lines of research that no one could follow without time, a secure position and the kind of reputation that allows one to take risks. His courage has earned its rewards, intellectual and other; the odd thing is that more than one of the programs, visionary at the beginning, has reached actuality just as Bill approaches his retirement.

5. UNIFYING PRINCIPLES AND THE STRUGGLE TO CREATE

In a famous critical essay on Tolstoy, published in 1953 under the charming title The Hedgehog and the Fox, Isaiah Berlin applied a quotation from the Greek poet Archilochus, "The fox knows many things but the hedgehog knows one big thing," to mark what he considered as "one of the deepest differences which divides writers or thinkers." On one side are those like Dante or Plato or Kafka, who relate everything to a single central vision; on the other are the Shakespeares, Aristotles and Goethes, who with the quickness of the fox "pursue many ends, often unrelated and even contradictory, related by no single or aesthetic principle; ... their thought is scattered or diffused, moving on many levels, seizing upon the essence of a vast variety of experiences and objects for what they are in themselves, without, consciously or unconsciously, seeking to fit them into, or exclude them from, any one unchanging, all-embracing ... unitary inner vision."

To Berlin most writers are easily categorized, but not Tolstoy. His genius for analyzing people and society seems to have been that of a mercilessly sharp-eyed fox, yet his lifelong quest, never truly fulfilled, was to find a coherent world view. This inner tension set Tolstoy apart from most other writers, Russian or European; in Berlin's view it was the key to an understanding of his special creative power.

The game of hedgehogs and foxes, of mentally placing colleagues on one side or other of the Berlin wall, is one that physicists, too, may enjoy. Often the choice is obvious; but Bill Fairbank, like Tolstoy though for a different reason, resists categorization. The very title of this book, *Near Zero: New Frontiers of Physics*, articulates the dilemma. On first glance "Near Zero" seems to express what Leonard Schiff once called "Fairbank's principle," the idea that any experiment which can be done can be done better at low temperature. Thus read, it would be a typical hedgehog's unifying principle; and when in 1965 Bill used the almost identical form of words "Near Zero: A New Frontier of Physics" as the title for his Richtmyer lecture to the American Physical Society, such may have been his thought. But over the years Bill Fairbank's vision, never narrow, has widened. I well remember his saying in a discussion about the gyroscope experiment after that lecture, that if ever we were to see a better approach to the experiment not involving low temperatures we should at once jettison our preconceptions and adopt it. This may seem ordinary common sense, but few strong-idea'd people have the strength seriously to reconsider a plan of action to which they have committed themselves.

Going further, the reader of this volume will discover that "Near Zero," as now interpreted, unites two distinct meanings. One is the usefulness of cryogenic techniques. The other is the separate and more elastic notion that new types of physics experiments become possible if several variables in the environment—pressure, magnetic field, electric field, gravitational acceleration, thermal noise, creep, and coeficients of expansion of materials, for example—are brought almost to zero. The two near zero principles overlap in that low temperatures aid in obtaining low pressure, low noise, low coefficients of expansion, and even low electric field; but not all variables are reduced by cryogenic operation (gravity is not), nor is this its only advantage. The low temperature world, as Fritz London saw and as Bill Fairbank has often emphasized, is a realm of long range order in momentum space. Quantum phenomena occur on a macroscopic scale. This fact provides the basis for many inventions, from Zimmerman and Silver's SQUID (Superconducting QUantum Interference Device) magnetometer to the idea due to Bill, Peter Selzer and myself of using a porous plug to control the operation of a helium dewar in space. Bill's genius has been to draw together almost the whole range of meanings in "near zero." Should one seek a low temperature physicist who was a hedgehog, one must pick not him but Walther Nernst, who did weave a whole career around one idea, the "third law of thermodynamics," the perception that the entropies of all materials asymptotically approach zero as the temperature approaches zero.

So much for "near zero." But notice, too, the evolution in Bill's thinking that has made us choose for a title not "a frontier of physics" but "new frontiers of physics." Frontier singular means a frontier of technique, applied to diverse ends; frontiers plural emphasizes the variety of territory opened up by the near zero principle (in its two senses) ... quantum phenomena, gravitation, surface physics, bio- and medical physics, superconducting accelerator technology, even new approaches to particle physics in the searches for quarks and monopoles.

A wider unifying principle in Bill's career is this: he is interested in physics. So, surely, is every physicist; but not in Bill's sense. Too often the contemporary academic as his career moves forward focuses ever more narrowly on some small region of thought, abandoning like Candide the effort to comprehend the world in favor of cultivating his garden. The pressures, internal and external, to retreat from the fray are enormous. The propaganda that ours is an age of specialization, that physicists produce their best work in their twenties, that progress in science is so rapid that no one can hope to keep up except on the narrowest of fronts, is intimidating enough. Add to it the formidable barrier, if one is an experimentalist, that research costs money, that money comes from funding agencies, that agencies reach decisions by peer review, which often means review by entrenched experts with little kindness for naive outsiders, and one may indeed wonder how even a man as gifted as Bill can contribute effectively over so many fronts.

It cannot be done alone. One needs the intellectual infrastructure. But before anatomizing Bill's success in collaboration we should pause to consider what physics the enterprise, as practised by an experimentalist on the West Coast of the United States in the latter half of the 20th century, truly is. This physics is not an abstract mental exercise. It is a strange amalgam, or rather an agglomerate, of great ideas, theoretical red herrings, history, mathematics, instrumentation, inventiveness, computers, departmental meetings, philosophical speculation, government property, teaching, technological applications, proposal writing, discussions with students, overhead rates, electronics design, cross-country travel, planning of buildings, preparation of viewgraphs, servo loops, salary structures, site visits, letters of recommendation, long distance telephone calls, reprints, preprints, international conferences, vacuum leaks, rumors about new results from MIT, explanations to deans about the nonarrival of funds, recalcitrant apparatus, false starts, sudden success. This hectic collage is not to be set against some Botticelli-like vision of the grace of physicists' lives in former ages. Galileo may not have been concerned with explanations to deans, he was with explanations to cardinals. Newton never became alarmed about competition from MIT; he did about competition from Hooke. The point is the actuality within which a contemporary interest in physics must flower.

To function well in this world one needs robustness of health and character, and a certain efficiency. Now efficiency is no obvious thing. Onlookers might expect the efficient physicist to be one whose desk is always tidy and who disposes of bureaucratic paperwork in a flash, for that person will have the most time free for research. Such paragons do exist; Leonard Schiff was an example; Bill Fairbank is not among them. Even Bill's greatest admirers would not claim for him a tidy desk; while I, who have more than once spent as many as five hours helping him write a single letter of recommendation, can testify that his handling of paperwork is not always speedy. Nevertheless, in his own way Bill is notably efficient. It is instructive to learn how.

First, in the interconnectedness of things it is a mistake sharply to separate the activities in which one is engaged into the two categories physics and bureaucracy. A good letter of recommendation, to take the example just given, is more than bureaucracy, and in Bill's hands, besides being an act of friendship, it can become a document of real scientific insight. In one recent instance I found that by the time Bill had completed the letter not only had he persuasively argued the claims of a colleague, but we had digressed fruitfully on three other topics and Bill had succeeded in teaching me the basics of a region of physics about which previously I had known nothing. Behind the seemingly disproportionate investment of energy lies a special kind of efficiency, the efficiency that kills several birds with the same stone. Or take budgets. To spend, as we sometimes did in the early days of the gyroscope experiment, the better part of two days going over an annual budget submission that might have been completed in twenty minutes seems an odd proceeding. Within the modest funding available, the only variables were quantities of helium, the question of whether two or three man-months of machine shop time should be included, and which three of a dozen necessary pieces of capital equipment should be purchased, all of them issues on which our decisions might change. Yet these agonizings produced some of our most concentrated thinking and often led to a far more coherent plan of research. The method is not one to recommend to other people, but it is Bill's and it works.

I emphasize the apparent wastefulness of those procedures because, although they occur in matters away from the center of Bill Fairbank's interests, they give a clue to his creativity. Efficiency in creation is subtle. The relevant time scale is not days or weeks but years or even decades. Bill once remarked to me that a man who has one new idea a year will in a career of fifty years have had fifty ideas, and that is an enormous output. A superficial observer of Bill's development up to the time of his leaving Amherst at the age of thirty-five might have questioned where all this was tending. Walter Gordy's judgment was not superficial. Bill's next decade was (as we now know) to be wonderfully productive, and it laid the foundations for the even deeper work that followed. Circumstances affected Bill's career growth, but the delay must be seen in part as establishing his wide outlook on physics and in part as drawing our attention to complexities in the creative process itself.

The account of mathematical creativity popularized eighty years ago by Henri Poincaré, and extended later by Jacques Hadamard, is this: The mathematician struggles for months to comprehend a difficult topic; he gets stuck; he gives up; then while he is away on vacation or at work in a remote field, sudden insight waves its magic wand and hey presto! out of air comes the answer. Another splendid chapter has been added to the story of the forward march of intellect. The picture is a romantic one. It humanizes the mathematician and harmonizes with our latent feeling that if only our unconscious would buckle down to its job we, too, would be as brilliant as Poincaré. It almost makes us want to take a vacation. But though the account has truth in it, it is not the whole truth. It does not say how long we should keep battering our heads against the wall before taking the vacation. Nor does its portrayal of the sovereign idea as springing forth full-grown and fully armed, like Athena from the head of Zeus, represent fact save in the rarest of instances. Missing from the picture are two fundamental "inefficiencies" inherent in much creative thinking, including Bill Fairbank's: repetition and roundaboutness.

Anyone who has discussed physics with Bill has been struck by his habit of repeating the same point or the same calculation over and over again. The practice is the more surprising given Bill's vigor and elasticity of mind, and some people find it exhausting. A waste of time, one thinks, except that among Newton's manuscripts are examples, seemingly equally futile, of a passage being written out almost without change ten or a dozen times. Perhaps the repetitions are a mechanism for concentrating attention; if so, the process ties in with Poincaré's description of the first phase of mathematical creativity, though with Bill Fairbank the image one forms is more one of strenuous hammering than of scholarly exploration. But there is a further point. Stick the discussion out with Bill and you will notice that eventually, possibly at the ninth time around, something new and valuable has been added. The hammering has done more than serve as a means of concentrating attention; a mental barrier has been smashed down. One is led to suspect a gap in Poincaré's account of creativity. True the final illumination may come in a moment of relaxation, but the dawning of a great idea probably depends on a precursion of smaller ideas won during the long initial struggle. In that struggle repetition seems inescapable.

Few physicists have been more efficient than Leonard Schiff. Schiff in particular is a man one would expect to reason in a direct, unwasteful way, straight to the point. For the orbiting gyroscope experiment, Schiff's most original scientific idea, we happen to possess a set of notes from which, as I show in paper VI.3(A), the path of his invention can be traced. Nothing could be more roundabout. Again Poincaré oversimplified. Whatever may be in mathematics, in physics almost never do creative ideas blossom abruptly. Think of the twelve years' groping between the Bohr atom and modern quantum mechanics. Or take Maxwell's greatest idea, the displacement current. Most textbooks credit Maxwell with a pretty· but straightforward piece of detective work, allegedly uncovering a discrepancy between two of the fundamental equations of electromagnetism. In truth the first recognition came circuitously from an awkward mechanical model of the ether, soon abandoned; and eight years, three papers and a fair part of the writing of the *Treatise on Electricity and Magnetism* were to go by before Maxwell, in a letter to William Thomson (Lord Kelvin), formulated his ultimate hypothesis that all electric currents, even those in apparently open circuits, are in reality closed.

Bill Fairbank, too, has emphasized how circuitous the routes to his ideas sometimes are. When he and Bascom Deaver wanted a method for detecting the quantization of magnetic flux in a superconductor, their first thought was to heat and cool a superconducting ring through its transition temperature and look for the flux jumps. That failed because the energy change for a single.loop would be much less than the kT noise at the transition temperature. The next thought was to have not one ring but a thousand, all making the transition together. Then it dawned on Bill and Bascom that a thousand rings in parallel equate to a cylinder. Finally another idea emerged. Instead of rapidly heating and cooling the superconducting cylinder through its transition temperature, why not mount it in an oscillating sample magnetometer? The approach, once defined, is so obvious that one is mystified by all the intervening steps. Equally roundabout have been some of the paths in the gyroscope and free fall experiments. As Bill has put it, "You have this idea for slowing down positrons; then you learn about something you supposedly knew called Liouville's theorem; that messes up everything, so you start over."

If I am right in describing intellectual creation, at least in some modes, as an innately wasteful process, then plainly one must be singularly efficient in what one chooses to be inefficient about. Good judgment on research topics is of the essence. Scientific judgments, however, are not formed in isolation. Physics itself is a process. It involves people, it involves things, it involves ideas, all changing with time. Bill Fairbank as a creative physicist is a man of ideas. If we are to understand his creativity, we need to ponder the complex intellectual milieu of present-day physicists, and (almost as elusive) the complex web of relationships in physics between things and ideas.

6. EXPERIMENTAL IDEAS, LARGE AND SMALL

The line of Bill Fairbank's career has been drawn through science to coincide with the rise of "big physics." The earliest particle accelerators were constructed in the 1930's while Bill was in high school, but they were only hints of what was to come. Not until after World War II, and partly as a result of it, did physicists demand and governments provide, first in high energy physics and then elsewhere, the kind of massive research funding that is the envy and astonishment of other scholars (though other scholars forget that the libraries in which they work also cost enormous sums of money to build and operate). The existence of big physics poses a dilemma. Without huge machines and research teams some vital investigations would be impossible; with them often comes a stifling of individual creative power. Bill, as we have seen, avoided big physics before coming to Stanford; since then his energies have been divided between small and large programs, but even in the large programs he has tended to resist too rapid an expansion of the structures of research. Other physicists avoid big physics altogether.

Even what one avoids affects one, however. Big physics has had many unforeseen consequences, among which one of the more curious has been the socioeconomic fact that because a training in theory costs less than a training in experiment, the huge increases in funding, though driven by the cost of experimentation, have resulted in a vast overproduction of theoretical physicists. This little-recognized truth, in conjunction with the growing difficulty of experimental work, has transformed the sociology of research for devotees of big physics and little physics alike. On one side experimentalists find it harder and harder to stay abreast of theoretical detail except in some narrowly bounded field. The pressure toward specialization, already great, is intensified. On another side is a phenomenon best described in terms of time constants and signal-to-noise ratios. Regrettably, not all of the theoretical outpourings are on a par with Einstein's 1905 paper on the photon. Genuine conceptual signals lie deeply buried in noise. As for time constants, with the time for executing an experiment being generally much longer than that for writing a theory paper, the experimenter may see theoretical opinion reverse itself two or three times over before his work is half done. Small wonder if he, like some rabbit on the freeway, tends to suffer "cognitive dissonance."

For some people the connection in science between things and ideas is simple. Ideas equate to theories, and experiments to things. From there it is but a step to the opinion expressed not long since by a physicist of distinction whose interests overlapped Bill Fairbank's and whose contributions were principally but not exclusively in the theoretical domain, that to have more than one theorist in a physics department is as foolish as having more than one rooster in a hen-run. The metaphor, though reassuring to some and congruent with my remarks about the current surplus of theoretical physicists, fails unless one is prepared to adopt, in opposition to its author, a radically feminist interpretation of the creative process. Later I will seek a truer representation of the relationship between theory and experiment; meanwhile we need to recognize the existence of certain offspring of cerebration best called *experimental ideas*.

A master of experimental ideas was C. V. Boys (1855-1944), inventor of, among other things, the fused quartz fiber for torsion suspensions and the earliest ultra high speed camera, and author in 1887 of the first critical analysis of the effects of the size of an apparatus on its performance. In this last, done for the purpose of carrying out an improved Cavendish experiment, Boys may be regarded as the originator of the "near zero" principle, and as a seer who, before big physics existed, proved once and for all that not all good physics would be big physics.

The Cavendish experiment measures the tiny gravitational attractions between two fixed masses and two masses suspended from a torsion balance. Success hinges on the elimination of disturbing forces, the worst of which, in Cavendish's apparatus of 1798 with its six-foot long balance arm, had come from air convection in the balance chamber occasioned by the long thermal time constant of the system. Boys found from dimensional arguments that convective effects scale with the size of the apparatus at somewhere between the fifth and seventh power of the lineal dimension L. Making the apparatus small would eliminate them. More surprising was the scaling law for the effect to be measured. What counts here (within certain limits) is not the magnitude of the force but the magnitude of the angular deflection it produces. In a system with fixed torsional period and source masses of a given size, the deflection from the gravitational forces *increases* in inverse proportion to L. For two reasons, therefore, smaller is better. Boys chose to work with a balance arm only 5/8 inch long. This decision, so contrary to most people's intuition, illustrates how experimental ideas may involve theory, though their objective is not theory *qua* theory but improved instrumentation. Of course, new intruments may open new theoretical arenas, as did the high speed camera when Boys applied it to the study of lightning and discovered that certain kinds of lightning stroke are initiated from the ground rather than the clouds.

Bill Fairbank also is a master of experimental ideas. Two examples, one small scale, one large scale, are the calorimeter for the λ-point experiment and the use of high-Q superconducting cavities in accelerator physics. In the λ-point experiment the goal Bill and his colleagues set themselves in response to Feynman's bet was to measure the specific heat of liquid helium, above and below the superfluid transition, with a temperature resolution of 10^{-6} K, three orders of magnitude beyond earlier work. There were two issues. First was thermometry. This proved surprisingly easy; an ordinary carbon resistance thermometer, properly ins'rumented, was good enough. Second was the need to devise a calorimeter in which the temperature of the liquid would stay uniform to 10^{-6} K, even in measurements above the λ-point where helium, being in the normal state, has poor thermal conductivity. Here Bill's perception matched that of Boys. He recognized, as no one before him had, that low temperature operation offers a fresh vision of calorimetry, at least as applied to liquid helium.

Before the work of Buckingham, Fairbank and Kellers, the typical calorimeter used in measuring the specific heat of liquid helium was a thin-walled copper can, suspended in an evacuated cavity, with heater and thermometer so disposed that temperature differences as large as a millikelvin might exist between different parts of the liquid. Such designs followed a pattern set by the calorimeters used in elementary physics laboratories for measuring specific and latent heats of water and other liquids near room temperature, where in a tradition going back to the 18th century the walls of the containing vessel were kept thin to make the correction for its heat capacity small. With liquid helium, which has a specific heat per unit volume two orders of magnitude less than that of water, one might suppose that the walls would have to be even thinner. At low temperatures, however, two other things change. The specific heat of copper falls off as T^3 in accordance with Debye's law, becoming even less than that of helium, while if the copper is sufficiently pure its thermal conductivity is actually higher at 2 K than at room temperature. Connecting these facts, Bill and his colleagues devised a sample chamber of small volume (about 1 cc) within which were a series of closely spaced copper fins such that no portion of the liquid was more than 0.5 mm from a high thermal conductivity wall. The chamber was filled with liquid while cold, permanently sealed off, and suspended in the dewar inside an evacuated can containing a heater, a thermometer and a mechanically operated heat switch. The high thermal conductivity of copper and small spacing between the fins allowed the temperature of the liquid to be gradually raised and lowered while staying uniform throughout the sample. Withal, despite the large mass of the sample chamber, its heat capacity was less than 10% of that of the helium, and remained sensibly constant over the narrow range of temperature under investigation. The subtlety in the design lay in its lack of subtlety. The problem of temperature equilibrium had been overwhelmed by intelligent brute force.

Though Bill and his colleagues devised the scheme of their λ-point apparatus from first principles, their work gains in power and perspective through being seen in the context of two centuries of invention spanning two realms of science, thermodynamics and statistical mechanics. Boys' paper was the pivotal one; behind it lay a series of investigations on scaling, including Kelvin's development of an improved marine compass (1881), William Froude's study of the scaling laws for ships' models in towing tanks (1870 on), and James Watt's examination in 1757 of the reasons for the unsatisfactory performance of a one-eighth size model of a Newcomen atmospheric steam engine. The theme connects with Daniel DeBra's remarks in paper VI.3(G) on the "near zero" principle in engineering. For Kelvin, as for Boys, the key to an improved instrument was a reduction in size or, more precisely, in the ratio of the compass' moment of inertia to its effective magnetic moment. Watt's investigation pointed in the opposite direction. The Newcomen engine worked by injecting a fine spray of cold water into the steam-filled cylinder to create a vacuum on each stroke. In the model engine, as contrasted with the full scale one, thermal losses stopped the operation after a few strokes because the time for heating and cooling the cylinder wall was short in comparison with the period of the stroke. When the need is not for equilibrium of temperature but for disequilibrium, larger is better. From there Watt went on to the idea of lagging the cylinder wall to keep it hot, and having a separate permanently cooled condenser for the steam. Thus was started the line to Carnot's theory of the ideal heat engine (1824) and thence to thermodynamics. Seldom has an experimental idea borne such fruit in scientific theory.

If Watt's finding proves that small is not always beautiful, so for different reasons does the more recent history of instrumentation, including even later progress on the λ-point experiment. Boys in the end concluded that he had gone too far in making the balance arm of his apparatus only 5/8 inch long. In practice twice as large would have been better. Had Boys reduced the dimensions still further he would have found, as did W. J. H. Moll and H. C. Burger thirty years later (1925) in another instrument, a continuous angular jitter of the suspended body, limiting the resolution of measurement. The cause, as G. A. Ising soon proved, was Brownian motion. So the year in which Heisenberg formulated the uncertainty principle was the year for establishing, a century after Robert Brown's studies of pollen on water, this classical but fundamental limit on instrument performance. By combining it with scaling considerations similar to those of Boys, H. A. Daynes and A. V. Hill in separate papers published in 1926 were able to fix optimum sizes and observation times for practical instruments. A later masterful investigation was P. M. S. Blackett's optimization (1950) of the design of the astatic magnetometer, which yielded an instrument capable of resolving 3×10^{-10} G in 30 sec, an advance of three orders of magnitude over anything preceding it. Blackett's remained the most precise magnetometer available until, in the mid 1960's, quantized magnetic flux and the Josephson effect led to the SQUID.

Examples from Bill's interest where molecular statistics influence the sizing of apparatus include bar detectors for gravitational radiation (paper VI.5) and John Lipa's new λ-point experiment (paper II.9). In the former, large size increases the instrument's radiation cross section, and so increases the signal energy by comparison with the kT (Brownian) noise in the bar, while low temperature reduces the noise. In the latter, with the goal of attaining a temperature resolution of 10^{-10} K rather than the 10^{-6} K of Buckingham, Fairbank and Kellers, the argument takes a surprising turn. The issues, as before, are thermometry and calorimeter design, but the root need at this level is for a better thermometer, which Lipa has provided by applying a SQUID to measure the temperature dependent magnetic susceptibility of a paramagnetic salt. In calorimeter design the task is to remove limitations on the experiment from gravity and from statistical fluctuations in the helium, and since fluctuations depend inversely on the free volume of helium there is a lower limit of about 1 cm on the spacing between the walls of the sample chamber. Intelligent brute force, so successful before, no longer works; instead, a cylindrical chamber much like the traditional calorimeter's is subjected to a system of temperature control of the utmost delicacy, with five nested stages, active or passive, between the chamber and the surrounding helium bath. The limitation from gravity is removed by operating the experiment on Shuttle. The total rethinking of a successful design illustrates the mental flexibility that is needed in carrying forward an experimental idea, a flexibility which Bill Fairbank outstandingly has, and encourages in others.

If the λ-point experiment is in two senses "little physics," the superconducting accelerator is preeminently "big physics." The idea, already described, is to exploit the low thermal losses in superconducting cavities to operate an electron linear accelerator in a continuous current mode rather than with the 0.1% duty cycle typical of SLAC. Continuous operation increases the signal strength; it also opens other prospects such as recirculating the beam to increase the energy of the electrons provided by an accelerator of given length. Success hinges, as in all big physics, on interconnections between technique and economics. In conventional linear accelerators the duty cycle is limited by the enormous thermal losses due to the low Q (2×10^4) of the copper cavities in the resonant structure. Continuous operation is ruled out on three grounds: the cavities would melt, the klystrons driving them could not deliver the requisite amount of power, and (not least) the total power for running the accelerator would

be monstrous. For SLAC, which now consumes 1 MW in normal operations, the requirement would be 1000 MW, more than the full output of
the Grand Coulee dam.

One consideration in the design of a superconducting accelerator is the
Carnot efficiency of the refrigerator which extracts heat from the resonant
structure. A typical figure for a Collins machine is about 0.1%, which
means that 1000 W must be expended to deliver 1 W of cooling power at
low temperature. To operate the accelerator with 100% rather than 0.1%
duty cycle means another factor of 1000. Hence if the machine is to run at
about the same power level as a comparable room temperature accelerator, the superconducting cavities should have Q's roughly 10^6 times higher
than those for copper cavities, that is, around 2×10^{10}. This calculation
critically affects the design. The lead cavities developed in 1964 had Q's of
4.5×10^8 at 4.2 K, rising nearly to 10^{10} at 1.8 K. Since cooling the accelerator structure from 4.2 K to 1.8 K only increases the refrigeration demand
by a factor of two, the net gain from cooling is to reduce the total power
demand by a factor of ten, a point of great economic significance. Also
the helium at 1.8 K is superfluid, which greatly simplifies the transport
of heat away from the accelerator structure. Everything pointed therefore
to superfluid operation, yet in the early 1960's distinct boldness of vision
was needed to conceive of an accelerator hundreds of feet long all at 1.8 K
linked vertically by an 80 foot line to a refrigerator continuously generating
superfluid helium. That boldness Bill Fairbank and Alan Schwettman had.
One of the grounds for it was, as I have already remarked, Bill's experience
in building his own refrigerator at Amherst College.

Sustained work by Michael McAshan, Todd Smith, John Turneaure and
others finally produced by 1978 a superconducting accelerator with five
20 foot sections, all superfluid, delivering average beam currents up to
450 μA and an energy after three stages of recirculation of 230 MeV. This
machine, after evaluation in nuclear physics experiments, is now being developed as a free electron laser by Schwettman, Smith and their colleagues.
The proposed U.S. National Continuous Electron Beam Accelerator Facility (CEBAF) in Virginia for intermediate energy physics is planned as a
superconducting accelerator.

7. THE EXPERIMENTALIST'S ENCOUNTER
WITH THEORY:
I. BRIDGING THE CHASM

The thought-world of a creative experimental physicist is not limited to
advances in instrument design. Within the realm of experimental ideas
come the uses, sometimes very unexpected, to which instruments are put.
The cloud chamber, invented by C. T. R. Wilson to study conditions of

cloud formation, and then applied by him and others to observe the interactions of high energy particles, offers a prime example. Another is Eötvös'
application of torsional gravity gradiometers to test the equivalence of gravitational and inertial mass. By replacing one of the two masses in his apparatus successively with others of different composition, and arranging to
turn the torsion head through 180° in the combined centrifugal and gravitational field of the rotating earth, Eötvös created a test of equivalence nearly
four orders of magnitude more exact than earlier pendulum experiments,
supplying Einstein thereby with one essential clue to the general theory
of relativity. Eötvös' work began in 1890 contemporaneously with that of
Boys on the Cavendish experiment. With equipment so similar, a comparison of the two men's achievements is illuminating. As an instrumentalist
Boys was far superior. His paper of 1895 is a model of critical thinking,
whereas Eötvös' technique, despite the praise that has been lavished on
it, had many deficiencies. Nevertheless, most reflective physicists would
judge Eötvös' leap of the imagination, so unexpected and so influential, to
surpass all Boys' ingenuity. Comparisons are odious, and anyone who in
his eightieth year can write, as did Boys, a witty and absorbing article on
soldering stands outside ordinary categories of praise and blame. That a
man of his insight should have missed such a pearl deepens one's respect
for Eötvös' originality.

Sometimes the experimentalist's ties to theory are far closer. One thinks
of Rutherford who, from his experimental idea of looking for high angle
α-particle scattering from thin gold plates, went on, after the work of Geiger
and Marsden, to formulate the theoretical idea of the atomic nucleus. Or
one thinks of Faraday whose method, in Maxwell's words, consisted in
a "constant appeal to experiment as a means of testing the truth of his
ideas, and a constant cultivation of ideas under the direct influence of
experiment." These are extreme cases. Bill Fairbank has lived in a different
era, where different methods apply, but critical examination will show that
even when, as at Amherst, his work has had a large engineering content,
Bill has always hitched his experimental wagon to one or more theoretical
stars.

Part of Bill's strength has been his gift for conversing with theoretical physicists. The seven who have contributed to this volume (John
Bardeen, Felix Bloch, Michael Buckingham, Remo Ruffini, Kip Thorne,
Dirk Walecka and Frank Yang) make the point; but beyond these Bill at
different phases in his career has had unusually fruitful relationships with
three great men, all now gone from us: Lars Onsager, Fritz London and
Leonard Schiff. Through seeing the uncommon qualities these three had in
common one can learn something about Bill. Each had an exceptionally
wide view of physics. Each was mature and had reached maturity before
the onset of the post-World War II overproduction of theoretical physicists.

Each, while competent in many fields, stood a little apart from current theoretical fads and fashions, except insofar as he originated themes that were to become fashionable later on. These were good mentors. For Bill, the ex-chemistry major, it is gratifying to note that two of them, London and Onsager, though usually regarded as physicists, held appointments in chemistry departments.

Of the three, Onsager's influence, extending over fifteen years from second sound to quantized flux, meant most. To hear from the oracle (as Bill did during a walk in Cambridge, England, in 1960) a disquisition on flux quantization in which Onsager remarked that when London advanced the hypothesis, he (Onsager) had at first concluded that it could be true only if *all* magnetic flux were quantized in the London units of hc/e, but that now he had changed his mind, quantization was probably restricted to superconductors and probably in units of $hc/2e$ ... to hear such brooding pronouncements from a master physicist is a vitalizing experience. The power of Onsager's thinking can only be felt now by recalling the deeply-forgotten skepticism among theorists at the time about the whole idea of flux quantization. The lines from Onsager to ^{3}He, to phase transitions, and *via* the Ising model to the logarithmic discontinuity in the specific heat of liquid helium near the λ-point, were also strong. Especially potent, as it seems to me, were Onsager's early discussions with Lane, Henry and Bill on second sound. By showing Bill that a good theorist, besides posing theoretical questions, may point to experimental solutions (may produce, in other words, "experimental ideas"), Onsager subtly imparted the inverse lesson that an experimentalist might dare to penetrate the arcana of theory.

The influence of London, cut short by his death, was strongest not at Duke but later, during Bill's first five years at Stanford. Obvious in the experiments on quantized flux, quantized rotation in helium, and the London moment, it had also a less expected dimension. William Little, taking the words of Job and Isaiah, has pictured Bill Fairbank as bursting on the Stanford scene like a whirlwind from the south, speaking fierce visionary language about long range order in momentum space. The image is a pleasing one, and I myself can recall some of the prophetic utterances wherein Bill, following London, invoked long range order as the basis for explaining not only superconductivity but the stability of living systems. Little's paper (paper III.9) describes his own brilliant development of London's vision in the idea of the organic superconductor, with its offshoots. For Bill Fairbank, biophysical interests issued in the work of Tad Day and John Philo on the magnetic susceptibility of biological molecules; then subsequently in the still-continuing work in medical physics. The prophetic word works in strange ways but does not lie idle.

Of Leonard Schiff, whose instincts were more rabbinic than prophetic, I can speak personally, having myself known him for eight years. Here the exchanges were more evenly balanced, with Bill influencing Leonard as well as Leonard Bill. Intellectually Schiff's most memorable quality was his odd combination of rigor and flexibility. He would work out a result with extreme speed and assurance, hold to it tenaciously, but if ever the smallest flaw in assumption were exposed he would instantly scrap the entire structure and start over. Robert Cannon tells of Leonard producing a large sheet of calculations on a difficult fluid dynamics problem which Cannon knew too well from having worked on it himself for months only to get the wrong answer. Leonard had reached the same point in a few days. The moment Cannon told him, without having said why, Leonard dropped the whole sheet in the wastebasket and asked for an account of the true theory. Schiff's work connected with Bill's directly in the free fall and gyroscope experiments; beyond there he interested himself in everything Bill did, and had, I fancy, a clearer eyed understanding than anyone of Bill's extraordinary powers. To have such a friend, trustworthy, self-giving, self-controlled, in command of his subject, quick but never in a hurry, masterly in administration, and with an honesty so direct as to be at times quite disconcerting, is among the rarest of human experiences. Few reactions are as telling as Bill's in continuing for several years after Schiff's death often to speak of him in the present tense.

Yet whatever Bill Fairbank has gained from discussions with individual theorists he remains his own man with his own outlook on physics. How does an experimentalist intelligently address issues in contemporary physical theory, with its elaborations and abstractions? The answer comes through seeing theory as a multileveled structure with connections into the experimental soil at every level.

Some years ago, in an article on theory and experiment in physics,[2] Ian Hacking and I analyzed from this view one of the most famous of physics experiments, Faraday's discovery of magnetooptical rotation. We identified six levels of theorizing, beginning with Faraday's initial speculation (1839) that there must be a connection between magnetism and light, and ascending to the electron theory of the Faraday and Zeeman effects advanced by Lorentz in 1895. Quantum mechanics added a seventh level. Faraday, who knew no algebra, could never have followed the details of Lorentz's theory, let alone quantum mechanics, yet he had good theoretical grounds for what he did. His apparatus comprised a block of extra-dense lead glass placed between the poles of an electromagnet, through which holes had been drilled to allow a beam of polarized light to travel parallel to the lines of magnetic force. The plane of polarization of the transmitted beam

---

[2] "Theory or Experiment, Which Comes First?" See also Ian Hacking, *Representing and Intervening* (Cambridge University Press, Cambridge, England, 1983).

underwent a rotation proportional to the optical path length and the intensity of the magnetic field. How, one may ask, did Faraday find his way from that first vague speculation to an experimental setup so specific, nonobvious and difficult to execute? The path was long, with many side turnings, but there were two crucial points: first, analogy with David Brewster's earlier discovery (1818) of induced birefringence in mechanically strained glass (which suggested that if magnetism does influence light, it might do so *via* a polarization effect); second, an understanding of the symmetries of the magnetic field (which suggested that lines of force might serve as axes of rotation). This, without a line of mathematical symbolism, was theoretical reasoning of a higher order.

Bill, whose grip on mathematical technique though superior to Faraday's is (as he would be the first to admit) in no way exceptional, is admirably deft at crossing the various bridges from experiment to theory. A personal recollection illustrates one method, the most elementary. Shortly after my arrival at Stanford, an idea occurred to me while reading Schiff's paper on the gyroscope experiment, which, if it had been correct, would have greatly simplified the experiment. Schiff's paper in hand, I went to Bill and said, "Look, I don't entirely understand where this equation comes from, but if you look at it, thus and such seems to be the case." Bill waxed enthusiastic, said that he hoped I was right (I was not), and then added, "Still, even if you aren't, you have the right idea; don't be afraid of looking at the equations just because you can't follow the derivations." Simple as this remark is, it points to the least appreciated of Bill's qualities, his realism.

Yes, realism. People who have met Bill only casually, and have heard him talking wild visionary schemes, or promising results "next week" which do not materialize for years, will smile at the word; but on the fundamental issue of whether an experiment can be done Bill is deeply realistic. Very few of the experiments he has started, however farfetched, have proved infeasible. The secret lies in a process I call *matching the numbers*, a process which requires one to master the art of extracting from theory the relevant numerical predictions, and of extracting from the confused mess of experimental issues the one or two that truly count, bringing these together in an effective scientific judgment. It sounds obvious. Most experimentalists estimate the sizes of effects they wish to measure. When, however, one leaves the realm of what Thomas Kuhn might call "normal" experimentation for wider adventure, great resourcefulness and flexibility of mind are required. Some new experimental idea is wanted, but the idea cannot be formed in a vacuum. The bridge (to revert to our metaphor) has to be erected from the theoretical and experimental ends simultaneously with only the sketchiest of surveys and design drawings. One only need examine some of the peculiar suggestions for experiments made by good physicists in the last twenty years to see how tricky the process is.

Examples of number-matching in Bill Fairbank's work are many. The λ-point and quantized flux experiments, discussed earlier, illustrate it, as does the course taken by Bill, along with Bill Hamilton, in 1968 when they began pursuing the detection of gravitational waves. Joseph Weber, as is well known, had pioneered this field much earlier at a time when to most physicists the venture, except for its technological interest, seemed quixotic at best, especially since one of the ideas was to detect radiation generated from a man-made source. In 1964 Hamilton began discussions with Bill about starting a similar program at Stanford, but Bill resisted. Then in 1967 Remo Ruffini at Princeton made a calculation in which the slowing down of the Crab pulsar was attributed to the loss of energy from gravitational radiation. Given thus a source of known power, Bill became interested. He and Hamilton calculated that a Weber bar, 150 feet long if aluminum, 80 feet long if lead, cooled to 1 mK, would have a cross section large enough and kT noise small enough for detecting the signal; and that a displacement monitor based on the high-Q cavities from the superconducting accelerator should be capable of resolving the motions of the bar. All these discussions took place (as I vividly remember) after the prodding from Hamilton but before Weber's sensational announcement of the possible existence of a source of gravitational radiation near the center of our galaxy. The stampede to prove Weber right or wrong had little effect on either of the Bills. Their inspiration, after Weber's basic work, was twofold: a bridge had been thrown between theory and experiment, and HEPL, with its unique resources for large scale cryogenics, seemed the ideal location for such an activity. The bridge was a rickety one, for the theory was wrong, the proposed apparatus too grandiose, and the displacement monitor not what would eventually be used; but it was a start. For Bill Fairbank there was also a start in the comprehension of the physics and astrophysics of gravitational wave sources.

If matching the numbers is one means of building bridges between theory and experiment, another is the more searching process which may by analogy be called *matching the ideas*. Too often theorists can make calculations and experimentalists can make measurements, but a chasm exists between the two activities. Finding points of theory that give meaning to experiment or phenomena to measure that bear on theory is a high art, impossible to codify, with few successful practitioners.

A master of the art was Heinz London. When one critically examines the contributions of the two London brothers to the understanding of superconductivity, it is hard to know which more to admire. After they had collaborated in producing the famous phenomenological equations of superconductivity in 1935, Fritz London elucidated the equations' deeper significance through his amazing conceptual discovery that whereas the flow of electrons in a superconducting wire is confined to a thin surface layer, their canonical momentum is (according to the equations) uniform throughout the wire. Thus were born the concept of long range order in momentum space and Fritz's inspired guess about macroscopic quantization. But it was Heinz who started construction work on the bridges that were to lead to the next stage of theorizing. One was the experimental study of superconductivity in thin metallic films, which he along with E. T. S. Appleyard, A. D. Misener and J. R. Bristow pioneered in 1938, fixing the penetration depths of actual superconductors from critical field measurements. Another, continued through the same period and crowned with success in 1939, was the search for alternating current losses in superconductors, already described. A brilliant, unforeseen byproduct of the latter work (and one that Bill Fairbank himself was to study further in his doctoral research) was the discovery of the anomalous skin effect in normal metals, which Heinz correctly attributed to an increase in electron mean free path at low temperatures.

In these two fields, high frequency and thin film superconductivity, Heinz London had created two of the three experimental methods that would lead to the Bardeen-Cooper-Schrieffer microscopic theory of superconductivity. (The third was the study of the isotope effect, the shift in superconducting transition temperature between different isotopes of the same metal.) Even that did not exhaust Heinz's skill in bridge-building. Another bridge started in 1935 was the application of thermodynamics to superconductivity. Heinz investigated the transition to the normal state which occurs in high magnetic fields. He discovered that unless a superconductor has a positive surface energy sufficient to counteract the field energy in the penetration layer, the state will split in high fields into a finely divided mixture of normal and superconducting regions. Since in most pure metals superconductivity does abruptly disappear at the critical field, he concluded that for them the surface energy is indeed positive. He conjectured, however, that some hysteresis effects just then observed in alloys might arise from negative surface energy. These ideas were afterwards entirely confirmed; during the 1950's they were elaborated into the distinction between type I and type II superconductors.

If Bill Fairbank has not quite reached Heinz London's standing as a matchmaker for ideas (few people have), he has his own special flair. His independent search for alternating current losses in superconductors exemplifies his instinct for the experiment that portends new theory, while his search for Fermi degeneracy in ^{3}He exemplifies his instinct for the wrenching theoretical issue. One of the best examples of Bill's feeling for theory occurs in my opinion in what has become the most controversial of all his experiments, the search for fractional charges. I still recall the day in 1965 when Bill dropped the remark that he was beginning to get interested in these ideas about quarks. My own reaction at the time was one of skepticism, not from belief in confinement, which had not yet been proposed, but from a distaste for the whole picture of particles within particles within particles. Swift's jingle,

> Larger fleas have smaller fleas upon their backs to bite 'em
> And smaller fleas have smaller fleas and so *ad infinitum*,

stuck in the mind. Yet whatever may be the final judgment of history on the quark concept and on Bill's experiment, few people now would disagree that of all the loud succession of trumpet flourishes that have sounded forth over the last thirty years from the 'massed band of high energy theorists, quarks and quantum chromodynamics have formed the one most deserving of attention. To have picked out this signal and this alone, and to have chosen to test it in the way he did, was excellent judgment on Bill's part. But that brings us to the larger question of choosing a research field.

8. THE EXPERIMENTALIST'S ENCOUNTER WITH THEORY: II. THE CHOICE OF A RESEARCH FIELD

The process of judging the scientific potential of a field of research engages issues of theory and experiment at a different level from those so far discussed. Here one pauses to admire the independence of mind that made Bill Fairbank, as far back as the University of Washington, define his own research field, and likewise at Yale, once committed to low temperature physics, choose a thesis topic of which Lane knew nothing. How few of us, in graduate school or after, do choose! Sheep seeking pasture, we follow some local bellwether into a field, and thereafter in the scramble for jobs and tenure cannot or dare not look further, all our future hanging on the luck of that first wandering. Yet the choice affects not only discovery but who we are. Think of Rutherford, the younger contemporary of Boys and Eötvös. Admirable as those two men were, his was a larger mind. It was so in part because radioactivity, his chosen field, offered more scope than experimental gravitation for intellectual growth.

In choosing a field the wise experimentalist will recognize that theorists, however adroit in calculations, may have no more discernment than he. They too are wandering souls. Nor will he bow to preemptive judgments like that of the later Rutherford, boasting in the 1930's that there were two kinds of physicists, nuclear physicists and stamp-collectors. To think that one's own subject and one's own style of reasoning are the only reality is the disease of the learned, sharply depicted by Leonhard Euler (who did not lack in learning) two centuries ago: "The natural philosopher and chymist will have nothing but experiments, ... the geometrician and logician nothing but arguments, and ... some ... devoted to the study of history and antiquity, would admit nothing as true but what you could prove by history or the authority of some ancient author."[3] It is not hard to identify 20th century counterparts to the attitudes Euler condemns. Yet choose a field one must. One suggestion, offered by Peter Kapitza in a passage Bill admires, is to go into previously unexplored regions, because that is where surprises lie. True, but only to a point. P. W. Bridgman gave forty years of ingenuity to high pressure physics with never a great discovery. Kapitza himself worked in two regions, low temperatures and high magnetic fields. In the former, which had already a long history, his discoveries were profound; in the latter, which was wholly new, they were not. Novelty alone is an insufficient criterion.

The basis for Kapitza's achievement in low temperature physics was technological. The field was rich but difficult to work; he, in building the first expansion engine helium liquefier, made it more accessible. Historians will see Kapitza's contribution as part of a wider movement of the 1920's and 1930's, where the two older styles of experimentation, the tradition of "string-and-sealing-wax" and the tradition of the instrument-maker, began to give way to a new kind of physics, rooted in engineering. Kapitza himself, while a professor of physics at Cambridge, had a degree in engineering from Leningrad. Though not in itself what would now be called "big physics," his work depended on big industry, in particular on the vast contemporary increase in the production of helium gas for use in airships. The power of what he had done is revealed in his ability to stay a leader in the field even after suffering a four year disruption of his career when, just after the liquefier's first run in 1934, the Soviet government forbad him to return to England from a visit to his parents in Russia. During the next fifteen years the three institutions with Kapitza liquefiers, Cambridge, Moscow, and Yale under Lane, all gained a competitive edge.

Reverting to Rutherford's heavy-handed claim (in which connection it is pleasant to recall that Kaptiza, though not a nuclear physicist, was Rutherford's favorite younger colleague), one is led to reflect on the importance of *timing* in scientific research. Fields come and go; and then surprisingly come back. Experimental gravitation is far more interesting now than it was in the 1890's. Spectroscopy was near the center of physics from 1910 to 1930, then faded, and then had a revival in the 1970's with the coming of the laser. Nuclear physics, the tidal wave created and ridden by

---

[3]L. Euler, *Letters to a German Princess on Different Subjects in Physics and Philosophy* (Paris, 1757–1759); English translation edited by Henry Hunter (London, 1802), Vol. ii, p. 3.

Rutherford,[4] had spent its first force by the time Pollard was advising Bill Fairbank to look elsewhere.

In weighing the scientific potential of a field of research, three interlocking questions, each of great subtlety, apply: Is the subject matter important? How accessible is it to experiment? To what extent is the theory understood? The pitfalls are many. Some grand topics (quantizing general relativity or interpreting the fine structure constant, for example) defy theoretical attack and seem inaccessible to experiment. Other topics are accessible but not grand. Bill Fairbank has said that the ideal field for an experimentalist is one where the theoretical issues are important and partially understood, but not so well understood that there is no room for surprise. Such were processes of electrical discharge in gases in 1873 when Maxwell, with his wonderful prescience, singled them out as phenomena which "when they are better understood will probably throw great light on the nature of electricity." Such also in different ways were radioactivity for Rutherford in 1897 and low temperature physics in the late 1940's when Bill Fairbank entered the field.

Viewing the physics of solids and liquids at low temperatures as it would have struck the more aware theoretical physicists in say 1948, the interest was that here manifestly was a quantum mechanical world but not one which like the atom was well understood. The same held for solid state physics in general, not just at low temperatures; but at low temperatures quantum effects stood out. Debye's T^3 law of specific heats, known for over thirty years, was one example; another, more recent, was the anomalous skin effect. Because large numbers of particles were involved one might at low temperatures expect to see pecularities of quantum statistics: Fermi degeneracy for particles with half-integer spin like electrons and ^{3}He atoms; Bose-Einstein condensation for those like ^{4}He with integer spin. Then there were superconductivity and superfluidity, phenomena that might or might not be susceptible of quantum mechanical explanation. How important all this might be remained uncertain. Landau, London and Onsager were enthusiastic; Gregory Breit, on the other hand, was less so. He told Bill that the nucleus and new particles were the center of physics; low temperatures were unlikely to yield answers to more than a few minor puzzles.

Whatever the case, neither solids nor liquids were simple systems. The challenge to theorists lay in meshing quantum theory with thermodynamics, hydrodynamics and intermolecular force laws. The challenge to experimenters lay in deciding what to do, not only which phenomena to measure

---

[4]C. D. Ellis: "You have been lucky, Rutherford, you have always ridden the crest of the wave." Rutherford (*laughing*): "Yes, but I made the wave ... (*after a pause*) at least to some extent I did."

but which techniques to apply. Should one, for example, stick to temperatures above 1 K, reached by pumping on liquid ^{4}He, or should one venture below 1 K, a region only accessible then by adiabatic demagnetization? In gauging Bill Fairbank's work in low temperatures during the ten years after 1946, we need to see both what he did and what he did not do. Other people were entering the field and there was much to explore. Data were being accumulated on the standard superconductive properties of metals and alloys, and on such properties of liquid helium as surface tension, film thickness and film flow. More sophisticated studies covered exotic phenomena like the de Haas-van Alphen effect, leading in combination with advanced theory to detailed mappings of the Fermi surfaces of single metallic crystals. None of those topics drew Bill. Setting aside the joint experiment with Lane and Henry on second sound, his chosen themes were two: at Yale ac losses in superconductors, at Amherst and Duke the study of ^{3}He by nuclear magnetic resonance. These were intelligent choices. Each explored a novel topic; each applied a technique Bill knew, but added fresh techniques to his repertoire; each, differently, established a bridge between theory and experiment.

Having sufficiently discussed the work on ac losses earlier, I concentrate here on ^{3}He. In 1947 when Bill went to Amherst, the deepest question in low temperature physics was whether superconductivity and superfluidity were or were not consequences of Bose-Einstein condensation. The uncertainties were many. Most radical was a theorem of Bloch's which to some minds seemed proof that no explanation of superconductivity could be developed within quantum theory. That aside, there were problems of quantum statistics. For free particles these were at least nominally understood, but no one was sure of their application to the strongly interacting systems of solids and liquids. Superconductivity added to the mystery. Suppose Bose-Einstein condensation were a factor, it would make sense in superfluidity because ^{4}He, having integral spin, is a Bose particle; but the electrons in superconductors ought to be Fermi particles. That issue, as is well known, was resolved in 1957 through Cooper's brilliant hypothesis of electron pairing, the basis of the BCS theory. Meanwhile for Bill and everyone else the questions hung in the air.

One line on the problem was to study liquid ^{3}He, since ^{3}He atoms like electrons should be Fermi particles. A test for superfluidity was obvious; Bill went deeper. If liquids do obey quantum statistics, the vital question was whether and where ^{3}He undergoes Fermi degeneracy. Fermi-Dirac statistics dictate that no more than two particles, paired with opposite spins, occupy the same state. As the temperature is lowered, more and more particles congregate in the lower states. With a fixed number of particles there will be a certain temperature T_F (the Fermi temperature) below which all states are filled; at higher temperatures particles will begin populating the higher states in numbers approximating to the classical Maxwell-Boltzmann law. Measurements of magnetic susceptibility supply a test. Above T_F the susceptibility χ will more or less obey the classical Curie formula, with χT a constant. Below T_F thermal effects are unimportant; the only contribution to susceptibility is the redistribution of spins from the slight splitting in energy levels for parallel spins; χ should become independent of T. This was the effect Bill went in search of. Given the ^{3}He, the questions were how to measure its very small magnetic susceptibility (Bill as we have seen fixed on a nuclear magnetic resonance technique), and what T_F might be.

For an ideal gas with density equal to that of liquid ^{3}He, the Fermi temperature was calculable; it was 5 K. Bill asked Onsager what T_F might be in the liquid; he thought for a moment and then said 0.5 K. Right or wrong this was a critical prognostication; it meant that mere pumping on ^{4}He would not provide a low enough temperature. Bill and his colleagues designed an apparatus in which the sample could be cooled by adiabatic demagnetization. On plotting χT as a function of temperature they found as expected a constant value down to a certain temperature, and then a curve falling off as $1/T$, corresponding to constant susceptibility. The measured Fermi temperature was 0.45 K. When Bill asked Onsager how he had known, the oracle smiled.

I have already described the impact of this work on Bill's career. More was to come. With King Walters, then a graduate student at Duke, Bill set out to measure how T_F would change in ^{3}He diluted with ^{4}He. They saw strange effects which they eventually attributed to a phase separation between two distinct ^{3}He-^{4}He solutions of different densities and concentrations. The critical temperature below which the phase separation took place was 0.8 K. With masterly ingenuity Bill and King Walters devised an apparatus with three tiny sample chambers, one above another, joined by fine capillary tubes, by means of which they could observe three NMR signals, one for the dilute solution of ^{3}He in ^{4}He, one for the concentrated solution, and one for the portion of liquid containing the boundary between the two. From the changes in position and amplitude of the three NMR signals they plotted the phase curve as a function of temperature. Afterwards D. O. Edwards and colleagues at Ohio State University extended the measurements to lower temperatures, finding that even near the absolute zero the concentration of ^{3}He in the denser phase remains finite ($\sim 6\%$). Edwards' result supplied the last link in the chain of reasoning that led up to Heinz London's invention of the ^{3}He-^{4}He dilution refrigerator.

As a science evolves, the relationships in it between theory and experiment change. No field remains long in Bill Fairbank's preferred state, where theoretical issues are focused sharply enough for their importance to be clear but not so sharply for there to be little chance of experimental surprises. In low temperature physics a turning point was the formulation of the BCS theory of superconductivity in 1957. I still recall a discussion from 1960 with an older colleague in London, not a low temperature physicist, who, when he heard of my burgeoning interest in the field, told me that there could not be much left to do or discover now that the problem of superconductivity was solved. The obituary was premature; nevertheless the style of research appropriate to the physics of condensed matter at low temperatures was changing. With such changes come the opportunity and the need for the alert experimentalist to rethink his career.

The direction of rethinking will be a matter of taste. Some physicists enjoy the precision of thought that comes with elaborating the applications of a well-established theory. With Bill a first stage of rethinking was to go not forward but back, to Fritz London and the unfinished business of macroscopic quantization. Taking on, rashly, the invidious task of rank-ordering physics experiments (where on a scale of 1 to 10 the measuring of an expansion coefficient might be a 1 and the discovery of the atomic nucleus a 10), one should in my judgment count the work of Bill Fairbank and his colleagues at Duke on Fermi degeneracy in ^{3}He as a very good 7, and the discovery of quantized magnetic flux in superconductors as an 8. As an experimental achievement the former was in some respects the more impressive. There Bill and his colleagues were on their own, whereas in flux quantization several groups were close behind, and indeed Doll and Näbauer in Germany had the result at the same time as Bascom Deaver and Bill, though without the vital calibration that would set quantization in half-London units. Nevertheless the judgment stands. Important as Fermi degeneracy in ^{3}He was, the demonstration that a single wave function suffices to describe the collective behavior of 10^{23} particles was more so, while the factor of two made this the decisive experiment on Cooper pairing.

How subtle are the interrelations of experiment and theory! The search for quantized flux only made sense in the context of a quantitative theory; without a numerical prediction Bill would have had no occasion to try it. Yet work based on Fritz London's radical intuitive leap from the macroscopic theory of superconductivity ended by confirming a theory London had no part in developing, which did not originally embody the prediction. The history of physics does not lack poignancy. Beyond the theoretical turning point, so well captured by Frank Yang in paper III.2, is another ground for ranking flux quantization higher among Bill's discoveries than Fermi degeneracy: its consequences reached further. Through Brian Josephson's insight it led to the Josephson effect and SQUID magnetometers. Through the work of Bill Hamilton and Blas Cabrera, with Bill Fairbank, it led to ultralow magnetic fields. The place of SQUID's and low field shielding in the later work of Bill and his colleagues is one of the grand themes of this book.

Things, ideas ... and people. To Bill Fairbank's preference for research fields in which theory is partially but only partially understood, there is a personal corollary, according to which the ideal field might be described as one where there are enough people to share and compete with, but no so many that competition is damaging. The place of competition in science deserves thought. No one who has run on an athletic field with Bill will doubt that he, like some of his co-workers, has the kind of competitive drive that athletes admire but physicists (despite the lure of Nobel prizes) rarely admit to. Yet touches of competition stimulate creative thought, and, if subject to civilizing constraints, add to the pleasure of research. There is a happy balance between Will Thompson's early advice to Bill to disregard popular opinion in choosing a career, and the effect of Feynman's bet in making Bill and his colleagues do a λ-point experiment that would astonish the low temperature world.

If enthusiasm for macroscopic quantization was one advance in Bill's thinking after the BCS theory, another was the gradual shift from low temperature to "near zero" physics. The roots of the latter were two, a wish to enter new regions of physics and the triumphal near zero experience of the λ-point experiment. How far Bill was consciously responding to the times is unclear. The process of applying low temperature techniques elsewhere began with the liquid helium bubble chamber, but came to be decisive in the free fall experiment. Bill has often told of a lecture in 1957 by Bryce DeWitt, in which DeWitt took a piece of chalk, tossed it in the air, caught it as it fell, and remarked that this was the only experiment we did or could do in gravitation. Later in that year Philip Morrison and Thomas Gold won the Gravity prize from the Babcock Foundation for an essay arguing that the nonexistence of antimatter in our universe might follow from there being a repulsive rather than attractive gravitational force between particles and antiparticles. So Bill conceived the idea of measuring the action of the earth's gravitational field on electrons and positrons. The experiment (a "near zero" experiment if ever there was one) was started in 1961 by Fred Witteborn; its ingenuities, successes and paradoxes are described in paper VII.2 by Lockhart and Witteborn. Curiously, Leonard Schiff's interest in gravitation also originated in Morrison and Gold's speculation, which Schiff, independently of meeting Bill, attempted to disprove by an argument based on applying the weak equivalence principle to virtual antiparticles. The tie of that particular paper of Schiff's to the gyroscope experiment appears in paper VI.3(A).

Twice already I have referred to the pressures that discourage physicists from changing fields. Bill, commenting on the experiment on Fermi degeneracy in ^{3}He, has remarked that there he plunged in with no knowledge of nuclear magnetic resonance, but with the advantage of having worked in both low temperatures and radar. Another of Bill's remarks is that since most people know only one experimental technique, someone who knows two has an advantage, while the advantage of anyone who has a command of three techniques is overwhelming. In truth, techniques are like languages: the more of them one knows, the easier it is to pick up others. Most people think that the first step in any venture is the hard one. In the physics venture the difficulty is the second step, to persuade the expert in some other field to let you join his group. Bill's diversity stems from the Rad Lab and curiosity. He was 24 when he and Jane, after being at the Rad Lab for only a few weeks, already knew (in Gordy's words) "everybody and everything that was going on there."

Returning to Archilochus, Isaiah Berlin, hedgehogs and foxes, one might characterize Bill as a hedgehog in the intensity with which his intellectual energies are concentrated on physics, a fox in the multiplicity of his interests within physics, and again as a hedgehog (or is it a bulldog?) in the tenacity with which, once he has set his mind to a problem, he worries it through to the end.

But there is one faculty that Bill has which has not been granted to hedgehogs or foxes or bulldogs, a purely human aptitude for conversation.

9. THE PERSONAL CONTEXT

Actors find that character often is established not in grand actions or grand emotions but through some tiny detail: a gesture, a trick of body position, a turn of phrase. Fairbank details come quickly to mind ... Bill, the once would-be optometrist, pushing his glasses up to his forehead to read a scientific paper more attentively ... Bill in a motel room in Phoenix, Arizona, at two in the morning, seeking to convince five collapsed colleagues that a superconducting motor and mechanical clutch, with superconducting bearings, is the ideally simple means for spinning up a gyroscope ... Bill in another motel room in Minneapolis on his first sight of a bathroom telephone at once making half a dozen long distance phone calls each beginning "Joe (or Henry or whoever), guess where I am" ... Bill, who never (well almost never) misses a plane, standing by his secretary's desk forty-four minutes before departure time from San Francisco airport, making one last call to his Air Force monitor while nervous traveling companions fret and fume ... Bill asking one to come back to the lab in the evening for one reason, only to spend the first three hours talking Monty-Python-like about something completely different .... And (a semi-apocryphal tale) Bascom Deaver, long after midnight, preparing to leave the basement laboratory of the old physics building at Stanford, seeing Bill standing in conversation at the nearer of two doors, deciding to crawl along behind the benches to the further door, colliding with Bill Hamilton in the process of effecting a similar escape, and in the ensuing scuffle finding himself with Hamilton eagerly welcomed to a discussion on physics lasting until 4:30 a.m.

Smile or not, it is no accident that six of the seven images that occur to an author during an afternoon's run, are of Bill engaged in conversation. Bill loves discussion. Occasionally he meets resistance in this, for few academics listen well; but even where the force of his enthusiasm dominates a discussion, Bill is no soliloquist. Many years ago Alan Schwettman remarked to me that Bill's ability to take part in a conversation is his greatest strength. The remark came as a surprise because I had put courage and imagination higher; but it has validity. Four aspects of a typical Fairbank conversation deserve attention: the fluidity of thinking, the mastery of numbers, the cheerful unconcern at exposing ignorance, and one further quality, most important of all, to be identified in a moment.

I noted earlier my own bewilderment in a first disucussion with Bill. Partly it was from being swept pell-mell into issues I knew nothing about; partly it was from the fluidity of Bill's thinking. Nothing seemed fixed; nothing indicated where we were going next. Some years later, after a four-hour discussion on the gyroscope experiment involving Bill, myself and two engineers, one of the engineers expressed his awe at a circumstance in which, as he put it, "the requirements changed by six orders of magnitude during the course of the meeting." But in this madness there was method; we were into the process of number-matching. In such discussions one impressive fact is Bill's memory for numbers. Working at the blackboard one may over a few hours need Planck's constant, the mass and the charge of the electron, the radius of the earth, the mass and diameter of the sun, the gravitational constant, the inductance of a 4 cm diameter metal ring, several conversion factors from SI to electrostatic or electromagnetic units, Boltzmann's constant, the black body radiation from a 5 1/2 inch diameter disk at 300 K, the mean free path in helium gas at 2 K and 10^{-3} torr, the conversion factor from ergs to electron volts, the specific heats and thermal conductivities of various materials at low temperatures, the bursting speed of steel, the binding energy of aluminum, the number of watts that makes a liter of liquid helium evaporate in an hour, and who knows what else. All these numbers Bill seems able to pluck out of the air. An amusing example occurred when in writing a brochure I decided to calculate the mass of a lump of neutron matter the size of a raindrop. Wishing to check my result I asked Bill's opinion as to the diameter of a raindrop; he replied that in chemistry he had learned that 1 cc was 33 drops from a pipette. So we got our number, which was 30,000 tons.

Memory is not the *sine qua non* of scientific greatness (Faraday's was woefully defective); it does help, and in the kinds of discussion Bill conducts it is all but essential. Naively one thinks that concepts are what count and that numbers can always be looked up in a handbook, but with four people at a blackboard one cannot be perpetually turning pages. This is one lesson one learns from Bill; another concerns ignorance.

We are all ignorant. We are all aware that Socrates' wisdom consisted in knowing that he knew nothing. And yet how hard it is to admit of any particular thing we do not know it. When I started working with Bill I was amazed at the naivety of the questions this famous professor would ask. How could he so expose himself? Later I saw a parallel with the quality that Maxwell found most praiseworthy in Faraday: Faraday's willingness in his writing to "shew us his unsuccessful as well as his successful experiments, and his crude ideas as well as his developed ones, [so that] the reader, however inferior to him in inductive power feels sympathy even more than admiration, and is tempted to believe that if he had the opportunity, he too would be a discoverer." Here we reach the heart of the matter, and the fourth of the four characteristic qualities of a Fairbank conversation. Somehow connected with Bill's fluidity of mind and unconcern at showing ignorance is an ability to ask and go on asking until out of the depths is projected a not-so-naive question which no expert would have posed. Curiosity gains its reward.

Enthusiasm, fox-like curiosity, hedgehog-like conversation, the urge to converse and the ability to pose a question: these are the personal resources of Bill the creative individual. He has chosen to apply them to physics, and except for the three years at the Rad Lab, itself a semiacademic organization, his entire professional life has been spent in academia. Yet if there is such a creature as the typical academic physicist (in truth there are several types), Bill is not it. In pondering his career one is led to reflect on the influence of character on destiny; and also on the degree to which the roots of character extend beyond professional boundaries into family and society. As someone who has spent half his life in England and half in the United States, I find interest in examining the similarities and differences of the two outstanding experimental physicists I have been privileged to work with: the quintessentially English Blackett and the quintessentially American Bill Fairbank.

Quintessential does not mean stereotypical. No one who knew Blackett or knows Bill Fairbank could carry away from meeting either man any impression except that of a uniquely memorable individual. Rather it is that each man's individuality belongs peculiarly to his own time, circumstance and country (which is obvious), and (not so obvious) that with all our rightful insistence on science being international, each man's style of doing physics is influenced by aspects of country and circumstance that have no connection with his professional training. The comparison is the more interesting because Blackett and Fairbank have so many affinities: enthusiasm, intensity, love of a challenge, independence of mind, and a wonderful power of stimulating other people.

Since this is an article about Bill Fairbank, not Blackett, I shall not pursue how one man, Blackett, could plausibly unite the traits of a physicist, a conservative naval officer, a left-wing Cambridge intellectual, and (in his own words) "an old-fashioned Victorian rationalist agnostic." My concern instead is with how Blackett's rationalist creed and his experience as a naval officer in World War I affected his physics. He learned physics from Rutherford. Central to his work was a very British (and not entirely Rutherfordian) pleasure in instrument design, as seen in his inventions of the counter-controlled cloud chamber, the curvature meter for particle tracks, the improved astatic magnetometer, and the Blackett bombsight. Like Bill he had a grip on theory; it was, as Norwood Russell Hanson established in the *Discovery of the Positron*, his work rather than C. D. Anderson's that definitively connected pair production with the Dirac equation. Yet even though Blackett knew how to use theory, as when in 1926 he applied special relativity to give for the first time the mathematical relationship between the length of tracks in cloud chambers and the lifetimes of high energy particles, theory was less part of his lifeblood than it has been of Bill's. Modifying his physics was a very Victorian faith in a different kind of science, philosophized by John Stuart Mill and practised by Francis Galton, a rationalistic Baconian process of collecting and classifying large masses of data. Not for nothing was *phenomenological* his favorite word; nor was it chance that in World War II he became the brilliant pioneer of operations research. Two remarks encapsulate his outlook: from the Victorian rationalist, "Make sure you gather plenty of data"; from the naval officer, "You should treat your research like a military campaign." To the pursuer of intellectual subleties it was a wonderful corrective to see Blackett's brutally direct way of smashing scientific problems.

Bill's background and ancillary experiences are very different. One senses in him the strength of a long family tradition, confronting and surviving the Depression. One senses also the impact of that uniquely American version of the *Wanderjahre*, in which young men (and nowadays young women) from the middle classes find opportunity, over several years rather than one, to move about and take on jobs that would be unthinkable for their European counterparts. It is hard to picture Blackett in his teens acting like Bill and Henry Fairbank as a door-to-door salesman of Real Silk hosiery, an experience of which Bill often and amusingly speaks in connection with the task of persuading government agencies to dispense funds for research. Such experiences can foster early scientific independence, as they did in Bill and Henry. Blackett, by contrast, who had obviously had trouble with Rutherford, once told me that he himself did not grow up until he left Cambridge for a year in 1924, against Rutherford's advice, to work with James Franck at Göttingen on spectroscopy. From a man who had fought in the battle of Jutland, and who was at 27 already well on the way to scientific fame, it was a notable admission.

In writing of national differences I cannot, in view of Anglo-American cliches about the Latin races, forbear quoting the words of an Italian physicist of great eminence, who some years ago said, "Fairbank is a fine physicist, but he is too excitable." Those who know Bill will know what was meant; on the other hand, Bill's capacity to become excited is part of what makes him a good physicist. Bill's grandfather was a Congregationalist missionary in India, whose convictions in part Bill shares. How far Bill has inherited the old man's character I have no means of knowing; but if Blackett regarded physics as a military campaign, one may suspect Bill at times of regarding it as a missionary endeavor. Bill, however, is one of those rare missionaries (the best) who through willingness to admit a lack of knowledge learns fully as much from his auditors as he imparts to them.

Mingling with the two traditions of independence and of a slightly old-fashioned but wholly admirable family loyalty, there is in Bill's character a third attribute that seems to me peculiarly American, the style of his inventiveness. It is an inventiveness that combines three things, eagerness for novelty, sharpness of ingenuity, and an enormous capacity for hard work. One remembers Edison's prescription for genius as 2% inspiration and 98% perspiration. Whether or not the proportions are correct, Bill has the capacity for both perspiration and inspiration, and shares some of Edison's inventive power. In contrast to the traditional American inventor, however, who functioned outside learned circles, Bill has linked his inventiveness to an understanding of fundamental science.

Bill's approach to physics embodies another and even higher virtue than the willingness to admit ignorance. Deeply ingrained in his being is a conviction which most people call optimism, but which I prefer to call hope. He expresses the conviction in various ways. Sometimes he speaks of the antimurphy law, sometimes of triumph in adversity. According to the antimurphy law, when one finally understands an experimental situation correctly, one will find three or four things miraculously coming together to make the impossible possible. As for adversity, the moment of disaster is the moment of opportunity; it is then that one can rethink a situation or an experiment and come up with something even better. Polyannaish as this may sound, it is impressive to observe how often setbacks have served as a spur to Bill's creative imagination. One recalls a comment by another great physicist who has overcome setbacks, S. Chandrasekhar, who, in a reminiscence about his close friend, E. A. Milne, chides Milne for a certain negativism and remarks that in order to succeed in science it is essential to cultivate an optimistic and positive outlook.

So entrenched are the antimurphy and antidisaster principles in Bill Fairbank, so well do they express his scientific personality, that I feel compelled to end this account of Bill's creativity with my favorite Fairbank story, one told by Alex Dessler, Bill's first graduate student. Alex was working in the laboratory at Duke late one night on his second sound experiment; he had just removed one of the precious glass dewars from his apparatus and was standing with his back to the door holding the dewar under his arm. In walked Bill to ask how things were going. Alex turned to answer, and as he did so brought the dewar round in a smooth arc intersecting the location of an iron bottle of compressed helium gas. There was stunning implosion as the dewar shattered into a hundred fragments. In the silence which followed Bill looked at Alex for about thirty seconds and then said, "Alex, I cannot think of a single reason why that was a good thing to have done."

CHAPTER II

Liquid and Solid Helium

(*Top*) Recalling Duke days: Dwight Adams, John Goodkind, Fred Kellers, Walter Gordy, G. King Walters, Michael Buckingham, Henry Fairbank and William Fairbank at the Near Zero conference, Stanford University, 1982. (*Bottom*) John Lipa and Talso Chui with the Stanford lambda point apparatus.

Liquid and Solid Helium: Introduction

George B. Hess

While at Amherst College, William Fairbank realized that even though ^{3}He occurs in only 5 parts in 10^7 of ^{4}He, a Navy blimp contains two liters of ^{3}He. If ^{3}He could be removed from large quantities of ^{4}He, enough could be obtained to perform significant experiments with the new isotope. Fairbank developed a technique for separating out this small quantity of ^{3}He during the liquefaction process of ^{4}He in a liquefier built at Amherst for this purpose. Although this heat flush method proved successful, reactor-generated ^{3}He from tritium made the heroic effort unnecessary.

Work with the newly more plentiful isotope proceeded at Duke, where Fairbank, Ard and Walters [1], using nuclear magnetic resonance, found the onset of Fermi degeneracy in liquid ^{3}He. Extension of this work to ^{3}He-^{4}He solutions by Walters and Fairbank [2] led to discovery of the isotopic phase separation. G. K. Walters gives an historical review of this work, and of recent developments in a new Fermi liquid system, spin polarized ^{3}He.

The NMR work at Duke was also extended to solid ^{3}He, for which spin ordering was expected at much lower temperatures than in the liquid phase. Dwight Adams and John Goodkind began work on this at Duke and continued as postdoctoral fellows at Stanford. Adams subsequently has played a leading role in solid helium research, and Goodkind has been a pioneer in submillikelvin research. In this volume they collaborate in a review of progress in this area from the Duke days through recent studies of the properties of the spin ordered phases.

Bruce Pipes continued the nuclear susceptibility measurements on solid ^{3}He at Stanford using a dilution refrigerator of his design. He and his coworkers review the use of isochoric pressure measurements to obtain the spin exchange parameters of solid ^{3}He and present the results of their research using this technique.

Fairbank collaborated in the construction at Duke of a prototype liquid helium bubble chamber. Below the lambda point the bubble chamber did not show charged particle tracks, but intriguing macroscopic bubble pairs appeared, which were conjectured to nucleate on quantized superfluid vortices. This inspired some optical experiments on liquid helium at Stanford, which led in several directions, and two papers here may be considered off-shoots. Pierce Webb and Hugh Griffiths report the previously unpublished results of Webb's thesis project on light scattering by ^{3}He in the neighborhood of the liquid-vapor critical point. Philip Marston, for his thesis research at Stanford, made an interferometric study of the surface profile of rotating liquid ^{4}He. He and D. S. Langley review experiments on bubble nucleation in superfluid ^{4}He, the theory of light scattering by bubbles, and the theory of bubble nucleation as applied to liquid helium.

In 1946 Henry and Bill Fairbank collaborated with C. T. Lane to detect the propagation of thermal waves, or second sound, predicted earlier by Tisza, in liquid ^{4}He. This was before a similar experiment by Peshkov in Moscow was known in the West. Henry Fairbank outlines the course of second sound research from that time, including subsequent measurements on ^{3}He-^{4}He mixtures, experiments on the propagation of second sound in crystals, and, more recently, observations of second sound in the A_1 phase of ^{3}He.

Bill Fairbank was strongly influenced by F. London during the brief period they were colleagues at Duke, particularly in his view that long range order in momentum is responsible for superfluidity, both in helium and in superconductors. First, London's identification of the lambda transition with Bose-Einstein condensation emphasized the importance of comparing the Fermi liquid case of ^{3}He. A more explicit consequence was London's prediction of fluxoid quantization and of the magnetization of rotating superconductors, which led to experiments described in chapters III and VI of this volume. In the case of liquid ^{4}He, one consequence of long range order is that the lambda transition should be sharp. Michael Buckingham, as a visiting theorist at Duke, was induced to participate in an experiment with Fairbank and graduate student C. F. Kellers to measure the specific heat of ^{4}He near the lambda point to very high resolution [3]. Their result, that the specific heat is indeed singular to within about a microkelvin of the transition and consistent with a logarithmic divergence, together with

Onsager's calculation of a logarithmic divergence for the two-dimensional Ising model, was a significant factor in stimulating interest in critical point singularities at the onset of the era of critical exponents. Buckingham, in his paper on the "shape" of the lambda transition, reviews the development of our understanding of critical singularities and scaling (and hyperscaling) laws and universality, and illuminates certain pitfalls along the way.

There has been much subsequent experimental work on the critical behavior of thermodynamic and dynamic properties at the lambda transition of liquid helium (see, for instance, [4]), which has come to be recognized as a uniquely favorable system. John Lipa reports here on a project at Stanford to attain yet more orders of magnitude of temperature resolution in measurements of the specific heat, and analyzes the design of such an experiment to be conducted in a near zero gravity environment in orbit.

An additional consequence of long range order in superfluid helium, stated explicitly by H. London [5], is that the superfluid in equilibrium in a rotating container should not participate in the rotation, provided the angular velocity is not too large. An experiment which demonstrates this effect is described by George Hess, in greater detail than it has been reported previously.

In the period following the earliest ^{3}He experiments, John Wheatley and his associates at Illinois and later La Jolla had a leading role in developing the techniques for cooling helium to the few millikelvin range, and in measuring the properties of liquid ^{3}He and ^{3}He-^{4}He mixtures in this range. Wheatley reviews progress in an area of his current interest, intrinsically irreversible heat engines.

References

[1] W. M. Fairbank, W. B. Ard and G. K. Walters, *Phys. Rev.* **95**, 566 (1954).

[2] G. K. Walters and W. M. Fairbank, *Phys. Rev.* **103**, 263 (1956).

[3] M. J. Buckingham and W. M. Fairbank, in *Progress in Low Temperature Physics*, edited by C. J. Gorter (North-Holland, Amsterdam, 1961), Vol. 3, Chap. 3.

[4] G. Ahlers, in *The Physics of Liquid and Solid Helium*, edited by K. H. Benneman and J. B. Ketterson (Wiley, New York, 1976), Part 1, Chap. 2.

[5] See F. London, *Superfluids* (Wiley, New York, 1954), Vol. 2, p. 144.

The ^{3}He Fermi Liquid and ^{3}He-^{4}He Phase Separation: Duke Origins

G. K. Walters

The 1950's were especially exciting years for the study of quantum fluids. In particular, small quantities of the rare ^{3}He isotope, obtained from β-decay of tritium gas, were just becoming available for research purposes from the Atomic Energy Commission. This afforded the first opportunity to explore the extent to which the unusual properties (*e.g.*, superfluidity) of liquid helium depend on quantum statistics (Bose-Einstein for the common ^{4}He isotope, Fermi-Dirac for ^{3}He), at a time when there were no satisfactory first-principles theories.

William Fairbank's nuclear magnetic resonance (NMR) studies, initiated upon his arrival at Duke in 1952, were to be perhaps the most revealing and definitive regarding the macroscopic quantum behavior of the ^{3}He Fermi liquid.

To place the Duke work in perspective, it is instructive first to review briefly early studies of the thermodynamic properties of liquid ^{3}He initiated in the late 1940's at the national laboratories where ^{3}He gas had first become available. ^{3}He was first liquefied in 1948 by Sydoriak, Grilly and Hammel at Los Alamos Scientific Laboratory [1]. These workers also located the ^{3}He critical point ($T_c = 3.34$ K, $P_c = 875$ torr) and the normal boiling point (~ 3.2 K) and determined approximate vapor pressures and liquid densities above 1.2 K. These properties were confirmed and the temperature range extended in 1950 by Abraham, Osborne and Weinstock at Argonne National Laboratory, and this group also was the first to determine the heat of vaporization and entropy of liquid ^{3}He above 1 K [2]. Their work suggested that the ^{3}He nuclear spins probably were not aligned above 1 K, but that interesting effects could be expected at lower temperatures. The Argonne group also demonstrated by melting pressure measurements that ^{3}He liquid is the stable condensed phase at absolute zero [3].

This was approximately the extent of experimental research pertinent to the Duke experiments that had been performed on liquid ^{3}He prior to 1952.

1.   THE ^{3}HE FERMI LIQUID

Fritz London, who remained an inspiring and enthusiastic supporter of Bill Fairbank's research on liquid ^{3}He until his untimely death in 1954, was the first to advance the hypothesis, in 1938, that the λ-transition at 2.2 K and superfluid properties of liquid ^{4}He might be intimately associated with the peculiar phenomenon of condensation of atoms into the state of momentum zero, characteristic of an ideal Bose-Einstein gas below a critical temperature

$$T_B = \frac{h^2}{2\pi m_4 k}\left[\frac{n}{2.612}\right]^{2/3} \quad , \tag{1}$$

where h and k are the Planck and Boltzmann constants, and m_4 and n the ^{4}He atomic mass and number density, respectively [4]. Indeed, the Tisza two-fluid model of superfluid helium based on considerations of the ideal Bose-Einstein gas enjoyed considerable success despite the worrisome question of how to incorporate interatomic forces and the circumstance that one is attempting to describe a liquid rather than a gaseous state [5].

Thus, the question naturally arose as to whether or not the properties of liquid ^{3}He might exhibit similarities to those of an ideal Fermi-Dirac gas of comparable density. In this connection it is noteworthy that, very shortly after ^{3}He was first liquefied, Sydoriak and Hammel showed that no ^{4}He-like λ-type transition exists in ^{3}He above 0.86 K [6].

In contrast to the characteristic thermal discontinuity expected of an ideal Bose-Einstein gas at T_B, the ideal Fermi-Dirac gas undergoes a continuous transition to the ordered state with decreasing temperature [7]. For ^{3}He, with nuclear spin 1/2, the ordered state at absolute zero according to this model would be one in which the lowest available energy levels are successively filled by paired ^{3}He atoms of opposite spin (much as conduction electrons in a metal do) until all the atoms are accommodated. The highest occupied energy level is defined as the Fermi energy E_F,

$$E_F = \frac{h^2}{8m_3} \left[\frac{3n}{\pi} \right]^{2/3} , \tag{2}$$

where m_3 is the ^{3}He atomic mass. E_F corresponds to a characteristic degeneracy temperature $T_F = E_F/k$. For an ideal Fermi-Dirac gas with density equal to that of liquid ^{3}He, $T_F \simeq 5$ K.

In 1950 Goldstein and Goldstein suggested the then relatively new method of nuclear magnetic resonance as a particularly direct means for probing properties of the liquid ^{3}He spin system [8]. Bill Fairbank and his associates performed the definitive experiments demonstrating antiparallel nuclear spin alignment in the liquid at temperatures below 1 K [9] beginning in 1952.

Actually, Fairbank's interest in the question of Fermi-Dirac degeneracy in liquid ^{3}He using nuclear magnetic resonance as a probe predated his 1952 arrival at Duke. He first discussed the question in 1947 with Lars Onsager who suggested that the degeneracy temperature observed with nuclear magnetic resonance techniques might be 0.5 K, which is remarkably close to the temperature 0.45 K later observed. At Amherst Fairbank had made extraordinary efforts to concentrate ^{3}He from natural well helium. However, it was with a 20 mL STP ^{3}He sample obtained from the Oak Ridge National Laboratory, subsequently concentrated to better than 99% purity by fractional distillation at Duke, that Fairbank initiated his now classic studies of the nuclear susceptibility of liquid ^{3}He.

The concept of the experiment was quite simple, yet definitive. The strength of the nuclear magnetic resonance signal for a spin one-half system such as ^{3}He is proportional to the difference in populations of spins aligned parallel and antiparallel to an externally applied static magnetic field B. For a classical system of non- (or weakly) interacting spins at constant density, this population difference is governed by the Boltzmann factor and results in a $1/T$ dependence (the Curie law) in signal strength as long as $\mu B \ll kT$, where μ is the nuclear spin magnetic moment [10]. However, at sufficiently low temperatures the spins of particles of an ideal Fermi-Dirac gas would, as noted above, be expected to pair off in an antiparallel alignment [7], hence causing the NMR signal strength to fall below the classical $1/T$ law and finally to become temperature independent for $T \ll T_F$.

Within a year after his arrival at Duke, with the help and collaboration of Walter Gordy, postdoctoral associate Hans Dehmelt, and Bill Ard, one of Gordy's graduate students, Fairbank had shown that the magnetic behavior of liquid ^{3}He was strictly classical down to 1.2 K [11]—the lowest temperature achievable by pumping on a liquid ^{4}He bath. In view of the earlier thermodynamic measurements [2,3,12], this was not a surprising result, but it was a direct demonstration that the spin degeneracy characteristic of an ideal Fermi-Dirac gas of the same density does not occur in the case of liquid ^{3}He in the expected temperature range.

I arrived at Duke to begin graduate studies in the fall of 1953, just after the NMR work above 1.2 K had been completed. I joined Fairbank's group immediately, as plans for extending the ^{3}He nuclear susceptibility measurements below 1.2 K were being formulated. By then Hans Dehmelt and Walter Gordy had withdrawn, leaving only Bill Ard and myself working with Fairbank on the project. I remember being very busy that year, with the usual first-year graduate student course load on top of the feverish research activity driven by Fairbank's infectious enthusiasm and boundless energy. Within about three or four months most of the parts for the new apparatus had been completed—the paramagnetic salt pill for cooling by adiabatic demagnetization, the ^{3}He sample chamber and thermal link to the salt, the vacuum housing and thermometry, the NMR spectrometer, *etc.*

About that time, when we were first ready to assemble the cryogenic apparatus, we were paid a visit by a prominent European low temperature physicist who, in assessing our experiment, was to enunciate what turned out to be the first of several unprophetic "triple negatives" that Bill Fairbank has had (and continues!) to endure over the course of his career. We had laid out the apparatus shown schematically in figure 1 on a table to provide an exploded view. After looking it over carefully and hearing a description of the proposed experiment, our visitor shook his head and pronounced that (a) we would never achieve temperatures below 1.2 K with the apparatus before him, but that, if we did, (b) the ^{3}He spin-lattice relaxation time would become so long at lower temperatures that we would be unable to detect the NMR signal, but that, if we did, (c) we would find no deviation from the classical $1/T$ behavior of the nuclear susceptibility. Events of the next few months were to prove our guest wrong on all counts, but at the time his evaluation was somewhat devastating to Bill Ard and me. After all, Fairbank had never done an experiment requiring such low temperatures, and had no previous experience with adiabatic demagnetization techniques. However, Fairbank's reaction was a mixture of anger, confidence and resolve, and as a consequence we all worked harder than ever. By spring of 1954—nine months after I arrived at Duke—the experiment was successfully completed, and the Fermi behavior of liquid ^{3}He directly demonstrated with an approximate magnetic degeneracy temperature of 0.45 K [13]. The original results are shown in figure 2.

The spin-lattice relaxation time was measured to be approximately 30 seconds over the temperature range from 0.23 K to 1.2 K, and shown to be determined primarily by interaction of the ^{3}He nuclei with surfaces within the sample cell. Surface spin-lattice relaxation effects in liquid ^{3}He were the subject of a number of later investigations [14], while the inherent

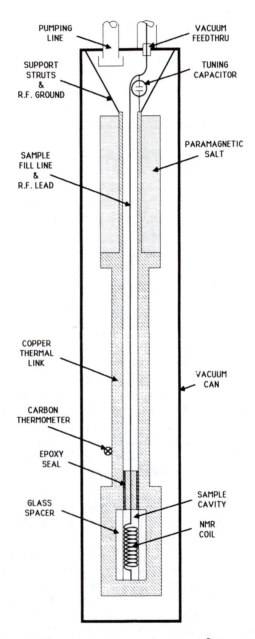

FIGURE 1. Schematic diagram of paramagnetic salt and ^{3}He sample cavity assembly used for studies of liquid ^{3}He below 1 K.

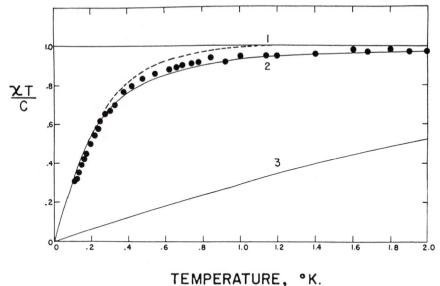

TEMPERATURE, °K.

FIGURE 2. Plot of $\chi T/C$ *versus* T. (χ = molar nuclear magnetic susceptibility of ^{3}He, T = absolute temperature, C = normalizing Curie constant.) The horizontal line 1 represents the classical Curie law expected from Boltzmann statistics, curve 3 represents an ideal Fermi-Dirac gas with the same density and atomic mass as liquid ^{3}He ($T_F = 5$ K), curve 2 represents an ideal Fermi-Dirac gas with a degeneracy temperature $T_F = 0.45$ K. The ●'s represent the experimental points normalized to curve 2 at 1.2 K [13]. The dashed curve represents the experimental data normalized to curve 1, the Curie curve, at 1.2 K. Later work showed that the proper normalization is as shown in the dashed curve.

bulk relaxation behavior was determined in 1959 by R. H. Romer [15]—in Fairbank's laboratory at Duke—and was shown to be in reasonable agreement with the theory of Bloembergen, Purcell and Pound [16].

All the NMR work described above was done in one of Gordy's laboratories using his 12 inch Varian electromagnet. Upon completion of that phase of the program, we set about establishing an independent NMR laboratory that would be dedicated to Fairbank's research. The centerpiece was a large new electromagnet and current-regulated power supply designed by Professor Harry Owen of the Duke Electrical Engineering Department, whom Fairbank had earlier commissioned for the task. The magnet yoke was fabricated by an outside contractor and delivered to Duke in late 1954, if memory serves. The magnet coils were mounted at Duke by Fairbank and his students amidst a sea of glyptol that ruined all our clothes. (Fairbank always wore a business suit and tie regardless of the task at hand.) After what seemed to me at the time to be an eternity—but in fact must have been no more than a few months—everything was up and running in the new laboratory. Figure 3 provides a view of the new experimental rig, including the original gas handling system we had moved from Gordy's lab and modified to permit measurements of the pressure dependence of the ^{3}He nuclear susceptibility.

Our first series of experiments in the new laboratory revealed that the characteristic temperature at which the liquid ^{3}He nuclear susceptibility deviates from the classical Curie behavior decreases with increasing pressure [17], as shown in figure 4. This is in the opposite direction from that which would be expected for an ideal Fermi-Dirac gas.

It was already clear, however, before these experiments were completed, that the ideal Fermi-Dirac gas model, even with T_F arbitrarily adjusted to 0.45 K to provide the best fit to the nuclear susceptibility data at the saturated vapor pressure, is unsatisfactory for liquid ^{3}He. This became quite apparent when Abraham, Osborne and Weinstock extended their measurements of the heat capacity down to 0.23 K [18].

In 1956 and 1957 Landau published his Fermi liquid theory [19], which is based on the Fermi-Dirac distribution function but introduces interactions between the atoms. Later experiments, most notably the elegant series in John Wheatley's laboratory at the University of Illinois on the temperature dependences of thermodynamic and transport properties, demonstrated that the Landau theory accounts very well for the properties of liquid ^{3}He above the superfluid transition temperature [20]. There still remains, however, the problem of establishing from first principles that liquid ^{3}He is properly represented by the system of elementary excitations, or quasi-particles, proposed in Landau's model. In any event, the experiments clearly demonstrated that the effect of Fermi statistics is manifested in a particularly direct and striking manner in the behavior of liquid ^{3}He.

Before I left Duke in 1957, Fairbank and I attempted to extend our studies to solid ^{3}He in order to test Pomeranchuk's suggestion that the spin ordering temperature would be much lower in the solid, with the consequence that adiabatic compression from liquid to solid at sufficiently low temperatures would result in cooling [21]. Indeed, this has since become the technique of choice for research on ^{3}He in the millikelvin range. Though we were unable to obtain reproducible results on the nuclear susceptibility of solid ^{3}He, we did show that when the solid was allowed to transform adiabatically into a liquid, heating occurred at the melting point for temperatures below about 0.4 K, and cooling occurred above this temperature [22], in agreement with Pomeranchuk's prediction. We also observed that the spin-lattice relaxation time near the melting pressure over the accessible temperature regime is markedly less in the solid than in the liquid phase ($\leq$ 0.5 sec *versus* > 30 sec), and this served as a convenient means for identifying the solid-liquid transition as the sample was allowed to warm

FIGURE 3. Photograph of the experimental apparatus used during 1955–56 for NMR measurements on liquid and solid ³He, and for studies of the ³He–⁴He phase separation.

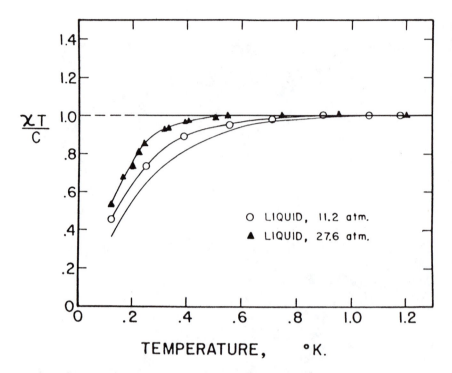

FIGURE 4. Plot of $\chi T/C$ *versus* T for liquid ^{3}He under pressure. (χ = molar nuclear magnetic susceptibility of ^{3}He, T = absolute temperature, C = normalizing Curie constant.) The horizontal line represents the classical Curie law. The lowest curve, without data points, indicates the behavior of the liquid ^{3}He susceptibility under saturated vapor pressure (same as dashed curve of figure 2).

at fixed density. This series of measurements suggested that the entropy of the solid is greater than that of the liquid below approximately 0.4 K and that a minimum in the melting pressure curve as predicted by Pomeranchuk should occur at that point, as was later confirmed by others. However, in our experiments, we could not rule out the possibility that friction was responsible for the heating observed below 0.4 K.

Fairbank and his students, notably Dwight Adams and John Goodkind, continued NMR research at Duke on liquid and solid ^{3}He until Fairbank's move to Stanford in 1959. At that time Horst Meyer assumed the leadership of the Duke low temperature group. The definitive work on the question of spin ordering in solid ^{3}He initiated at Duke in the late 1950's is described in this volume in the paper by Adams and Goodkind and the paper by Pipes, McQueeney, Van Degrift and Bowers.

Low temperature physics at Duke has continued to flourish through the years under the joint leadership of Horst Meyer and Bill's brother

Henry, who moved from Yale in 1962. It is worthy of note that the exciting discovery in 1972 of the superfluid phases of liquid ^{3}He was made
jointly by former students of Meyer and Henry Fairbank—Robert Richardson and David Lee, respectively—and their student at Cornell, Douglas
Osheroff [23].

2.   THE ^{3}HE-^{4}HE PHASE SEPARATION

Perhaps personally the most frustrating, and ultimately the most rewarding, research in which I was involved as a student at Duke was the NMR
study of ^{3}He-^{4}He solutions. After our work showing that the characteristic
spin degeneracy temperature of pure liquid ^{3}He decreases with the application of pressure, Bill Fairbank suggested that we study the spin ordering
properties of ^{3}He diluted by ^{4}He. This study eventually led us to the discovery that liquid ^{3}He-^{4}He mixtures separate into ^{3}He-rich and ^{4}He-rich
phases below about 0.8 K [24].

At the time this project was initiated in 1955, neither of us was aware
that the phase separation had been predicted in 1954 by Prigogine et al.
[25] and in 1955 by Chester [26] on fundamental theoretical grounds, and
even earlier by Henry Sommers in an elegant and convincing analysis of
vapor pressure data of liquid ^{3}He-^{4}He mixtures above 1.3 K [27]. Thus
what we anticipated to be a rather straightforward extension of nuclear
susceptibility measurements to lower ^{3}He densities led us initially into all
sorts of problems of nonreproducibility and apparently nonsensical results.
However, before long it occurred to Fairbank that a phase separation could
be the source of confusion in our experiments. To test for this, we moved
the sample to a region of the magnet where the field gradient was large, so
that the NMR line width and shape were governed by the magnetic field
distribution over the sample volume. Indeed, significant line shape changes
were observed as ^{3}He-^{4}He samples were allowed to warm after initial cooling of the sample to about 0.2 K by adiabatic demagnetization. However,
we still could not make much sense out of the data, which continued to
be irreproducible. It later became clear that the problem was a result of
surface tension effects at the boundary separating the two liquid phases.
As the sample warmed, the phase boundary moved through an inadvertent
constriction caused by bonding resin that had found its way into the sample chamber prior to curing. Surface tension prevented the sample from
achieving thermal equilibrium as the temperature rose; rather we would observe nonreproducible and discontinuous changes in signal size and shape
whenever the block caused by surface tension was broken. Indeed, before
this problem was identified, we for a short while believed that the data
suggested that there was no phase separation, and Fairbank presented a
paper to that effect in New Orleans in late 1955.

All the experimental problems were solved and the phase separation clearly established when the old sample cell of figure 1 was replaced by the new one shown in figure 5a. The new cell was divided into three vertically arranged sections, connected by small holes. When this container was placed in a magnetic field with strong gradient from top to bottom, the solutions in each of the three sections came to resonance sequentially, at constant excitation frequency, as the current in the magnetic field coils was swept linearly in time. Thus the resonance line was split into three separate components, corresponding respectively to magnetic resonance of the ^{3}He in each of the three sections. Since the strength of the NMR signal is proportional to the number of ^{3}He nuclei in the vicinity of the resonance coil, changes in relative strengths of the three resonance peaks as a function of temperature gave a measure of the relative ^{3}He concentrations in the three sections of the sample cell [28]. Representative three-peaked resonance signals are shown in figure 5b, and signal amplitudes as a function of temperature are plotted in figure 6 for 40% and 60% solutions of ^{3}He in ^{4}He.

The original phase diagram determined by analysis of the data in figure 6 is indicated by the data points in figure 7. The solid curve is the currently accepted phase diagram based on more accurate subsequent work of a number of investigators [29,30]. The agreement is quite good for the lower ^{3}He concentration branch, but our results deviate substantially from the accepted phase separation line at higher ^{3}He densities. The error in our work evidently was a consequence of the crudeness of the method, together with assumptions that had to be made in data reduction. We invoked the law for perfect solutions, and assumed the ^{3}He susceptibility per atom to be the same in solution as in pure ^{3}He at the same temperature. It was later shown that in fact the susceptibility exhibits the temperature dependence which one would calculate from equation (2) assuming an effective (Landau quasi-particle) mass of 2.5 m_3 [31]. We did point out the merging of lambda and phase separation lines below 0.8 K.

Special note must be made of the 1965 demonstration by Edwards et al. of the nonzero ($\sim 6\%$) solubility of ^{3}He in ^{4}He at 0 kelvin [31]. This important discovery opened the way to dilution refrigeration as the modern method for achieving and maintaining temperatures down to the 10 mK range.

3.   THE ^{3}HE FERMI LIQUID—REVISITED

The exciting era of the ^{3}He Fermi liquid began to fade more than a decade ago, giving way to the new frontiers of superfluid ^{3}He and solid ^{3}He phases at ultra low temperatures. However, recent technological advances are now opening the door to yet one more intriguing area of Fermi liquid research, spin polarized liquid ^{3}He [32].

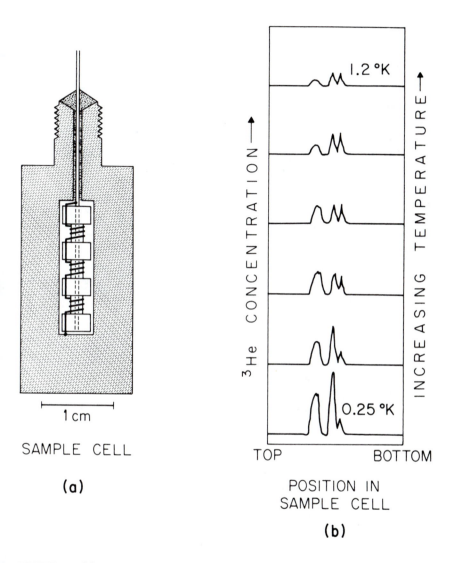

SAMPLE CELL

(a)

POSITION IN
SAMPLE CELL

(b)

FIGURE 5. (a) Three-section sample chamber used for phase separation measurements. Coarse hatch represents copper. Close hatch represents bakelite. Solid black represents insulating epoxy bonding resin. (b) Representative nuclear magnetic resonance signals *versus* temperature for 40% ³He in ⁴He, taken in three-section cavity shown in figure 5a. Temperature increases from 0.25 K at upper left to 1.2 K at lower right. The right peak corresponds to the bottom section, the central peak to the middle section, and the left peak to the top section of the cavity, as described in the text.

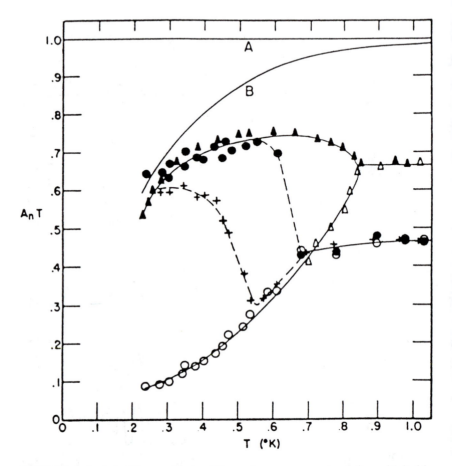

FIGURE 6. A plot of the amplitude of the nuclear resonance signal A_n multiplied by the temperature T *versus* T for each of three peaks shown in figure 5b. The peaks were normalized at 1.2 K to the value for pure ^{3}He. Curve A represents the Curie law for pure ^{3}He. Curve B represents the measured nuclear susceptibility for pure ^{3}He. The triangles represent 60% ^{3}He; the circles and +'s represent 40% ^{3}He. As the temperature decreases in the 60% sample the susceptibility follows approximately the Curie law until the temperature reaches 0.83 K. There the values for the top section (filled triangles) and the bottom section (open triangles) diverge rapidly due to phase separation. The 40% solution is represented by • top cavity, + middle cavity, o bottom cavity. As the temperature decreases below 0.67 K, the signal (•) representing the top section increases rapidly, corresponding to the phase boundary moving downward through the top section. As the temperature is further decreased, the phase boundary moves downward into the central section, causing the signal (+) in this section to rise rapidly. This occurs at 0.54 K. Once the phase separation occurs, the ^{3}He concentrations in the two phases are independent of the initial mixture. Hence the entire phase diagram can in principle be mapped out with a single initial mixture.

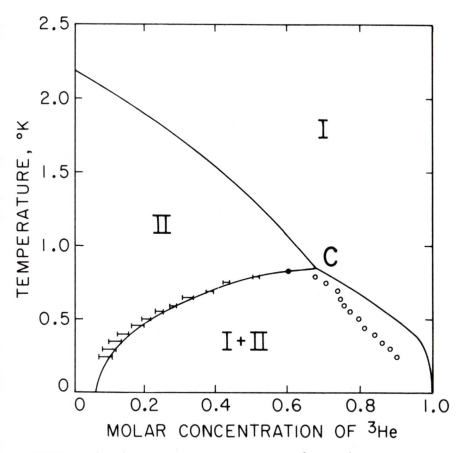

FIGURE 7. The phase separation and lambda lines of ^{3}He and ^{4}He solutions. The data points (⊢ and o) are from the original paper of Walters and Fairbank [24]. The solid lines represent the more accurate descriptions resulting from later work [29,30].

It has long been recognized that high polarization of the nuclear spins would profoundly alter the macroscopic thermodynamic and transport properties in liquid ^{3}He, but until recently there appeared to be no feasible approach to exploring this regime experimentally.

Spin polarizing liquid ^{3}He—forcing or "tricking" the nuclei into parallel rather than the antiparallel alignment characteristic of Fermi systems— would have the effect that only one-half of the spin states in the Fermi sea would be populated, raising the Fermi energy E_F in consequence. Thus, we are discussing the possibility of attaining a macroscopic excited state of the Fermi liquid, conceptually analogous to the situation that would result if one decoupled the electron spins in the conduction band of a metal and forced them into parallel alignment.

Lhuillier and Laloë have recently investigated theoretically the properties of an ensemble of nuclear spin polarized ^{3}He atoms, and quantitatively predicted the changes that can be expected in transport properties, liquid-vapor and liquid-solid phase diagrams, and solubility of ^{3}He in ^{4}He in a spin polarized ^{3}He sample [33].

Two methods have been proposed for achieving, at least partially, the spin polarized state of liquid ^{3}He. In 1979 Castaing and Nozières pointed out that rapid melting of solid ^{3}He, spin polarized in a strong magnetic field at very low temperatures, would result in a transient spin polarized liquid for approximately the nuclear spin-lattice relaxation time (several hundred seconds) [34]. Several groups have recently initiated experimental efforts along these lines.

The other approach is based on an extension of the optical pumping method for spin polarizing low density ^{3}He gas developed some years ago by Colegrove, Schearer and Walters [35], which can be described briefly as follows. A weak electrical discharge is maintained in a helium gas sample at a pressure of a few torr, to produce a steady state population of the metastable 2^3S_1 state (ground state of orthohelium) of about 10^{10} to 10^{11} per cm^3. Atoms in this state can be spin polarized by optically exciting transitions to the 2^3P states with circularly polarized 1083 nm resonance radiation, the pumping cycle being completed by spontaneous decay from the 2^3P states back to one of the 2^3S_1 Zeeman sublevels. A weak magnetic field is applied along the pumping light axis to provide a unique axis of quantization. The pumping light introduces net angular momentum into the atomic system by virtue of the selection rule $\Delta m = +1$ (for right-hand circular polarization) on the change in magnetic quantum number in the $2^3S_1 \to 2^3P$ transitions.

This process produces electron spin polarization of the 2^3S_1 atoms; the hyperfine interaction couples the polarization to the nuclear spins. Collisions involving excitation transfer between 2^3S_1 atoms and ground state atoms,

$$^3\text{He}(2^3S_1) \uparrow + {}^3\text{He}(1^1S_0) \downarrow \to {}^3\text{He}(1^1S_0) \uparrow + {}^3\text{He}(2^3S_1) \downarrow \quad , \qquad (3)$$

are effective in transferring the optically induced nuclear spin polarization (indicated by the arrows) from the metastable atomic system to the much more highly populated ground state system. Spin angular momentum is well conserved in this process, so that ground state ^{3}He atoms emerging from reaction (3) have in general each gained a unit of angular momentum at the expense of their respective metastable 2^3S_1 collision partners, which in turn can again be optically pumped and recycled through reaction (3). As a result, the entire system of ground state atoms, typically 10^6 to 10^7 times more dense than the 2^3S_1 atoms being directly pumped, become nuclear spin polarized. The attainable steady state polarization is determined by balancing the angular momentum input from the pumping radiation against the angular momentum losses due to spin relaxation processes in both the ground and metastable (2^3S_1) states.

Until quite recently, the only available sources of 1083 nm pumping radiation were helium discharge lamps. Under optimum conditions such lamps can produce ^{3}He nuclear spin polarizations up to about 20–25% at room temperature for gas pressures in the vicinity of 1 torr [35].

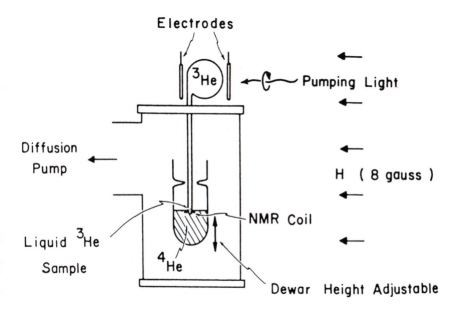

FIGURE 8. Schematic diagram of apparatus used to spin polarize liquid ^{3}He by diffusive exchange with the optically oriented saturated vapor above it [36].

In 1967, McAdams and Walters demonstrated that liquid ^{3}He could be spin polarized by optically pumping the vapor above it, relying on diffusion to transfer the polarization from gas to liquid phase [36]. Figure 8 is a schematic diagram of the experiment. Steady state spin polarization of nearly 0.1% was achieved at 0.93 K in a 20 mm^3 liquid sample in an applied magnetic field of 8 gauss. This corresponds to polarization enhancement by a factor $> 10^3$ over the thermal equilibrium value, but is far below the polarization attainable in gas samples at room temperature, and much too small for interesting quantum fluid experiments. The primary loss of polarization was traced to very rapid spin disorientation as a result of collisions of the ^{3}He atoms with the container walls in the gas just above the liquid surface [37]. Not perceiving a clear means for markedly reducing polarization losses at the cold container surfaces, and recognizing the severe intensity limitations of the discharge lamps used for optical pumping, we reluctantly decided to shelve the project. However, solutions to both of these problems appear to have emerged in recent years, so the experiment is currently being resurrected.

In 1975, Barbé, Laloë and Brossel reported that ^{3}He nuclear spin relaxation at low temperature cell walls could be dramatically inhibited by solid H_2 coatings [38]. Hopefully this technique will go a long way toward solving the wall relaxation problem in transfer of polarization from gas to liquid.

The problems associated with limitations in optical pumping radiation intensity have been overcome only very recently with the development of cw color center lasers [$(F_2^+)^*$ in NaF] tunable over the 1083 nm pumping band [39,40]. Laser optical pumping of ^{3}He was first reported by Leduc, Trénec and Laloë in 1980 [41], and only in recent weeks that group has achieved 70% polarization in ^{3}He gas at room temperature [42] with a cw laser power of about 300 mW [40]. A similar laser developed at Rice [39] has been used to optically orient the electron spins in a ^{4}He(2^3S) beam used for surface physics experiments [43], with 66% polarization so far realized [44], and also to produce 80% polarization of ^{4}He(2^3S) atoms in a flowing helium afterglow apparatus that is the source of the Rice spin polarized free electron beam [45].

Thus it would seem that the time has finally arrived for research on spin polarized ^{3}He Fermi liquids, more than 25 years after Bill Fairbank's pioneering work at Duke. Looking back, it seems that so much was accomplished in such a short period at Duke. Those were good years for Bill, and exciting and rewarding ones for his students. The good years continue at Stanford, and I know that Bill's devotion to physics and his bold, imaginative leadership are as inspirational to his present colleagues and students as they were to us at Duke 25 years ago.

References

[1] S. G. Sydoriak, E. R. Grilly and E. F. Hammel, *Phys. Rev.* **75**, 303 (1949). Until this time there was a question as to whether ^{3}He would liquefy at all. F. London and O. K. Rice, *Phys. Rev.* **73**, 1188 (1948), had predicted that it "cannot exist in a liquid phase at any temperature, at least not for normal pressure," and that it was "questionable whether solutions of much more than 1% ^{3}He could exist in a liquid phase." However, de Boer and Lunbeck, *Physica* **14**, 410 (1948), predicted, by means of a quantum theory of corresponding states, vapor pressures and critical temperature and pressure that agreed well with subsequent measurements. On

the experimental side, Henry Fairbank and collaborators, *Phys. Rev.* **74**, 345 (1948), on the basis of measurements of the vapor pressure difference between ^{4}He and solutions enriched to 0.16 percent ^{3}He, had calculated vapor pressures of pure ^{3}He using ideal solution concepts, and had predicted a normal boiling point of 2.9 K, not far from the 3.2 K Sydoriak *et al.* measured.

[2]   B. M. Abraham, D. W. Osborne and B. Weinstock, *Phys. Rev.* **80**, 366 (1950).

[3]   B. Weinstock, B. M. Abraham and D. W. Osborne, *Phys. Rev.* **85**, 158 (1952).

[4]   F. London, *Nature* **141**, 643 (1938); *Phys. Rev.* **54**, 947 (1938).

[5]   L. Tisza, *Nature* **141**, 913 (1938); *Compt. Rend.* **207**, 1035, 1186 (1938); *J. Phys. Rad.* **1**, 164, 350 (1940); *Phys. Rev.* **72**, 838 (1947). Fritz London's book *Superfluids II* (John Wiley and Sons, 1954) contains a comprehensive review of the properties of an ideal Bose-Einstein gas, and of the two-fluid model of liquid helium.

[6]   S. G. Sydoriak and E. F. Hammel, *Proceedings of the International Conference on the Physics of Very Low Temperatures* (Massachusetts Institute of Technology, 1949), p. 42.

[7]   A brief description of the properties of an ideal Fermi-Dirac gas, and Landau's Fermi liquid theory, with comparisons to properties of liquid ^{3}He, may be found in J. Wilks, *An Introduction to Liquid Helium* (Clarenden Press, Oxford, 1970).

[8]   L. Goldstein and M. Goldstein, *J. Chem. Phys.* **18**, 538 (1950).

[9]   Previously, Hammel, Laquer, Sydoriak and McGee, *Phys. Rev.* **86**, 432 (1952), had demonstrated the absence of nuclear ferromagnetism in liquid ^{3}He by means of static magnetic susceptibility measurements.

[10]   See, for example, F. Reif, *Statistical Physics* (McGraw-Hill, 1965), p. 166.

[11]   W. M. Fairbank, W. B. Ard, H. G. Dehmelt, W. Gordy and S. R. Williams, *Phys. Rev.* **92**, 208 (1953).

[12]   B. Weinstock, B. M. Abraham and D. W. Osborne, *Phys. Rev.* **89**, 787 (1953).

[13]   W. M. Fairbank, W. B. Ard and G. K. Walters, *Phys. Rev.* **95**, 566 (1954).

[14]   G. Careri, in *Helium Three*, edited by J. G. Daunt (Ohio State University Press, 1960), p. 26; F. J. Low, H. E. Rorschach and H. A. Schwettman, *ibid.*, p. 29; G. K. Walters, *ibid.*, p. 37; F. J. Low and H. E. Rorschach, *Phys. Rev.* **120**, 1111 (1960).

[15] R. H. Romer, *Phys. Rev.* **117**, 1183 (1960).

[16] N. Bloembergen, E. M. Purcell and R. V. Pound, *Phys. Rev.* **73**, 679 (1948).

[17] G. K. Walters and W. M. Fairbank, *Phys. Rev.* **103**, 263 (1956); W. M. Fairbank and G. K. Walters, *Proceedings of the Ohio State Conference on Liquid Helium Three* (Ohio State University Press, 1957), p. 205. These papers also report the first measurements of liquid ^{3}He density and isothermal compressibility as a function of pressure, the results of which were confirmed and extended by Sherman and Edeskuty, *Ann. Phys. (New York)* **9**, 522 (1960). Thomson, Meyer and Adams, *Phys. Rev.* **128**, 509 (1962), later extended the susceptibility measurements over a wider range of temperatures and pressures.

[18] B. M. Abraham, D. W. Osborne and B. Weinstock, *Phys. Rev.* **98**, 551 (1956).

[19] L. D. Landau, *Sov. Phys.–JETP* **3**, 920 (1957); *ibid.*, **5**, 101 (1957); *ibid.*, **8**, 70 (1959). See also A. A. Abrikosov and I. M. Khalatnikov, *Repts. on Progress in Phys.* **22**, 329 (1959), and [7].

[20] For reviews of experimental tests of Landau's theory of Fermi liquids, see J. C. Wheatley, in *Quantum Fluids*, edited by D. F. Brewer (North-Holland, 1966), p. 183; and J. C. Wheatley, in *Progress in Low Temperature Physics*, edited by C. J. Gorter (North-Holland, 1970), Vol. VI, p. 77.

[21] I. Pomeranchuk, *Sov. Phys.–JETP* **20**, 919 (1950).

[22] W. M. Fairbank and G. K. Walters, *Proceedings of the Ohio State Conference on Liquid Helium Three* (Ohio State University Press, 1957), p. 220.

[23] D. D. Osheroff, R. C. Richardson and D. M. Lee, *Phys. Rev. Lett.* **28**, 885 (1972); D. D. Osheroff, W. J. Gully, R. C. Richardson and D. M. Lee, *Phys. Rev. Lett.* **29**, 920 (1972). For a complete account of the discovery and properties of the superfluid phases of liquid ^{3}He, see D. M. Lee and R. C. Richardson, in *The Physics of Liquid and Solid Helium, Part II*, edited by K. H. Bennemann and J. B. Ketterson (John Wiley and Sons, 1978), p. 287.

[24] G. K. Walters and W. M. Fairbank, *Phys. Rev.* **103**, 262 (1956); W. M. Fairbank and G. K. Walters, *Proceedings of the Ohio State Conference on Liquid Helium Three* (Ohio State University Press, 1957), p. 226.

[25] I. Prigogine, R. Bingen and A. Bellemans, *Physica* **20**, 633 (1954).

[26] G. V. Chester, *Proceedings of the Paris Conference on the Physics of Low Temperatures, 1955* (Centre National de la Recherche Scientifique and UNESCO, 1956), p. 385.

[27]   H. S. Sommers, *Phys. Rev.* **88**, 113 (1952).

[28]   More sophisticated versions of this method for probing relative concentrations of a specific constituent as a function of position in a sample have been exploited in recent years in the development of NMR tomography. See, for example, P. C. Lauterbur, *Nature* **242**, 190 (1973).

[29]   See K. W. Taconis and R. de Bruyn Ouboter, in *Progress in Low Temperature Physics*, edited by C. J. Gorter (North-Holland, 1964), Vol. IV, p. 38, for a detailed description of work prior to the 1965 discovery [30] of the 6% solubility of ^{3}He in ^{4}He at absolute zero. A more recent review by G. Ahlers, in *The Physics of Liquid and Solid Helium, Part I*, edited by K. H. Bennemann and J. B. Ketterson (John Wiley and Sons, 1976), describes later studies of the tricritical point, where the lambda line intersects the phase separation line.

[30]   D. O. Edwards, D. F. Brewer, P. Seligman, M. Skertic and M. Yagub, *Phys. Rev. Lett.* **15**, 773 (1965).

[31]   D. L. Husa, D. O. Edwards and J. R. Gaines, *Phys. Lett.* **21**, 28 (1966).

[32]   The spin polarization of a spin one-half system such as ground state ^{3}He is defined as
$$P = \frac{n_+ - n_-}{n_+ + n_-} \quad ,$$
where n_+ and n_- are the densities of spin-up (magnetic quantum number $m = +1/2$) and spin-down ($m = -1/2$) atoms, respectively.

[33]   C. Lhuillier and F. Laloë, *J. Physique* **40**, 239 (1979).

[34]   B. Castaing and P. Nozières, *J. Physique* **40**, 257 (1979).

[35]   F. D. Colegrove, L. D. Schearer and G. K. Walters, *Phys. Rev.* **132**, 2561 (1963). This work extended the earlier demonstration of optically-induced electron spin polarization of metastable ^{4}He(2^3S) atoms by Colegrove and Franken, *Phys. Rev.* **119**, 680 (1960), and Schearer, *Advances in Quantum Electronics* (Columbia University Press, 1961), p. 239.

[36]   H. H. McAdams and G. K. Walters, *Phys. Rev. Lett.* **18**, 436 (1967); H. H. McAdams, *Phys. Rev.* **170**, 276 (1968). I believe the first suggestion of the interesting properties expected of spin-polarized ^{3}He was put forth in the 1967 paper.

[37]   W. A. Fitzsimmons, L. L. Tankersley and G. K. Walters, *Phys. Rev.* **179**, 156 (1969).

[38]   R. Barbé, F. Laloë and J. Brossel, *Phys. Rev. Lett.* **34**, 1488 (1975).

[39]   K. W. Giberson, C. Cheng, F. B. Dunning and F. K. Tittel, *Appl. Optics* **21**, 172 (1982).

[40] G. Trénec, P. J. Nacher and M. Leduc, *Optics Comm.* (to be published).

[41] M. Leduc, G. Trénec and F. Laloë, *J. Physique* **41-C7**, 75 (1980).

[42] P. J. Nacher, M. Leduc, G. Trénec and F. Laloë (private communication).

[43] T. W. Riddle, M. Onellion, F. B. Dunning and G. K. Walters, *Rev. Sci. Instr.* **52**, 797 (1981).

[44] K. W. Giberson, C. Cheng, M. Onellion, F. B. Dunning and G. K. Walters, submitted to *Rev. Sci. Instr.*

[45] P. J. Keliher, R. E. Gleason and G. K. Walters, *Phys. Rev. A* **11**, 1279 (1975); *Rev. Sci. Instr.* **50**, 1 (1979).

The Quest for
Nuclear Spin Ordering in Solid ^{3}He

E. D. Adams and J. M. Goodkind

The availability of ^{3}He, beginning in the early 1950's, fostered an enormous burst of activity in low temperature physics. William Fairbank's involvement in the field began even before this time with his heroic project at Amherst College to concentrate ^{3}He by distillation from enormous quantities of ^{4}He. The earliest experiments were performed on the liquid state, but it was only a matter of increasing the pressure to produce the solid [1–3]. The liquid was of interest because it provided the Fermi particle counterpart to the ^{4}He base fluid. As such it was expected that the Fermi statistics would force the spins to align with equal numbers in opposite directions in much the same manner as for the conduction electrons in a metal.

The solid at first appeared to be of interest because of the suggestion by Pomeranchuk [4] that the spins would not align until very much lower temperatures than in the liquid. He argued that the only force tending to align the spins in the solid state would be the magnetic force between the dipole moments of the nuclei. This led him to the fascinating conclusion that, over a wide range of temperatures, solid in equilibrium with liquid would have more entropy than liquid. This has proven to be the case and provides one of the techniques for refrigeration to millikelvin temperatures. However, Primakoff [5] pointed out that there should be a quantum mechanical exchange interaction in solid ^{3}He which would dominate the dipole-dipole interaction. In this case, solid ^{3}He became of fundamental importance as an extremely clean and simple paradigm for understanding the phenomenon of magnetism, which has been a puzzle since the time of the ancient Greeks.

A fundamental understanding of magnetism was not possible until the discovery of the electron with its spin and magnetic moment and the development of quantum theory. The electron magnetic moments provide the observed magnetic field when they are aligned. However, the interaction between the moments which was required to align them at the high Curie temperatures of magnetic materials needed to be thousands of times larger than the magnetic dipole-dipole force of classical electromagnetism. The only known force of sufficient magnitude was the Coulomb force between electrons, but this would not exert a torque on the spin. The riddle was resolved when Dirac and Heisenberg showed how the Pauli exclusion principle led to an *apparent* spin dependent force between electrons which interact, explicitly, only *via* the Coulomb field. This is called the "exchange interaction" and it will appear whenever the wavefunctions of interacting Fermi particles overlap. It was shown by Heisenberg that it could be represented in the form $J(\mathbf{S}_i \cdot \mathbf{S}_j)$, where J is the "exchange integral" and the $\mathbf{S}$'s are spin operators for the particles i and j.

Unfortunately, magnetic materials are not simply a collection of electrons. Though the magnetism results from the moments of the electrons, the electrons are for the most part bound in atoms. Magnetism usually results from electrons in unfilled inner shells, and, when the magnetic material is a metal, the interaction between the inner shell electrons is mediated by the conduction electrons. Thus, even though the Heisenberg Hamiltonian represents the fundamental cause of magnetism, real magnetic materials are too complicated to be understood in terms of that Hamiltonian alone. Magnetism is still an unsolved problem.

Solid ^{3}He, by contrast, can be considered as a pure collection of identical Fermi particles. The Coulomb interaction is replaced by the much weaker van der Waals force so that the exchange interaction will be correspondingly weaker and the magnetic ordering temperature correspondingly lower. X-ray diffraction reveals a well-defined body-centered cubic crystal so that the wavefunctions of the atoms are indeed localized. Thus the overlap must be restricted to nearby neighbors. Therefore the Heisenberg Hamiltonian was expected to describe the system well, and it was even hoped that the exchange integral could be computed from first principles. It appeared that solid ^{3}He would provide a clean test of the fundamental idea behind our understanding of magnetism.

The initial experiments by Fairbank and colleagues used nuclear magnetic resonance (NMR) to study the susceptibility and the spin relaxation times, τ_1 and τ_2 [1–3]. At the time of the first measurements, knowledge of the phase diagram was incomplete (the melting pressure below the minimum had not been measured). *In situ* pressure measuring devices had not come into use, so that sample density and pressure were poorly known. Some early samples of ^{3}He contained large, unknown quantities of ^{4}He which could produce unexpected effects, perhaps because of phase separation in these solid mixtures (unknown at that time). As a result of these factors which were not understood at the time, the first susceptibilities were not consistent, at times deviating first above, then below, the Curie value at temperatures as high as 0.3 K.

From the outset there has been a close interaction between theory and experiment. Paralleling the early experiments were developments in the theory by Primakoff *et al.* [5,6]. These workers realized that the large zero-point motion in the solid would produce sufficient overlap of wavefunctions to lead to a sizable exchange interaction. First calculations of the interaction gave values in the range −0.1 to −0.3 K for the low density solid. This would produce an antiferromagnetic transition at about $T_c = 0.1$ K, with anomalies in such quantities as specific heat and susceptibility. However, the susceptibility data of the Duke group [2], which now extended to 70 mK, failed to show any departure from Curie's law for the low density solid. From these results, which were confirmed by specific heat data of Edwards *et al.* [7], it was concluded that an upper limit could be placed on T_c of 10–20 mK.

Soon new susceptibility data from Duke were available to 50 mK which suggested an antiferromagnetic interaction [8]. However, even this temperature was not low enough, given the accuracy of the measurements, to permit definite conclusions to be drawn.

With the sought-after ordering now placed below the range of ready experimental accessibility, experimentalists turned to ways of investigating the exchange interaction which did not require observing anomalies near T_c. During the early 1960's several groups applied NMR techniques to measure the spin relaxation times and diffusion [3,9,10]. Although the determination of the exchange energy J from these quantities is somewhat indirect, this approach has the great advantage of being applicable in the convenient temperature range of 0.3 to 1.2 K. The most extensive work was done by Meyer *et al.* [10]. From the measurements were extracted values of the exchange interaction J defined in terms of a Heisenberg Hamiltonian,

$$H = -2\sum_{i<j} J\mathbf{I}_i \cdot \mathbf{I}_j - \sum_i \boldsymbol{\mu}_i \cdot \mathbf{B} \quad , \tag{1}$$

where $\mathbf{I}$ is the nuclear spin operator, μ the magnetic moment, and $\mathbf{B}$ the magnetic field. Theoretical calculations assumed that only a simple interaction involving two-particle nearest-neighbor exchange was important [11]. The experiments gave a largest value of $|J|$ for the solid near melting of about 0.7 mK. This would correspond to a transition temperature of $T_c \approx 2$ mK. The strength of the interaction was found to decrease rapidly as the density was increased.

Although the anticipated transition temperature had now been pushed even lower, various groups continued to look for evidence of spin ordering in bulk properties. The first success in this endeavor was made by Adams and colleagues at the University of Florida [12]. Their method made use of a sensitive capacitive pressure gauge for measuring the thermodynamic constant-volume pressure $P_V(T)$. This method proved to be advantageous because the leading term in the high temperature expansion for the pressure goes as T^{-1} and has the quantity $d \ln |J|/d \ln V \approx 18$ as a coefficient,

$$P_V(T) = \frac{3NJ}{V} \frac{\partial \ln |J|}{\partial \ln V} \left[\left(\frac{J}{kT} \right) - \frac{1}{2} \left(\frac{J}{kT} \right)^2 + \cdots \right] \quad , \qquad (2)$$

where N is the number of particles and k is Boltzmann's constant. Thus at temperatures as high as 200 mK, the temperature dependence of $P_V(T)$ is dominated by the exchange contribution. Measurements of the Florida group extended over most of the bcc phase and gave values of $|J|$ which agreed well with those obtained from NMR relaxation times. Figure 1 shows J's obtained by various workers.

With the accurate values of $|J|$ provided by the thermodynamic data and the agreement with the NMR results, the major remaining question was whether the interaction was ferromagnetic or antiferromagnetic, $i.e.$, whether $J > 0$ or $J < 0$. There was some meager evidence that $J < 0$. However, the answer to this question would require additional susceptibility data. Unfortunately such data must be very precise and go to temperatures fairly near T_c in order to detect departure from the Curie law. This may be seen from the expansion for χ,

$$\chi = \frac{N\mu^2}{VkT} \left[1 + 4 \left(\frac{J}{kT} \right) + \cdots \right] \quad , \qquad (3)$$

where the leading term is the Curie law. In spite of this difficulty, several groups mounted efforts to measure the susceptibility [13–15]. It was not until 1969 that the first data showing a convincing departure from Curie's law were reported. As it turned out, three groups reported results almost simultaneously [13–15]. These all showed a decrease in χ below the Curie value, which was interpreted as indicative of antiferromagnetism. The value of the Curie-Weiss constant θ defined by $\chi = C/(T - \theta)$ with $\theta = 4 J$ was consistent with the previous exchange constants.

At this point it was felt that a satisfactory picture of nuclear spin ordering in solid ^{3}He was emerging. All of the experimental results could be interpreted consistently within the Heisenberg model using nearest-neighbor

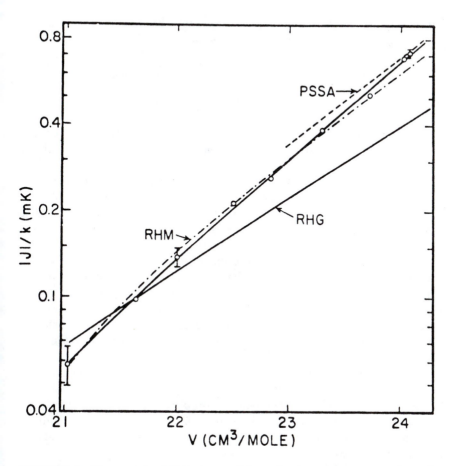

FIGURE 1. The nearest-neighbor two-particle exchange energy *versus* molar volume [12].

two-particle exchange (the HNN model). It appeared that all that remained was to cool below the expected temperature $T_N = 2$ mK (for the melting solid) to observe the antiferromagnetically ordered solid. However, a common feature of the experiments was that in all cases only the lowest order term in J (in series expansions in powers of T^{-n}) had been determined. The agreement between experiments was not too surprising since effective-field theories give the same leading terms in the series. The experiments had not really provided a crucial test of the HNN model, which theorists assured would be correct.

Beginning about 1970, several experiments were performed which demonstrated that the HNN model did not provide an adequate description of solid ³He magnetism. These experiments involved conditions of either high magnetic field or very low temperatures such that additional terms were needed in the series expansion of the partition function

$$
\begin{aligned}
\log Z \;=\;& \log 2 + \tfrac{3}{2}\left(J_{xx}/kT\right)^2 - \tfrac{1}{2}\left(J_{xxx}/kT\right)^3 + \cdots \\
& + \tfrac{1}{2}\left(\mu B/kT\right)^2\left[1 - 4J_{xzz}/kT + \cdots\right] \quad .
\end{aligned}
\tag{4}
$$

The J's are various moments involving either the exchange, x, or the Zeeman, z, part of the Hamiltonian. In the HNN model, the J's should all be identical.

The first experiment to identify J's other than J_{xx} was the high-field determination of the thermodynamic pressure [16], shown in figure 2. This experiment gave $J_{xzz} = -0.33$ mK, whereas $J_{xx} = -0.7$ mK (the J of the HNN model). These results were contrary to expectations of the HNN model and indicated that it would have to be discarded or modified in some major way. This experiment was followed by several others which confirmed the inability of the HNN model to explain observations [17,18].

In 1974, the first study of the ordered solid was reported [18]. Since then a number of results have been obtained which are beginning to provide an understanding of this system. Halperin et al. [18] measured the latent heat of solidification and from this deduced the melting curve $P(T)$ and the entropy $S(T)$. Their results extended through the ordering temperature, which turned out to be 1 mK, not 2 mK as expected from the HNN model. The entropy underwent a very abrupt drop at the transition, which has since then been verified as, indeed, first order. Dundon and Goodkind [19] made conventional measurements of the specific heat, which agreed qualitatively with the entropy of Halperin et al.

The latent heat technique was applied by Kummer et al. [20] to study the ordering in applied fields up to 1.2 T. They found that the transition temperature decreased slightly for fields up to 0.4 T. Above this field, the character of the ordering changed. There was no longer an abrupt decrease in entropy and the transition temperature increased with increasing field. The phase diagram which they deduced is shown in figure 3 (some recently discovered features have been included also). It is now generally believed that there are two ordered phases, as shown on this diagram. This phase diagram has now been extended to fields as high as 7.0 T by Godfrin et al. [21]. The high field phase boundary shows an increasing upward curvature. On the basis of current theoretical models [22], this line is expected to bend over with a maximum field for the ordered phase of $B \approx 16$ T at $T = 0$.

Two types of experiments have given information on the magnetization through the transition. Using a SQUID, Prewitt and Goodkind [23] determined the static magnetization to well below T_N. In the ordered phase

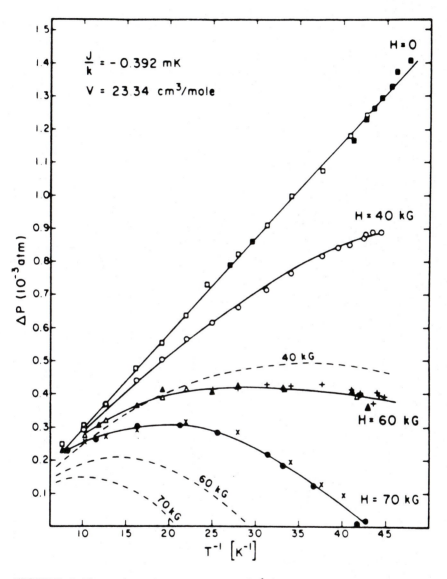

FIGURE 2. Thermodynamic pressure *versus* T^{-1} for various magnetic fields. The dashed lines show the behavior expected from the HNN model [16].

the magnetization had only about 1/3 of its value just above T_N. This provided fairly convincing evidence that the ordering was antiferromagnetic. Adams *et al.* [24] applied a thermodynamic analysis to the melting pressure in applied fields to deduce the magnetization. The data for $B < 0.4$ T confirmed the drop in magnetization at T_N. For fields

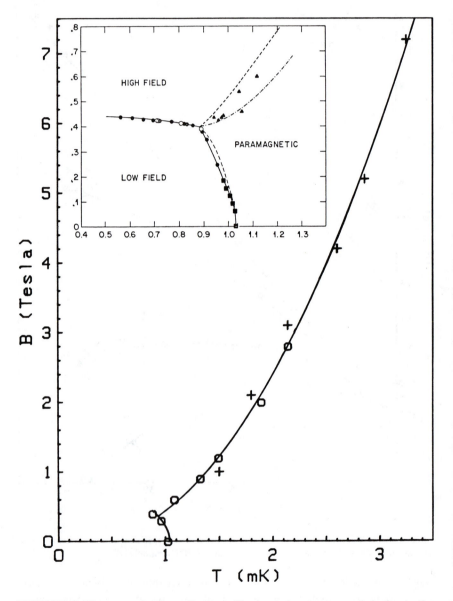

FIGURE 3. The magnetic phase diagram. The inset shows an expanded view in the region of the "triple point" [20,21,30].

above 0.4 T, the magnetization continued to increase through the ordering temperature.

Prewitt and Goodkind [25] have extended their measurements to several different magnetic fields up to 0.58 T. Their results are shown in figure 4 in the form H/M versus T. For all fields below 0.4 T, they found the magnetization to decrease abruptly at T_N, as in their previous experiment. However, for fields greater than 0.4 T, there was a well-defined temperature at which the magnetization rose above the Curie-Weiss value. This temperature increased with increasing field and was interpreted to be on the second order phase boundary reported earlier by Kummer et al. [20]. Within the high field phase the magnetization leveled off at about 1/2 the saturation magnetization, confirming the evidence for a high magnetization previously deduced from the melting pressure [24].

For fields of 0.42 and 0.46 T, the magnetization showed a cusp at still lower temperatures. This signaled another phase boundary which joined the two previously found at about 0.4 T and 0.9 K in a "triple point" (see figure 3) [26]. The existence of this phase boundary and its location, separating a low field phase from a high field phase, are now well established [27].

We will now discuss recent NMR experiments which have changed rather drastically our view of the ordered phases. NMR allows one to examine the ordered phases on a microscopic scale and to obtain information not available through the study of bulk properties. The magnetic structure cannot be deduced from the observed NMR spectrum; however, structures inconsistent with the observed spectrum can be eliminated. Two NMR studies of the ordered phases were reported simultaneously in 1980 [28,29]. Adams et al. [28] used Pomeranchuk cooling with the drawback that they could cool only slightly below T_N and their NMR coil contained a polycrystalline mixture of ordered and disordered solid. They were able to go to fields as high as 3.0 T, well into the high field phase. Osheroff et al. [29] used nuclear cooling which allowed much better control of temperature and crystal growth. They studied single crystals at temperatures well below T_N.

Both experiments showed a large shift in the NMR frequency above the Larmor frequency in the low field phase. The prevailing theoretical view at the time was that this phase was simple cubic antiferromagnetic. Such a cubic magnetic structure would have very little anisotropy in the dipole energy and no frequency shift. The observed large shift allowed this structure to be excluded immediately.

Osheroff et al. [29] obtained detailed spectra of the resonances in the low field phase, one such example being shown in figure 5. From the form of the antiferromagnetic spectra, they were able to arrive at important conclusions concerning the symmetry of the magnetic sublattices. Based on the symmetry of the low field phase, they proposed the structure shown

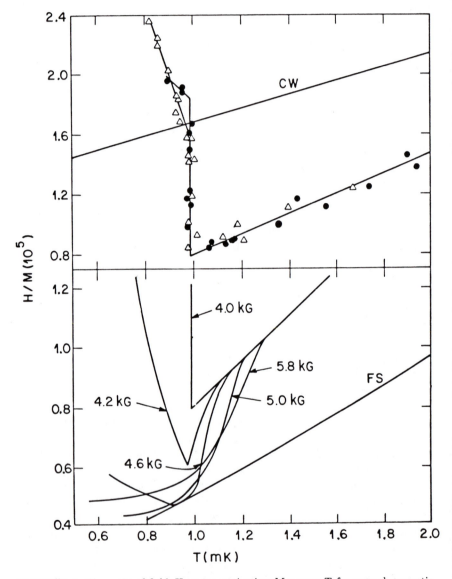

FIGURE 4. The ratio of field H to magnetization M *versus* T for several magnetic fields. The lines CW and FS are the Curie-Weiss law with $\theta = -2.5$ mK and the free-spin behavior, respectively. The circles and triangles are for fields of 0.8 and 0.4 T, respectively [25].

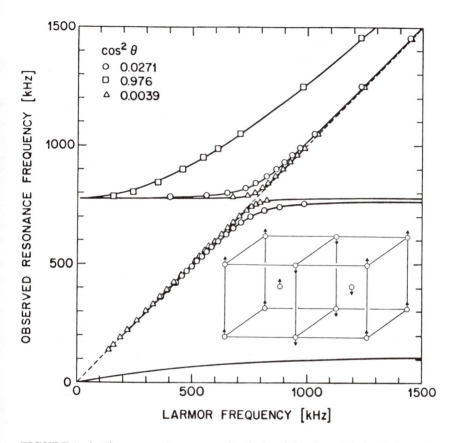

FIGURE 5. Antiferromagnetic resonance for the low field phase. The inset shows the proposed structure [29].

in the inset of figure 5. This consists of pairs of planes normal to the cube axis with all spins up, the next pair of planes having all spins down.

Instead of the large shifts discussed above, the spectra of Adams *et al.* for fields greater than 0.4 T showed a much smaller positive shift of only about 3 kHz independent of field [28]. Such a shift does not of itself imply an ordered antiferromagnetic state. A highly magnetized sample would produce a shift in frequency, because of the demagnetizing field, which would depend on sample geometry. For a sample in the shape of a long cylinder with its axis along the field, as used by Adams *et al.*, the observed 3 kHz shift would correspond to a magnetization of about 0.5 the saturation magnetization M_s. Godfrin *et al.* [21] have obtained $M = 0.7\ M_s$ while the data of Prewitt and Goodkind [25] correspond to $M = 0.45\ M_s$. These data suggest that there is an ordered state at high fields, but do not provide enough information to allow its structure to be deduced. The observed magnetization is consistent with that expected in the spinflopped normal antiferromagnetic phase predicted by Roger *et al.* [22].

Recently Osheroff and Cross [30] have extended the previous measurements of Osheroff *et al.* [29] to fields of about 0.6 T. Osheroff and Cross used the very different NMR spectra in the low field and high field phases to plot the location of the boundary between these two phases. From the behavior of the signal near and on crossing this phase boundary, they have concluded that it is first order. Osheroff and Cross also studied the boundary between the high field and paramagnetic phases. They found that samples held near this phase boundary showed two different demagnetization shifted lines. Again, from the behavior of the high field resonance line near the boundary, they concluded that this line was first order as well. This is unexpected since the previously cited magnetization [23,25] and entropy [20] suggest a second order transition.

Almost all the work in recent years has been on solid in equilibrium with the liquid, except that of Goodkind and colleagues [19,23,25]. Quite recently [31] the density dependence of various quantities, such as the constant-volume pressure and magnetization, through T_N has been reported. A surprising result is that T_N showed the same volume dependence, $T_N \propto V^{18}$, as J in the HNN model. The current molecular-field model which best fits the data uses various multiple-exchange parameters which might be expected to have different volume dependences [22]. In fact, in this model $J = 0$.

In spite of the 25 year history of this field and the major advances of recent years, many questions remain to be answered. On the experimental side, the magnetic lattice structure of the ordered phases must be confirmed and the phase diagram in field, temperature, and pressure space must be completely determined. More subtle questions such as the magnitude and volume dependence of the exchange parameters or the properties near the intersection of the phase boundaries will also be investigated. They will probably lead to still other questions which are not yet obvious. On the theoretical side it must be determined if multiparticle exchanges in mean field approximation provide a qualitatively correct description of the system or if a new approach is required. In addition, it must be determined if the exchange interaction can truly be calculated microscopically from first principles. Only then will we know if the apparent simplicity of this system and of the notion of exchange have provided truly fundamental knowledge.

References

[1]   W. M. Fairbank and G. K. Walters, in *Proceedings of the Symposium on Solid and Liquid* ^{3}*He*, Columbus, Ohio (1957), p. 220.

[2]   E. D. Adams, H. Meyer and W. M. Fairbank, in *Helium 3*, edited by J. G. Daunt (Ohio State University Press, Columbus, 1960), p. 57.

[3]   J. M. Goodkind and W. M. Fairbank, *ibid.*, p. 52.

[4]   I. I. Pomeranchuk, *Zh. Eksp. Theo. Fiz. [Sov. Phys.-JETP]* **20**, 919 (1950).

[5]   H. Primakoff, *Bull. Am. Phys. Soc.* **2**, 63 (1957).

[6]   N. Bernardes and H. Primakoff, *Phys. Rev. Lett.* **2**, 290 (1959); *Phys. Rev.* **119**, 968 (1960); N. Bernardes, in *Helium 3*, edited by J. G. Daunt (Ohio State University Press, Columbus, 1960), p. 115.

[7]   D. O. Edwards, J. L. Baum, D. F. Brewer, J. G. Daunt and A. S. McWilliams, in *Helium 3*, edited by J. G. Daunt (Ohio State University Press, Columbus, 1960), p. 126.

[8]   A. L. Thomson, H. Meyer and P. N. Dheer, *Phys. Rev.* **132**, 1455 (1963). William Fairbank had moved to Stanford in 1959, with Horst Meyer becoming the leader of the low temperature research at Duke.

[9]   J. M. Goodkind and W. M. Fairbank, *Phys. Rev. Lett.* **4**, 458 (1960); H. A. Reich, in *Helium 3*, edited by J. G. Daunt (Ohio State University Press, Columbus, 1960), p. 63; R. Garwin and A. Landesman, *Phys. Rev.* **133**, A1503 (1964); M. G. Richards, J. Hatton and R. P. Giffard, *Phys. Rev.* **139**, A19 (1965).

[10]   R. C. Richardson, E. Hunt and H. Meyer, *Phys. Rev.* **138**, A1326 (1965).

[11]   R. L. Garwin and A. Landesman, *Physics* **2**, 107 (1965); J. H. Hetherington, W. J. Mullin and L. H. Nosanow, *Phys. Rev.* **154**, 175 (1967).

[12]   M. F. Panczyk, R. A. Scribner, G. C. Straty and E. D. Adams, *Phys. Rev. Lett.* **19**, 1102 (1967); M. F. Panczyk and E. D. Adams, *Phys. Rev.* **187**, 321 (1969).

[13]   H. D. Cohen, P. B. Pipes, K. L. Verosub and W. M. Fairbank, *Phys. Rev. Lett.* **21**, 677 (1968); P. B. Pipes and W. M. Fairbank, *Phys. Rev. Lett.* **23**, 520 (1969).

[14]   J. R. Sites, D. D. Osheroff, R. C. Richardson and D. M. Lee, *Phys. Rev. Lett.* **23**, 836 (1969).

[15]   W. P. Kirk, E. B. Osgood and M. Garber, *Phys. Rev. Lett.* **23**, 833 (1969).

[16]   W. P. Kirk and E. D. Adams, *Phys. Rev. Lett.* **27**, 392 (1971).

[17]   D. D. Osheroff, W. J. Gully, R. C. Richardson and D. M. Lee, *Phys. Rev. Lett.* **29**, 920 (1972).

[18]   W. P. Halperin, C. N. Archie, F. B. Rasmussen, R. A. Buhrman and R. C. Richardson, *Phys. Rev. Lett.* **32**, 927 (1974).

[19]   J. M. Dundon and J. M. Goodkind, *Phys. Rev. Lett.* **32**, 1343 (1974).

[20]   R. B. Kummer, E. D. Adams, W. P. Kirk, A. S. Greenberg, R. M. Mueller, C. V. Britton and D. M. Lee, *Phys. Rev. Lett.* **34**, 517 (1975); R. B. Kummer, R. M. Mueller and E. D. Adams, *J. Low Temp. Phys.* **27**, 319 (1977).

[21]   H. Godfrin, G. Frossati, A. S. Greenberg, B. Hebral and D. Thoulouze, *Phys. Rev. Lett.* **44**, 1695 (1980).

[22]   M. Roger, J. M. Delrieu and J. H. Hetherington, *Phys. Rev. Lett.* **45**, 137 (1980).

[23]   T. C. Prewitt and J. M. Goodkind, *Phys. Rev. Lett.* **39**, 1283 (1977).

[24]   E. D. Adams, J. M. Delrieu and A. Landesman, *J. Phys. Lett. (Paris)* **39**, 190 (1978).

[25]   T. C. Prewitt and J. M. Goodkind, *Phys. Rev. Lett.* **44**, 1699 (1980).

[26]   The proper phase transition terminology to apply to this point will not be clear until the nature of the various transitions is fully established.

[27]   D. D. Osheroff, in *Proceedings of the 1981 International Conference on Low Temperature Physics LT16 Part III*, edited by W. G. Clark (North-Holland, Amsterdam, 1982), p. 1461.

[28]   E. D. Adams, E. A. Schubert, G. E. Haas and D. M. Bakalyar, *Phys. Rev. Lett.* **44**, 789 (1980).

[29]   D. D. Osheroff, M. C. Cross and D. S. Fisher, *Phys. Rev. Lett.* **44**, 792 (1980).

[30]   D. D. Osheroff and M. C. Cross, presented in [27].

[31]   For recent work' on solid ^{3}He, see *Proceedings of the 1981 International Conference on Low Temperature Physics LT16*, edited by W. G. Clark (North-Holland, Amsterdam, 1981).

Measurements of the Dependence of the Isochoric Pressure of Solid ^{3}He on Temperature, Magnetic Field and Density

P. B. Pipes, D. F. McQueeney,
C. T. Van Degrift and W. J. Bowers, Jr.

When W. M. Fairbank and his coworkers began the study of nuclear magnetism in solid ^{3}He in the 1950's there was little reason to believe that magnetic effects could be observed above 10^{-7} K where nuclear dipole interactions would become important. By 1960 it was clear, both from the experimental work of Fairbank and others and from theoretical work, that nuclear exchange would dominate the magnetic properties of solid ^{3}He just as it did in liquid ^{3}He [1]. This meant that, although one must study solid ^{3}He near zero temperature, interesting magnetic effects should be experimentally accessible.

In the period from 1960 to 1970 there was growing experimental and theoretical evidence that solid ^{3}He was a simple Heisenberg magnet with a single exchange parameter. It was predicted on the basis of indirect measurements (*e.g.,* nuclear spin relaxation times) that a nuclear antiferromagnetic transition would occur somewhere near 1 mK. All this seemed to be confirmed when, in 1969, Fairbank and Pipes [2] and two other groups [3,4] reported that the Weiss constant, θ, deduced from nuclear susceptibility measurements was negative and had a value of about −3 mK at a molar volume of 24 ml/mole. Figure 1 shows some of the data of Fairbank and Pipes. Pressure measurements at zero magnetic field, though not giving the sign of θ, were in accord with the magnitude [5].

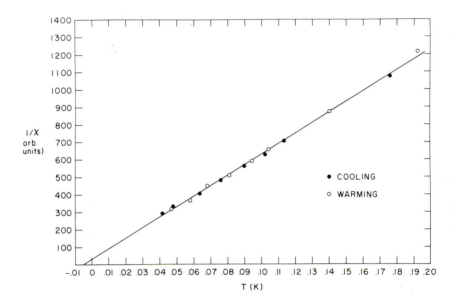

FIGURE 1. The inverse nuclear susceptibility of solid ^{3}He as a function of temperature for a molar volume of 24.2 ml/mole [2].

The confidence that magnetism in solid ^{3}He could be explained rather simply was shaken in 1971 when Kirk and Adams measured the magnetic field dependence of the spin contribution to the isochoric pressure [6]. Although the sign of the Weiss constant was confirmed, the magnitude needed to fit these experiments appeared to be about 1/2 the accepted value. Guyer pointed out that this discrepancy, if real, was model independent [7]. Further evidence that solid ^{3}He is not a simple Heisenberg antiferromagnet has been provided by numerous experiments near the magnetic transition at 1 mK [8,9,10].

More recent nuclear susceptibility measurements [11,12] have confirmed the original estimates of θ. The discrepancy with the results of Kirk and Adams has remained a mystery and, since the precision of their measurements was not high, recent theoretical attempts to explain the magnetic phase diagram of solid ^{3}He have used the more numerous susceptibility measurements as part of the data base [13]. In an attempt to clarify

the situation we have undertaken high precision measurements of the isochoric pressure of solid ³He as a function of temperature, magnetic field
and density [14].

The main experimental difficulty in these experiments is ascertaining the
temperature of the sample in the presence of a large magnetic field. Our
solution to this problem is shown in figure 2. The upper portion of the sample chamber consists of a single, 38 cm long, coin silver rod, partially split
into two parallel arms. One arm is attached to the mixing chamber of a

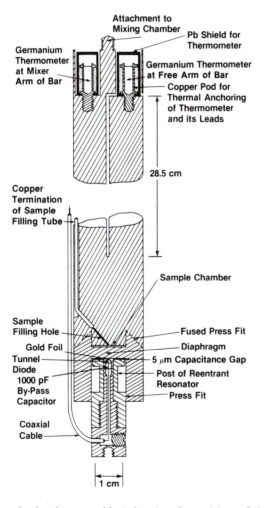

FIGURE 2. Sample chamber assembly indicating the positions of thermometers and
the details of the pressure transducer.

dilution refrigerator and supports an auxiliary germanium thermometer while the other, free, end supports the primary germanium thermometer. Both thermometers are enclosed in lead shields and located where the magnetic field is always less than 0.06 T. The calibration accuracy of the primary thermometer is 0.3 mK in the range $0.023 \text{ K} \leq T \leq 1.000 \text{ K}$. Extensive measurements of these thermometers in conjunction with heaters placed on each arm of the sample assembly allowed us to determine that the primary thermometer registered the temperature of the sample chamber to within the calibration accuracy.

The sample pressure is monitored by a capacitance transducer which utilizes a pulsed tunnel diode oscillator as a readout device [15]. Typical precision in the measurement of pressure changes is 1 Pa. Calibration of the pressure transducer was done with a dead weight tester.

Measurements have been completed at three molar volumes, 23.834, 24.163 and 24.371 ml/mole. Essentially all experimental control and data acquisition are done by a minicomputer based measurement system and each molar volume requires about 30 days of continuous running. The temperature is slowly scanned up and down between 29 mK and 500 mK at three magnetic fields, 0.0, 6.0 and 8.0 tesla. Over 2000 data points are acquired at each molar volume.

In order to compare our results with other experiments we have fit our data to a standard expression containing both phonon and magnetic terms. The functional form we used for magnetic pressure at each molar volume is

$$
P_M(T,H) = P_0 + A_{xx}\frac{1}{T} + A_{xxx}\frac{1}{T^2} + A_{xzz}\frac{H^2}{T^2}
$$
$$
+ A_{xxzz}\frac{H^2}{T^3} + A_{xzzzz}\frac{H^4}{T^4} \quad .
$$
(1)

The subscripts on the coefficients conform to the notation of Guyer [16]. Table 1 shows the results of weighted least squares fits of our data to

TABLE 1. PARAMETERS OF EQUATION (1) PERTAINING TO SPIN INDUCED PRESSURE CHANGES. The units are combinations of kPa, mK, and tesla as required by equation (1). The uncertainties are given in parentheses.

v(ml/mole)	P_0	A_{xx}	A_{xxx}	A_{xzz}	A_{xxzz}	A_{xzzzz}
23.834(0.004)	3780.7(0.4)	6.29(0.06)	−3.5(0.7)	−2.51(0.06)	7(4)	0.5(0.3)
24.163(0.006)	3498.9(0.4)	10.00(0.10)	−7.7(0.8)	−3.42(0.08)	19(6)	0.8(0.4)
24.371(0.005)	3336.8(0.4)	13.30(0.12)	−12.2(0.9)	−3.97(0.06)	26(5)	1.1(0.3)

equation (1). The residual errors are randomly distributed with standard deviations of 0.85, 0.57 and 0.90 Pa at 23.834, 24.163 and 24.371 ml/mole, respectively.

Three coefficients (A_{xxx}, A_{xxzz} and A_{xzzzz}) are only weakly determined by the data because of the small contribution they make to the pressure and correlations with each other and the dominant terms. We, therefore, chose to set the values for A_{xxx} to agree with preliminary, very low temperature, zero-field measurements of Wildes et al. [17], adjusted for consistency with our values of A_{xx}. We also forced A_{xzzzz} to obey a molar volume dependence similar to that of the dominant terms, and then deduced values for A_{xxzz} as well as A_{xx} and A_{xzz}.

It is found that the molar volume v dependence of the dominant terms is fit well by the following expressions,

$$A_{xx} = [7.95(0.01)] \left(\frac{v}{24}\right)^{33.67(0.50)} \tag{2}$$

and

$$A_{xzz} = -[2.92(0.04)] \left(\frac{v}{24}\right)^{20.7(2.0)} . \tag{3}$$

The numbers in parentheses are the uncertainties. This sort of fit for A_{xxzz} cannot be justified because of the lack of precision in that coefficient.

Again using Guyer's notation [16] we can write the magnetic contribution to the pressure predicted by an effective spin Hamiltonian as

$$
\begin{aligned}
P_M(T, H, v) \;=\; & P_0(v) + \tfrac{3R}{2}\tfrac{\delta J_{xx}^2}{\delta v}\tfrac{1}{T} + \cdots \\[6pt]
& + \tfrac{R}{2}\left(\tfrac{\mu H}{kT}\right)^2 \left[4\tfrac{\delta J_{xzz}}{\delta v} + 12\tfrac{\delta J_{xxzz}^2}{\delta v}\tfrac{1}{T} + \cdots\right] ,
\end{aligned}
\tag{4}
$$

where R is the gas constant, μ is the nuclear magnetic moment of ^{3}He, k is the Boltzmann constant, and J_{xx}, J_{xzz} and J_{xxzz} are the exchange parameters in temperature units. Thus we see that $|J_{xx}|$ and J_{xzz} can be obtained by integration of equations (2) and (3). Performing those integrations we get

$$|J_{xx}| = [0.664(0.004)] \left(\frac{v}{24}\right)^{17.34(0.25)} \text{mK} \tag{5}$$

and

$$J_{xzz} = -[0.320(0.018)] \left(\frac{v}{24}\right)^{21.7(2.0)} \text{mK} . \tag{6}$$

These are plotted in figure 3 where comparison is made with other measurements

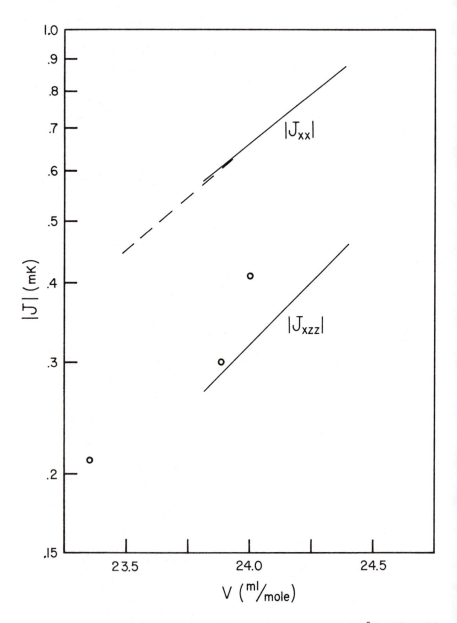

FIGURE 3. Molar volume dependence of exchange constants in solid ^{3}He. The solid lines are from this work. The dashed line is from the work of Panczyk and Adams on J_{xx} [5] and the open circles are from Guyer's analysis of the work of Kirk and Adams on J_{xzz} [6,16].

Figure 3 shows that our results for $|J_{xx}|$ agree quite well with the work of Panczyk and Adams [5]. Since many of the data points used to determine $|J_{xx}|$ were taken at high magnetic field, this agreement confirms the precision of our temperature and pressure measurements in a magnetic field. It is also clear from equation (6) and figure 3 that the indication of the work of Kirk and Adams that J_{xzz} is negative and $|J_{xzz}| < |J_{xx}|$ is borne out by our more accurate results. Also a crude estimate of $|J_{xxzz}|$ can be obtained from A_{xxzz}. At $v = 24.2$ ml/mole we get $|J_{xxzz}| = 0.500(0.100)$ mK.

Roger, Delrieu and Hetherington [13] have shown that the magnetic phase diagram of solid ^{3}He can be explained quite well using an effective spin Hamiltonian with two exchange parameters: triple exchange, J_t, and planar four spin exchange, K_p. Their best fit values for J_t and K_p predict a Weiss constant, $\theta = -2.80$ mK, at $v = 24.25$ ml/mole. This value for θ also fits the nuclear susceptibility data [2,3,4,11,12] quite well. At that molar volume we get $J_{xzz} = -0.401(0.016)$ mK, which gives $\theta = 4\ J_{xzz} = -1.604(0.064)$ mK. The inconsistency noted by Guyer [7] persists.

At the Second Symposium on Liquid and Solid Helium Three, held at Ohio State University in 1960, W. M. Fairbank said the following in an invited talk [1]:

> And finally let me say one word about the solid as a medium for further low temperature research. Experiments thus far seem to indicate that it is a worthy partner for liquid ^{4}He and ^{3}He. The experiments are interesting, difficult and often confusing, just as in the liquid. The addition of the spin to solid ^{3}He has made it a much more fertile field of research than solid ^{4}He.

Little did he realize how true those words would still ring more than 20 years later.

Acknowledgments

We would like to thank D. G. Wildes, J. Saunders and R. C. Richardson for the use of their unpublished results in our data analysis. Also, useful discussions with L. I. Zane and J. H. Hetherington are acknowledged. This work was supported in part by the National Science Foundation.

References

[1] W. M. Fairbank, in *Helium Three: Proceedings of Second Symposium on Liquid and Solid Helium Three*, (Ohio State University Press, Columbus, 1960) p. 47.

[2] P. B. Pipes and W. M. Fairbank, *Phys. Rev. Lett.* **23**, 520 (1969).

[3]   W. P. Kirk, E. B. Osgood and M. Garber, *Phys. Rev. Lett.* **23**, 833 (1969).

[4]   J. R. Sites, D. D. Osheroff, R. C. Richardson and D. M. Lee, *Phys. Rev. Lett.* **23**, 836 (1969).

[5]   M. F. Panczyk and E. D. Adams, *Phys. Rev.* **187**, 321 (1969).

[6]   W. P. Kirk and E. D. Adams, *Phys. Rev. Lett.* **27**, 392 (1971).

[7]   R. A. Guyer, *Phys. Rev.* **A9**, 1452 (1974).

[8]   W. P. Halperin, C. N. Archie, F. B. Rasmussen, R. A. Buhrman and R. C. Richardson, *Phys. Rev. Lett.* **32**, 927 (1974).

[9]   R. B. Kummer, R. M. Mueller and E. D. Adams, *J. Low Temp. Phys.* **27**, 319 (1977).

[10]   H. Godfrin, G. Frossati, A. Greenberg, B. Hebral and D. Thoulouze, *J. Physique Coll.* **41**, C7/125 (1980).

[11]   T. C. Prewitt and J. M. Goodkind, *Phys. Rev. Lett.* **39**, 1283 (1977).

[12]   D. M. Bakalyar, C. V. Britton, E. D. Adams and Y. C. Hwang, *Phys. Lett.* **A64**, 208 (1977).

[13]   M. Roger, J. M. Delrieu and J. H. Hetherington, *J. Physique Coll.* **41**, C7/241 (1980).

[14]   C. T. Van Degrift, W. J. Bowers, Jr., P. B. Pipes and D. F. McQueeney, *Phys. Rev. Lett.* **49**, 149 (1982).

[15]   C. T. Van Degrift, *Physica* **108B**, 1361 (1981).

[16]   R. A. Guyer, *J. Low Temp. Phys.* **30**, 1 (1978).

[17]   D. G. Wildes, private communication.

Critical Opalescent
Light Scattering from ^{3}He

J. Pierce Webb and Hugh Griffiths

1. INTRODUCTION

The experimental work reported in this paper was stimulated by the widespread revival of interest in critical phenomena that began more than two decades ago. The specific heat measurements of W. M. Fairbank, M. J. Buckingham and C. F. Kellers [1] at the lambda point of ^{4}He, in which they verified the logarithmic divergence predicted by Onsager's solution to the Ising problem, was a key catalyst in reviving interest in this field.

Light incident on an inhomogeneous dielectric medium is scattered by fluctuations in the dielectric constant of the medium. These fluctuations in the bulk dielectric constant ϵ are related to fluctuations in the number density n and to the molecular polarizability α through the Lorentz-Lorenz equation

$$\frac{\epsilon - 1}{\epsilon + 2} = \frac{4\pi n\alpha}{3} \quad . \tag{1}$$

In the immediate vicinity of a critical point, thermodynamic fluctuations which are normally microscopic increase dramatically. This marked growth of density fluctuations, both in amplitude and in the spatial range over which fluctuations are correlated, enhances the scattering of light, which is the phenomenon of critical opalescence.

Additionally, critical scattering becomes anisotropic as density fluctuations become correlated over distances comparable to the wavelength of the light being scattered.

The growth of isothermal density fluctuations as the critical point is approached results from a vanishing of the thermodynamic restoring force which promotes density homogeneity. This restoring force is given by the partial derivative of the chemical potential μ with respect to density ρ at constant temperature, $(\frac{\partial \mu}{\partial \rho})_T$. Since $(\frac{\partial \mu}{\partial \rho})_T = \frac{1}{\rho^2} \frac{1}{\kappa_T}$, the thermodynamic restoring force vanishes as the isothermal compressibility κ_T diverges. Thus a measurement of the light scattered by a fluid in its critical region provides a very direct and important technique for investigating the isothermal compressibility. Since the forward scattering (at zero degree scattering angle) is directly proportional to $\rho^2 \kappa_T$, a measurement of forward scattering in the critical region as a function of temperature and density, for example, could map out the functional behavior of the divergence of κ_T as the critical point is approached along some locus. κ_T (or $\rho^2 \kappa_T$) is expected to diverge as $\kappa_T = Ct^{-\gamma}$ when the critical point is approached along the critical isochore from $T > T_c$, where $t = (\frac{T}{T_C} - 1)$, the reduced temperature. γ is defined as the critical index characterizing the divergence of the susceptiblity.

In this experiment we measured the intensity of light scattered from ^{3}He while very slowly warming the sample at constant pressure through the critical region. About 0.3 mW from a helium-neon laser was incident on the cell; cell temperature was recorded continuously, as was the light scattered at 45° and 135°. A succession of passes was made at different pressures, all less than P_c, so these results apply only in the one-phase region above the critical point. Sample density at the beginning of a pass began at $\rho > \rho_c$, then decreased as the sample warmed to $\rho < \rho_c$. The scattering intensity increased monotonically as the molar density decreased toward the critical density, then decreased monotonically thereafter on that isobaric pass. The scattering maximum on each pass occurred at the temperature T_{max} where the isothermal compressibility was a maximum. The asymptotic divergence of κ_T was evaluated along the locus of these maxima.

The small anisotropy in the scattered light could be used to calculate forward (zero degree) scattering from 45° and 135° data using equation (3) in the section on data and results.

According to Debye, anisotropic scattering results from correlations of the density fluctuations, and these correlations ultimately are related to the intermolecular potential that determines the microscopic distribution of individual atoms. According to this interpretation, the asymmetry data can be used to investigate the nature of the intermolecular potential function through the Debye ℓ parameter [2,3].

The Debye ℓ parameter is defined using the interatomic potential energy $\phi(r)(a \leq r \leq \infty)$, where a is the radius of the repulsive hard-core part of the potential such that a van der Waals type equation of state can be derived from $\phi(r)$ in a mean field approximation (far from T_c). Then,

$$\ell^2 = \frac{\int_{2a}^{\infty} r^2 \phi(r) d^3 r}{\int_{2a}^{\infty} \phi(r) d^3 r} \quad . \tag{2}$$

2. NEAR ZERO: SOME PERSPECTIVE ON THE STANFORD LOW TEMPERATURE LIGHT SCATTERING FACILITY

This light scattering experiment was first suggested in 1964 when not much was known about liquid-gas critical points. Michael Fisher could say [4] "the accurate experimental measurement of the isothermal compressibility of a gas near its critical point is not easy and the classical prediction that κ_T should diverge as $(T - T_c)^{-1}$ along the critical isochore does not seem to have been properly tested." Fisher goes on to quote the theoretical prediction that the divergence of κ_T is described by the exponent $\gamma = 1.25$. (The latest value based on the renormalization group technique is $\gamma = 1.2412 \pm .0011$ [5].)

Even less was known about the critical point of ^{3}He, or quantum fluids in general. And less still was known about the tricritical point in ^{3}He-^{4}He mixtures. The Stanford light scattering facility was designed and built with the intention of ultimately looking at both systems.

Besides looking at the lowest temperature liquid-gas critical point (3.31 K), this experiment approached zero in another way. The intensity of scattering from fluctuations in a fluid is proportional to $(\epsilon - 1)^2$ and to kT. With an index of refraction less than 1.03, compared to more typical fluids with refractive index greater than 1.3, and with T_c approximately two orders of magnitude below room temperature critical points, this translates into scattering intensities between four and five orders of magnitude lower than ordinary critical opalescence experiments. By going toward zero scattering, we avoided the problem of multiple scattering that clouds the interpretation of so many critical opalescence experiments [6]. (We estimated that the fraction of detected photons that had been multiply scattered was only $\sim 2 \times 10^{-4}$, even at the peak critical scattering.) The smallness of the ^{3}He sample used helped reduce gravitationally induced density gradients over the thickness of the sample, i.e., that submillimeter thickness illuminated by the focused laser beam.

More to the point, a cryogenic apparatus, with its "near zero" heat leaks, provided superb thermal isolation for the sample cell and very high resolution thermometry.

The scattering facility was designed around the "near zero" scattering anticipated and the need to maintain good thermal equilibrium between

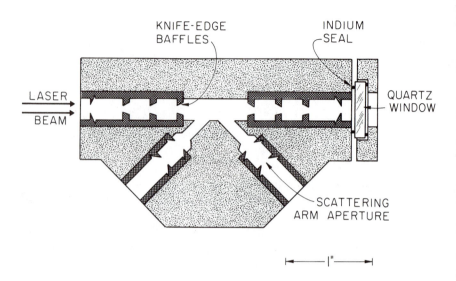

FIGURE 1. Sample cell: cross section of scattering cell (top view), including sample cavity, exit ports for scattered radiation, and one (typical) window and window mount.

the illuminated sample of helium actually observed and the cell containing it, where thermometers and heating resistors were mounted. The cell is illustrated in figure 1. Stray light was expected to be the great enemy in the light scattering facility; to eliminate stray light, 2.5 mm i.d. knife-edged apertures were brazed into the OFHC copper cell along the laser beam axis, and a separately mounted set of 1.675 mm i.d. baffles preceded the cell to further remove stray laser light. These served both to reduce background and to minimize heating of the cell. Background scattering levels were sensitive to alignment adjustments of ~ 75 μm and 0.3 milliradians. This quality of alignment could be achieved only by mounting and aligning the cell and prebaffle at room temperature, where they were accessible and visible, and then developing an unusually rigid mounting whose contraction and distortion under the stresses associated with cooldown were less than 100 μm, while still maintaining an adequately small heat leak. The light scattering facility built to achieve this is illustrated schematically in figure 2. Our laser was mounted in a robust gimbal assembly on a bench milling machine rigidly attached to the massive lathe bed shown. The bottom flange of the dewar assembly rested on the same lathe bed. A fused quartz column 60 cm high and a compressed stack of ~ 500 stainless steel shims dusted with silicon carbide grit (used in a sandwich configuration to isolate the cell from the compressing screws) supported the cell rigidly from the bottom flange. All other connections to the cell (such as the gas heat

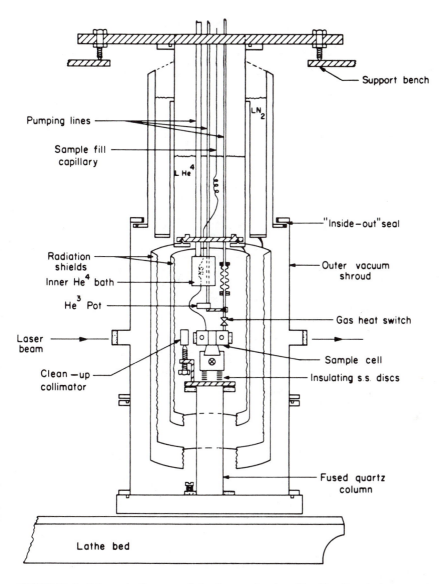

FIGURE 2. Schematic diagram of the dewar, showing "fixed" cell and cleanup collimator mounted on the rigid quartz column and horizontal dovetail and vertical motion leafspring mounts, the "floating" main cryogen reservoir part of the dewar, the removable outer vacuum shroud with "inside-out" o-ring seal, and the removable radiation shields.

switch mounted on bellows) were flexible. With the main part of the dewar held up at the top by the supporting bench as shown, with the central portion of the outer vacuum shroud lifted completely out of the way, and with the schematically indicated radiation shields removed, the cell and prebaffle were accessible for careful alignment with the focused laser beam; also, the thermal contraction of the quartz/stainless steel stack mount was small enough and geometrically simple enough so it could be reasonably well predicted and compensation made. The radiation shields, consisting of half cylinders of OFHC copper and a squirrel cage of copper flanges and bus bars, were assembled in place and heat sunk; then the outer vacuum shroud was lowered into place and sealed, displacing the top reservoir part of the dewar but not affecting the cell assembly.

Furthermore, the cell and prebaffle interiors were blackened, the antireflection-coated windows on the heat shields were canted to reflect light into blackened radiation dumps, low-level light paths were isolated with tubes of "coffin paper," *etc.*

In its complexity, our light scattering facility was quite representative of Fairbank-inspired research, for William Fairbank is legendary for the sophistication of the equipment he has used to pursue frontiers in physics. However, he also deserves to be better known for his uncanny ability to complete his experiments with no more sophistication than is necessary. This skill was exemplified when the low temperature light scattering facility was buttoned up for its first major run.

All was in readiness—indium squeezed in place under heat sinking clamps and fluid sealing flanges, radiation shields erected in place, pump-down started— when a small vacuum leak was detected in the inner, pumped ^{4}He refrigerator. Disassembly, replacement of the refrigerator, resealing and reassembly would have been at least a 4 month operation. Inspired by his uncanny intuition, Fairbank climbed onto the shoulder-high shelf that gave access to the top of the dewar and emptied an entire can of vacuum leak sealant down the ^{4}He pumping tube. The leak sealant crept down the pumping tube, past the overlapped radiation baffles, into the ^{4}He refrigerator pot, and was sucked into the offending crack. At this point, liquid nitrogen was introduced, and the sealant either cured or froze in place as the system contracted! The system was kept cold for 6 months, all of the ^{3}He data reported here were taken, and the leak never reappeared until the dewar was warmed up.

3.   APPARATUS AND MEASUREMENT LIMITATIONS [7]

The cell thermometer-resistor formed one arm of a capacitively balanced ac null bridge. The temperature resolution was less than 2×10^{-5} K. The uncertainty in our measurement of relative cell temperature was less than

5×10^{-5} K. The sample cell was maintained at liquid helium temperature from six weeks before any data were taken until the experiment was completed, so thermal coupling between resistor core and cell should have been invariant [7].

Warming rates of the cell for temperatures within ± 0.001 degrees of T_{max} ranged from 7 to 20×10^{-7} K per second, and the reduced temperature t was increasing at less than 6×10^{-7} sec^{-1}. A solution to the idealized heat flow equation of a ^{3}He-like sample in a uniformly warming cylindrical cell showed that the temperature at the center of the cell lagged the wall temperature by < 0.0001 K for these warming rates. This is an upper bound. Actually, sample homogeneity was further promoted by the extremely gentle stirring action occasioned by the slow evolution of gas as sample density decreased on each isobaric warming pass.

The ^{3}He sample was precleaned by passage through a sintered copper sponge mounted in an external liquid helium dewar, serving as a dust filter and ensuring that no condensible gas could pass on to freeze and block the filling capillary or form frozen gas snowflakes in the cell. Large Toepler pumps served as variable-volume reservoirs to receive the ^{3}He evolved from the sample cell (at constant pressure) as the sample warmed. ^{3}He system pressure could be read to ± 0.1 torr, and a practised operator could manually regulate Toepler pump volume to keep this pressure constant to ± 0.2 torr ($\pm 0.025\%$), with occasional pressure excursions of $\sim \pm 0.5$ torr. A short-term fluctuation of ± 0.2 torr at the reservoir had no discernible effect upon cell temperature or light scattering. Gas was continuously evolving from the sample chamber through a long capillary tube. Thus the absolute pressure at the Toepler pumps was certainly different from the cell pressure and hence was not used as a thermodynamic variable in our analysis. Pump pressure was a control parameter related to and controlling cell pressure; once a pump pressure was chosen, only its short-term constancy was critical.

Our ^{3}He sample contained a ^{4}He impurity of $0.4\% \pm .08\%$ as measured immediately after scattering runs were completed, leading to an uncertainty in T_c that had to be resolved. T_c could be determined with greater precision from our own scattering data than could be achieved transferring "absolute" temperatures or estimating (impurity-induced) displacement of the critical temperature from values measured in other labs.

Incident laser light was modulated by a tuning fork light chopper and was detected by EMI 9558A photomultipliers coupled to PAR lock-in amplifiers. The input attenuating networks in these two amplifiers were calibrated separately to intercompare data measured on separate ranges. Laser output was monitored to normalize data taken on different runs.

The delicate angular asymmetry data were recorded only for two passes on the same night's run. We estimate the uncertainty in the relative measurement of the scattered light on the two passes measured during one night to be less than 1%. An additional uncertainty of $\sim \pm 2\%$ was estimated between different nights' runs.

Background scattering, as measured with a cell full of ^{3}He more than one degree below T_c, was almost negligibly small compared to critical opalescent scattering near T_c; the small background was subtracted.

4. DATA AND RESULTS

The basis of our analysis is the observation that since forward scattering diverges like κ_T, then the reciprocal forward scattering should go to zero as $\frac{1}{\kappa_T}$ when T approaches T_c. The intensity $I(k)$ of the scattered light in the direction of the wave vector k is given by equation (3) developed by Fisher [4,8]:

$$I(k)\alpha\chi(k) = \frac{I_0 t^{-\gamma}}{(1 + k^2\xi^2)^{1-\frac{\eta}{2}}} \quad , \tag{3}$$

where I_0 is a constant,

$$\xi = \xi_0 t^{-\nu} \quad , \tag{4}$$

and

$$k = \left(\frac{4\pi}{\lambda}\right) \sin\left(\frac{\theta}{2}\right) \quad . \tag{5}$$

$\chi(k)$ in equation (3) is the wave-vector-dependent generalized susceptibility $\chi(k) = (\frac{\partial \rho}{\partial \mu})_T$. The divergence of χ near the critical point is characterized by the critical exponent γ; ν is the critical exponent describing the divergence of the correlation length ξ. If $\nu = 0.5$ and $\eta = 0$, one obtains the Debye formulation with $\xi_0^2 = \frac{\ell^2}{6}$. The small index η introduced by Fisher [4] measures departure from the classical Ornstein-Zernike prediction ($\eta_{oz} = 0$).

In our analysis η was set equal to 0, 0.1 and 0.2, and ν was varied from $\nu = 0.6$ ($\sim \frac{\gamma}{2}$) through $\nu = 0.5$ (classical value) to $\nu = 0.45$ (slightly less than classical value). T_c was determined from trial values of T_c^* every tenth of a millidegree ranging over 3.3196 K $\leq T_c^* \leq 3.3211$ K. T_c unambiguously lay above 3.3196 K (where boiling was observed on another run), and very near and slightly above 3.3203 K, where $45°$ reciprocal scattering extrapolated to zero when plotted against temperature. Scattering data from $45°$ and $135°$ were extrapolated (separately) to forward scattering for each set of trial parameters using equations (3) through (6), and the reciprocal forward scattering was plotted against T. For the "correct" trial parameters, the two reciprocal scattering curves had to be identical and extrapolate to the trial T_c^* when reciprocal scattering equaled zero, as shown in figure 3 for the set $\eta = 0$, $\nu = 0.5$ and $T_c^* = 3.3207$ K.

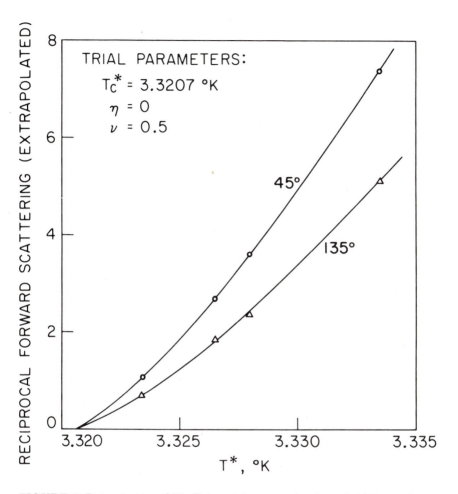

FIGURE 3. Determination of T_c^*. Data points represent reciprocals of the maximum scattering intensities (minus background) extrapolated to forward scattering. The trial parameters for the extrapolations illustrated are $\eta = 0$, $\nu = 0.5$ and $T_c^* = 3.3207$ K. Extrapolated reciprocal intensities derived from 45° and 135° scattering are deliberately plotted in different units to separate the two curves. The reciprocal scattering curves should go to zero and intersect at 3.3207 K; the relative magnitudes of the two curves are deliberately distorted by this plot.

(A scaling factor was used to separate these two curves, but the extrapolation to the trial T_c^* is clearly shown.) Over the considerable range of ν and η described, $T_c^* = 3.3207 \pm .0003$ K. Figure 4 shows how gamma was derived from log-log plots of forward scattering *versus* t, where the same values of T_c^* were used to calculate t and forward scattering. The values of γ derived from extrapolated 45° scattering were between 1.24 and 1.27 for

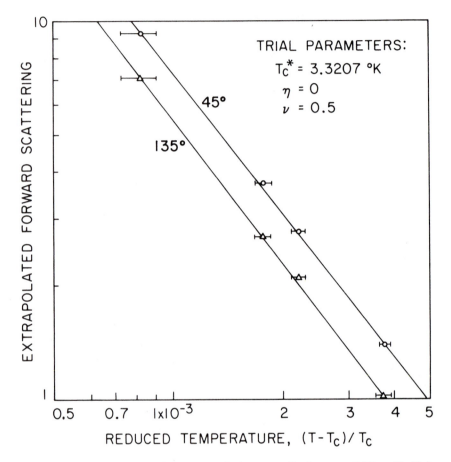

FIGURE 4. Asymptotic divergence of the generalized susceptibility, $(\partial\rho/\partial\mu)_T = \rho^2\kappa_T$. Data points represent maximum scattering intensities extrapolated to forward scattering. Trial parameters used here to calculate the extrapolated intensities and reduced temperatures t for our slightly impure ^{3}He sample are $\eta = 0$, $\nu = 0.5$ and $T_c^* = 3.3207$ K. Each point represents the peak scattering observed on an isobaric pass through the critical region. The horizontal uncertainty bars are associated with a ±0.0003 degree uncertainty in the assignment of the critical temperature T_c^*; they do not represent a random uncertainty. The left-hand ends of the error bars correspond to $T_c^* = 3.3210$ K and the right-hand ends to $T_c^* = 3.3204$ K.

all values of $0 \le \eta \le 0.2$ and $0.45 \le \nu \le 0.6$ for $T_c^* = 3.3207$ K. Extrapolated 135° scattering showed $1.26 \le \gamma \le 1.28$ for the same T_c^* over the same range of η and ν.

The largest source of uncertainty in a determination of γ is the assignment of T_c. Our extrapolated scattering *versus* t data were less consistent when the T_c^* limits were used: $T_c^* = 3.3204$ K led to values for γ

consistently differing by ~ 0.05 when derived from 45° *versus* 135° data, *i.e.*, $\gamma(45°) \sim 1.30$ and $\gamma(135°) \sim 1.35$. Log-log plots of extrapolated scattering *versus* t for the same range of ν and η were more rounded, less linear for $T_c^* = 3.3210$ K; these data are also consistent with $\gamma = 1.25 \pm .03$ or $.04$, giving an excellent fit to all of the data taken at both scattering angles and over the entire range of T_{max} covered. This was true over the full range of ν and η described.

Reduced correlation lengths ξ_0 were calculated from the angular asymmetry data by means of the expression:

$$\frac{W(45°, T_1)/W(135°, T_1)}{W(45°, T_2)/W(135°, T_2)} \equiv \frac{(I(45°)/I(135°))_{T_1}}{(I(45°)/I(135°))_{T_2}} , \qquad (6)$$

where $W(\theta, T)$ is the apparent output voltage from the photomultiplier system, and $I(\theta)$ is given by equation (3). The data for two scattering maxima at temperatures T_1 and T_2 were obtained on the same night and analyzed as above to cancel out uncertainty in gain and geometric parameters. The double ratio of Fisher's expressions was expanded to 4th order in $k\xi$, inverted, and solved for ξ_0 in terms of the trial parameters and the data, giving a different value of ξ_0 for each night's pair of runs and each set of trial parameters. Some values of $\xi_0(\eta, \nu, T_c^* = 3.3207$ K) are listed in table 1.

We found that the value of ξ_0 and its temperature dependence were strongly influenced by the choice of trial value for ν. For $\nu = 0.45$ we found agreement to within $\pm 4\%$ for ξ_0 from three different sets of data, but consider this to be fortuitous rather than real and do not feel that these data conclusively demonstrate any significant superiority for any trial value of ν in the range examined. For ^{3}He we deduce a reduced correlation length $\xi_0 \sim 8.2$ Å, corresponding to a Debye ℓ parameter of 20 Å. The value of η, which is small, could not be determined accurately.

Our value for ξ_0 can be compared with the much smaller values found for the classical gases. For argon $\xi_0 \simeq 2.2$ Å [9], for nitrogen $\xi_0 \simeq 2.45$ Å [9], and for xenon $\xi_0 \simeq 1.8$ Å [10]. The smallest value of ξ_0 listed in table 1 is ~ 3.4 Å, still significantly higher than for the classical gases. This value and others in the table are consistent with Ohbayashi and Ikushima's value [8] of $\xi_0 = 4.8 \pm 2$ Å calculated with $\nu = 0.59$ and $\eta = 0.15$.

5. DISCUSSION

The critical exponent $\gamma = 1.25 \pm .04$ determined here is unambiguously larger than the values $\gamma = 1.14 \pm .05$ determined from light scattering by Ohbayashi and Ikushima [8] and probably larger than the values

$\gamma = 1.18 \pm .04$, $1.16 \pm .02$, and $1.19 \pm .01$ determined from capacitive techniques by Meyer *et al.* [11,12,13]. The latest theoretical value [5] based on the renormalization group technique is $\gamma = 1.2412 \pm .0011$.

TABLE 1. REDUCED CORRELATION LENGTHS FOR $T_c = 3.3207$ K.

	Run 8	Run 7	Run 3
$\dfrac{[I(45°)/I(135°)]_{T_1}}{[I(45°)/I(135°)]_{T_2}}$	1.10693	1.05552	1.00517
t_1:	8.071×10^{-4}	17.47×10^{-4}	8.197×10^{-3}
t_2:	21.83×10^{-4}	38.40×10^{-4}	10.34×10^{-3}
$\ell = (6)^{1/2}\xi_0$:	18.4 Å ± 1	20.4 Å ± 2	25.6 Å ± 25
$\dfrac{\Delta\xi_0}{\xi_0}$ (For 1% uncertainty in asymmetry data)	$\pm 5\%$	$\pm 10\%$	$\pm 100\%$
$\underline{\xi_0\,(\eta, \nu)}$			
$\xi_0\,(0, 0.45)$	11.2 Å	11.9 Å	11.6 Å
$\xi_0\,(0, 0.50)$	7.5 Å	8.3 Å	10.4 Å
$\xi_0\,(0, 0.6)$	3.44 Å	4.13 Å	5.97 Å
$\xi_0\,(0.1, 0.5)$	7.74 Å	8.59 Å	10.7 Å
$\xi_0\,(0.1, 0.6)$	3.55 Å	4.25 Å	6.13 Å
$\xi_0\,(0.15, 0.59)$	3.89 Å	4.63 Å	6.57 Å
$\xi_0\,(0.2, 0.5)$	8.01 Å	8.86 Å	11.1 Å
$\xi_0\,(0.2, 0.6)$	3.66 Å	4.38 Å	6.31 Å

Probably the largest difference between the experiments reported in this paper and those of [8,11,12,13] is that we did not seal off our sample chamber. This had several fundamental consequences:

(1)   We did not have to measure pressure or density accurately as a thermodynamic variable; we did not have to worry about pressure measured

in the lab differing from pressure in the cell, or about pressure at the top of the cell differing from that at the sampled plane. This last can be particularly difficult to deduce because the hydrostatic head includes highly variable densities in the highly compressible fluid near the critical point. Our pressure measurement was used only as a control, and we did not have to know it absolutely as long as it did not change.

(2)   Our sample was stirred extremely gently but continuously, as the open cell evolved fluid while it slowly warmed. We did not attempt to "bring" the cell to an equilibrium point to take a measurement, but let it drift as slowly as possible toward what was identified as a datum point only after the scattering maximum was passed. Thus the degree of equilibrium within our sample may have been different from that in the other experiments.

(3)   By analyzing only data from scattering maxima, we let nature pick the natural locus of approach toward the critical point.

(4)   We did not approach T_c as closely as the other experimenters: Ohbayashi took data to $t < 2 \times 10^{-4}$, and Meyer's most recent measurements extended to $t < 10^{-5}$, whereas in our closest approach $t \sim 8 \times 10^{-4}$.

In terms of the Debye ℓ parameter ($\ell = 6^{\frac{1}{2}} \xi_0$), we note that the intermolecular attractive potential is unusually broad and shallow in ^{3}He, owing to the addition of the (positive) zero-point energy to the (negative) London-type attractive potential. It was of interest whether this broad attractive potential would manifest itself through the Debye formalism using critical opalescence data, as well as how the nonclassical ($\gamma \neq 1.00$) region was affected by the quantum nature of ^{3}He [11]. Our unusually large value of ℓ (~ 20 Å) compared to ~ 6 Å for nitrogen and argon [9] does show that the Debye formalism reflects this effect of ^{3}He's zero-point energy. It is of interest to compare our results for ℓ with those found in ^{4}He, since the attractive potential is presumably the same for the two isotopes but the zero-point energy should be reduced by the square root of the reciprocal ratio of the atomic masses relative to ^{3}He. Ohbayashi and Ikushima [14] found that $\ell \sim 8.8$ Å for ^{4}He, which is significantly less than our best estimate for ^{3}He, but in reasonable agreement with the minimum value in table 1 with $\nu = 0.6$.

Acknowledgments

It is a pleasure to thank W. M. Fairbank, who inspired the techniques used in pursuing zero in this work. We are also grateful to Martin Edwards for stimulating discussions and useful suggestions. This research was supported in part by the Air Force Office of Scientific Research.

References

[1]   W. M. Fairbank, M. J. Buckingham and C. F. Kellers, in *Proceedings of the 5th International Conference in Low Temperature Physics*, edited by J. R. Dillinger, (Univ. of Wisconsin Press, Madison, Wisconsin, 1958), p. 50; C. F. Kellers, Ph.D. Thesis, Duke University, 1960; M. J. Buckingham and W. M. Fairbank, in *Progress in Low Temperature Physics*, edited by C. J. Gorter, (North-Holland Publishing Co., Amsterdam, 1961), Vol. 3, p. 80.

[2]   P. Debye, *J. Chem. Phys.* **31**, 680 (1959).

[3]   B. Chu, *Molecular Forces*, based on the Baker Lectures of Peter J. W. Debye, (Interscience Publishers, New York, 1967).

[4]   M. E. Fisher, *J. Math. Phys.* **5**, 944 (1964).

[5]   D. Z. Albert, *Phys. Rev. B* **25**, 4810 (1982).

[6]   D. S. Cannell, *Phys. Rev. A* **16**, 431 (1977).

[7]   For more experimental details, see also Julian Pierce Webb, Ph.D. Thesis, Stanford University, 1968. The original data have been extensively reworked for publication in *Near Zero*. Although the original conclusions stand essentially unchanged, they can be stated with more confidence after extensive computer variation of the parameters T_c, ν and η used in the analysis.

[8]   K. Ohbayashi and A. Ikushima, *J. Low Temp. Phys.* **19**, 449 (1975).

[9]   H. Brumberger, in *Critical Phenomena*, edited by M. S. Green and J. V. Sengers (National Bureau of Standards, Washington, D.C., 1966), Misc. Publication 273, p. 116.

[10]   M. Giglio and G. B. Benedek, *Phys. Rev. Lett.* **23**, 1145 (1969).

[11]   B. Wallace, Jr., and H. Meyer, *Phys. Rev. A* **2**, 1563 (1970).

[12]   R. P. Behringer, T. Doiron and H. Meyer, *J. Low Temp. Phys.* **24**, 315 (1976).

[13]   C. Pittman, T. Doiron and H. Meyer, *Phys. Rev. B* **20**, 3678 (1979).

[14]   K. Ohbayashi and A. Ikushima, *J. Low Temp. Phys.* **15**, 33 (1974).

Bubbles in Liquid ^{4}He and ^{3}He: Mie and Physical Optics Models of Light Scattering, and Quantum Tunneling and Spinodal Models of Nucleation

Philip L. Marston and Dean S. Langley

1. INTRODUCTION

As with ordinary liquids, the temperature of liquid helium can be lowered by reducing the pressure of the vapor above the liquid. When this is done for He I, the normal phase of ^{4}He, bubbles rise to the surface and normal boiling occurs. For He II, the low temperature phase of ^{4}He, thermal transport properties of the superfluid inhibit normal boiling, and evaporation takes place only at the free surface. Consequently, visual observation of bubbles in He II is not a common event. They can be created by expanding He II in a sealed piston without direct contact with vapor [1-3]. Transient bubbles can also be created in He II in response to large amplitude ultrasonic waves [4]. We begin this paper with a review of experiments by W. M. Fairbank and coworkers [1-5] in which bubbles were generated by these processes in He I and He II. In subsequent sections, we describe recent models which give insight into the optical and nucleation properties of bubbles in the helium liquids.

2. PHENOMENA RELATED TO BUBBLES IN ^{4}HE

Fairbank and coworkers developed the first liquid helium bubble chambers for the purpose of recording the tracks of ionizing radiation. The project was motivated by high energy physics experiments which required a ^{4}He target nucleus. In both the prototype and the final version of the bubble chamber, tracks of bubbles were produced in He I following a sufficient expansion of the chamber. The tracks revealed the path of radiation, as is the case for bubble chambers containing ordinary liquids.

The prototype bubble chamber was also expanded at temperatures *below* the lambda transition temperature, $T_\lambda = 2.172$ K, so that the chamber contained He II. The expanded liquid was insensitive to radiation and no bubble tracks were formed. It is now generally agreed that bubble tracks in ordinary liquids nucleate in response to the "thermal spikes" produced by radiation [6,7]. It has been pointed out by several workers (*e.g.*, [4]) that the unusual thermal transport properties of He II should partially inhibit the local rise in temperature produced by radiation; consequently, the absence of *radiation-induced bubbles* is plausible. However, almost every expansion of the He II in the prototype chamber resulted in production of *radiation-insensitive pairs of small bubbles* [1]. A photograph of these pairs is reproduced in figure 1. The individual bubbles have been estimated to be between 0.07 and 0.3 mm diameter and are separated by 0.3 to 0.7 mm.

FIGURE 1. Bubble pairs in superfluid helium in prototype bubble chamber.

One conjecture regarding these observations [1,3] was that the pairs may have nucleated on the cores of vortex lines or rings. Vortices in He II have a superfluid circulation which is quantized in units of h/m, where h is Planck's constant, and m is the mass of a ^{4}He atom [8]; they can persist in a metastable state [9,10]. The nucleation of bubbles on vortices is plausible since there should be a localized region of low pressure at the vortex core. To test this nucleation hypothesis, Edwards, Cleary and Fairbank [3] photographed the expansions of a bubble chamber in which vortices were deliberately introduced by two methods. In the first, the entire chamber could be rotated; however, no clear correlation between rotation and bubble nucleation was evident. In the second, expansion of the chamber forced He II through a constriction. On subsequent expansions (approximately 100 minutes later), unpaired bubbles nucleated in regions some distance from the constriction, provided the liquid was maintained at a temperature well below T_λ (e.g., 1.7 K). Other observations suggested that the liquid flowing through the constriction produced turbulence and possibly vortex rings. The relative absence of bubbles for those expansions in which the constriction was not present suggests that the constriction introduced a persistent vorticity or turbulence which served as a nucleation site for bubbles.

Subsequent bubble experiments performed in Fairbank's laboratory by one of us, Philip Marston, in collaboration with Douglas Greene, were directed toward understanding the nucleation and dynamics of ultrasonic cavitation in He II [4] and in He I [5]. In experiments on He II at $T = 2.09$ K the absolute pressure amplitude of sound waves was measured, and from it the negative pressure extremum, $-\xi$, associated with the nucleation of bubbles growing to visible size was inferred. The empirical tensile strength ξ was estimated to be 60 kPa (+0, −30 kPa); however, there were several events where ξ may have been as large as 120 kPa. (Note that 100 kPa = 1 bar ≃ 0.99 atm.) Classical homogeneous nucleation theory [6,7] predicts an intrinsic tensile strength $\xi \simeq 480 \pm 20$ kPa for the conditions of the experiment. The empirical ξ, though well below the theoretical ξ, is the largest observed for He II (for reviews see [4,6 and 11]).

One reason why the intrinsic tensile strength appears to be difficult to reach may be that persistent metastable vortices act as nucleation sites. The origin of vortices in all of the experiments is not certain; however, there is now evidence [12] that ultrasonic transducers, when driven at large amplitudes in He II, produce quantized vortex lines. Another source of nuclei would affect both He I and He II: expansion of the tiny bubble (radius ≃ 15 Å) which surrounds free electrons produced by the interaction of cosmic rays with ^{4}He. The ξ associated with the unstable expansion of an "electron bubble" has been estimated to be 0.16 MPa for $T \lesssim 2.5$ K [4]. Unfortunately, technical problems prevented the measurement of ξ for $T \gtrsim T_\lambda$ in the experiments of Marston [4].

Space does not permit us to give a detailed account of the dynamics of ultrasonic cavitation bubbles in He I and II. Bubbles in either phase [4,5] are acted on by Bjerknes forces which are a particular manifestation of acoustic radiation pressure, which also affects bubbles in ordinary liquids [6]. An unusual resonance phenomenon [5,6,11,13], associated with evaporation and condensation, should affect the dynamics of cavitation in He I.

To complete this summary of work on bubbles and vortices, we describe
the detection by Marston and Fairbank [9,10] of large metastable vortices in
He II. These experiments relied not on detecting bubbles but on observing
the dimple at the free surface which manifests the low pressure of the vortex core. The dimple was present after the rotation of the He II container
was stopped, and also when the container was rotating under conditions
such that the energetically favored state of the liquid should have been a
uniform distribution of singly quantized vortex lines, which gives rise to
a parabolic profile [8]. For the creation of a dimple, the container had to
undergo a rotational history in which a critical velocity was exceeded, and
had to be at a temperature at which the superfluid fraction was large. Vortex depressions were detected by generating a contour map of the surface
through optical interferometry. Figure 2 is an example of an optical fringe
pattern of a large vortex. The energetically favored state would have been a
parabolic profile with a central depth of 17 μm. Instead, there was a central
depression which left only a thin film of He II on the container's surface.
Figure 2 was taken after the metastable condition had persisted for 222 sec.

3. LIGHT SCATTERING BY BUBBLES IN HE II

Light scattering may provide the best techniques for investigating bubbles
in He II, especially in experiments involving ultrasonic cavitation. The
usual acoustic indicators of cavitation are, in some cases, unreliable for
He II. For example, Marston [4] found that when the driven frequency
ω of a resonator was varied, the ω which maximized the driven pressure
amplitude (ω_M) differed from the ω which maximized the production of
cavitation-like noise. Maximum production of visible bubbles required that
ω be ω_M, as is to be expected. Furthermore, the onset of subharmonics can
be attributed to the generation of vortex lines instead of to cavitation [12].
   For scattering by bubbles in He II there is only a small difference between
the refractive index n_o of the outer medium (liquid) and the index n_i of
the inner medium (vapor). Since their ratio $n_o/n_i = 1.03$ is close to unity,
the scattering should be weak except at certain angles. (To appreciate
the problem, view an ice cube surrounded by water; this has $n_o/n_i \simeq$
$1.33/1.31 \simeq 1.02$.) An exact solution for scattering from dielectric spheres
[14] was given by Mie in 1908, but only recently has his work been applied
to the description of how bubbles in ordinary liquids and glass scatter
light [15–18]. Effects of diffraction have been observed to be significant
even for large bubbles in water [18,20] and silicone oil [19]. Approximate
models have been developed which give insight into certain features of
the scattering by spherical and nonspherical bubbles. A physical optics
approximation is used to describe certain effects of diffraction [18–21]. The
procedure is (i) to compute (*via* ray optics) wave amplitudes in an exit

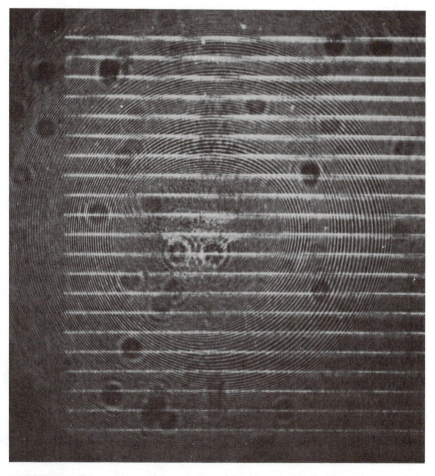

FIGURE 2. Fringe pattern from a metastable vortex in He II. Measurements of the radius of the transition between the region close to the vortex (where the surface curves downward) and the outer region (where the surface curves upward) yield a circulation $\simeq$ 1900 h/m. The lines in the superposed grid have a spacing of 0.5 mm.

plane in contact with the bubble, and (ii) to allow this wave to diffract to the far field where the distance from the bubble's center $R \gg xa$. Here a is the bubble's radius and x is the size parameter defined below.

Figure 3 illustrates, for a bubble in He II, several rays which affect the total scattering when the scattering angle $\phi = 15°$. The angles of incidence and refraction of each class of ray are denoted by θ_ℓ and ρ_ℓ. The impact parameters are denoted by b_ℓ. When θ_0 exceeds the critical angle of incidence $\theta_c = \sin(n^{-1})$, the $\ell = 0$ ray is totally reflected unless tunneling and effects of curvature are significant [20,21]. The corresponding condition on

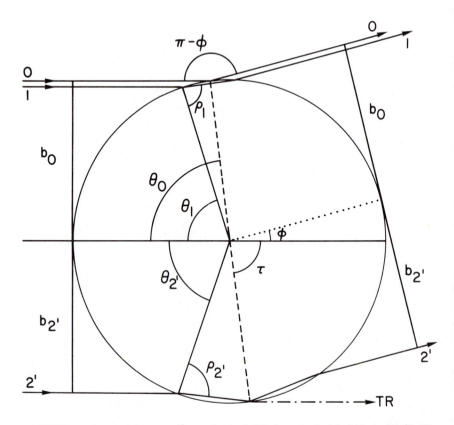

FIGURE 3. Rays with scattering angle ϕ of 15° for spherical bubble in He II. The number ℓ of internal chords is shown on the left.

ϕ is $\phi \leq \phi_c$, where $\phi_c = 180° - 2\,\theta_c \simeq 27.72°$ is the critical scattering angle. Let I_j denote the normalized scattering intensity defined as follows: The actual j-polarized intensity for $R \gg a$ is the incident j-polarized intensity multiplied by $I_j(a/R)^2/4$. For the electric vector perpendicular to the scattering plane, $j = 1$; for the parallel case, $j = 2$. This normalization is appropriate for bubbles since geometric optics predicts that if the intensity of the $\ell = 0$ reflection could be considered by itself, $I_j(\phi \leq \phi_c) = 1$ due to total reflection. Both the physical optics approximation and the Mie theory depend on the size parameter x, which may be written ka where $k = 2\pi/\lambda_o$, λ_o being the wavelength of light in the outer medium. For red light $x = 1000$, 100 and 5 gives $a = 100$, 10 and 0.5 μm, respectively. In figure 3 the ray denoted TR is a forward directed tunneling ray which reduces the reflected ray's amplitude when the size parameter x is small [21].

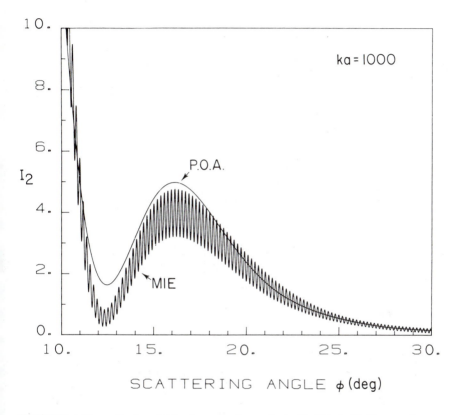

FIGURE 4. Normalized scattering intensity for a spherical bubble in He II as predicted by a physical optics approximation and by Mie theory.

Figure 4 compares results of the two theories. The Mie theory exhibits a fine structure with an angular quasi-period $\Omega_f \simeq \arcsin[\lambda_o/(b_0 + b_{2'})] \simeq 0.2°$ which is due to the interference of the 0 and 2′ rays. The physical optics approximation includes the interference and diffraction of only the $\ell = 0$ and 1 rays. In contrast to the geometric prediction, $I_j(\phi_c) = 1$, the physical optics approximation and Mie theory both give $I_j(\phi_c) \simeq 0.25$ (when fine structures in the Mie I_j are neglected). This reduction in intensity is due to diffraction which is significant near ϕ_c because the reflection coefficient of the $\ell = 0$ ray varies rapidly as θ_0 approaches θ_c. The broad maximum at 16° and minimum at 12.5° are due to effects of diffraction and interference; they are part of a coarse oscillation which has a quasi-period Ω_c discussed below. The physical optics approximation and the coarse oscillation concept are useful for x as small as 25 for bubbles in water [15] and glass [16]. Figure 5 shows that bubbles in He II with $x = 100$ should not have coarse

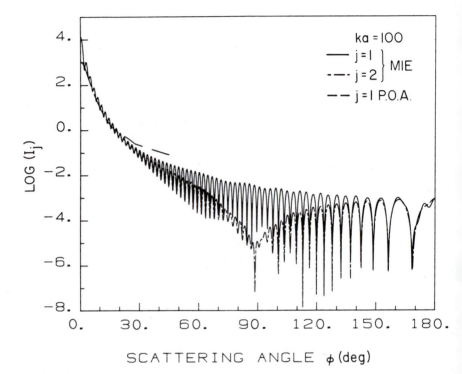

FIGURE 5. Modeled scattering as in figure 4 except that $x = 100$ and the physical optics approximation is for $j = 1$.

oscillations. As $\phi \to 0°$, the physical optics approximation underestimates the total scattering since it fails to include forward diffraction [16,18].

Other features in the Mie scattering may be explained by ray optics provided x is sufficiently large. The $j = 2$ reflection coefficient of the $\ell = 0$ ray vanishes when θ_0 is equal to Brewster's angle $\theta_B \equiv \arctan(n^{-1})$. The corresponding scattering angle $\phi_B = 180° - 2\theta_B \simeq 91.7°$ [16,18]. The resulting curves displayed in figures 5 and 6 show broad minima near ϕ_B and an alteration of the fine structure. For $x = 5$ (figure 7) a broad minimum is also present in I_2 in the general vicinity of ϕ_B but now $I_j < 1$ for all ϕ. For $x \lesssim 10$ the total reflection of light is expected to be inhibited by electromagnetic tunneling [21].

Far-field I_j curves like those displayed in figures 4–7 can be used to predict which rays can be resolved in an imaging system [18]. It is advisable that photographic or photoelectric detectors make use of scattering in the region where it is strongest, which is $\phi \lesssim \phi_c - \Omega_c$ for $x \gtrsim 100$. An approximation for Ω_c in [18] gives $\Omega_c \simeq 235x^{-1/2}$ degrees when $n = 1.03$; consequently, when $x = 1000$, $\phi_c - \Omega_c \simeq 20°$. If the source light is excluded

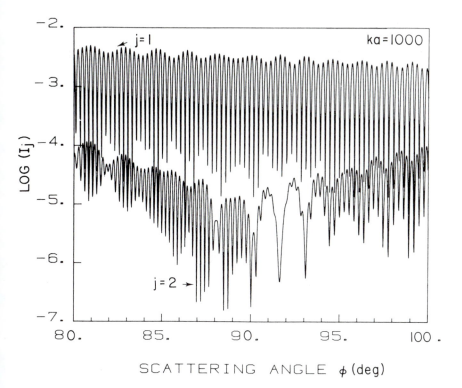

FIGURE 6. Normalized Mie intensities for ϕ near $90°$ and $x = 1000$. Except for the fine structure, $I_2 \ll I_1$ due to Brewster's minimum in the amplitude of the reflected ray.

from the camera, the bubbles in He II appear bright as in figure 1. If the camera accepts light from the source while rejecting much of the scattered light, the bubbles appear dark on a bright background [4,5]. Coarse and fine structures in the scattering can be observed by focusing the camera on infinity [18,20]. Figure 8 shows the $j = 2$ scattering from a bubble with $x \simeq 1100$ in water where $\phi_c \simeq 83°$. The region on the right is dark since these $\phi > \phi_c$. Spheroidal bubbles and other nonspherical bubbles will also have a critical scattering angle [16], but its value may depend on the bubble's orientation.

4. QUANTUM TUNNELING AND SPINODAL MODELS OF BUBBLE NUCLEATION IN ⁴HE AND ³HE

Since He II tends to wet solid surfaces completely, we expected that the intrinsic strength ξ would be observable *provided* vortices, electron bubbles,

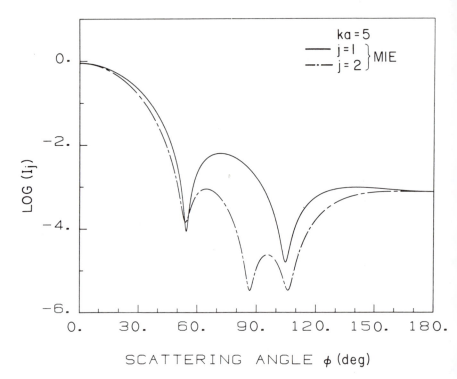

FIGURE 7. Normalized Mie intensities for a small bubble ($a \simeq 0.5$ μm) in He II.

and other heterogeneous nuclei were completely absent. When suitable precautions are taken, observed ξ for classical liquids [6] agree with predictions of the theory for homogeneous nucleation where the nucleation

FIGURE 8. Far-field scattering of monochromatic light from an 83 μm radius bubble rising through water. The left edge corresponds to $\phi \simeq 71°$ and the right edge to $\phi \simeq 86°$. The fine structure is visible and has a spacing $\simeq 0.3°$. The first coarse maximum and minimum are visible near the center and the left, respectively.

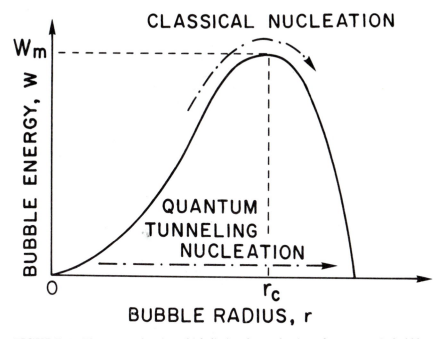

FIGURE 9. The energy barrier which limits the nucleation of macroscopic bubbles in liquids under tension. Thermally induced voids can overcome the barrier and grow to terminate the state of tension; however, at sufficiently low temperatures the rate of tunneling may be significant if the tension is large.

rate J [events/(cm^3 sec)] and ξ are [4,7]

$$\xi = [\gamma kT \, \ln(\beta/J)]^{-1/2} - p_v \quad , \tag{1}$$

$$J = \beta \exp(-W_m/kT) \quad , \tag{2}$$

where k is Boltzmann's constant, p_v is the vapor pressure, $\gamma = 3/16\,\pi\sigma^3$, σ is the surface tension, and the preexponential factor β is thought to depend only weakly on T. Nucleation occurs when there is sufficient thermal energy available to overcome the energy barrier (figure 9): $W(r) \simeq 4\pi\sigma r^2 + (4\pi/3)r^3(p - p_v)$, where r is the radius of a fluctuating void, p is the pressure in the liquid, $W_m = W(r_c)$, and r_c is the critical radius.

When equations (1) and (2) are applied to ³He and ⁴He, and T is allowed to approach the absolute zero, for a given J the predicted ξ diverges while for a given p, the predicted J vanishes. The quantum properties of the fluid must be taken into account. We have investigated two approaches for estimating ξ.

The first method makes use of an investigation by Lifshitz and Kagan [22]. They calculate that at temperatures less than a "quantum temperature" T^*, tunneling through the energy barrier is important in first order phase transitions. Application of their theory to bubble nucleation gives [23]

$$\xi(T < T^*) \simeq \left\{ \frac{135\pi^2\sigma^4(6m)^{1/2}}{16\hbar a^{3/2} \ln(N\omega/J)} \right\}^{2/7}, \qquad (3)$$

where m is the atomic mass, $a = N^{-1/3}$, $N = A/v$, where A is Avogadro's constant, v is the molar volume of the liquid, and $\omega \simeq \hbar/ma^2$ is an estimate of the "attempt frequency" associated with zero-point oscillations. It is customary to take $J = 1$ cm^{-3} sec^{-1} as an estimate of experimental conditions; ξ is only weakly dependent on J. Akulichev and Bulanov [23] find that $T^* \simeq 0.31$ K and $\xi(T < T^*) = 1.48$ MPa for ^{4}He.

We have evaluated equation (3) for both ^{4}He and ^{3}He. For ^{4}He with $\sigma = 37.3$ dyne/cm and v evaluated at $p = 0$, we find $\xi \simeq 1.66$ MPa. It is more realistic, however, to evaluate v at $p = -\xi$. For example, the first pv relation given below gives $v(-\xi)/v(p = 0) \simeq 1.5$ and equation (3) gives $\xi \simeq 1.57$ MPa. For ^{3}He we use data for $v(p = 0)$ and take $\sigma = 15.5$ dyne/cm; equation (3) gives $\xi \simeq 0.56$ MPa. We are not aware if the tensile strength of ^{3}He has been measured for any T; however, it should be less than this estimate for all finite T.

The second method of estimating ξ makes use of a limiting condition for thermodynamic stability, first stated by J. W. Gibbs and named by van der Waals the "spinodal condition" [7]: $(\partial p/\partial v)_T = 0$ when $p = -\xi$. In practice, the condition is difficult to test since pvT data at negative pressures are unavailable. A pv relation is available for ^{4}He at $T = 0$ kelvin which may be extrapolated to negative pressures. This equation (3.1 of [24]), obtained by applying Green's functions methods to the Lennard-Jones potential of ^{4}He, has the form $p(y) = k\rho_0 y^2[2B(y-1) + 3C(y-1)^2]$, where $y = \rho/\rho_0$, ρ is the atomic number density, $\rho_0 = \rho(p = 0) \simeq 0.02183$ Å^{-3}, $B = 14.5 \pm 2.5$ K, and $C = -1.2 \pm 7.8$ K. At the spinodal, $y = y_s \equiv [(9C - 3B) + (9B^2 - 6BC + 9C^2)^{1/2}]/12C$ and $\xi = -p(y_s) = 1.35 \pm .32$ MPa where the uncertainty is due to that of C. The $p(y)$ relation can also be fitted to certain positive-pressure data by taking $B = 13.65$ K and $C = 7.67$ K (equation (2) of [25]); we find that these give $\xi = 0.92$ MPa. We conclude that the zero degree pv relations from [24] and [25] predict strengths for ^{4}He similar to that of equation (3). For comparison, the van der Waals equation, when fitted to the critical constants of ^{4}He, gives $\xi \simeq 6$ MPa at the absolute zero.

In summary, classical theories of homogeneous nucleation fail since no thermal energy is available to overcome the energy barrier to bubble formation. The theory of Lifshitz and Kagan allows for tunneling through this barrier and predicts tensile strengths $\xi = 1.6$ MPa and 0.6 MPa for ^{4}He and ^{3}He, respectively. An application of the equation of state of Whitlock *et al.* to compute the spinodal (Gibbs' thermodynamic limit on stability $(\partial p/\partial v)_T = 0$) yields a ξ of 1.4 MPa for ^{4}He.

Acknowledgments

This work was supported in part by the Office of Naval Research. P. Marston was an Alfred P. Sloan Research Fellow during this work.

References

References [1-5] describe experiments performed at Duke and Stanford Universities.

[1] W. M. Fairbank, J. Leitner, M. M. Block and E. M. Harth, in *Problems of Low Temperature Physics and Thermodynamics* (Pergamon Press, New York, 1958), Vol. 1, p. 45.

[2] E. M. Harth, M. M. Block, W. M. Fairbank, M. J. Buckingham, G. G. Slaughter and M. E. Blevins, *CERN Symposium* **2**, 22 (1956).

[3] M. H. Edwards, R. M. Cleary and W. M. Fairbank, in *Quantum Fluids*, edited by D. F. Brewer (North-Holland, Amsterdam, 1966), p. 140.

[4] P. L. Marston, *J. Low Temp. Phys.* **25**, 383 (1976).

[5] P. L. Marston and D. B. Greene, *J. Acoust. Soc. Am.* **64**, 319 (1978).

References [6] and [7] review the nucleation and dynamics of bubbles; [6] also reproduces photographs (from reference [4]) of bubbles in He II.

[6] R. E. Apfel, in *Ultrasonics, Methods of Experimental Physics*, edited by P. D. Edmonds (Academic, New York, 1981), Vol. 19, p. 383.

[7] V. P. Skripov, *Metastable Liquids* (Wiley, New York, 1974).

Reference [8] summarizes several papers concerned with vortices and other superfluid phenomena. The creation and detection of large vortices is described in references [9] and [10]. Line 17 of p. 419 of reference [10] should read "similar in appearance to figure 4."

[8] R. B. Hallock, *Am. J. Phys.* **50**, 202 (1982).

[9] P. L. Marston and W. M. Fairbank, *Phys. Rev. Lett.* **39**, 1208 (1977); 1497 (E) (1977).

[10] P. L. Marston and W. M. Fairbank, in *Quantum Fluids and Solids*, edited by S. B. Trickey, E. D. Adams and J. W. Dufty (Plenum, New York, 1977), p. 411.

References [11-13] are related to the investigation of sonic cavitation in helium.

[11]   R. D. Finch and E. A. Neppiras, *Ultrasonics International 1973 Conference Proceedings* (1973), p. 73

[12]   C. W. Smith, M. J. Tejwani and D. A. Farris, *Phys. Rev. Lett.* **48**, 492 (1982).

[13]   P. L. Marston, *J. Acoust. Soc. Am.* **66**, 1516 (1979).

The following group of papers are related to light scattering. References [15-18] adapt Mie theory (reference [14]) for use with bubbles, and test the approximations introduced in references [19] and [21]. References [18-20] describe experiments.

[14]   G. Mie, *Ann. Phys. (Leipzig)* **25**, 377 (1908).

[15]   D. L. Kingsbury and P. L. Marston, *J. Opt. Soc. Am.* **71**, 358 (1981).

[16]   D. L. Kingsbury and P. L. Marston, *Appl. Opt.* **20**, 2348 (1981).

[17]   P. L. Marston and D. S. Langley, *J. Opt. Soc. Am.* **72**, 456 (1982).

[18]   P. L. Marston, D. S. Langley and D. L. Kingsbury, *Appl. Sci. Res.* **38**, 373 (1982); D. S. Langley and P. L. Marston, *Appl. Opt.* **23**, 1044 (1984).

[19]   D. S. Langley and P. L. Marston, *Phys. Rev. Lett.* **47**, 913 (1981).

[20]   P. L. Marston, *J. Opt. Soc. Am.* **69**, 1205 (1979); **70**, 353 (E) (1980).

[21]   P. L. Marston and D. L. Kingsbury, *J. Opt. Soc. Am.* **71**, 192 (1981); **71**, 917 (1981).

The following are related either to phase transitions or to equations-of-state near absolute zero.

[22]   I. M. Lifshitz and Yu. Kagan, *Sov. Phys.–JETP* **35**, 206 (1972).

[23]   V. A. Akulichev and V. A. Bulanov, *Sov. Phy. Acoust.* **20**, 501 (1975).

[24]   P. A. Whitlock, D. M. Ceperley, G. V. Chester and M. H. Kalos, *Phys. Rev. B* **19**, 5598 (1979).

[25]   P. R. Roach, J. B. Ketterson and C. W. Woo, *Phys. Rev. A* **2**, 543 (1970).

Second Sound:
A Historical Perspective

Henry A. Fairbank

For the invitation to present this paper, I wish to thank the organizers of the Near Zero conference and book; and for my long acquaintance with William Fairbank, it is appropriate that I thank my father and mother. I have known Bill since he was 21 months old in the small town of Hobson, Montana (population 200), where I first met him.

Second sound was the subject of the first paper that Bill and I published together, and it was Bill's first low temperature physics paper. However, it was not the first time that we collaborated; I think there were several instances of that in Montana. Bill reminded me of one such collaboration which I had forgotten. Someone dug a well on our Hobson property. We were intrigued by this, so we decided that we would try to dig a well, too. In the true fashion of experimentalists, we picked out a location where it was easiest to dig, not necessarily where it was most useful. It turned out to be in the path across a field. The path was used the next day by a minister, who fell into our well. But not every experiment is successful the first time.

I will skip on quickly to Whitman College, where Bill and I both did our undergraduate work. Bill was a chemistry major, and he had a good friend, Jane Davenport, who was also a chemistry major. In my sophomore year Carroll Zimmerman came to Whitman as a physicist and revived the interest in physics among many of us; there was also Ivar Highberg in math, who was one of our more inspirational teachers. I decided to major in physics. Bill and Jane, after completing chemistry majors, spent an extra year at Whitman to complete physics majors. It was a very interesting situation. The physics population at Whitman was just beginning to increase at that time, but in my last year, and Bill's postgraduate year, we had one class in which there were only three members. It was a seminar class with Jane, Bill and me. It was a little awkward, but fun. Of course it was embarrassing for Bill and me, as you can understand, because Jane was the smartest of the three. Bill and Jane went on to the University of Washington and then to the Radiation Laboratory at MIT for several years. I went to Yale and Los Alamos, and we ended up together at Yale in 1945. I was an instructor and brother Bill was a delayed graduate student. Jane was busy becoming the mother of William M. Fairbank, Jr., who is the author of another paper in this volume.

Yale was an exciting place at that time in low temperature physics. Professor C. T. Lane had built one of the few helium liquefiers outside of Europe, a Kaptiza expansion engine of the Mond laboratory design. Lars Onsager was becoming very interested in liquid helium, superconductivity, the de Haas-van Alphen effect, and so forth. So it was not surprising that Bill decided to do his thesis in low temperature physics with Lane on the microwave properties of superconducting tin cavities. This was a time when many exciting problems in low temperature physics were waiting to be explored, many of which were conceptually very simple. The properties of superfluid ^{4}He had yet to be fully unraveled. ^{3}He was not yet around, but we were eager to find its properties and those of ^{3}He–^{4}He mixtures at low temperatures.

Fritz London had written some important papers in which he anticipated the quantum nature of liquid helium [1]. He suggested that the properties were a reflection of a degenerate Bose system. Tisza produced a phenomenological model based on London's ideas and predicted the existence of a new type of wave motion in superfluid helium [2]. Landau's theory became available to us in the fall of 1945, and he likewise predicted a new wave mode in liquid helium which he called second sound [3]. So our first project that year was to see if we could produce and detect second sound waves. First, let us recall what second sound really is. Landau coined the term when he showed that there must be two types of sound in He II. However, they are quite different in origin. First sound is the conventional pressure or density wave obeying the equation

$$\frac{\partial^2 p}{\partial t^2} = u_1^2 \frac{\partial^2 p}{\partial x^2}$$

(1)

for a plane wave, where p is the pressure, t is the time, x is the displacement and u_1 is the velocity of the sound wave.

On the other hand, second sound is a temperature or entropy wave obeying the equation

$$\frac{\partial^2 T}{\partial t^2} = u_2^2 \frac{\partial^2 T}{\partial x^2} \quad , \tag{2}$$

where T is the temperature and u_2 is the velocity of the second sound wave, given by

$$u_2^2 = \frac{\rho_s}{\rho_n} \frac{T_0}{C_0} S_0^2 \quad , \tag{3}$$

where ρ_s and ρ_n are the densities of the superfluid and normal components, respectively, T_0 is the ambient temperature, C_0 is the specific heat, and S_0 is the entropy [3].

Clearly any type of heater providing a temperature pulse or periodic temperature fluctuation in superfluid helium will generate second sound. Following Landau's 1941 paper [3], Shalnikov and Sokolov made unsuccessful attempts to generate second sound with a piezoelectric transducer. That they failed was not surprising since the temperature fluctuation associated with a piezoelectric transducer radiates second sound with only a millionth of the intensity of ordinary sound [4]. Later in 1944 Peshkov [5] successfully used an electric heater to generate second sound with the velocity predicted by Landau.

The report of the 1944 experiment had not reached us at Yale, and we were unaware of Peshkov's results. So Lane, Bill and I set out to detect second sound [6] with no knowledge that it had already been observed.

To produce the liquid helium we used Lane's Kapitza-type liquefier which is described in Alan Schwettman's paper entitled "RF Superconductivity and its Applications." The liquefier was a device about five feet tall and a foot in diameter. It was mounted about a foot and a half above the floor, with a spigot at the bottom from which the liquid helium could be transferred. We put a small glass dewar about an inch in diameter underneath the liquefier, and all of our experiments on second sound were done next to the floor in this flask with about 200 cc of liquid helium. The experimental problems with second sound were smaller than the experimental problems of liquefying the helium. The liquefier was tempermental; it had about ten degrees of freedom and at the time Lane, Bill and I were the only technicians who could "run" it. So we would spend all day getting the liquid helium (and hoping a leak would not develop) in order to perform a short three quarters of an hour of measurements on second sound.

Figure 1 shows a schematic drawing of the apparatus used. It is very simple. An electric heater at the bottom of the liquid column generates second sound. The fluctuations in temperature at the surface of the liquid,

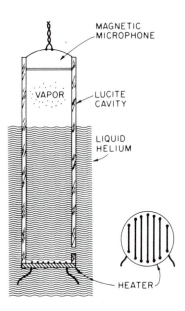

FIGURE 1. Schematic diagram of the second sound apparatus used by Lane, Fairbank and Fairbank [7].

by producing periodic evaporation, generate first sound in the vapor, which we detected with a hearing aid placed at the top of this vapor column. This method was suggested by Onsager. We measured the velocity of the second sound by observing the resonances in the liquid column as the level of the helium fell due to evaporation. A resonance occurred every time the liquid level dropped by one half-wavelength, and was observable by the amplitude of the first sound in the vapor. The method worked well. Figure 2 shows some of the first results taken at 1.57 K [7]. One can see the peaks arising from second sound resonance in the liquid. The peaks are modulated by first sound resonances in both the liquid and the vapor. The second sound wavelength and velocity can be measured with good precision, and the results of these early measurements are shown in figure 3.

An interesting and controversial question in 1946 was what happens to the velocity of second sound as T approaches zero. On the basis of his phenomenological two-fluid theory, Tisza [2] predicted that the second sound velocity should go to zero as $T \rightarrow$ zero. Landau [3], however, insisted that the second sound velocity should rise at the lowest temperatures and approach $u_1/\sqrt{3}$, where u_1 is the velocity of first sound. This result followed from Landau's characterization of He II as being made up of elementary excitations which he called rotons and phonons. At the lowest temperatures only the phonons, quantized longitudinal sound waves, are excited.

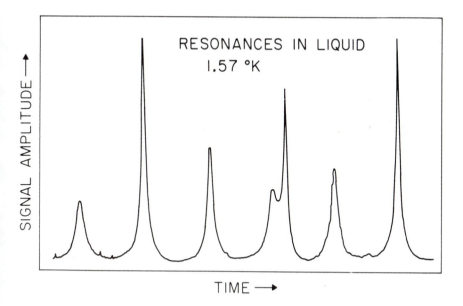

FIGURE 2. Second sound resonances at 1.57 K as registered on a recorder from a hearing aid microphone in the vapor above the liquid surface [7].

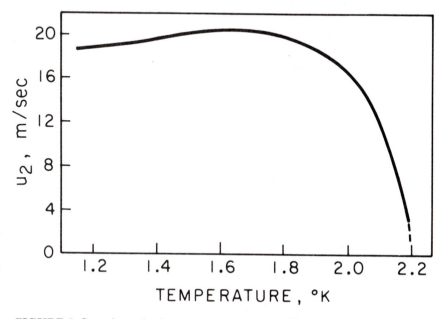

FIGURE 3. Second sound velocity *versus* temperature [7].

At higher temperatures the excitations consist of a mixture of phonons and a new kind of excitation which Landau called rotons. Thus, at the lowest temperature, in Landau's theory second sound is a density wave in a phonon gas. As in the case of first sound in an ideal gas, the velocity of the wave is just the particle's velocity divided by the square root of three. When measurements at lower temperatures were made, Landau's predictions were confirmed [8]. However, at the very lowest temperatures below about 0.5 K where the phonon mean free path becomes longer than the wavelength of the second sound, a heat pulse is propagated ballistically rather than as a second sound wave, with the leading edge arriving with the speed of first sound.

The next problem of interest was what happens when you add ^{3}He to ^{4}He. London [1] had proposed that the superfluid properties of ^{4}He were associated with a Bose-Einstein condensation to the ground state similar to that predicted for a Bose-Einstein ideal gas. The superfluid portion represented the atoms condensed into the ground state. ^{4}He, having two neutrons, two protons and two electrons, obeys the Bose-Einstein statistics, while ^{3}He, with only one neutron, obeys the Fermi statistics. It was suggested by Pollard [9] that ^{3}He would not partake in the superfluid properties of ^{4}He. To verify this was one of the exciting problems of 1946. A considerable research effort in our laboratory demonstrated that ^{3}He moves with the normal fluid and not the superfluid and can be separated from ^{4}He by the application of a temperature gradient in which the ^{3}He moves with the normal fluid toward the colder regions [10]. This is called heat flush, and was used by Bill Fairbank in a liquefier built at Amherst College to separate ^{3}He from ^{4}He. With 100% separation 1 liter of ^{3}He could be extracted from 10^7 liters of ^{4}He (obtained from commercial well sources). Fortunately, the production of ^{3}He from decaying tritium (for the hydrogen bomb) provided relatively inexpensive ^{3}He in reasonable quantities, and made this somewhat heroic method unnecessary.

Although the original interest in the heat flush experiments with ^{3}He emphasized the difference in the statistics between the two isotopes, it is now clear that any foreign particle in ^{4}He will associate with the normal component. So a superleak filter is an excellent way of ridding ^{4}He of all impurities.

Second sound measurements in ^{3}He-^{4}He mixtures provide an excellent means of measuring the large effects of even small concentrations of ^{3}He on the superfluid properties of ^{4}He. Results from our early work at Yale are shown in figure 4 [11–13]. Several features of the data are of interest. Note that the velocity goes to zero at the lambda transition temperature. T is shifted to lower temperatures with increasing ^{3}He concentration. Unlike the pure ^{4}He case, the velocity goes toward zero in the low temperature range. Finally, below 0.8 K the ^{3}He-^{4}He mixtures are seen to separate into

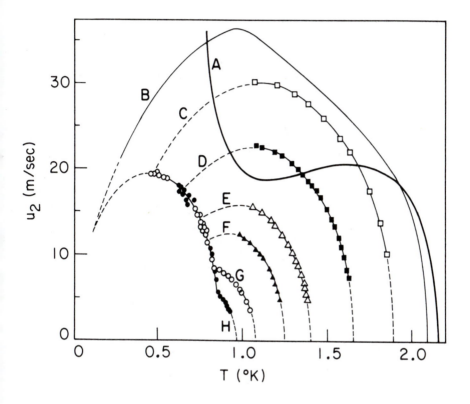

FIGURE 4. The velocity of second sound *versus* temperature. Curve A, pure ^{4}He. Curve B, 4.30% ^{3}He. Curve C, 18.4% ^{3}He. Curve D, 31.4% ^{3}He. Curve E, 43.9% ^{3}He. Curve F, 50.5% ^{3}He. Curve G, 59.4% ^{3}He. Curve H, 63.9% ^{3}He [13].

two phases. The existence of this phase separation was first seen by Walters and Bill Fairbank [14] at Duke University (see paper in this book entitled "The ^{3}He Fermi Liquid and ^{3}He-^{4}He Phase Separation: Duke Origins"). The behavior in the dilute region is nicely understood on the theory of Pomeranchuk [15] in which he considers the ideal-solution theory to hold. He finds that

$$u_2^2 = \frac{\rho_s}{\rho_n} \frac{T}{C} \left[\left(S_0 + \frac{k\epsilon}{m_4} \right)^2 + \frac{k\epsilon C}{m_4} \right] \tag{4}$$

$$\rho_n = \rho_{no} + \frac{\mu\epsilon\rho}{m_4} \tag{5}$$

$$C = C_0 + \frac{3}{2}\frac{k\epsilon}{m_4} \quad , \tag{6}$$

where S_0 is the entropy per gram of pure ^{4}He, m_4 is the mass of the ^{4}He atom, k is the Boltzmann constant, and C, ρ_s and ρ_n are the heat capacity per gram, superfluid density and normal component density of the solution. C_0 and ρ_{no} are corresponding quantities for pure ^{4}He. ϵ is the mole fraction N_3/N_4.

Below about 0.05 K the ^{3}He contribution to the entropy, specific heat and normal component density becomes much larger than that of ^{4}He, and u_2 approaches $\frac{5}{3}\frac{kT}{\mu}$, which is just the value of the sound velocity in an ideal gas of particles of mass μ. (μ turns out to be about 2.0 to 2.5 m_3).

When the ^{3}He interaction forces are taken into account, the results are slightly modified, and very low temperature second sound measurements in dilute solutions have been useful in the calculation of these interaction potentials.

Is the second sound phenomenon unique to superfluid ^{4}He and ^{3}He-^{4}He mixtures? Clearly it might be expected to exist in any superfluid in which countercurrent flow of normal and superfluid components could occur with the accompanying temperature changes. The light isotope of helium, ^{3}He, even though a spin 1/2 particle obeying Fermi statistics, does indeed (because of a pairing interaction) become superfluid below about 3 mK. In both the A and B phases the expected velocity is quite small and the attenuation so large as to make detection unlikely. However, in a very narrow region of the phase diagram, the A_1 phase, which exists in the presence of a magnetic field, Liu [16] predicted and Corruccini and Osheroff [17] experimentally verified that second sound could be propagated. The A_1 superfluid phase is apparently made up of pairs of ^{3}He atoms lined up parallel to the strong applied magnetic field, the fraction of aligned pairs being temperature dependent. Thus a second sound (temperature wave) in the A_1 phase with countercurrent normal and superfluid component flow would also involve a spin density wave. The transmitter and receiver were vibrating, porous membranes which couple only to the normal component. An additional NMR coil placed in the middle of the cell detected the spin wave associated with this hybrid mode. The measured velocity is found to be directly proportional to the reduced temperature $(1 - T/T_{A_1})$ and reaches a maximum value of about 1.4 m/sec at T_{A_2}; T_{A_1} and T_{A_2} are upper and lower temperature limits of the A_1 phase. Again, second sound measurements provide a useful tool in understanding the details of this superfluid.

From the early days of second sound investigations an interesting question was whether second sound could be propagated in solids under suitable conditions. In liquid ^{4}He at about 0.5 K second sound is essentially a density wave in the gas of longitudinal phonons. As the temperature is low-ered further and the phonon-phonon mean free path becomes long, heat pulses are propagated ballistically and true second sound no longer exists. So in a dielectric solid perhaps appropriate conditions could be found for second sound. Attempts by Ward, Wilks and others to produce second sound in high conductivity crystals such as sapphire and diamond were un-successful [18,19]. But in 1966 Ackerman, Bertman, Fairbank and Guyer [20,21], following the suggestion of Guyer and Krumhansl [22], found a temperature and pressure range of solid ^{4}He where second sound clearly exists. Subsequently it has been found in ^{3}He crystals [23] and in NaF [24,25] and Bi [26] crystals.

The conditions for second sound propagation are straightforward. If a temperature pulse or wave is generated by a heater in the solid, pure sec-ond sound would propagate if the phonon-phonon nondissipative (normal process) interactions were frequent (short mean free path) and the phonon dissipative type interactions (*e.g.*, phonon-phonon umklapp scattering and phonon scattering by impurities and imperfections) were infrequent (long mean free path). This turns out to uniquely favor solid helium, since it is possible to make high quality single crystals of helium which are isotopically and chemically extremely pure. The umklapp scattering process virtually disappears below about 1 K and the nondissipative normal process inter-actions are still frequent to a few tenths of a degree kelvin. Liquid helium can be made chemically pure very easily since all chemical impurities can be frozen out and heat flush techniques (mentioned earlier) can reduce ^{3}He impurities in ^{4}He to a negligible level. It seems clear that attempts to gen-erate second sound in many other solids have been thwarted by isotopic, chemical or imperfection scattering of the phonons.

It is instructive to see the behavior of a heat pulse transmitted in a solid helium single crystal in different temperature ranges [21,27,28]. Be-low about 0.2 K phonon-phonon interactions are infrequent and a burst of phonons generated at a resistive film heater at one end of a crystal is propagated ballistically with the leading edge of the received (drawn out) pulse arriving with the velocity of the highest velocity phonons in the solid. At higher temperatures the received pulse sharpens up into a well-defined second sound pulse with the characteristic second sound velocity. Above about 1 K the pulse again broadens out and takes on the characteristic shape of a heat pulse transmitted in an ordinary dissipative solid by diffu-sion of phonons. The spreading of a transmitted pulse in the second sound region of temperature can be used to measure the normal process mean free path. This turns out to be proportional to T^{-3}.

This is a limited report on second sound, obviously slanted toward work in which Bill and I have had involvement. The larger story is impressive and important, but will not be told here.

References

[1]   F. London, Nature 141, 643 (1938); Phys. Rev. 54, 947 (1938).

[2]   L. Tisza, Nature 141, 913 (1938); Compt. Rend. 207, 1035, 1186 (1938); J. Phys. et Radium 1, 164, 350 (1940).

[3]   L. Landau, J. Phys. (U.S.S.R.) 5, 71 (1941); 8, 1 (1944); 11, 91 (1947).

[4]   E. Lifshitz, J. Phys. (U.S.S.R.) 8, 110 (1944).

[5]   V. Peshkov, J. Phys. (U.S.S.R.) 8, 381 (1944); 10, 389 (1946).

[6]   C. T. Lane, H. A. Fairbank, H. Schultz and W. M. Fairbank, Phys. Rev. 70, 431 (1946).

[7]   C. T. Lane, H. A. Fairbank and W. M. Fairbank, Phys. Rev. 71, 600 (1947).

[8]   D. De Klerk, R. P. Hudson and J. R. Pellam, Phys. Rev. 93, 28 (1954).

[9]   E. Pollard and W. L. Davidson, Applied Nuclear Physics (John Wiley and Sons, New York, 1942), p. 183.

[10]   C. T. Lane, H. A. Fairbank, L. T. Aldrich and A. O. Nier, Phys. Rev. 73, 256 (1948).

[11]   E. A. Lynton and H. A. Fairbank, Phys. Rev. 80, 1043 (1950).

[12]   J. C. King and H. A. Fairbank, Phys. Rev. 93, 21 (1954).

[13]   S. D. Elliott and H. A. Fairbank, in Proceedings of the Fifth International Conference on Low Temperature Physics, edited by J. R. Dillinger (University of Wisconsin Press, Madison, 1958), p. 180; H. A. Fairbank, Nuovo Cim. (Series X) 9, 325 (1958).

[14]   G. K. Walters and W. M. Fairbank, Phys. Rev. 103, 262 (1956).

[15]   I. Pomeranchuk, Sov. Phys.–JETP 19, 42 (1949).

[16]   M. Liu, Phys. Rev. Lett. 43, 1740 (1979).

[17]   L. R. Corruccini and D. D. Osheroff, Phys. Rev. Lett. 45, 2029 (1980).

[18]   J. C. Ward and J. Wilks, Phil. Mag. 43, 48 (1952).

[19]   E. W. Prohofsky and J. A. Krumhansl, Phys. Rev. 133, A1403 (1963).

[20]   C. C. Ackerman, B. Bertman, H. A. Fairbank and R. A. Guyer, Phys. Rev. 16, 789 (1966).

[21]   C. C. Ackerman and R. A. Guyer, Ann. Phys. (N. Y.) 50, 128 (1968).

[22]   R. A. Guyer and J. A. Krumhansl, Phys. Rev. 148, 778 (1966).

[23]   C. C. Ackerman and W. C. Overton, Jr., Phys. Rev. Lett. 22, 764 (1969).

[24] T. F. McNelly, S. J. Rogers, D. J. Channin, R. J. Rollesson, W. M. Goubau, G. E. Schmidt, J. A. Krumhansl and R. O. Pohl, *Phys. Rev. Lett.* **24**, 100 (1970).

[25] H. E. Jackson, C. T. Walker and T. F. McNelly, *Phys. Rev. Lett.* **25**, 26 (1970).

[26] V. Narayanamurti and R. C. Dynes, *Phys. Rev. Lett.* **28**, 1461 (1972).

[27] K. H. Mueller and H. A. Fairbank, in *Low Temperature Physics–LT13*, edited by K. D. Timmerhaus, W. J. O'Sullivan and E. F. Hammel (Plenum Press, New York, 1974), p. 90.

[28] K. H. Mueller and H. A. Fairbank, in *Proceedings of the 12th International Conference on Low Temperature Physics 1970* (Academic Press of Japan, Tokyo, 1971), p. 135.

The Shape of the Helium λ-Transition

M. J. Buckingham

1. INTRODUCTION

Exactly twenty five years before the Stanford meeting in his honor, William Fairbank presented a contributed paper at the Spring 1957 meeting of the American Physical Society in Washington D.C. The paper was entitled "Specific Heat of Liquid ^{4}He near the λ-Point" by W. M. Fairbank, M. J. Buckingham and C. F. Kellers, and was the first published result [1] of an experiment that was to prove something of a milestone.

The experiment demonstrated that it was possible to achieve for thermodynamic properties a resolution previously limited to fields such as spectroscopy. The temperature interval $\Delta T = T - T_\lambda$ could be brought as "near zero" as 10^{-6} K. Compared with other systems, liquid helium does have very favorable properties; not only are there special reasons for great theoretical interest in its transition, there are special characteristics that make its λ-transition, out of all cooperative transitions, the one most amenable to precise high resolution measurement. In all other cases there are effects due to impurities, strains, imperfections, gravity, etc., which tend to smear out what would otherwise be sharp features; for the λ-transition all such effects are either absent or very small.

The preliminary results presented in Washington showed already the essential "shape" of the specific heat; it is that of a logarithmic singularity symmetric above and below T_λ, except that the high temperature branch is reduced by a constant; this shift gives the characteristic asymmetric cusp shape, like that of the Greek letter "λ," for which the transition is named. Later we will discuss more quantitatively the shape of this and other thermodynamic singularities in terms of critical parameters and exponents and the theoretical viewpoint of universality. Twenty five years ago, understanding of the nature of cooperative transitions was still relatively primitive. It will be a purpose of this paper to sketch a somewhat personal view of some aspects of the evolution of what has since then become a good understanding.

The pattern of this sketch will be much guided by hindsight, but we will not neglect the false leads and errors which influence the often tortuous path of advance. Valid insights are frequently at first entangled in false ones and the latter themselves can often prove of value as a challenge or as a stimulus to probe in an otherwise unlikely direction. There can be few fields of physics in which there are more or better examples of creative, ingenious, seminal and often widely embraced ideas which, in spite of major contributions to advance and in spite of appearance, are nevertheless eventually seen to be false. "Truth," of course, has no monopoly as a source of inspiration in any field, but at least in fundamental physics one eventually arrives at very firm ground, frequently amenable even to "proof"; looking back with hindsight from there, the "false" often appears foolish and naive, the "true" obvious. Usually only the perspective of hindsight finds its way into the textbooks. There, as the established viewpoint, it becomes so "sanitized" that the valuable lessons apparent from the superseded or erroneous views are forgotten.

Before continuing in section 3 with the story specific to the shape of the helium λ-transition, we discuss in section 2 the nature of statistical order and its relation to phase transitions in general. In section 2.1 we discuss the breaking of symmetry and long range order. Section 2.2 briefly recalls some aspects of critical phenomena, introduces the critical exponents and emphasizes how all the asymptotically significant information is contained in the correlations of the fluctuations. This aspect is developed further in section 2.3 where the enigmatic status of the "hyperscaling" equations is reviewed. Resolution of a long-standing dilemma is seen to depend on the interplay between thermodynamics and geometry and the subtlety of exchange between the singular and the regular. The opportunity is taken in section 2.4 to probe also the tangled story of the classical theory of the correlation function. Section 2.5 concludes section 2 with a summary list of some important aspects of order relevant to the shape of the λ-transition.

2. SYMMETRY AND ORDER; STABILITY AND TRANSITION

2.1 Phase Transitions and Long Range Order

2.1.1 Symmetry and Stability

A system undergoes a thermodynamic transition when, as its temperature is reduced (or the value of some other parameter changes), it cannot both remain stable and retain its symmetry. The system has no choice but to abandon the latter; stability is then preserved at the expense of symmetry. This resulting state involves a broken symmetry and a long range ordering of the property whose stability was threatened.

Most frequently the symmetry that is lost is homogeneity. Each region or volume element, which has the same properties as any other before the transition, comes to possess a particular one of two (or sometimes more) distinct values for the properties. Each of these distinct values characterizes a "phase" and various macroscopic regions of space are occupied by different phases, which of course are in equilibrium with each other.

There is an element of mystery in the existence and emergence of long range order in a system whose microscopic parts only interact with short range forces. It is important therefore to emphasize that this widespread phenomenon is a purely *statistical* effect. It does not depend on any feature of the rules of quantum or classical mechanics (or any other rules such as, for some models, "lattice statistics") which determine the microscopic states of the system. What these do determine is the nature of the properties that can become long range ordered and hence the nature of the state, whether a crystal, an anti-ferromagnet, a superfluid or merely a fluid. It is in quantum but not in classical mechanics that the peculiar type of order can develop that is necessary for the superfluid state and hence for the helium λ-transition. Furthermore in this case it is not the symmetry of homogeneity that is broken, but a type of gauge symmetry. A single quantum state becomes macroscopically occupied, which implies a long range order of the phase of the complex wavefunction.

2.1.2 A Phase Transition

To clarify these ideas let us consider a simple and familiar example of a phase transition, the condensation of a fluid, but looked at from a perhaps unfamiliar point of view.

Suppose a fixed volume V contains a fixed number $\rho_o V$ of atoms, so that the overall number density is ρ_o. At a high temperature the atoms form a uniform fluid homogeneously[1] filling the volume V; a volume element at position r, say, has then a density with an instantaneous value $\rho(r)$ which fluctuates about the mean value ρ_o. Another volume element at r', a distance R from the first, has density $\rho(r')$, similarly fluctuating about the mean ρ_o. The distribution function for each of these fluctuating values is a narrow Gaussian and the particular values at r and r' are uncorrelated.

Suppose now that the temperature is reduced. At a certain temperature, T_{svp}, the pressure has fallen to correspond to the saturated vapor pressure, and when the temperature is further reduced below this value, the distribution of the density has become qualitatively different: the system has lost its homogeneity. It is now in a two-phase state; part of its volume is liquid (or perhaps solid) and part vapor. The densities at positions r and r' are no longer near the mean value ρ_o. Instead they fluctuate close to one or the other of two possible values neither of which *depends in any way* on the mean ρ_o. The two possible values are determined solely by the temperature and are the liquid density ρ_ℓ and the vapor density ρ_v, the first much greater and the second much less than ρ_o. It is merely the numerical fraction of the whole volume constituting each phase that ensures that the overall mean density is ρ_o as it must always be. Except for this fraction, no property depends on the particular value of ρ_o.

Whatever the value of ρ_o, it will be true that in some parts of the volume $\rho(r)$ is near the value ρ_ℓ, in the remaining parts near ρ_v. The distribution function for values of the density, instead of being a Gaussian about the value ρ_o, has become the sum of two separate Gaussians, one about ρ_ℓ and one about ρ_v. Furthermore the values at r and r' are no longer uncorrelated. For separations R up to values a fraction of the linear dimension of the whole volume V, the values $\rho(r)$ and $\rho(r')$ are overwhelmingly likely to belong to the *same* Gaussian; that is, both r and r' are in the same phase, either liquid or vapor. It is this property of being in the same phase—having the same large departure from the mean ρ_o—that is the property with long range order.

2.1.3 One Bit of Order

This simplest of cases exemplifies the way a symmetry is broken when a thermodynamic transition of first order takes a system into a two-phase state. The two phases (or more) in equilibrium, with the same temperature and pressure (and chemical potential, magnetic field, *etc.*) have quite different values for one or more of their extensive properties—density, entropy,

---

[1] The homogeneous symmetry would be compromised in an earth-bound laboratory, because of gravity. Our example should be thought of as in an inertial frame, approximated within an orbiting satellite.

magnetic moment, *etc.* The two-phase state also provides a typical example of long range order. Having seen how the values of the density at two positions are highly correlated over separations R up to long range, let us be very precise about what in fact is correlated at long range. It is just *one bit of information*, the information as to which of the two possible values to choose, ρ_ℓ or ρ_v. Long range order arises because the volume elements at r and r' make the *same choice*, even for a separation of long range.

Only because of its capacity for long range order can a volume element of the system at a particular point fit the requirements the particular boundary conditions (in our example the value of ρ_o) have set. These conditions could at first sight be satisfied by every element "trivially" satisfying them, with $\rho = \rho_o$ everywhere, as indeed happens in the one-phase, symmetry unbroken state. When this simple answer is not available (because it would be unstable), the conditions cannot be satisfied locally and a much more subtle way of doing so must be invoked. But this way, the global way, requires long range order so that each element can "know" what it must do. Both the possession of long range order and the loss of symmetry, are inevitable features of the global solution.

2.2 Critical Phenomena

2.2.1 Correlation and Fluctuations

Another quite different type of correlation effect that can also become long ranged is central for the helium λ-transition and for critical phenomena in general, although it is not necessary for the first-order type of transition discussed above. It is again a *purely statistical effect* and is involved with the *onset* of symmetry breaking rather than the presence of broken symmetry.

Let us consider the correlation between the fluctuations at the two points r and r', where the fluctuation $\delta\rho(r)$ is the deviation of the density $\rho(r)$ from the "local mean." In the liquid, the local mean is ρ_ℓ, so $(\rho(r) - \rho_\ell)$ is the fluctuation. If we continue with the example from the last section we see that in a one-phase state, the local mean density everywhere is just the same as the overall mean ρ_o, so that for $T > T_{svp}$, $\delta\rho(r) = \rho(r) - \rho_o$. On the other hand for $T < T_{svp}$, $\delta\rho(r)$ is $(\rho(r) - \rho_\ell)$ or $(\rho(r) - \rho_v)$ in the liquid or vapor phase, respectively. Thus in the symmetry-broken state the local mean invokes in its definition just the one bit of information that is available in the long range order. Ordinarily these fluctuations in a one-phase or a two-phase state will be described by a normal distribution and the mean square deviation, $\langle \delta\rho(r)^2 \rangle$, will be proportional to the compressibility and to N^{-1}, where there are N atoms in the volume element concerned. That is, the distribution function for the fluctuations is a narrow Gaussian, of width proportional to $N^{-1/2}$ and the distributions at r and r'

are uncorrelated except at microscopic separation between r and r'. When circumstances are *not* ordinary we shall see that the exponent in $N^{-1/2}$ becomes modified.

Let us define a correlation function

$$C(R) = \frac{\langle \delta\rho(r)\delta\rho(r')\rangle}{\{\langle \delta\rho(r)^2\rangle\langle \delta\rho(r')^2\rangle\}^{1/2}} \tag{1}$$

which, except near a boundary, will depend only on the separation R between r and r'. As defined, the dimensionless function $C(R)$ is unity at $R = 0$ and zero at $R \to \infty$; ordinarily it decreases exponentially to zero over a correlation length R_c, where R_c is a few times the mean interparticle spacing. That is, within microscopic distances, correlation of the fluctuations "ordinarily" decays to zero whether in a one- or two-phase state. But what is meant by "ordinarily"? It means not in the critical region where the N dependence above becomes abnormal.

Near the critical point of the fluid in our example, or near a cooperative transition in general, the correlation length R_c ceases to be microscopic; in fact it must become infinitely large. It diverges in such a way that, if ρ_o equals the critical density ρ_c, then as the temperature approaches the critical temperature T_c, R_c diverges as $(T - T_c)^{-\nu}$, where ν is the critical exponent for the correlation length. The space integral of $C(R)$, which according to statistical mechanics is just the compressibility, is ordinarily finite. However the compressibility, and thus the integral too, become infinite at the critical point. Furthermore the magnitude of $C(R)$ is everywhere bounded (in fact $C(R) \leq 1$ for all R), so the integral can only diverge because of the contributions from greater and greater distances as the critical point is approached. This does not, of course, prevent the correlations from decreasing with R; it only means that they do not decrease rapidly enough for the integral to converge. At the critical point $C(R)$ decreases, say as R^{-m}, where m is the "critical point correlation function exponent," which is approximately unity for three-dimensional systems, but exactly $1/4$ for the two-dimensional Ising model. Usually instead of m we employ the exponent η, which is defined in d dimensions by $\eta = m - d + 2$. It also is $1/4$ for the two-dimensional Ising model, but is approximately zero for three-dimensional systems.

In the neighborhood of the critical point the distribution of the fluctuations thus becomes completely different from the "ordinary" one. No longer is the distribution narrow nor is it Gaussian; nor indeed are the fluctuations at r and r' uncorrelated. This new structure of the fluctuations and their correlations now contains the information which determines the asymptotic nature of the thermodynamic singularity, and in liquid helium the asymptotic shape of the λ-transition.

2.2.2   Critical Exponents

We cannot here delve far into the details of critical phenomena, but refer the reader for background to the comprehensive volumes edited by Domb and Green [2] and to the excellent 1967 reviews by Fisher [3] and Heller [4] for accounts, respectively, of the theoretical and experimental aspects of the field, before the impact in the early 1970's of the renormalization group method. An account of that impact and a wide ranging discussion of its wider implications may be found in the book by Pfeuty and Toulouse [5], while the field theory connection is developed in the review by Kogut [6].

Let us recall the four principal critical exponents α, β, γ and δ. At the critical density ρ_c, the specific heat of a fluid diverges as $|T - T_c|^{-\alpha}$; below T_c the order parameter vanishes as $(T_c - T)^\beta$. The inverse compressibility vanishes, at $\rho = \rho_c$, as $(T - T_c)^\gamma$, while on the critical isotherm, at $T = T_c$, it vanishes as $|\rho - \rho_c|^{\delta-1}$. These exponents are not independent; thermodynamic stability requires that they obey certain inequalities, such as $\beta(\delta + 1) \geq 2 - \alpha$; and within our present context of universality, discussed in section 2.2.6, only two exponents are actually independent. Then we have the scaling equations:

$$2 - \alpha = \beta(\delta + 1) = 2\beta + \gamma \quad . \tag{2}$$

Furthermore, we can write equations which connect these exponents with the correlation exponents ν and m already introduced, and d, the dimensionality of space:

$$\alpha = 2 - \nu d \; ; \quad \delta = 2(d/m) - 1 \; ; \quad \gamma = (d - m)\nu \quad . \tag{3}$$

These are the so-called hyperscaling equations. They derive from inequalities originally proved, respectively, by Josephson [7] (the form $\nu d \geq 2 - \alpha$), Gunton and Buckingham [8,9] ($m \geq 2d/(\delta + 1)$) and Fisher [10,11] ($(2 - \eta)\nu \geq \gamma$, i.e., $(d - m)\nu \geq \gamma$). The hyperscaling equations (as opposed to the inequalities) have a checkered history and have often been under attack. We will return to this subject in section 2.3.3 where we will see that the attacks were often misdirected.

2.2.3   The Phase-Set Number

In our example of the gas-liquid transition the order parameter is the density difference $\rho - \rho_c$, and the way its magnitude vanishes as the temperature approaches T_c from below $(T \to T_c^-)$ defines the exponent β. Although there is a continuum of values for this *magnitude*, there is only, as we have seen, one bit of information that is carried by the long range order. This one bit can be represented as a choice between ± 1. In a more general case, when the order parameter can be a vector in an n-dimensional space, the set of equilibrium phases forms a continuum instead of just two. Again the vanishing of the magnitude as $T \rightarrow T_c^-$ defines β, but now the phase information that is long range ordered is more than one bit. It is the information needed to specify the direction of the n-dimensional vector. Thus the phase can be identified by the angle of a unit vector in n-dimensional space or, equivalently, by the points on the surface of a unit-radius hypersphere. Our gas-liquid example corresponds to the value $n = 1$ for the phase-set number n, in which case the "unit one-dimensional sphere" degenerates to the two points ± 1. For $n = 2$, we have an ordinary angle around a unit circle. This case applies to the liquid helium λ-transition for which the "phase" angle χ really is a "phase angle," since the order parameter is a complex number, $\Psi = |\Psi| \exp i\chi$, representing the macroscopically occupied quantum state. The same is true for superconductivity where the phase angle χ can be manipulated electronically with exquisite sensitivity, as discussed in the paper by Bascom Deaver entitled "Fluxoid Quantization: Experiments and Perspectives" elsewhere in this book.

2.2.4 Thermodynamics *versus* Geometry

The equations (3) connect the properties of the correlations with those of the thermodynamic singularity. The correlation parameters should be a reflection of the long range correlation of the fluctuations in the critical state, while the relation between the thermodynamic coefficients and the fluctuations should be purely statistical and not at all a geometric matter. Thus it is at first sight surprising that the dimension d should appear so prominently in the relations (3).

This is however only appearance, although one that has proved a long standing source of confusion. It arises because of an intrusion of geometry by way of the description of the correlation function in terms of length R, rather than the thermodynamically appropriate number of particles or volume, R^d. It is easy to see that in those latter terms, instead of the exponents ν and m we would have had the combinations $c_1 = \nu d$ and $c_2 = m/d$. The thermodynamic exponents above can then be expressed in terms of c_1 and c_2 so that we obtain, instead of (3),

$$\alpha = 2 - c_1 \; ; \quad \delta = 2c_2^{-1} - 1 \; ; \quad \gamma = c_1(1 - c_2) \; . \tag{4}$$

These are now the proper *thermodynamic* relations, with no cognizance of the geometric dimensionality d, as there should not be. To complete the set of exponents, equations (4) can, using the scaling conditions (2), be augmented with $\beta = \frac{1}{2}c_1 c_2$.

Of course the geometric dimensionality of a system is important in determining its actual thermodynamic properties. After all, even the monatomic perfect gas in d dimensions has its specific heat at constant volume given by $C_V/R = \frac{1}{2}d$ per mole, where R is the gas constant. Dimensionality determines the number of neighbors an atom has, as well as interaction and symmetry effects, *etc.* Thus the value of d will be crucial in determining the specific thermodynamic structure, and hence the exponent values, that a given system can have. Similarly crucial is the phase-set number n, discussed in the previous section.

Nevertheless, the values of d and n should not intrude into the formal thermodynamic relations nor their connection with the fluctuations, and indeed they do not when properly expressed. The information determining all details of the thermodynamics of any equilibrium state of a system, including the critical state, can be described, in the parameter space ("Gibbs space") of its extensive variables, in terms of the relationship between the entropy surface and its tangent plane at the point representing that state. In particular, the highly anomalous relationship at the critical state reflects [12] the details of the singularity structure. As stated at the end of section 2.2.1, this information is also contained in the structure of the fluctuations and their correlations; the translation between the two information sources is embodied in the equations (4). We shall show in section 2.3.1 that, for example, for an "ordinary" state, $c_2 = 1$, and that the compressibility for a finite sized system with N atoms behaves in general, as $N \to \infty$, as N^{1-c_2} regardless of d. We postpone until that section further discussion of the d-independent derivation of the equations (4) and of the status of the hyperscaling equations.

2.2.5   Scaling and Renormalization

We have seen how, as the critical point is approached, the correlation length R_c becomes longer and longer, diverging on the approach from high temperatures $(T \to T_c^+)$ as $(T - T_c)^{-\nu}$ or $t^{-\nu}$, where t is the dimensionless temperature interval $(T - T_c)/T_c$. Once R_c has become very large compared with the range of the interactions and any other microscopic length parameters, the specific details of such microscopic features become less relevant since they are more and more thoroughly averaged over. The changes that occur when R_c, already very large, increases by a further factor of say 10^ν, correspond to those for the temperature interval t, already very small, decreasing by a factor of 10. Similarly along other thermodynamic paths the variables scale in an appropriate way. But these changes are just the ones representing the asymptotic form of the singularity.

Only the effective interactions between large cells or blocks, of size R_c^d containing $N_c \sim t^{-\nu d}$ atoms, can be essential in determining these asymptotic changes. They cannot depend on the microscopic details which have been more and more "forgotten" as the block size increases. Only the most general features such as the dimensionality of space and the tensor character of the order parameter survive as the details are lost. The formalization of such scale transformations and the resulting dependence solely on these general topological properties is the basis of Wilson's renormalization group theory [13]. The scaling properties, even without this formalization, imply connections between the thermodynamic variables such that the asymptotic singular part of the free energy is a homogeneous function, a function only of certain particular combinations. Let us write B for the symmetry breaking field which is the thermodynamic conjugate of the order parameter (in ferromagnetism, B is the magnetic field). Then such a combination is $B/t^{\beta\delta}$ and the singular term in the free energy can be written as

$$g(t, B) = t^{2-\alpha} K(B/t^{\beta\delta}) , \quad \text{for} \quad t > 0 \tag{5}$$

or

$$g(t, B) = B^{1+1/\delta} Q(t/B^{1/\beta\delta}) , \quad \text{for} \quad B > 0 , \tag{6}$$

or similar expression in other domains [12]. The functions K and Q are nonzero and regular for small values of their arguments. The specific heat, the magnetization derivative $(\partial M/\partial T)_B$ and the susceptibility are given respectively by $\partial^2 g/\partial t^2$, $\partial^2 g/\partial B \partial t$ and $\partial^2 g/\partial B^2$. They can easily be seen to diverge for $B = 0$, $t \to 0$, as $t^{-\alpha}$, $t^{-(1-\beta)}(= t^{1-\alpha-\beta\delta})$ and $t^{-\gamma}(= t^{2-\alpha-2\beta\delta})$, respectively; also the susceptibility diverges as $B^{-1+1/\delta}$ on the critical isotherm $t = 0$. Only two of the exponents are independent, with $2-\alpha = \beta(\delta+1) = 2\beta+\gamma$, etc., as we have seen already in section 2.2.2. Although these scaling relations are important, the quantitative methods of the renormalization group approach are required to progress towards the calculation of actual numerical values for the exponents.

2.2.6   Universality

It was stated in section 2.2.4 that while the dimensionality d and the phase-set number n should not be involved in the expression of general thermodynamic relations, they are crucial in determining the actual singular thermodynamic properties of a given system. A "principle of universality" can be defined, which asserts that the values of d and n are the *only* attributes of the systems which are relevant, at least as far as the dominant asymptotic behavior at the critical singularity is concerned, and all other attributes, such as the details of the microscopic interactions, play only a secondary, or irrelevant, role. The fundamental basis of universality and the characteristics of "relevance" are among the major clarifications that have been achieved from the viewpoint of the renormalization group. Although the strength of some particular microscopic interaction may well be what determines the actual *location* of the critical singularity in the domain of the thermodynamic variables, it has no bearing on its asymptotic shape. The latter is characterized by the values of the critical exponents and certain amplitude ratios and is the same for all members of a universality class.

Thus the gas-liquid, order-disorder and three-dimensional Ising model transitions are all members of the universality class $d = 3$, $n = 1$ and should have the same exponent values even though, at the level of microscopic description, they appear to have little in common. Our case of principal interest, that of the liquid helium λ-transition, is in the class with $d = 3$, $n = 2$ and therefore has a different set of exponent values. Of course the gas-liquid critical point for helium, like that of any other fluid, is in class $d = 3$, $n = 1$. Figure 1 shows both transitions.

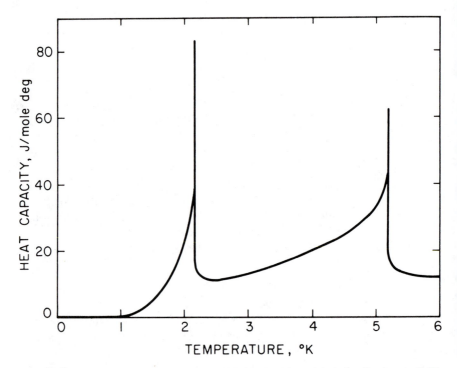

FIGURE 1. The heat capacity of liquid helium at its critical density ($\rho_c \sim 0.07$ gm/cm^3). Transitions from two different universality classes are displayed; the λ-transition, on the left, is in the class $d = 3$, $n = 2$; the critical transition, on the right, has $d = 3$, $n = 1$ (after Moldover and Little [54]).

2.3   Critical Information

2.3.1   Critical Decoding

As we have emphasized more than once already, the correlation of fluctuations in a critical state carries the information for the shape of the associated thermodynamic singularity. It is of some importance therefore to examine more closely the question of "decoding" this information.

As an example of the fluctuation analysis [14], let us recall that the compressibility coefficient χ (multiplied by $k_B T$, where k_B is Boltzmann's constant) is V^{-1} times the mean square deviation from the mean V of the volume occupied by a fixed number N of atoms. It is also the sum over all space of the correlation function describing these density fluctuations. In an ordinary, stable, noncritical thermodynamic state, these quantities approach, as $N \to \infty$, a finite limit which gives the value of the thermodynamic coefficient and of the correlation sum. At the critical point, however, where χ becomes infinite the correlation sum also diverges as $N \to \infty$. In fact it diverges with increasing N as N^{1-c_2}.

We have already remarked on the highly anomalous relationship between the entropy surface in the thermodynamic Gibbs' space and its tangent plane at the critical state. The surface is abnormally "flat" there and the fluctuations enormous; the free energy increase associated with a slight change of the order parameter from the critical value is abnormally small. Instead of being quadratic in this change, as it would be for a normal state, it is proportional to the $(1 + \delta)$th power. But being quadratic (i.e., $\delta = 1$) is a necessary condition for the distribution function to be a normal Gaussian and for its width to be proportional to $N^{-1/2}$. For the $(1 + \delta)$th power the distribution is non-Gaussian and the root mean square width can easily be shown to be proportional to $N^{-1/(1+\delta)}$; this gives the result that $1 - c_2 = 1 - 2/(1 + \delta)$, or, in the form given in equations (4), $\delta = 2c_2^{-1} - 1$. Note that, whether normal or abnormal, neither the result nor the argument depends in any way on the dimensionality. It is only when the statements are expressed in terms of *distance* and correlation length that the geometry and hence the dimensionality appears to become involved, as in the equations (3).

A similar analysis can be made of the correlations of the energy fluctuations in the critical state; this yields the first of equations (4) while extension to near-critical states yields the third. These arguments hinge on the anomalous free energy changes being dominated by the singular part of the free energy rather than any regular (nonsingular) contributions. When this ceases to be true, C_v remains finite and the exponent α describing the singular part becomes negative and but one of a bifurcated pair of values, the other arising from the regular part of the free energy, which generates an exponent value $\alpha = 0$.

2.3.2   Bifurcating Dominance

Thus when the free energy dominance passes from the singular to a regular contribution, the exponents "stick" at the "regular" values, that is, at the "classical" values $\alpha = 0$, $\delta = 3$, given below in equation (9). The exponents characterizing the "singular" and the "dominant" part can be identified by means of a subscript S or D, respectively. When the singular part dominates there is only one exponent value $\alpha = \alpha_D = \alpha_S$, *etc.*, but when dominance passes to the regular contribution, α and α_S separate, with $\alpha_S < \alpha_D = \alpha = 0$ and $\delta_S < \delta_D = \delta = 3$. Thus the exponent values bifurcate at the change of dominance, while at the precise point of degeneracy it can be shown that $\alpha = 0$ represents a logarithm, while beyond, $\alpha = 0$ represents a finite discontinuity. While the actual thermodynamic coefficients are numerically dominated by the D exponents, their higher derivatives (such as $\partial C_V(T)/\partial T$) remain determined by the S exponents. Any singular part, however "weak" and nondominant, nevertheless contributes a divergence to a sufficiently high derivative, then numerically outweighing any regular term. The case of dominance passing to a regular contribution and the exponents bifurcating is particularly relevant for systems with dimensionality $d > 4$, which are discussed further below.

The significance of "dominance" can be quite subtle, particularly for the correlation function. Whereas the exponents α_D and δ_D are greater than the corresponding singular exponents α_S and δ_S, the opposite is the case for the exponents ν and η. Thus $\nu_S > \nu_D$, and the range of the correlations, while becoming infinite for both, is greater for the "singular" than the "dominant" and increasingly so as the critical point is more closely approached. The "singular," not the "dominant," part is what actually determines the longest range correlations. This provides us with the key for resolving [14] the long standing dilemma concerning the status of the hyperscaling equations.

2.3.3 Hyperscaling Dilemma

The hyperscaling equations (3) have always been something of an enigma. They follow from renormalization group analysis and are obeyed as equalities in some exactly known cases, for example the ideal Bose gas at constant density (for which $d = 3$, $\alpha = -1$, $\delta = 5$, $\gamma = 2$, $\nu = 1$ and $\eta = 0$) and the two-dimensional Ising model ($d = 2$, $\alpha = 0$, $\delta = 15$, $\gamma = 7/4$, $\nu = 1$ and $\eta = 1/4$). Even the case of first order transitions, which can be brought [15] within the ambit of the renormalization group approach and scaling theory, provides another example, using a suitably generalized interpretation by Fisher and Berker [16] of the exponents (any d, $\alpha = 1$, $\delta = \infty$, $\gamma = 1$, $\nu = 1/d$ and $\eta = 2 - d$; $m = 0$). On the other hand, the equalities can also fail, for example in the limit of large dimension d. In that limit the mean-field theory applies, and the exponents take the classical values given below in equations (9) and (10) ($d \to \infty$, $\alpha = 0$, $\delta = 3$, $\gamma = 1$, $\nu = 1/2$ and $\eta = 0$).

The hyperscaling relations establish the important link between the thermodynamic and the correlation exponents but because of some failure of the equality and poor agreement with some early numerical estimates, their status has been much questioned, for example by Stell [17], Domb [18] and

Fisher [19]. It is the explicit appearance of d that has particularly provoked suspicion. Thus Fisher [19] has introduced the exponent ω^* (the "anomalous dimension of the vacuum") to generalize Josephson's equation, the first of (3). He defined it by $d - \omega^* = (2 - \alpha)/\nu$ and suggested that, for large d, $\omega^* = d - 4$.

To blame the explicit d-dependence of the hyperscaling relations is, we believe, to attack the wrong target. As we saw in equations (4) the purely thermodynamic statements need not involve d at all! That the first two relations (3) are not necessarily deficient is in any case indicated by the result of eliminating d between them; the outcome, still linking the thermodynamic and correlation exponents, but of course now independently of d, is just the third of equations (3) in the form $(2 - \eta)\nu = \gamma$, originally due to Fisher [10] himself and apparently always valid, even when the first two equalities are not.

2.3.4 Resolution

The dilemma can be resolved if more care is taken with the significance of "dominance." When a regular part of the changes in free energy does dominate the singular part (in particular when $\alpha < 0$), the two distinct exponent sets should be retained because both are important. Each set separately obeys all the equations (3), with $\alpha_D \geq \alpha_S$ and $\delta_D \geq \delta_S$. However, as we saw above, the opposite inequality applies for the correlation exponents, $\nu_S \geq \nu_D$ and $\eta_S \geq \eta_D$. What reflects the quality of functional dominance as far as the correlations are concerned, if more than one finite-ranged term is involved, is that term with the *longer range*, namely the larger ν. This term may not, and indeed usually does not, yield the larger contribution to the convergent integral for the thermodynamic coefficient (since m is also larger). Nevertheless it must do so for the higher moments of the integral, because of its longer range; it thereby reflects the fact that sufficiently high derivatives of the thermodynamic coefficients will be dominated by the "singular," not the "dominant," contribution.

Thus if we were to revert to the usual notation, which fails to retain and identify both sets of exponents, we would have to put $\nu = \nu_S$ and $\eta = \eta_S$ but $\alpha = \alpha_D$ and $\delta = \delta_D$ and so destroy the hyperscaling equalities. We would merely be left with the originally proved inequalities (namely $\nu d \geq 2 - \alpha$ and $m \geq 2d/(\delta + 1)$) resulting now from the two sets of relations

$$\nu d = \nu_S d \geq \nu_D d = 2 - \alpha_D = 2 - \alpha \tag{7}$$

and

$$m = d - 2 + \eta = d - 2 + \eta_S \geq d - 2 + \eta_D = 2d/(\delta_D + 1) = 2d/(\delta + 1) \quad . \tag{8}$$

Rather than the inequalities, however, it is more useful to retain all the information in the equalities (3) for both exponent sets.

When bifurcation occurs the usual homogeneity conditions break down, but in a very simple way. The free energy is the sum of two separately homogeneous terms, the D term representing a regular contribution.[2] It is ironic that these circumstances are completely contrary to conventional wisdom, since the homogeneity condition fails but hyperscaling *equalities* hold! Our resolution is thus the opposite of Fisher's [19] attempt to preserve homogeneity at the cost of compromising hyperscaling by introducing the anomalous dimension of the vacuum ω^*.

For the ordinary critical point, the universality parameters are d and n and, as discussed in more detail in the next section, the value $d = 4$ is the special and unique case in which the classical (regular) thermodynamic and correlation exponents can be self-consistent while the homogeneity condition is maintained. It is therefore the only place where the bifurcation of exponents can take place and where the ϵ-expansion method can have an unperturbed starting point. The renormalization group results [20,5] do indeed show that, for $d > 4$, the stable fixed point becomes that of the "trivial" Gaussian model for which the expected bifurcation and dominance exchange does occur. While obvious enough from the present point of view, exactly why this switch must take place suddenly at $d = 4$ is not so apparent from within the renormalization group approach itself. As Wilson [21] said in his 1979 article in *Scientific American*:

> It is not entirely surprising that the critical exponents should converge on the mean-field values as the number of spatial dimensions increases ... As the dimensionality increases, the physical situation begins to resemble more closely the underlying hypothesis of mean-field theory. It remains a mystery, however, why $d = 4$ should mark a sharp boundary above which the mean-field exponents are exact.

2.4 A Classical Conspiracy

2.4.1 Prejudice Incorporated

In section 2.2.4 we saw that, although it is customary to talk of correlation length rather than volume, it is the latter (or equivalently any other measure of size, such as N, the number of particles) which is really the appropriate statistical concept. At a critical point, where the correlation of the fluctuations becomes long ranged, we have the already defined m, in R^{-m}, as the exponent identifying the decrease with R of the correlation function. The usually employed $\eta = m - d + 2$ was not a fortunate choice as the convention for this exponent, since it incorporates a prejudice for

---

[2]What has been referred to as a "regular" contribution is actually [12] the Legendre transform of a function possessing a convergent Taylor expansion and will, in fact, involve branch points and exponents that are rational fractions.

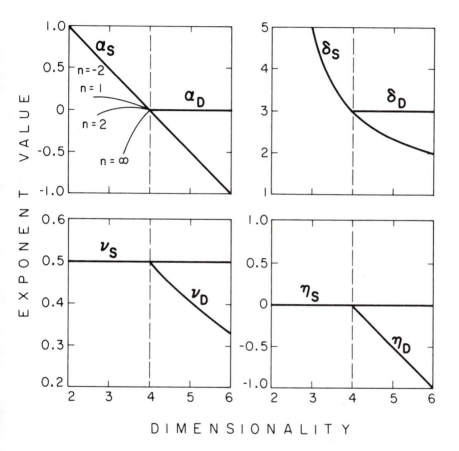

FIGURE 2. Restoration of hyperscaling by exponent bifurcation. A bifurcation of exponent values as a function of dimensionality d takes place at $d = 4$, as shown here for α, δ, ν and η. A regular contribution to the free energy dominates the singular one for $d > 4$, when it generates exponent values shown with subscript D. Those with subscript S arise from the singular term for all $d > 2$. The values represented by the heavy lines are for the "trivial" Gaussian fixed point, for which the phase-set number n equals -2. The expressions for the exponents are: $\alpha_S = 2 - d/2$; $\alpha_D = 0$ $(d > 4)$; $\delta_S = (d+2)/(d-2)$; $\delta_D = 3$ $(d > 4)$; $\nu_S = 1/2$; $\nu_D = 2/d$ $(d > 4)$; $\eta_S = 0$; $\eta_D = 2 - d/2$ $(d > 4)$. For $d > 4$ the same values apply for all n, according to renormalization group calculations, but for $d < 4$ the "nontrivial" fixed point takes over with results for α_S shown by light lines for several values of n. Note that the hyperscaling *equalities* ($\nu_S d = 2 - \alpha_S$, $\nu_D d = 2 - \alpha_D$, *etc.*; see text) apply in all cases, although this would not be so in the usual notation. In that, the exponents (without subscripts) take the value of the larger of the bifurcated pair in each case; this destroys hyperscaling for $d > 4$, since α and δ equal α_D and δ_D whereas ν and η equal ν_S and η_S, respectively.

a spurious d-dependence in what, as we have seen, should be a statistical not a geometric concept. The convention arose as the specification for a departure from the prediction $m = d - 2$ of the classical Ornstein-Zernike theory. It is, however, only this comparison theory which brings in the explicit d-dependence, not the concept itself.

The story of the classical theory of the correlation function is a fascinating example of the tangling of the threads referred to in the introduction. It is a case in which a fertile and creative idea, in many ways long before its time, contributes positively and suggestively but at the same time, aided and abetted by the camouflage of convention and prejudice, contributes to confusion and distortion. In this convoluted case, the concept itself is at the same time both penetrating and misleading; furthermore it yields predictions which are numerically nearly correct but, as it finally turns out, correct for the wrong reason!

2.4.2 Classical Correlations

The Ornstein-Zernike theory [22,3,10] involves the simple and appealing idea of representing the correlation function by an integral equation in terms of a supposedly simpler "direct" correlation function. The theory recognizes that the influence at a distance R caused by a fluctuation in a volume element at r can be seen as the sum of a direct contribution and an indirect one, the latter caused by the influence at r' due to this fluctuation in turn producing its effect, to be then integrated over all r'.

The integral over space of this "direct" function converges, even at the critical point, and the theory makes the (plausible, but in fact unwarranted and usually incorrect) assumption that its Fourier transform, in terms of wave number k, is analytic and possesses a Taylor expansion in k^2 at the long wavelength limit $(k \rightarrow 0)$. This leads to the result $\nu = \gamma/2$ and to that already mentioned above, namely $m = d - 2$. Now it has long been clear, from the results of light scattering and other experiments [4], that the value of m at the critical point of fluids is indeed very close to unity, the value given by the Ornstein-Zernike theory for three dimensions. Because of this success (and no doubt also because the venerable theory was published as long ago as 1914) it became established as the classical theory of correlations and was bracketed with the classical theories of the thermodynamic transition, such as that of van der Waals.

2.4.3 Quantum Links

Another significant but potentially misleading thread links the Ornstein-Zernike theory with quantum field theory. There the connection is with a field which has quanta with rest mass M corresponding to R_c^{-1} (times h/c, Planck's constant divided by the velocity of light). At the critical point itself, where $R_c \rightarrow \infty$, the field quanta have zero rest mass, like the photon. The Ornstein-Zernike theory describes the bare, noninteracting field and the solution for $M = 0$, $d = 3$, corresponds to the Coulomb field, for which indeed the potential falls as r^{-1}, *i.e.*, exponent $m = 1$, $\eta = 0$. Actual correlation fields, however, are not noninteracting and so possess in general a nonzero anomalous dimension η; for real fluids the exponent m, while nearly unity, is not exactly so. Of course real fields are not noninteracting in quantum field theory either, and it is that domain from which the renormalization group originated, in attempts to cope with those interactions. The great developments and great insights that have followed the adaptation of the renormalization group by Wilson [13] to critical phenomena, have been not only in statistical mechanics; the insights gained there have provided major feedback to field theory itself [5,6]. Let us return our discussion now, however, to the thermodynamics, rather than the correlation field, of the critical transition.

2.4.4 Classical Exponents

The classical critical transition for fluids is exemplified by the famous equation of state put forward by J. D. van der Waals in his dissertation at Leiden University in 1873. This was only four years after the discovery by Andrews [23] of "the continuity of the gaseous and liquid states of matter." Van der Waals' theory, even more venerable than that of Ornstein and Zernike, began the theoretical analysis of critical phenomena. Even the concept of universality was foreshadowed by his "law of corresponding states" which played a major role in the turn of the century attempts to liquefy the gases and in the beginnings, particularly in the Netherlands, of low temperature physics. In 1907 came the Weiss theory of magnetism and in the 1930's the Bragg-Williams theory of order-disorder transformations in binary alloys. These and all mean-field theories share with that of van der Waals the same fatal weakness; they embody the implied existence of Taylor expansions, this time for the thermodynamic properties near the critical point (an assumption, like that for the Ornstein-Zernike theory, again plausible, but unwarranted and usually incorrect). The assumptions of valid Taylor expansions in the two cases, while apparently similar, are actually completely independent. It is quite possible for either one to apply and the other not. In fact, worse than independent, they are in general *inconsistent* with each other; furthermore, as proved [8] long ago, it is *impossible* for both to apply to the same three-dimensional system!

The mean-field assumptions lead to thermodynamic exponent values which are the well-known "classical" ones

$$\alpha = 0, \quad \beta = \frac{1}{2}, \quad \gamma = 1, \quad \delta = 3 \quad , \tag{9}$$

while the "classical" correlation exponents are the Ornstein-Zernike ones, with $\gamma = 1$, namely

$$\nu = \frac{1}{2}, \quad \eta = 0 \quad . \tag{10}$$

2.4.5 Classical Inconsistency

In sections 2.2.4 and 2.3.1 we saw that the correlation of fluctuations in the critical state requires the exponents c_1 and c_2, regardless of d, to be determined solely by α and δ, respectively, as in equations (4). The correlation exponents ν and m are then given by $\nu = (2 - \alpha)/d$ and $m = 2d/(\delta + 1)$; these and the consequent relation for η are displayed in the first row of table 1. From the classical values $\alpha = 0$, $\delta = 3$, we then have the result that for a classical transition in d dimensions the value for the exponent ν is $2/d$ while that for m is $\frac{1}{2}d$. We can call these the *classical thermodynamic* values. Having called the Ornstein-Zernike theory "classical," however, convention has already pre-empted as the classical values $\frac{1}{2}$ for ν and $d - 2$ for m (*i.e.*, $\eta = 0$), as given by equations (10). For our present purpose we call these latter the *classical convention* values. The two sets of classical values are shown in the second and third rows of table 1. The fourth row gives the classical thermodynamic values for three dimensions while the fifth lists the only case, namely $d = 4$, in which the two classical values are self-consistent. The table also includes some other nonclassical cases.

It is hardly surprising, with the inbuilt semantic inconsistency, that there has been much confusion concerning the correlation function exponents. Furthermore, reliably accurate and even exact results have sometimes provided misleading and even completely false clues.

The classical value zero for α happens to apply exactly (but representing a logarithm) to the two-dimensional Ising model and very closely to the helium λ-transition, while its value is only about 0.1 for the real gas-liquid critical point in three dimensions. Thus in all of these cases (see last rows of table 1), none of which is a *classical* system, the actual value of ν which is $(2 - \alpha)/d$ is close to the "classical thermodynamic" value $2/d$ while not being at all close to the "classical convention" value $\frac{1}{2}$. However, the contrary appears to apply in the case of the other exponent η, for which the classical "convention" fares rather well. For the binary fluids and gas-liquid transitions which are the cases best studied, scattering experiments show that η is very close to the zero value of the Ornstein-Zernike theory. This apparent agreement with the classical "convention" arises, however, for quite the wrong reason from the classical point of view—it is because the classical value 3 for δ is in fact quite wrong; δ is approximately 5, just the value needed (see again row 6 of table 1 and equations (3)) in three dimensions to give $\eta = 0$!

TABLE 1. CORRELATION EXPONENTS FOR d DIMENSIONS. Confusion has arisen because of two independent applications of the epithet "classical." These are distinguished here as "thermodynamic" and "convention." Consistency with fluctuation theory requires the correlation exponents v and m to be given in terms of the thermodynamic exponents α and δ by the first row in the table. Thus for the "thermodynamic" classical case ($\alpha = 0$, $\delta = 3$), this gives the second row. However, the "classical" Ornstein-Zernike theory gives the classical "convention" in the third row, inconsistent (except for $d = 4$) with the "thermodynamic." Exponent η is defined as $\eta = m - d + 2$ so that in any dimension, d, η is zero by "convention." Observed results for three-dimensional systems show approximate agreement with the classical "thermodynamic" v and the classical "convention" η. However this agreement is not because the system is classical but (row 6) because α is numerically small (so $v \sim \frac{2}{3}$) and δ is nearly 5 (so $\eta \sim 0$).

Correlation Exponents	v	m	η
From correlation of critical fluctuations:	$(2 - \alpha)/d$	$2d/(\delta + 1)$	$2 - d\dfrac{(\delta - 1)}{(\delta + 1)}$
Classical "thermodynamic"	$2/d$	$\frac{1}{2} d$	$2 - \frac{1}{2} d$
Classical "convention"	$\frac{1}{2}$	$d - 2$	0
Classical "thermodynamic", $d = 3$	$\frac{2}{3}$	$\frac{3}{2}$	$\frac{1}{2}$
Classical consistent, $d = 4$	$\frac{1}{2}$	2	0
Nonclassical, $d = 3$, $\alpha = 0$, $\delta = 5$	$\frac{2}{3}$	1	0
Ising, $d = 2$, $\alpha = 0$, $\delta = 15$	1	$\frac{1}{4}$	$\frac{1}{4}$
Fluid critical transition (experimental)	$0.63 \pm .01$	$1.03 \pm .03$	$\leqslant .05$
Helium λ-transition (experimental)	$0.675 \pm .001$	?	?

2.5 Summary of Long Range Order

To conclude this look at the nature of order in macroscopic equilibrium systems, we summarize some important features discussed in the sections above.

- Long range order is a statistical not a quantum effect.
- It arises under thermodynamic conditions in which there exists a set (more than one) of phases in equilibrium with each other.
- The phase-set exists as a consequence of the spontaneous symmetry-breaking necessary to ensure stability.
- In the long range ordered state, every volume element departs grossly from the symmetry-mean to a value close to that of one member

of the phase-set. The symmetry-mean value is unstable and inaccessible; symmetry may be maintained globally, but it is broken locally.

- In contrast, in the disordered one-phase state (usually at higher temperature) symmetry is preserved locally in the mean in every volume element.
- What physical property is ordered depends on the particular system; it is the "order parameter," a value for which characterizes each member of the phase-set.
- The order parameter can be a purely quantum effect, as it is for the superfluid states.
- The information in the long range order determines which phase a particular volume element is in, what value its order parameter has. The choice is correlated over the whole volume of the system, however large, and is determined by the specifics of boundary conditions.
- The local value of the order parameter may vary with position but it is a macroscopic ("hydrodynamic") variable, not a microscopic one.
- The members of the equilibrium phase-set can be represented by the points on the surface of an n-dimensional hypersphere, where n is the phase-set number.
- Two numbers characterize the universality class of a critical transition; they are n and d, where d is the dimensionality of geometrical space.
- Condensation, order-disorder in binary crystals, the Ising model, *etc.*, correspond to the phase-set number n having the value $n = 1$; this set has only two discrete phases, representable as $+1$ and -1.
- For $n > 1$ there is a continuum of phases; these can be identified by their angle on the hyperspherical surface.
- For $n = 2$ a single angle χ, with period 2π, describes the phase-set, representable in this case by the points on a unit circle.
- This $n = 2$ case describes the superfluids and characterizes the helium λ-transition.
- Helium at the critical density has two transitions in different classes, the λ-transition in the class with $d = 3$, $n = 2$, and the gas-liquid critical transition with $d = 3$, $n = 1$.
- Because of the nonexistence of a real symmetry-breaking field for the helium λ-transition, most of the critical exponents are not accessible to experiment. (See section 3.)
- Partly for the same reason, those that are accessible, α and ν, present opportunity for the greatest experimental precision and thus for stringent test of the hyperscaling equation $\nu d = 2 - \alpha$.

3.   THE SUPERFLUID TRANSITION

3.1   Superfluid Order

3.1.1   London Equations

As with so much of low temperature physics, the theory of the helium λ-transition begins with Fritz London. The strange flow properties of the fluid became known [24,25] in 1937–38, and very soon London proposed [26] that the transition occurring about 2.2 degrees above absolute zero was essentially related to the condensation phenomenon characteristic of the ideal Bose-Einstein gas. A few years earlier, Fritz London and his brother Heinz had put forward their macroscopic electromagnetic equations for the phenomenological theory of the superconducting state. The new proposal was not unrelated.

Developments in the understanding of cooperative phenomena resumed after the interruption of the second world war. They were stimulated particularly by growing awareness of the dramatic and little understood properties displayed by the superconductors and liquid helium. There was as yet no microscopic theory, but powerful and elegant phenomenological theories had been developed by London and were embodied in the two small volumes [27] of his masterpiece *Superfluids*. London's theory emphasized that the superfluid phases were macroscopic quantum states, which involved long range order in *momentum* space, rather than in ordinary coordinate space as in all familiar ordered phases.

A fundamental assertion of the London theory was that at low speeds the superfluid motion is irrotational, curl $p_s = 0$, where p_s is the momentum field of the superfluid. This gives the well-known irrotational hydrodynamics of superfluid helium, as well as the famous London equation for superconductivity.

This type of behavior for a system in thermodynamic equilibrium is highly singular. In contrast, standard thermodynamic consideration of rotational equilibrium seemingly shows [28] that, whether or not a system is solid, it should in equilibrium possess solid-like motion. It should have a velocity field $v = \Omega \times r$ for which curl $v = 2\Omega$, far from irrotational. Here Ω, taken to be vanishingly small (to allow a focus on linear effects), is a constant vector representing the angular velocity of the containing "reservoir," with which the system is in equilibrium. This theorem would imply for example that if any substance, frozen solid and in rotational equilibrium, were reversibly melted (without the application of any mechanical forces, or redistribution of its moment of inertia) its *state of motion* could not change. For this to be true, the substance could not obey London's irrotational equation; if it did, a change of motion would have to appear spontaneously.

3.1.2   The Wrong Dilemma

We are now faced with a dilemma. It seems that a choice is being forced between London's irrotational motion and thermodynamic equilibrium motion with constant rotation.

This is not where the real dilemma lies, however. The "solid-like rotation" theorem is another "classical" inheritance. It is as vulnerable as the classical critical transition and the classical correlation theory discussed in section 2. Here we can employ the word "classical" not only in the sense of *established by tradition*, but also in the sense of *non-quantum-mechanical*.

The fatal flaw of the classical theories discussed in section 2 was the unwarranted assumption of the existence of convergent Taylor expansions; the two theories made assumptions that were independent of, even inconsistent with, each other. In the proof of the "solid-like rotation" theorem a tacit presumption (once more, plausible, but in general unwarranted and incorrect) is made that motion in two volume elements separated by a macroscopic distance is statistically independent. In effect the proof depends on the absence of long range momentum correlations. It is the existence of just such correlation, however, that forms the basis of London's theory and causes the highly singular response to magnetic fields and rotation in superconductors, and to rotation in superfluid helium.

It is ironic that it is not the *existence* but the *specific form* of long range correlation that undermined the classical theories of critical phenomena. Both the Ornstein-Zernike and the van der Waals' theories involve long range correlation in the specific forms given in table 1; both were seen to be wrong for three-dimensional systems. It is clear, however, that classical mechanics (or indeed any sort of mechanics, as we saw in section 2.1.1) has no objection to long range order as such. Classical mechanics permits order of any configurational or *static* property, but not of *motion*; that is vehemently forbidden by the classical equipartition theorem, but not of course by quantum mechanics.

3.1.3   Motion a Point of Order

The nature of the equilibrium response of a system to attempts to change its motion is best seen in terms of the Fourier transforms, as shown by Schafroth [29,30]. Let us deal with the more familiar magnetic case, and suppose that M_k is the magnitude of the equilibrium (diamagnetic) magnetization produced by the kth Fourier component, B_k, of an imposed magnetic field $B(r)$. The linear response coefficient χ_k is just the ratio M_k/B_k. For London's equation this susceptibility ratio at long wavelengths diverges as k^{-2}. In contrast the diamagnetic susceptibility of a normal metal would correspond to a coefficient which has a small constant value for $k = 0$. Furthermore, in *classical mechanics* the equipartition theorem shows that momentum correlations actually vanish in equilibrium and that the diamagnetic coefficient, far from diverging, is not even a finite constant—it is identically zero! Unlike the static phenomenon of paramagnetism, equilibrium diamagnetism does not exist in classical mechanics.

London's theory both for superconductors and for superfluid helium requires a highly singular response, dependent on a correlation of the momentum of particles over arbitrarily long range. It encompasses as essentially equilibrium properties the conservation (and quantization) of magnetic flux and the existence of persistent current. For example, a superconducting ring in a very small magnetic field would carry in equilibrium a charge-transporting circulating current proportional to that field and exactly sufficient to cancel the imposed flux. The question of whether persistent current states really can be equilibrium states or merely ultra-long-lived nonequilibrium ones, was the real dilemma confronting London's theory, not the false one described in section 3.1.2. Can it be true that the equilibrium motion-response function in the long wavelength limit, $k \to 0$, really does change from being identically zero in classical mechanics, to infinity? A group of theorists in Australia thought not and in 1954 were developing a theory with less drastic demands on long range motion correlation.

The nature of the real dilemma was stated very clearly by Casimir [31] in an article in the 1955 volume commemorating the 70th birthday of Niels Bohr:

> If London's explanation of persisting currents is valid, we have to consider such stable wave functions extending over a mile or so of dirty lead wire. This idea of macroscopic quantum states has also been considered in connection with liquid helium. If London's equation on the other hand would only hold in cells of the order of for instance 10^{-4} cm, then we would still find (the diamagnetic properties) with sufficient accuracy, but the existence of persisting currents would remain unexplained. So we are faced with the following dilemma: either we have to abandon London's idea altogether and must look for a separate explanation of the infinite conductivity or we have to accept some sort of coherence extending over very large dimensions.

3.2 Limited Range Order

A theory that modified London's with the idea of a large but finite correlation length for momentum correlations, was advanced [32] in 1955 by the Sydney group, J. M. Blatt, S. T. Butler and M. R. Schafroth. They suggested that the diamagnetic property of superconductors should be characterized by a response function $\{k(k + \mu)\}^{-1}$, still singular but less violently so than the London form, k^{-2}, and containing a constant $\mu = \Lambda^{-1}$, determined by the finite correlation length Λ. London's theory would be regained if the correlation length were taken as infinite. Schafroth [33] had shown earlier that the ideal Bose-Einstein gas of charged (but non-Coulomb-interacting) particles effectively obeys the London equation, so that in 1955 the following were the cases of exactly known behavior of the equilibrium diamagnetic response function χ_k in the $k \to 0$ limit: In classical mechanics χ_k vanishes; in "normal" quantum mechanics, χ_k approaches a small constant; in the unphysical ideal Bose-Einstein gas, $\chi_k \to k^{-2}$. The finite correlation length theory assumed the intermediate behavior, $\chi_k \to k^{-1}$.

In liquid helium the angular momentum response to equilibrium in a rotating container plays the role of the diamagnetic response in superconductors, with very similar consequences. As Casimir was to imply in the statement quoted above, persistent currents and superfluid flow must in this theory be metastable, not equilibrium properties, and this aspect was emphasized in the publication entitled "Nonequilibrium Nature of the Superfluid State" by Butler and Blatt [34].

The equilibrium properties were of course also implied by the theory. Their evaluation, however, required extensive computation and had to await completion of the SILLIAC computer. As a sobering reminder of the speed of computer development, it is interesting to recall now that this machine, the first fully-engineered electronic computer to be built in Australia, was the Sydney version of the ILLIAC machine at the University of Illinois. When commissioned, SILLIAC was one of the fastest machines in the world and could carry out, on twelve decimal digits, half a million operations *per minute*. To do so it used 2,800 electronic vacuum tubes, with 40 cathode ray tubes for its fast access memory, consumed 35 kilowatts and, of course, had to be programmed in machine language. The first scientific calculation carried out on SILLIAC, on July 5th, 1956, was in fact that for the equilibrium properties of liquid helium. The results of the computations were reported [35] in the September 1956 issue of *Il Nuovo Cimento*, with due acknowledgement to the just commissioned SILLIAC. The computations involved clusters of 10^7 atoms or more, and yielded results for the specific heat and superfluid density, calculated essentially from first principles but incorporating the assumption of a finite correlation length, $\Lambda \sim 10^{-5}$ cm. The results showed an agreement with the then existing experimental results that greatly improved on any obtained by previous theoretical approximations.

Because of the limited correlation length, the theory did not yield a singularity in the variation of the properties as a function of temperature and hence did not give a true thermodynamic transition. It gave instead a smoothed out "quasi-transition." The value of the specific heat, although a continuous and differentiable function of temperature, did nevertheless change extremely suddenly, falling from a maximum value by a factor of two within a millidegree interval above the λ-point; this occurred after the value had risen steadily, for some two thousand times that interval, as the quasi-transition was approached from lower temperatures.

3.3 Theoretical and Experimental Connections

It happened that I was visiting the University of Sydney Physics Department at just this time, on leave from Duke University where I had been doing theoretical research for the previous two years in close collaboration with the experimental low temperature physics group led by Bill Fairbank.

I had made arrangements to go to Duke in 1954 to work as a post-doctoral fellow with Fritz London in the chemistry department. His sad and untimely death on March 30 that year was a great loss to the world and an irreplaceable loss to physics and chemistry. In his correspondence with me, London had told me of the exciting experimental work being done in the physics department low temperature laboratories. In a letter written only a few days before his death, London had told me of the deviations from the Curie law below 0.5 K then being indicated by measurements of the nuclear magnetic susceptibility of liquid ^{3}He. This he suggested as something that I might be interested in looking into, as indeed I later did.

In a letter informing me of the tragedy, Bill Fairbank asked me still to come to Duke and suggested that I change from chemistry to the physics department.

After having worked with the intrepid Fairbank for two years, I had a different perspective from that of the Sydney theorists, who saw their predicted quasi-transition more as a theoretical oddity than as an experimentally decisive opportunity. It was not hard to conclude, however, that with good design it would be quite feasible to improve greatly the experimental resolution of specific heat measurements near the λ-point. Bill Fairbank was equally enthusiastic to carry out quickly the crucial test of the quasi-transition theory. Immediately on my return to North Carolina in late September we set to work, soon to be joined by C. F. Kellers, on a "crash program" to do the experiment and to try to obtain the qualitative result by the end of the year.

3.4 Transition Class

3.4.1 Sharp Enough

The experiment was successful and the first report [1] of the results, the one mentioned in the introduction, was published in April 1957. It described a logarithmic singularity, sharp to nearly a microdegree. Immediately it was clear that the quasi-transition concept and the finite correlation length theory were untenable. The demise of the theory is actually witnessed in the literature [36] even before the Washington meeting. In the *Bulletin of the American Physical Society* for the January 1957 New York meeting, the cryogenics section displays the evidence of the theory withdrawn.

The unequivocal conclusion from these preliminary findings was confirmed in more detail as measurements were further refined in the next few years [37,38, 39], particularly by C. F. Kellers [40] as reported in his thesis. The results were consistent with the specific heat diverging as the logarithm of the temperature interval $T - T_\lambda$, with the same coefficient above and below the transition at T_λ. They were reported in the form of the specific heat at the saturated vapor pressure, C_s, in joules per gram degree, given by

$$C_s = 4.55 - 3.00 \log_{10} |T - T_\lambda| - 5.20\Delta \quad , \qquad (11)$$

where $\Delta = 0$ for $T < T_\lambda$ and $\Delta = 1$ for $T > T_\lambda$. This expression described the results for four decades of the temperature interval $|T - T_\lambda|$, from $10^{-2} T_\lambda$ to $10^{-6} T_\lambda$. Figure 3, adapted from one of the original publications, illustrates how very sharp the transition is. The "steep" descent from its maximum, calculated for the quasi-transition, would have noticeably finite slope on the scale of the central illustration. On the scale of the right illustration it would be reduced to being indistinguishable from a horizontal straight line across the figure at about the value 15 for C_s.

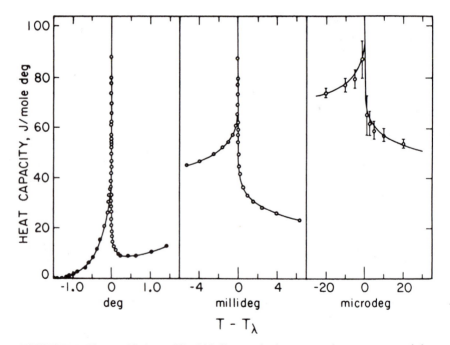

FIGURE 3. The specific heat of liquid helium under its saturated vapor pressure (after Buckingham and Fairbank [38]).

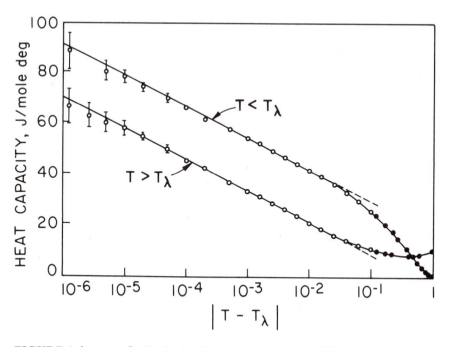

FIGURE 3 (*continued*). The broken lines represent equation (11).

3.4.2   Real Class

In the general discussion of long range order in section 2.1.1 we recalled that its microscopic mechanics and symmetries determine for a system what properties and therefore what extensive thermodynamic variables it may have. But we also saw that, given the existence of any such property, its possible manifestation in long range order is a purely statistical phenomenon. No property that exists at all is inherently immune from a potential threat to break its symmetry and disrupt the symmetry-mean manner in which it normally distributes itself. One must therefore be suspicious of the denial of the potential of symmetry-breaking to any property merely because it does not exist in classical mechanics. At least with hindsight, it is not hard to see that the quasi-transition theory was doomed, in spite of its ingenuity, because of its denial of this principle.

Real matter possesses in general a finite, although usually small, diamagnetic as well as paramagnetic response coefficient. Either property, the one a motion, the other static, may be involved in symmetry-broken long range ordered states; the first becomes a superfluid, the second a ferromagnet. Exactly solvable but somewhat unphysical models exist for both systems. These are, respectively, the ideal Bose gas, as first realized by London, and the "spherical" model, originally introduced by Berlin and Kac [41]. The relationship between the condensation of the ideal Bose gas and the λ-transition in real liquid helium can be said to be similar to the relation between the spherical model and a real ferromagnet. Furthermore the two models can be shown [42] to share the same set of critical exponents, identified now as the universality class $d = 3$, $n = \infty$.

The helium λ-transition is, then, a genuine "mainstream" cooperative transition and, as indicated already in section 3.6, is in universality class $d = 3$, $n = 2$. As stated in the introduction, it is an extremely favorable system for experimental high resolution measurements. This carries with it a serious penalty, however. One reason the helium transition is so favorable is that there exists no possible interference from a symmetry-breaking field conjugate to the order parameter Ψ introduced in section 2.2.3. Comparably precise measurements would not be possible on fluid critical transitions in an earthbound laboratory, because gravity is effectively conjugate to the order parameter in that case. The very nonexistence of such a symmetry-breaking field means, however, that there is nothing to manipulate in experimental analogy to the magnetic field for a ferromagnet, and therefore there is no access to the thermodynamic exponents, other than α and β. The *correlation* exponent ν is accessible, however, as was shown by Josephson [43], through measurement of the manner in which the superfluid density ρ_s vanishes as $T \to T_\lambda$. Thus although the other equations remain hidden, the hyperscaling equation $\nu d = 2 - \alpha$ can be tested, and with much greater accuracy than would be possible for any other system.

3.4.3   Log-Jam

One of the recognized turning points in the field of critical phenomena was the conference held at the National Bureau of Standards in April 1965. The meeting was hosted by Melville Green, now so sadly missed. In a contribution to the recent volume dedicated to the memory of Mel Green, Cyril Domb [44] states:

> The conference at NBS in 1965 can reasonably be termed the founding conference of critical phenomena, since for the first time all the different strands of the subject were woven into a coherent fabric.

This is true, but it is also true that some of the strands were very tangled. One of the knots, which had its heyday at that time, was the ubiquitous logarithm. In 1965 the only "realistic" theoretical model with known critical properties, albeit only in two dimensions and in zero symmetry-breaking field, was Onsager's Ising model. The only real system with its asymptotic behavior accurately measured was the helium λ-transition, again with only zero symmetry-breaking field accessible. For both of these the specific heat diverged with the logarithm of the temperature interval; the former exactly, the latter within measurement accuracy over four orders of magnitude in relative temperature interval.

With only these two cases confidently known, it is not surprising that many people thought that perhaps all cooperative transitions might have logarithmic specific heat singularities. It had been clear for many years that the coexistence curve for fluids was much better described by the Guggenheim [45] "law" $\beta = \frac{1}{3}$ than the classical $\beta = \frac{1}{2}$, but otherwise no reliable, high resolution information was available. The keynote lecture at the NBS conference was delivered by Uhlenbeck [46] who emphasized the importance of the failure of the classical critical point description, and, after recalling the widespread $\beta = \frac{1}{3}$ behavior and the logarithmic specific heats, said:

> If there is such an universal, but nonclassical behavior, then there must be an universal explanation which means that it should be largely independent of the nature of the forces. The only corner where this can come from is I think the fact that the forces are not long range. The Onsager solution gives I think a strong hint. It may well be so that away from the critical points the classical theories give a good enough description but that they fail close to the critical point where the substance remembers so to say Onsager. I think that to show something like this is the central theoretical problem. One can call it the reconciliation of Onsager with van der Waals.

These remarks were prophetic indeed; they also reflect the widely held view of the time about the "universality" of the logarithm. In fact, it was to take until nearly the end of the decade before the logarithm was convincingly displaced from its privileged position. As far as I know, it was a report from our own laboratory in Western Australia by Edwards, Lipa and Buckingham [47] that first clearly stated that the logarithmic form was ruled out by experiments. These were measurements of the specific heat very near the critical point of xenon and later [48, 49], much more precisely, of CO_2. One whole section of the NBS conference was devoted to "Logarithmic Singularities" and logarithms arose in several other segments as well; amongst other speakers, both Bill Fairbank and I presented papers.

Fairbank [39] reviewed the helium λ-transition specific heat results and compared them with those for some magnetic systems. He also related [39] an amusing exchange between Blatt and Feynman concerning the conceptual difference between zero and 10^{-70} in reduced momentum correlations across the width of the universe! As it happens, I possess a copy of a letter (dated 17 September, 1956) written by John Blatt to Robert Schafroth at the time of my Sydney visit mentioned above, and it refers to the same exchange. Blatt, displaying his unwavering confidence in the quasi-transition theory, reported that he had lunch with Feynman

> ... and beat him down considerably from his initial position, although
> he is still far from convinced ... I also have a $10 bet with Feynman
> that C_V of liquid He, measured with $> 10^{-5}$ °K resolution, will be
> continuous.

My own paper [50] in the same segment, although somewhat preten-
tiously entitled "The Nature of the Cooperative Transition," did contain
some significant premonitions of later developments. It is another example
of the entangling of insight and error and of the fact that while an idea too
late is useless, one that is premature tends to stumble over unseen hazards
and imaginary logs. The paper purported to show that the critical fluctuations at a classical transition (one with the classical exponents, given by
equations (9) of section 2.4.4) would prevent it from being self-consistent
unless it had a logarithmically infinite, rather than a discontinuous, specific
heat. It did this by considering the description of the singularity from the
point of view of a sequence of divisions of the system. An infinite system
was divided into cells or blocks containing a larger and larger number of
atoms, increasing by a factor for each higher order in the sequence. This
led to a recursion relation which was solved for the logarithmic specific
heat. The argument was that, once the cells were large enough,

> the problem has ceased to depend on any detailed property of the
> interaction between atoms, or of any finite number of atoms and has
> taken an asymptotic form ... Thus any thermodynamic singularity
> may be regarded as an example of a generalized near neighbour lattice
> problem.

The further analysis, however, had several flaws, one an embarrassingly
elementary error of thermodynamics. Carried through correctly, the argument should have led to the result that α is $2 - \nu d$, not that it represents a
logarithm, an example indeed of how prejudice can intrude its expectations,
even at the expense of simple logic!

3.4.4   Fluid Logs

The first measurements attempting to resolve the fluid specific heat
anomaly were made in 1962 by Bagatskii, Voronel and coworkers in the
USSR [51,52]. They interpreted their measurements on argon and oxygen as indicating the same type of logarithmic singularity as shown by the
helium λ-transition. Fisher [53] analysed their results and compared them
with numerical computations for the Ising model lattice-gas in three dimensions. The latter had indicated a larger value for the specific heat exponent
α than for α', where α and α' apply to the high temperature and low temperature branches, respectively. The Russian workers had suggested that
both branches may become logarithmic with the same coefficient, nearer to
T_c than about 0.1 percent; however, as Fisher pointed out, this speculation
put undue weight on the few uncertain experimental points closest to T_c.

In his analysis Fisher was able to find a type of agreement within about 10 percent between the computations and the observed results. This was based on $\alpha \sim 1/5$ for $T > T_c$, and a logarithmic divergence, $\alpha' = 0$, for $T < T_c$.

At the NBS meeting Moldover and Little [54] reported higher resolution specific heat measurements for the critical transition of helium. These results also suggested a logarithmic $\alpha' = 0$ form for $T < T_c$ but a larger exponent for $T > T_c$. None of these early measurements, however, was at all adequate for a real determination of the exponent values, as a review in 1967 by Kadanoff et al. [55] concluded. Any value for α between 0 and 0.3 or 0.4 and for α' between 0 and 0.25 would be consistent with all the data available at that time. Similarly, specific heat measurements on magnetic and other systems were quantitatively ineffective, although usually described in terms of logarithmic behavior.

However, as we have already stressed, precision was much more readily obtainable for the helium λ-transition. By 1966 measurements of the superfluid density ρ_s near the λ-point had been made, using two different methods, by Clow and Reppy [56] and by Tyson and Douglass [57]. Both obtained results indicating that ρ_s/ρ vanishes with an exponent ζ equal to 2/3 as the λ-point is approached. In 1968 measurements of u_2, the velocity of second sound, were made near T_λ by Pearce, Lipa and Buckingham [58] in Australia and by Tyson and Douglass [59] in Chicago. Since u_2^2 is proportional to ρ_s/C_p near $T = T_\lambda$, the exponent for u_2^2 should be $\zeta + \alpha'$, so for the first time there was an opportunity to test directly the consistency of the asymptotic forms of two thermodynamically equal quantities. While all the independent results were indeed fully consistent with thermodynamics, they illustrated well the numerical significance of terms additional to the asymptotically dominant ones.

Tyson and Douglass [57] had reexamined the Duke specific heat results and concluded that the specific heat exponent should be described as $\alpha = -0.014 \pm .016$, consistent with the logarithmic form, $\alpha = 0$ and equation (11), but hinting at a slightly negative value. They reported their own ρ_s/ρ measurements as giving $\zeta = 0.666 \pm .006$. Thus these results, $\alpha = 0$ and $\zeta = 2/3$, would satisfy, to within one percent, the condition $\zeta d = 2 - \alpha$. This is expected from the hyperscaling equation since the exponent ζ in fact equals the correlation length exponent ν. At first the exponent ζ had been incorrectly identified [56] with 2β, by analogy with the ideal Bose gas, for which ρ_s is $|\Psi|^2$, the square of the order parameter. In the real superfluid, however, the operational definition of ρ_s is a hydrodynamic one and, as Josephson [43] pointed out, consistency requires its identification with $2\beta - \eta\nu$, not just 2β. Using the hyperscaling equations (3), this requires $\zeta = (d - 2)\nu$ so, in three dimensions, $\zeta = \nu$.

3.4.5 Scaled

As the 1960's advanced, the earlier indication that the exponent α is greater than α' for the fluids gave way to support for the scaling condition $\alpha = \alpha'$. One reason for the initial discrepancies, in both experimental and numerical results, is illustrated in figure 4. Any effect, such as that of

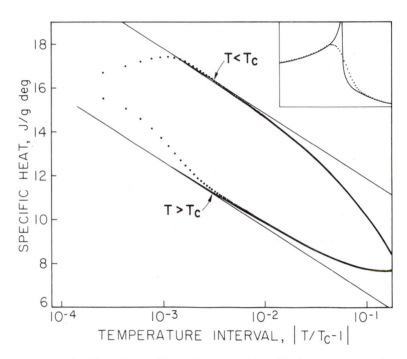

FIGURE 4. The effect of smoothing a sharp transition. The low temperature branch has a larger magnitude than the high and this asymmetry has the result that any smoothing, experimental or numerical, appears to increase the exponent α and decrease α'. This deception is only enhanced by the departures from the asymptotic form away from the transition. In this illustration a simple logarithmic singularity (given by equation (11) plus a constant) is shown by the straight lines. The heavy line is the same function multiplied by $(T/T_\lambda)^3$ (to simulate departures from the asymptote), while the dotted line is that product smoothed by convolution with a simple Gaussian; the inset shows a temperature interval of 16×10^{-3} around the singularity on a linear temperature scale (with vertical scale reduced by a factor 2.5). Note how the departures from the logarithm, $\alpha = \alpha' = 0$, produce an upward curvature, simulating $\alpha > 0$ for the lower $T > T_c$ branch, and a downward curvature $\alpha' < 0$ for the $T < T_c$ branch.

gravity, magnetic field, impurities, numerical approximation, *etc.*, which causes a smoothing of the sharpness of a transition, while maintaining the total integrated area, has the result of apparently increasing the $T > T_c$ branch, exponent α, and decreasing α'. This misleading effect is a direct consequence of the asymmetry, the magnitude of the heat capacity in the two-phase being substantially larger than in the one-phase branch. Furthermore, a second effect conspires to enhance the deception, since on moving away from the transition at T_c, the sign of the departures from the asymptotic form happens to reinforce this tendency of the smoothing near T_c.

Improved experimental and numerical results in the late 1960's and at the turn of the decade led to accumulating empirical evidence for the validity of scaling and universality. Some problems remained in obtaining consistency between specific heat and equation of state data, as shown in the careful analysis by Barmatz, Hohenberg and Kornblit [60] of experimental results for xenon and carbon dioxide and for magnetic systems. The dominant events of this period, however, were in the theoretical domain, where the great breakthrough resulting from the renormalization group approach was taking place. As far as the shape of the helium λ-transition is concerned, it was, meanwhile, being subjected to intense experimental probing, principally by Guenter Ahlers at the Bell Laboratories.

3.5   Final Shape

3.5.1   Refinement

The very extensive work by Ahlers and his coworkers is well documented [61], and here we will only mention the results. By accurate measurements, not only at the vapor pressure but under higher pressure and with admixtures of the isotope ^{3}He, Ahlers was able to demonstrate convincingly that singular terms additional to the dominant ones were numerically important in practice. He showed that by allowing for these confluent singularities in the form predicted theoretically [62], it was possible to determine in great detail properties satisfying all scaling requirements fully self-consistently and over a wide range of conditions. The conclusion reached was that the value for the helium λ-shape exponent is actually less than zero, the final result being $\alpha = \alpha' = -0.026 \pm .004$, with the ratio of the amplitude for the high and low temperature branches given by $A/A' = 1.112 \pm .022$. The most accurate calculations, using the renormalization group approach, are by Le Guillou and Zinn-Justin [63], who very effectively employ Borel transformation methods. They obtain, for the universality class $d = 3$, $n = 2$, the result $\alpha = -0.007 \pm .006$. One striking result from Ahlers' work is the empirical confirmation (recalling that $\zeta = \nu$) of the hyperscaling equation $\nu d = 2 - \alpha$, already confirmed at the one percent level with $\alpha = 0$, $\zeta = 2/3$. Ahlers' new experimental result was that $\zeta = 0.6749 \pm .0007$; this value confirmed agreement with hyperscaling and the above α value at the 0.1 percent level.

3.5.2   Nearer Zero

The last word on the nature of the λ-transition has by no means been said. Even the detailed experiments by Ahlers could not penetrate nearer to zero than about 10^{-6} in the relative temperature interval. John Lipa, who has been a member of the Fairbank group at Stanford since his Ph.D. days with us in Western Australia, is now developing new techniques of thermometry which should permit measurements as close as 10^{-10} in the space shuttle where the effects of gravity are reduced. Some details of this development are described elsewhere in this book so we conclude with the expectation that still more remains to be learned from the shape of the helium λ-transition. Hopefully this will be reported at a Stanford meeting, called perhaps "Nearer Zero," to honor Bill Fairbank's 70th birthday in 1987.

Acknowledgments

I wish to thank Cyril Edwards for helpful comments on the manuscript and Robert Penny for preparing figure 4. I also wish to record my thanks to Bill Fairbank for his inspiration and unquenchable enthusiasm over so many years.

References

[1]   W. M. Fairbank, M. J. Buckingham and C. F. Kellers, *Bull. Am. Phys. Soc., Ser. II* **2**, 183 (1957).

[2]   *Phase Transitions and Critical Phenomena*, edited by C. Domb and M. S. Green (Academic Press, New York, 1972–77), Vols. 1–6.

[3]   M. E. Fisher, *Rept. Progr. Phys.* **30**, 615 (1967).

[4]   P. Heller, *Rept. Progr. Phys.* **30**, 731 (1967).

[5]   P. Pfeuty and G. Toulouse, *Introduction to the Renormalization Group and to Critical Phenomena* (Wiley, London, 1977).

[6]   J. B. Kogut, *Rev. Mod. Phys.* **51**, 659 (1979).

[7]   B. D. Josephson, *Proc. Phys. Soc.* **92**, 269 and 276 (1967).

[8]   J. D. Gunton and M. J. Buckingham, *Phys. Rev. Lett.* **20**, 143 (1968).

[9]   M. J. Buckingham and J. D. Gunton, *Phys. Rev.* **178**, 848 (1969).

[10]   M. E. Fisher, *J. Math. Phys.* **5**, 944 (1964).

[11]   M. E. Fisher, *Phys. Rev.* **180**, 594 (1969).

[12]   See M. J. Buckingham, in *Phase Transitions and Critical Phenomena*, edited by C. Domb and M. S. Green (Academic Press, New York, 1972), Vol. 2, Chapter 1.

[13]   K. G. Wilson, *Phys. Rev.* **B4**, 3174 and 3184 (1971).

[14]   M. J. Buckingham, *Aust. J. Phys.* **36**, 683 (1983).

[15]   B. Nienhuis and M. Nauenberg, *Phys. Rev. Lett.* **35**, 477 (1975).

[16]   M. E. Fisher and A. N. Berker, *Phys. Rev.* **B26**, 2507 (1982).

[17]   G. Stell, *Phys. Rev. Lett.* **20**, 533 (1968) and **24**, 1343 (1970); *Phys. Rev.* **B1**, 2265 (1970), and in *Critical Phenomena*, edited by M. S. Green (Academic Press, New York, 1971), Enrico Fermi course No. 51, p. 188.

[18]   C. Domb, *Phys. Rev. Lett.* **20**, 1425 (1968).

[19]   M. E. Fisher, in *Renormalization Group in Critical Phenomena and Quantum Field Theory*, edited by J. D. Gunton and M. S. Green (Temple University, Philadelphia, 1973) and in *Collective Properties of Physical Systems*, edited by B. and S. Lundqvist (Academic Press, New York, 1974), 24th Nobel Symposium.

[20]   K. G. Wilson and M. E. Fisher, *Phys. Rev. Lett.* **28**, 540 (1972).

[21]   K. G. Wilson, *Sci. Am.* **241**, 155 (1979).

[22]   L. S. Ornstein and F. Zernike, *Proc. Acad. Sci. Amsterdam* **17**, 793 (1914).

[23]   T. Andrews, *Phil. Trans.* **159**, 579 (1869). Reprinted in *Cooperative Phenomena Near Phase Transitions*, edited by H. E. Stanley (MIT Press, Cambridge, Mass. and London, 1973), p. 101.

[24]   P. Kapitza, *Nature* **141**, 74 (1938).

[25]   J. F. Allen and A. D. Misener, *Nature* **141**, 75 (1938).

[26]   F. London, *Nature* **141**, 643 (1938); *Phys. Rev.* **54**, 947 (1938).

[27]   F. London, *Superfluids, Vol. I: Macroscopic Theory of Superconductivity* (Wiley, New York, 1950) (republished by Dover, New York, 1961) and *Superfluids, Vol. II: Macroscopic Theory of Superconductivity* (Wiley, New York, 1954) (republished by Dover, New York, 1961).

[28]   See for example: L. D. Landau and E. M. Lifshitz, *Statistical Physics* (Pergamon, London, 1958), 2nd edition (London and New York, 1969).

[29]   M. R. Schafroth, *Helv. Phys. Acta* **24**, 645 (1951).

[30]   M. R. Schafroth and J. M. Blatt, *Nuovo Cim.* **4**, 786 (1956).

[31]   H. B. G. Casimir, in *Niels Bohr and the Development of Physics*, edited by W. Pauli (Pergamon, London, 1955), p. 127.

[32]   J. M. Blatt, S. T. Butler and M. R. Schafroth, *Phys. Rev.* **100**, 481 (1955).

[33]   M. R. Schafroth, *Phys. Rev.* **96**, 1149 (1954).

[34]   S. T. Butler and J. M. Blatt, *Phys. Rev.* **100**, 495 (1955).

[35]   S. T. Butler, J. M. Blatt and M. R. Schafroth, *Nuovo Cim.* **4**, 674 and 676 (1956).

[36]   *Bull. Am. Phys. Soc., Ser. II* **2**, 64 (1957).

[37]   W. M. Fairbank, M. J. Buckingham and C. F. Kellers, *Proceedings of the 5th International Conference on Low Temperature Physics* (Madison, Wisconsin, 1957), p. 50.

[38]   M. J. Buckingham and W. M. Fairbank, in *Progress in Low Temperature Physics*, edited by C. J. Gorter (North-Holland, Amsterdam, 1961), Vol. 3, Chapter III.

[39]   W. M. Fairbank and C. F. Kellers, in *Critical Phenomena*, edited by M. S. Green and J. V. Sengers (National Bureau of Standards, Washington, 1966), NBS Misc. Publ. No. 273, p. 71.

[40]   C. F. Kellers, Thesis (Duke University, 1960).

[41]   T. H. Berlin and M. Kac, *Phys. Rev.* **86**, 821 (1952).

[42]   J. D. Gunton and M. J. Buckingham, *Phys. Rev.* **166**, 152 (1968).

[43]   B. D. Josephson, *Phys. Lett.* **21**, 608 (1966).

[44]   C. Domb, in *Perspectives in Statistical Physics*, edited by H. J. Raveche (North-Holland, Amsterdam, 1981), p. 175.

[45]   E. A. Guggenheim, *J. Chem. Phys.* **13**, 253 (1945).

[46]   G. E. Uhlenbeck, in *Critical Phenomena*, edited by M. S. Green and J. V. Sengers (National Bureau of Standards, Washington, 1966), NBS Misc. Publ. No. 273, p. 3.

[47]   C. Edwards, J. A. Lipa and M. J. Buckingham, *Phys. Rev. Lett.* **20**, 496 (1968).

[48]   J. A. Lipa, C. Edwards and M. J. Buckingham, *Phys. Rev. Lett.* **25**, 1086 (1970).

[49]   J. A. Lipa, C. Edwards and M. J. Buckingham, *Phys. Rev.* **A15**, 778 (1977).

[50]   M. J. Buckingham, in *Critical Phenomena*, edited by M. S. Green and J. V. Sengers (National Bureau of Standards, Washington, 1966), NBS Misc. Publ. No. 273, p. 95.

[51]   M. I. Bagatskii, A. V. Voronel and V. G. Gusak, *Zh. Eksp. Teor. Fiz.* **43**, 728 (1962) (*Sov. Phys.–JETP* **16**, 517 (1963)).

[52]   A. V. Voronel, Yu. R. Chashkin, V. A. Popov and G. G. Simkin, *Zh. Eksp. Teor. Fiz.* **45**, 828 (1963) (*Sov. Phys.–JETP* **18**, 568 (1964)).

[53]   M. E. Fisher, *Phys. Rev.* **136**, A1599 (1964).

[54]   M. R. Moldover and W. A. Little, in *Critical Phenomena*, edited by M. S. Green and J. V. Sengers, (National Bureau of Standards, Washington, 1966), NBS Misc. Publ. No. 273, p. 79.

[55]   L. P. Kadanoff *et al.*, *Rev. Mod. Phys.* **39**, 395 (1967).

[56]   J. R. Clow and J. D. Reppy, *Phys. Rev. Lett.* **16**, 887 (1966).

[57]   J. A. Tyson and D. H. Douglass, *Phys. Rev. Lett.* **17**, 472 (1966).

[58]   C. J. Pearce, J. A. Lipa and M. J. Buckingham, *Phys. Rev. Lett.* **20**, 1471 (1968).

[59]   J. A. Tyson and D. H. Douglass, *Phys. Rev. Lett.* **21**, 1308 (1968).

[60]   M. Barmatz, P. C. Hohenberg and A. Kornblit, *Phys. Rev.* **B12**, 1947 (1975).

[61]   See G. Ahlers, *Rev. Mod. Phys.* **52**, 489 (1980) and references given there.

[62]   F. Wegner, *Phys. Rev.* **B5**, 4529 (1972).

[63]   J. C. Le Guillou and J. Zinn-Justin, *Phys. Rev.* **B21**, 3976 (1980).

High Resolution
Heat Capacity Measurements
Near the Lambda Point

J. A. Lipa

1. INTRODUCTION

Ever since the pioneering experiments of Buckingham, Fairbank and Kellers [1] on the heat capacity singularity at the lambda point of helium, this unique transition has occupied a special place in the field of cooperative phenomena. As early as 1934 Keesom and Keesom [2] had established the characteristic lambda-like shape of the heat capacity curve, clearly indicating an intimate relationship between this transition and those observed elsewhere, for example, at Curie and Néel points, in binary alloys, and at gas-liquid critical points. William Fairbank's group at Duke University showed that the divergence of the heat capacity at T_λ was very close to logarithmic and, further, that the transition was extremely sharp, with little sign of rounding until $t = |1 - T/T_\lambda| \sim 10^{-6}$. A logarithmic behavior of the heat capacity near a cooperative transition had also been found by Onsager [3] in his exact solution of the two-dimensional Ising problem. However, the details of the relationship between this model and the lambda transition were not clear. Nevertheless, this first high resolution experimental study of a cooperative transition was a great stimulus to further work in the field, some of which is described below. To date, no sharper cooperative transition has been discovered, leaving the lambda transition unchallenged as the premier testing ground of the latest theoretical ideas in this field.

Elsewhere in this volume Buckingham has described the early theoretical ideas which led to a new understanding of cooperative transitions. With the development of the Renormalization Group (RG) approach by Wilson [4] in the early 70's, these ideas were put on a much firmer footing. For the first time the predicted exponent values for a wide variety of systems were reasonably close to the experimental values. More recently, sophisticated perturbation techniques have been applied to improve the precision of these estimates to a level comparable to that available from experiment [5]. Also, the functional representation of thermodynamic parameters in the experimentally accessible region has been refined to include so-called "confluent singularity" terms. The extended representation of a variable $X(t)$ can be written as

$$X(t) = A_x t^{\lambda_x}(1 + a_x t^{\Delta_x} + \ldots) + \ldots \quad , \qquad (1)$$

where the exponents λ_x and Δ_x characterize the leading singularity and the confluent term, respectively. Recent estimates of these exponents have uncertainties of the order ± 0.005 and ± 0.01 respectively. The amplitudes, A_x and a_x, which are system dependent, are far more difficult to estimate, although certain universal ratios have been predicted.

The existence of the confluent singularity term causes difficulties when high precision tests of the RG predictions for the leading exponent are attempted. Since in general the coefficient a_x may be of the order of unity, the value of the confluent term can easily amount to several standard deviations of the data obtained in accurate experiments in the range $10^{-3} < t < 10^{-6}$. Thus statistical analyses of high quality data must take account of this term when the parameters of the leading term are extracted. This problem is much more severe for systems exhibiting significant rounding of the transition out to values of t as large as 10^{-4} or 10^{-5}, for example a fluid at its critical point or a ferromagnet at its Curie point. In these systems the range of data available for curve fitting is restricted, which significantly degrades the statistical accuracy of the value found for λ_x and possibly introduces systematic errors due to the neglect of higher order terms in equation (1). At present it appears that the lambda transition offers the best possibility of minimizing these difficulties because of the very low level of intrinsic rounding. In principle this transition can be resolved to $t \lesssim 3 \times 10^{-8}$ on earth, and to at least 10^{-11} in space. Measurements of any thermodynamic property of interest in this region would introduce a new era of investigation of cooperative phenomena, potentially extending the range of observations beyond the present-day level of $t \sim 10^{-6}$ by a larger factor than has been achieved since Andrews [6] discovered the critical point in 1863. Thus we have a unique opportunity to perform tests of the RG predictions with results that could have repercussions for the whole field of cooperative phenomena and perhaps other fields where the same theoretical techniques are applied.

In this paper we explore some of the possibilities for increasing the resolution of the lambda point, and describe the high resolution work that is in progress at Stanford. We begin by briefly reviewing the main RG predictions and related experimental results at the lambda point.

2. BACKGROUND

The most advanced tests of the RG predictions for cooperative transitions are currently experiments performed near the lambda transition of helium. This transition has a number of experimental advantages not found elsewhere; since the system is superfluid below the transition, thermal relaxation times are short, and even above the transition temperature the thermal conductivity diverges. Also, since it is a fluid system, the problems with strains and crystal defects encountered with solids are avoided. The transition temperature itself is conveniently accessible. It is in a region where high resolution thermometry is well developed and where the advantages of superconductivity can be readily applied. In contrast to the behavior near a critical point, the divergence of the compressibility is very weak, minimizing gravity rounding from this effect; and no special care is needed in setting the sample density. These advantages have facilitated the collection of a wide body of accurate data extending to $t \sim 10^{-6}$ on which tests of the RG predictions have been made. To date, the data sets most useful for accurate exponent determination were obtained by Ahlers and coworkers; these include isobaric thermal expansion coefficient data [7], heat capacity measurements at the vapor pressure [8] and superfluid density measurements [9]. Also of major importance are the heat capacity measurements of Gasparini and Moldover [10] as a function of ^{3}He concentration, x. Many other experiments have been reported, but generally these have either been analyzed in a restricted way, or are of lower accuracy. These experiments allow the determination of the exponents α characterizing the leading singularity of the heat capacity, and ζ, that for the superfluid density:

$$C_p = \frac{A}{\alpha} t^{-\alpha} (1 + D t^\Delta + \ldots) + B \quad , \tag{2}$$

$$\frac{\rho_s}{\rho} = A_\rho t^\zeta (1 + D_\rho t^\Delta + \ldots) \quad , \tag{3}$$

where ρ and ρ_s are the total and superfluid densities, respectively. In addition, information on the leading coefficients and other parameters in these expressions is obtained.

TABLE 1. COMPARISON OF OBSERVED AND PREDICTED VALUES OF LAMBDA POINT PARAMETERS.

Parameter	$\alpha\,(=\alpha')$	A/A'	x	D/D'	$B - B'$	ζ	D'/D_ρ
RG predictions	−0.008 ±.003	1.03	0.521 ±.006	1.17	0	0.669 ±.002	0.15 .18
ρ_s data	—	—	0.5 ± 0.1	—	—	0.6749 ±.0007	−0.17
Isobaric expansion data	−0.026 ±.004	1.112 ±.022	(=0.5)	1.29 ±.25	(=0)	—	±0.04
^{3}He-^{4}He mixture data	−0.024	1.088	(=0.5)	—	(=0)	—	—

The RG calculations give predictions [5] for the exponents α, ζ and Δ, the coefficient ratios A/A', D/D' and D'/D_ρ, and the difference $B - B'$, where the primed quantities refer to equation (2) below the transition. Also, the universality hypothesis [11] predicts that the above quantities will be independent of "irrelevant" variables, in this case the pressure and the ^{3}He concentration, and scaling [12] predicts that $\alpha = \alpha'$. In table 1 we list the theoretical predictions for the above quantities along with the experimental results. Quantities in parentheses represent constraints.

The most precise determination of a leading exponent reported to date is from the superfluid density data. This is because equation (3) does not contain a parameter equivalent to B in equation (2), easing the curve fitting task, and also because ρ_s has been determined from very precise second sound velocity measurements. The agreement with theory shown in table 1 is very good, and recently Ahlers [13] has reported that even better agreement can be obtained if equation (3) is modified by replacing A with $A_0 + A_1(\rho)t$. In this case, with $\Delta = 0.5$, he obtains $\zeta = 0.6716 \pm .0004$, which is within the range of uncertainty of the theoretical estimate. At face value this appears to be very strong support for the RG predictions, but on closer examination there are some remaining difficulties. These stem from the use of second sound to determine ρ_s. The basic relationship for the conversion is

$$\frac{\rho_s}{\rho_n} = \frac{U_2^2 C_p}{S^2 T} + \cdots \quad , \qquad (4)$$

where U_2 is the second sound velocity, $\rho_n = \rho - \rho_s$, and S is the entropy. This equation is based on the two-fluid model of He II, and is assumed reliable to the order of 0.1%. Independent verification [14] by alternative measurements of ρ_s shows agreement to a level of about 1% over a restricted temperature range, but no measurements yet exist which fully support this assumption, nor have dispersion measurements been made near T_λ. More important, even if equation (4) can be shown to be correct to the required accuracy, the determination of ζ from the velocity of second sound inevitably requires a knowledge of the heat capacity singularity *via* the quantity C_p. The velocity data of Greywall and Ahlers were converted to ρ_s using a logarithmic form for C_p, with $\alpha' = 0$. We have reanalyzed these data using the best fit function to our recently obtained data [15] with $\alpha' = -0.013$, and find a shift of ζ of 0.005, more than ten times the statistical uncertainty quoted by Ahlers [13]. Another serious problem occurs when we try to work backwards by using the Josephson scaling relation [16] to obtain α' from ζ; the values obtained are inconsistent with the input values used in equation (4). For example, if the input α' is -0.013, then the output value is -0.030, in clear disagreement with experiment. These difficulties with the analysis of the second sound data imply that the determination of ζ must be subordinated to the prior determination of α' to a much greater extent than has been acknowledged previously.

Precise determinations of the heat capacity exponent have been made by three different methods: direct heat capacity measurements at the vapor pressure, isobaric thermal expansion coefficient measurements as a function of pressure along the lambda line, and heat capacity meaurements as a function of ^{3}He concentration. The results obtained are summarized in table 2. A quick look at the table reveals an interesting effect. All measurements at the vapor pressure give exponent values close to -0.017, while those at higher pressures or in the mixtures give values close to -0.025. Some aspects of this effect have been discussed by Gasparini and Gaeta [18]. They demonstrated that while the exponent differences are not large compared with the standard errors, the vapor pressure heat capacity data [8,10] are clearly incompatible with the optimum values of the universal parameters from the thermal expansion data [7]. When the constraints $\alpha = \alpha' = -0.026$, $A/A' = 1.11$ and $D/D' = 1.11$ were applied to the heat capacity data, marked systematic deviations were observed. In figure 1 we have plotted the deviations of Ahlers' data from the best fit function with these constraints, and also from the best fit unconstrained function. The difficulty clearly can be seen, indicating a possible breakdown of universality along the lambda line. The possibility of large uncontrolled systematic errors in the data seems to be discriminated against by the consistency of the many experimental results listed in table 2. If, for some as yet undetermined reason, there is a weak breakdown of universality at the lambda transition, further difficulty is encountered with the superfluid density results. Since the exponent ζ was obtained from data taken over a range of pressures, consistency requires that the conversion from U_2 to ρ_s be performed using the expansion coefficient result for α', exacerbating the problems described earlier in this section.

TABLE 2. EXPERIMENTAL DETERMINATIONS OF THE HEAT CAPACITY EXPONENT.

Investigators	Reference	Parameter	α	Range of t
Buckingham, Fairbank, and Kellers	[1]	C_{sat}	$0.0 \pm .05$	$10^{-6} \rightarrow 10^{-3}$
Ahlers	[8]	C_{sat}	$0.000 \leq \alpha \leq -0.025$	$10^{-6} \rightarrow 3 \times 10^{-3}$
Ahlers	[7]	C_{sat}	$-0.0163 \pm .0017$	$10^{-6} \rightarrow 3 \times 10^{-3}$
Mueller, Ahlers, and Pobell	[7]	Expansion coefficient	$-0.026 \pm .004$	$3 \times 10^{-5} \rightarrow 3 \times 10^{-3}$
Gasparini and Moldover	[10]	C_{sat}	$-0.0198 \pm .0037$	$10^{-6} \rightarrow 3 \times 10^{-3}$
Gasparini and Moldover	[10]	Mixtures	-0.025	$3 \times 10^{-6} \rightarrow 3 \times 10^{-3}$
Takada and Watanabe	[17]	C_{sat}	-0.017	$10^{-5} \rightarrow 2 \times 10^{-3}$
Takada and Watanabe	[17]	Mixtures	-0.024	$10^{-5} \rightarrow 3 \times 10^{-3}$

It is unfortunate that the various high resolution measurements show small but persistent departures from the predictions, weakening the support for universality and the RG results in general. It would appear to be of very high priority to attempt a resolution of these discrepancies between theory and experiment, and it is clear that only the lambda transition has significant potential for improving the experimental results, because of the unique properties of this system. By performing new measurements at higher resolution, perhaps approaching $t \sim 10^{-11}$ in space, it will be possible simultaneously to probe much deeper into the asymptotic region and to extend the dynamic range of the measurements. Improved thermometry will also allow more accurate measurements over the whole range, further increasing our knowledge of the asymptotic form of the singularity.

3.   TRANSITION BROADENING

In this section we will first discuss the transition broadening effects encountered on earth and then examine the situation expected in space. As is well

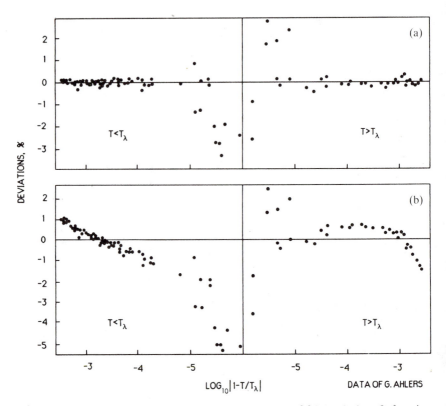

FIGURE 1. Deviations of the heat capacity data of Ahlers [8] from the best fit functions (a) with $\alpha = -0.016$, and (b) with the constraints $\alpha = -0.026$, $A/A' = 1.11$ and $D/D' = 1.11$ (after Gasparini and Gaeta [18].)

known, the gravitational rounding is due almost entirely to the finite slope of the lambda line, since the compressibility is only very weakly divergent. All impurities except ^{3}He are frozen out, and the concentration of this isotope can easily be reduced below one part in 10^8, reducing renormalization of the heat capacity to a negligible level. Thus we expect this transition to be extremely sharp and limited primarily by the competition between the slope of the lambda line as a function of pressure and finite size effects. In a free-flying spacecraft additional effects to be considered are thermal fluctuations, self-gravitation and gravity gradient.

3.1 Lambda Transition Broadening on Earth

Considering first the effect of gravity alone, the heat capacity of a sample of helium of height h reaches a finite maximum at a reduced temperature displacement t_m below the lambda point given by

$$t_m = \frac{\rho g h}{T_\lambda} \left(\frac{dP}{dT}\right)_\lambda^{-1} \quad , \tag{5}$$

where g is the acceleration due to gravity, and $(dP/dT)_\lambda$ is the slope of the lambda line. Between the lambda "point" and the temperature given by equation (5) the helium is in a two-phase region with co-existing helium I and helium II, and the heat capacity curve is highly distorted. Outside this region we can expand the deviation from the ideal heat capacity, C_0, in the form

$$\Delta C_g = C_g - C_0 = \frac{A t_m}{2t} + \frac{A}{6}\left(\frac{t_m}{t}\right)^2 + \cdots \quad , \tag{6}$$

where C_g is the heat capacity in a gravity field and A is the coefficient of the assumed logarithmic singularity in C_0. A similar formula can be developed for small but nonzero α. It can be seen from equations (5) and (6) that the perturbation due to gravity is in first order directly proportional to the height of the sample, dictating that this quantity be minimized in any high resolution experiment.

On the other hand, the divergence of the superfluid healing length below the transition dictates a large sample for an accurate representation of C_0. An exact calculation of this effect is not yet available, but an approximate idea of the magnitude can be obtained as follows. Near the transition, the healing length, r, diverges as

$$r = r_0 t^{-2/3} \quad , \tag{7}$$

where $r_0 \approx 2 \times 10^{-8}$ cm. For a horizontal disc-shaped sample, a fraction of the volume $2r/h$ is within one healing length of the boundaries. If the relative distortion of the specific heat in this surface layer is y (< 1), then the relative error in the measurement of the total specific heat is

$$\frac{\Delta C}{C_0} \approx \frac{2ry}{h} \quad . \tag{8}$$

Thus, for example, if we assume $y = 20\%$ and require that the distortion of the measurement be less than 1%, then we must have

$$h > 4 \times 10^{-7} t^{-2/3} \text{ cm} \quad . \tag{9}$$

A straightforward minimization of these competing effects leads to an optimum height of 0.3 mm and a maximum resolution $t \sim 5 \times 10^{-8}$ assuming a 1% distortion. However, we can clearly do better than this because equation (6) can be used to correct for the gravity effect. At some point it will probably also be possible to correct for the finite size effect in the region where it is small, but here this will be ignored. A slightly more complex calculation indicates that the limit set by the *uncertainty* in the gravity effect is near $h \approx 0.48$ mm and $t \sim 2.8 \times 10^{-9}$. While it is an exciting prospect to contemplate measurements at this resolution, it must be pointed out that a number of difficulties exist. First, the lambda point is not clearly marked in heat capacity measurements; instead, the most prominent feature is the maximum, C_m, located an interval t_m below the point of interest. As the resolution is increased beyond t_m the extrapolation to T_λ becomes less certain. Second, the shape of the heat capacity curve in the neighborhood of C_m will be distorted somewhat by the finite size effects at the He I/He II interface. Third, in the region of long correlation length there may be an appreciable gravitational correction due to the pressure gradient which is explicit in the free energy, in addition to the implicit contribution we have calculated above. In spite of these potential problems, it appears that there are reasonable grounds for expecting that a resolution of $t \sim 10^{-8}$ may be attainable.

3.2   Broadening in Space

In a spacecraft with no external forces and far from all sources of gravitational fields, the lambda transition would be broadened by at least three effects. First, the finite size effect described above would still occur, dictating a large sample size for small rounding. With the same assumptions as above we can easily show that for a spherical sample a nominal 1% correction occurs when

$$t = 4.6 \times 10^{-10} \, r_s^{-3/2} \quad , \tag{10}$$

where r_s is the radius of the sample. The transition will be completely smeared out when r_s equals the correlation length, giving a second relation

$$t = 2.8 \times 10^{-12} \, r_s^{-3/2} \quad . \tag{11}$$

Second, Goldstein [19] has pointed out that temperature fluctuations ultimately will wash out the transition. This effect occurs at

$$t = 8.7 \times 10^{-13} \, r_s^{-3/2} \quad , \tag{12}$$

somewhat less than the previous limits. The third factor is due to the self-gravitation of the helium, generally an exceedingly small effect, which nevertheless sets an ultimate limit on the sharpness of the transition. For a self-gravitating sphere of helium in hydrostatic equilibrium it is easy to show that the pressure difference ΔP between the surface and the center is given by

$$\Delta P = \frac{2\pi}{3} \rho^2 G r_s^2 \quad , \tag{13}$$

where G is the gravitational constant. Since the transition broadening is

$$\Delta t = \left(\frac{dP}{dT}\right)_\lambda^{-1} \Delta P \quad,$$ (14)

we obtain the result

$$\Delta t = 2.6 \times 10^{-16} \, r_s^2 \quad.$$ (15)

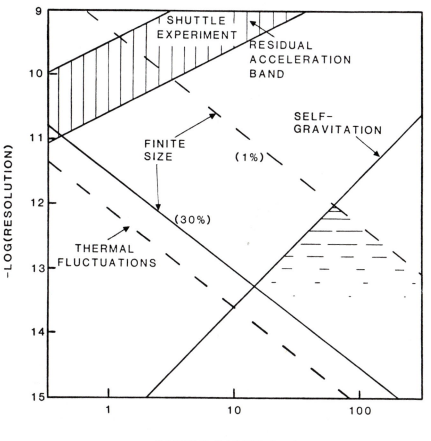

SAMPLE RADIUS (cm)

FIGURE 2. Comparison of the factors limiting the ultimate resolution of the lambda transition in space.

The above expressions are plotted in figure 2, which shows that the maximum possible resolution is $t \sim 10^{-12}$ with the application of minor corrections to the data, and $t \sim 10^{-13}$ if more extensive correction terms are used. These levels represent the best that can be achieved in any presently envisaged system, and extend four to five orders of magnitude beyond that which can be achieved on earth.

TABLE 3. RESOLUTION AND OPTIMUM SAMPLE SIZE *versus* RESIDUAL ACCELERATION.

g/g_0	t_0	a_0 (cm)
1	5.0×10^{-8}	0.088
10^{-1}	1.3×10^{-8}	0.22
10^{-2}	3.2×10^{-9}	0.56
10^{-3}	7.9×10^{-10}	1.40
10^{-4}	2.0×10^{-10}	3.5
10^{-5}	5.0×10^{-11}	8.8
10^{-6}	1.3×10^{-11}	22.2

In more practical situations, effects other than self-gravitation often set an upper limit on the sample size. For example, for a free-flying spacecraft in low earth orbit which is maintained drag-free to a level of 10^{-8} g or better, we must also consider tidal forces due to the gravity gradient. On board the shuttle the residual acceleration competes with the finite size effect to limit the resolution. This situation is similar to that described in the previous section except that the acceleration is now a small, variable quantity. The results of the optimization calculations for this case are summarized as a function of reduced acceleration in table 3. In this table g_0 is the acceleration on earth, a_0 is the optimum sample size, and t_0 is the corresponding temperature resolution. In the calculations we assumed that only the upper limit on the acceleration was known and no corrections were made to the data. The values quoted for the resolution correspond to the point at which a 1% distortion occurs. If the acceleration during a flight experiment were recorded and were sufficiently stable, corrections for this effect could be made, allowing a somewhat higher resolution. It can be seen from the table that in a shuttle environment of 10^{-4} g a resolution of 2×10^{-10} could be obtained, about two orders of magnitude higher than that possible on earth.

4. HIGH RESOLUTION EXPERIMENTS

For the past few years we have been developing a new high resolution thermometer to probe the temperature zone extremely close to the lambda transition which is accessible in a micro-gravity environment. This program is continuing and we hope to perform an experiment in the latter part of this decade on the shuttle. In this section we outline the development work to date, which has resulted in a thermometer with a resolution approaching the thermal fluctuation limit and an advanced thermal control system for use in the experiments. We have used these devices to perform heat capacity measurements up to the gravity limit on earth, and plan to measure other parameters in the near future.

4.1 Thermometer Development

At the start of our program the highest resolution thermometers suitable for studying the lambda transition were carbon and germanium resistors. With these devices it is possible to resolve to $t \sim 10^{-7}$ at the lambda point when the measuring power is about 10^{-8} W. Higher resolution can be obtained with increased dissipation, but with present readout technology a power level of about 10^{-2} W would be needed to resolve down to $t \sim 10^{-10}$. Clearly this is a high level for any low temperature experiment, but especially for heat capacity measurements in which we must measure power levels of the order of 10^{-12} W! This simple estimate underlines the importance of developing a completely different concept in high resolution thermometry if we wish to resolve the lambda transition to the limit imposed by gravity and beyond.

The new thermometer we have developed uses a superconducting readout system and paramagnetic salt as the sensitive element. This device has an exceptionally low intrinsic dissipation level ($< 10^{-17}$ W) and a very high sensitivity due to the use of a SQUID magnetometer as a detector. Similar devices have been built [20] for use below 1 K, but their noise characteristics have not been reported. We built a thermometer optimized for use near the lambda point of helium and provided it with considerably improved shielding to minimize the effect of external magnetic fields. The construction of this device has been described elsewhere [21]. A schematic view of the thermometer is shown in figure 3.

We initially made measurements of the basic sensitivity of the device for comparison with calculations and to allow projections for its ultimate performance. We obtained sensitivities of up to 400 Φ_0/K/gauss at 4 K, where Φ_0 is the quantum of magnetic flux, within a factor of two of the expected values. Since the magnetometer has a noise level of about 10^{-4} Φ_0 in a 1 Hz bandwidth, we estimated a resolution of $t \sim 1.2 \times 10^{-10}$ with a trapped

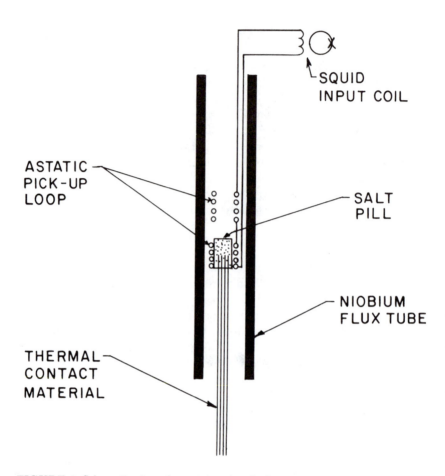

FIGURE 3. Schematic view of paramagnetic salt thermometer.

field of 500 gauss, the maximum field used so far. In order to measure a resolution of this magnitude, it is necessary to carefully control the thermal environment of the thermometer during the observations. This has been achieved with feedback systems using carbon or germanium thermometer detectors. By careful control of the salt thermometer environment we have been able to observe the noise level during brief intervals of high thermal stability. The lowest noise so far observed is equivalent to a resolution of 3.3×10^{-10} at 2 K, about a factor of three worse than the original estimate. More recent calculations have indicated that this noise contribution may be due to thermal fluctuations causing random heat transfer between the calorimeter and the temperature sensor. It can be shown that the mean-squared noise due to thermal fluctuations in our system is given by

$$\langle \Delta T^2 \rangle = \frac{2kT^2\tau_0}{c_e\tau} \quad , \tag{16}$$

where k is Boltzmann's constant, c_e is the heat capacity of the sensing element, τ is the integration time, and τ_0 is the thermometer relaxation time for heat transfer to the helium sample. At the lambda point we obtain

$$\delta t_{rms} = \frac{\langle \Delta T^2 \rangle^{1/2}}{T_\lambda} \approx 4 \times 10^{-10} \quad , \tag{17}$$

which is close to the noise we observe.

One difficulty with the operation of our thermometer at higher resolution is due to the presence of $1/f$ noise from the SQUID which limits the usefulness of integration in the low frequency regime. However, if we increase the trapped field until thermal fluctuation noise dominates the SQUID noise, integration can again be used since the thermal noise is in principle white. We have integrated down to a 0.01 Hz bandwidth and obtained a very marked improvement in sensitivity, but the thermal control system then in use did not provide the stability required for accurate measurements. Nevertheless it appears relatively straightforward to obtain sensitivities of the order of 10^{-11} under narrow bandwidth conditions. It can be seen from equation (16) that thermal fluctuation noise can also be reduced by minimizing the ratio τ_0/c_e which contains the only physical design variables. It appears possible to obtain a further reduction of an order of magnitude in this quantity before construction of the device becomes prohibitively difficult. This would reduce δt_{rms} by a factor of about three.

With the availability of such high thermal resolution it is important to establish the null stability of the thermometer. This has been done by observing the location of T_{max}, the temperature of the heat capacity maximum, as a function of time during multiple scans through the lambda transition. With the apparatus free from external disturbance, no detectable drift in T_{max} was seen over periods up to 2 hours to the limit of our resolution of this quantity, which was about 2×10^{-8} K using an analog data collection system.

4.2 Heat Capacity Measurements

In 1981 we completed some preliminary measurements of the heat capacity singularity at the lambda transition with a sample height of 0.3 mm [22]. In this experiment we achieved a measurement accuracy of about 5% down to $t \sim 5 \times 10^{-9}$. This can be compared with previous experiments which achieved an accuracy of about 3% with a resolution of $t \sim 10^{-6}$. Although we improved the resolution of the lambda transition by about two orders of magnitude, our results did not span a wide enough range of temperature to allow accurate estimates of the parameters in equation (2).

More recently we have improved the thermal control system by adding a second stage of isolation and reducing the heat leak to the calorimeter. These changes have allowed us to achieve much better reproducibility of the measurements and to operate at lower heating rates. Also the brass calorimeter was replaced with a copper container, significantly reducing temperature gradients in the system. New data obtained with this apparatus are shown in figure 4. These data indicate that our earlier measure-

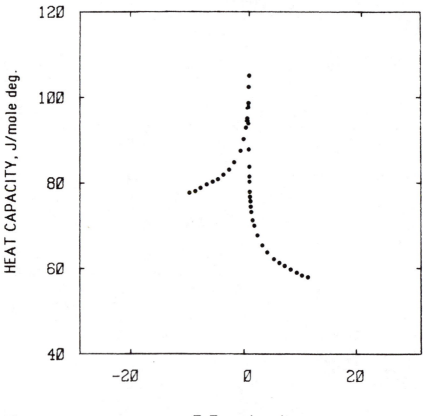

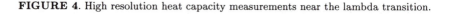

FIGURE 4. High resolution heat capacity measurements near the lambda transition.

ments very close to the transition were somewhat high. We believe this to be due to the changes in thermal gradients in the calorimeter as the helium switched between the normal and superfluid states. We have reexamined the original data in the light of the new measurements and find some residual heating rate dependence of the results for $t < 10^{-7}$ which was previously overlooked. In figure 5 we plot the new data on a semilogarithmic scale

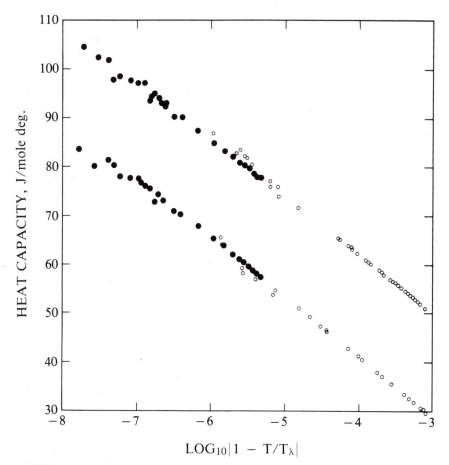

FIGURE 5. High resolution heat capacity measurements plotted on a semilogarithmic scale: ● : Lipa and Chui, o : Ahlers [8].

along with that of Ahlers [8]. It can be seen that an exponent value close to zero is indicated for the whole range $3 \times 10^{-8} < t < 10^{-3}$. When our data outside the two-phase region is corrected for the effect of gravity, the best fit exponent is $\alpha = -0.016 \pm .010$ for $5 \times 10^{-8} < t < 5 \times 10^{-6}$. If we also include the data of Ahlers in the curve fitting, we obtain $\alpha = -0.015 \pm .002$

for $5 \times 10^{-8} < t < 10^{-4}$. It must be remembered, however, that possible systematic differences between the two experiments could give errors larger than the statistical uncertainty in this last result.

The main factor limiting the accuracy of the above measurements was stray heat leak control. A new thermal control system designed to minimize this problem is shown in figure 6. It contains four stages of thermal isolation and includes a thermal shield completely surrounding the calorimeter which can be controlled to within a few microkelvins. Inside this shield is a final isolation stage which is controlled using a paramagnetic salt thermometer as a sensor. This system is capable of improving the thermal stability of the calorimeter by at least two orders of magnitude. We have demonstrated the excellent level of decoupling between the stages and the stability of the inner isolation stage. The system is now being modified to allow heat capacity measurements [23] and to provide a lambda point reference temperature for further measurements of thermometer null stability.

5. PROSPECTS FOR EXPERIMENTS IN SPACE

In section 3 we estimated the potential resolution available at the lambda point in various low gravity situations. On the shuttle a resolution approaching 2×10^{-10} can be achieved, the limit being set primarily by acceleration noise. In this section we describe a method of obtaining high resolution heat capacity information in such an environment based on extrapolating our ground-based measurements to the flight situation. First, we consider the uncertainty in the heat leak from the surroundings during the measurements. To date we have been able to measure the heat capacity of an 8 mg sample of helium with an accuracy of $\pm 2\%$ and a resolution of about 5×10^{-9}, using heating rates of the order 3×10^{-10} K/sec. This implies an upper limit on the stray heat leak of no more than 4×10^{-12} W. For a 3.5 cm diameter sample of helium, this uncertainty implies a lower limit of 2×10^{-12} K/sec in the heating rate to give a similar accuracy in the heat capacity data. This presents no problem for measurements below T_λ, but immediately above T_λ some distortion might be encountered due to the rapid variation of the sample relaxation time with temperature. This difficulty can be avoided by taking advantage of the improved stability of the new thermal control system described above. With this system a reduction of two orders of magnitude in the stray heat leak is anticipated.

The limitations set by sensor noise are still under evaluation. However, our observations indicate that we are close to achieving the required resolution. All noise measurements to date indicate that $1/f$ noise is not important down to a bandwidth of 0.01 Hz. This means that we should be able to maintain an accuracy of about 2% to a resolution of 5×10^{-10}

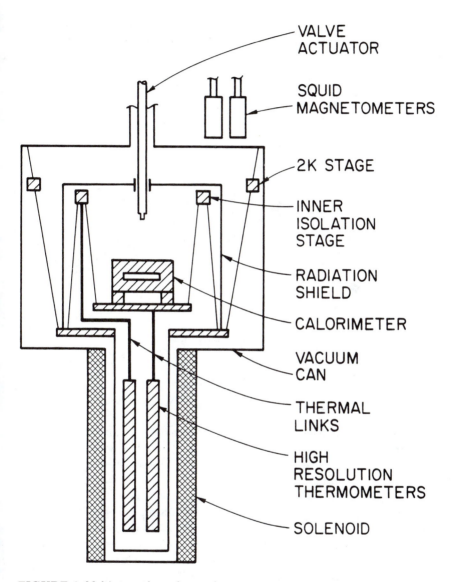

FIGURE 6. Multi-stage thermal control system.

by increasing the integration time to 100 sec and reducing the heating rate to 3×10^{-12} K/sec, an easily obtained value. Further improvement to resolve to the shuttle limit can be achieved by reducing the heating rate by a further factor of three, which is well within the capability of the new thermal control system, and by averaging a number of scans through the transition. This approach has the advantage of allowing the rejection of a certain amount of data that might be corrupted by periods of high acceleration, without losing all information in a given range of temperature.

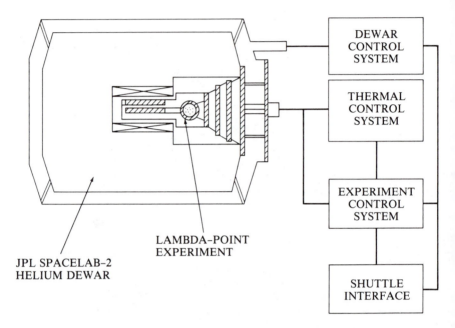

FIGURE 7. Schematic view of main components of lambda point flight experiment.

The apparatus needed to perform the experiment is similar to that shown in figure 6. For space, the major changes needed are to cage the system for launch, to completely automate the system, and to increase the sample size to the optimum for flight. Figure 7 shows a schematic view of the proposed system. The heat capacity apparatus will be housed in a flight-qualified helium dewar, such as the Jet Propulsion Laboratory superfluid helium facility. Possibly other experiments will be housed in the same dewar. The dewar control system manages the helium venting and dewar temperature, and records the information from a local three-axis accelerometer. The experiment is controlled by two systems: a thermal control system which takes care of the thermal isolation system set points as a function of experiment mode, and a main experiment control system which selects the mode, logs the data, and communicates with the shuttle aft flight deck. In operation, three primary modes are envisaged. The first mode involves coarse setup of the thermal control system and calibration. Next there will be a wide range heat capacity collection mode used to bridge the gap between ground-based data and the highest resolution measurements. The third mode will be used for high resolution measurements, and will be activated only during periods of low acceleration.

The experiment we have briefly described would lead to a detailed representation of the heat capacity curve through the lambda transition to a resolution of 2–3 $\times 10^{-10}$. This information will be invaluable for performing new tests of the RG theory of cooperative transitions, and for the first time will probe the nature of cooperative phenomena at resolutions below 10^{-8}. The completion of an experiment of this type will lead to a number of other possibilities. Of particular interest will be studies of the superfluid density, the normal fluid thermal conductivity, and second sound damping, which involve other predictions of the RG theory. Since the healing length of the superfluid component reaches truly macroscopic dimensions very close to the transition, we also have a unique opportunity to probe the spatial variation of ρ_s near walls, vortex cores and the He I/He II interface using well-defined trajectories for electrons or ions to give new insight into the nature of these boundary phenomena. Experiments designed to reach the ultimate resolution of 10^{-12} or 10^{-13} must await an improved platform, such as the space station, and the next generation of instrumentation.

Acknowledgments

I wish to thank the PACE committee of NASA for its support in funding the high resolution thermometry program, and the Ames Research Center Fund for Independent Research for a grant to complete the heat capacity measurements. I am extremely grateful to T. C. P. Chui for assistance with the data collection.

References

[1]   M. J. Buckingham and W. M. Fairbank, in *Progress in Low Temp. Phys.*, edited by C. J. Gorter (North-Holland, 1961), Vol. 3, p. 80; and C. F. Kellers, Thesis, Duke University (1960).

[2]   W. H. Keesom and A. P. Keesom, *Physica* **2**, 557 (1935).

[3]   L. Onsager, *Phys. Rev.* **65**, 117 (1944).

[4]   K. G. Wilson, *Phys. Rev. B* **4**, 3174 (1971).

[5]   J. C. LeGuillou and J. Zinn-Justin, *Phys. Rev. B* **21**, 3976 (1980); and E. Brezin, J. C. LeGuillou and J. Zinn-Justin, *Phys. Lett.* **47A**, 285 (1974).

[6]   By 1869, six years after his discovery of the critical point of CO_2, T. Andrews had determined the shape of an isotherm within 0.2° of the transition, or with $|1 - T/T_c| < 7 \times 10^{-4}$. See *Phil. Trans.* **159**, 575 (1869).

[7]   K. J. Mueller, G. Ahlers and F. Pobell, *Phys. Rev. B* **14**, 2096 (1976).

[8]   G. Ahlers, *Phys. Rev. A* **3**, 696 (1971).

[9]   D. S. Greywall and G. Ahlers, *Phys. Rev. A* **7**, 2145 (1973).

[10]   F. M. Gasparini and M. R. Moldover, *Phys. Rev. B* **12**, 93 (1975).

[11]   L. P. Kadanoff, in *Phase Transitions and Critical Phenomena*, edited by C. Domb and M. S. Green (Academic, New York, 1976), Vol. 5A, p. 1.

[12]   B. Widom, *J. Chem. Phys.* **43**, 3892 (1967).

[13]   G. Ahlers, *Physica (Utrecht)* **107B**, 347 (1981).

[14]   J. A. Tyson, *Phys. Rev.* **166**, 166 (1968); J. R. Clow and J. D. Reppy, *Phys. Rev. Lett.* **16**, 887 (1966).

[15]   J. A. Lipa and T. C. P. Chui, *Phys. Rev. Lett.* **51**, 2291 (1983).

[16]   B. D. Josephson, *Phys. Lett.* **21**, 608 (1966).

[17]   T. Takada and T. Watanabe, *J. Low Temp. Phys.* **41**, 221 (1980).

[18]   F. M. Gasparini and A. A. Gaeta, *Phys. Rev. B* **17**, 1466 (1978).

[19]   L. Goldstein, *Phys. Rev.* **135**, A1471 (1964).

[20]   R. P. Giffard, R. A. Webb and J. C. Wheatley, *J. Low Temp. Phys.* **6**, 533 (1972).

[21]   J. A. Lipa, B. C. Leslie and T. C. Wallstrom, *Physica (Utrecht)* **107B**, 331 (1981).

[22]   J. A. Lipa, *Physica (Utrecht)* **107B**, 343 (1981).

[23]   Recent heat capacity measurements with this system are described in [15].

Rotation of Superfluid Helium

George B. Hess

1. INTRODUCTION

One of the ideas which William Fairbank brought to Stanford in 1959 was to look for nonrotation of superfluid helium in a rotating container. This was based on the notion, advocated particularly by Fritz London [1], that frictionless superfluid flow could not be due to absence of interactions, but must be a macroscopic quantum state established by the interactions. Then, in suitable circumstances, the thermodynamic equilibrium state can be a state of flow relative to the confining walls. In particular, in a sufficiently slowly rotating container, the equilibrium state of the superfluid is not to rotate. I began working on this shortly after, and in 1966 we succeeded in observing the effect. Two brief reports were published [2]. In this paper I describe the experiment in more detail and review some subsequent work.

Conceptually this experiment goes back to Landau, who used it to introduce the two-fluid model: "To investigate this problem [of helium II at finite temperature] we shall consider helium in an axial-symmetric vessel rotating around its axis at a constant angular velocity ... During the rotation of the vessel containing helium the superfluid part remains stationary as has already been pointed out. It can be said that the superfluid liquid is not capable of rotation." [3] At the conference LT1, H. London remarked on the desirability of testing the uniqueness of this irrotational state by cooling a rotating container of liquid helium through the lambda transition (T_λ), and looking for transfer to the container of the angular momentum of that fraction of the liquid which becomes superfluid [4]. This would be analogous to the Meissner experiment in establishing the reversibility of the superconducting state with respect to an applied magnetic field. However London recognized a difficulty: Bijl, de Boer and Michels [5] had already pointed out that the product of the experimental maximum velocity of frictionless flow v_c and the width of the flow channel d is close to $\hbar/m$, where m is the mass of the helium atom. Applied to the rotating case, this suggests a condition for irrotational response in a vessel of radius R rotating at angular velocity ω:

$$\omega \leq \omega_0 = \hbar/mR^2 \quad . \tag{1}$$

The proposal of quantized vortices of circulation $2\pi\hbar/m$ by Onsager [6] and by Feynman [7] put this condition on an explicit foundation. For angular velocities less than

$$\omega_{c_1} = (\hbar/mR^2) \ln(R/a) \approx 15\,\omega_0 \quad , \tag{2}$$

the superfluid state with one vortex on the cylinder axis has greater free energy than the superfluid rest state. (Here $a \approx 10^{-8}$ cm is the radius of the vortex core.) For larger angular velocities the equilibrium number of vortices increases, asymptotically approaching ω/ω_0, and the equilibrium angular momentum soon becomes indistinguishable from its classical value.

Beginning with Osborne [8], and Andronikashvili and Kaverkin [9], a number of experimenters looked for equilibrium nonrotation of the superfluid by measuring the curvature of the free surface [10]. In all cases the classical curvature was found, indicating rotation of the whole liquid: not a surprising result because the achievable sensitivity always required $\omega > 10^3$ ω_0. Direct measurements of angular momentum transfer were made by Hall [11] and by Lane, Reppy and coworkers [12,13]. In the final form of their experiment, Reppy and Lane [13] used a 2.5 cm diameter container of liquid helium hanging in vacuum from a Beams-type magnetic suspension by a 1 m long capillary. This container was given an impulsive angular acceleration, and its subsequent free rotation history provided a record of the entrainment of the helium. It was found that in most runs, within a random error of about 5%, the whole liquid came into rotation, although occasionally the superfluid remained metastably at rest. The lowest rotational speed reported in this experiment was 500 ω_0. The resolution in final angular momentum was of order 20 L_0, where $L_0 = (\rho_s/\rho)M_{He}(\hbar/m)$ is the angular momentum induced in the superfluid by a single quantized vortex along the axis, ρ_s/ρ is the superfluid fraction, and M_{He} is the mass of helium in the vessel. It is clear that a substantial advance in technique was still required in order to observe equilibrium nonrotation. Other work prior to 1966 is reviewed in [14].

The experiment to be described below used the following general procedure. A container of liquid helium is set into rotation on a nearly frictionless magnetic bearing and then cooled to well below T_λ. As part of the helium, initially in classical (rigid) rotation, becomes superfluid, it should give up its angular momentum along with its enthalpy to the container. However this cannot be observed because cooling must be done with exchange gas and the gas exerts large background torques. Instead, measurements of angular momentum transfer are made on the return transition. The exchange gas is pumped out, and then the rotor is heated to its original temperature with light. In the final state the helium must quickly return to rigid rotation with the rotor. Thus, if the superfluid state had less than the classical angular momentum, the rotor would slow down (by $-\Delta\omega$). To be explicit, angular momentum conservation requires

$$I_R\Delta\omega = L_S - \frac{1}{2}\frac{\rho_s}{\rho}M_{He}R^2\omega \qquad (3)$$

where I_R is the moment of inertia of the rotor, and L_S is the angular momentum and $(\rho_s/\rho)M_{He}$ the mass of the superfluid in the cold state. The last term will be referred to as the classical angular momentum associated with the superfluid mass.

This is the basic experiment. Several variations, in which the angular velocity of the rotor is changed while in the cold state, will also be discussed.

2.   EXPERIMENTAL DETAILS

Several design factors contributed to the necessary improvement in sensitivity. (a) Use of the Beams ferromagnetic suspension (following Walmsley, Lane and Reppy [12,13]). This not only provides sensitive integration of torques, but is very robust compared to a torsion fiber and obviates the need to rotate the apparatus. (b) Operation of the whole suspension at low temperature, so that the rotor can be compact and isothermal. (c) Use of a permanently sealed sample. (Bill Fairbank knew from his lambda point specific heat experiment [15] that there is no problem in silver soldering a cap on a container of liquid helium.) Although the vessel must then be heavy compared to the helium inside, it is the elimination of *in situ* filling that makes the experiment practical. (d) Use of a small radius vessel, made easier by (b) and (c) above, puts the time scale of the experiment in a practical range. Both ω_0^{-1} and the viscous relaxation time vary as R^2.

The actual parameters of our rotating vessel are summarized in figure 1. The moment of inertia is $I_R = 3.68 \pm .07$ mg cm^2, of which one part in 535 ± 30 is contributed by the 7 mg of helium inside.

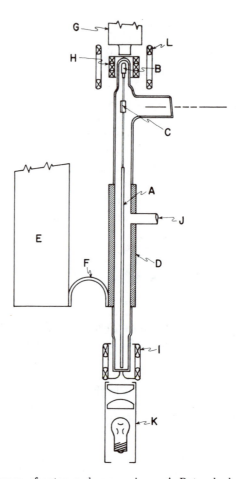

FIGURE 1. Diagram of rotor and suspension. A–Rotor body. B–Ferrite slug. C–Mirror. D–Copper chamber. E–1.1 K Liquid helium reservoir. F–Thermal shunt. G–Magnet pole. H–Height sensing coils. I–Induction motor coils. J–Pumping line. K–Heating light assembly. L–Horizontal field trim coil. The body is a stainless steel tube of inside radius $R = 0.0443$ cm and wall thickness 0.020 cm. It is sealed with silver solder at the bottom and has a short stainless steel stem and cap on top. The rotor is completed by a ferrite slug attached to the cap for magnetic suspension, and a front-aluminized mirror glued to the stem. The entire rotor is 17.6 cm long and has a mass of 1.04 g.

This helium content was determined by heat capacity measurements over the temperature range 1.7 to 2.5 K, where it gives the dominant contribution. An identical but empty rotor was substituted to determine the contribution (3 to l0%) of the rotor, bomb and addenda. After a small vapor volume correction [16], the difference was compared to published specific heat data for liquid helium [15–17], leading to the estimate $M_{He} = 7.30 \pm .25$ mg. Further, a discontinuity in specific heat was found at $T = 5.02 \pm .01$ K, which was identified as departure from the liquid-vapor coexistence region. This implies a density of 0.098 g/cm^3 [18] and gives $M_{He} = 6.71 \pm .25$ mg.

Next I describe the magnetic suspension and other elements of the environment of the rotor during the experiment. The rotor is suspended within a copper chamber with glass end sections, attached to a pumped ^{4}He reservoir (normally) at a temperature of about 1.1 K. The suspension electromagnet is directly above this chamber, attached to a 4.2 K liquid helium bath. The magnet current is controlled by the signal from a differential transformer surrounding the ferrite slug, which provides a sensitive measure of the height of the rotor. The rotor is set into rotation by an induction motor, consisting of a triad of coils surrounding the lower end of the body. A step and vernier attenuator for the 3 phase current allows fine adjustment of the motor torque over a wide range.

The chamber enclosing the rotor is pumped continuously with a diffusion pump. For cooling, helium exchange gas is bled into the pumping line at a rate sufficient to give a pressure of about 20 millitorr. Constant angular velocity is maintained during cooling by manually adjusting the induction current. After the gas supply is shut off, evacuation to the point where gas torques are no longer observable requires about 1/2 minute. The heat flux out of the rotor was monitored by recording the temperature difference across the shunt connecting the rotor chamber to the 1.1 K reservoir. The relationship is nearly linear in the relevant range. This heat flux monitor provides a way to measure heat inputs to the suspended rotor which result from various operations, through the heat removed on subsequent admission of exchange gas. The background heat leak, probably from eddy currents induced by the suspension, was found to be 6 μW. After evacuation, our protocol is to outgas the chamber surrounding the rotor by heating it (at 18 mW) to about 2.4 K for 20 sec. This procedure delivers $1.4 \pm .6$ millijoules to the rotor, and also produces a counterclockwise angular impulse of magnitude comparable to that we wish to measure. However repetition produces no further effect, so the outgassing is apparently effective. It is inferred from these measurements that in our standard procedure (4 minutes between evacuation and heating) with the reservoir at 1.1 K, the rotor is at $1.61 \pm .06$ K just before heating. Thus the superfluid fraction is $\rho_s/\rho = 0.82 \pm .03$.

The rotor is heated by light from a flashlight bulb, focused by condensing lenses on the bottom of the rotor body. The bulb is enclosed in a superconducting lead tube for magnetic shielding. Under standard conditions of 1/3 rated power, this light supplies 1.6 mW to the rotor (heating it to T_λ in 11 sec) and also delivers 2.1 mW to the chamber. The gas evolved in the standard operations was measured with a helium mass spectrometer. Provided the chamber has been previously outgassed, the gas flux associated with heating the rotor is about 10^{-3} of that due to the outgassing. Torques associated with this small amount of gas should be completely negligible.

The period of rotation is measured with an autocollimator, which observes the mirror and provides a trigger pulse to a counter once per revolution. Although the damping torque is small, corresponding to a rotational decay time of order 10^5 sec under standard vacuum conditions, there are additional torques acting on the rotor. A counterclockwise torque of about 5×10^{-8} dyne-cm is thought to result from residual gas streaming. These two can be balanced by a weak motor drive. There is also a conservative torque, apparently due to a permanent horizontal component of magnetization of the ferrite slug. This can be reduced to about 10^{-4} dyne-cm by trimming the horizontal components of magnetic field at the slug. Since the corresponding angular potential energy is small compared to the kinetic energy at experimental angular velocities, it has only a small effect on the uniformity of rotation.

A more serious problem is coupling of pendulum modes to the rotation, which is believed to result from the same horizontal magnetization, coupling *via* tilt to the vertical support field. (There are many possible second order coupling mechanisms, but all others are considerably smaller.) The two types of pendulum modes have respectively a period of 0.703 sec with nodal point slightly above the top of the rotor, and a period of 0.15 sec with nodal point one-third up from the bottom of the rotor. The high frequency modes are well damped and are only observed by intentional driving, but the low frequency modes have high Q and are invariably excited. This results in a modulation of the observed period of rotation τ at the difference between the pendulum frequency and the nearest multiple of the rotation frequency. The amplitude of this modulation can approach 1 percent. However if τ is chosen to be very close to a multiple of 0.703 sec, this modulation averages out over the rotation cycle, so that the amplitude as well as the frequency of the beat goes to zero. Thus the effect can be ignored if we make measurements at this set of discrete rotation periods, and we have done so in all but a few cases.

3.   VORTEX MODEL CALCULATIONS

The equilibrium number and configuration of vortices is that which minimizes $E - \omega L_s$, where E is the energy (essentially kinetic). This is easily calculated for small numbers of vortices in a circular cylinder [19–21], for which relevant configurations have high symmetry, and $L_s(\omega)$ can be evaluated for the equilibrium configurations [21] (see figure 4). Havelock [22] has examined the low-ω stability limit for a single ring of vortices. For more than eight vortices, stable configurations have been found by a numerical relaxation procedure by Stauffer and Fetter [23], and more extensively by Campbell and Ziff [24].

4. EXPERIMENTAL RESULTS

The procedure for cooling and heating at constant angular velocity has been described above. Figure 2 shows a portion of the rotation record including

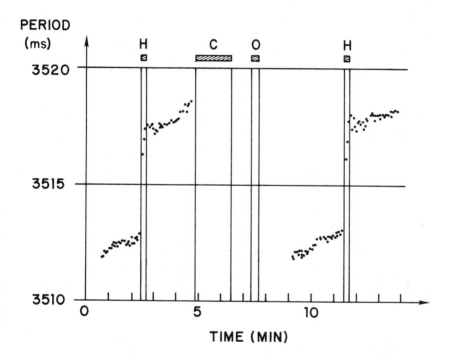

FIGURE 2. Sample of the free rotation record. The points are successive periods of rotation. C–Cooling (exchange gas in). O–Outgassing chamber. H–Heating light on. During cooling the motor was used to maintain a nearly constant rotation period of about 3.51 sec.

two successive heatings. The change in period on heating is estimated by extrapolating a free-rotation record of about 2 minutes from either side. The values of $\Delta\omega$ obtained in 117 trials are plotted against angular velocity in figure 3. The first point to note is that $\Delta\omega$ is (with one exception)

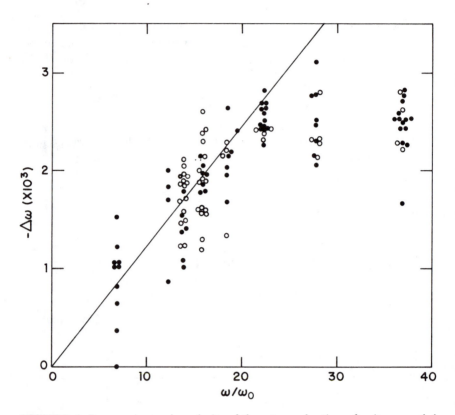

FIGURE 3. Decrease in angular velocity of the rotor on heating, after it was cooled in rotation at the same angular velocity, given by the abscissa. Open circles: clockwise rotation; solid circles: counter clockwise rotation. All of these measurements were made very close to one of eight discrete values of ω, but some points have been offset horizontally to avoid overlap. The diagonal line is the result expected if the superfluid has zero angular momentum.

always nonzero, and negative,[1] which implies that the superfluid does in fact have less than classical angular momentum. Also there does not appear to be any systematic difference between clockwise and counterclockwise rotation. This is confirmation that gas torques, which are predominantly unidirectional at low pressures, do not make an important contribution.

The average angular momentum of the superfluid, calculated by equation (3) for all trials at the same angular velocity, is plotted against ω in figure 4. Classical angular momentum (rigid rotation) is indicated by the diagonal dashed line and would correspond to $\Delta\omega = 0$. The heavy line

---

[1]The sign of ω is defined to be positive for each trial.

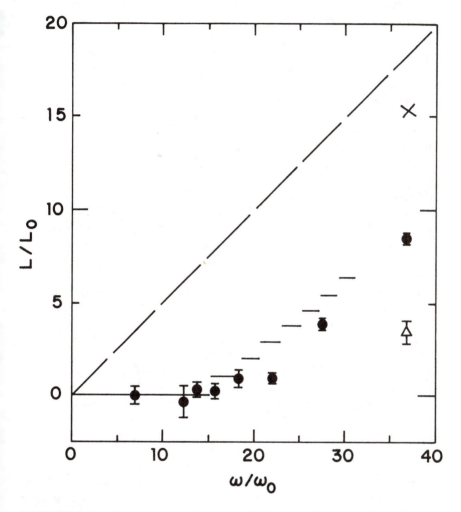

FIGURE 4. Angular momentum of the superfluid after cooling in steady rotation, as a function of angular velocity (solid circles). The reduced units are defined in the text. Each point is the average of 10 to 23 trials (except the point at $\omega/\omega_0 = 12.3$ which is 4 trials) and error bars represent the standard deviation of the mean. The heavy horizontal line segments are the prediction of the vortex model with $\ln(R/a) = 15.3$. The diagonal dashed line is the classical angular momentum. Results of other procedures are also shown (see text). The triangle is the average angular momentum after cooling under conditions expected to produce no vortices and then accelerating to the indicated angular velocity. The cross is the result of cooling at $\omega/\omega_0 = 60$ and then decelerating to the indicated angular velocity.

segments are the predictions of vortex theory, and give the angular momentum of the stable configuration of $0, 1, 2 \ldots$ rectilinear vortices over the range of ω for which that configuration gives the lowest free energy [21].

The experimental results are seen to be in reasonable agreement with the equilibrium vortex model. The experimental scatter is too large (standard deviation $\approx 1.3\ L_0 = 1.2 \times 10^{-6}$ dyne-cm-sec) to permit determination of the exact number of vortices present in a given trial, or to resolve the step structure in the average angular momentum. This scatter may originate entirely from external perturbations, or may include real variability in the number of vortices, possibly including vortices which extend only part of the length of the helium column and terminate on the wall [25].

More precisely, the occurrence of a given angular momentum is shifted to perhaps 10 per cent larger ω than predicted, as would be expected if the vortex had larger energy. However the vortex parameters are well established by ion experiments [26], and the deviation must probably be interpreted as a small departure from equilibrium. It might be reduced considerably by a new value of M_{He} within the present uncertainty of that parameter.

In order to check that the angular momentum deficit is in fact proportional to the superfluid mass, a series of measurements was made using exactly the same procedure, but with the helium reservoir at temperatures higher than 1.1 K. Figure 5 shows the result at a rotation period of 5.6 sec, where no vortices are expected. The solid line represents the predicted effect, proportional to $\rho_s(T)$, assuming that the superfluid has no angular momentum, and involves no adjustable parameters. The agreement is excellent. In particular, for initial temperatures above T_λ, $\Delta\omega$ is zero within experimental error. Similar measurements at a rotation period of 2.1 sec are shown in figure 5b. Vortices are present at this speed and the solid line, proportional to $\rho_s(T)$, is normalized to the full set of data at 1.61 K. The scatter in this run is somewhat larger than usual and agreement with the expected temperature dependence is fair.

In an additional series of experiments the rotor was cooled at an angular velocity ω_1, then its rotation speed was changed to ω_2 while in the exchange gas stage, and finally it was heated at ω_2 to determine the angular momentum of the superfluid. One such procedure is to cool at rest, or at sufficiently low ω_1, where no vortices are expected, and look for the angular momentum of vortices generated mechanically during acceleration to or rotation at ω_2. A procedure which is probably equivalent is to cool in reverse rotation, since any vortices of opposite sense should become unstable and be lost rapidly as the rotation is reduced to zero and reversed. The average result of 11 trials using both procedures and with $2\pi/\omega_2 = 2.1$ sec is given by the triangle in figure 4. The resulting angular momentum is considerably less than the equilibrium value, but greater than zero; apparently 3 or 4 vortices have entered the superfluid. Eight trials were made with

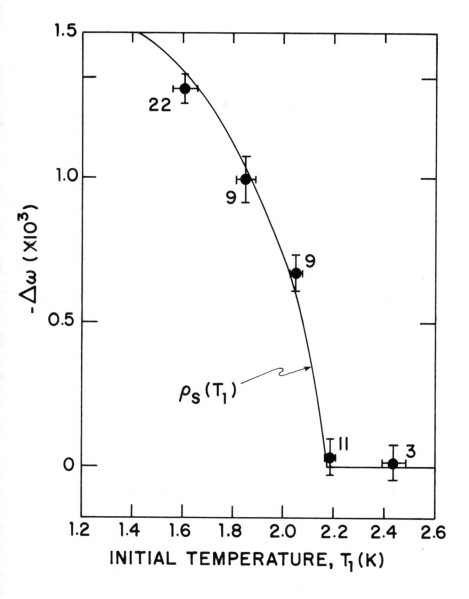

FIGURE 5a. Change in angular velocity on heating as a function of initial temperature T_1, for steady rotation at $\omega/\omega_0 = 14$. Numerals are the number of trials averaged and the error bars represent the standard deviation of the mean. The same heat input is used in all trials. The solid line is proportional to the superfluid density at T_1 and is normalized to give the expected $\Delta\omega$ if no vortices are present.

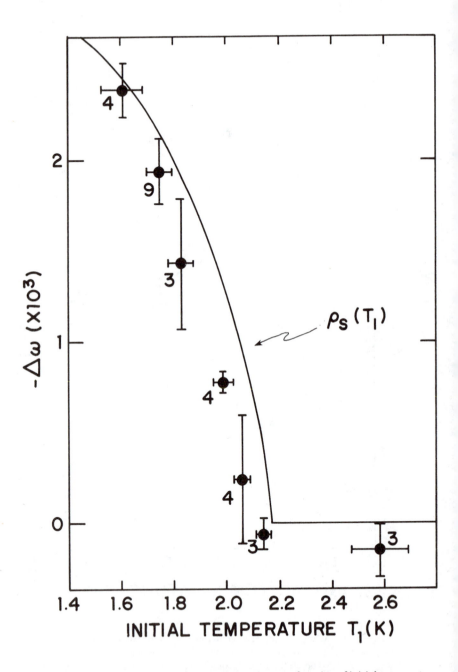

FIGURE 5b. Change in angular velocity on heating as a function of initial temperature T_1, for steady rotation at $\omega/\omega_0 = 37$. The solid line is proportional to the superfluid density at T_1 and in this case is normalized to the full set of data at $T = 1.61$ K.

$\omega_1 < 8\,\omega_0$ and ω_2 in the range 12 to 17 ω_0. The mean angular momentum corresponds to about one vortex, but is consistent with zero.

At what angular velocity might one expect a vortex loop to be generated at the wall and expand into the interior of the helium? A reasonable guess is at a tangential velocity equal to the critical velocity for vortex generation in axial flow. The most sensitive measurements of dissipation in pure superfluid flow in a tube comparable to our rotor body remain those of Kidder and Fairbank [27]. They found a critical velocity of 0.085 cm/sec at 1.57 K, with a resolution in pressure gradient which corresponds to generation of about two vortex lines per sec per cm of tube length. This critical velocity corresponds to $\omega_{crit} = 24\,\omega_0$, and would give an offset from equilibrium rotation consistent with our observation of several vortices at 37 ω_0.

Nineteen trials were made in deceleration, $\omega_2 < \omega_1$. With one exception, only one to three trials were done for each combination of angular velocities, and scatter is large. One possible outcome of slow deceleration is that the number of vortices remains unchanged until there is no longer a metastable configuration for that number, at which point the angular momentum will be close to the classical value [21]. On the other hand, it is possible that the instability will result in loss of additional metastable vortices and the final angular momentum will be smaller. Experimentally, the final angular momentum is found to be less than classical (by about 3 vortices on average) but well above the equilibrium value. The cross in figure 4 represents the average of five trials cooled at $\omega_1 = 60\,\omega_0$ and heated at $\omega_2 = 37\,\omega_0$, and illustrates this trend.

Finally, seven trials were made in which the rotor was cooled and heated at 14 ω_0, but while in the exchange gas stage was accelerated to between 48 and 66 ω_0, rotated at that speed for about a minute, and then decelerated. The intention was to verify mechanical generation of vortices. Our interpretation of the previous two procedures suggests many vortices (of order 20) will be generated at the maximum angular velocity, and enough of these to provide nearly classical angular momentum might be retained after deceleration. In fact, the final angular momentum corresponded to only zero to three vortices.

Two possible explanations come to mind. First, the relatively large decrease in ω, going well beyond the stability limit for the initial number of vortices, may drive a stronger instability, which may carry nearly all of the vortices to the wall. Second, vortices formed mechanically may be different from those formed by cooling in rotation, such that they are more easily lost on reduction of rotation speed. Specifically, the former, created by a three-dimensional process, may be more likely to extend only part of the length of the cylinder, with one or both ends remaining pinned to the side wall. In this case the energy barrier opposing contraction may be small compared to the barrier keeping a rectilinear metastable vortex from the wall.

We deemed it not feasible to make more extensive measurements over a two-parameter field, so long as multiple measurements under the same conditions had to be averaged to obtain the desired resolution.

5. MODIFICATIONS OF THE APPARATUS

Following these measurements, the apparatus was rebuilt with several modifications. Most important, the servoed ferromagnetic suspension was replaced by a passive superconducting suspension. For this purpose the ferrite slug at the top of the rotor was replaced by a chemically polished niobium sphere of 0.36 cm diameter, and the iron-core magnet was replaced by a small coil of Nb-Ti wire (1 cm diameter × 0.25 cm high) located just below the sphere. A persistent current is trapped in this coil to levitate the rotor.

Torques acting on the new suspension were found to be quite similar to those on the old. In particular, the rotor had a horizontal component of magnetization which produced an angular potential well of about the same depth as in the ferromagnetic case. However in this case the magnetization, apparently due to trapped flux in the niobium sphere, changed magnitude and direction on each cooling of the rotor from room temperature. This was a serious inconvenience, as the nulling process is very tedious.

The rotational noise was not noticeably improved. Use of ^{3}He exchange gas did not significantly affect the amount of outgassing or the magnitude of resulting torques. Several trials were made of the standard procedure, and the decrease in angular velocity on heating was observed with about the expected magnitude. In view of the lack of improvement in noise level and experimental convenience, no systematic data collection was undertaken. (This work was done partly at Stanford and partly at Virginia.)

6. SUBSEQUENT EXPERIMENTS
ON ROTATING HELIUM

I am not aware of any more recent studies of equilibrium rotation of bulk helium by measurement of angular momentum. (Reppy and coworkers [28] did elegant experiments on persistent currents in porous media and in thin helium films.) There have been very important advances, however, in studying the appearance and the distribution of vortices by means of second sound attenuation and especially by negative ion trapping.

6.1   Second Sound

The earliest support for the quantized vortex model came from observations by Hall and Vinen of extra attenuation of second sound propagating perpendicular to the axis of rotation of a resonant cavity, and their analysis in terms of scattering of rotons by vortex lines [29]. This experiment was refined by Bendt and Donnelly to detect the threshold ω_c for appearance of a ring of free vortices in an annular resonator rotating about its axis [30,31]. The experiment depends on mechanical generation of vortices on isothermal changes in angular velocity, not only to introduce free vortices, but to establish equilibrium circulation around the annulus. However, only a small hysteresis was found, provided the apparatus was once rotated above a certain critical velocity. Apparently this served to introduce a population of pinned vortices which facilitated subsequent approach to equilibrium.

This work was extended by Shenk and Mehl [32], who studied isothermal rotation at three temperatures (1.4 to 1.8 K) and included measurements to larger angular velocities. They found somewhat more vortices than are expected if $a \sim 1$ Å. Further measurements [33] of the threshold in the range $T_\lambda - T = 0.2$ to 0.01 K show that ω_c decreases by about a factor of 2 approaching T_λ across this range. This is attributed to a much stronger divergence of $a(T)$ than is expected. Curiously, metastable nonequilibrium states were found to be more stable above ~ 2.1 K than below. Possibly pinned vortices are lost near T_λ.

6.2   Ion Trapping

The discovery by Rayfield and Reif [26] that ions accelerated in liquid helium may create and become bound to quantized vortex rings provided the most direct means for measuring the parameters of the quantized vortex, at least at low temperatures. Ion trapping on vortices in rotating helium was then studied extensively, especially by Donnelly and coworkers [34], and provided important information on both the ions and the vortex.

In 1969 Packard and Sanders [35] succeeded in adapting this technique to the detection of individual vortices in a rotating cylindrical vessel. Vortices, if present, were charged with negative ions (electron bubbles) in the lower part of the cylinder. The trapped ions were then drawn up the vortex lines by appropriate electric fields and extracted through the meniscus, where they were detected by a unique helium vapor proportional counter. Measurements were made on slow angular acceleration at constant temperature (1.2 K), and often showed clear stepwise increases in the ion current corresponding to the introduction of successive vortices. If the acceleration is reversed and the rotation speed reduced, the signal decreases but there is not a well-developed step structure. On a second acceleration from rest without heating above T_λ, the signal reappears at about the same threshold, but with large fluctuations and no sign of steps. This history dependence is interpreted to mean that segments of lost vortices remain pinned at walls and can reenter the interior of the cylinder under favorable circumstances, but probably not in an ideal rectilinear configuration.

These data can be compared with our results for isothermal acceleration. In the figures in [35], the first vortex appears at $\omega/\omega_0 = 25 \pm 4$ and the fourth vortex at $\omega/\omega_0 = 34 \pm 6$. (The numbers are slightly higher for the vessel of radius 0.087 cm than that of radius 0.05 cm.) This is quite comparable to our observation of 3–4 vortices at $\omega/\omega_0 = 37$.

DeConde and Packard [36] employed a similar apparatus to study the threshold for appearance of the first vortex on slow cooling (through T_λ to 1.2 K) in steady rotation. They used a circular cylinder of radius 0.159 cm, as well as several cylinders of elliptical or rectangular cross section. In all cases they found agreement within 10% with equilibrium theory [37]. The threshold for the circular cylinder was $\omega/\omega_0 = 16.2 \pm .7$.

Finally in 1974 Williams and Packard [25] used the ion trapping technique to obtain images of the vortex array, that is, of the intersections of vortices with the meniscus. This was accomplished by accelerating electrons extracted through the meniscus into a phosphor screen, and transferring the image on the screen by a coherent fiber optics bundle to room temperature optics. The techniques have been described in detail recently [38]. Early images showed irregular patterns of vortices. After additional efforts, including reduction of mechanical disturbances, stable arrays of 1 to 11 vortices were obtained, in close correspondence to the calculated patterns [39]. These observations were made near $T = 0.1$ K. On acceleration from rest successive vortices appeared, but always fewer than the equilibrium number (e.g., 4 instead of 23 at $\omega/\omega_0 = 54$ in one trial). On deceleration the number may exceed equilibrium, confirming the interpretation of previous experiments.

These experiments establish the existence and configuration of quantized vortices in a rotating cylinder in considerable detail, although data on the dynamics remain somewhat sparce. Given the distribution of vortices, one can calculate the superfluid velocity field and angular momentum. In fact all stationary configurations of a given number of vortices at given ω have essentially the same angular momentum [24], so in a sense a measurement of angular momentum gives no information beyond counting the vortices. What the angular momentum experiment does uniquely is establish the existence of a regime of nonrotation of the superfluid directly and independent of the vortex model.

Acknowledgments

The experimental work at Stanford University was supported in part by the Army Research Office, Durham, and by the Advanced Research Projects Agency through the Center for Materials Research, Stanford. Further work was supported by the Center for Advanced Studies at the University of Virginia.

References

[1]   F. London, *Superfluids* (John Wiley and Sons, Inc., New York, 1954), Vol. 2, especially pp. 142–5 and 200.

[2]   G. B. Hess and W. M. Fairbank, *Phys. Rev. Lett.* **19**, 216 (1967); and in *Proceedings of the 10th International Conference on Low Temperature Physics*, edited by M. P. Malkov (Viniti, Moscow, 1967), Vol. 1, p. 423. See also G. B. Hess, Ph.D. thesis, Stanford University (1967), unpublished.

[3]   L. D. Landau, *J. Phys. USSR* **5**, 71 (1941), p. 78.

[4]   H. London, *Report on an International Conference on Fundamental Particles and Low Temperatures*, Cambridge, July 1946 (The Physical Society, London, 1947), Vol. 2, p. 48.

[5]   A. Bijl, J. de Boer and A. Michels, *Physica* **8**, 655 (1941).

[6]   L. Onsager, *Nuovo Cimento Suppl.* **6**, 249 and 269 (1949).

[7]   R. P. Feynman, in *Progress in Low Temperature Physics*, edited by C. J. Gorter (North-Holland Publishing Co., Amsterdam, 1955), Vol. 1, Chap. 2.

[8]   D. V. Osborne, *Proc. Roy. Soc. (London)* **A63**, 909 (1950).

[9]   E. L. Andronikashvili and I. P. Kaverkin, *Zh. Eksperim. i. Teor. Fiz.* **28**, 126 (1955) [*Sov. Phys.–JETP* **1**, 174 (1955)].

[10]   For instance, R. J. Donnelly, *Phys. Rev.* **109**, 1461 (1958); R. R. Turkington, J. B. Brown and D. V. Osborne, *Can. J. Phys.* **41**, 820 (1963); R. Meservey, *Phys. Rev.* **133**, A1471 (1964). See also P. L. Marston and W. M. Fairbank, *Phys. Rev. Lett.* **39**, 1208 (1977).

[11]   H. E. Hall, *Phil. Trans. Roy. Soc. (London)* **A250**, 359 (1957).

[12]   R. H. Walmsley and C. T. Lane, *Phys. Rev.* **112**, 1041 (1958); J. D. Reppy, D. Depatie and C. T. Lane, *Phys. Rev. Lett.* **5**, 541 (1960); J. D. Reppy and D. Depatie, *Phys. Rev. Lett.* **12**, 187 (1964).

[13]   J. D. Reppy and C. T. Lane, *Phys. Rev.* **140**, A106 (1965).

[14]   E. L. Andronikashvili and Yu. G. Manaladze, *Rev. Mod. Phys.* **38**, 567 (1966) or in *Progress in Low Temperature Physics*, edited by C. J. Gorter (North-Holland Publishing Co., Amsterdam, 1966), Vol. 5, Chap. 3.

[15]   M. J. Buckingham and W. M. Fairbank, in *Progress in Low Temperature Physics* (North-Holland Publishing Co., Amsterdam, 1961), Vol. 3, Chap. 3.

[16]   R. W. Hill and O. V. Lounasmaa, *Phil. Mag.* **2**, 143 (1957).

[17]   A. C. Kramers, J. D. Wasscher and C. J. Gorter, *Physica* **18**, 329 (1952).

[18]   M. H. Edwards and W. C. Woodbury, *Phys. Rev.* **129**, 1911 (1963).

[19]   W. F. Vinen, *Proc. Roy. Soc. (London)* **A260**, 218 (1961).

[20]   A. L. Fetter, *Phys. Rev.* **138**, A429 (1965); **152**, 183 (1966).

[21]   G. B. Hess, *Phys. Rev.* **161**, 189 (1967).

[22]   T. H. Havelock, *Phil. Mag.* **11**, 617 (1931).

[23]   D. Stauffer and A. L. Fetter, *Phys. Rev.* **168**, 156 (1968).

[24]   L. J. Campbell and R. M. Ziff, *Phys. Rev.* **B20**, 1886 (1979).

[25]   G. A. Williams and R. E. Packard, *Phys. Rev. Lett.* **33**, 280 (1974).

[26]   G. W. Rayfield and F. Reif, *Phys. Rev.* **136**, A1194 (1964).

[27]   J. N. Kidder and W. M. Fairbank, *Phys. Rev.* **127**, 987 (1962).

[28]   J. R. Clow and J. D. Reppy, *Phys. Rev. Lett.* **19**, 289 (1967) and *Phys. Rev. A* **5**, 424 (1972); M. H. W. Chan, A. W. Yanof and J. D. Reppy, *Phys. Rev. Lett.* **32**, 1347 (1974).

[29]   H. E. Hall and W. F. Vinen, *Proc. Roy. Soc. (London)* **A238**, 204 and 215 (1956).

[30]   P. J. Bendt and R. J. Donnelly, *Phys. Rev. Lett.* **19**, 214 (1967); P. J. Bendt, *Phys. Rev.* **164**, 262 (1967).

[31]   A. L. Fetter, *Phys. Rev.* **153**, 285 (1967).

[32]   D. S. Shenk and J. B. Mehl, *Phys. Rev. Lett.* **27**, 1703 (1971).

[33]   I. H. Lynall, D. S. Shenk, R. J. Miller and J. B. Mehl, *Phys. Rev. Lett.* **39**, 470 (1977).

[34]   R. J. Donnelly, W. I. Glaberson and P. E. Parks, *Experimental Superfluidity* (University of Chicago Press, Chicago, 1967), and references therein.

[35]   R. E. Packard and T. M. Sanders, Jr., *Phys. Rev. Lett.* **22**, 823 (1969) and *Phys. Rev. A* **6**, 799 (1972).

[36]   K. DeConde and R. E. Packard, *Phys. Rev. Lett.* **35**, 732 (1975).

[37]   A. L. Fetter, *J. Low Temp. Phys.* **16**, 533 (1974).

[38]   G. A. Williams and R. E. Packard, *J. Low Temp. Phys.* **39**, 553 (1980).

[39]   E. J. Yarmchuk, M. J. V. Gordon and R. E. Packard, *Phys. Rev. Lett.* **43**, 214 (1979).

Intrinsically Irreversible Heat Engines

John C. Wheatley

I would like to begin by saying that the remarkably simple and beautiful experiments on liquid ^{3}He and on ^{3}He-^{4}He solutions done at Duke University by William Fairbank, King Walters and others in the decade of the fifties helped lay the foundations for a field which is very active today. King Walters will describe those days better than I. But I want to add how important those early observations of Fermi-like behavior in liquid ^{3}He and of the phase separation phenomenon in ^{3}He-^{4}He solutions have turned out to be and how they provided something secure on which I could build my own experimental work in the field of liquid helium.

My contribution to this volume differs from most of the others, yet I feel that it is still appropriate. Certainly one of the themes of the Near Zero conference is the importance to physics of attaining temperatures near absolute zero. Although there is no longer any problem in reaching exceedingly low temperatures, in some sense the motivation for my present far-from-complete work is to reexamine old ideas and investigate new ones to see if it might not be possible to attain low temperatures *more simply*, and in doing so to open up the temperature region "near zero" to even more widespread discovery and invention. In this work I am only following Carnot's admonition, written in the last paragraph of his book [1], to the effect that there is more to the subject of heat engines [2] than the efficient use of heat and work; that convenience, economy, and above all simplicity are essential components of the solution to any practical problem. I like to think that some of the hallmarks of Bill Fairbank's contributions to physics are both the ingenuity and the essential simplicity of his approach to the solution of practical measurement problems.

1. INTRODUCTION

Only heat engines which work on a long time scale can have the Carnot efficiency. If power, or work (or heat) per unit time, is the important quantity, then irreversibilities are unavoidable. For example, a Carnot-like engine with a finite thermal resistance between working fluid and reservoirs has the diminished efficiency at optimal power output [3,4] of $1 - (T_C/T_H)^{1/2}$ rather than $1 - (T_C/T_H)$, where T_C and T_H are the temperatures of the cold and hot reservoirs.

An example of an engine which has Carnot's efficiency in the limit of long periods is the Stirling engine [5]. Rev. Robert Stirling invented this reciprocating engine in 1816, eight years before Carnot's book was published. I think of Stirling's engine as being intrinsically reversible. Stirling introduced into the engine what I call a second thermodynamic medium in addition to the primary thermodynamic medium (air for Stirling's engine). Engineers call this second medium a heat regenerator. It plays the role of a continuously distributed heat source and sink internal to the engine, and makes possible an idealized cycle having Carnot efficiency in which all processes are locally isothermal. The Stirling engine, like all intrinsically reversible engines, has two separately controlled elements, in this case a power piston and a displacer which are phased appropriately with respect to one another to produce a useful output.

The intrinsically irreversible heat engines I discuss here use the irreversible process of thermal conduction to achieve the necessary phasing between temperature changes and motion of a primary medium and therefore have only one moving mechanical element. Stirling's second thermodynamic medium is also an essential element of such engines, but now the thermal contact with it is neither very good nor very bad. Indeed, no useful output is obtained either in the reversible isothermal limit, where the engine runs very slowly, or in the reversible adiabatic limit, where the engine runs very fast. That is why I call such engines "intrinsically irreversible." It is also necessary for the geometrical symmetry between primary and secondary media to be broken in order to achieve a useful thermodynamic result.

If an external source works on such an engine then a temperature difference is produced. If the primary medium is a thermodynamically active fluid with a very low Prandtl number (the dimensionless ratio of viscosity to thermal conductivity), so that effects of viscosity are negligible, then I think that very large temperature differences should ensue. As it is possible to design engines in which thermodynamic effects are emphasized over frictional effects, I am encouraged to pursue a scientific study of intrinsically irreversible engines with the hope that there will be practical consequences, especially for the simple attainment of low temperatures. Essential to the simplicity are the need for only one moving element and the quality that thermal contact to the second medium *must* not be good.

I refer to the present class of engines as intrinsically irreversible but *functionally* reversible to distinguish them, for example, from refrigerators which employ the irreversible Joule-Thomson expansion of a fluid to produce low temperatures. The Joule-Thomson, or perhaps better the Linde-Hampson, cooling method is characterized by simplicity and effectiveness and indeed was the means employed by Kamerlingh Onnes to liquify helium. The present engines are functionally reversible in the sense that they can be prime movers as well as refrigerators, just depending on the temperatures spanning the engine. As prime movers, in one form they are the famous Sondhauss tubes in acoustics [6]. In another form familiar to all low temperature experimentalists they lead to the "Taconis oscillations" in a tube inserted through a temperature gradient into a vessel containing liquid helium.

The possibility of using the irreversible process of thermal conduction as part of a method to produce cold was first proposed by Gifford and Longsworth and then developed by them in a series of papers [8–11]. They called their device a "pulse tube refrigerator" and described its operation in terms of a concept called "surface heat pumping." It operated at low frequency (*ca.* 1 Hz) and used pressure changes of several atmospheres. I was reminded of the Gifford and Longsworth papers last year when I ran across them again while browsing in the library at the Kamerlingh Onnes laboratory. Just at that time I was also studying a paper by Ceperley [12] on a traveling wave acoustical engine. The Ceperley engine is essentially a novel Stirling engine. It requires, as for all Stirling engines, regions of high thermal contact between primary and secondary media, and it suffers badly from effects of viscous dissipation. However, combining acoustical techniques with thermodynamically irreversible processes seemed like an eminently sensible thing to do. So on my return to Los Alamos a year ago, and with the help of my colleagues, Greg Swift and Al Migliori, I began studying intrinsically irreversible acoustical heat engines experimentally. The practical qualities of simplicity, high frequency (for a heat engine), automatic provision of a "flywheel" by the inertia of the gas, absence of moving seals, and dependence on *poor* thermal contact were immediately obvious. Only more recently have I appreciated the generality of the concepts.

2. THE PRINCIPAL THERMODYNAMIC EFFECTS

An apparatus in which the principal thermodynamic effects can be observed is shown in figure 1. I want to emphasize that from a fundamental point

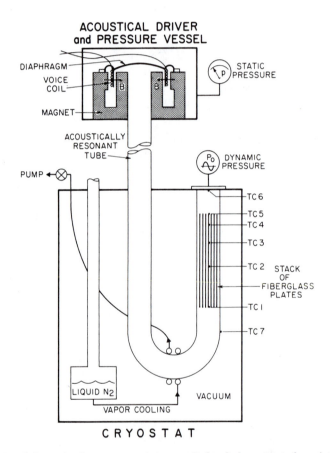

FIGURE 1. Schematic of apparatus using acoustical techniques to study an intrinsically irreversible heat engine, using ^{4}He gas near room temperature as primary medium and fiberglass plates as secondary medium. TC1 through TC7 are thermocouple sensors, p is the static pressure, and P_0 is the amplitude of the dynamic pressure.

of view the use of acoustical techniques in this apparatus is peripheral. An acoustical compression driver (loud speaker) excites a resonant acoustical tube in the form of a U which is closed at one end. The U shape is used to eliminate natural convective heat transfer. The second thermodynamic medium, on which our attention will be focused, is a stack of 19 fiberglass plates in the right half-tube. These are instrumented by epoxying five chromel-constantan thermocouples TC1 to TC5 to the central plate with the junctions located more or less as shown. The tube contains ^{4}He gas at an average pressure p that we can control. At the closed end of the tube the dynamic pressure amplitude P_0 can be measured accurately. Both the driver and the closed end of the tube, whose temperature is measured by thermocouple TC6, are near room temperature. The walls of the straight sections of the U-tube are 1 mm thick fiberglass and are insulated with superinsulation, the whole U-tube being in vacuum. The bottom of the U-tube is copper whose temperature, measured using TC7, can be adjusted by controlling the flow of N_2 boiloff gas from a liquid N_2 bath. The open space above and below the second medium is designed to have a high thermal resistance.

This form of the apparatus, with gas cooling at the bottom of the U, is necessary to reduce the heat input to the bottom end of the second medium to a controllably small value. The tube is operated at resonance. Although heat generated in the second medium itself by viscous effects is small, owing to its proximity to the pressure antinode, heat generated viscously in other parts of the tube and heat from the room temperature region are transported toward the second medium, in part by acoustical streaming. In a practical cooling engine it will be necessary to play a trick to eliminate this problem, but for the present purposes of illustrating the principles the arrangement shown is satisfactory. As the length of the second medium is short compared to the radian length of the acoustical wave, whose frequency is typically 400 Hz, the same results would presumably be achieved ideally in another shorter apparatus with an oscillating piston located close to the second medium. Acoustical techniques are used here primarily for practical reasons.

In this apparatus we can measure temperature and static and dynamic pressure. Two types of experiments give readily interpretable results. In one, with the apparatus initially at a uniform temperature (usually ambient), the acoustical power is suddenly applied for a short time and the thermocouple response measured. In such a measurement the initial time rate of change of temperature T_i is proportional to the heat flux into the surface of the second medium. In a second experiment the temperature distribution along the second medium is observed for essentially adiabatic conditions under the action of continuous wave acoustical power.

Typical results of the sudden application of acoustical power are shown in figure 2. They show the essence of the thermodynamic effects. Before acoustical power is applied, we have $(grad_z T) = 0$ along the plates. The ^{4}He pressure is 4.90 bar and the sound frequency is adjusted to resonance, near 400 Hz. At time $t = 0$ acoustical power is applied with amplitude

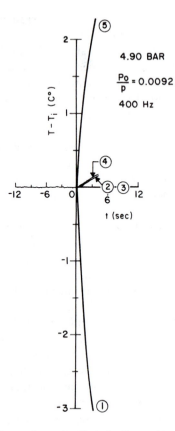

FIGURE 2. Essential thermodynamic effects in the engine. Shown are superimposed chart recordings of thermocouple temperature changes *versus* time. There is no initial longitudinal temperature gradient: $(grad_z T) = 0$. At $t = 0$ the acoustical power is suddenly switched on. For $t > 0$ the acoustical power is constant. The numbers refer to the thermocouple locations given on figure 1.

ratio $P_0/p = 0.0092$. This corresponds to only about 4% of a typical power level and was adjusted for demonstration purposes to allow all five thermocouple responses to be shown with the same sensitivity. Each of thermocouples 2, 3, and 4, those within the plates, had very nearly the same response, a heating with T constant but not large. Thermocouple 5, located at the edge of the plates closest to the closed end, heats rapidly while 1, located at the edge of the plates closest to the driver end, cools even more rapidly.

These observations are consistent with the following time-averaged energy flows in the gas toward the closed end. Within the plates but far from the ends, the energy flow toward the closed end increases linearly with the distance z from the closed end. Then, from conservation of energy, the time-averaged heat flow per unit length into the plates will be constant, as observed. There are sudden changes in the energy flow at the ends of the plates, the energy flow in the gas being smaller at both ends outside the plates. Then, at TC1, the energy flow toward the closed end is greater within the plates than it is outside them, so heat must be extracted from them, their temperature dropping as observed. At TC5 the energy flow toward the closed end is greater within than outside the plates, so the plates must absorb heat, their temperature rising as observed. The thermodynamic action corresponding to the heating at TC5 and the cooling at TC1 is a central feature of the irreversible engine.

The second experiment which can be done reasonably well is to study the adiabatic temperature distribution. Here the acoustical power is applied continuously, the temperature at TC7 is adjusted by controlling the flow of cold nitrogen gas, and the temperatures at TC1 and TC5 are observed. As I pointed out in describing the apparatus, the thermal resistance of the spaces adjacent to the stack of plates is high, so the response of the plates is nearly adiabatic. Typical results of this experiment for ^{4}He gas at $p = 1.90$ bar and a dynamic pressure ratio $P_0/p = 0.04$ are shown in figure 3, where the absolute temperatures of the five thermocouples are shown *versus* their longitudinal positions on the stack. The data are obtained as follows. We start with the whole stack at ambient temperature, that of TC6, about 287 K. The acoustical power is turned on. There is an immediate response. In a few minutes a temperature distribution is developed, with T_5, the uppermost thermocouple in the stack, hotter than T_0 and with T_1, the lowest thermocouple, cooler than T_0.

As shown in figure 3, we find a variety of equilibrium curves, very similar in shape, but different in intercept, depending on how T_7 is manipulated. The temperature difference (T_5-T_1) is typically in the range 100–120 K. A good way to describe this distribution is "rigid." It is established quickly. The general temperature level can be changed slowly by changing T_7, but this control is very "soft." The distribution is insensitive, at a level of 5 to 10%, to variations of static pressure over a factor of 10 (from 0.5 to 5 bar), to variations of dynamic pressure P_0/p over a factor of 5 (and thus in power over a factor 25), and to small variations in frequency (factor of 2 to 3). It also does not depend strongly on either T_6 or T_7. The dashed line on figure 3 is calculated using the adiabatic equation of state of ^{4}He gas, with T being the temperature at z, V being the volume included between the closed end of the tube and the point z, and a forced fit at TC5. A relationship like this between the adiabatic equation of state of the gas and the temperature distribution in absence of heat flows was proposed by Gifford and Longsworth [9].

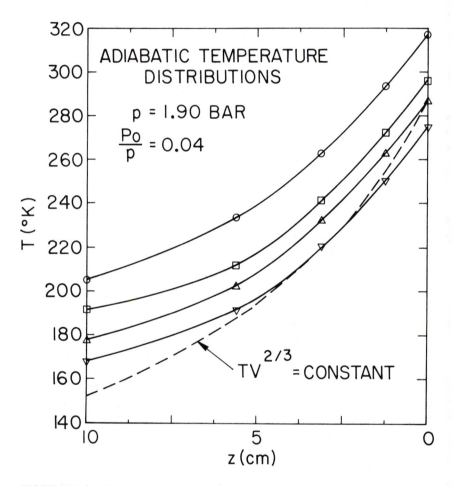

FIGURE 3. Adiabatic temperature distributions as obtained from dynamic equilibrium values of the five thermocouple outputs for $p = 1.90$ bar and $P_0/p \simeq 0.04$. These four curves were obtained for four different values of T_7, figure 1, the lowest curve, being for the lowest value of T_7. The point $z = 0$ corresponds to the end of the plates where TC5 is located. The line labeled $TV^{2/3} = $ constant is based on theoretical arguments presented in section 3, where T is the temperature at z and V is the gas volume included between z and the closed end of the tube.

3. UNDERSTANDING THE INTRINSICALLY IRREVERSIBLE ENGINE WITH IDEAL GAS PRIMARY WORKING SUBSTANCE

The concept of phase is important in heat engines. As shown in figure 4a, the phasing in an internal combustion engine is provided by the correct timing of the ignition or injection of fuel into the cylinder with respect to the motion of the piston. The resulting indicator diagram (or pV plot) has an enclosed area corresponding to the work done in a cycle. Proper phasing is achieved in a Stirling engine, figure 4b, by moving the piston P and the displacer D independently to maximize the area of the indicator diagram. In each of the above, two quantities had to be controlled and caused to act in proper time sequence to achieve the desired result.

In figure 4c I consider a piston and cylinder enclosing a gas with only one moving element, the piston, and with only reversible processes considered. If τ is the thermal relaxation time of gas in the cylinder and ω is the angular frequency of piston motion then the two possibilities for reversible processes are either isothermal ($\omega\tau \ll 1$) or adiabatic ($\omega\tau \gg 1$). Starting at the same state (p, V) and letting the piston oscillate produces no included area on the indicator diagram in either extreme. But if, as suggested in figure 4d, the thermal contact is too poor to allow isothermal processes yet not so poor as to make the processes adiabatic, then the indicator diagram will have a net included area. In this case, and for a small sinusoidal variation of volume with $\omega\tau < 1$ so that the processes are not quite isothermal, the indicator diagram will appear as shown. During expansion the gas will be generally lower in temperature than its surroundings, and hence at a lower pressure than the isotherm; during compression the gas will be generally higher in temperature and hence at a higher pressure than the isotherm. Thus with only one moving element, the piston, and irreversibility, there is a finite included area on the indicator diagram. But that is *not enough* to produce an interesting thermodynamic effect.

Excepting the region very near to the piston itself, the above irreversibility just leads to the absorption of heat by the walls of the cylinder. What is necessary to produce something interesting is to make the symmetry of the second medium and of the space occupied by the primary medium different. Why this is important can be understood rather easily by considering energy flow in the gas. Refer to figure 5a, which shows a cylinder with walls at constant temperature containing a gas confined by an oscillating piston. Excluding kinetic energy of mass motion the energy flow in the gas is the enthalpy flow. What is of interest is the time-averaged energy flow $\bar{H}$, taken

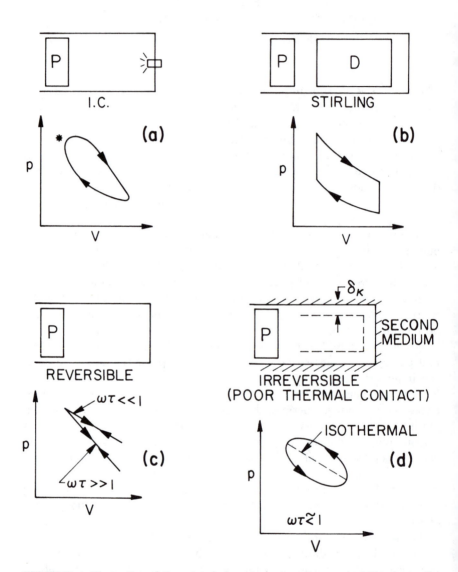

FIGURE 4. Illustration of the role of phase in heat engines. p and V refer to the pressure and volume of gas in the cylinder. (a) refers to an internal combustion (I.C.) engine with heat introduction at the (*). (b) shows a Stirling engine with piston P and displacer D moved individually in an articulated cycle. (c) describes reversible isothermal ($\omega\tau \ll 1$) and reversible adiabatic ($\omega\tau \gg 1$) processes; ω is angular frequency and τ is thermal relaxation time. (d) shows the behavior for an intermediate thermal contact with thermal penetration depth δ_κ comparable to the cylinder radius.

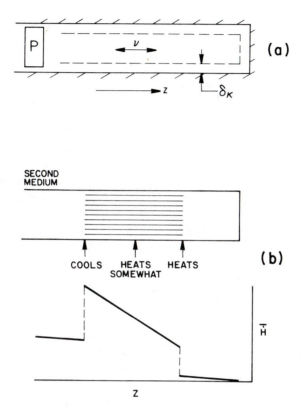

FIGURE 5. (a) Geometry used for the calculation of energy flow in the gas. z and v are local gas position and velocity. δ_κ, equation (2) in text, is the thermal penetration depth. (b) Time-averaged energy flows as a function of z for $(grad_z T) = 0$ when a stack of plates is introduced into the cylinder of (a).

positive in the direction along z toward the closed end. If $\bar{H}$ depends on x then $-d\bar{H}/dx$ is, by conservation of energy, the time-averaged heat flow per unit length into the walls of the cylinder. The time-averaged energy flow can be written schematically

$$\bar{H} = \overline{\overline{Av\rho h}} = \overline{\overline{Av\rho c_p \delta T}} \quad , \tag{1}$$

where the double bar indicates both a time and a spatial average, A is the cross-sectional area, ρ is mass density, c_p and h are specific heat and enthalpy per unit mass, and δT is the temperature deviation from the isothermal surroundings. For isothermal processes the temperature deviation δT is zero; for adiabatic processes v and δT have a $\pi/2$ phase difference, so that their time-averaged product is zero. Hence for irreversible processes of both types $\bar{M}$ is zero.

Let us define a penetration depth

$$\delta_\kappa = (2\kappa/\omega)^{1/2} \tag{2}$$

where $\kappa = K/\rho c_p$ is the thermal diffusivity, K being the thermal conductivity and ω the angular frequency of the piston motion. In the limit when the Prandtl number is zero (zero viscosity) and δ_k is small compared to the distance between solid walls, the energy flow can be calculated simply, with the result that $\bar{M} \propto v_0 P_0 \Pi \delta_k$, the proportionality constant being of order unity. From (1) the area A is replaced by the product of the perimeter (at z) Π and the penetration depth δ_κ, the velocity v by the velocity amplitude (at z) v_0, and the quantity $\rho c_p \delta T$ by P_0, the amplitude of the dynamic pressure variation. (For ideal gases $(\partial T/\partial p)_s = (\rho c_p)^{-1}$.) Expressing v_0 in terms of the compressibility, the distance $|z|$ from the closed end, the angular frequency ω, and P_0, and putting in the numerical constants valid for ^{4}He gas, one obtains

$$\bar{H} = \frac{1}{4}\omega \delta_\kappa p \left(\frac{P_0}{p}\right)^2 \Pi z \quad . \tag{3}$$

The energy flow depends on both perimeter, or surface area per unit length, and distance from the closed end. For a geometry such as that of figure 5b, where a set of plates has been inserted into the cylinder, the dependence of energy flow on $|z|$ is as shown. Moving away from the closed end there is a sudden increase where the plates begin, a steady increase over the region of the plates, and then a sudden decrease where the plates end. $\bar{H}$ does not drop to zero outside the plates, as of course work is being performed on the gas in the plates and beyond. (There are no *sudden* changes in $\bar{H}$, but rather rapid changes over the distances of the same order as the amplitude of the fluid motion.) These are exactly the qualities expected from the experimental data shown in figure 2. Absent the symmetry breaking geometrical change, only heat rejection to the second medium would result.

If the above is somewhat unfamiliar, then for purposes of understanding consider an analogous magnetic system. Imagine a magnetic body with a system of spins (the primary medium) and a lattice (the second medium) coupled by a thermal resistance. Let the magnetic body move in a sinusoidal fashion at angular frequency ω in an inhomogeneous field so that temperature changes will result. If τ is the spin-lattice relaxation time, then there is no *average* heat consequence either in the isothermal ($\omega\tau \ll 1$) or the adiabatic ($\omega\tau \gg 1$) limits. But in between, for $\omega\tau \sim 1$, we know that the magnetization and field will be phased so that, on the average, heat is rejected to the lattice. There is, however, no useful thermodynamic result, only the direct conversion of work into heat at the lattice temperature. This corresponds to the situation in the cylinder of figure 5a before the plates are introduced. But now suppose that the "spins" and the "lattice" are in *different* bodies coupled by some thermal resistance and let the symmetry between them be broken by, say, limiting the size of the "spins." Furthermore, let the "spins" be moved in an oscillatory fashion with respect to the "lattice" in the inhomogeneous magnetic field. Then the end of the "spins" in the weaker field will cool while the end in the stronger field will heat. A useful thermodynamical effect has been produced which will be optimal in the vicinity of $\omega\tau \sim 1$.

The above considerations were based on a starting condition of uniform temperature with $(grad_z T) = 0$. Suppose we let the engine run continuously so that a steady temperature difference is achieved. In a practical engine this temperature difference will be limited by, for example, external heat flows, internal heat generation due to viscosity, longitudinal conductivity of the various thermodynamic media and of the container, and heat transport due to natural or forced convective motions of the fluid. The ideal fluid has no viscosity but does have thermal conductivity; it has zero Prandtl number. In a typical fluid the effect of viscosity need not be overwhelming, since the refrigeration effect at the end of the plates is proportional to the plate length while the overall viscous heating effect is proportional roughly to the cube of the plate length; I think that the two can be adjusted appropriately by choice of geometry and frequency so that viscous effects will not be large. For the ideal fluid in the ideal engine in the absence of external heat flows, the temperature gradient will continue to develop as acoustical power is applied until heat flow to the second medium drops to zero.

To understand this situation, consider a particle of fluid oscillating simple harmonically with amplitude x_0 at a point situated a distance z from the closed end of a cylinder with one open end. As the particle moves toward the closed end its temperature rises; as it moves away its temperature falls. Initially $(grad_z T)$ is zero, and a given fluid particle tends to give heat to the walls on compression and take heat from the walls on expansion. But as the engine runs and T increases toward the closed end, the temperature difference driving the heat flow to the second medium decreases. Finally at some limiting temperature gradient in the wall $(grad_z T)_{lim}$ the heat flow to the wall will reach zero and we will have the condition

$$x_0|(grad_z T)|_{lim} = \delta T_0 \quad , \tag{4}$$

where x_0 is the amplitude of the particle motion and δT_0 is the amplitude of the adiabatic temperature change of the fluid particle. For adiabatic processes in an ideal monatomic gas $\delta T_0/T = -(2/3)\delta V/V$ and $P_0/p = -(5/3)\delta V/V$.

Since the pressure and pressure changes are uniform even when the temperature of the gas is non-uniform, the amplitude of the fractional volume change $\delta V/V$ of a given fluid particle at z is the same as the amplitude of the total volume change $A(z)x_0/V(z)$ beyond z, where $A(z)$ is the cross-sectional area at z and $V(z)$ the volume included between the point z and the closed end. As a consequence, from (4)

$$|grad_z T|_{lim} = \frac{2}{3} \frac{T}{V(z)/A(z)} \quad . \tag{5}$$

This is equivalent to

$$T(z)V(z)^{2/3} = \text{constant} \tag{6}$$

independent of pressure amplitude, pressure and frequency. This result, which was derived by Gifford and Longsworth [9] in a similar way, gives a reasonable accounting of the results in figure 3. The "rigid" temperature distribution found there is essentially independent of the various experimental parameters and seems to be correlated with the adiabatic equation of state of the gas.

Heat flows into plates at $(grad_z T) = 0$ have been studied extensively by us and compared with a thermoacoustical theory by Rott [14], with excellent agreement where comparison is appropriate. The behavior of the engine in the "adiabatic limit" is reasonable. We have not yet dealt quantitatively with the qualities of the engine in the presence of both substantial heat flows and a temperature gradient, nor have we attempted prime mover operation. Hence we are not yet able to describe the engine's efficiency under prescribed conditions. Since for a prime mover heat must be absorbed at the hot end and rejected at the cold end, the prime mover and refrigerator functions of this engine are separated by the temperature distribution corresponding to the adiabatic limit. Only if the temperature gradient exceeds that given by equation (5) will the gas on compression be cooler than its immediate surroundings so that heat can be absorbed. For refrigerator operation heat transfer to the hot reservoir and from the cold reservoir can be *via* regions of constant temperature. But for prime mover operation some of the overall temperature difference available must be used up to produce suitably large temperature gradients to achieve the necessary direction of heat transfer. This could be looked upon as an undesirable quality, but then it could also mean greater simplicity of the necessary heat exchangers. And simplicity after all is one of the most important practical characteristics of these engines.

4.   GENERAL REMARKS

The engine as we have described it really does not depend on acoustical concepts. But since simplicity is such an important quality of intrinsically irreversible engines, and since acoustical techniques are so simple and elegant and the relatively high frequency useful, future study of an acoustical engine is very desirable. But there are problems to be overcome. One is concerned with viscous heating. To reduce viscous heating to a small value relative to the thermodynamic cooling effect, it is essential that the second medium be placed near a pressure antinode and that the ratio of the length of the second medium to the radian length of the acoustical wave be small. Even then the heating effect of the remainder of the open resonant tube is excessive. A way must be found to deal with this problem. A second problem is associated with the effects of acoustical streaming [3]. In this well-known effect an oscillatory flow in the presence of boundaries leads to a steady circulational or forced convective flow. As a consequence we can expect that $|(grad_z T)|$ will be decreased. I suspect this is the reason we have not been able to achieve the full adiabatic temperature difference in experiments like that which produced figure 3. No doubt even more interesting problems will appear as we go more deeply into this subject.

The general configuration of the second medium that we have used, as in figure 5b, could be changed by letting it extend fully to the closed end of the tube. In that case one might want to use the last bit to make contact with the hot reservoir. But suppose that we chose instead to lengthen the open space at the end of the tube. Now in the way we have operated the engine, if τ_p is the thermal time constant in the region of the plates and τ_0 is the thermal time constant in the open region, then we normally have $\omega\tau_0 \gg 1$ and $\omega\tau_p \simeq 1$. Processes in the open space are essentially adiabatic while in the region of the plates thermal contact is just poor. The energy flow curve of figure 5b applies. However, if for the same geometry we decrease ω until the thermal penetration depth in the open tube is comparable to the radius, then we will have $\omega\tau_0 \sim 1$ and $\omega\tau_p \ll 1$. The time-averaged energy flow in the open space goes up; that in the plates decreases toward zero. The *signs* of the discontinuities in $\tilde{H}$ at the ends of the plates change; the regions of heating and cooling are interchanged. This latter condition is the one that probably applied to the experiments of Gifford and Longsworth. In any case, more than geometrical configuration is important in these engines.

I have said before that I think the concepts presented here are quite general, not limited to any particular working substance. Whether or not they are used in any practical case probably will be related to the importance of simplicity. The essential requirements seem to be (a) both primary and secondary thermodynamic media, (b) relative motion between primary and secondary media, (c) an irreversible process, preferably thermal, to provide suitable phasing between motion and temperature changes, and (d) a breaking, either continuously or discontinuously, of the relative symmetry between primary and secondary media. The primary medium must be thermodynamically active, that is, its entropy at constant temperature should depend sufficiently strongly on some externally applied parameter such as pressure or magnetic field. The secondary medium should have an adequate heat capacity.

The general principle which determines the steady state operation of this type of engine has not been stated. It probably should be consistent with the principle of minimum rate of production of entropy [15], at least for the present experiments. Thus, for the ideal engine with no external heat input and where the thermodynamic effects are due to thermal conduction and viscous effects are negligible, as the engine operates and the temperature gradient develops the rate of production of entropy decreases. In the dynamic equilibrium state, when heat transfers to the second medium cease, the rate of production of entropy has been reduced to zero.

The intrinsically irreversible heat engine is certainly a conceptually and scientifically interesting object. Whether or not it will lead to the hoped-for solutions to practical problems in low temperature technology has yet to be shown. The limiting temperature distribution of equation (6) should apply provided that sources of heat can be eliminated and that $|(grad_z T)|$ is not so large as to lead to an excessive heat transfer within the second medium. From equation (6) factors of 2, 4 and 8 reductions in temperature require, respectively, factors of 2.8, 8 and 27.6 in volume ratio. It is also possible in principle to provide stages of temperature reduction, as is common in cooling using more conventional cooling methods. Or one might combine an irreversible acoustical engine with a Joule-Thomson engine or with, say, an intrinsically irreversible magnetic engine. For the present, a detailed scientific understanding of intrinsically irreversible heat engines and the problems and peculiarities of their acoustical manifestations must precede practical developments.

Acknowledgments

I wish to acknowledge here the many important contributions to this work made by my colleagues Greg Swift and Al Migliori, starting at the earliest stage. I also wish to acknowledge the work of Tom Hofler, with whom the measurements shown in figure 3 were made. I am indebted to the U.S. Department of Energy for the support of this work, both through the Basic Energy Science Program and through Institutional Supporting Research and Development funds at the Los Alamos National Laboratory.

References

[1]   S. Carnot, *Reflections on the Motive Power of Fire, and on Machines Fitted to Develop that Power*, (1824), translated and edited by

R. H. Thurston, edited by E. Mendoza (Peter Smith Publisher, Inc., Gloucester, Mass., 1977), p. 59.

[2]   In this paper the term "heat engine" refers to any apparatus in which the concepts of heat, work, energy and temperature are important. Heat engines may be classified according to their function. A heat engine whose function is to transfer heat from one reservoir to another by the performance of external work is called a refrigerator or heat pump. A heat engine whose function is to perform external work by absorbing heat from one reservoir and rejecting heat to a lower temperature reservoir may be called a prime mover. Here I am mainly concerned with refrigerators.

[3]   F. L. Curzon and B. Ahlborn, *Am. J. Phys.* **43**, 22 (1975).

[4]   B. Andresen, R. S. Berry, A. Nitzan and P. Salamon, *Phys. Rev.* **A15**, 2086 (1977). References 3 and 4 introduce the effect of irreversibility on optimizing power out from intrinsically reversible engines.

[5]   G. Walker, *Stirling Engines* (Clarendon Press, Oxford, 1980). This is primarily an engineering text. It contains some interesting historical material.

[6]   Lord Rayleigh, *The Theory of Sound* (Dover Publications, New York, 1945), Vol. II, 2nd edition, p. 230.

[7]   K. W. Taconis, J. J. M. Beenakker, A. O. C. Nier and L. T. Aldrich, *Physica* **15**, 733 (1949).

[8]   W. E. Gifford and R. C. Longsworth, *Adv. in Cryogenic Eng.* **10**, 69 (1965).

[9]   W. E. Gifford and R. C. Longsworth, *Adv. in Cryogenic Eng.* **11**, 171 (1966).

[10]   R. C. Longsworth, *Adv. in Cryogenic Eng.* **12**, 608 (1967).

[11]   W. E. Gifford and G. H. Kyanka, *Adv. in Cryogenic Eng.* **12**, 619 (1967). References 8–11 give a good general discussion of the phenomena which occur in a pulse tube refrigerator, though I am inclined to explain the phenomena from a somewhat different viewpoint than the authors.

[12]   P. H. Ceperley, *J. Acoust. Soc. Am.* **66**, 1508 (1979).

[13]   J. Lighthill, *J. of Sound and Vibration* **61**, 391 (1978). This is an excellent general reference to the phenomenon of acoustical streaming.

[14]   N. Rott, *J. Appl. Math. and Phys.* **25**, 619 (1974). Calculation of thermoacoustical effects for $(grad_z T) = 0$ for any Prandtl number in the approximation that the penetration depths are small compared to the spacings between solid walls.

[15]   G. Nicolis and I. Prigogine, *Self-Organization in Nonequilibrium Systems* (John Wiley, New York, 1977), p. 42.

CHAPTER III

Superconductivity

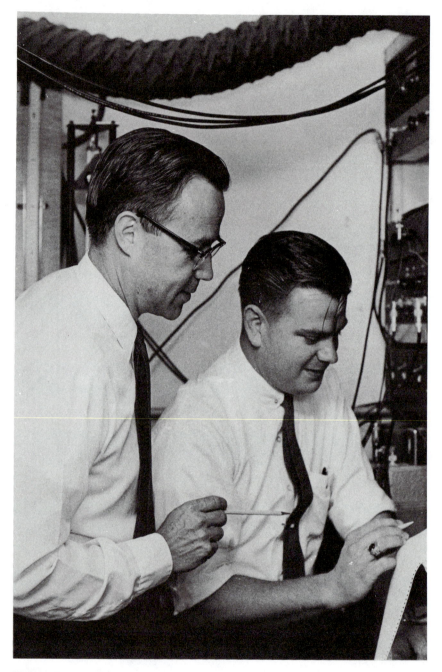

William Fairbank and Bascom Deaver, Jr., with quantized flux apparatus, Stanford University, 1961.

Superconductivity: Introduction

Bascom S. Deaver, Jr.

As a graduate student, William Fairbank had become involved with fundamental questions about superconductivity, and his thesis research [1] led to subsequent applications of superconducting microwave cavities to superstable oscillators and to superconducting electron accelerators (see chapter IV). As Walter Gordy has described in his paper in chapter I, he and Fairbank had collaborated at Duke on an experiment to measure the superconducting energy gap. Also at Duke, Fairbank became familiar with London's concept of superconductivity as a macroscopic quantum effect involving long range order in momentum and a wave function describing many particles in the same quantum state having a phase that was coherent over macroscopic distances. Fairbank devised experiments to test some of London's predictions arising from these concepts, particularly fluxoid quantization and the magnetization induced in rotating superconductors, and undertook those experiments immediately after arriving at Stanford University. The subsequent experimental verification of fluxoid quantization gave strong support for London's ideas about macroscopic quantum effects and also, through the appearance of the factor 2 in the quantum, gave direct evidence for pairing of electrons in superconductors in agreement with the then-newly-developed microscopic theory of superconductivity of Bardeen, Cooper and Schrieffer.

C. N. Yang and Nina Byers showed that fluxoid quantization in superconductors can be derived from general principles of quantum mechanics without additional assumptions. In the first paper of this chapter of *Near Zero*, Yang describes the fundamental connection between fluxoid quantization and the evolving understanding of the concept of phase in quantum mechanics and also its role in his ideas about off-diagonal long range order [2]. Felix Bloch used the idea of an order parameter of the Ginzburg-Landau type to discuss fluxoid quantization [3], he clarified ideas about the momentum ordering in superconductors [4], and he showed the very basic connection between fluxoid quantization and the Josephson effects [5]. He comments briefly on these topics in the next paper in this chapter.

Bascom Deaver describes in his paper the original fluxoid quantization experiments at Stanford and reviews many other experiments and phenomena involving the ubiquitous quantity ϕ_0, the fluxoid quantum, in superconductivity. Many of the subsequent experiments involve Josephson effects and SQUID's. There is an interesting interplay between the fundamental studies of fluxoid quantization in rings containing Josephson junctions (SQUID's) and the use of supersensitive electronic devices based on SQUID's to investigate fundamental properties of superconductors.

Fairbank recognized immediately that an implication of the existence of fluxoid quantization is the possibility of producing a zero magnetic field region within a superconducting cylinder or shell entirely in the zero quantum state. Blas Cabrera describes the progress that has been made in achieving nearly zero field regions and some of the fundamental experiments that can be carried out in them. He also describes experiments to determine the ratio h/m from measurements on rotating superconducting rings with such precision that he can detect relativistic corrections to the electron mass in superconductors. At the Near Zero conference Walter Gordy noted that if these corrections are well enough characterized then this measurement may indeed provide one of the most precise determinations of h itself.

As a graduate student working with Fairbank, Allen Goldman performed the first experiments at Stanford University on the Josephson effects. In his *Near Zero* paper he describes the use of the Josephson effects as a probe to measure the spatial and temporal variations of the order parameter in superconductors. Only quite recently has the theory been extended to describe the kind of nonequilibrium phenomena which these experiments can explore.

When he was a graduate student working with Fairbank, Edward Wilson undertook experiments to explore the nonlocal behavior of superconductors as evidenced by the magnetic field penetrating within a hollow superconducting shield. He describes those experiments here.

As a colleague of Fairbank's at Stanford, W. A. Little often discussed with him London's concept of superconductivity and also some of London's speculations about long range order in organic and biological molecules. Here Little describes the stimulus that these ideas provided for his thinking about and predictions of the possibility of organic superconductors with high transition temperatures. He reviews some of the developments which have ensued in the physics of low dimensional materials and the remarkable progress that is being made in achieving superconductivity in organic materials.

References

[1] W. M. Fairbank, Ph.D. Thesis, Yale University (1948); *Phys. Rev.* **76**, 1106 (1949).

[2] Yang comments on these topics more extensively in C. N. Yang, *Selected Papers 1945–1980 with Comments* (W. H. Freeman Co., San Francisco, 1982).

[3] F. Bloch and H. E. Rorschach, *Phys. Rev.* **128**, 1697 (1962).

[4] F. Bloch, *Phys. Rev.* **137A**, 787 (1965).

[5] F. Bloch, *Phys. Rev. Lett.* **21**, 1241 (1968); F. Bloch, *Phys. Rev. Lett.* **B2**, 109 (1970).

Flux Quantization,
A Personal Reminiscence

Chen Ning Yang

In 1948–1949 Onsager [1] speculated that in a superfluid there might be quantized vortices with a quantum of circulation around the vortex line given by

$$\oint pdq = nh \quad .$$ (1)

Very soon after that London discussed [2] the possibility that in superconducting rings the magnetic flux might be quantized. It may be useful to quote from this discussion where flux quantization appeared in the literature for the first time:

> We note that in order for ψ to be a single valued function as required by quantum mechanics, it is necessary that the moduli of χ fulfill a kind of quantum condition:
>
> $$\langle \chi \rangle = \oint = \bar{p}_s \cdot ds = Kh$$ (2)
>
> where K must be an integer. This means that there exists a universal unit for the fluxoid:
>
> $$\Phi_1 = hc/e \simeq 4 \cdot 10^{-7} \text{gauss} \cdot \text{cm}^2 \quad .$$ (3)

London's discussions were very interesting but confusing. In particular they seem to be based on the requirement of a single valued wave function and are independent of whether the ring is superconducting or not. Indeed the confusion led Onsager to make the following comments [3] in the session on superconductivity at the 1953 Kyoto–Tokyo international conference:

> Now the question is: at what stage in this sequence of operations does the flux become integral? A tenable theory is that as soon as we have trapped it, it becomes integral. This line of thinking is akin to Dirac's suspicion that when nature has an opportunity to make such a unit, (g), she would not fail to take advantage of that opportunity. Alternatively, we may observe that the electromagnetic field, which always acts on electric charges e, may have such a structure that it can only be probed by charges e. The philosophy seems acceptable; but if it is right we still have to find the correct mathematical formulation. In a recent conversation, Professor Frohlich brought up the question whether such a principle might arise out of cooperation in many particle systems. If we have to invoke not just the cooperation between the electrons in a superconductor but a conspiracy between all the elementary charges in the universe, then we are still dealing with a fundamental law of nature which we do not yet understand.

Onsager was evidently groping for the meaning of flux quantization in terms of possible new structures of electromagnetism. After the 1961 experimental discovery of flux quantization, Onsager wrote a short paper [4], in which he referred back to these comments in the following sentence:

> London's result inspired the suggestion that the quantization of flux might be an intrinsic property of the electromagnetic field.

He then argued that the suggestion was untenable:

> The notion that the electromagnetic field itself might be subject to a similar condition seems untenable now, for singly charged bosons exist (deuterons) and a condition imposed on the electromagnetic field ought to be equally compatible with all charged particles.

Indeed, whether flux quantization in superconducting rings, if found, is a new fundamental law of nature or not was a very confusing question.

I was a participant at the 1953 conference but that was before I had serious interests in superfluidity or superconductivity, and I don't remember even having attended the session where Onsager made those comments. Eight years later, in the spring of 1961, through the arrangements of Felix Bloch and Leonard Schiff, I spent a couple of months at Stanford. The department was then still in the old Spanish styled building which I thought was very beautiful but not very convenient. Soon after my arrival William Fairbank told me that with Bascom Deaver he was looking for quantized magnetic flux in superconducting rings. He asked me if they did find quantized flux whether or not that would be a new principle of physics. I could not answer.

Bill's question stimulated my interest, and with Nina Byers I began to work on this subject. We soon got thoroughly confused. A few weeks later, Bill and Bascom said they had found quantized flux. Bill showed me a preliminary version of figure 1, which at that time had much fewer

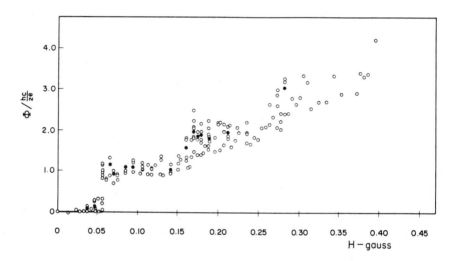

FIGURE 1. Trapped flux (in units of the flux quantum) in a hollow tin cylinder as a function of the magnetic field in which the cylinder was cooled. Open circles are individual data points. Solid circles represent the average of all data points at a particular value of applied field.

data points. (This diagram later appeared in figure 1 of their paper [5] in the *Physical Review Letters*.) Bill claimed that the experimental data in this figure showed steps in quantized units. I drew a straight line through the points and said that I was not convinced of any quantized steps at all!

Around that time, Nina and I finally began to understand that there might be quantization of flux in superconductors just because of the already known physical principles. We were excited about this possibility and spent considerable time working out the details. In the meantime, Bill and Bascom made measurements on a second specimen and produced experimental results which convinced even theorists like Nina and me that they had discovered quantized magnetic flux.

Their beautiful experiment was published in the *Physical Review Letters* referred to above. Immediately following it was a paper by Nina Byers and me giving our understanding of flux quantization. Allow me to quote from the beginning paragraphs of this paper.

The previous discussions leave unresolved the question whether quantization of the flux is a new physical principle or not. Furthermore, sometimes the discussions seem to be based on the assumption that the wave function of the superconductor in the presence of the flux is proportional to that in its absence, an assumption which is not correct. We shall show in this Letter that (i) no new physical principle is involved in the requirement of the quantization of magnetic flux through a superconducting ring, (ii) the Meissner effect is closely related to the requirement that the flux through any area with a boundary lying entirely in superconductors is quantized, and (iii) the quantization of flux is an indication of the pairing of the electrons in the superconductor.

And from a paragraph near the end:

It is interesting to notice that the existence of the variation of the energy levels of the electrons in P with the flux Φ, even when there is no magnetic field in P, is based on the same principle as the experiment proposed by Aharonov and Bohm.

Many years later I discussed with Bill why he was able to see steps in their first result while I was not. His answer was very revealing to me. He said that to an experimental physicist each point in his plot has a personality. He gives it, implicitly, almost subconsciously, a weight. The plot thus says many more things to him than to a theorist.

The discovery of quantized magnetic flux was a milestone in the physics of superconductivity. For me it led to my searching for a precise meaning of phase coherence, especially in fermion systems. That produced a later paper in which I introduced the term ODLRO (off-diagonal long range order), a paper of which I have been very fond.

In preparing my talk for the Near Zero conference, I thought about the concept of "phase". It is really a fundamental concept with many, many facets. The following is a partial list of some milestones in our evolving understanding of its meaning and importance:

Bose-Einstein Condensation	(1924)
Dirac's Monopole	(1931)
London's Equation	(1935)
BCS Theory	(1956)
Aharonov-Bohm Experiment	(1959)
Quantized Flux	(1961)
Josephson Effects	(1962)
Quantized Hall Effect	(1981)

Furthermore the concept of a *gauge field*, which is so popular in elementary particle physics today, is in fact misnamed. It should have been called a *phase field*. The change from "gauge" to "phase" historically [6] required the insertion of an $i = \sqrt{-1}$, as first pointed out by London. In retrospect, it certainly was not an accident that London was also the person who first speculated about quantized flux!

At the Near Zero conference at Stanford in March of 1982 I directed to Bill Fairbank these concluding remarks:

> Bill, it was in 1958, I believe, that I first met you when I went to Duke for a visit, and you showed me your unbelievably accurate experiment on the specific heat of He near the λ-point. I have, ever since that time, been an admirer of yours, not only for your beautiful and elegant experiments, not only for your pioneering achievements in so many developments, not only for your independence in thinking, in clear physical terms, about fundamental phenomena, but also for your stubborn refusal to be brow-beaten by theorists. I wish you continued success in future years in bringing to all of us great excitements in our explorations of the secrets of nature.

Acknowledgments

This work was supported in part by the National Science Foundation under grant PHY 81-09110.

References

[1] L. Onsager, remark at a low temperature conference at Shelter Island, 1948 (unpublished); see F. London, *Superfluids* (Wiley, 1954), Vol. II, footnote 10, p. 151. See also V. L. Ginsberg, *Dok. Akad. Nank.* **69**, 161 (1949), and L. Onsager, *Nuovo Cim. Suppl.* **6**, No. 2, 249 (1949).

[2] F. London, *Superfluids* (Wiley, 1950), Vol. I, p. 152,

[3] L. Onsager, in *Proc. International Conference of Theoretical Physics* (Science Council of Japan, 1954), p. 936.

[4] L. Onsager, *Phys. Rev. Lett.* **7**, 50 (1961).

[5] B. S. Deaver, Jr., and W. M. Fairbank, *Phys. Rev. Lett.* **7**, 43 (1961).

[6] F. London, *Z. Phys.* **42**, 375 (1927). For a short history, see C. N. Yang, *Annals of N. Y. Acad. Sci.* **294**, 86 (1977). See also C. N. Yang, "Hermann Weyl's Contribution to Physics," to appear in a book on Hermann Weyl, edited by K. Chandrasekharan (Springer).

Flux Quantization: Retrospective

Felix Bloch

Like any old man I like to reminisce about the past, and it gives me a special pleasure to look back on flux quantization because it was one of the first among many other links between William Fairbank's work and my own interests.

Before we even knew each other, however, there already existed a link between us in the person of Fritz London, and my retrospective would not go back far enough if I did not speak about him first. I met him when I was a student in Zurich and he worked as a postdoctoral student with Walter Heitler on their famous theory of covalent bonds. The events in Europe brought both London and me some years later to the United States, where I landed at Stanford and he at Duke University. Bill has often spoken to me about his fond memories of London as a colleague, and I fully shared his admiration of this man, both as a person and as a physicist. After London's untimely death, Duke established the London Memorial Lecture, which I was invited (surely thanks to Bill) to deliver in 1958. I was a most ungrateful guest of Duke because I used the opportunity to pry Bill loose from there in order to have him with us at Stanford. He might not have paid any attention to my seducement had it not been for the far too favorable opinion of me which London had conveyed to him; and certainly there were more weighty reasons for him to fall for the temptation. Be that as it may, I was very happy when he came here to join the research in low temperature physics that had been started one year earlier by W. A. Little.

Fairbank came with very clear ideas about what he was going to do, which brings me once more back to Fritz London. Together with his brother, London had long before developed the phenomenological theory of superconductivity, and with fine intuition had recognized it in his own words to be "a quantum phenomenon on a macroscopic scale." Guided by this recognition, he suggested already in 1948 that the magnetic flux through a superconductive ring might be quantized in units hc/e. To most physicists, however, including myself, his arguments did not appear to be very convincing, because he based them on the wave function of a single electron running freely around the ring. Later one would have called it, with Landau and Ginzburg, "the order parameter," and London's idea might have been more widely acceptable under that name.

Well, Bill Fairbank was among the few who were open-minded enough to take it seriously, so seriously, in fact, that he decided to look for the effect. And that is what he had planned and started just after he came to Stanford. When he told me how he would proceed, I was still doubtful about London's suggestion, and felt that it took a lot of courage to undertake such a delicate experiment on the basis of a mere hunch. Now we all know that Bill is a very courageous man, and this time his courage seemed to border on recklessness. He went so far as to risk the future career of a promising graduate student, Bascom Deaver, by proposing the experiment to him for his Ph.D. thesis. I am not sure that Bascom fully realized what he was getting into, but I am sure he has no regrets about having accepted the proposal.

In fact, it did not take very long before Bill showed me the first data, which already looked pretty good. But in my customary skeptical attitude I remained unconvinced until they had collected more data and there could be no doubt anymore about the reality of quantized flux. Fortunately, this did not cause an appreciable postponement of the publication; an independent and simultaneous investigation by Doll and Näbauer in Munich had led to the same conclusion, so that a major delay would have given them sole priority.

Even in a field which has been as rich in surprises as superconductivity, the existence of quantized flux ranks as a major discovery. It was the first and most direct demonstration of a quantum effect on a macroscopic scale or, in other words, of a long range order that distinguishes the superconductive from the normal state of a metal. Contrary to London's prediction, however, the flux was found to be quantized not in units hc/e but $hc/2e$. This says clearly that the carriers of electricity in a superconductor have twice the elementary charge, and thereby confirms the presence of Cooper-pairs, which is one of the cornerstones in the theory of Bardeen, Cooper and Schrieffer.

A few weeks after the discovery of flux quantization Frank Yang paid one of his frequent visits to Stanford. Like myself, he was greatly impressed and also felt that such a striking fact called for more than the suggestive arguments of London. But, unlike myself, he was able to show in a paper with Nina Byers that it could be understood on entirely general grounds by ultimately tracing it back to gauge invariance and thus to a fundamental feature of quantum mechanics.

Bascom Deaver's article describes some small and not so small steps to which flux quantization has since been leading. I hope he will forgive me if I intrude somewhat on his topic in telling about two later visits by Yang. The first was shortly after the appearance of Josephson's papers on his new effect. Yang told me that he could not understand it, and asked whether I could. In all honesty I had to confess that I could not either, but we made a deal that whoever of us first understood the effect would explain it to the other. It took me a long time before I finally realized that by considering a barrier in a superconducting ring, one could interpret the Josephson effect by essentially the same general considerations which Byers and Yang had introduced in their treatment of flux quantization. So, in keeping with our deal, I told Yang about it on his next visit and he was very pleased. But then I could not help saying to him, "Frank, you should not only have understood the effect right away, but you should have predicted it before Josephson."

With this little story I end my retrospects and now look forward to the prospects of flux quantization to be presented in the next paper.

Fluxoid Quantization: Experiments and Perspectives

Bascom S. Deaver, Jr.

1. BACKGROUND AND INITIAL QUANTIZED FLUX EXPERIMENTS

In the fall of 1959 William Fairbank joined the faculty at Stanford University. The previous spring, having completed two years of course work as a part-time graduate student, I had gone to George Pake, who was then a professor at Stanford and whose friendship and judgement I valued highly, and asked him what kind of research I should pursue. He said I should work for Fairbank, who would be arriving in a few months, and that he would write to him and arrange it. And so it was that, as his first graduate student at Stanford, I came to be sitting in Fairbank's office that fall as he described to me a series of incredible experiments he proposed to undertake.

One was an experiment to measure the quantized states of liquid helium in a rotating container, an experiment that George Hess completed a few years later and describes in detail in his paper entitled "Rotation of Superfluid Helium" in this book. Another was a search for quantized vortex lines in rapidly rotating liquid helium using optical techniques, an investigation that was subsequently undertaken by Pierce Webb. Still another was a measurement of the magnetization of a rotating superconductor which was predicted by Fritz London. Morris Bol carried out this experiment and observed what is now usually called the "London moment." (See section 22 of this paper, and the paper entitled "Rotating Superconductors and Fundamental Physical Constants" by B. Cabrera, S. B. Felch and J. T. Anderson.) Fairbank described an experiment to investigate another prediction of Fritz London, that a quantity London called the "fluxoid" would be quantized. This experiment I found enormously appealing and I asked to be allowed to work on it.

London did not live to know about the microscopic theory of Bardeen, Cooper and Schrieffer or of the existence of electron pairs, but he already understood and described in an article [1] in *Physical Review* in 1948 and in his beautiful monograph [2] on superconductivity the essential, universal characteristic of the superconducting state, namely the existence of a many particle condensate wave function $\Psi = \psi e^{i\chi}$ which maintains phase coherence over macroscopic distances. The most basic implication of the existence of a phase factor is exhibited in the simple case of a superconducting ring for which the single valuedness of Ψ requires that the phase return to itself on going once around the circuit, thus

$$\oint \nabla \chi \cdot ds = n2\pi \quad n = 0, 1, 2 \ldots \quad . \tag{1}$$

A consequence of this requirement is readily exhibited by calculating the current density in the state Ψ, which is

$$\mathbf{j}_s = \frac{n_s e}{2m}(\hbar \nabla \chi - 2e\mathbf{A}) \quad , \tag{2}$$

where we have already included the fact that the supercurrent $\mathbf{j}_s$ is carried by pairs of electrons with density $|\Psi|^2 = \frac{n_s}{2}$, where n_s is the number density of superconducting electrons and $\mathbf{A}$ is the magnetic vector potential. Using equation (2) to express $\nabla \chi$ in terms of $\mathbf{j}_s$ and $\mathbf{A}$ in equation (1) leads to the fact that

$$\oint \frac{m}{n_s e^2} \mathbf{j}_s \cdot ds + \oint \mathbf{A} \cdot ds = n\Phi_0 \quad , \tag{3}$$

where $\Phi_0 = \frac{h}{2e}$. This is the quantity London called the "fluxoid" and whose quantization he predicted [1,2]. However, without knowing that the current was carried by pairs of electrons, he anticipated a quantum that was twice as large as the observed value Φ_0.

By writing the current density in terms of the pair velocity $\mathbf{v}_s$ so that $\mathbf{j}_s = n_s e \mathbf{v}_s$, we arrive at another useful expression for the fluxoid,

$$\oint \frac{m\mathbf{v}_s}{e} \cdot ds + \oint \mathbf{A} \cdot ds = n\Phi_0 \quad . \tag{4}$$

Written in terms of the pair canonical momentum, $\mathbf{p}_s = 2m\mathbf{v} + 2e\mathbf{A}$, equation (4) becomes

$$\oint \mathbf{p}_s \cdot d\mathbf{s} = nh \quad , \tag{5}$$

in analogy to the corresponding condition for an atom. London discussed extensively the concept of superconductivity as a macroscopic quantum effect as implied by this result.

Fairbank understood London's concept of the superconducting state and his prediction of fluxoid quantization. A consequence of the London theory [2] is that in a superconductor the magnetic field $\mathbf{H}$ obeys the equation $\nabla^2 \mathbf{H} = \mathbf{H}/\lambda^2$, and thus the field $\mathbf{H}$ and the current density $\mathbf{j}$ accompanying it decay exponentially with penetration depth λ upon entering the superconductor. Fairbank recognized that for a superconducting ring whose thickness t is much greater than λ the total flux trapped in the ring would be quantized in units of Φ_0. This result follows immediately from equation (3) since for a thick ring a path can be chosen deep enough within the superconductor that $\mathbf{j}_s$ is zero. Then

$$\Phi = \oint \mathbf{A} \cdot d\mathbf{s} = n\Phi_0, \quad t \gg \lambda \quad . \tag{6}$$

It was these quantized values of trapped flux that he proposed to measure. Little did I realize as I talked to Fairbank that working with him on this experiment would be a continuing thrill of a lifetime.

There was a very large laboratory in the basement of the old Physics Corner in part of which W. A. Little and his group, including Ron Parks, Bob Johnson and Marcel LeBlanc, were already at work. Dwight Adams and John Goodkind, who came with Fairbank from Duke University, were soon to take their places there as research associates working on nuclear magnetic resonance in ^{3}He. Some rooms along one side were available for graduate students, and in one of those, along with George Hess and Pierce Webb in adjacent ones, I undertook my research. Slightly later Allen Goldman began work down the hall in a room with W. O. Hamilton, who was completing his research with George Pake. After clearing out several decades of accumulated junk, or treasures, in the room assigned to me I began work in earnest on creating an apparatus for our measurement.

Fairbank was in the lab regularly discussing the experiment. But not only this experiment. In fact, he had a constant stream of ideas, and he talked about them with everyone, Nobel laureate and graduate student alike. He had some ideas about using a superconducting sphere supported losslessly in the magnetic field of a persistent current loop as an accelerometer and as a gyroscope for some experiments in general relativity. (See papers entitled "The Stanford Relativity Gyroscope Experiment.") He talked about measuring the free fall of a single positron to observe the sign of the gravitational force on an antiparticle. (See paper entitled "Near Zero Electric Field: Free Fall of Electrons and Positrons and the Surface Shielding Effect.")

His insights into physics are amazing. His explanations of the most abstruse concepts make them seem simple, and he has an incredible ability to identify the fundamental principles in even the most complicated experiment. His energy and his enthusiasm for physics are truly stimulating, as everyone who works with him knows.

Then, as now, there was a continual flow of eminent visitors through the Stanford Physics Department. Fairbank would routinely bring his visitors to the lab to meet and talk with the graduate students—another feature of his personality which added to the excitement of working with him.

Fairbank's original concept of the quantized flux experiment was of a thousand small, identical loops encircled by a thousand small pickup coils, all in series. By trapping flux in the loops, warming through the transition temperature, and observing the voltage pulse from the pickup coil array, it should be possible to measure the flux quantum. Then came the realization that a tiny cylinder is equivalent to many loops in the amount of magnetic energy stored, this being the crucial feature for being able to detect the signal, and a cylinder is easier to couple to a pickup coil. After seeing Foner's paper [3] on a vibrating sample magnetometer, I proposed to use this technique and we agreed on it for our initial experiments. (Subsequently Francis Everitt pointed out to me that many years earlier Blackett [4] had used a technique almost exactly like the one we used.)

The measurement scheme is shown in figure 1. A tin cylinder 13 μm i.d. and 1.0 cm long was vibrated at 100 Hz with an amplitude of about 1 mm, and the emf induced in two pickup coils encircling the ends of the cylinder was measured to determine the magnetic flux in the cylinder. The cylinder was formed by electroplating Sn onto #56 copper wire (13 μm o.d.), and, to protect the Sn and to make a more robust sample, electroplating a thick copper coating on top of the Sn to make a solid rod $\sim$ 80 μm o.d. There were many days of winding coils of #56 copper wire, making cylinders, constructing lockin amplifiers and fighting noise, but finally it all came together in workable form as shown in figure 2.

Fairbank's optimism was certainly needed as we began our first runs. Everything seemed to go wrong. Samples broke, electronics malfunctioned and vacuums leaked, but he stoutly maintained that this was a good sign because one could not truly succeed without first having failed soundly.

While I was working through various experimental problems we got word of other experiments in progress to search for quantized flux. On a visit to Leiden in the summer of 1960, Fairbank visited C. J. Gorter who discussed cautiously an experiment [5] to observe quantized increments of flux

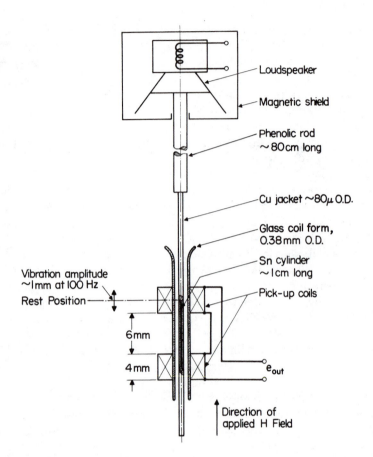

FIGURE 1. Scheme used by Deaver and Fairbank for quantized flux experiments.

emerging from a small cylinder as it was warmed through the superconducting transition temperature by observing the voltage pulses in a surrounding coil. As far as I know there were no published results from these
experiments.

There was even more cause for concern when at the March 1961 meeting
of the American Physical Society we heard in a paper by Mercereau and
Vant-Hull [6] that in their experiment there was no observable evidence
for quantized flux. They used a thin, superconducting ring about one
millimeter in diameter with pickup loops immediately above and below it.
One segment of the ring was periodically heated and cooled through the
transition temperature. The emf induced in the pickup coils was observed
as a magnetic field applied through the loop was slowly varied. Minima in
the amplitude of the emf were expected whenever the applied field produced

FIGURE 2. Apparatus used by Deaver and Fairbank. Ambient field was maintained near zero with large three-axis Helmholtz coils. Field was applied with small coils around dewar. The photograph at the beginning of this chapter shows Fairbank and Deaver with this apparatus in 1961.

an integral number of Φ_0 in the ring. This very nice technique should certainly work in principle, although for such a large ring it requires a detector with excellent energy sensitivity.

Still later that spring Fairbank received word from Joseph Reynolds at Louisiana State University that he was also undertaking an experiment to detect quantized flux. He had originated an extremely clever technique. With superconductor coated on the outside of a bismuth wire about 1 μm diameter he was planning to detect the trapped field by measuring the resistance of the wire. Because of the very large magnetoresistance of bismuth there should be plenty of sensitivity. However, about the time he was beginning his experiments the helium liquifier at LSU broke down. This, coupled with travel plans for the summer, prevented his carrying out the experiments.

An entry in my lab notebook on May 3, 1961, shows the first definitive signature of quantized flux, namely a change in sign of the induced flux when the applied field in which the cylinder was cooled was changed by a tiny amount. This was a complete change in character of the response of the cylinder to an applied field from completely ejecting all the applied flux and trapping no flux (the $n = 0$ state) to having an induced flux in the same direction as the applied flux and a discrete amount of trapped flux (the $n = 1$ state) as the applied field was changed by only a very small increment. This change was completely convincing, and Fairbank recognized it the first time it occurred.

That day began almost six weeks of frantic activity and sleepless nights. We soon accumulated enough data to know certainly that the flux was quantized. However, because of the way the sample was constructed we were not certain of the actual size of the cylinder and thus of the measured quantum. After several weeks of measurements we removed the sample, made x-ray photographs on which the Sn could be precisely located within the copper, and obtained careful dimensional measurements. Then we found that within our precision the quantized value was h/2e. Fairbank immediately recognized that the possibility of this value for the quantum had been mentioned to him by Lars Onsager at a conference in Cambridge, England, in 1959.

After many measurements on a single cylinder, we were convinced of the validity of the result. However, Felix Bloch suggested that data on a second cylinder would make the evidence much stronger.

Our second cylinder had much thinner walls than the first and the transition temperature was depressed about 0.2 K from the T_c of bulk tin by what we would now call the proximity effect. Upon seeing this reduced T_c, Fairbank recognized the cause and brought down a book in which he showed me a graph [7] of the depression of T_c of Sn films electroplated on the surface of a bulk nonsuperconducting metal. The data indicated that a layer about 0.6 μm thick in contact with the copper would not be superconducting at the temperature where we made our measurements. With this correction to the diameter of the cylinder, the data from this cylinder gave even better agreement with the value h/2e. When these measurements were completed about the middle of June 1961 the results were much more convincing and we submitted a paper to *Physical Review Letters* [8].

As shown in figure 3, we made two measurements that showed the quantized states. To discuss them it is convenient to refer to equation (6) with the total flux Φ written in terms of the induced flux in the cylinder, $\Phi_i = i_s L$, where L is the inductance of the cylinder, and the applied flux Φ_x, so that

$$i_s L + \Phi_x = n\Phi_0 \quad . \tag{7}$$

For one of the measurements the cylinder was cooled through its transition temperature T_c with a flux Φ_x applied. As T_c was passed, a current i_s

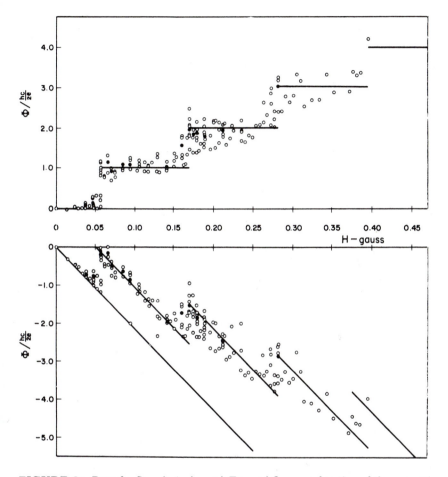

FIGURE 3a. Data for Sample 1. (upper) Trapped flux as a function of the magnetic field in which the sample was cooled below the superconducting transition temperature. The open circles are individual data points. The solid circles represent the average value of all data points at a particular value of applied field including all the points plotted and additional points which could not be plotted due to severe overlapping. Approximately 200 data points are represented. The lines are drawn at multiples of $hc/2e$. (lower) Net flux before turning off the applied field as a function of the applied field. The line through the origin is the diamagnetic calibration to which all runs have been normalized. The other lines are translated vertically by successive steps of $hc/2e$.

flowed in the cylinder to put it in the quantized state of lowest energy for that particular value of Φ_x. The lower graphs in figures 3a and 3b are plots of the values of this induced flux, which is proportional to the equilibrium magnetization of the cylinder. Those data show a sharp jump at a particular Φ_x (corresponding to $H \sim 56$ mG and 70 mG in figures 3a and 3b,

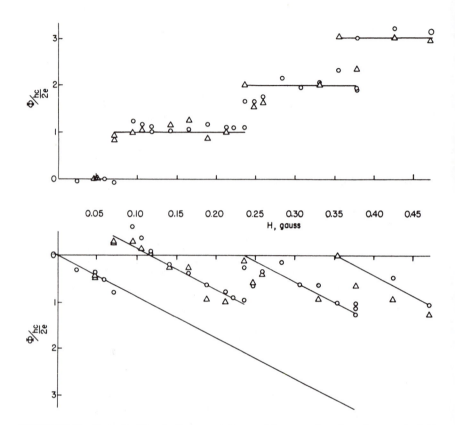

FIGURE 3b. Data for Sample 2. (upper) Trapped flux as a function of magnetic field in which the sample was cooled below the superconducting transition temperature. The circles and triangles indicate points with oppositely directed applied fields. Lines are drawn at multiples of $hc/2e$. (lower) Net flux before turning off the applied field as function of the applied field. The line through the origin is the diamagnetic calibration to which all runs have been normalized. The other lines are translated vertically by successive steps of $hc/2e$.

respectively), showing that the cylinder changes from an induced current to expel the excess flux to one that makes up the odd $\sim \frac{1}{2}\Phi_0$ to reach the nearest quantized state. It continues as a sawtooth function with period $h/2e$ as expressed by equation (7).

Then, for the second measurement, the applied flux in which the cylinder was cooled was turned off and the trapped flux $\Phi = n\Phi_0$ was measured, giving the graphs at the top of figures 3a and 3b, showing quantized flux steps.

In June 1961 there was a conference on superconductivity as part of the dedication ceremonies for the new IBM Research Center at Yorktown

Heights which Bill Little was attending. We gave him a copy of our data to discuss with people there. At that meeting Doll and Näbauer (about whose experiment we had not known) announced also that they had been making similar measurements using a torsion fiber to determine the torque on a suspended cylinder containing trapped flux. They had observed quantized steps and a value that within their precision was 0.4 h/e. The results of both our experiment [8] and theirs [9] appeared in the same issue of *Physical Review Letters* in July 1961 and have been discussed more fully in subsequent reports [10,11].

The success of these experiments gave very powerful support for the concept of superconductivity as a macroscopic quantum state as envisioned by London, and furthermore made strong contact with the microscopic theory since the appearance of the factor of two in the quantum was taken as direct evidence for electron pairs. However obvious this last point seems now, it was surprisingly difficult to understand at that time.

2. POSSIBILITY OF PRODUCING A ZERO MAGNETIC FIELD REGION

Another consequence of fluxoid quantization recognized by Fairbank, and on which he commented in the initial letter [8], is that by cooling a superconducting cylinder in a sufficiently low applied magnetic field it should be possible to create a region of exactly zero magnetic field, since the cylinder would pass into the $n = 0$ quantum state and exclude all the flux within it. This observation was the basis for an extensive effort initiated by Bill Hamilton [12] and pursued by Blas Cabrera [13,14] to achieve a nearly zero field region within a superconducting shield. These nearly zero field regions are now playing crucial roles in several fundamental experiments discussed in this book. (See paper entitled "Near Zero Magnetic Fields with Superconducting Shields.")

3. THEORETICAL DISCUSSIONS OF FLUXOID QUANTIZATION

Following the first quantized flux experiments there were numerous theoretical papers [15–42]. C. N. Yang, who during the course of our experiments was visiting at Stanford, together with Nina Byers showed [15] that quite generally quantum mechanics requires that the free energy of a collection of electrons, confined to an annular region within which there is a flux Φ, be a periodic function of Φ. For most normal conductors the amplitude of that variation is vanishingly small. However, for electrons constrained by the BCS pairing condition for a superconductor there are extremely deep minima of the free energy for values of flux equal to integral multiples of Φ_0.

Onsager [16] discussed the connection of the factor of two with the pairing of electrons. Bardeen [18] calculated the reduced value of the flux to be expected in cylinders having walls thin with respect to the penetration depth, because it is the fluxoid that is quantized and not the flux itself. Felix Bloch and H. Rorschach [33] discussed the relationship of the stability of persistent currents to fluxoid quantization.

Microscopic theory and various phenomenological theories were used. All related the factor of 2 in Φ_0 to pairing of electrons. And there were several detailed calculations of the magnetization expected for small cylinders [17–19,25,28,37,38].

4. LITTLE-PARKS EXPERIMENT

At Stanford, Bill Little was also thinking about fluxoid quantization. He realized that, as a result of the periodic variation of the free energy of a superconducting cylinder with the flux within it, the transition temperature should also be a periodic function of the flux, and he used the BCS theory to calculate the variation. By measuring the resistance at the superconducting transition temperature of tiny cylinders ($\sim 1\ \mu$m diameter) as a function of magnetic field applied along the axis, he and Ron Parks [43–46] demonstrated this periodic dependence in a very direct and elegant way. An example of their data is shown in figure 4. Experiments of this type have been extended and refined [47–51], and there have been corresponding improvements in the detailed theoretical treatments of the data, including an early paper by Tinkham [35] and a more recent one by Fink and Grünfeld [52].

5. QUANTIZED VORTICES

Michael Tinkham [35,53] used the concept of fluxoid quantization and the Ginsburg-Landau theory not only to interpret the Little-Parks experiments but also, in a more novel application of these ideas, to calculate the critical field of a thin superconducting film in a perpendicular magnetic field. He found that near T_c the screening currents flow in a uniform array of vortices each containing a single flux unit Φ_0. He noted the close connection of this behavior to that of type II superconductors for which such a vortex array had been predicted in 1957 by Abrikosov [54], and he made clearer the fundamental role of fluxoid quantization in determining the magnetic properties of superconductors.

Type II superconductors [55] are so well understood now that it is difficult to appreciate how unusual these ideas were in 1962–1963. In contrast

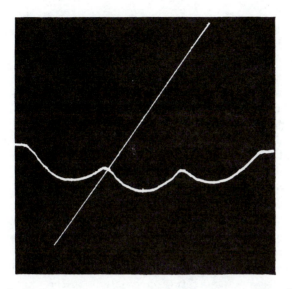

FIGURE 4. Oscilloscope traces showing variation of resistance of Sn cylinder (1.4 μm diameter, 375 Å thick) at its superconducting transition temperature as a function of magnetic field. Straight line is magnetic field variation. Zero is at the center of the picture (Little and Parks [43]).

to type I superconductors for which the coherence length ξ is much greater than the penetration depth λ and which are perfectly diamagnetic in fields up to H_c, the critical field, at which superconductivity is completely destroyed, Abrikosov identified as type II those materials for which $\lambda > \frac{1}{\sqrt{2}}\xi$. Their magnetic behavior he calculated to be quite different. A type II material is perfectly diamagnetic up to a field H_{c_1} at which flux begins to enter the material in isolated flux lines each containing a single quantum Φ_0. At the center of each line is a normal core of radius $\sim \xi$ surrounded by a current vortex with current density decreasing exponentially into the superconductor with characteristic distance λ. The field within the core is $H_{c_2} \sim \frac{\Phi_0}{\pi\xi^2}$, while $H_{c_1} \sim \frac{\Phi_0}{\pi\lambda^2}$.

As the applied field is increased, the density of flux lines increases, and when it approaches H_{c_2} the lines are arranged in a triangular lattice. When the applied field reaches H_{c_2}, the normal cores overlap and bulk superconductivity is destroyed. Flux penetration allows superconductivity to exist at much higher fields in type II than in type I materials and makes possible high field magnets.

Experimental evidence for the existence of quantized vortices was obtained shortly after Tinkham's paper by Parks and Mochel [56,57] and Parks, Mochel and Surgent [58,59], who measured the resistance of superconducting strips 1–5 μm wide near T_c in a perpendicular applied magnetic field. They interpreted a kink in the resistance *versus* field curve as the field for which a vortex just fit into the bridge.

The reality of the Abrikosov vortex array was shown most vividly in 1967 by Essmann and Träuble [60] who used iron particles to decorate the flux lines and electron microscopy to produce photographs of lattices consisting of individual quanta as well as other more complicated flux structures. By now this subject has been extensively studied by a variety of techniques, giving information about both the microscopic properties of isolated vortices [61] and flux structures and flux motion, which are the subject of a very useful monograph by Huebener [62].

The study of vortices in superconducting films is currently of great interest for the study of the Kosterlitz-Thouless transition because the very general theories of phase transitions in two-dimensional systems are now believed to apply to thin, dirty superconducting films [63]. The superconducting transition in such films is expected to be characterized by a critical temperature T_{KT} lying below the usual BCS T_c. As T_{KT} is approached from lower temperature, fluctuations in the phase of the order parameter give rise to bound vortex-antivortex pairs which at T_{KT} become unbound. The best experimental evidence for this transition is provided by ac impedance measurements and dc I-V data by Hebard and Fiory [64]. The current state of this very active topic has been reviewed by Hebard and Fiory [65] and by Mooij [66].

The motion of flux lines is central to the understanding of many phenomena in superconductors and is discussed in section 13 of this paper.

6.   QUANTIZED FLUX MEMORY ELEMENT

Returning to a more nearly chronological account of the early experiments, I would like to mention the application of quantized trapped flux to a superconducting memory element originated at Stanford by Dumin and Gibbons [67] in 1962. In this early version of a superconducting microcircuit they formed a Pb loop by sequential evaporation of layers of Pb, insulator and Pb. Above this loop and insulated from it, a Sn strip was deposited. By measuring the critical current of the Sn strip, which sensed the magnetic field in the loop, they were able to detect the quantized flux states of the loop.

7.   MODULATED INDUCTANCE MAGNETOMETER

Immediately following the initial quantized flux measurements we were very eager to have a more sensitive magnetometer for further experiments, so we began development of the device [68] shown schematically in figure 5. A flux change in the pickup coil A produced a persistent current in the

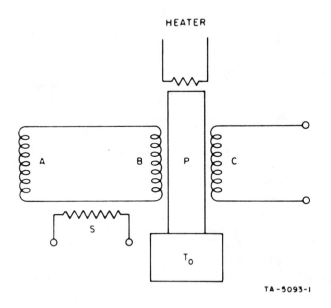

HEATER

FIGURE 5. Modulated inductance magnetometer. P is a superconducting rod or hollow cylinder, T_0 a heat sink with $T_0 < T_c$, S a heater to eliminate unwanted persistent currents in superconducting circuit A-B. Heater on P cycles it through T_c at 10–100 kHz.

superconducting circuit consisting of coils A and B. A superconducting rod within coils B and C was alternately switched between the supercon- ducting and normal states by heating and cooling it at frequencies up to 10 kHz, and the resulting modulated flux induced a voltage across the coil C proportional to the current in A-B and thus to the original flux change. Subsequently a hollow superconducting cylinder was used as the modulator because it could be cycled at ~ 100 kHz [69,70]. For small fluxes it was just as effective as a solid cylinder since the quantized flux condition required that the entire flux be ejected from within it as it switched into the superconducting state.

A magnetometer using a mechanically modulated inductance achieved by vibrating a superconducting plane near a long superconducting mean- der line was developed by Opfer [71] for use with a gravity wave detector at Stanford. An analysis of the operation of modulated inductance magne- tometers (in effect parametric amplifiers) was carried out by John Pierce [69]. His analysis is useful also in understanding the operation of SQUID magnetometers, particularly ones operating in the nonhysteretic mode (see section 12), since they can be described in terms of a modulated kinetic inductor [72].

8. SUPERCONDUCTING SUSCEPTOMETER

During the development of these new magnetometers, Fairbank pointed out the great potential for new magnetic measurements provided by the combination of an extremely sensitive magnetometer with a superconducting solenoid operated in the persistent mode to give stable applied fields and a superconducting shield to provide the necessary isolation from external magnetic disturbances. Fairbank, Bol and I worked on the concept of such a system [73], and two instruments were later constructed for measuring the magnetic properties of materials [69,70]. Although the modulated inductance detectors used with these first instruments were very sensitive, SQUID magnetometers based on the Josephson effects, which emerged soon afterwards, are much more sensitive and they were used for all subsequent systems. The Josephson effects and SQUID's are discussed briefly below, and the evolution of superconducting susceptometers to their current highly developed state and their increasing use for chemical, biological and medical applications are discussed in chapter IV.

9. EXPERIMENTS WITH
THERMALLY CYCLED CYLINDERS

In 1964 Alvin Kwiram and I were both research associates working with Fairbank. We worked together on experiments [74,75] using thermally cycled hollow cylinders, and were able to set some limit on the time required to establish the quantized state as well as obtain a much more precise value for Φ_0. The technique we used is indicated in figure 6a. With this device immersed in liquid helium, the cylinder was cyclically heated through the transition temperature by passing an alternating current through the copper-gold film. The amplitude of the alternating voltage (which was at twice the frequency of the heater current) induced in a pickup coil around the cylinder, plotted as a function of applied field along the axis of the cylinder, had the form shown in figure 6b. This curve, which is proportional to the equilibrium magnetization of the cylinder, shows the quantized flux states, and from the periodic variation and the cross-sectional area of the cylinder the fluxoid quantum was determined. The average slope of the curve is due to the Meissner effect of the walls of the cylinder.

A few years later at the University of Virginia Hugh Henry and I [76] carried out a variation of this experiment to study the quantized states of a pair of concentric superconducting cylinders with an internal heater. By plotting the amplitude of the voltage from a pickup coil around the

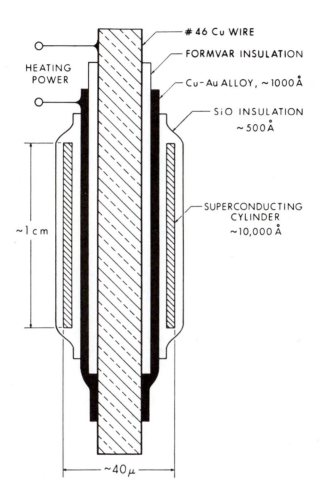

FIGURE 6a. Superconducting cylinder with coaxial internal heater. Pickup coil was wound around cylinder to detect flux changes.

pair of cylinders as a function of applied field along the axis, we measured their equilibrium magnetization. A comparison of the data with the calculated magnetization (figure 7) shows excellent agreement with the expected doubly periodic behavior of these coupled macroscopic quantum systems.

10.   THE PERSISTATRON

Immediately after the first quantized flux experiments, a topic of great interest was the possibility of inducing transitions between quantized states without heating the ring above its transition temperature. Various schemes

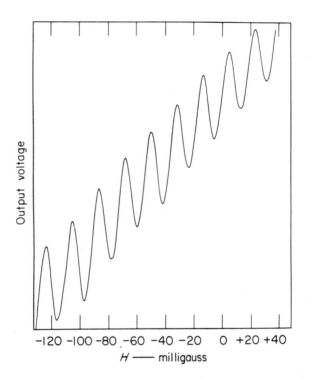

FIGURE 6b. Amplitude of voltage induced in coil around an indium cylinder of type shown in figure 6a as a function of axial magnetic field as the cylinder was thermally cycled at 20 kHz (Kwiram and Deaver [75]).

were considered, including using phonon drag by flowing an acoustic wave along an arc of the ring to push on the normal electrons to try to induce a change in the pair current. Another idea was to study the flux states of a persistatron, a superconducting memory element devised by Buckingham and Fairbank [77]. In this device a pair of current leads were attached at two closely spaced points along the periphery of a superconducting ring. A current from a pulse applied to these leads divides between the arcs of the loop in inverse proportion to their inductances until the critical current of the shorter arc is approached, whereupon the impedance of the short arc increases and the excess current flows around the larger arc. Upon removal of the applied current, a persistent current remains trapped in the ring. Reapplication of the same current pulse does not change the persistent current, but a larger pulse or one of the opposite sign does.

Before these ideas progressed very far Josephson made his famous predictions [78] and a flurry of activity began to study them. Thereafter Josephson effects provided a basic mechanism for changing the state of the ring.

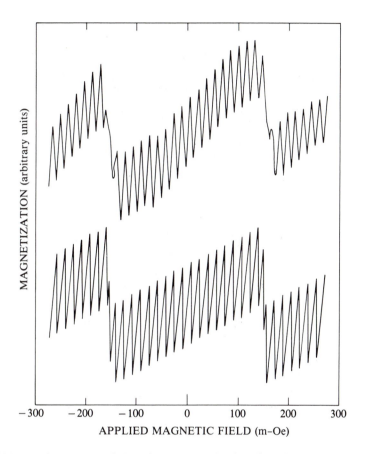

FIGURE 7. Comparison of plot of experimental value of equilibrium magnetization (top) and calculated magnetization for pair of concentric superconducting cylinders (H. L. Henry and B. Deaver [69]).

11. THE JOSEPHSON EFFECTS

The Josephson effects [78,79] provide remarkable demonstrations of the macroscopic quantum nature of the superconducting state, and superconducting rings containing one or more Josephson junctions make possible extremely precise studies of fluxoid quantization. In fact, Bloch [80] has emphasized the close fundamental relationship between fluxoid quantization and the Josephson effects.

I will not be able to discuss much of the vast amount of work that has been and continues to be done on this subject, but I will mention a few basic phenomena because of their direct relationship to fluxoid quantization and describe some applications that play such a crucial role in subsequent experiments on fluxoid quantization and in so much of the other research described in this book.

For two weakly coupled superconductors which exchange electrons through tunneling or through some barrier that prevents full phase coherence, Josephson showed that there is a pair current described by the following equations:

$$j = j_c(z) \sin \phi(z) \quad , \tag{8}$$

$$\frac{\partial \phi}{\partial t} = \frac{2e}{\hbar} V \quad , \tag{9}$$

$$\frac{\partial \phi}{\partial z} = - \left(2\mu_0 \frac{ed}{\hbar} \right) H \quad . \tag{10}$$

Here

$$\phi = \chi_2 - \chi_1 - \left(\frac{2e}{\hbar} \right) \int_1^2 \mathbf{A} \cdot d\mathbf{s} \tag{11}$$

is the gauge invariant phase difference between the macroscopic wave functions on the two sides of the junctions, V is the voltage across the junction, j_c is the critical current density, $d = \lambda_1 + \lambda_2 + \ell$ with ℓ being the separation between the two superconductors and λ_1, λ_2 the respective penetration depths, H is the magnetic field applied normal to the current and in the plane of the junction, and z is distance along the width of the junction. (See upper part of figure 8a.)

For a junction of sufficiently small width w (*i.e.*, $w \ll (h/2\mu_0 ed j_c)^{1/2}$), the current density is essentially uniform across the junction. However, with an applied field H it follows from equations (8) and (10) that the current density varies sinusoidally across the junction. Integrating this current density to obtain the total current gives for the maximum zero-voltage current

$$I_c \propto \left| \frac{\sin \pi H/H_0}{\pi H/H_0} \right| \quad , \tag{12}$$

where $\mu_0 H_0 w d = \Phi_0$. As indicated in figure 8, for values of the applied field for which the flux within the area wd is an integral multiple of Φ_0, the net current is zero. The first experimental evidence for the Josephson effects was reported in 1963 by Anderson and Rowell [81,82] who found a dc supercurrent with this characteristic behavior in an applied field.

The ac Josephson effect follows from equations (8) and (9), which give for an applied steady voltage V_0 across the junction an oscillating pair current with frequency

$$\nu_J = \left(\frac{2e}{\hbar} \right) V_0 \quad . \tag{13}$$

The voltage across the junction can be pictured as a number ν_J of flux quanta Φ_0 crossing the junction each second. This picture becomes more

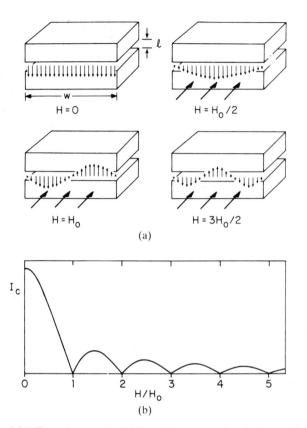

FIGURE 8. (a) Effect of magnetic field on the current distribution and the critical current of a Josephson junction. (b) Variation of critical current with applied field.

precise for a wide junction consisting of a thin film bridge in "Josephson vortices," characteristic of a tunnel junction (figure 8a), become Abrikosov vortices, characteristic of a type II superconductor, flowing across the junction. (See section 13.)

The first experimental evidence for the existence of the ac Josephson effect was obtained by Shapiro [83] in 1963. He applied, in addition to a steady voltage V_0, an oscillating voltage at microwave frequency ν, which results, through mixing in the junction, in a series of current steps in the dc I-V curve at discrete voltages $V = n\left(\frac{h}{2e}\right)\nu$.

By far the most accurate determinations [84–88] of the fluxoid quantum have been through experiments of this type as done originally by Parker, Taylor and Langenberg [85] using accurately known frequencies and voltages. Determinations made in these experiments are shown in figure 9.

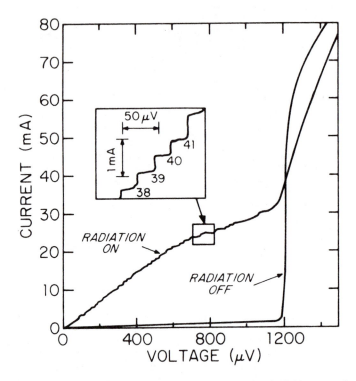

FIGURE 9. Microwave-induced supercurrent steps on the dc *I-V* curve of a Sn-Sn oxide-Sn tunnel junction at $T = 1.2$ K with $\nu = 10$ GHz (W. H. Parker, D. N. Langenberg, A. Denenstein and B. N. Taylor [84]).

Daniel Bracken [89] working with Fairbank at Stanford compared equivalent voltage steps on junctions of Pb and of Sn and showed that the two voltages were identical to within a few parts in 10^9.

12. MACROSCOPIC QUANTUM INTERFERENCE AND SQUID'S

Early 1964 was the beginning of an incredibly beautiful series of experiments on the Josephson effects and macroscopic quantum phenomena reported [90–102] by various members of the Ford Research Laboratories. Using two Josephson junctions in a superconducting ring (figure 10) in the configuration that has now become so familiar as the dc SQUID (Superconducting Quantum Interference Device), Jacklevic, Lambe, Silver and Mercereau [90] first demonstrated macroscopic quantum interference.

The behavior of a superconducting ring containing a Josephson junction can be readily calculated by applying the quantized fluxoid condition,

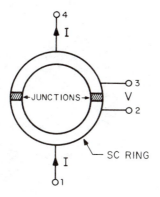

FIGURE 10a. Superconducting ring containing two Josephson junctions (interferometer configuration).

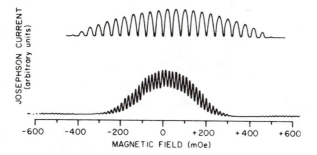

FIGURE 10b. Maximum supercurrent *versus* magnetic field for configuration of (a) (R. C. Jacklevic, J. Lambe, J. E. Mercereau and A. H. Silver [96]).

equation (4). For an otherwise thick ring, the first term is zero everywhere along a path within the ring except across the junction. There, using the definition of ϕ from equation (11) and $\nabla\chi$ from equation (2), we find that

$$\phi = \frac{2\pi}{\Phi_0} \int_1^2 \frac{m\mathbf{v}_s}{e} \cdot d\mathbf{s} \tag{14}$$

across the link. Since $\oint \mathbf{A} \cdot d\mathbf{s} = \Phi$ is the total flux within the ring, we can write for equation (4) applied to the ring with one junction

$$\frac{\Phi_0}{2\pi}\phi + \Phi = n\Phi_0 \quad . \tag{15}$$

Similarly for a ring with two junctions, evaluating equation (4) gives

$$\frac{\Phi_0}{2\pi}(\phi_2 - \phi_1) + \Phi = n\Phi_0 \quad . \tag{16}$$

For a ring with two identical Josephson junctions the total current is

$$i = i_c(\sin\phi_1 + \sin\phi_2) \quad , \tag{17}$$

which can be written as

$$I = 2i_c \sin\left(\frac{\phi_1 + \phi_2}{2}\right)\cos\left(\frac{\phi_1 - \phi_2}{2}\right) \quad . \tag{18}$$

The last factor can be expressed in terms of the total flux using equation (16), giving

$$I = 2i_c \sin\frac{(\phi_1 + \phi_2)}{2}\cos(\pi\Phi/\Phi_0) \quad . \tag{19}$$

The angle $\frac{\phi_1 + \phi_2}{2}$ varies with applied current I; however, the maximum value I_c is

$$I_c = 2i_c|\cos\pi\Phi/\Phi_0| \quad , \tag{20}$$

which is a periodic function of the total flux in the ring with period Φ_0, giving the characteristic interference pattern.

If the field is applied to the junction as well as within the loop, a diffraction-modulated interference results (figure 10b) since i_c in equation (20) is of the form given by equation (12).

About this time Allen Goldman undertook the first experiments at Stanford on the Josephson effects. He demonstrated for the first time the trapped flux states of rings with two Josephson junctions [103].

Using a superconducting ring with a single Josephson junction, Silver and Zimmerman first studied the quantized states of the ring and transitions between them. They reported their experiments in an extremely comprehensive paper [102] in 1967 in which they also described the use of the ring as a magnetometer in the configuration (figure 11) now universally known as an rf SQUID.

Some properties of this device are readily obtained by applying the quantized fluxoid condition to the ring, which, as mentioned above, gives equation (15) as the relationship between the total flux in the ring and the phase difference ϕ across the junction. What is usually of interest is the current i in the ring, or the total flux Φ as a function of externally applied flux Φ_x. Since the total flux is $\Phi = iL + \Phi_x$ where L is the inductance of the ring, and the current through the junction is $i = i_c \sin \phi$, we can eliminate ϕ from equation (15) and find

$$i_c \sin \left(2\pi \Phi / \Phi_0 \right) = \left(\Phi - \Phi_x \right) / L \tag{21}$$

and

$$\frac{i}{i_c} = \sin 2\pi \left[\left(iL + \Phi_x \right) / \Phi_0 \right] \quad . \tag{22}$$

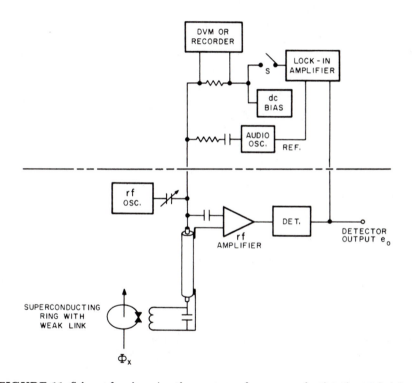

FIGURE 11. Scheme for observing the response of a superconducting ring containing a single Josephson junction. Components below the dotted line were used in initial experiments to study behavior of ring. With the addition of components above the line, the circuit becomes a practical magnetometer.

Some examples of the variations calculated from these equations are shown in figure 12. A parameter $\beta = 2\pi i_c L / \Phi_0$ determines the behavior. For $\beta < 1$ there is a continuous transition between states (the case

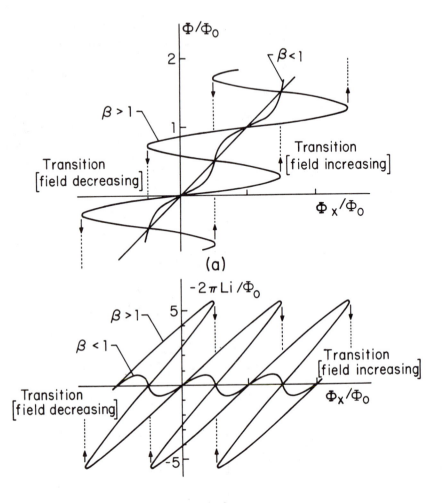

FIGURE 12. Total flux Φ and current i as function of external flux Φ_x for ring containing a Josephson junction. Curves are shown for two values of $\beta = 2\pi i_c L/\Phi_0$.

for nonhysteretic SQUID operation). For $\beta > 1$, when $i = i_c$ there is a discontinuous switch in current to a lower quantized state and an abrupt flux change in the ring (the case for the more common hysteretic SQUID).

The ac response can be determined from these characteristics by assuming $\Phi_x = \Phi_x(t)$ and has been widely studied in experiments with rf and microwave SQUID's. A typical kind of measurement is made using the scheme originated by Silver and Zimmerman [102] (bottom

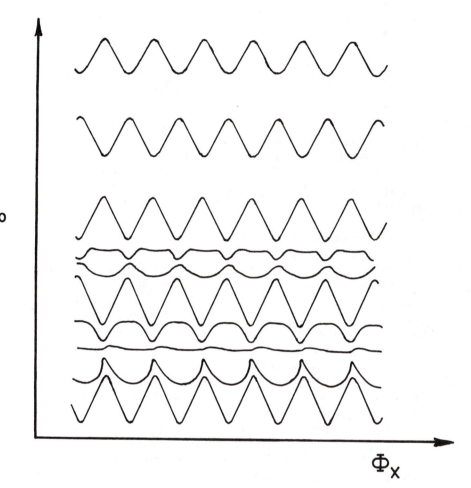

FIGURE 13. Response of rf SQUID showing periodic variation of detected tank voltage (see figure 11) with applied flux Φ_x through the ring for various levels of rf drive current. Period is Φ_0 (D. A. Vincent [104]).

part of figure 11) to plot the detector output e_0 as a function of applied flux Φ_x for a SQUID operating at $\sim$ 30 MHz, giving data like that shown in figure 13, again illustrating the ubiquitous periodicity with Φ_0.

SQUID's, of course, have been used for a multitude of demonstrations of fluxoid quantization and macroscopic quantum effects, and as magnetometers and amplifiers form the basis for an ever increasing variety of superconducting electronic devices [105]. SQUID's are discussed briefly in chapter IV, and throughout the book experiments are described in which they play crucial roles.

13. FLUX MOTION AND PINNING IN SUPERCONDUCTORS

Previously (in section 5) I mentioned that magnetic flux enters type II materials and also thin films of type I as individual quantized vortices each contaning one flux quantum. Any current flowing at the vortex produces a Lorentz force that causes the vortex to move. However, in most materials there are grain boundaries, defects or other inhomogeneities at which the energy of a vortex is lower than in the homogeneous material, so the vortices are pinned at these sites. Through interactions among the vortices the whole lattice can be immobilized.

Direct experimental confirmation of the pinning of flux in localized quantized units in a type II superconductor, as predicted by Abrikosov [54], was obtained by Zimmerman and Mercereau [94] in 1964 in one of the first applications of a dc SQUID. By cooling a Nb wire in a magnetic field and then passing the wire through the ring of a two-junction interferometer, they observed the emergence along the length of the wire of individual flux lines with magnitude Φ_0.

If the pinning force is strong compared to the force due to a transport current, there can be random flux motion due to thermally activated hopping of the vortices. This gives rise to flux "creep" which can cause a slow decay of the field in solenoids.

If the pinning is weak compared to the driving force, the vortices move in a relatively steady flow at a terminal velocity determined by viscous drag. Theoretical explanations of the drag forces involve several mechanisms, including dissipation arising from normal current in the vortex cores calculated by Bardeen and Stephen [106], from a finite relaxation time of the order parameter as treated by Tinkham [107], and from a diffusion mechanism of relaxation of the order parameter treated by Gorkov and Kopnin [108].

The first basic studies of flux flow in type II materials was carried out by Kim, Hempstead and Strnad [109]. Very recently Poon and Wong [110] have shown that in nearly ideal type II amorphous bulk superconductors all three kinds of dissipation are required to explain the observed flux flow resistance.

An ingenious experiment to study flux flow in thin films is the dc transformer of Giaever [111] which consists of two superimposed superconducting films separated by a thin insulating layer. A current in one film creates a flux flow voltage across it. An equal voltage appears across the second film even though no current passes through it. The interpretation of the result is that flux lines common to both films experience a force due to current in the primary and as they move are dragged across the secondary at the same rate they move in the primary film. As the separation between the films is increased the interaction between the two flux arrays is weakened and there is slippage between them [112].

The current-voltage characteristics of small wires, thin film strips and some microbridges are determined by flux flow. In a microbridge this phenomenon is closely related to the Josephson effect, since for single vortices carrying flux Φ_0 across a bridge at a rate of ν per second the voltage is just $V = \nu\Phi_0$, which is the Josephson frequency relation, equation (12). In a very long Josephson tunnel junction there can be isolated vortices each corresponding to a 2π change in phase difference $\phi(z)$ along the junction, and an accompanying circulating current superimposed on the uniform current flow through the junction. (See figure 8a, case for $H = H_0$.) The position of the vortex can be sensed and used as the basis for a flux shuttle logic element for a superconducting computer [113]. As the vortex moves it constitutes a soliton, a subject on which there is currently much active research in long Josephson junctions [114]. Very complex phenomena can occur in junctions, and for some values of the junction parameters chaotic motions are possible, so such junctions provide an experimental arena for investigation of chaos as well [115].

14. OTHER MEASUREMENTS OF Φ_0

During the period 1965–1970 there were several new measurements of the fluxoid quantum using a variety of techniques including another very clever use of flux motion. Van Ooijen [116] at Philips Research Laboratories observed the entry of flux into a hollow superconducting tin cylinder mounted inside the coil of a resonant circuit. As a magnetic field along the axis of the cylinder was continuously increased, the entry of flux into the cylinder by the incoherent flow of discrete flux units produced a noise voltage in the tank circuit. From measurements of the average square of the noise voltage and calculations from a shot-noise model, he obtained a value for the flux increment in good agreement with Φ_0.

At Stanford Stuart Spence [117], working with Bill Little, observed flux quantization in 2 μm diameter indium cylinders deposited on quartz fibers. Each fiber was hung as a pendulum with a weight on one end, and the flux trapped upon cooling in an axial magnetic field was found from a measurement of the static deflection of the fiber in a transverse magnetic field.

A novel measurement of the fluxoid quantum was made by Blas Cabrera [118] working in my lab during his senior year at the University of Virginia.

He used a double junction interferometer in the form of a Clarke SLUG [119] (figure 14), with a precisely measured insulation thickness t and separation ℓ between the contacts. The critical current of the SLUG is a periodic function of the current I in the wire. From a measurement of this period, and a calculation of the flux through the area $\ell\,(\lambda_1 + \lambda_2 + t)$, the quantum was determined within a few percent.

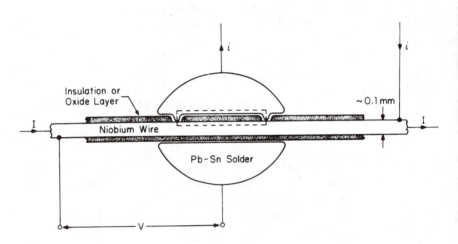

FIGURE 14. Cross-sectional view of Clarke SLUG showing two discrete weak links and indicating the area (within the dashed line) into which the magnetic field from I penetrates (Cabrera, Williams and Deaver [118]).

Using techniques of electron microscopy, Lischke [120,122,123], Boersch and Lischke [121] and Wahl [124–126] have observed interference patterns from pairs of electron beams encircling superconducting cylinders with trapped flux in them. Through the dependence of the pattern on the enclosed flux (Aharonov-Bohm effect—see section 19) they measured the quantized states and determined the value of Φ_0 with an accuracy of a few percent. They also observed that when several quanta were trapped in a tapered cylinder, the flux escaped in discrete steps of Φ_0 along the length.

15. DETAILED MEASUREMENTS OF QUANTIZED STATES OF A CYLINDER

With the advent of SQUID magnetometers it was possible to make magnetic flux measurements with immensely improved sensitivity. Continuing our research on fluxoid quantization at the University of Virginia, William Goodman [127,128] used rf SQUID's for some detailed studies of the quantized flux states of cylinders, as reported in 1970. For his first measurements he essentially repeated the initial quantized flux experiments but with much improved precision. An example of his data for trapped flux *versus* magnetic field in which the cylinder was cooled is shown in figure 15. The flat steps are $n\Phi_0 \pm \frac{1}{2}\%$.

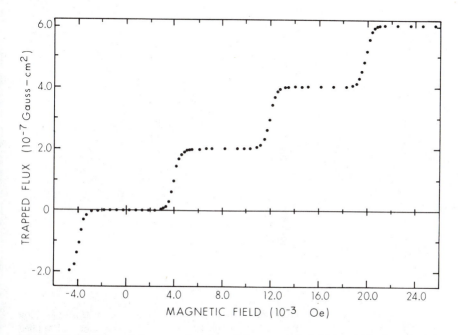

FIGURE 15. Trapped flux as a function of the magnetic field in which the cylinder was cooled below its transition temperature. These data were taken with a tin cylinder 56 μm i.d. and 24 mm long with walls about 500 nm thick (Goodman and Deaver [127]).

In these data, as in the original quantized flux experiments, there were points along a continuous curve between quantized values. It was known, of course, that flux could leak out of the cylinder as a quantized flux line through a normal core surrounded by a current vortex in the wall. However, John Pierce [129], working with us at Virginia, was the first to suggest that the cylinder might break up into a domain-type structure. For more detailed flux measurements to examine this possibility, Goodman [128] used a SQUID with an extremely small pickup loop to map the trapped flux as a function of position along the cylinder. Examples of these data are shown in figure 16.

For most values of the applied field the entire cylinder was in the same quantized state, with flux maps like those for 6.30 mOe and 14.28 mOe in

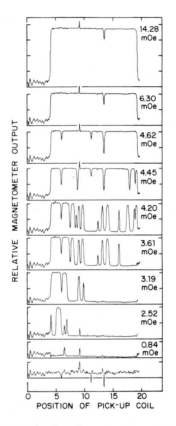

FIGURE 16. Flux trapped in a small superconducting cylinder as a function of position along the length of the cylinder. The curves are labeled with the values of applied field in which the cylinder was cooled through its transition temperature. The bottom curve is for $T > T_c$. It shows randomly distributed magnetic impurities in the cylinder (Goodman, Willis, Vincent and Deaver [128]).

figure 16, corresponding to points along the flat steps in figure 15. However, for some values of applied field a mixed state occurs with bands along the cylinder in states differing by one flux quantum. A distribution of bands of alternating quantum number separated by pinned vortices can be pictured as a quantized flux line weaving in and out of the wall of the cylinder.

For values of applied field corresponding to points between steps in figure 15, the quantized bands are distributed in such a way that the average flux in the cylinder is between quantized levels. Since the data in figure 15 were obtained with a large diameter SQUID ring that coupled to flux along most of the cylinder, it responded to the average flux. This interpretation is substantiated by integrating a series of flux maps like those in figure 16 and finding a series of average flux values that almost exactly reproduce the data in figure 15.

16. TEMPERATURE DEPENDENCE OF FLUX FOR THIN-WALLED CYLINDER

For a cylinder with wall thickness $t < \lambda$ there is a current throughout the walls, and the first term in equation (3) cannot be neglected as it could for thick-walled cylinders. In this case the trapped flux can be less than the full value $n\Phi_0$, and was first shown by Bardeen [18] to be

$$\phi = \frac{n\Phi_0}{1 + \frac{2\lambda^2}{rt}} \qquad (23)$$

for a cylinder of radius r and wall thickness t. This result and expressions for flux trapped in cylinders of arbitrary wall thickness have been obtained from microscopic theory, Ginsburg-Landau theory, and various models [17,19,25,28,29,37,38,52].

As shown by Mercereau and Crane [130], for a long, slender cylinder with walls so thin that the current density can be assumed to be constant, we can obtain equation (23) readily from equation (3). For a cylinder of length ℓ, radius r and wall thickness $t < \lambda$ so that $\mathbf{j}_s$ is nearly uniform, and assuming no applied flux so that the total flux $\oint \mathbf{A} \cdot d\mathbf{s} = i_s L$, where L is the inductance of the cylinder, the condition of fluxoid quantization, equation (3), can be written

$$\frac{m}{n_s e^2} i_s \frac{2\pi r}{\ell t} + i_s L = n\Phi_0 \quad . \qquad (24)$$

Solving for i_s and calculating the trapped flux $\Phi = i_s L$ explicitly, using $\lambda^2 = m/\mu_0 n_s e^2$ and $L = \mu_0 \frac{\pi r^2}{r\ell}$, gives immediately equation (23).

From equation (23) it is evident that for sufficiently thin-walled cylinders the trapped flux will differ from $n\Phi_0$, even at very low temperature. For any cylinder, but particularly for ones with $t < \lambda(0)$, the trapped flux Φ will depend on temperature because of the temperature dependence of the penetration depth $\lambda(T)$, or equivalently because of the variation of $n_s(T)$ since $\lambda^2 \propto [n_s(T)]^{-1}$. This variation of the flux with T corresponds to a reversible interchange between electromagnetic angular momentum and kinetic angular momentum of the pairs, since the total angular momentum (fluxoid) remains constant so long as T is not varied so much that there is a change in quantum number n. This is an adiabatic perturbation of a single quantum state.

Mercereau and Crane [131] and Hunt and Mercereau [132] verified the constancy of the fluxoid by observing the reversible change of flux with

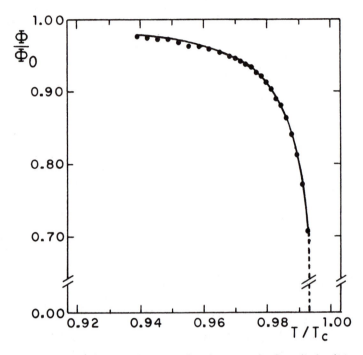

FIGURE 17. Trapped flux as a function of temperature for Sn cylinder (14 μm i.d., 500 nm walls, 1 cm long) in a state with $n = 1$. The curve is a fit to a calculation for a long cylinder. The dashed line indicates a jump to the $n = 0$ state (Goodman, Willis, Vincent and Deaver [128]).

temperature in thin film rings containing 10^5–10^7 flux quanta, and they found agreement with the variation expressed by equation (23).

Using the same apparatus that we used for mapping the domain structure, Goodman [128] measured the temperature variation of the flux in a thin-walled, small cylinder with single flux quantum trapped, and found a reversible variation $\phi(T)$ in good agreement with that calculated including the effects of wall thickness (figure 17).

Measurements of the temperature variation of trapped flux at the single quantum level in small crystals of the layer compounds $NbSe_2$ and TaS_2 intercalated with pyridine have been used by Finley [133] at Virginia to determine the penetration depth of these highly anisotropic materials.

17. FLUX MEASUREMENTS ON RINGS CONTAINING A WEAK LINK

For a thick-walled superconducting ring $\Phi(\Phi_x)$ is not a very interesting function, since Φ remains constant at a quantized value and the screening flux $i_s L$ exactly cancels the applied flux, until, of course, the critical current is reached. However, for a ring containing a weak link, the current through the link is determined by the phase difference ϕ across it. For an ideal Josephson junction, *e.g.*, a tunnel junction or very small microbridge, the current-phase relation is $i_s(\phi) = i_c \sin\phi$, but in general it may differ from this form for other kinds of weak coupling.

For a ring containing a weak link, as we have discussed previously (equation (13)), the condition of fluxoid quantization can be expressed as

$$\frac{\Phi_0}{2\pi}\phi + \Phi = n\Phi_0 \quad , \tag{25}$$

where the first term in equation (3) contributes only across the link and is $\phi = \frac{2\pi}{\Phi_0}\int_1^2 \frac{m}{n_s e^2}\mathbf{j}_s \cdot d\mathbf{s}$, and Φ is the total flux within the ring. In terms of an applied flux Φ_x and the induced flux iL, where L is the inductance of the ring,

$$\Phi = iL + \Phi_x \quad . \tag{26}$$

The dependence $i_i(\phi)$ of the current through the weak link on ϕ implies through equation (25) a corresponding variation $\Phi(\Phi_x)$. In a ring containing a Josephson junction with $i_s(\phi) = i_c \sin\phi$, the variation of Φ and of the induced flux $i_s L$ are shown in figure 12.

A conceptually simple experiment permits a quite direct determination of the current-phase relation $i_s(\phi)$ for any weak link using the technique shown schematically in figure 18a. A pickup loop surrounding the ring containing the weak link is connected to a SQUID magnetometer and used to measure the total flux Φ, as an external flux Φ_x applied with an extremely long solenoid is varied. By equation (25), ϕ is proportional to Φ. From equation (26), for a ring of known L with the measured $\Phi(\Phi_x)$, the current $i_s(\Phi_x)$ is calculated; and, eliminating Φ_x, $i_s(\phi)$ is determined.

Note that for rings with sufficiently small inductance $L < \Phi_0/2\pi i_c$, corresponding to $\beta < 1$ in figure 12, no discontinuous transitions are possible, so the ring can be stably phase-biased and the current determined over the whole range 0–2π.

In their pioneering paper in 1967, Silver and Zimmerman [102] reported measurements made by this technique. More recently, Jackel, Buhrman and Webb [134] have used it for a direct measurement of the current-phase relation of Nb point contacts, and, taking careful account of thermal fluctuations, find excellent agreement with $i_c \sin\phi$.

At the University of Virginia we have used a closely related technique. By applying a small amplitude alternating flux in addition to the steady external flux bias, it is possible by measurement of the alternating voltage across the link to determine the dynamic impedance $Z(\phi)$ of phase-biased weak links. Assuming that $Z(\phi) = R(\phi) + jw\mathcal{L}(\phi)$, where the pair current is represented by the inductance $\mathcal{L}(\phi)$ and the quasiparticle current by $R(\phi)$, Rifkin [135] has obtained data for Nb point contacts. He found an inductance (figure 18b) in good agreement with that expected for an ideal Josephson junction, $viz.$ $\mathcal{L} = \phi_0/2\pi i_c \cos\phi$ including the expected negative inductance branch, which is accessible only because of the constraint of fluxoid quantization. Since $\mathcal{L}(\phi) = \frac{h}{2e}\left(\frac{\partial i_s(\phi)}{\partial \phi}\right)^{-1}$, the data also determine the current-phase relation, and since they give the form expected for the Josephson inductance also give excellent agreement with $i_c \sin\phi$.

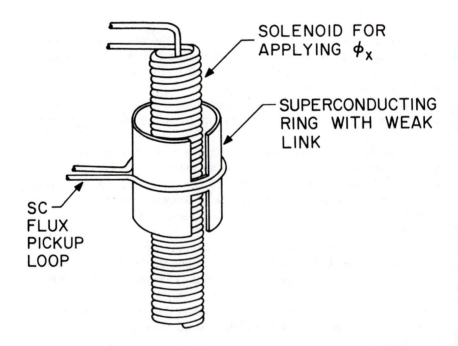

FIGURE 18a. Scheme for measuring total flux Φ as a function of applied flux Φ_x to determine current-phase relation of weak link.

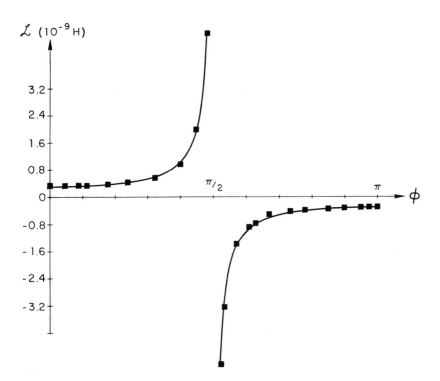

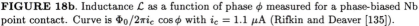

FIGURE 18b. Inductance $\mathcal{L}$ as a function of phase ϕ measured for a phase-biased Nb point contact. Curve is $\Phi_0/2\pi i_c \cos\phi$ with $i_c = 1.1\ \mu$A (Rifkin and Deaver [135]).

18.   QUANTIZED FLUX STATES OF A TWO-TURN RING

Now I would like to examine another parameter in the fluxoid expression, namely the path along which the integrals are taken.   Consider a two-turn superconducting ring (see figure 19a).   For a thick ring as usual we can neglect the first term in equation (3), and since for the second term the path goes twice around the hole we have $2\Phi = n\Phi_0$. Thus we expect the quantized states of this ring to be in units of $\Phi_0/2$. That is, if on passing once around a flux Φ_0 a pair picks up 2π of phase, then on passing twice around $\Phi_0/2$ the phase advance is 2π and the condition of single valuedness is preserved.   In general for an N turn loop, $\Phi = n\Phi_0/N$, and since $\Phi = iL + \Phi_x$, the induced flux $iL = \frac{n\Phi_0}{N} - \Phi_x$.

With the configuration shown in figure 18a but with the weakly linked ring replaced by a two-turn loop, we used a SQUID to measure the induced flux in the ring [76].   The induced flux is a sawtooth function, just as for a

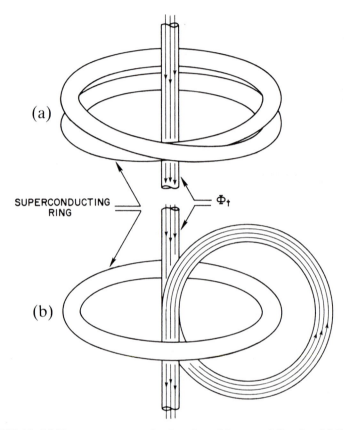

(a)

SUPERCONDUCTING RING

Φ_t

(b)

FIGURE 19. (a) Two-turn superconducting ring with trapped flux Φ_t. (b) Same ring untwisted to be a single turn. Flux passes twice through it.

single turn ring (*e.g.*, lower part of figure 3b), but now with period $\Phi_0/2$ and with flux changes between states of $\Phi_0/2$.

Yang has given an interesting interpretation of this result. He noted that if the original two-turn ring is untwisted without letting the trapped flux cross the superconductor, the flux passes twice through the resulting single turn ring (figure 19b). Since the quantized values for the single turn ring are $n\Phi_0$, it follows that the quantized states for the two-turn ring are $n\Phi_0/2$.

19.   AHARONOV-BOHM EFFECT

The result described in the previous section can also be considered to be an experimental demonstration of the winding number dependence of the Aharonov-Bohm (A-B) effect [136,137], about which I would like to say a bit here because of its close connection with fluxoid quantization.

In 1957 Aharonov and Bohm [136] called attention to the fact that the phase difference, between the wave functions of charged particles passing around an inaccessible region containing magnetic flux, depends on the enclosed flux even though the particles were never in the magnetic field. For the two-slit interference experiments with free electrons the phase difference between the two paths is changed by 2π for each change of h/e of the magnetic flux, within the paths, and thus the interference pattern is shifted by a full fringe. Considerable controversy ensued over their interpretation of the role of the vector potential in quantum mechanics; however, the experimental situation seems relatively clear.

The first experiments [138–141] designed specifically to examine the effect were carried out by diffracting electron beams around magnetic whiskers or solenoids. There were objections that the wide extent of the electron wave functions and the nonzero field far from the solenoid might invalidate the results. Very recently a group [142] in Japan has used electron holography and small toroidal ferromagnets for an elegant demonstration of the A-B effect that seeks to remove these objections.

Peshkin [143] has redirected attention to the fact that experiments using superconductors, which have wave functions confined to well-defined paths, overcome some of the questions that arise in experiments with free electrons, and that experiments that demonstrate fluxoid quantization can be regarded as confirming the A-B effect. In fact, in their paper Byers and Yang [15] had commented on this relationship.

The first experiment using superconductors to demonstrate the A-B effect with care taken specifically to confine the flux within the ring was done by Jacklevic et al. [91] with a dc SQUID.

Immediately after the first quantized flux experiments Onsager [16] pointed out that, as a consequence of fluxoid quantization, by simply closing a superconducting ring around an inaccessible region of space and measuring the current that flowed in the ring, it is possible to determine, modulo Φ_0, the flux enclosed by the ring. Thus the type of experiment that Al Kwiram and I did on thermally cycled rings (section 9) is a nice confirmation of the A-B effect, particularly when the flux is applied with a long solenoid within the ring, as was done later by Willis [144] at the University of Virginia.

Very recently Bernido and Inomata [145] have used a Feynman path integral approach, and have emphasized the topological nature of the A-B effect by noting that the phase difference between the two wave functions depends not only on the magnetic flux within the inaccessible region but also on the winding number N, the number of times the region is encircled. They questioned how one could devise an experiment for detecting this topological shift. The experiment done by Hugh Henry on a two-turn loop (section 18) is just such an experiment and demonstrates explicitly the effect for $N = 2$, as Donaldson and I have pointed out [146].

20.   THERMALLY INDUCED FLUX IN A SUPERCONDUCTING RING

Thermally induced magnetic flux in a bimetallic superconducting ring is a consequence of fluxoid quantization, or at least the constancy of the fluxoid, that has been studied experimentally only recently [147]. Garland and Van Harlingen [148] and Gal'perin, Gurevich and Kosub [149] independently proposed that thermoelectric currents would be observable in rings composed of two different superconductors with one junction at a higher temperature than the other. The essence of their arguments is contained in the following simple calculation.

The thermal gradient ∇T will induce quasiparticle currents along the two superconductors since $j_n = L_V \nabla T$, where L_V is the transport coefficient. At each point in the bulk superconductors there is an equal and opposite pair current to maintain $\mathbf{B} = 0$ except at the surface, where the cancellation is not complete and the net current induces a flux within the ring. We can find the flux Φ by requiring that in the bulk $j_n + j_s = 0$. This implies $j_s = -L_V T$, and since $j_s = n_s e v_s$ there is a pair velocity $v_s = -L_V \nabla T / n_s e$. If the transport coefficients of the two superconductors are different, there will be a net supercurrent encircling the ring and a flux Φ within the ring, which can be calculated using these pair velocities in evaluating the fluxoid equation (4) to find

$$\frac{m}{e^2} \left(\frac{L_{V_2}}{n_{s_2}} - \frac{L_{V_1}}{n_{s_1}} \right) \nabla T + \Phi = n\Phi_0 \quad . \tag{27}$$

Soon after the above proposals there were several experiments [150–152] with bimetallic rings. In each case a flux was observed that was proportional to ∇T and that diverged as T_c was approached from below. But the flux was much larger than that calculated from the theories.

Recently Van Harlingen, Heidel and Garland [153] have used a toroidal ring with a SQUID detector for very careful measurements of the effect and they find indeed a large induced flux differing from their calculated values by a factor of 10^5. Although several explanations have been advanced [154–156], this discrepancy has not yet been resolved. In any case it appears reasonably certain that the effect follows from the constraint of fluxoid quantization and represents an experimental situation in which a driving force specifically on the normal electrons causes a superfluid response.

A related phenomenon has been predicted by Falco [157] who has calculated that a magnetic flux will be induced in a ring composed of dissimilar materials when an acoustical wave is propagated through them.

21. FLUX INDUCED IN ROTATING RINGS

Finally in this pattern of discussing parameters that enter the fluxoid, I would like to mention the effects of rotation on a superconducting ring. Again a simple calculation can be used to derive a generally valid result. Consider a hollow superconducting cylinder of radius r, with walls thick with respect to the penetration depth, rotating with angular speed ω about its axis. Since at equilibrium the normal electrons move with the lattice ions, there will be a normal current density

$$j_n = (n_{ion} - n_n)\, ev_n = n_s er\omega \quad , \tag{28}$$

as viewed from the lab. Since in the bulk superconductor $j_n + j_s = 0$, then $j_s = -n_s er\omega$ from which $v_s = r\omega$. Evaluating equation (4) for the fluxoid along a path deep within the superconductor using this pair velocity gives

$$\frac{2m}{e}\omega\pi r^2 + \Phi = n\Phi_0 \quad . \tag{29}$$

For $n = 0$,

$$B = \left(\frac{2m}{e}\right)\omega \quad . \tag{30}$$

This is a quite general result first derived by London [2] who predicted that a spatially uniform magnetic field of this magnitude would be generated within the body of any rotating superconductor. This effect has become known as the "London moment." The effect occurs because it is the fluxoid that is constant. In the bulk superconductor the pairs move in synchronism with the lattice ions and normal electrons, giving $j_n + j_s = 0$, but at the surface within the penetration depth they rotate slightly slower, giving rise to a current that produces the London moment.

As I said at the beginning of this paper, the measurement of the London moment was the topic of Morris Bol's thesis [158] with Fairbank. By now there are a number of experiments [159–165] that give excellent agreement with the London result. The London moment is the basis for determining the orientation of the spin axis of the precision gyroscope in the relativity experiment described by Everitt and coworkers in the papers entitled "The Stanford Relativity Gyroscope Experiment."

The most precise measurement of the London moment, one of the best observations of the quantized flux states of a ring, and the most precise determination of h/m for electrons in a superconductor are already results of an extremely elegant experiment done by Blas Cabrera, Sue Felch and John Anderson [166] at Stanford. This experiment is certainly a pinnacle of achievement in measurements involving fluxoid quantization. It is described in detail in the paper entitled "Rotating Superconductors and Fundamental Physical Constants."

To measure h/m they use a rotating ring with a closely coupled pickup loop around it to determine for a particular quantum state n of the ring the rotational speed ω_n in which the flux Φ is zero in accord with equation (29). For two such speeds then from equation (29),

$$\frac{h}{m} = 4\pi r^2 \left(\omega_n + \omega_{n-1} \right) \quad , \tag{31}$$

so the ratio h/m can be determined to high accuracy from measurements of rotational speed and ring geometry.

At the time of the high accuracy determination of h/e using the Josephson effects it was suggested [87] that h/m could be determined in this way, which in effect balances the magnetic flux from n flux quanta against the flux from the London field. The first experiments were, in fact, done in 1965 by Zimmerman and Mercereau [97], and the most precise results prior to the Stanford experiments were reported by Parker and Simmonds [164] in 1970.

22. FLUXOID QUANTIZATION IN NONSUPERCONDUCTING SYSTEMS

There are now several experimental results that can be interpreted as manifestations of fluxoid quantization in nonsuperconductors. In fact, London's derivation [1,2] and the arguments used by Byers and Yang [15] and by Bloch [80,81] to discuss the periodicity with Φ of the free energy and of the currents in a gas of electrons confined to an annular region encircling a magnetic flux Φ are very general, and apply not just to superconductors. For systems of electrons with the pair correlations of superconductivity, the dominant period is $h/2e$ and the fluxoid is quantized in these units. However, as emphasized recently by Imry [168], for other systems with discrete states and phase coherence around the ring, even in the presence of scattering if it preserves phase memory, these arguments lead to macroscopic quantum effects including fluxoid quantization in normal conductors with period h/e.

Using the Little-Parks [43] technique of measuring the resistance of a small hollow cylinder as a function of magnetic flux applied along the axis, Shablo et al. [169] have observed periodicity of the resistance of Al cylinders with period $h/2e$ at temperatures well above the superconducting transition temperature. This result was anticipated theoretically [170–172], as was the possibility [173,174] that in a very small normal metal cylinder with long mean free path there might be fluxoid quantization with period h/e.

More surprising is the result of Sharvin and Sharvin [175], who also use the Little-Parks technique with a cylinder formed of highly disordered Mg, which has never been found to be superconducting, and find a resistance variation with period $h/2e$. This result was predicted by Al'tshuler, Aronov and Spivak [176] who calculated that in disordered metals the Aharonov-Bohm effect would be manifested in oscillations of the kinetic coefficients with magnetic flux within the cylinder but with period one half of that for the ordinary A-B effect with free electrons.

Recent experiments [177–179] with essentially two-dimensional conductors formed by the inversion layer in semiconductors have shown that the Hall conductance of these systems can exhibit precisely quantized values in units of e^2/h. Laughlin [180] and Imry [168] have discussed the relationship between this quantized Hall effect and other macroscopic quantum effects, and Imry concludes that these systems should also exhibit fluxoid quantization with period h/e.

The quantized rotational states of superfluid ^{4}He described by George Hess in his paper entitled "Rotation of Superfluid Helium" are another example of macroscopic quantum effects, and his experiments are the helium analog of the original quantized flux experiments in superconductors.

23. COMPUTERS, THE MAGNETIC MONOPOLE, MACROSCOPIC QUANTUM TUNNELING, AND THE FUTURE

In closing I want to mention several of the newest involvements of fluxoid quantization.

In satisfying the desire for the fastest possible computing speeds superconductivity appears to offer great potential, and the development of the superconducting computer is proceeding vigorously [181,182]. The fundamental basis for the memory in this computer is the storage of quantized persistent currents in superconducting rings. Josephson devices including multijunction SQUID's provide the high speed and low dissipation for logic and memory functions [183–185].

Certainly one of the most exciting events at the Near Zero conference was the report on the experiment to detect magnetic monopoles being done by Blas Cabrera [186], who has observed an event that appears to have all the right characteristics to be a monopole. In a talk entitled "From Fluxoid Quantization to Magnetic Monopoles" which he gave in 1982 at the University of Virginia, he mentioned various ways that fluxoid quantization is involved in the low field shield, in the SQUID detector, and in the upcoming experiment for recording the passage of a monopole through the shield by the resulting pair of localized vortices with two quanta each at the entry and exit locations.

For the future there are certainly some basic questions. Are there other systems with periodic variations of the free energy with flux? Are there, in addition to $h/2e$ and h/e, other possible periodicities corresponding to other correlations? Are there interesting dynamics of vortices in thin amorphous films? Are there interesting phenomena in arrays of magnetically-coupled and phase-coupled rings?

It has recently been argued that quantum tunneling between metastable fluxoid states in a SQUID may constitute evidence that a collective variable describing the macroscopic state of a complex system, in this case the flux Φ in the ring, obeys the superposition principle of quantum mechanics [187]. The effects of dissipation on the tunneling rate have been calculated and it appears that tunneling should be observable at very low temperatures [188]. In fact, there is already evidence for macroscopic quantum tunneling in current-biased Josephson junctions [189, 190]. It has been suggested very recently that a time varying field imposed on an rf SQUID ring may produce photo assisted tunneling between fluxoid states, a striking phenomenon because it involves a transition between two macroscopically distinguishable states caused by absorption of a single photon [191].

Macroscopic quantum tunneling and other basic tests of quantum mechanics using fluxoid states are intriguing possibilities for future experiments.

In addition to the monopole search, there are several other experiments at Stanford including the h/m measurements, the relativity gyroscope, and the search for the electric dipole moment of ^{3}He in which fluxoid quantization plays crucial roles. These and other experiments in Bill Fairbank's lab manifest a remarkable synergism that results from combining temperatures near zero, magnetic fields near zero, and sensitivity to signals near zero for new fundamental experiments at this frontier of physics.

Acknowledgments

The work on fluxoid quantization and related topics carried out at the University of Virginia has been supported annually by the National Science Foundation and the Office of Naval Research and is currently funded under NSF grant DMR-82-04365 and ONR contract N00014-81-K-0400.

References

[1]   F. London, *Phys. Rev.* **74**, 562 (1948).

[2]   F. London, *Superfluids, Vol. 1: Macroscopic Theory of Superconductivity* (John Wiley and Sons, New York, 1950). Reissued by Dover Publications, Inc., New York, 1961.

[3] S. Foner, *Rev. Sci. Instr.* **30**, 548 (1959).

[4] P. M. S. Blackett, *Lectures on Rock Magnetism* (Weizmann Press, 1956), Appendix I, Blackett and Sutton.

[5] C. J. Gorter, *Ned. T. Natuurk* **27**, 269 (1961).

[6] J. E. Mercereau and L. L. Vant-Hull, *Bull. Am. Phys. Soc.* **6**, 121 (1961).

[7] E. Burton, H. Grayson-Smith and J. Wilhelm, *Phenomena at the Temperature of Liquid Helium* (Reinhold Publishing Corp., New York, 1940), p. 120.

[8] B. S. Deaver, Jr., and W. M. Fairbank, *Phys. Rev. Lett.* **7**, 43 (1961).

[9] R. Doll and M. Näbauer, *Phys. Rev. Lett.* **7**, 51 (1961).

[10] B. S. Deaver, Jr., Ph.D. Thesis, Stanford University, 1962.

[11] R. Doll and M. Näbauer, *Z. Physik* **169**, 526 (1962).

[12] W. O. Hamilton, *Revue de Physique Appliquée* **5**, 41 (1970).

[13] B. Cabrera and W. O. Hamilton, *The Science and Technology of Superconductivity*, edited by W. O. Gregory, W. N. Mathews, Jr., and E. A. Edelsack (Plenum Press, New York, 1973), Vol. II, p. 587.

[14] B. Cabrera, Ph.D. Thesis, Stanford University, 1975.

[15] N. Byers and C. N. Yang, *Phys. Rev. Lett.* **7**, 46 (1961).

[16] L. Onsager, *Phys. Rev. Lett.* **7**, 50 (1961).

[17] J. M. Blatt, *Phys. Rev. Lett.* **7**, 82 (1961).

[18] J. Bardeen, *Phys. Rev. Lett.* **7**, 162 (1961).

[19] J. B. Keller and B. Zumino, *Phys. Rev. Lett.* **7**, 164 (1961).

[20] W. Brenig, *Phys. Rev. Lett.* **7**, 337 (1961).

[21] K. Maki and T. Tsuneto, *Progr. Theoret. Phys. (Kyoto)* **27**, 228 (1962).

[22] A. Bohr and B. R. Mottelson, *Phys. Rev.* **125**, 495 (1962).

[23] G. Lüders, *Z. Naturforsch* **17a**, 181 (1962).

[24] W. Weller, *Z. Naturforsch* **17a**, 182 (1962).

[25] H. J. Lipkin, M. Peshkin and L. J. Tassie, *Phys. Rev.* **126**, 116 (1962).

[26] W. Weller, *Phys. Lett.* **1**, 222 (1962).

[27] E. Merzbacher, *Amer. J. Phys.* **30**, 237 (1962).

[28] F. Schwabl and W. Thirring, *Nuovo Cimento* **25**, 175 (1962).

[29] V. L. Ginsburg, *Sov. Phys.–JETP* **15**, 207 (1962).

[30]   M. Peshkin and W. Tobocman, *Phys. Rev.* **127**, 1865 (1962).

[31]   D. Bohm, in *Proceedings of the 8th International Conference on Low Temperature Physics (LT8), London, 1962*, edited by R. O. Davies (Butterworth, Washington, 1963), p. 109.

[32]   L. N. Cooper, H. J. Lee, B. B. Schwartz and W. Silvert, in *Proceedings of the 8th International Conference on Low Temperature Physics (LT8), London, 1962*, edited by R. O. Davies (Butterworth, Washington, 1963), p. 126.

[33]   F. Bloch and H. E. Rorschach, *Phys. Rev.* **128**, 1697 (1962).

[34]   C. N. Yang, *Rev. Mod. Phys.* **34**, 694 (1962).

[35]   M. Tinkham, *Phys. Rev.* **129**, 2413 (1963).

[36]   M. Peshkin, *Phys. Rev.* **132**, 14 (1963).

[37]   D. H. Douglass, *Phys. Rev.* **132**, 513 (1963).

[38]   L. P. Rapaport, *Sov. Phys.–JETP* **18**, 1003 (1964).

[39]   G. Wentzel, *Physics* **49**, 679 (1963).

[40]   B. B. Schwartz and L. N. Cooper, *Rev. Mod. Phys.* **36**, 280 (1964).

[41]   D. A. Uhlenbrock and B. Zumino, *Phys. Rev.* **133A**, 350 (1964).

[42]   F. Bloch, *Phys. Rev.* **137A**, 787 (1965).

[43]   W. A. Little and R. D. Parks, *Phys. Rev. Lett.* **9**, 9 (1962).

[44]   W. A. Little and R. D. Parks, in *Proceedings of the 8th International Conference on Low Temperature Physics (LT8), London, 1962*, edited by R. O. Davies (Butterworth, Washington, 1963), p. 129.

[45]   W. A. Little and R. D. Parks, *Phys. Rev.* **133A**, 97 (1964).

[46]   W. A. Little, *Rev. Mod. Phys.* **36**, 264 (1964).

[47]   L. Meyers and W. A. Little, *Phys. Rev. Lett.* **11**, 156 (1963).

[48]   L. Meyers and W. A. Little, *Phys. Rev. Lett.* **13A**, 325 (1964).

[49]   R. P. Groff and R. D. Parks, *Phys. Rev.* **176**, 567 (1968).

[50]   R. Meservey and L. Meyers, *Phys. Lett.* **26A**, 367 (1968).

[51]   L. Meyers and R. Meservey, *Phys. Rev.* **B4**, 824 (1971).

[52]   H. J. Fink and V. Grünfeld, *Phys. Rev.* **B22**, 2289 (1980).

[53]   M. Tinkham, *Rev. Mod. Phys.* **36**, 268 (1964).

[54]   A. A. Abrikosov, *Sov. Phys.–JETP* **5**, 1174 (1957).

[55]   M. Tinkham, *Introduction to Superconductivity* (McGraw Hill, 1975), provides excellent discussions of type II superconductors, fluxoid quantization and the Little-Parks experiment using Ginsburg-Landau theory.

[56] R. D. Parks and J. M. Mochel, *Phys. Rev. Lett.* **11**, 354 (1963); *Rev. Mod. Phys.* **36**, 284 (1964).

[57] J. M. Mochel and R. D. Parks, in *Proceedings of the 9th International Conference on Low Temperature Physics (LT9), Columbus, Ohio, 1964*, edited by J. G. Daunt, D. O. Edwards, F. J. Milford and M. Yaqub (Plenum Press, New York, 1965), p. 571.

[58] R. D. Parks, in *Proceedings of the 9th International Conference on Low Temperature Physics (LT9), Columbus, Ohio, 1964*, edited by J. G. Daunt, D. O. Edwards, F. J. Milford and M. Yaqub (Plenum Press, New York, 1965), p. 34.

[59] R. D. Parks, J. M. Mochel and L. V. Surgent, Jr., *Phys. Rev. Lett.* **13**, 331 (1964).

[60] U. Essmann and H. Träuble, *Phys. Lett.* **24A**, 526 (1967).

[61] G. B. Donaldson, D. J. Brassington and W. T. Band, *J. Phys. F* **5**, 1726 (1975).

[62] R. P. Huebener, *Magnetic Flux Structures in Superconductors* (Springer-Verlag, Berlin, 1979).

[63] M. R. Beasley, J. E. Mooij and T. P. Orlando, *Phys. Rev. Lett.* **42**, 1165 (1979).

[64] A. F. Hebard and A. T. Fiory, *Phys. Rev. Lett.* **44**, 291 (1980); A. F. Hebard and A. T. Fiory, *Phys. Rev. Lett.* **50**, 1603 (1983).

[65] A. F. Hebard and A. T. Fiory, *Physica* **109** & **110B**, 1637 (1982).

[66] J. E. Mooij, in *Advances in Superconductivity*, edited by B. Deaver and J. Ruvalds (Plenum Press, New York, 1983), p. 435.

[67] D. J. Dumin and J. F. Gibbons, *J. Appl. Phys.* **34**, 1566 (1963).

[68] B. S. Deaver, Jr., and W. M. Fairbank, in *Proceedings of the 8th International Conference on Low Temperature Physics (LT8), London, 1962*, edited by R. O. Davies (Butterworth, Washington, 1963), p. 116.

[69] J. M. Pierce, in *Proceedings of a Symposium on the Physics of Superconducting Devices, University of Virginia, Charlottesville, 1967* (National Technical Information Service, Springfield, VA), AD661848, p. B1.

[70] B. S. Deaver, Jr., and W. S. Goree, *Rev. Sci. Instr.* **38**, 311 (1967).

[71] J. E. Opfer, *Revue de Physique Appliquée* **5**, 37 (1970).

[72] B. S. Deaver, Jr., and D. A. Vincent, in *Methods of Experimental Physics*, edited by R. V. Coleman (Academic Press, New York, 1974), Vol. II, p. 267.

[73] M. Bol, B. S. Deaver, Jr., and W. M. Fairbank, U. S. Patent No. 3,454,875, July 8, 1969 (Filed Sept. 3, 1963).

[74]   A. L. Kwiram and B. S. Deaver, Jr., *Phys. Rev. Lett.* **13**, 189 (1964).

[75]   A. L. Kwiram and B. S. Deaver, Jr., in *Proceedings of the 9th International Conference on Low Temperature Physics (LT9), Columbus, Ohio, 1964*, edited by J. G. Daunt, D. O. Edwards and M. Yaqub (Plenum Press, New York, 1965), p. 451.

[76]   H. L. Henry and B. S. Deaver, Jr., *Bull. Am. Phys. Soc.* **15**, 1353 (1970); H. L. Henry, Ph.D. Thesis, University of Virginia, 1970.

[77]   M. J. Buckingham, in *Proceedings of the 5th International Conference on Physics and Chemistry (LT5), Madison, Wisconsin, 1957*, edited by J. R. Dillinger (University of Wisconsin Press, Madison, 1958), p. 229.

[78]   B. D. Josephson, *Phys. Lett.* **1**, 251 (1962); *Rev. Mod. Phys.* **36**, 216 (1964); *Adv. Phys.* **14**, 419 (1965).

[79]   B. D. Josephson, in *Quantum Fluids*, edited by D. F. Brewer (North-Holland, Amsterdam, 1966), p. 174.

[80]   F. Bloch, *Phys. Rev. Lett.* **21**, 1241 (1968); *Phys. Rev.* **B2**, 109 (1970).

[81]   P. W. Anderson and J. M. Rowell, *Phys. Rev. Lett.* **10**, 230 (1963).

[82]   J. M. Rowell, *Phys. Rev. Lett.* **11**, 200 (1963).

[83]   S. Shapiro, *Phys. Rev. Lett.* **11**, 80 (1963).

[84]   W. H. Parker, D. N. Langenberg, A. Denenstein and B. N. Taylor, *Phys. Rev.* **177**, 639 (1969).

[85]   W. H. Parker, B. N. Taylor and D. N. Langenberg, *Phys. Rev. Lett.* **18**, 287 (1967).

[86]   B. N. Taylor, W. H. Parker and D. N. Langenberg, *Rev. Mod. Phys.* **41**, 375 (1969).

[87]   R. F. Dzinba, B. F. Field and T. F. Finnegan, *IEEE Trans. on Instr. and Meas.* **I.M.23**, 264 (1974).

[88]   V. Kose, *IEEE Trans. on Instr. and Meas.* **I.M.25**, 483 (1976).

[89]   T. D. Bracken, Ph.D. Thesis, Stanford University, 1971.

[90]   R. C. Jacklevic, J. J. Lambe, A. H. Silver and J. E. Mercereau, *Phys. Rev. Lett.* **12**, 159 (1964).

[91]   R. C. Jacklevic, J. J. Lambe, A. H. Silver and J. E. Mercereau, *Phys. Rev. Lett.* **12**, 274 (1964).

[92]   J. E. Zimmerman and A. H. Silver, *Phys. Lett.* **10**, 47 (1964).

[93]   J. Lambe, A. H. Silver, J. E. Mercereau and R. C. Jacklevic, *Phys. Lett.* **11**, 16 (1964).

[94]   J. E. Zimmerman and J. E. Mercereau, *Phys. Rev. Lett.* **13**, 125 (1964).

[95]   R. C. Jacklevic, J. J. Lambe, A. H. Silver and J. E. Mercereau, in *Proceedings of the 9th International Conference on Low Temperature Physics (LT9), Columbus, Ohio, 1964*, edited by J. G. Daunt, D. O. Edwards and M. Yaqub (Plenum Press, New York, 1965), p. 446.

[96]   R. C. Jacklevic, J. J. Lambe, J. E. Mercereau and A. H. Silver, *Phys. Rev.* **140A**, 1628 (1965).

[97]   J. E. Zimmerman and J. E. Mercereau, *Phys. Rev. Lett.* **14**, 887 (1965).

[98]   A. H. Silver and J. E. Zimmerman, *Phys. Rev. Lett.* **15**, 888 (1965).

[99]   J. E. Zimmerman and A. H. Silver, *Phys. Rev.* **141**, 367 (1966).

[100]   J. E. Zimmerman and A. H. Silver, *Solid State Commun.* **4**, 133 (1966).

[101]   A. H. Silver, R. C. Jacklevic and J. J. Lambe, *Phys. Rev.* **141**, 362 (1966).

[102]   A. H. Silver and J. E. Zimmerman, *Phys. Rev.* **157**, 317 (1967).

[103]   A. M. Goldman, in *Proceedings of the 9th International Conference on Low Temperature Physics (LT9), Columbus, Ohio, 1964*, edited by J. G. Daunt, D. O. Edwards and M. Yaqub (Plenum Press, New York, 1965), p. 421.

[104]   D. A. Vincent, Ph.D. Thesis, University of Virginia, 1970.

[105]   *Future Trends in Superconductive Electronics, Charlottesville, 1978*, edited by B. S. Deaver, Jr., C. M. Falco, J. H. Harris and S. A. Wolf, AIP conference proceedings (American Institute of Physics, New York, 1978), No. 44.

[106]   J. Bardeen and M. J. Stephen, *Phys. Rev.* **140A**, 1197 (1965).

[107]   M. Tinkham, *Phys. Rev. Lett.* **13**, 804 (1964).

[108]   L. P. Gorkov and N. B. Kopnin, *Sov. Phys.–JETP* **37**, 183 (1973); **38**, 195 (1974).

[109]   Y. B. Kim, C. F. Hempstead and A. R. Strnad, *Phys. Rev. Lett.* **12**, 145 (1964); *Phys. Rev.* **139**, A1163 (1965).

[110]   S. J. Poon and K. M. Wong, *Phys. Rev.* **B27**, 6985 (1983).

[111]   I. Giaever, *Phys. Rev. Lett.* **15**, 825 (1965).

[112]   J. W. Ekin, B. Serin and J. R. Clem, *Phys. Rev.* **B9**, 912 (1974).

[113] P. W. Anderson, *Phys. Today* **23**, 29 (1970); T. A. Fulton, R. C. Dynes and P. W. Anderson, *Proc. IEEE* **61**, 28 (1973); T. A. Fulton and L. N. Dunkelberger, *Appl. Phys. Lett.* **22**, 232 (1973).

[114] N. F. Pedersen, in *Advances in Superconductivity*, edited by B. Deaver and J. Ruvalds (Plenum Press, New York, 1983), p. 151.

[115] R. F. Miracky, J. Clarke and R. H. Koch, *Phys. Rev. Lett.* **50**, 856 (1983).

[116] D. J. Van Ooijen, *Phys. Lett.* **14**, 95 (1965); *Philips Res. Rep.* **22**, 219 (1967).

[117] S. T. Spence, Ph.D. Thesis, Stanford University, 1967.

[118] B. Cabrera, J. Williams and B. S. Deaver, Jr., *Bull. Am. Phys. Soc.* **13**, 1691 (1968).

[119] J. Clarke, *Phil. Mag.* **13**, 115 (1966).

[120] B. Lischke, *Phys. Rev. Lett.* **22**, 1366 (1969).

[121] H. Boersch and B. Lischke, *Z. Physik* **237**, 449 (1970).

[122] B. Lischke, *Z. Physik* **237**, 469 (1970).

[123] B. Lischke, *Z. Physik* **239**, 360 (1970).

[124] H. Wahl, *Optik* **28**, 417 (1969).

[125] H. Wahl, *Optik* **30**, 508 (1970).

[126] H. Wahl, *Optik* **30**, 577 (1970).

[127] W. L. Goodman and B. S. Deaver, Jr., *Phys. Rev. Lett.* **24**, 870 (1970).

[128] W. L. Goodman, W. D. Willis, D. A. Vincent and B. S. Deaver, Jr., *Phys. Rev.* **B4**, 1530 (1970).

[129] J. M. Pierce, *Phys. Rev. Lett.* **24**, 874 (1970).

[130] J. E. Mercereau and L. T. Crane, *Phys. Lett.* **7**, 25 (1963).

[131] J. E. Mercereau and L. T. Crane, *Phys. Rev. Lett.* **12**, 191 (1964).

[132] T. K. Hunt and J. E. Mercereau, *Phys. Rev.* **135**, A944 (1964).

[133] J. J. Finley and B. S. Deaver, Jr., *Solid State Commun.* **36**, 493 (1980).

[134] L. D. Jackel, R. A. Buhrman and W. W. Webb, *Phys. Rev.* **B10**, 2782 (1974).

[135] R. Rifkin and B. S. Deaver, Jr., *Phys. Rev.* **B13**, 3894 (1976).

[136] Y. Aharonov and D. Bohm, *Phys. Rev.* **115**, 485 (1959).

[137] Y. Aharonov and D. Bohm, *Phys. Rev.* **123**, 1511 (1961).

[138] R. G. Chambers, *Phys. Rev. Lett.* **5**, 3 (1960).

[139] H. A. Fowler, L. Marton, J. A. Simpson and J. A. Suddeth, *J. Appl. Phys.* **32**, 1153 (1961).

[140] H. Boersch, H. Hamisch, K. Grohmann and D. Wohlleben, *Z. Physik* **165**, 79 (1961).

[141] G. Möllenstedt and W. Bayh, *Phys. Blatte* **18**, 299 (1962).

[142] A. Tonomura, T. Matsuda, R. Suzuki, A. Fukuhara, N. Osakabe, H. Umezake, J. Endo, K. Shinagawa, Y. Sugita and H. Fujiwara, *Phys. Rev. Lett.* **48**, 1443 (1982).

[143] M. Peshkin, *Phys. Rev.* **A23**, 360 (1981).

[144] W. D. Willis, M. S. Thesis, University of Virginia, 1971.

[145] C. Bernido and A. Inomata, *Phys. Lett.* **77A**, 394 (1980).

[146] B. S. Deaver, Jr., and G. B. Donaldson, *Phys. Lett.* **89A**, 178 (1982).

[147] C. M. Falco and J. C. Garland, in *Nonequilibrium Superconductivity, Phonons and Kapitza Boundaries*, edited by K. E. Gray (Plenum Press, New York, 1981), p. 521.

[148] J. C. Garland and D. J. Van Harlingen, *Phys. Lett.* **47A**, 423 (1974).

[149] Y. M. Gal'perin, V. L. Gurevich and V. I. Kosub, *Sov. Phys.–JETP* **39**, 680 (1974).

[150] N. V. Zavaritskii, *JETP Lett.* **19**, 126 (1974).

[151] C. M. Pegrum, A. M. Guénault and G. R. Pickett, in *Low Temperature Physics–LT14*, edited by M. Krusius and M. Vuorio (North-Holland, Amsterdam, 1975).

[152] C. M. Falco, *Solid State Commun.* **19**, 623 (1976).

[153] D. J. Van Harlingen, D. F. Heidel and J. C. Garland, *Phys. Rev.* **B21**, 1842 (1980).

[154] G. R. Pickett, *Phys. Rev. Lett.* **47**, 134 (1981); *J. Low Temp. Phys.* **51**, 561 (1983).

[155] D. J. Van Harlingen, *Physica* **109** and **110B**, 1710 (1982).

[156] V. L. Ginsburg, G. F. Zharkov and A. A. Sobyanin, *J. Low Temp. Phys.* **47**, 427 (1982).

[157] C. M. Falco, *Phys. Rev.* **B14**, 3853 (1976).

[158] M. Bol, Ph.D. Thesis, Stanford University, 1965.

[159] A. F. Hildebrandt, *Phys. Rev. Lett.* **12**, 190 (1964).

[160] A. F. Hildebrandt and M. Saffren, in *Proceedings of the 9th International Conference on Low Temperature Physics (LT9), Columbus, Ohio, 1964*, edited by J. G. Daunt, D. O. Edwards and M. Yaqub (Plenum Press, New York, 1965), p. 459.

[161] C. A. King, J. B. Hendricks and H. E. Rorschach, in *Proceedings of the 9th International Conference on Low Temperature Physics (LT9), Columbus, Ohio, 1964*, edited by J. G. Daunt, D. O. Edwards and M. Yaqub (Plenum Press, New York, 1965), p. 466.

[162] M. Bol and W. M. Fairbank, in *Proceedings of the 9th International Conference on Low Temperature Physics (LT9), Columbus, Ohio, 1964*, edited by J. G. Daunt, D. O. Edwards and M. Yaqub (Plenum Press, New York, 1965), p. 471.

[163] N. F. Brickman, *Phys. Rev.* **184**, 460 (1969).

[164] W. H. Parker and M. B. Simmonds, *Precision Measurements and Fundamental Constants I*, National Bureau of Standards (U.S.) Publication 343, 243 (1970).

[165] K. Oide and H. Hirakawa, *J. Phys. Soc. Japan Lett.* **43**, 1087 (1977).

[166] B. Cabrera, S. B. Felch and J. T. Anderson, in *Precision Measurements and Fundamental Constants II*, edited by B. N. Taylor and W. D. Phillips, National Bureau of Standards (U.S.), Special Publication 617 (1981).

[167] B. Cabrera, H. Gutfreund and W. A. Little, *Phys. Rev.* **B25**, 6644 (1982).

[168] Y. Imry, in *Anderson Localization—Proceedings of the 4th Taniguchi International Symposium on the Theory of Condensed Matter* (Springer, Berlin, 1982), p. 198.

[169] A. A. Shablo, T. P. Narbut, S. A. Tyurin and I. M. Dmitrenko, *JETP Lett.* **19**, 246 (1974).

[170] I. O. Kulik, *Sov. Phys.–JETP* **31**, 1172 (1970).

[171] I. O. Kulik and K. V. Mal'chuzhenko, *Sov. Phys.–Solid State* **13**, 2474 (1972).

[172] L. Gunther and Y. Imry, *Solid State Commun.* **7**, 1391 (1969).

[173] I. O. Kulik, *JETP Lett.* **11**, 275 (1970).

[174] E. N. Bogachek, G. A. Gogadze and I. O. Kulik, *Phys. Stat. Sol. B* **67**, 287 (1975).

[175] D. Y. Sharvin and Y. V. Sharvin, *JETP Lett.* **34**, 273 (1981).

[176] B. L. Al'tshuler, A. G. Aronov and B. Z. Spivak, *JETP Lett.* **33**, 94 (1981).

[177] K. von Klitzig, G. Dorda and M. Pepper, *Phys. Rev. Lett.* **45**, 494 (1980).

[178] D. C. Tsui and A. C. Gossard, *Appl. Phys. Lett.* **38**, 550 (1981).

[179] M. A. Paalanen, D. C. Tsui and A. C. Gossard, *Phys. Rev. B* **25**, 5566 (1982).

[180] R. B. Laughlin, *Phys. Rev. B* **23**, 5632 (1981).

[181] *IBM J. Res. Develop.* **24** (1980) is devoted entirely to Josephson Computer Technology.

[182] H. Zappe, in *Advances in Superconductivity*, edited by B. Deaver and J. Ruvalds (Plenum Press, New York, 1983), p. 51.

[183] H. H. Zappe, *Appl. Phys. Lett.* **27**, 432 (1975).

[184] S. M. Faris, W. H. Henkels, E. A. Valsamakis and H. H. Zappe, *IBM J. Res. Develop.* **24**, 143 (1980).

[185] P. Guéret, A. Moser and P. Wolf, *IBM J. Res. Develop.* **24**, 155 (1980).

[186] B. Cabrera, *Phys. Rev. Lett.* **48**, 1378 (1982).

[187] A. J. Leggett, *Prog. Theor. Phys. Suppl.* **69**, 80 (1980).

[188] A. O. Caldeira and A. J. Leggett, *Phys. Rev. Lett.* **46**, 211 (1981).

[189] R. F. Voss and R. A. Webb, *Phys. Rev. Lett.* **47**, 265 (1981).

[190] L. D. Jackel, J. P. Gordon, E. L. Hu, R. E. Howard, L. A. Fettar, D. M. Tennant, R. W. Epworth and J. Kurkijärvi, *Phys. Rev. Lett.* **47**, 697 (1981).

[191] S. Chakravarty and S. Kivelson, *Phys. Rev. Lett.* **50**, 1811 (1983).

Near Zero Magnetic Fields
with Superconducting Shields

Blas Cabrera

1. INTRODUCTION

In the 1961 article announcing the discovery of flux quantization at Stanford [1] is the following statement:

> Below a certain value of applied field, the total cross section of the cylinder acts as a perfect diamagnet, excluding all the flux, and no flux is trapped when the applied field is turned off. (We believe this provides a way of obtaining a truly zero magnetic field region.)

Thus, as has been the case on many occasions throughout his distinguished career, William Fairbank foresaw important applications of a discovery from the earliest moments.

Conventional magnetic shielding with ferromagnetic materials has been used since the nineteenth century. Today, with the advent of modern high-mu materials, nested shields with remanent fields approaching a microgauss are produced with drifts of order 0.1 μG per minute. The discovery of superconductivity in 1911 opened the way for a totally new type of magnetic shield. Because of truly zero dc resistance, the field inside a superconducting shield remains constant under a changing externally applied field. The induced shielding currents do not decay as they do for a normal conductor such as copper, thus arbitrarily slow changes in external fields are shielded perfectly. The next step in sophistication was to reduce the ambient field as the shield was cooled below its superconducting transition temperature to obtain low absolute fields as well, possibly reaching the zero flux quantum state as suggested by Bill Fairbank in 1961.

Shortly after Fairbank's suggestion, W. O. Hamilton, also at Stanford, proposed an expansion technique for achieving this aim in a large shield. In the middle 1960's measurements were made by Hamilton and D. K. Rose on expandable superconducting shields constructed of niobium sputtered on Kapton sheeting [2]. These were cooled in a folded configuration and then opened, acting as a flux pump to reduce the ambient field inside the shield.

Elsewhere, successful work was reported by Brown using expandable shields made of pure lead foil [3], and by Hildebrandt using thick rigid lead shields which were slowly cooled in an open configuration and achieved field attenuation through the Meissner effect [4].

When I began my graduate studies at Stanford under Bill Fairbank, all reported research on superconducting shielding had used fluxgate magnetometers with resolutions no better than a microgauss. Brown had reported absolute fields at this resolution limit. Working closely with George Hess during the summer of 1968 and with Bill Hamilton until his departure from Stanford in 1970, we decided to build a SQUID magnetometer modeled after a successful instrument which had just been completed by Art Hebard for his fractional electric charge search. (The two-point niobium foil SQUID was based on a design developed by M. R. Beasley.) The expected resolution was better than a nanogauss. Because of continuing adhesion problems with niobium sputtered on Kapton, we decided to adopt the Brown technique using pure lead foil in an expandable configuration. By the mid-1970's we had developed a technique for producing ultra low magnetic shields with remanent fields approaching 10 nG over a useful volume [5]. We use the term "ultra low magnetic shielding" to indicate absolute fields below the conventional high-mu shielding limit of a microgauss.

My original motivation for developing the ultra low magnetic shielding was for an experiment to precisely determine h/m_e using rotating superconducting rings (see paper entitled "Rotating Superconductors and Fundamental Physical Constants" in this volume), although Fairbank and Hamilton were also interested in its application to the relativity gyroscope experiment and the ^{3}He nuclear gyroscope. In fact, I first used the technique, as part of my thesis work, for precise magnetic charge measurements of the niobium sphere fractional electric charge candidates from the Fairbank-Hebard-LaRue-Phillips experiments (see paper entitled "A Superconductive Detector to Search for Cosmic Ray Magnetic Monopoles" in this volume).

In 1974 the technique was applied to work on the ^{3}He nuclear gyroscope, under development to perform a precise electric dipole moment measurement on the ^{3}He nucleus (see paper entitled "The ^{3}He Nuclear Gyroscope and a Precise Electric Dipole Moment Measurement" in this volume). To date, this research remains the most demanding application of the ultra low field technology.

Between 1975 and 1979, working together with Frank van Kann until his departure in 1977, we adapted the technique for use on the relativity gyroscope experiment [6] (see papers entitled "The Stanford Relativity Gyroscope Experiment" in this volume). During this work using the absolute magnetometer, the cumbersome two-point contact SQUID which I had made during my thesis work was replaced with a commercial SHE unit. In addition, significant improvements were made in the expansion and cooling techniques.

A brief description of the technique is presented below, together with a summary of our work on absolute magnetometry using SQUID sensors and its application to the design of other instruments which use the ultra low field environment. Prospects for future improvements are discussed in closing.

2. MAKING AN ULTRA LOW MAGNETIC FIELD SHIELD

The technique uses expandable cylindrical superconducting shields as mechanical flux pumps to reduce the magnetic field below ambient levels. The cylinders, made of 70 μm thick lead foil (containing 0.2% tin), are tightly folded in accordion fashion as shown in figure 1a. As these are slowly cooled through their superconducting transition temperature they trap the ambient magnetic field. The field penetrates the foil as an array of supercurrent vortices, each encircling one quantum of magnetic flux. Because of the Meissner effect, an ideal superconductor would spontaneously exclude all of the ambient magnetic field. However, in practice it is very difficult to observe large Meissner effects because defects in the superconducting material produce strong flux pinning sites which prevent flux motion. Once the shield is superconducting, any changes in the external magnetic field are exactly canceled by supercurrents on the outer surface of the shield. Therefore, by mechanically expanding one of these tightly folded cylinders while maintaining it superconducting, a region of lower magnetic field is obtained, as shown in figure 1b. Field reduction factors of 0.1 to 0.01 are obtained. During the expansion the supercurrent vortices remain strongly pinned; thus the magnetic field lines are tied to fixed locations along the shield surfaces. A single high-mu shield is used around the outside of the

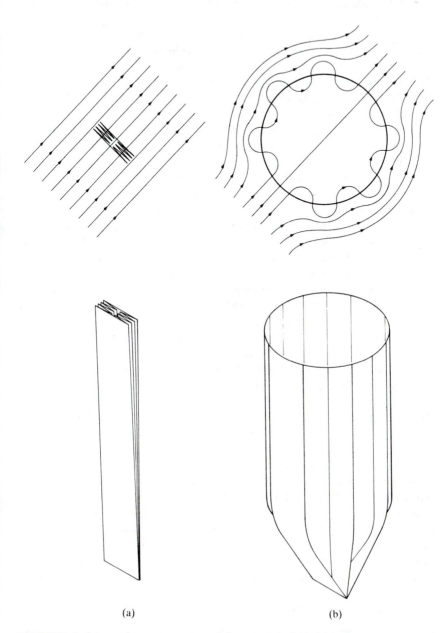

(a)   (b)

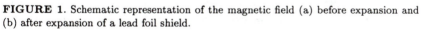

FIGURE 1. Schematic representation of the magnetic field (a) before expansion and (b) after expansion of a lead foil shield.

dewar, providing an ambient field of several milligauss outside of the superconducting shields.

To obtain the lowest magnetic fields a bootstrap technique is used on a sequence of identical prefolded shields (as in figure 1a). Each new tightly folded shield is mounted in a vacuum-insulated tube and lowered inside the most recently expanded shield which remains submerged in the liquid helium bath (4.2 kelvin). The temperature inside the cooling tube is then slowly reduced until the new shield has become superconducting (7.2 kelvin). There is no longer any need for the outer shield. Now that the new shield has become superconducting, magnetic field changes are exactly canceled by induced supercurrents in its surfaces. The old shield is torn away and the cooling tube is raised out of the dewar, leaving behind the new shield now submerged in the bath. Then the new shield is expanded using a spherically shaped mechanical plunger. This shield is now in the same configuration as the previous shield but contains a lower magnetic field inside. One cycle has been completed. After three or four such identical cycles, field levels below 100 nG are obtained routinely.

Typically, each cycle is completed in one week; thus the production of a final shield requires about six weeks. Because of the lengthy lead time and because warming the final shield above its superconducting transition requires repeating the entire six-week expansion procedure, these shields once produced are kept perpetually submerged in a liquid helium bath. Efficient dewar systems are used to reduce maintenance costs. Several shields presently in use at Stanford are more than three years old.

Magnetic field profiles obtained in a final shield are shown in figure 2. Three component measurements along the axis of a 20 cm diameter × 100 cm long shield are shown, and correspond to the magnetic field averaged over a 5 cm diameter loop (see discussion on the magnetometer). Over a 15 cm length this shield provides an ultra low field environment of about 20 nG. Less than 30 flux quanta thread the cylinder in this region.

As seen in figure 2, an increase in the field is always found near the bottom of the shields where the cylinder is closed and thus never expanded. In addition, an exponential increase in all field components is found approaching the open top of the shield. This rapid increase results from the external magnetic field leaking in through the open top. Assuming that there are no sources of field inside the shield, the leading term in a series solution of Laplace's equation within the shield predicts field attenuation of the form

$$\mathbf{B}(r,\theta,z) = \mathbf{B}(r,\theta)e^{-k'_{11}z/a} \tag{1}$$

where $k'_{11} = 1.8412$ is the first zero of the derivative of J_1, a is the shield radius, and z is measured in from the top of the shield. Thus for each radius in from the open end, all field components are attenuated by at least

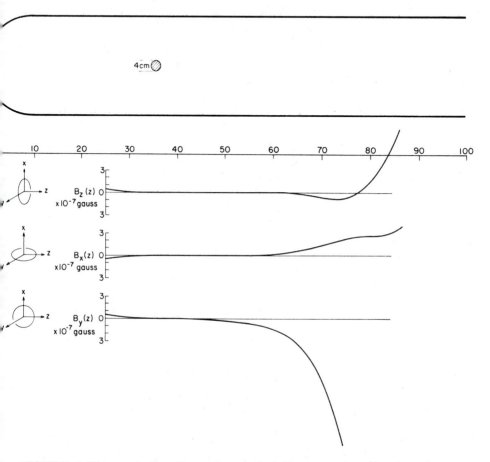

FIGURE 2. Three mutually orthogonal magnetic field component profiles along the axis of a 20 cm diameter × 100 cm long superconducting shield.

a factor of $e^{-1.8412} = 0.159$. The magnetic field lines resulting from the leakage of a transverse external field have been calculated along a portion of the shield and are shown in figure 3. The plot is oriented along the symmetry plane of the magnetic field which is parallel to the externally applied field.

Next we discuss the instrument used to measure the ultra low magnetic field.

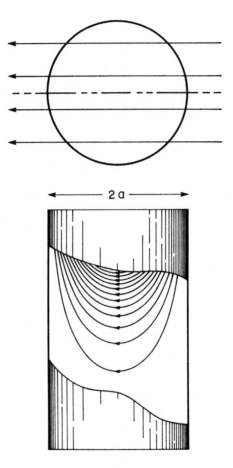

FIGURE 3. Exponentially attenuated magnetic field resulting from the leakage of an external magnetic field through the open top of a cylindrical superconducting shield.

3. ABSOLUTE MAGNETOMETRY

Constructing magnetometers capable of measuring ultra low magnetic fields has been the most difficult part of the work. It has been relatively straightforward to obtain lower magnetic fields by simple modification of the expansion technique; however, it has proven much more difficult to make instruments with background magnetic contamination levels below the residual fields in the shields.

The instrument used for measuring the magnetic field profiles of figure 2 is shown in figure 4. It is capable of measuring absolute fields at the nanogauss level. It has a flip coil assembly made almost entirely of fused

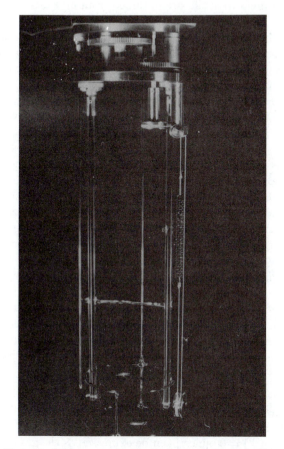

FIGURE 4. Photograph of nanogauss absolute magnetometer. This magnetometer was also used as the prototype detector in a search for a flux of cosmic ray supermassive magnetic monopoles (see paper entitled "A Superconductive Detector to Search for Cosmic Ray Magnetic Monopoles" in this volume.)

quartz and housed in an aluminum vacuum can. Rotation is achieved using a dacron string wrapped around a quartz pulley, and the tension in the string is maintained with a fused quartz spring. Four turns of 0.005 cm diameter niobium wire are wound inside the 2 mm diameter quartz tubing which had been formed into a 5 cm diameter circular coil form. The twisted-pair niobium leads come out through the axis and go up to a SQUID sensor some 35 cm above the coil. As the flip coil is rotated, the ambient magnetic flux threading it changes. Since the total flux through the superconducting circuit must remain constant, a compensating supercurrent change is induced in the circuit and detected with the SQUID. Absolute magnetic field measurements are obtained from an exact 180° rotation of the flip coil.

This same instrument has been used more recently as the prototype detector in a search for a flux of cosmic ray magnetic monopoles (see paper entitled "A Superconductive Detector to Search for Cosmic Ray Magnetic Monopoles" in this volume).

Learning to make an absolute magnetometer capable of measuring the field levels in the ultra low field shields has taught us how to construct apparatus for other experiments which fully utilize the advantages of the ultra low field environment. The magnetic fields associated with the apparatus must be reduced to levels comparable to those within the shield. As expected, ferromagnetic contamination must be kept at a minimum both within the bulk and on all surfaces. Measurements have been made and tabulated on the typical remanent magnetization of many candidate construction materials [6]. At these low field levels nearly all available materials exhibit detectable residual fields; however, these stable fields have not been the biggest concern in the work. Unexpectedly, much larger magnetic fields are found associated with nonsuperconducting metals.

These fields arise from thermal gradient induced currents. It is well known that in an isotropic normal conductor a temperature gradient will generate an emf. For many pure crystalline metals the thermopower alpha is isotropic. In this case, we can express the current in the material as

$$\mathbf{j} = -(\sigma/e)\boldsymbol{\nabla}\mu - \sigma\alpha\boldsymbol{\nabla}T \tag{2}$$

where μ is the electrochemical potential, σ is the electrical conductivity, and $\alpha(T)$ is the thermopower. In the steady state div $\mathbf{j} = 0$ and from equation (2), curl $\mathbf{j} = 0$ (since $\mathbf{j}$ can be expressed as the gradient of a scalar function). Thus $\mathbf{j} = 0$ throughout the sample and no magnetic fields are generated.

However, if the thermopower is different along different directions in the sample, then equation (2) must be replaced by the tensor equation

$$j_i = -\frac{1}{e}\sigma_{ij}\nabla_i\mu - \sigma_{ij}\alpha_{jk}\nabla_k T \quad . \tag{3}$$

In this case we are led to the conclusion that we must have circulating currents. In two dimensions, for example, the circulating current in the x-y plane will be proportional to the temperature gradient. From equation (3),

$$j_{cir} \propto \bar{\sigma}(\alpha_x - \alpha_y)\nabla T \quad , \tag{4}$$

where $\bar{\sigma}$ is the average electrical conductivity, and α_x and α_y are the thermopowers in the x and y direction, respectively [7]. Thus, in general, a magnetic field is produced by a temperature gradient along an anisotropic normal conductor.

The key contribution, made by Peter Selzer as part of his thesis work on anisotropic crystalline tin [7], was that the metal need not have an anisotropic crystalline structure, but that it is sufficient for it to have been stressed to exhibit an anisotropy in the thermopower. Construction materials are typically rolled or extruded, leading naturally to such anisotropies. Near 4 K in an OFHC (oxygen-free high conductivity) copper test piece (isotropic in crystalline form), Selzer found a generated field of 0.1 gauss cm/K. Later, we found similar coefficients for aluminum and for lead above its superconducting transition temperature of 7.2 K. Thus even small temperature gradients along metallic structural parts of an apparatus can easily produce fields many orders of magnitude larger than the ambient magnetic field inside the ultra low magnetic field shields. This effect has caused the dominant contamination fields throughout the research.

We have largely overcome this problem by eliminating normal conductors entirely from the sensitive area of an apparatus. Thus only clean insulating materials such as glasses or ceramics are used for structural members. The quartz flip coil assembly in figure 4 is an example.

Thermally induced fields also limit the lowest absolute fields attainable with the expandable shield technique. As the lead is cooled through its transition temperature, any temperature gradient will generate fields which are then trapped in the superconducting phase. Since these fields are self-generated, subsequent expansions do not produce further field reductions. We have produced our best shields by cooling in a vacuum with a small helium exchange gas pressure (about 10^{-5} torr), thus reducing the thermal gradient along the shields. Once the shield is superconducting, the thermopower disappears and magnetic fields are no longer generated by thermal gradients.

4. CONCLUSIONS

We have developed a technique for generating ultra low magnetic fields over usable volumes with expandable superconducting shields. Several experiments have successfully used field levels approaching 10 nG. Thermally induced magnetic fields have produced the greatest noise sources, but extensive use of nonconducting materials reduces the contamination fields to acceptable levels.

Future work with all fused quartz structures will allow experiments in fields approaching a nanogauss. Further improvements will be more difficult as the remanent magnetization of nearly all materials becomes the dominant problem. But as I have learned from Bill Fairbank over the years, never say that anything is impossible!

Acknowledgments

It is a particular pleasure to acknowledge the seed ideas suggested by Bill Fairbank and Bill Hamilton. So many others at Stanford also have contributed ideas and collaborated in the development of the near zero magnetic field technique that I do not have space to acknowledge them individually. Suffice it to say that Bill Fairbank has always maintained a relentless enthusiasm which rubbed off on us all.

The work described here has been funded over the years by the Air Force Office of Scientific Research, the Office of Naval Research, the Army Research Office, the National Science Foundation, and the National Aeronautics and Space Administration.

References

[1]   B. S. Deaver and W. M. Fairbank, *Phys. Rev. Lett.* **7**, 43 (1961).

[2]   W. O. Hamilton, *Rev. Phys. Appli.* **5**, 41 (1970).

[3]   R. E. Brown and W. M. Hubbard, *Rev. Sci. Instr.* **36**, 1378 (1965); R. E. Brown, *Rev. Sci. Instr.* **39**, 547 (1963).

[4]   A. F. Hildebrandt, *Rev. Phys. Appli.* **5**, 49 (1970).

[5]   B. Cabrera and W. O. Hamilton, *The Science and Technology of Superconductivity* (Plenum, 1973), Vol. II, p. 587; B. Cabrera, Ph.D. Thesis, Stanford University, 1974.

[6]   B. Cabrera and Frank van Kann, *Acta Astronautica* **5**, 125 (1978).

[7]   P. M. Selzer, Ph.D. Thesis, Stanford University, 1974.

Rotating Superconductors and Fundamental Physical Constants

B. Cabrera, S. B. Felch and J. T. Anderson

1. INTRODUCTION

In 1961 flux quantization was discovered at Stanford by Bascom Deaver and William Fairbank [1] (also independently by Doll and Näbauer in Germany). Then in 1964 Morris Bol and Fairbank verified the existence of the London moment, a spontaneous magnetic moment developed by a superconductor when set into rotation. (Again, independent experiments were performed by Hildebrandt et al. and by King et al. [2].) Several years later Bill Fairbank, in discussions with George Hess and other group members, realized that a logical next step was the determination of h/m_e (Planck's constant divided by the free electron mass), which could be done using a rotating superconducting ring. The measurement is based on balancing an integral number of flux quanta $n(hc/2e)$ against the London moment flux $(2mc/e)\omega S$, where S is the area bounded by the ring and ω the angular velocity. Experiments based on this technique by Zimmerman and Mercereau, and later by Parker and Simmonds [3] at a higher resolution of 400 parts per million (ppm), have been in agreement with the accepted value of h/m_e for the free electron at rest obtained by other techniques.

Technological advances in superconductive devices and in high precision dimensional metrology during the intervening decade now allow a significant improvement in these measurements. The experiment described here is aimed at a resolution approaching 1 ppm, for the first time allowing investigation of relativistic corrections to the electron mass within a conducting lattice. A quantitative theory by B. Cabrera, H. Gutfreund and W. A. Little at Stanford has predicted corrections of 100–200 ppm in many superconducting materials [4]. If the experimental results are in agreement with this theory, the data can be corrected for the relativistic shift, and an independent determination of h/m would be obtained at the 1 ppm level or better. This ratio of fundamental physical constants is particularly interesting, since

$$\alpha^2 = \frac{2R_\infty}{c} \left(\frac{h}{m_e} \right) \quad , \tag{1}$$

where α is the fine structure constant of atomic physics. The Rydberg constant R_∞ and the speed of light c are the two best known constants, known to better than 10 ppb. Thus, a measurement of h/m leads directly to a determination of α.

We will discuss a simple theoretical derivation for flux quantization and the London moment. Then we will describe the apparatus and summarize recent progress.

2. FLUX QUANTIZATION, LONDON MOMENT AND THE h/m MEASUREMENT

The theoretical basis for the determination of h/m can be simply understood using the nonrelativistic Ginzburg-Landau supercurrent density [5]

$$\mathbf{j} = \frac{e^* \hbar}{2im^*} (\psi^* \boldsymbol{\nabla} \psi - \psi \boldsymbol{\nabla} \psi^*) - \frac{e^{*2}}{m^* c} \psi^* \psi \mathbf{A} \quad , \tag{2}$$

where the Cooper pair mass m^* and charge e^* are equal to $2m$ and $2e$, respectively, and ψ is the superconducting order parameter with a phase $\phi, \psi = |\psi| e^{i\phi}$. Equation (2) looks nearly identical to the standard Schrödinger current density equation for a single electron in a magnetic field, where $\psi^* \psi$ would be the probability of finding the electron at a particular location. Now we interpret $\psi^* \psi$ as the density of Cooper electron pairs which form a many-particle coherent wave with phase ϕ.

To describe the rotating superconducting ring, we transform equation (2) into the ring's rotating frame using the Galilean relation $\mathbf{v}' = \mathbf{v} - \boldsymbol{\omega} \times \mathbf{r}$. The canonical momentum $\mathbf{p} - \frac{e}{c} \mathbf{A}$ becomes $\mathbf{p} - \frac{e}{c} (\mathbf{A} + \mathbf{A}_\omega)$, where we define an effective vector potential due to the rotation as $\mathbf{A}_\omega = \frac{mc}{e} \boldsymbol{\omega} \times \mathbf{r}$. Then

$$\mathbf{j} = \frac{e\hbar}{m} |\psi|^2 \boldsymbol{\nabla} \phi - \frac{2e^2}{mc} |\psi|^2 (\mathbf{A} + \mathbf{A}_\omega) \quad , \tag{3}$$

where $\mathbf{A}_\omega$ appears on equal footing with the magnetic vector potential $\mathbf{A}$.

Before proceeding, we can easily show that the current density $\mathbf{j}$ vanishes inside a thick superconductor, by taking the curl twice on both sides of equation (3) ($n_s = 2|\psi|^2$ remains constant to a high degree throughout the superconductor):

$$\nabla \times (\nabla \times \mathbf{j}) = -\frac{e^2 n_s}{mc} \nabla \times \mathbf{B} \quad , \tag{4}$$

where $\nabla \times (\nabla \times \mathbf{A}_\omega) = 0$. Using Maxwell's equation $\nabla \times \mathbf{B} = \frac{4\pi}{c}\mathbf{j}$ and taking $\nabla \cdot \mathbf{j} = 0$ for a time-independent current, we obtain

$$\nabla^2 \mathbf{j} - \frac{1}{\lambda^2}\mathbf{j} = 0 \quad , \tag{5}$$

where

$$\lambda = \left(\frac{mc^2}{4\pi e^2 n_s}\right)^{1/2} \tag{6}$$

is the London penetration depth, typically 300–500 Å. Solutions for the supercurrent density $\mathbf{j}$ in equation (5) fall exponentially to zero over a characteristic distance λ into the superconductor. For a semi-infinite slab of superconductor the exact solution is $\mathbf{j}(z) = \mathbf{j}(0)e^{-z/\lambda}$.

By integrating equation (3) around a closed path Γ contained within the wire of a superconducting ring which rotates at angular velocity ω, we obtain

$$\frac{4\pi\lambda^2}{c} \oint_\Gamma \mathbf{j} \cdot d\boldsymbol{\ell} = n\left(\frac{hc}{2e}\right) - \int_{S_\Gamma} \mathbf{B} \cdot d\mathbf{S} - \frac{2mc}{e}\boldsymbol{\omega} \cdot \mathbf{S}_\Gamma \quad , \tag{7}$$

where $n_s = 2\psi^*\psi$ and S is the area bounded by Γ. We have set the line integral of $\nabla\phi$ around a closed path equal to an integral multiple n of 2π to satisfy the physical requirement that the order parameter ψ be single-valued.

Considering a nonrotating thick ring ($\omega = 0$) and taking any path Γ everywhere many λ away from the wire surface where $\mathbf{j} = 0$, we find from equation (7) that the total flux through the area bounded by the ring must be an integer number of ϕ_0:

$$\int_{S_\Gamma} \mathbf{B} \cdot d\mathbf{S} = n\phi_0 \quad , \tag{8}$$

where

$$\phi_0 = hc/2e = 2.07 \times 10^{-7} \, \text{gauss cm}^2 \tag{9}$$

is the flux quantum of superconductivity (see figure 3 for recent experimental measurements).

The rotation of a thick, singly-connected superconductor can also be analyzed using equation (7). Again $\mathbf{j}$ vanishes and $n = 0$, since there are no holes in the order parameter ψ (neglecting trapped flux). Now the flux through any path Γ is

$$\int_{S_\Gamma} \mathbf{B} \cdot d\mathbf{S} = -\frac{2mc}{e}\boldsymbol{\omega} \cdot \mathbf{S}_\Gamma \quad . \tag{10}$$

Taking an arbitrary infinitesimal path, we find that $\mathbf{B}$ is uniform and constant throughout the interior of the superconductor and given by

$$\mathbf{B} = -\frac{2mc}{e}\boldsymbol{\omega} \quad . \tag{11}$$

A uniform magnetization is produced throughout the bulk and accounts for the observed London magnetic moment, which is proportional to spin speed. The readout of the superconducting relativity gyroscope experiment is based on accurately monitoring the London moment of a spinning superconducting sphere (see paper entitled "The Stanford Relativity Gyroscope Experiment (C): London Moment Readout of the Gyroscope" in this volume.)

The measurement of h/m is also derived from equation (7). We do not require a thick ring, and here for simplicity we assume the only source of $\mathbf{B}$ to be the supercurrent $\mathbf{j}$. Then for each number n of flux quanta threading the ring, there always exists a unique frequency ω_n such that $\mathbf{j} = 0$ and simultaneously $\mathbf{B} = 0$. We obtain the simple relation

$$h/m = 4\Delta\omega S \quad , \tag{12}$$

where $\Delta\omega = \omega_n - \omega_{n-1}$ is the angular velocity difference between successive nulls and S is the cross-sectional area bounded by the ring. Thus, by measuring a macroscopic area and a frequency difference, a high precision determination of h/m is possible.

The theoretical analysis for the relativistic corrections [4] to equation (12) can be summarized by replacing m with an effective mass m' (not to be confused with the solid state effective mass), where

$$m' = m\left[1 + \frac{\langle T \rangle}{mc^2} - \frac{e\Phi_{in}}{mc^2}\right] \quad . \tag{13}$$

The first parenthetical term is due to the mass-velocity shift and is given by the expectation value of the electron kinetic energy $\langle T \rangle$ averaged over Fermi surface states. It is a factor of 5 or more larger than the second correction term, which depends on the average bulk electrostatic potential Φ_{in}. Net mass increases of 100 ppm or larger are predicted for several superconductors.

3. THE EXPERIMENT

The measurements are made with a cryogenic helium gas bearing constructed entirely of fused quartz (see figure 1). The rotor is a sphere of radius 25 mm truncated 1 mm above its equator. A 15 μm wide by 40 nm thick superconducting ring is deposited onto the equatorial plane using photoresist masking techniques (see figure 2). Gas bearings in the flat top and spherical bottom of the housing support the rotor with clearances of 15 μm. Helium spin up channels are located in the sides of the housing, and spin speeds above 100 Hz in both directions are possible. The rotor spin speed is measured using a fiber optic readout mounted through the bearing on top. The intensity of the reflected light is modulated by an encoder pattern, etched into the rotor top surface, as it passes continuously under the fiber optic bundle.

FIGURE 1. Photograph of all-quartz helium gas bearing. The areas of both interchangeable rotors are known to 2 ppm.

The magnetic flux is measured using a superconducting pickup loop placed in a groove cut into the housing. This loop is connected to a SQUID

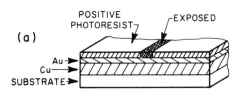

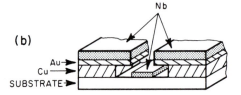

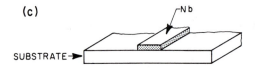

FIGURE 2. Deposition of a precision ring. (a) Successive layers (0.5 μm of copper, 100 nm of gold and 10 μm of positive photoresist) form a band around the rotor equator. A fine line of resist is exposed with a focused He-Cd laser. (b) The resist is then developed, and the gold and copper are chemically etched, forming an all metal mask after the removal of the remaining resist. The superconductive material (here niobium) is then deposited. (c) Finally, the mask is lifted off using a chemical copper etch.

magnetometer and is capable of resolving 0.01 of a flux quantum through the rotating ring in a several hertz bandwidth. As the rotor spin speeds are slowly varied, the ring is continuously modulated through its transition temperature at 1 to 10 Hz using a second micro-fiber optic line, which shines an infrared laser directly onto the superconducting line. The overall temperature of the bearing and rotor is stabilized by feedback control on heaters in the spin up gas lines.

An important part of the experiment is the use of an ultra low magnetic field shield around the bearing (see paper entitled "Near Zero Magnetic Fields with Superconducting Shields" in this volume). The shield used for these measurements provides an absolute magnetic field environment of 50 nG and is essential for obtaining the required stability in the SQUID detection circuit.

4.   THE MEASUREMENTS

Recent preliminary measurements on a niobium wire ring taken with this apparatus have demonstrated sufficient resolution to determine $\Delta\omega$ to several ppm [6]. Figure 3 shows the quantization of magnetic flux when the ring, held stationary, is biased within several mK of its transition temperature. This 5 cm diameter ring represents the largest area for which flux quantization has been directly verified. Determination of the frequency spacing $\Delta\omega$ is accomplished by counting the integer number of magnetic field nulls over a range of spin speeds. A change in frequency of 1 Hz corresponds to passage through 70 nulls. In figure 4 raw data over a 1 Hz range are shown. A single slope and the null spacing $\Delta\omega$ were determined by a least squares fit to equation (7).

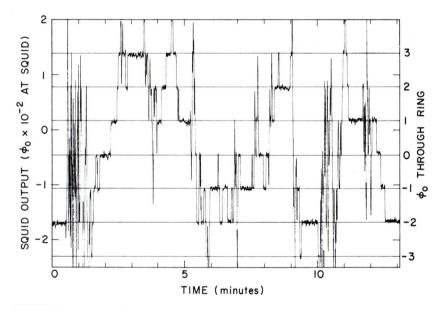

FIGURE 3. Thermally induced transitions between quantum states. These clearly demonstrate flux quantization in this 5 cm diameter ring.

These very encouraging initial data demonstrate an rms resolution of better than 1% of the frequency interval between successive nulls. Thus, for spin speeds between ±50 Hz, covering 7000 nulls, the same 1% resolution between successive nulls will result in a statistical error for $\Delta\omega$ of 2 ppm. Since improvements in this accuracy to better than 0.1 ppm seem relatively straightforward, for the foreseeable future the dominant error

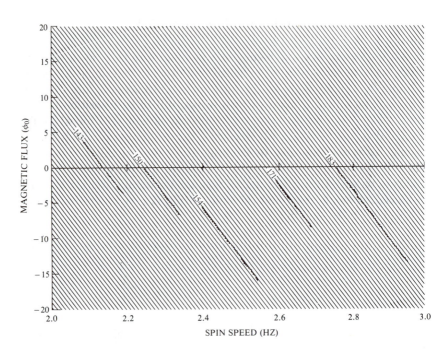

FIGURE 4. Flux *versus* spin speed data used to determine $\Delta\omega$.

in the experimental determination of h/m through equation (12) is the determination of the area S.

We are completing the deposition of a precise niobium ring. Its thickness (40 nm) and width (15 μm) on the equatorial plane of the rotor are chosen so that all possible cross-sectional areas bounded within the superconductor have the same value to a resolution of 2 ppm. Thus, determining the area of the ring to this accuracy requires only measurements on the equatorial plane of the bare rotor prior to the deposition.

The area of a plane figure cannot be determined by any number of diametral measurements, since there exist an infinite number of constant width figures other than a circle, all having the same circumference but varying in area by as much as 10%. To avoid this ambiguity in area, the radius of the figure as a function of angle must be measured with respect to a fixed point. The uncertainty in diametral measurements is eliminated by using the out-of-roundness data for the rotor to correct one absolute diameter measurement to a true mean diameter.

A computer-aided Talyrond roundness measuring system, available to us through the Stanford relativity gyroscope program (see paper entitled "The Stanford Relativity Gyroscope Experiment (B): Gyroscope Development" in this volume) is used to determine both the rotor profile and its least squares reference circle with a precision of better than 5 nm. If matched by corresponding accuracy in the diametral measurement, the area could be determined to 0.1 ppm. Sub-ppm diameter measurements are very difficult, although significant advances appear likely in the near future. The National Bureau of Standards, Gaithersburg, has been assisting us with the diameter measurements using a horizontal diameter measuring machine based on the Hewlett Packard laser interferometer. Two of our precision rotors have been measured to 1 ppm along a known diameter, providing a 2 ppm uncertainty ($1\ \sigma$) in the determination of their areas after the roundness corrections have been made [7].

Since the area measurements were made at 20 °C and the $\Delta\omega$ data are taken at 4 K, a correction must be made for the thermal expansion of fused quartz. A total change of 17 ppm is expected. Direct measurements on samples from the original rotor blanks will be made in a separate apparatus. Again, the absolute diameter determination will dominate the final error.

5. CONCLUSIONS

Initial cryogenic measurements of the induced magnetic flux from a rotating superconducting ring (London moment), made with the precision helium gas bearing, have demonstrated an improvement in resolution by two orders of magnitude over previously reported observations. Flux changes from angular frequency changes as small as a few mHz were detected. We also clearly observed flux quantization in our 5 cm diameter ring. To our knowledge this represents the largest area in which quantization has been observed. To date, our combined measurements determine h/m to 3 parts per thousand. The value is in agreement with the accepted value (known to several tenths of a part per million from other experiments not using superconductivity) and is limited by our preliminary wire ring's area uncertainty. Using this same apparatus with minor improvements, we will next make measurements on a precisely deposited niobium ring, and we expect to reach a resolution of 2 ppm on the determination of $\Delta\omega$. Combining this result with our area determination corrected to 4 K, we expect an overall uncertainty of 2 ppm on the right side of equation (12).

By comparing this value with the presently accepted value of h/m for the free electron at rest, we will be investigating for the first time relativistic corrections to the electron mass within the superconducting lattice. Detailed condensed matter calculations based on the relativistic theory of Cabrera, Gutfreund and Little are expected to match the experimental accuracy. Various elemental superconductors will be measured to verify these calculations.

If the theory is in good quantitative agreement with the measurements, we will turn the argument around and obtain a corrected independent value for h/m at the several ppm level. We will also obtain an independent value for α through equation (1). Further improvements which reach sub-ppm levels would provide a significant input datum for the next least squares adjustment of the fundamental physical constants.

Acknowledgments

We greatly appreciate many useful discussions with Bill Fairbank and Bill Little and their continued encouragement. The helium gas bearing was primarily worked out by George Hess while still at Stanford, and constructed by Don Davidson. The authors also wish to thank the Stanford relativity gyroscope group and, in particular, Francis Everitt for technical assistance and much encouragement.

This work has been supported by a precision measurement grant from the National Bureau of Standards and a grant from the National Science Foundation.

References

[1]   B. S. Deaver and W. M. Fairbank, *Phys. Rev. Lett.* **7**, 43 (1961); R. Doll and M. Näbauer, *Phys. Rev. Lett.* **7**, 51 (1961).

[2]   M. Bol and W. M. Fairbank, in *Proceedings of LT9*, 471 (1964); A. F. Hildebrandt, *Phys. Rev. Lett.* **12**, 190 (1964); A. F. Hildebrandt and M. M. Saffren, in *Proceedings of LT9*, 459 (1964); C. A. King, J. B. Hendricks, and H. E. Rorschach, in *Proceedings of LT9*, 466 (1964).

[3]   J. E. Zimmerman and J. E. Mercereau, *Phys. Rev. Lett.* **14**, 887 (1965); W. H. Parker and M. B. Simmonds, *PMFC-I*, NBS Spec. Publ. **343**, 243 (1970).

[4]   B. Cabrera, H. Gutfruend and W. A. Little, *Phys. Rev. B1* **25**, 6644 (1982).

[5]   See, for example, M. Tinkham, *Introduction to Superconductivity* (Kreiger, New York, 1980).

[6]   B. Cabrera, S. Felch and J. T. Anderson, in *Precision Measurement and Fundamental Constants II*, edited by B. N. Taylor and W. D. Phillips (National Bureau of Standards (U.S.)), Spec. Publ. 617 (in press).

[7]   B. Cabrera and G. J. Siddall, *Precision Engineering* **3**, 125 (1981).

The Study of
Superconducting Order Parameter Dynamics:
An Application of
the Josephson Effect

A. M. Goldman

1. INTRODUCTION

Flux quantization experiments dramatically demonstrated the importance of long range phase coherence in the description of the superconducting state, an idea originally proposed by London as an integral part of his phenomenological theory of the Meissner-Ochsenfeld effect. The most striking experimental demonstration of the phase coherence of the superconducting state is that the maximum dc Josephson current in a thin-film tunneling junction exhibits a Fraunhofer-like dependence on magnetic field.

The understanding of superconductivity which emerged from the theoretical and experimental ferment of the late 1950's and early 1960's did not deal adequately with spatial and temporal variations of the order parameter. Equilibrium fluctuation effects seemed to be unobservable, at least according to the Ginzburg criterion [1] which seemed to assure the applicability of mean-field theory at temperatures very close to T_c. Furthermore, the possibility of driving a superconductor out of equilibrium was not considered seriously at all and would have to wait a decade before receiving any attention.

Although the analogy between the macroscopic wavefunction or order parameter of the superconducting state and the order parameters of other phase transitions was appreciated in the early 1960's, there appeared to be no obvious way to carry out experiments on the spatial and temporal behavior of the order parameter with the elegance that was possible in the study of other systems.

The difficulty in identifying experiments was a fundamental one having to do with the unusual character of the superconducting order parameter. In magnetic systems, for example, the order parameter M has a conjugate thermodynamic field which couples to it. As a consequence it is possible to measure a static magnetic susceptibility and to determine the spatial and temporal behavior of the order parameter in magnetic neutron scattering experiments where the wave vector and frequency-dependent structure factor $S(q, \omega)$ is proportional to the angle and energy dependence of the inelastic magnetic scattering cross section. Because the order parameter of a superconductor is off-diagonal in number space, there is no thermodynamically conjugate field which induces superconductivity in a normal metal in the manner that a laboratory magnetic field induces a finite magnetization in a paramagnet [2]. Correspondingly, there is also no obvious counterpart of magnetic neutron scattering in the superconducting case.

What had not been appreciated in the superconducting case is that within the spirit of mean-field theory one can treat the internal interaction within a material as an effective field acting on a given region of the material. This internal interaction is long range (on the order of the coherence length ξ), and gives rise to the well-known proximity effect in a clean contact between a superconductor and a normal metal. The same type of physical effect, although weaker, occurs across a tunneling barrier, and it is this type of coupling that gives rise to the Josephson effects between two superconducting electrodes [3]. Even in a tunneling junction between a superconductor and a normal metal, there is a small induced order parameter on the normal side. Furthermore, there is an "excess" supercurrent across the junction, which is a measure of the generalized superconducting pair-field susceptibility $\chi(q, \omega)$ [4]. Similar considerations are relevant when both electrodes are superconducting, with the excess pair current for nonzero voltages also being related to the susceptibility.

The study of the wave vector and frequency-dependent pair-field susceptibility of the superconducting order parameter $\chi(q, \omega)$ and its relationship to an excess tunneling current in a Josephson junction is by now established as perhaps one of the most sensitive probes of superconducting order parameter dynamics. The development of this subject has proceeded in several stages. Experimental attention was initially directed at establishing whether the proposed connection between pair tunneling and the excess current existed [5]. Then the technique was used to study in detail the dynamics of superconducting fluctuations in the normal state. What followed was the experimental discovery of a way to extend the measurements into the superconducting state and the discovery of propagating charge imbalance waves [6].

2. THEORY

The results of several types of calculations show that the imaginary part of the susceptibility $\chi''(q, \omega)$ is proportional to an excess current in the dc I-V characteristic of a tunneling junction in which the superconductor under study is incorporated as one electrode of the junction, with the second electrode a metal with a much higher transition temperature. The former will be described by unprimed variables and the latter by primed. (See

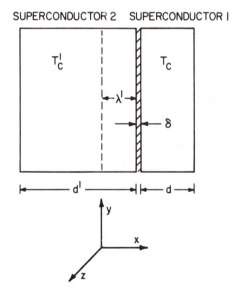

FIGURE 1. Schematic of tunnel junction configuration used to measure the pair-field susceptibility, for $T \sim T_c \ll T_c'$. The magnetic field is in the z direction. The pair field is modulated in the y direction and current flow is in the x direction.

figure 1.) The variables ω and q, the frequency and wave vector, respectively, are related to the dc voltage across the junction and the magnetic field H applied in the junction plane through the relations $\omega = 2eV/\hbar$ and $q = (2e/\hbar)(d/2 + \lambda')H$, where d is the thickness of the film whose susceptibility is being determined and λ' is the magnetic penetration depth of the electrode with the higher transition temperature. Simple expressions are found if for the electrode of interest $d \ll \xi(T)$, where $\xi(T)$ is the coherence length, and for the reference electrode $\lambda' \ll d'$.

The easiest way to see the connection between the generalized susceptibility and the excess current due to pair tunneling is to consider the linear response model of Scalapino [4], in which the coupling between the two superconductors is described by an effective Hamiltonian of the form

$$H_1 = C \, \exp(-i\omega t) \int_A d^2\mathbf{x} [\exp(i\mathbf{q} \cdot \mathbf{x})] \widehat{\Delta}(\mathbf{x}, t) + h.c. \quad , \qquad (1)$$

where $\widehat{\Delta}(\mathbf{x}, t)$ is the pair-field operator

$$\widehat{\Delta}(\mathbf{x}, t) = g\psi_\downarrow(\mathbf{x}, t)\psi_\uparrow(\mathbf{x}, t) \quad . \qquad (2)$$

Here the quantities $\widehat{\psi}_\uparrow$ and $\widehat{\psi}_\downarrow$ are field operators for spin-up and spin-down quasiparticles, g is the superconducting coupling constant of the fluctuating electrode, and $\mathbf{q}$ is the wave vector of the tunneling coupling, which is given by C. The effective Hamiltonian implies an average pair transfer current

$$\langle I_1 \rangle = (4eC/\hbar d)Im\{ \exp(-i\omega t)\langle \widehat{\Delta}(-\mathbf{q}, t) \rangle \} \quad . \qquad (3)$$

The commonly understood situation when $T > T_c$ is for $\langle \widehat{\Delta}(-\mathbf{q}, t) \rangle_0 = 0$ which results in vanishing pair transfer or Josephson current. Here $\langle \ldots \rangle_0$ denotes an average with respect to an isolated superconductor. However, because equation (3) involves an average with respect to a superconductor coupled to a reference electrode, there is actually a contribution to the pair current from an "induced" order parameter, which is the linear response to the coupling $H_I(t)$ given by

$$\widehat{\Delta}(-\mathbf{q}, t) = (1/i\hbar) \int_{-\infty}^{t} dt' \langle [\widehat{\Delta}(-\mathbf{q}, t'), H_I(t')] \rangle_0 \quad . \qquad (4)$$

This induced order parameter can in turn be written as

$$\widehat{\Delta}(-\mathbf{q}, t) = C \int_A d^2\mathbf{x} \, \exp(-i\mathbf{q} \cdot \mathbf{x}') \int_{-\infty}^{\infty} dt' \, \exp(i\omega t')\chi(\mathbf{x} - \mathbf{x}', t - t') \quad , \qquad (5)$$

where $\chi(\mathbf{x} - \mathbf{x}', t - t')$ is the response function

$$\chi(\mathbf{x} - \mathbf{x}', t - t') = (i/\hbar)\theta(t - t')\langle [\widehat{\Delta}^+(\mathbf{x}', t'), \widehat{\Delta}(\mathbf{x}, t)] \rangle_0 \quad , \qquad (6)$$

and $\theta(t - t')$ is the unit step function. Carrying out the various integrals, one obtains for the pair transfer current the expression

$$\langle I_1 \rangle = (4e|C|^2 A/\hbar d)\chi''(q,\omega) \quad , \tag{7}$$

which is second order in the coupling.

When $T < T_c$, there is a contribution to the pair transfer current which is the usual Josephson tunneling current, first order in C. This must be augmented by an additional contribution to $\langle I_1 \rangle$ associated with $\chi''(q,\omega)$, which is second order in C. This is an additional feature of pair tunneling evidently not anticipated by Josephson and which may be interpreted as a second order Josephson effect. This contribution to the tunneling current can also be viewed as a kind of proximity effect in which the tunneling interaction induces an order parameter in the normal state or increases the order parameter relative to its equilibrium value in the superconducting state. In other words, the pair transfer Hamiltonian characterizing the Josephson coupling of an asymmetric junction provides an external field which couples to the order parameter. The linear response of the order parameter to this coupling determines the characteristic susceptibility associated with the phase transition and results in an excess current due to pair tunneling in the junction proportional to $\chi''(q,\omega)$.

3. EXPERIMENT

The crossed-film superconducting tunnel junctions of the configuration Al-Al$_2$O$_3$-Pb needed for these experiments were prepared in a conventional manner in an oil-free vacuum chamber capable of maintaining a pressure of less than 5×10^{-8} torr. Since important results are associated with the superconducting transition, care was taken to mask the edges of the unprimed film. Quality checks on junctions included testing for zero-field steps of the I-V characteristic, which if found are evidence of weak shorts, study of the magnetic field dependence of the zero-voltage current which will exhibit a well-defined Fraunhofer pattern if the insulator is uniform, and selecting for study only junctions in which the product of the leakage conductance and the normal-state tunneling resistance was 10^{-3} or lower.

Measurements were carried out in an apparatus designed to give substantial isolation from electromagnetic noise and at the same time allow for substantial temperature regulation. Electrical leads entering the low temperature apparatus all passed through π-section rf filters. The experimental environment was shielded from the earth's field down to the 10^{-4} G level. Sample temperatures were regulated to $\pm 10^{-5}$ K electronically.

The excess current due to a pair tunneling was determined by subtracting the quasiparticle current from the total measured tunneling current. The

appropriate quasiparticle current in early work was initially determined by
an extrapolation to low voltages of an expression derived from the BCS theory parameterized by fitting over a range of voltages for which the excess
current due to pair tunneling was small. A typical excess current *versus*
voltage characteristic resulting from such an analysis is shown in figure 2.
Actually, the excess current may be significant with the Al electrode in the

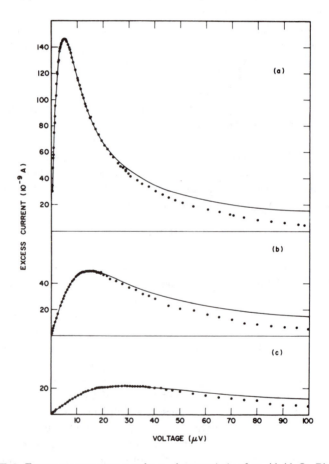

FIGURE 2. Excess current *versus* voltage characteristic of an Al-Al$_2$O$_3$-Pb junction.
Curves (a)–(c) correspond to $T = 1.8473$, 1.933 and 2.099 K, respectively. The solid lines
are computed using equation (8). The deviations at higher voltages can be corrected
with a more detailed analysis of the quasiparticle current.

normal state even for voltages greater than 200 μV. Thus very precise determinations of the excess current *versus* voltage characteristic required a fit

of the total measured current to a model containing a sum of contributions from the quasiparticle, excess and leakage currents, respectively. A set of parameters of the excess current *versus* voltage characteristic, the normal tunneling resistance R_N, the leakage conductance G_L, and the energy gap in Pb, $\Delta_{Pb}(0)$ were thus determined for $T > T_c$. These same parameters were then used to compute the background quasiparticle and leakage current below T_c. Such a procedure was essential for the analysis of the regime below T_c for which the functional form for the excess current was in doubt.

3.1 The Normal State $(T > T_c)$

Measurements in this regime are a critical test of dynamical behavior of a superconducting order parameter governed by a diffusive time-dependent Ginzburg-Landau equation. When $d < \xi(T)$, the imaginary part of the pair-field susceptibility is given by

$$\chi''(q,\omega) = \frac{2m\xi^2(0)}{\hbar e} \frac{\omega/\Gamma_0}{(1 + q^2\xi^2(T))^2 + \omega^2/\Gamma_0^2} \quad , \tag{8}$$

where $\Gamma_0 = (8/\pi)(k_B T_c/\hbar)\epsilon$ is the relaxation frequency of the order parameter, and $\epsilon = (T - T_c)/T_c$ is a reduced temperature. More complicated expressions are relevant for the case of thick films. It should be emphasized, as was pointed out by Scalapino [4], that $1/\Gamma_0$ is not the pair lifetime but is the time for a perturbation of the order parameter to diffuse a coherence distance $\xi(T)$.

Figure 3 contains a plot of the peak voltage of the excess current *versus* voltage characteristic as a function of temperature. This is in good agreement with equation (8) except near T_c. There have been a number of discussions of the physical origins of the disagreement of experiment and theory for this regime. Film inhomogeneity and disorder, finite size effects, critical behavior and circuit effects have all been treated or have been the subject of speculation. The subject is not quantitatively resolved at this time.

3.2 The Superconducting State $(T < T_c)$

Measurements of $\chi''(q,\omega)$ below T_c must be made at finite, nonzero values of q because of the necessity of suppressing the usual Josephson effects which are first order in the coupling constant. Below T_c three main features are found in the excess current *versus* voltage characteristic. Just below T_c a shoulder develops on the low voltage side of the main peak. As T is reduced further, this moves to lower voltages. The second feature is a sharpening of the main peak and its movement to higher rather than lower voltages as T is reduced. At lower temperatures the peak again broadens.

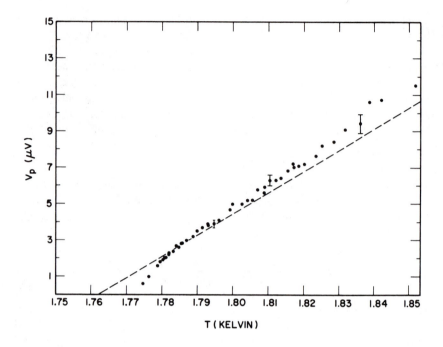

FIGURE 3. Peak voltage $V_p = \frac{\hbar}{2e}\Gamma_0$ *versus* T for an Al-Al$_2$O$_3$-Pb junction.

The third feature is the appearance of a temperature-dependent secondary peak, which occurs at a voltage given by $eV = \Delta(T)$, where $\Delta(T)$ is the energy gap of the fluctuating electrode (Al in this instance). The general behavior described above is documented in figure 4.

It is somewhat more revealing to study the structure factor $S(q,\omega)$ rather than $\chi''(q,\omega)$, since from the former one can ascertain the character of the dynamics without a detailed model. The structure factor is the space and time Fourier transform of the order parameter correlation function, whereas the susceptibility is the transform of the response function. The two are related through the fluctuation-dissipation theorem

$$S(q,\omega) = (\hbar/\pi)[\exp(\hbar\omega/k_BT) - 1]^{-1}\chi''(q,\omega) \quad . \tag{9}$$

In the limit $\hbar\omega/k_BT \ll 1$ one finds

$$S(q,\omega) = (k_BT/\pi)\chi''(q,\omega)/\omega \quad . \tag{10}$$

Since χ'' is proportional to the excess current and $\omega = 2eV/\hbar$, $S(q,\omega)$ is proportional to the quotient of the excess current and the voltage at which it is measured. Typical results are shown in figure 5. With the frequency

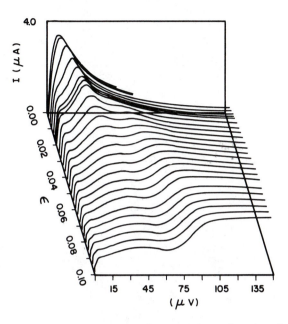

FIGURE 4. Excess current *versus* voltage characteristic as a function of $\epsilon = (T_c - T)/T_c$ for a junction with $T_c = 1.155$ K and $R_N = 0.053$ Ω.

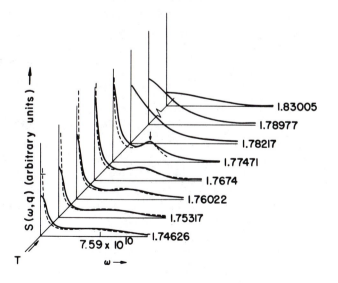

FIGURE 5. Structure factor I_{ex}/V in arbitrary units *versus* T and ω. The magnetic field is 125 Oe. Curves are shown only for $\omega > 0$ portion and are symmetric about $\omega = 0$. $T_c(H) = 1.780$ K. Dashed lines are from [7].

scale used, the aluminum gap peak is not seen. Above T_c the structure factor is a Lorentzian centered at $\omega = 0$. Below and near T_c, $S(q,\omega)$ has a peak at a nonzero frequency in addition to the usual one at the origin. Usually a peak in the dynamical structure factor at nonzero frequency implies the existence of a propagating mode, which is the new feature below T_c in addition to the diffusive mode found both above and below T_c.

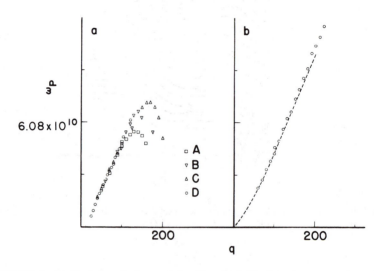

FIGURE 6. (a) Position of the peak in $S(q,\omega)$ of Al plotted *versus* q in units of magnetic field at several temperatures. The points A through D correspond to $T = $ 1.771, 1.767, 1.764 and 1.753 K, respectively. (b) Frequencies of peaks in $S(q,\omega)$ at fixed $\epsilon(H) = 6.32 \times 10^{-3}$. The dashed line is from [7].

An approximate dispersion relation for the propagating mode can be determined by measuring the wave vector dependence of the finite frequency peak of the structure factor, where the wave vector is set by the magnetic field. Figure 6 shows the result for an Al film. In figure 6a the dispersion relation is plotted at fixed T and in figure 6b at fixed reduced temperature $\epsilon(H,T)$. The bending of the curves in figure 6a is a consequence of the approach to the superconducting-normal phase boundary with increasing H. Because an applied magnetic field shifts T_c as well as sets the wave vector, the physical character of the mode is best revealed by the plot at fixed ϵ. The curve in figure 6b corresponds to $\epsilon(H,t) = 6.32 \times 10^{-3}$. For this sample $\xi(0) = 5.46 \times 10^{-6}$ cm and $T_c(H) = 1.780$ K. The apparent propagation velocity is 1.74×10^6 cm/sec. The dashed line in figure 6b was computed from the theory of Schmid and Schön [7] which provided the first successful explanation of the propagating mode.

4. DISCUSSION

The fundamental origin of the excess pair current is the Josephson coupling between the strong superconductor and the weakly superconducting or normal film. Although the results of measurements of $\chi''(q, \omega)$ are related to the spectrum of spontaneous thermal fluctuations within the weak superconductor, they are in no sense due to these fluctuations. The excess current being a Josephson current, it depends on the existence of a phase correlation across the barrier. This does not exist with the thermal fluctuations, which therefore do not contribute to the excess current. In a measurement of excess conductivity within a film above T_c, on the other hand, only correlations within the film are important, and to these thermal fluctuations do contribute. In fact, thermal fluctuations will act to smear out well-defined features in the excess current [8].

For $T < T_c$, the situation is somewhat more complicated due to the fact that the order parameter has a nonzero equilibrium average value. The fluctuations separate into parts associated with the magnitude of the order parameter (the longitudinal or L-mode) and the phase (the transverse or T-mode). A key feature of the theory is that these fluctuations are coupled to the transport equation for the quasiparticle disequilibrium. The resultant equations for $\chi''(q, \omega)$ are more complicated than those derived from the time-dependent Ginzburg-Landau equations valid for a gapless superconductor below T_c or for $T > T_c$.

There is also a profound physical distinction between the modes. In the L-mode the absolute value of the order parameter changes and the disequilibrium of the quasiparticle distribution function of δf_E^L is an odd function of energy measured from the Fermi level. Since the number of Cooper pairs is proportional to $|\Delta|^2$, relaxation in the L-mode is a Cooper pair condensation. Here $\int dE N_1(E)\delta f_E^L = 0$, resulting in the number of quasiparticles remaining constant but their energy distribution changing.

The T-mode is a charge-imbalance mode [9]. The order parameter phase varies and the disequilibrium of the quasiparticle distribution function δf_E^T is an even function of energy as measured from the Fermi surface. Since spatial variations of the phase are supercurrents, relaxation processes in this mode relate to the interconversion between normal and supercurrents. As quasiparticle number changes, this mode can be excited by quasiparticle injection. The T-mode can also be described in the language applied to describe branch or charge imbalance. In particular, the propagating T-mode is the high frequency limit of the charge imbalance dynamics. The low frequency limit is relevant to the description of electric field penetration into a superconductor in the presence of current flow across a normal superconductor interface.

In the language of a two-fluid model, the propagating T-mode corresponds to a counterflow of normal and supercurrents. The motion of the superfluid actually generates the dispersion law. The T-mode, when it propagates, is a collisionless mode in that its frequency $\omega > 1/\tau_E$, where τ_E is the phonon inelastic scattering time. However, ω is still less than the reciprocal of the impurity scattering time. There is a temperature window of finite width below T_c in which the mode propagates. This window is a consequence of the fact that only near T_c is the superelectron density low enough for the required counterflow of normal fluid to be small enough to permit the mode not to be overdamped. The counterflowing normal fluid, as a consequence of impurity scattering, is the source of the damping of the mode. As T drops below T_c, the superelectron density grows and the damping increases. When magnetic impurities are present, the T-mode propagates only in the gap regime which develops at a temperature below T_c [10], a result qualitatively consistent with theory [11].

Acknowledgments

The development of this subject is the result of the labors of a number of graduate students including J. T. Anderson, R. V. Carlson and F. E. Aspen. Dr. A. M. Kadin collaborated in the development of a complete macroscopic theory of the pair-field susceptibility experiment, and contributed greatly to the author's understanding of this subject. Stimulating discussions have been held with R. A. Ferrell, D. J. Scalapino, L. Nosanow, P. Hohenberg, J. W. Halley, Jr., C. E. Campbell, R. Orbach, O. Entin-Wohlman, P. A. Lee, S. R. Shenoy, G. Schön, A. Schmid, H. Schmidt, I. O. Kulik and V. P. Galaiko. The work was supported by the National Science Foundation under grant NSF/DMR80-06959.

References

[1]   V. L. Ginzburg, *Fiz. Tverd. Tela* **2**, 2031 (1960) [*Sov. Phys.–Solid State* **2**, 1824 (1960)].

[2]   V. Ambegaokar, in *Proceedings of the International Conference on the Science of Superconductivity*, Stanford, USA, 26–29 August 1969, edited by F. Chilton (North-Holland, Amsterdam, 1971), p. 32.

[3]   R. A. Ferrell, *J. Low Temp. Phys.* **1**, 423 (1969).

[4]   D. J. Scalapino, *Phys. Rev. Lett.* **24**, 1052 (1970).

[5]   J. T. Anderson, R. V. Carlson and A. M. Goldman, *J. Low Temp. Phys.* **8**, 29 (1972).

[6]   R. V. Carlson and A. M. Goldman, *J. Low Temp. Phys.* **25**, 67 (1976).

[7]   A. Schmid and G. Schön, *Phys. Rev. Lett.* **34**, 941 (1975).

[8]   A. M. Kadin and A. M. Goldman, *Phys. Rev. B* **25**, 6701 (1982).

[9]   C. J. Pethick and H. Smith, *Ann. Phys. (NY)* **119**, 133 (1979).

[10]   F. E. Aspen and A. M. Goldman, *J. Low Temp. Phys.* **43**, 559 (1981).

[11]   O. Entin-Wohlman and R. Orbach, *Ann. Phys. (NY)* **116**, 35 (1978);
G. Schön and V. Ambegaokar, *Phys. Rev. B* **19**, 3515 (1979).

Local and Nonlocal Effects in the Near Zero Magnetic Field Penetrating into Superconducting Tin Film Cylinders

Edward G. Wilson

1. INTRODUCTION

As had been the case in the detailed theory of the anomalous skin effect in normal metals, careful studies of the electrodynamics of superconductors showed that local theory was not sufficient to describe the experimental results completely. This paper presents experimental data and supporting theoretical analysis which demonstrate nonlocal effects in the near zero magnetic field penetrating into superconducting tin film cylinders, which is an area not previously explored in detail.

The London theory of superconductivity [1] set the supercurrent density $\mathbf{J}$ proportional to the magnetic vector potential $\mathbf{A}$ at each point inside a superconductor. This completely local formulation provided the first model which predicted the screening of magnetic fields from inside a superconductor. The Pippard [2] and the BCS [3] theories provide very similar generalizations of the London local relationship between $\mathbf{J}$ and $\mathbf{A}$. They are called nonlocal theories because $\mathbf{J}(\mathbf{r})$ is proportional to a weighted average of $\mathbf{A}(\mathbf{r}')$ within a volume roughly equal to the cube of a characteristic length (ξ, the coherence length), centered at **r**. In Pippard's theory the length ξ is introduced phenomenologically, but in the later BCS theory ξ appears as the distance over which the quantum wave function describing a Cooper pair extends coherently. (A Cooper pair is two conduction electrons bound by an interaction through the phonons into a superconducting quasiparticle.) Thus the nonlocal averaging of **A** is done mathematically by integrating over a volume of roughly ξ^3, but it may be viewed as being done physically by a quasiparticle of the same size. The coherence length ξ is shortened from its pure bulk value of ξ_0 by any impurities and irregularities which reduce the mean free path ℓ, following the relation $1/\xi = 1/\xi_0 + 1/\ell$. As long as ξ exceeds the London penetration depth λ_L, the penetration of a magnetic field into a superconductor exhibits nonlocal effects, *i.e.*, effects that cannot be predicted from a local theory. For $\xi < \lambda_L$, however, the Pippard theory becomes approximately local, and it reduces for $\xi \ll \lambda_L$ to a London-type theory with an effective $\lambda = \lambda_L \sqrt{1 + \xi_0/\ell}$.

Consider tin, for which $\xi_0 = 230$ nm and $\lambda_L = 34$ nm. In a pure bulk sample, tin should be highly nonlocal, showing effects dramatically different from the London theory. These include a depth D_0 at which the penetrating field is zero, and a small field opposite in sign to the outside field at depths greater than D_0. Sommerhalder and Thomas [4] showed that the nonlocal theories predict similar effects for the magnetic field penetrating to the interior of a hollow cylinder. Subsequently, Drangeid and Sommerhalder [5] reported observing a reversed field during a single low temperature run on one of the twenty-five tin film cylinders which they produced and tested. The magnitude of the reported reversed field is, however, at least 300 times greater than predicted by the theory.

Since the reversed field appears *only* if $\xi > \lambda_L$, William Fairbank suggested that observing it as a function of film thickness would provide perhaps the most direct way of observing the effects of ξ itself as a function of distance. With his guidance and support I made 15 tin cylinders by vacuum deposition, and investigated them for field reversal and other evidence of nonlocal behavior. None of the films showed field reversal (probably because of impurities and pinholes), but two films showed clearly nonlocal effects, in that they are much better modeled by the nonlocal than by the local theory. I also discovered a mechanism (motion of trapped flux) which could explain the implausibly large field reversal reported by Drangeid and Sommerhalder [5], referenced above.

2. EXPERIMENTAL APPARATUS AND PROCEDURES

The two most significant problems with the experimental project of looking for nonlocal effects in superconducting films result from the impossibility of producing films which are continuous pure crystals. The first problem is the presence of impurities and crystalline imperfections which shorten the mean free path and thus move the film toward the local limit. Impurities are minimized by evaporating in as low a pressure as possible, and as rapidly as possible.

The second problem is the presence in almost all evaporated films of small openings or pinholes. During an experiment on the film in the super-conducting state, these holes allow the externally applied magnetic field to extend directly into the interior, thus swamping the small field that penetrates through the superconductor. The holes are usually due to dust particles on the substrate, although some may occur due to thermal stresses [6]. To reduce pinholes caused by dust particles, I brushed the film softly with a camel's hair brush after depositing half the desired thickness of tin. After a second backsputtering treatment, the second half was deposited. This procedure moved dust particles from their holes in the first half of the film to new locations, so that they did not make continuous holes through both halves of the film. It produced completely pinhole-free films on flat substrates in initial tests, but only reduced, rather than eliminated, pinholes in cylindrical films. Pinholes were observed by putting a tiny light bulb inside the cylinder in a dark room, and films that were badly pinholed were not run at low temperatures.

The film thicknesses were determined during deposition by a Sloan thickness monitor, and calibration was verified on one film by weighing before and after deposition. The disagreement was 8%.

Proceeding from the outside inward, the experimental setup included a series of concentric cylinders consisting of a room temperature μ metal shield, a superconducting lead shield, a superconducting field coil, the 6 inch long by 0.43 inch diameter tin film sample and the magnetometer pickup coil.

The primary measurements taken during a run were the change in the applied field (ΔH_a), and the resulting change in the field inside the center of the tin cylinder (ΔH_i), as a function of temperature, thereby making the ~ 0.05 mG background field invisible to the experiments. The measured values $\Delta H_i / \Delta H_a$ are the experimental values to be compared with H_i / H_a in the theoretical equations below. The field ΔH_i was measured with a SQUID magnetometer with a minimum observable field at 16 Hz bandwidth of 2×10^{-8} G.

Because field transmission ratios $\Delta H_i / \Delta H_a$ down to 10^{-9} were measured in these experiments, the magnetometer and the superconducting transformer leads were shielded with a structure of molded 2 mil lead foil. The shielding was tested by running a lead foil cylinder in place of the usual tin film cylinder. The results placed an upper limit of $\Delta H_i / \Delta H_a = 5 \times 10^{-10}$ for leakage, which is well below the smallest measureable signals obtained from the experiments on tin film cylinders. This test was successfully repeated at the end of the experiments.

3. THEORY

The solution to the Pippard equation for the (infinitely long) hollow superconducting cylinder derived by Sommerhalder and Thomas [4] is given by:

$$\frac{H_i}{H_a} = \frac{\frac{1}{2K_0} + \sum_{n=1}^{\infty} (-1)^n \left[\left(\frac{n\pi}{D}\right)^2 + K_n \right]^{-1}}{\frac{RD}{4} + \frac{1}{2K_0} + \sum_{n=1}^{\infty} \left[\left(\frac{n\pi}{D}\right)^2 + K_n \right]^{-1}} \quad , \tag{1}$$

with

$$K_n = \frac{3}{2\lambda_L^2 \left(\frac{n\pi}{D}\xi_0\right)\left(\frac{n\pi}{D}\xi\right)^2} \cdot \left\{ \left[\left(\frac{n\pi}{D}\xi\right)^2 + 1 \right] \tan^{-1}\left(\frac{n\pi}{D}\xi\right) - \left(\frac{n\pi}{D}\xi\right) \right\} \tag{2}$$

where H_a is a uniform magnetic field applied externally to the cylinder and parallel to its axis, H_i is the resulting field inside the cylinder, D is the superconducting wall thickness, and R is the cylinder radius.

Equations (1) and (2) give H_i/H_a *versus* D for $T = 0$ kelvin. Although films can be made in a range of thicknesses D, the films of different D cannot be made identical in other properties, so it is not possible to generate experimental data for H_i/H_a *versus* D while holding the other parameters fixed. Instead, for each film, data were taken as a function of temperature, so temperature dependence must be inserted into (1) and (2) in order to compare the theory with experiments. Various suggestions have been made about how to insert temperature dependence into the Pippard theory [7,8,9]. The prescription chosen here is

$$\lambda(T) = \lambda_L \sqrt{1 - (T/T_c)^4} \quad , \tag{3}$$

which is the simplest and also gives a good approximation to the BCS theory for $T > T_c/2$ [10], the temperature range of interest in these experiments.

The ratio of the film thickness to the temperature-dependent penetration depth $[D/\lambda(T)]$ was a more convenient choice for the independent variable than the temperature T. Figure 1 gives a set of curves for $\ln|H_i/H_a|^1$ computed from the Pippard theory for a tin cylinder of radius 0.55 cm (the radius of our cylinders), with $D = 1000$ nm, $\xi_0 = 230$ nm, and $\lambda_L = 34$ nm. Values for which H_i/H_a is negative are indicated by dotted lines. For $\xi = \xi_0$

---

[1]labeled $\ln|\Delta H_i/\Delta H_a|$ to conform with experimental notation.

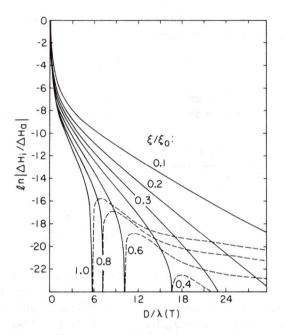

FIGURE 1. Pippard theory for magnetic field penetration into a hollow superconducting tin cylinder as a function of the wall thickness D divided by the temperature-
dependent penetration depth $\lambda(T)$. Here $\lambda_L = 34$ nm, $\xi_0 = 230$ nm, and D is held
fixed at 1000 nm. The solid lines represent positive H_i/H_a values and the dotted lines
represent negative H_i/H_a values. H_i/H_a is labeled $\Delta H_i/\Delta H_a$ to conform with the
experimental notation in figures 2a and 2b. Zero temperature gives $D/\lambda(0) = 29.41$ and
the critical temperature gives $D/\lambda(T_c) = D/\infty = 0$.

the initial decay is considerably faster than the London theory predicts, and
the negative extremum is the largest in magnitude.

For $\xi \leq 0.2\ \xi_0$ the Pippard solution is approximately reduced to the
London limit with an effective $\lambda = \lambda_L \sqrt{1 + \xi_0/\ell}$, and no negative H_i/H_a
values appear. This mean free path-dependent modification of the penetra-
tion depth, together with the temperature prescription in (3), was used to
generate London theory best fits to compare with Pippard theory best fits
to the experimental data. The London theory predicts for a semi-infinite
superconducting cylinder (with penetration depth λ) of wall thickness D
and radius R:

$$\frac{H_i}{H_a} = \left[\frac{R}{2\lambda} \sin h\left(\frac{D}{\lambda}\right) + \cos h\left(\frac{D}{\lambda}\right)\right]^{-1} , \qquad (4)$$

where

$$\lambda = \frac{\lambda_L \sqrt{1 + \xi_0/\ell}}{\sqrt{1 - (T/T_c)^4}} . \qquad (5)$$

Even our best films had some small pinholes. Thus in order to match theory with experiment a model of the leakage of H_a through pinholes was needed. A homogeneous external field of magnitude H_a and direction parallel to the surface of a superconducting sheet leaks through a hole in that sheet according to the following relation (see [11]):

$$\frac{H_i}{H_a} \propto \left(\frac{a_0 + \lambda(T)}{z}\right)^3 , \tag{6}$$

where $a_0 + \lambda(T)$ is the temperature-dependent effective hole radius and z is the distance away from the hole. In the case of a cylinder the falloff will be somewhat faster. The signal seen by the pickup coil from one pinhole will be the average of the pinhole field over the area of the coil, and for many holes it will be the sum of such averages. The first, temperature-independent term is dominant except very close to T_c and thus provides a satisfactory approximation for the effect of the pinholes.

4. RESULTS AND ANALYSIS

The best fit of both the Pippard and the London theories to each set of experimental data was obtained by minimizing the reduced chi-square (χ^2) with the normalized coherence length ξ/ξ_0 and β as adjustable parameters, where β is a constant added to H_i/H_a to approximate pinhole contributions. $\lambda_L = 34$ nm and $\xi_0 = 230$ nm, appropriate for tin, were used for this fit. Both the Pippard and the London theories can be expressed in terms of ξ/ξ_0. Finding ξ/ξ_0 determines the film's mean free path ℓ by means of the relationship $1/\xi = 1/\xi_0 + 1/\ell$.

The curve fitting procedure indicated that the fifth film, which produced data sets F5-1 and F5-2 (figures 2a and 2b), was the most nonlocal film studied. Since the Pippard and the London theories are the same in the local limit, but differ increasingly as the mean free path becomes long (the nonlocal regime), it is not surprising that these data sets also show the largest distinction between the best Pippard and the best London fits. The superiority of the Pippard fit is clear from visual inspection of the graphs, as well as from the much smaller chi-square values for the Pippard fits in the two cases.

Data set F5-2 was taken about six months after set F5-1. (During the six months the film was stored in an imperfect vacuum.) The main difference between the two occurs at the smallest measureable $\Delta H_i/\Delta H_a$ values, where the slope of F5-1 becomes steeper, while the slope of F5-2 begins to flatten out. Probably the difference resulted from one or more small pinholes opening up (or growing) in the film during the six-month interval. The best fit Pippard curves corroborate the hypothesis, since the ξ/ξ_0 value

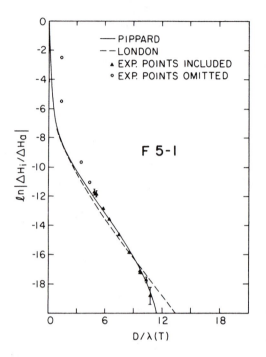

FIGURE 2a. Magnetic field penetration into a hollow superconducting cylinder as a function of $D/\lambda(T)$ for film F5-1.

differs between the curves by only 1%, while the pinhole term for F5-2 is 300% larger than for F5-1.

The size of the triangles representing the fitted data points was chosen to represent error bars of ±10%. Error bars of ±15% or greater are marked explicitly. The circles represent data points which were not used in the theory fits because neither theory can be made to fit them well. These points are always the ones closest to the transition temperature, and reasons why they cannot be fitted are discussed later.

Table 1 contains a summary of the experimental results compared to the best theoretical fits for each of the data sets that were judged interesting enough to fully analyze. Standard statistical techniques were used [12]. The results are described in detail in my thesis [9]. Clearly, films F5 and F12 are modeled significantly better by the Pippard theory than by the London theory, and therefore show nonlocal effects.

The data for the other three films could not be well fitted by either theory even after excluding the points nearest T_c. And even the best two

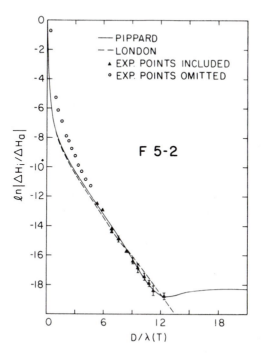

FIGURE 2b. Magnetic field penetration into a hollow superconducting cylinder as a function of $D/\lambda(T)$ for film F5-2.

films could not be well fitted if the points near T_c were not excluded. (The exclusion range was 25 mK to 40 mK below T_c, except for F12, which had points excluded down to 99 mK below T_c.) It is obvious to the eye that all the excluded data points indicate a broadening of T_c in that the penetrating magnetic field is not attentuated as rapidly as theory predicts as T drops below T_c. (See figures 2a and 2b; also figures 5.4 through 5.8 in [9].) Three explanations were considered.

The omitted pinhole terms do predict an increasing pinhole leakage as T approaches T_c, which would tend to cause the observed effect. However, pinholes of the size range indicated by the low temperature measurements would provide a leakage increase (as T approached T_c) that would be much too small. Fluctuation effects also cause a broadening of the superconducting transition, and significant fluctuations effects have been observed at 40 mK and farther below T_c, but only under special conditions [13] which do not exist in the films considered here.

TABLE 1. SUMMARY OF THE EXPERIMENTAL RESULTS COMPARED TO THE BEST THEORETICAL FITS. For each experimental data set the best fit values of ξ/ξ_0 and β are given for the Pippard and the London theories, together with the corresponding minimum (reduced) chi-square values. $P(\chi^2, \mu)$ is the probability that an artificial data set constructed from the best-fit theoretical curve with Gaussian errors would have a χ^2 as large as or larger than the χ^2 for the real data set.

Data Set	D (μm)	T_c (K)	Number of Data Points	Theory Used	ξ/ξ_0	$\beta \times 10^{-9}$	Minimum χ^2	P
F5-1	0.83	3.820 ± .002	10	Pippard	0.499 ± .027	4.07 ± 20	0.98	0.46
				London	0.587 ± .027	<0.07	4.03	<0.001
F5-2	0.83	3.8169 ± .0003	11	Pippard	0.505 ± .019	12.2 ± 6.1	1.20	0.28
				London	0.567 ± .027	<0.14	3.54	<0.001
F-7	0.75	3.7805 ± .0005	14	Pippard	0.451 ± .010	42.8 ± 4.4	28.4	<0.001
				London	0.503 ± .016	36.5 ± 4.4	25.7	<0.001
F13-1	1.02	3.7623 ± .00015	11	Pippard	0.442 ± .015	<0.032	13.9	<0.001
				London	0.463 ± .019	<0.40	16.7	<0.001
F13-2	1.02	3.7594 ± .0002	15	Pippard	0.482 ± .007	<0.20	32.8	<0.001
				London	0.548 ± .016	<0.040	46.1	<0.001
F12	0.68	3.7343 ± .0003	17	Pippard	0.216 ± .007	4.90 ± 8.1	0.71	0.79
				London	0.220 ± .007	1.82 ± 6.6	1.23	0.23
F11	0.85	3.7825 ± .0005	22	Pippard	0.101 ± .001	<0.028	5.52	<0.001
				London	0.102 ± .001	<0.014	5.69	<0.001

The most likely explanation for the broadening is the presence of various regions in a film with differing local T_c values. Since the overall T_c values for our different films (see table 1) vary over a range of 80 mK, it is plausible that individual films have regions whose T_c values vary from each other over roughly 40 mK. Further, some of the poor (badly pinholed) films also showed obvious spatial variation in appearance, and when such an optically varied film with few pinholes was measured in the superconducting state, it showed sharp changes of slope in $\ln |H_i/H_a|$ *versus* T. This indicated that different regions had distinctly different T_c values. The "good" films gave smooth curves, but that is not inconsistent with a spatially varying T_c as long as the differing regions are small and interact through long range ordering to give an overall but broadened "effective" T_c.

Furthermore, spatial variations in T_c have been observed by other workers using different methods. Measurements on a tin film microbridge found a range of about 25 mK in T_c values [14], which is comparable to the range hypothesized to explain the broadening in our tin films.

A model which would take account of multiple T_c regions would be much more complicated than the models used here, and could not be properly determined by our system of measurement anyway. Thus it was appropriate to omit data close to T_c. Further, the effects of the broadening can be seen in the graphs even in some of the data points which were used, in that the fitted data points at temperatures just below the cutoff are always above the best fit curve. And the effect is stronger in data sets which have implausibly large minimum χ^2 values. Thus, broadening of the superconducting transition due to spatial inhomogeneities appears to be our major source of discrepancy between theory and experiment.

After the superconducting state measurements were completed, the normal resistivity (ζ) of the films was measured at 4.2 K using a four terminal hookup. At 4.2 K (and below) the phonon contribution to the resistivity was negligible, so the mean free path (ℓ) could be calculated by $\zeta\ell = 1.05 \times 10^{-11}\Omega$ cm^2 for tin [15]. Then a predicted value for ξ/ξ_0 was calculated from these normal state data using $1/\xi = 1/\xi_0 + 1/\ell$. The results are displayed in figure 3, where each film is represented by a point, with the ξ/ξ_0 from superconducting measurements plotted *versus* ξ/ξ_0 from normal state measurements. There is obviously a high positive correlation, which validates our superconducting measurements and analysis.

During one run on film F13, an apparent reversed field was observed inside the film, but the effect did not appear again in several subsequent runs. The reversed signal was also quite nonlinear (H_i/H_a should be constant at a fixed temperature), hysteretic, and too large to be the reversed penetrating field predicted by the theory. The explanation of this spurious field reversal is of interest because it very likely accounts for the field reversal reported by Drangeid and Sommerhalder [5]. The explanation we propose is that if

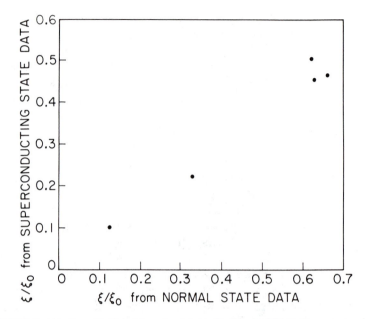

FIGURE 3. ξ/ξ_0 for each film determined from the Pippard fit to the superconducting data, *versus* ξ/ξ_0 for the same film determined from the normal state resistivity.

magnetic flux is pinned in the film at one or more "weak spots" (pinholes and other crystal imperfections), this flux may be pushed around (to other weak spots) by turning up H_a outside the film. This pinned flux motion could result in an apparent positive, *negative* or zero H_i response, depending on how the pinned flux was linked to the magnetometer pickup loop inside the film. And pinned flux motion would obviously give a nonlinear and hysteretic signal, just as we saw in our "reversed field" run on F13.

The fact that earlier in this run we saw evidence of flux being driven into the tin film cylinder increases our confidence in the pinned flux motion explanation of the reversed field. In addition, the fact that, after the film was warmed above T_c (which would allow any pinned flux to escape) and then cooled again, the reversed field could not be observed supports this explanation.

This same effect of pinned flux motion is very likely the explanation of the field reversal reported by Drangeid and Sommerhalder for the following reasons. (a) Their reversed field is at least 300 times greater than predicted by theory (more than 10^4 too big for $\xi/\xi_0 = 0.4$). (b) They mentioned observing nonlinearities in H_i/H_a which they ascribed to trapped flux in their films, and their figure 3 shows that there was considerable nonlinearity present in the reversed field signal. (c) They report that the film which showed field reversal at 2.88 K was subjected to some "high field runs" while being cooled to that temperature. (d) They do not report doing a subsequent (after warming above T_c) measurement on the film which showed field reversal.

5. CONCLUSIONS

We have observed two films (F5 and F12) whose magnetic field penetration behavior in the superconducting state is well modeled by the nonlocal theory, and poorly modeled by the local theory. We believe this is the first unambiguous demonstration of the effects of the nonlocal averaging over the coherence length on magnetic field penetration into superconductors.

Acknowledgments

This work was supported in part by the Air Force Office of Scientific Research.

References

[1] F. and H. London, *Physica* **2**, 341 (1935).

[2] A. B. Pippard, *Proc. Roy. Soc. A* **216**, 547 (1953).

[3] J. Bardeen, L. N. Cooper and J. R. Schrieffer, *Phys. Rev.* **108**, 1175 (1957).

[4] R. Sommerhalder and H. Thomas, *Helv. Phys. Acta* **34**, 29, 265 (1961).

[5] K. E. Drangeid and R. Sommerhalder, *Phys. Rev. Lett.* **8**, 363 (1962).

[6] P. Scharnhorst, *Surface Science* **15**, 380 (1969).

[7] A. B. Pippard, *op. cit.*, 558 (1953).

[8] J. Halbritter, *Z. Physik* **243**, 201 (1971).

[9] E. G. Wilson, Ph.D. Thesis, Stanford University (1976), p. 15.

[10] J. Halbritter, *op. cit.*, 214 (1971).

[11] B. Cabrera, Ph.D. Thesis, Stanford University (1975), p. 23.

[12] S. L. Meyer, *Data Analysis for Scientists and Engineers* (John Wiley and Sons, Inc., New York, 1975), p. 367.

[13] J. T. Anderson, Ph.D. Thesis, University of Minnesota (1971).

[14] W. J. Skocpol, M. R. Beasley and M. Tinkham, *J. Low Temp. Phys.* **16**, 149 (1974).

[15] J. L. Olsen, *Electron Transport in Metals* (Interscience Publishers—John Wiley and Sons, New York, 1962), p. 84.

Organic Superconductivity: The Duke Connection

W. A. Little

I would like to start by reminiscing a little about some of the early days when William Fairbank first came to Stanford in 1959. I had come the previous year as an assistant professor. My responsibility was to install the helium liquefier and a helium pumping system and to set up a laboratory in the old physics building, so that there would be some low temperature facilities available for Bill when he arrived from Duke.

About halfway through that first year we had most of the equipment working and we started on our own rather modest program of research. This was directed to the production of very low temperatures by the adiabatic demagnetization of various salts and metals. It was before the time that ^{3}He was generally available, so it was much more difficult to reach temperatures which are readily attainable now. In thinking back, the work that we were doing was very mundane physics. We were interested in heat transfer, the problems of the Kapitza resistance, various problems of phonon physics, and some work on superconductivity. At that time it was customary to produce large magnetic fields for adiabatic demagnetization by using water-cooled or kerosene-cooled magnets. It was a very unpleasant business! Seeking to avoid these problems I had thought of using a superconducting magnet. In the literature I had come across some work on niobium which indicated that niobium could carry superconducting currents at relatively high fields, that is, above 6 kilogauss; and late in 1959 we built a solenoidal magnet. I believe it was one of the first superconducting magnets that was ever built for a useful purpose. It was wound of niobium wire which the manufacturer had failed to anneal, and as a result it was able to carry a lot more current than we anticipated. It produced a field of about 8 kilogauss. We did not understand why, but it worked, so we used it. However, my opinion of superconductivity as a field of research at that time was rather low, because the superconductive properties were so dependent upon the presence of impurities and dislocations in the metal. It appeared to me to be a dirty field!

It was then that Bill burst on the scene—like a whirlwind, coming in from the South—as an evangelist preaching as he went. He was totally overwhelming. I did not know what he was saying, but he talked, and he talked and he talked! I remember many times thinking I would never get home. As I was leaving the laboratory at 5:30 p.m., Bill would catch me in the corridor and walk with me to my car. We would get to the car and then he would put his foot in the door and talk, and talk, while I got hungrier and hungrier! Eventually I got away—to a cold dinner! But what Bill was talking about was important. He brought with him the concept of "long range order" in superfluids and the idea of the "order parameter." That was not how he said it, as the message was not so clear at that time, but in retrospect that was what he was trying to tell me.

I remember one story which really hit home. This was about a beautiful experiment that he and Michael Buckingham had done on the lambda point of helium. This experiment was inspired by an argument due to Feynman who pointed out that the shape of the lambda transition and its sharpness should be dependent upon the size of the sample no matter how big, "even if it were the size of the sun." (I think this is how Bill put it, with his foot in my car.) I thought about that a great deal. I think I understand it better now, from the work on phase transition by T. D. Lee and C. N. Yang [1], but to me at that time it appeared to be an indication that the helium at one point "knew" what the helium somewhere far distant from it was doing. This was my first introduction to the concept of long range order. But there was much more to it. Bill told us of discussions he had had with Fritz London about condensation in momentum space, order in momentum space, and the existence of long range order in helium and in superconductors.

At the Near Zero conference we have heard of the connection between these ideas and the experiments on flux quantization. The ideas left their mark; but in addition Bill brought to my attention other ideas of Fritz London's, which had evolved from his work on the covalent bond, super-conductivity, and very large organic molecules [2]. As a result I became interested in their possible biological implications, and in the summer of 1962 Tad Day and I started a study of a number of problems in biology, in particular, questions relating to the stability of the genetic code, the origin of life, and the character of enzymatic activity. In the course of that study, Tad brought to my attention a very beautiful piece of work by J. Cairns [3]. This was an autoradiograph of DNA. The striking thing about this was that the DNA was in the form of a closed loop. (See figure 1.) We later learned that if this is labeled with ^{32}P then, on the average, the decay of a single phosphorous atom (the recoil energy of which is sufficient to sever the chain) destroyed the biological activity of the molecule. This immediately brought to my mind what Bill had said concerning London's thinking about the possibility of there being long range order in large organic molecules, specifically the existence of order the whole way round the DNA molecule. This got me interested in the problem of superconductivity in organic molecules.

FIGURE 1. Autoradiograph of DNA showing it to be in the form of a closed loop. (See [3].)

In order to investigate this possibility I considered for a model a system which was based on a simplified picture of the DNA structure: a polymer analogous to the sugar-phosphate helix with polarizable side chains, analogous to the base pairs. Using the simplest form of the Bardeen-Cooper-Schrieffer (BCS) theory I then attempted to calculate the transition temperature for a hypothetical system like this. The idea was that the pairing interaction would come about through the interaction of electrons in the polymer or "spine," by the polarization of the side chains. These side chains would give rise to an interaction analogous to the phonon-mediated, electron-electron interaction of superconductors. I found that this hypothetical material should become superconducting, at least in theory, and that the superconducting transition temperature should be enormously high, of the order of 2000 K [4]. (See figure 2.) This led me to think that London was absolutely right and that superconductivity might exist in certain biological materials.

Let me hasten to say that in spite of these arguments, which we still believe to be valid, there is absolutely no evidence of superconducting phenomena in any of these biological materials nor do I believe a need exists

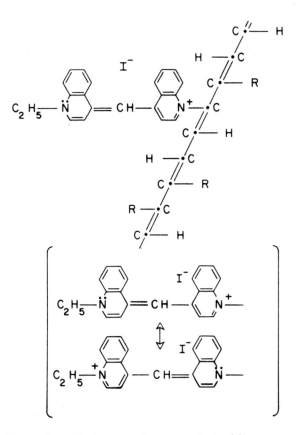

FIGURE 2. Proposed model of an organic superconductor [4].

for such phenomena to explain their biological function. However, it was by this bizarre route that we were led to the idea of a new mechanism in superconductivity, now known as the exciton mechanism, and to the study of complex organic compounds which were totally different from those which had been studied for superconductivity prior to this. The net result has been the development of a new area of physics and materials science, and an associated and coordinated growth of activity in chemistry in the synthesis of new and exciting materials. This new chapter in physics has involved the design, study and theoretical analysis of organic metals and more recently organic superconductors [5]. It has created a new generation of physicists and chemists who can communicate with one another and work together.

I would like to illustrate the developments in this field by sketching first the work in low dimensional physics which has evolved from it. The study of superconductivity in linear chain systems led to discoveries of new phenomena in limited dimension solid state physics. From the study of fluctuations in these systems has come an understanding of the role of noise in superconducting devices.

A linear chain system is at once simple and complex. It is simple because it is one dimensional, and the dynamics are thus simplified, but because of the role of fluctuations it becomes complicated. Let us ignore the fluctuations for the moment and focus on the simple dynamic aspects. Consider the energy of an electron confined to move in a one-dimensional chain, and ignore the interactions. One obtains the typical energy *versus* momentum relationship shown in figure 3a. The electrons interact with the ions of the

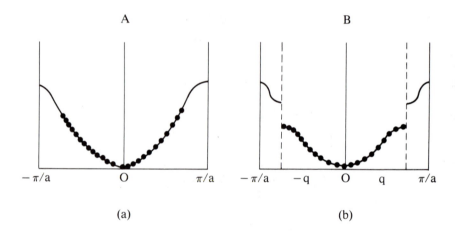

(a) (b)

FIGURE 3a and 3b. (a) Energy-momentum relation for free electrons. (b) Energy-momentum relation for electrons in a lattice which has undergone a Peierls' transition.

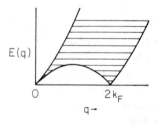

FIGURE 3c. Particle-hole energy as a function of momentum for a one-dimensional system.

lattice. If the lattice is distorted, a periodic potential is created in which the electrons move, and this produces a gap in the energy spectrum. If one now introduces this perturbation so as to put a gap at the Fermi surface, then the states which are filled will be depressed in energy while the states which are not filled will be raised in energy (figure 3b). The net effect is to produce a state of lower energy with a distorted structure. This is the Peierls' transition and it occurs in many linear chain systems.

However, the electron-electron interaction modifies the result. The modification can be understood from perturbation theory. Consider the particle-hole excitations in such a one-dimensional system. There are two regions where one can get low energy excitations (see figure 3c): a region near $q = 0$, involving small momenta, and another near twice the Fermi momentum. This is much simpler than in three dimensions, where one can scatter at any angle from the Fermi surface to another point on the Fermi surface with very small energy changes. It is clear from perturbation theory that these low energy denominators give rise to strong effects in these regions. Often the result is a pairing of the electrons. Different types of pairs can be formed. An electron can be paired with another electron of the opposite spin to give singlet superconductivity; an electron with spin up can be paired with another of spin up to give triplet superconductivity; an electron can be paired with a hole to give either a spin zero charge-density-wave state or a spin one state resulting in a spin-density-wave. All of these types of pairings can be described by a BCS-like expression for the temperature at the transition to the new ordered state,

$$kT_c \simeq \hbar\omega \exp(1/\lambda) \quad . \tag{1}$$

The effective coupling constant λ for each of the paired states is given by different combinations of g_1 and g_2, where g_1 corresponds to the interaction for scattering across the Fermi sea, and g_2 that for scattering with small momentum transfer (see figure 4). Which state occurs depends on the relative strengths of g_1 and g_2. One obtains a very simple phase diagram, at least in that region where the electron-electron interaction is weak relative to the hopping probability between elements of the chain. This region we refer to as the region of "g"-ology, where the g-coupling constants play a dominant role. (See figure 5.) Many compounds are well represented by it. In another part of the diagram, which has been referred to as "U-t"opia, the Coulomb interaction U is large compared with the transfer integral t, and there the phase diagram presumably is more complicated. An impressive amount of elegant work has been done in developing this theoretical structure and in correlating the behavior of compounds with it.

Equally impressive have been the achievements of the synthetic chemists, who have developed a number of extraordinarily ingenious ways of synthesizing new materials. When we first proposed the idea of making this

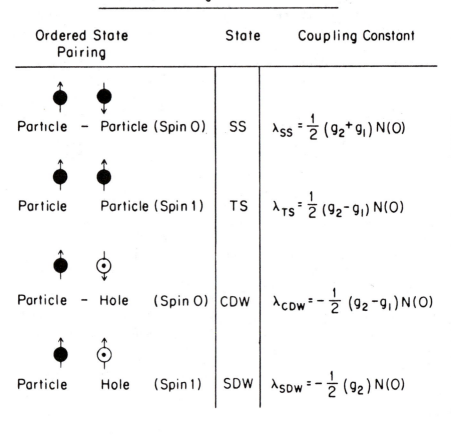

MEAN FIELD T_C IN I.D. SYSTEMS

Ordered State Pairing		State	Coupling Constant
Particle – Particle (Spin 0)		SS	$\lambda_{SS} = \frac{1}{2}(g_2 + g_1)N(O)$
Particle Particle (Spin 1)		TS	$\lambda_{TS} = \frac{1}{2}(g_2 - g_1)N(O)$
Particle – Hole (Spin 0)		CDW	$\lambda_{CDW} = -\frac{1}{2}(g_2 - g_1)N(O)$
Particle Hole (Spin 1)		SDW	$\lambda_{SDW} = -\frac{1}{2}(g_2)N(O)$

$$k T_C = h\omega \exp\left\{\frac{1}{\lambda_i}\right\}$$

FIGURE 4. Expressions for mean field critical temperature in I.D. systems. Each paired state [singlet superconductivity (SS), triplet superconductivity (TS), charge-density-wave (CDW) and spin-density-wave (SDW)] has a mean field transition temperature given by a BCS-like expression (bottom). The value of λ for each is a different function of the coupling constants g_1 and g_2, as shown in the right hand column of the table. Here $N(O)$ is the density of states at the Fermi surface.

complex organic superconductor, some of our most gracious colleagues in chemistry said that what we were asking for was at least 50 years ahead of the state-of-art of chemistry and that there was no hope of ever being able

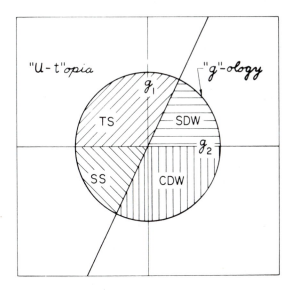

FIGURE 5. Phase diagram of a one-dimensional system as a function of the coupling constants g_1 and g_2.

to realize it. Great progress has nevertheless been made in the course of the last 10 years.

One method of attack has been that used by Gerhard Wegner [7] and his group in Freiberg in West Germany, and by Ray Baughman [8] at Allied Chemicals in this country. What they have done is to produce single crystals of poly-diacetylene. The monomer molecules are designed such that the unit cell dimensions of the crystal structure of the monomer are identical to that which the compound would have if it were polymerized. One starts with a clear single crystal of the basic material, irradiates it with x-rays, ultraviolet, or heat, and thus initiates polymerization. Because the unit cells of the monomer and polymer are identical, the single crystal monomer ends up as a single crystal polymer (figure 6). Thus one evades the fundamental problem of producing a polymer of such great complexity, namely the fact that polymers which are conjugated in the manner proposed in our model (figure 2) are insoluble in nearly all solvents. Having them in solution is essential in order to carry out the polymerization by conventional means.

These materials are very interesting in their own right. They start out being totally colorless, but upon exposure to light polymerize and become metallic looking. I might add that they are not good conductors; we do not understand exactly why. They are extraordinarily strong materials because of the near perfection of the molecular structure. The rate of polymerization can be controlled by the choice of end groups on the diacetylenes.

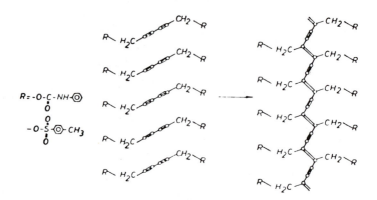

FIGURE 6. Solid state polymerization of diacetylenes.

This has created a novel use for them as "labels" for integrating the time-temperature history of a material [9].

As is well known, many organic polymers such as mylar and polyethylene are among the best insulators that we have. One finds, however, that they become conducting if they are doped, with bromine or with iodine for example, introducing carriers into the bands that were empty, or holes in the bands that were previously filled. Even polyacetylene when it is doped changes in conductivity by eight orders of magnitude. (See figure 7.) Starting out as a colorless material, it becomes transformed, upon doping, into a material which looks very much like aluminized mylar [10]. This is an organic metal and can be formed with traces of bromine, iodine, or various other compounds in it.

On another topic, it was brought to our attention about ten years ago [11] while we were searching for a material able to carry current along a spine-like structure with polarizable organic ligands, that there are a number of platinum compounds which meet this criterion [12]. These form linear chains with the d_{z^2} orbital of one metal atom in the chain overlapping the d_{z^2} orbital of the next. Oxidation gives a linear chain compound with a partially filled band which becomes an extremely good metal. If one looks at it under a polarizing microscope, one finds that for light polarized along the direction of the chain the crystal appears metallic, because conduction can occur along this direction, but a crystal oriented at right angles to the plane of polarization is transparent, because there is no lateral conductivity.

This is just one of a large class of materials which have been discovered in the course of these investigations which have enormously anisotropic electrical, thermal and other properties. Among these are two of the most extensively studied, TCNQ and TTF, which when combined form a

FIGURE 7. Polyacetylene doped with bromine to form an organic metal $(CHBr_{0.05})_x$.

structure in which the individual molecules stack on top of one another, again forming a linear chain compound [13]. (See figure 8.) These linear chain compounds exhibit a number of phase transitions. In studying their conductivity one finds that as one approaches a temperature of about 60 K the conductivity becomes enormously high [14]. At one time it was thought that this might signify the existence of superconductivity, but it does not; it is just that the anisotropy is so high that one has to be extraordinarily careful in making the measurements. The material undergoes a Peierls' transition near 60 K and becomes insulating below it. The study of these materials with various subtle changes of the organic structures and modification of these with selenium have given a number of compounds of extraordinarily interesting properties. Several members of a particular class of compounds $(TMTSF)_2X$ undergo a transition into the superconducting state [15]. So we now know that there are compounds which are organic superconductors, and we know of polymers, the polymer of SN_x for example, which also superconduct [16].

I think this topic is appropriate for the *Near Zero* volume because while our efforts have been directed to a search for a high temperature organic superconductor, and we continue to believe there is a possibility of

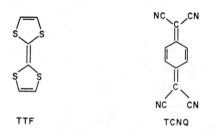

TTF   TCNQ

FIGURE 8a. Construction of highly conductive anisotropic organic metals. The molecules of TCNQ and TTF and various modifications of these are the basis for a series of highly conductive organic metals.

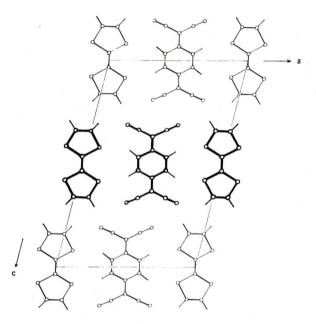

FIGURE 8b. The compound TCNQ-TTF consists of separate one-dimensional stacks of TCNQ and TTF molecules arranged in the crystal as shown.

obtaining high temperature superconductivity, thus far the superconductivity we have observed has been "Near Zero"! In conclusion, I want to repeat that it was largely Bill Fairbank's influence and his enthusiasm and constant interest in the fundamental problems of superconductivity that started us on the road to organic superconductivity. The legacy of all this work has been a new group of physicists and chemists who have done great things, and will continue to do so in the future, I am sure. For this we owe a debt of thanks to Bill Fairbank.

Addendum

The discovery of superconductivity in $YBa_2Cu_3O_{6.9}$ above 95 K has been announced by M. K. Wu et al. [17]. The remarkable high transition temperature of this material, together with its structure of one- and two-dimensional components, and the complete absence of an isotope effect on T_c as reported by R. J. Cava et al. [18] at the Materials Research Society meeting in Anaheim in April 1987 are strong indications that the electronic or "excitonic" interaction which we proposed for the organic superconductor [4] is operative here. This connection is made in papers presented at the MRS meeting by W. A. Little, J. P. Collman and J. T. McDevitt [19] and submitted to Physical Review Letters [20].

References

[1]   C. N. Yang and T. D. Lee, *Phys. Rev.* **87**, 404 (1952); T. D. Lee and C. N. Yang, *Phys. Rev.* **87**, 410 (1952).

[2]   F. London, *Superfluids* (Dover Publications, Inc., New York, 1961), p. 8.

[3]   J. Cairns, *J. Mol. Biol.* **6**, 208 (1963).

[4]   W. A. Little, *Phys. Rev.* **134**, A 1416 (1964).

[5]   For a recent review of the field see *Chemica Scripta* **17**, No. 1–5 (1981).

[6]   A simplified outline of this is given by H. Gutfreund and W. A. Little, in *Highly Conducting One Dimensional Solids*, edited by J. T. Devreese, R. P. Evrard and V. E. van Doren (Plenum, New York, 1979), p. 305.

[7]   G. Wegner and W. Schermann, *Colloid and Polymer Sci.* **252**, 655 (1974).

[8]   R. H. Baughman, in *Contemporary Topics in Polymer Science*, edited by E. M. Pearce and J. R. Schaefgen (Plenum, 1977), Vol. 2, p. 205; R. H. Baughman, *J. Appl. Phys.* **43**, 4362 (1972).

[9]   G. N. Patel, A. F. Preziosi and R. H. Baughman, U.S. Patent 3,999,946, Dec. 28, 1976.

[10]   M. Akhtar, C. K. Chiang, M. J. Cohen, J. Kleppinger, A. J. Heeger, E. J. Louis, A. G. MacDiarmid, J. Milliken, M. J. Moran, D. L. Peebles and H. Shirakawa, *Annals N. Y. Acad. Sci.* **313**, 726 (1978).

[11]   J. Collman, *J. Polymer Science* **C**, Polymer Symposia No. 29 (J. Wiley and Sons, 1970), p. 136.

[12]   K. Krogmann, *Angewandte Chemie* **81**, 10 (1969).

[13]   T. E. Phillips, T. J. Kistenmacher, J. P. Ferraris and D. O. Cowan, *J. Chem. Soc. Chem. Sommun.*, 471 (1973); T. J. Kistenmacher, T. E. Phillips and D. O. Cowan, *Acta Cryst.* **B30**, 763 (1974).

[14]   L. B. Coleman, M. J. Cohen, D. J. Sandman, F. G. Yamagishi, A. F. Garito and A. J. Heeger, *Solid State Commun.* **12**, 1125 (1973).

[15]   M. Ribault, G. Benedek, D. Jerome and K. Bechgaard, *J. Physique Lett.* **41**, L-397 (1980); D. Jerome, *Chemica Scripta* **17**, 13 (1981).

[16]   R. L. Greene, G. B. Street and L. J. Suter, *Phys. Rev. Lett.* **34**, 577 (1975).

[17]   M. K. Wu *et al.*, *Phys. Rev. Lett.* **58**, 908 (1987).

[18]   R. J. Cava *et al.*, *Materials Research Society Proceedings*, Anaheim, 1987.

[19]   W. A. Little, J. P. Collman and J. T. McDevitt, *Materials Research Society Proceedings*, Anaheim, 1987.

[20]   J. P. Collman, J. T. McDevitt and W. A. Little, "Possible Role of the Excitonic Interaction in the New High T_c Superconductors," submitted to *Phys. Rev. Lett.*, 1987.

CHAPTER IV

Applications of
Superconductivity

(*Top*) Alan Schwettman, Michael McAshan, Todd Smith and John Turneaure in the superconducting accelerator tunnel. (*Bottom*) John Madey making an adjustment on the free electron laser.

Applications of Superconductivity: Introduction

Bascom S. Deaver, Jr.

At Duke, and in his early years at Stanford, William Fairbank focused his research on basic questions about liquid helium, superconductivity and magnetism, topics within the traditional purview of the low temperature physicist. Thereafter his attention has been directed increasingly to a broad range of fundamental questions in physics encompassing topics in particle physics, symmetry principles and gravitation, all of which are discussed in subsequent chapters of this book. Invariably the experiments he has devised in these diverse areas have depended crucially on low temperature. He has continually espoused the advantages of the low temperature environment—low thermal noise, mechanical and thermal stability, and the possibility of exploiting the phenomena of superfluidity and superconductivity.

The papers in this chapter are concerned specifically with some applications of superconductivity. They are grouped into two general categories: (a) rf superconductivity and the superconducting accelerator, (b) superconductive electronics and magnetic measurements. Papers in the first category describe a series of accomplishments whose origins can be traced to Fairbank's thesis research on superconducting microwave cavities. Alan Schwettman describes the research at Stanford as it evolved from the early work by John Pierce on high-Q cavities to the superconducting electron accelerator with its advantages of cw operation, high beam current and excellent energy resolution. The unique properties of the Stanford accelerator are discussed by Todd Smith and colleagues in the next paper.

In conjunction with the accelerator there were two other major developments of quite different types. One is a technique for obtaining more efficient localized irradiation of cancerous tumors by means of a beam of pions, as described by Peter Fessenden and coauthors. The second is the free electron laser, which converts the kinetic energy of an intense high energy electron beam into coherent photons by passing the beam through a spatially periodic magnetic field. This remarkable device was conceived by John Madey who is author of the paper about it in this chapter.

An outstanding example of what can be achieved by obtaining near zero energy loss is the research on superstable oscillators based on extremely high-Q superconducting microwave cavities. In his paper John Turneaure describes these devices and the host of fundamental measurements that can be made with them.

The last four papers in this chapter deal with superconductive electronics, a term that has come to denote the collection of analog and digital devices, sensors and signal processors based largely on the Josephson effects, fluxoid quantization and the nonlinear conductance of tunnel junctions. A ubiquitous element in superconductive electronics is the SQUID (Superconductive Quantum Interference Device), which consists of a superconducting ring containing either one or two Josephson junctions. As a sensor of magnetic flux changes and as an amplifier at relatively low frequencies, it is unexcelled. SQUID's are the key element in a family of ultrasensitive devices including magnetometers, magnetic gradiometers, accelerometers, gravimeters and gravity gradiometers.

The majority of the experiments described in the remaining chapters of this book use SQUID's in one of these roles. The gravity wave detector will require the ultimate quantum-limited sensitivity of the device, *i.e.*, sensitivity to as near zero energy change as possible. (See paper by Michelson *et al.* entitled "Near Zero: Toward a Quantum-Limited Resonant-Mass Gravitational Radiation Detector.")

John Wikswo and Mark Leifer describe the use of SQUID magnetometers and gradiometers to study biomagnetism through measurements of the magnetic fields from the human heart and brain as well as fields from individual nerves.

Fairbank recognized the efficacy of using superconducting devices in concert. He conceived a susceptometer for measuring the magnetic properties of weakly magnetic materials by using the stable field environment made possible by a superconducting persistent-current magnet and a superconducting shield within which the superb sensitivity of a superconducting magnetometer can be exploited fully. A series of graduate students working with him at Stanford developed this instrument into one with uniquely high sensitivity and resolution. Concurrently, similar instruments were developed at other research labs and by instrument companies.

John Pierce built the first susceptometer at Stanford using a modulated inductance magnetometer. Edmund Day then developed a SQUID susceptometer and was the first to detect nuclear magnetic resonance with a SQUID. Subsequently John Philo produced an instrument with resolution nearly a factor of 100 better than that of any other susceptometer and with very fast response. He used it to study reaction kinetics following photo excitation of a hemoglobin solution with a laser pulse. In their paper in this chapter, Day and Philo describe the use of superconducting susceptometers for new measurements in biophysics.

Josephson devices and SQUID's are also the basis for very high speed (Δt near zero) and low loss (dissipation near zero) digital devices, including picosecond samplers, fast analog to digital converters with exceptional resolution, and logic and memory elements. In conjunction with rapidly evolving techniques for fabricating superconducting microcircuits, these devices provide the components for extremely high speed digital systems including the current pinnacle of superconducting technology, the Josephson computer. The current status of superconducting digital electronics is described by James Hollenhorst, who as a student at Stanford working with Robin Giffard developed a microwave SQUID for use in a low temperature gravity wave detector.

RF Superconductivity and Its Applications

H. A. Schwettman

1. EARLY HISTORY

H. London pointed out in 1934 [1] that resistive losses should be observable in superconductors at very high frequencies. London's prediction was based on a two-fluid model of superconductivity and has a simple electrical circuit analog illustrated in figure 1. At low frequencies the electrical current in a superconductor will be carried by the superconducting component of the fluid which is represented in our figure by a simple inductance since it exhibits only the inertial properties of the electron. At very high frequencies, however, this path has appreciable impedance, and therefore part of the current will be driven through the normal conducting component of the fluid where one has not only the inertial properties of the electron but also the resistive properties.

It was six years before that idea was put to a test. It was tested by London himself [2], who immersed an ellipsoidal tin sample in a 1.5 GHz rf field and made calorimetric measurements of the power dissipated in the sample. He made two important observations in the experiment. The first related to the superconducting properties of tin. At the transition temperature, the power losses dropped measurably and then gradually vanished as the temperature approached zero. That qualitatively confirmed his own prediction. But London also observed at that time a very interesting

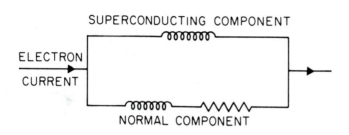

FIGURE 1. Circuit analog for electrical conduction in a two-fluid model of superconductivity. At very high frequencies part of the current is driven through the normal conducting component of the fluid.

phenomenon related to the normal state properties of tin. The rf conductivity of tin, instead of rising dramatically as one proceeds from high temperature to very low temperature, rose to quite a modest value and then became independent of temperature. This behavior he attributed correctly to the fact that the electron mean free path in the metal finally exceeded the classically calculated skin depth in the material. This low temperature conduction phenomenon is now known as the anomalous skin effect.

About this time World War II interrupted research in many areas of physics. But the tremendous development in microwave techniques during the war created for all of those who returned and entered physics research a wonderful opportunity. One of those persons was William Fairbank. Not everyone could easily enter the field which H. London had established, because it required a combination of two things: knowledge of microwave techniques, and knowledge of low temperature techniques. There were, in fact, in those early days after World War II, only three groups involved in rf superconductivity. Pippard [3] at Cambridge started making measurements at 1.2 GHz using a hairpin resonator. Bill Fairbank [4], who went to Yale after the war to become a graduate student, undertook measurements at 9.4 GHz using a resonant cavity. Maxwell, Marcus and Slater [5] at MIT proceeded with measurements, also using a resonant cavity, at 24 GHz.

2. THE FAIRBANK THESIS

It has been an interesting experience for me to read Bill Fairbank's thesis [6]. I had read the published paper earlier but, as you all know, there is much more interesting content in the thesis than can ever be found in the published scientific paper. What he set out to do was to reproduce the measurements London made, but at a much higher frequency, in the hopes of confirming the superconducting state and normal state behavior that London observed, and of obtaining information about the frequency dependence. He made measurements from nitrogen temperature, 75 K roughly, where the Q of the X-band cavity was about 5,000, to helium temperature, 4.2 K, where the Q was 23,000, and then ultimately to a minimum temperature of 1.26 K, where the Q exceeded one million. That is a very large variation in the Q, and it was one of the special problems that had to be dealt with in the experiment. Bill's technical sophistication was already evident in these early days when he was a graduate student. In figure 2, taken from Bill's thesis, one can see the microwave apparatus

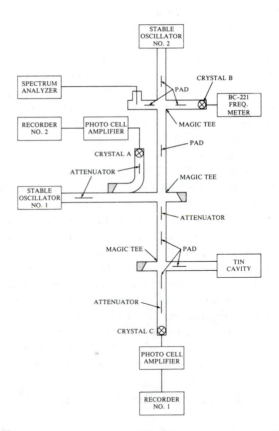

FIGURE 2. Block diagram of the microwave measurement system.

used to measure the cavity Q. Bill measured the power reflected from the cavity as a function of frequency. To make good measurements on a cavity where the Q reached a million required great stability of the oscillator, so he began work on stable oscillators, for which he found use many times in later research. In fact, for his thesis experiment Bill made two stable oscillators, one to check the frequency of the other. Characteristically, Bill has always had good foresight, and this was just the first example. One of the oscillators did not work, so it was very good that he had two.

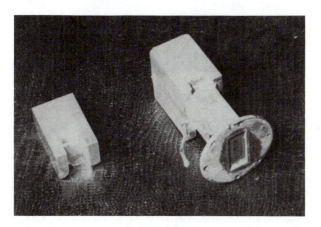

FIGURE 3. Photograph of the tin cavity. On the left is shown one of the two sections from which a cavity is made. On the right is shown the assembled cavity sealed with a resonant glass window.

Figure 3 is a photograph of the superconducting cavity. It was fabricated from very high purity tin, and it was molded in two halves and pressed into shape. Bill very carefully polished the die so that the inner surface of the cavity would be as smooth as possible. To exclude helium from the cavity during measurement, he soldered those two halves together, and coupled the cavity to the external waveguide with a metal-to-glass seal. Figure 4 shows something even more remarkable than the cavity itself. This is a photograph of the entire apparatus, and the first thing that will catch your attention is its size. It is somewhat smaller than most of the apparatus Bill has built since he came to Stanford. But as a graduate student he had to make liquid helium himself, so even one liter was challenge enough. In this same photograph you will see that the rectangular cavity is suspended from the top plate by a specially built waveguide. Helium was at such a premium in those days that one had to insure that the heat leak would be as small as humanly possible. Therefore, Bill's technique in fabricating the waveguide was to start with a piece of cylindrical glass pipe and form it into a rectangular shape using a series of tapered graphite dies. He found that he could not get a nice, smooth surface on the inside of the waveguide using the graphite dies; therefore, he made it slightly undersized and then carefully ground and polished the surface. In fact, I remember very well a

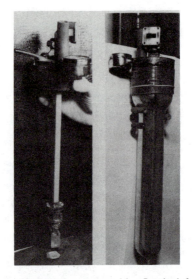

FIGURE 4. Photograph of helium dewar assembly. On the left one can see the specially built glass waveguide with the tin cavity attached at the bottom.

conversation Bill and I had one evening at the laboratory (about 4 o'clock in the morning) when we were commiserating about some of our own disasters enroute to our Ph.D. theses. I had told him a story of mine and he told me a story of his. It seems that it had taken Bill a few weeks to form one of these glass waveguides, and another few weeks to learn how to polish one, and then another week to actually polish the one he was going to use. After that he had to silver it just as you would a dewar. When he had it all finished, he held it proudly, turned in the room, knocked the fragile glass waveguide into the side of a table and broke it. Bill claims that he never told Jane about the disaster when he went home that night. He said it was too traumatic to tell anyone.

After rebuilding the waveguide, Bill completed his experiment. His measurements of the normal state properties of tin extending from nitrogen temperature to helium temperature confirmed the anomalous behavior first observed by H. London. The rf surface resistance approached a limiting value at the lowest temperatures, no matter how large the dc electrical conductivity became, and only the transition from the normal state to the superconducting state produced smaller losses. In figure 5 we show the measured surface resistance (the inverse of the rf conductivity) as a function of temperature in the superconducting state. At the superconducting transition temperature, the measured Q of the cavity was 2.3×10^4, while at the lowest temperatures achieved the Q-value exceeded 10^6. This work represents one of the earliest experiments producing information about the

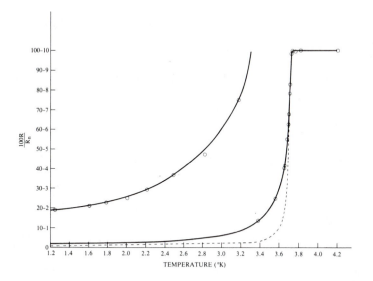

FIGURE 5. Graph of the measured surface resistance of superconducting tin as a function of temperature. The surface resistance is normalized to its normal state value at the transition temperature.

temperature dependence and the frequency dependence of microwave losses in superconductors.

3. THE ACCELERATOR APPLICATION

During the 1950's work in the field of rf superconductivity expanded dramatically. More laboratories had low temperatures available to them, and more research groups had become sophisticated in microwave techniques. In a review article by Maxwell [7], published in 1964, there is reference to fifty such papers. And let me just point out that in this decade of great activity the best Q-value that was achieved in a superconducting cavity was 5×10^6, a value higher by no more than a factor of three than Bill had achieved in his Ph.D. thesis work. In fact, Maxwell, in that same review article, made an observation which he had to correct in a note added in proof; he said that an order of magnitude improvement in Q from 5×10^6 might be possible but higher Q's were unlikely.

The note added in proof in Maxwell's article, of course, referred to Bill's work at Stanford in 1961 with John Pierce and Perry Wilson [8]. In 1961 people began to think about the possibility of building superconducting accelerators. A linear accelerator made of superconducting microwave cavities could be operated continuously instead of being pulsed as were all room temperature linacs. There were several groups which became active at essentially the same time: a group at Harwell [9], a collaboration between the group at CERN and the group at Lausanne [10], and the Stanford effort.

Of course, the situation was absurd; they were all crazy to think about the possibility of a superconducting accelerator, and the reason was clear. Let us take SLAC as an example. SLAC operates at an energy gradient of 6.6 MeV per meter. It operates at a duty factor of one part in one thousand; therefore, if you want to maintain the same energy gradient but operate the accelerator continuously at helium temperature, you must first improve the losses by a factor of a thousand in order to go from a duty factor of 10^{-3} to a duty factor of unity. But those losses will occur at helium temperature and therefore you must pay the Carnot efficiency, which is another factor of a hundred, and then if you actually want to remove that heat you must build a practical refrigerator that is likely to have an efficiency of 10%. Thus, you need approximately a factor of a million improvement in the conductivity of the walls to operate the superconducting accelerator with the same input power. You might do with 10^5 improvement, but even that improvement implies a Q-value exceeding 10^9. Furthermore, the rf field levels required are an appreciable fraction of the critical magnetic field of a superconductor such as lead. At 6.6 MeV the peak rf magnetic field in a typical accelerator cavity is approximately 250 Oe.

Let me characterize the situation all at once. It would be reasonable to predict that you will never achieve the Q that you need; but if you do, you will never achieve the field that you need; but if you do, you will never be able to make a superfluid refrigerator of the scale that you need for an accelerator.

Bill specializes in attacking problems that involve multiple negative predictions. The first entry in the data book, dated June 21, 1961, is a tentative sketch of the apparatus. By the summer of 1962, Bill working with John Pierce and Perry Wilson had very interesting results. Already in their initial experiments the cavity Q was higher than they had anticipated, and thus the oscillator was not stable enough to stay on resonance. The rf was pulsed repetitively and only once in a while was the oscillator frequency actually on the resonant frequency of the cavity. Nonetheless they were able to make Q measurements in their cavity, and achieved values as high as 3×10^8, still not high enough to imagine being able to build a superconducting accelerator, but really quite dramatic progress, as indicated in the note added in proof in Maxwell's article.

It was in the autumn of 1962 that I came to Stanford. Bill has often told me about a phenomenon that occurs with new graduate students and postdocs that I now recognize in myself, looking back on that time. Every time a new postdoc or a new student works on an experiment, absolutely nothing happens with the experiment for a year or two. The reason for this is that the person has to make the experiment over in his own image. In the first Stanford experiment the cavity was not immersed in liquid helium. I decided if you were going to get high fields you were going to have to have improved thermal contact, so we completely rebuilt the cavity, completely rebuilt the cryogenic system, and rebuilt the oscillator. Finally a year and a half later, in the summer of 1964, we made measurements [11] with the new system. At low power levels the Q-value of the electroplated lead cavity was measured as a function of temperature. As can be seen in figure 6, the Q had risen already to 5×10^9 at 1.85 K, and was indeed still rising.

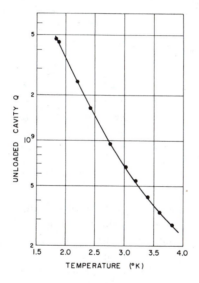

FIGURE 6. The unloaded Q of a superconducting S-band cavity as a function of temperature. The TE_{011}-mode cavity was fabricated from copper and electroplated with lead.

Furthermore, we succeeded in making measurements as a function of field strength. In figure 7 you can see that even up to the 200 gauss level we had maintained Q's in excess of 10^9.

Finally in the summer of 1965 we built a three-cell accelerator structure and accelerated electrons in a superconducting cavity for the first time [12]. This began an accelerator program that culminated in the machine whose characteristics are described in the paper by Todd Smith et al., entitled "Unique Beam Properties of the Stanford Superconducting Recyclotron," which follows in this volume.

The high Q-values achieved not only established the feasibility of the superconducting accelerator, but also provided an opportunity to make a detailed comparison of the measured surface resistance with that calculated from the theory of Mattis and Bardeen [13]. Up to that time

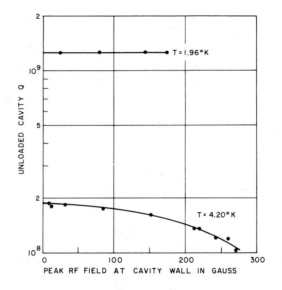

FIGURE 7. The unloaded Q of a superconducting TE_{011}-mode cavity as a function of the peak rf magnetic field at the cavity wall. In an accelerator cavity a peak magnetic field level of 200 gauss corresponds to an accelerating gradient of approximately 5 MV/m.

all of the microwave data had been fit to an empirical formula given by Pippard. Only with accurate measurements of the surface resistance spanning several orders of magnitude was it possible to distinguish between the T^4 dependence given in the Pippard formula and the exponential temperature dependence given in the Mattis-Bardeen theory. In the course of John Turneaure's thesis research, he and I made a detailed comparison between experiment and theory for both superconducting tin and superconducting lead at frequencies from 2856 MHz to 12 GHz. These were the first detailed comparisons of experiment and theory at microwave frequencies [14].

In 1967 we became interested in the possibility of fabricating cavities from niobium. Among elemental superconductors niobium has the highest transition temperature and the highest critical magnetic field, and thus was the favored superconductor for the accelerator application. The very first encouraging results with a niobium cavity were obtained in the spring of 1968 by Turneaure working with Ira Weissman of Varian Associates [15]. Finally in 1970 Turneaure and N. T. Viet [16] achieved peak magnetic field levels in superconducting niobium X-band cavities of 1000 Oe, peak electric field levels of 70 MV/m and Q-values of 10^{10}. The expectation that these remarkable results could be reproduced in accelerators turned out to be premature.

4. ELECTRON LOADING LIMITATIONS

The problems in extrapolating from X-band cavity performance to accelerator performance had little to do with superconductivity. Electron loading phenomena which were strongly suppressed in the small, high frequency test cavities dominated the performance of large, low frequency accelerator structures. The first of these electron loading phenomena is electron multipacting. Multipacting in rf cavities is an electron conduction phenomenon in which growth, or amplification, of the electron current is provided by secondary emission. The phenomenon occurs only in certain regions of the cavity and then only at specific field levels because of the strict dynamics requirement that an electron emitted from the cavity wall must return to that same point in some multiple of rf periods. Furthermore, it must return with an energy at which the secondary emission coefficient is greater than one. The nature of electron multipacting in superconducting cavities and the role it plays in limiting cavity performance were finally explained in 1977 by Lyneis, Schwettman and Turneaure [17], and a very effective method of suppressing this phenomenon was suggested in 1979 by Klein and Proch [18].

The second electron loading phenomenon that has limited performance of superconducting cavities is electron field emission. Electron field emission has been particularly important in limiting cavity performance at frequencies below 2000 MHz. Although a number of attempts have been made to explain the observed electron loading phenomena, these have had rather limited success and the situation has remained complex and confusing.

While on sabbatical at CERN in 1981 I had an opportunity to review the Stanford data of the early 1970's and the subsequent literature on electron field emission loading; the possibility of constructing a rather clear and simple picture of the observed phenomena emerged. Although definitive and unambiguous experiments have been difficult to achieve in this field, two general conclusions concerning electron field emission in superconducting cavities seem clear:

(1) Electron loading grows exponentially with the electric field level as illustrated in figure 8; and

(2) if interpreted in terms of Fowler-Nordheim field emission, the electron loading is strongly enhanced with an apparent enhancement of the field (β_a) that is roughly proportional to f^{-1}. This dependence, first pointed out by Lyneis [19], is the most perplexing experimental observation in the field, and it seems likely that it is connected to the general experience that field levels achieved in superconducting cavities increase monotonically with frequency. The monotonic frequency dependence of achieved field level for single-cell cavities and for short multi-cell structures is illustrated in figure 9.

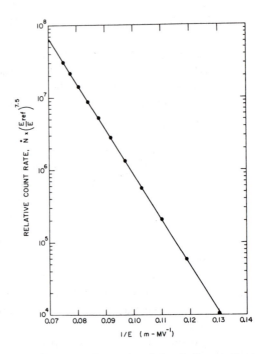

FIGURE 8. Modified Fowler-Nordheim plot of the photon count rate *versus* rf electric field level in a superconducting L-band cavity. The photons are produced by field emitted electrons that have been acccelerated in the cavity fields.

In discussing electron field emission loading, it is important to make a distinction between two classes of enhanced emission. The possibility of such a distinction was made clear in the early Stanford experiments on helium ion sputter processing [20]. As shown in figure 10, the graph of the apparent enhancement factor β_a as a function of accumulated processing (as measured by the accumulated radiative dose) consists of two regions. Initially the enhancement factor decreases rapidly with processing, while subsequently the enhancement factor decreases slowly. It was shown that the initial rapid decrease corresponded to removal of a few monolayers of material by helium ion sputtering. The distinction we want to make then is between "surface" enhancement and "bulk" enhancement. One can imagine several physical mechanisms for "surface" enhancement of emission and for "bulk" enhancement of emission, but for our present purpose it is not essential that we specify mechanisms. We can view this distinction as being purely empirical.

In the early Stanford work [21] it appears there are two sets of cavity data for which "surface" enhancement of field emission is small. The first set consists of data obtained with cavities that have been processed by

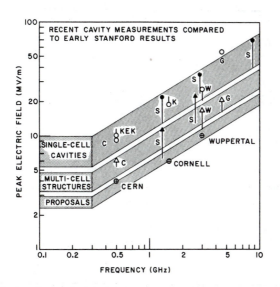

FIGURE 9. The maximum electric field levels achieved in single-cell cavities and multi-cell structures *versus* frequency. The data plotted here include the early Stanford experiments (S) and the more recent work at CERN (C), Karlsruhe (K), Genova (G), Wuppertal (W), and KEK.

helium ion sputtering, and the second set consists of data obtained with cavities that have been UHV fired and then immediately and permanently sealed. These specially selected data seem particularly important to our understanding of electron field emission phenomena in superconducting cavities, and therefore they are given in table 1. Although the average value of the apparent field enhancement is less at higher frequency, it is difficult to argue that β_a is proportional to f^{-1}. In fact, values of β_a equal to 380 (within 5% of the maximum value) are observed at 1300 and 2856 MHz, the two highest frequencies for which β_a-values were determined. A more plausible interpretation of this selected data set can be provided by making a statistical argument based on the assumption that there are characteristic emitters with β_a equal to 400, the maximum value observed, which cover the surface of these niobium cavities at some surface density n_A.

A word of explanation is in order. When β_a-values are determined, it is assumed that the emitting site is located in the region of the cavity where the electric field is a maximum. If, in fact, the emitting site is located at a point where the cavity field is one-half of the maximum value, then the value of β_a assigned will be one-half of the actual value. For a fixed surface density n_A, the probability of finding one emitter in the highest field region of the cavity will decrease as the frequency of the cavity increases simply because the high field surface area "at risk" decreases. The surface areas

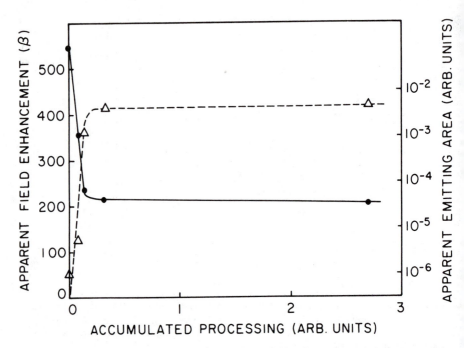

FIGURE 10. The dependence of the apparent field enhancement and the apparent emitting area on accumulated helium sputter processing. The apparent field enhancement and the apparent emitting area were determined by analysis of a modified Fowler-Nordheim plot. The accumulated helium sputter processing was estimated from the accumulated radiation dose measured external to the cavity.

"at risk" in the Stanford experiments referred to above are indicated in table 1.

For a large ensemble of cavities, of course, one would find at every frequency some cavities with β_a equal to 400, but the mean value of β_a determined experimentally would decrease monotonically with increasing frequency. Shown in table 1 are six measurements of β_a at 2856 MHz and five measurements at 1300 MHz. For an "actual" enhancement of 400 and a surface density of $0.03/\text{cm}^2$, one can calculate that 50% of the cavities at 1300 MHz should have $\beta_a > 365$, while 50% of the cavities at 2856 MHz should have $\beta_a > 225$. The experimentally determined mean values are approximately 330 and 230 at 1300 MHz and 2856 MHz, respectively. Although the statistics of the determination of n_A are poor, varying n_A by $\pm 50\%$ would substantially degrade the fit to the data. The statistical explanation given above is also consistent with the fact that field emission is not observed in X-band cavities. Finally, we have recently looked at the total of the CERN temperature maps [22], where emitters can be counted directly, and found a value of n_A comparable to the value determined here.

TABLE 1. PERMANENTLY SEALED AND SPUTTER PROCESSED CAVITIES.

Frequency (MHz)	Treatment	E_p (MV/m)	β_a	High Field Area (cm²)
8600	Permanently sealed	70	No emission	0.2
8600	Permanently sealed	63	No emission	0.2
8600	Permanently sealed	40	No emission	0.2
8600	Permanently sealed	51	No emission	0.2
2856	Permanently sealed	30	120	2
2856	Permanently sealed	21	380	2
2856	Permanently sealed	30	80	2
2856	Permanently sealed	35	210	2
2856	Permanently sealed	13	250	2
2856	Permanently sealed	19	250	2
1300	Permanently sealed	22	380	10
1300	Permanently sealed	11	330	10
1300	Permanently sealed	11	310	10
1300	Sputter processed	18	200	10
1300	Sputter processed	16	340	10
430	Permanently sealed	17	400	600
430	Sputter processed	17	400	600

How then can it be that for most cavities the maximum electric field achieved is proportional to frequency? The explanation for this might be found in the monotonic dependence of the mean value of β_a on frequency, but it is more likely the result of rf processing and the fact that such processing yields reduced "surface" enhancement of the field. Let us clarify one point about "surface" enhancement. In the early Stanford experiments on helium processing, we found β_a was 550 initially but after sputtering it reached a plateau value smaller than the initial value by a factor of $\sim$ 2.5. An interesting question is whether the mechanisms for "surface" enhancement themselves lead to β_a-values of 550, or whether the "surface" enhancement is built on top of the "bulk" enhancement and the "surface" mechanisms provide only an incremental enhancement by a factor of ~ 2.5. Very recently a helium processing experiment on the 500 MHz four-cell structure at CERN [23] has provided a definitive answer to this question. The initial β-value of 900 was reduced by helium processing to 450. Throughout the different stages of processing, electron emission, as observed by temperature mapping, originated from one site. Thus the "surface" enhancement appears to be built on the "bulk" enhancement factor, and the observed value of the "bulk" enhancement (~ 450) is remarkably consistent with the value determined from the early Stanford data.

Let us turn now to a discussion of processing. We mean by processing any mechanism which reduces field emission from a given site at a rate which increases monotonically with emitted current. Since the field emission current increases exponentially with field level, one expects that the rate of processing will also increase exponentially. The emitted current and thus the processing rate are negligible for an apparent field $\beta_a E_p$ of 4×10^9 V/m, while for an apparent field of 7×10^9 V/m one expects the emitted current to cause self destruction of the emitter. At some apparent field level greater than 4×10^9 V/m, where the emitted current is negligible, and rather near the breakdown field level of 7×10^9 V/m, there is a threshold field E_{th} where processing is effective and comfortable from an experimental point of view. Let us imagine this threshold field is 5.6×10^7 V/m. Since it is narrowly bounded from above and below, it cannot differ much from this value.

Let us now consider a processing experiment in a single-cell cavity at say 500 MHz. Initially we expect that the β_a-value would be approximately 1000 which implies that the first indication of field emission loading would appear at $E_p = 4 \times 10^9$ V/m divided by 1000. For a single-cell cavity where the accelerating field is half the peak electric field, loading would appear at an accelerating gradient of 2 MV/m. By the time we get to 2.8 MV/m, where $\beta_a E_p$ is 5.6×10^7 V/m, the loading has become substantial and processing is effective. Let us now hold the electron field emission loading approximately constant, and increase the field only as the effective processing of the emission sites permit. If processing is terminated for any reason and we then make a measurement of β_a and E_p, we will find to good approximation:

$$\beta_a E_p = 5.6 \times 10^7 \text{ V/m}. \tag{1}$$

In table 2, we have compiled from the experiments at Karlsruhe [24], CERN [25] and KEK [26] a few measurements of E_p and β_a, and calculated the product $\beta_a E_p$. The fact that the experimentally determined product $\beta_a E_p$ is nearly constant for different cavities measured at different laboratories and different frequencies is explained by the simple argument given above.

TABLE 2. 500 MHz CAVITY RESULTS.

Reference	E_p (MV/m)	β_a	$\beta_a E_p$ (MV/m)
Karlsruhe [24]	7.9	710	5.6×10^9
CERN [25]	9.3	600	5.6×10^9
	4.1	1350	5.5×10^9
KEK [26]	10.0	570	5.7×10^9

This argument, however, does not account for the fact that the maximum field level achieved in the very best cavities increases monotonically with frequency. This monotonic increase is more likely related to the expected decrease in the *mean value* of β_a with increasing frequency.

5. CONCLUSIONS

Although the simple picture of field emission loading described above captures some measure of the truth, it is best viewed as guidance for a program of research. An important fact to recognize is that the density of emitters on a niobium surface is remarkably small. Only one electron field emitter is found in 30 cm^2. If that density does not sound remarkably small, we can instead estimate the fractional surface area that is emitting significant numbers of electrons. An area of 3×10^{-10} cm^2 is a generous estimate for a single emitter and thus the fractional area is approximately one part in 10^{11}, a value that should qualify as remarkably small. The small density of emitters implies that metallurgical improvements that reduce the density even by a factor of two, or "surgical" procedures that remove existing emitters, could yield significant improvement in the field level achieved in single-cell cavities or in short multi-cell structures.

A second important fact is that the emitters on a niobium surface appear to have a well-defined field enhancement value of approximately 400. It would be interesting and helpful if we could characterize a niobium surface by a distribution that describes the fractional emitting area as a function of the field enhancement factor β. It is clear that the emitting area at $\beta = 400$ can only be the first prominent feature encountered in this distribution as the field level is increased. However, even with the limited data available it appears that no other important feature of the distribution occurs for β greater than 200.

How much progress will actually be made in controlling electron field emission and increasing the operating gradient in superconducting accelerators is not yet clear. However, simple helium sputter processing to eliminate "surface" enhancement of the apparent electric field should yield β-values of 400. Thus the accelerating cavities should support peak electric fields of 1.4×10^7 V/m without breakdown and 1.0×10^7 V/m without appreciable emission. For an accelerating field that is one-half the peak field this implies an energy gradient of 7 MeV per meter without breakdown and 5 MeV per meter without appreciable emission. Operation of a superconducting accelerator cw at this level is an important objective that is now clearly achievable.

References

[1]　H. London, *Nature* **133**, 497 (1934).

[2]　H. London, *Proc. Roy. Soc.* **A176**, 522 (1940).

[3]　A. B. Pippard, *Proc. Roy. Soc.* **A191**, 371 (1947).

[4]　W. M. Fairbank, *Phys. Rev.* **76**, 1106 (1949).

[5]　E. Maxwell, P. M. Marcus and J. C. Slater, *Phys. Rev.* **74**, 1234 (1948).

[6]　W. M. Fairbank, *The Surface Resistance of Superconducting Tin at 10,000 Megacycles*, Ph.D. Thesis, Yale University, 1948.

[7]　E. Maxwell, *Progress in Cryogenics* (Heywood, London, 1964), Vol. IV, p. 124.

[8]　W. M. Fairbank, J. M. Pierce and P. B. Wilson, in *Proceedings of the Eighth International Conference on Low Temperature Physics*, London, 1962 (Butterworths, Washington D.C. , 1963), p. 324.

[9]　A. P. Banford and H. G. Stafford, *J. Nucl. Energy, Part C* **3**, 287 (1961).

[10]　A. Susini, "Initial Experimental Results Concerning Superconductive Cavities at 300 Mc/s," *CERN Internal Report 63-2*, MCS Division (1963); J. Rüfenacht and L. Rinderer, *Z. Angew. Math. u. Physik* **15**, 192 (1964).

[11]　H. A. Schwettman, P. B. Wilson, J. M. Pierce and W. M. Fairbank, in *International Advances in Cryogenic Engineering*, edited by K. D. Timmerhaus (Plenum Press, New York, 1965), **10**, p. 88.

[12]　H. A. Schwettman, P. B. Wilson and G. Y. Churilov, in *Proceedings of the Fifth International Conference on High Energy Accelerators*, Frascati, Italy, 1965, edited by M. Grilli (CNEN, Rome, 1966), p. 690.

[13]　D. C. Mattis and J. Bardeen, *Phys. Rev.* **111**, 412 (1958).

[14]　J. P. Turneaure, *Microwave Measurements on the Surface Impedance of Superconducting Tin and Lead*, Ph.D. Thesis, Stanford University, 1966.

[15]　J. P. Turneaure and I. Weissman, *J. Appl. Phys.* **39**, 4417 (1968); I. Weissman and J. P. Turneaure, *Appl. Phys. Lett.* **13**, 390 (1968).

[16]   J. P. Turneaure and N. T. Viet, *Appl. Phys. Lett.* **16**, 333 (1970).

[17]   C. M. Lyneis, H. A. Schwettman and J. P. Turneaure, *Appl. Phys. Lett.* **31**, 541 (1977).

[18]   U. Klein and D. Proch, in *Proceedings of the Conference on Future Possibilities of Electron Accelerators*, Charlottesville, Virginia, 1979, edited by J. S. McCarthy and R. R. Whitney (Dept. Physics, U. Virginia, Charlottesville, 1979), p. N1.

[19]   C. M. Lyneis, in *Proceedings of the Workshop on RF Superconductivity*, held at Karlsruhe, KfK 3019, 119 (1980).

[20]   H. A. Schwettman, J. P. Turneaure and R. F. Waites, *J. Appl. Phys.* **45**, 914 (1974).

[21]   J. P. Turneaure, private communication; C. M. Lyneis, Y. Kojima, J. P. Turneaure and N. T. Viet, *IEEE Trans. Nucl. Sci.* **NS-20**, 101 (1973); I. Ben-Zvi, private communication.

[22]   CERN data kindly provided by H. Lengeler.

[23]   W. Weingarten, private communication.

[24]   W. Bauer, private communication.

[25]   W. Weingarten, private communication.

[26]   Y. Kojima, private communication.

Unique Beam Properties of the Stanford Superconducting Recyclotron

T. I. Smith, C. M. Lyneis, M. S. McAshan, R. E. Rand, H. A. Schwettman and J. P. Turneaure

1. INTRODUCTION

By the last half of the 1960's, work on the properties of superconductors at microwave frequencies had progressed to the point that a Stanford team undertook the design of a superconducting linear accelerator, taking advantage of the remarkable properties of superconducting metals and superfluid helium to obtain performance generally superior to that of conventional linear accelerators, particularly with regard to beam quality and duty cycle.

Today that machine exists, and the combination of beam emittance, energy resolution, beam current, duty factor and beam stability that is achieved in the superconducting accelerator is unmatched in the world. The superconducting linear accelerator (SCA) together with its energy multiplying recirculation system (forming the superconducting recyclotron, SCR) now provides beams with energies from 20 MeV to 230 MeV. In the future the upper limit could exceed 750 MeV without major changes in the recirculation system.

In this paper we describe the quality and versatility of the accelerator and its recirculation system. We begin by describing three experiments that have been carried out using the SCA. Each of these experiments exploited particular properties of the accelerator, and none of them would have been feasible at any other existing facility. We then give a brief description of the linac and the recirculation system, and outline some of the advantages gained in using low temperature techniques.

2.   THREE EXPERIMENTS

The first of the experiments is the study of the fission modes of ^{24}Mg by Sandorfi et al. [1]. In this experiment an electron beam of 26 to 40 MeV is incident on a ^{24}Mg target and the back-to-back fission fragments are detected in coincidence. In order for the fission fragments to get out of the target with acceptable energy loss, the target must be extremely thin. The thin target implies that the number of events per electron in the beam will be low. Consequently, for a reasonable counting rate, the fission detectors must subtend a large solid angle (to detect as many of the events as possible), and the average beam current must be large (to generate many events). On the other hand, in order to keep the probability of uncorrelated events from accidentally appearing as coincidences in the detectors, peak beam currents must be kept as low as possible. The simultaneous desire for high average currents and low peak currents obviously requires that the beam duty factor (defined as the fraction of the time that the beam is on) be as large as possible.

In the actual experiment the target thickness was only 33 μg/cm^2. The average beam current ranged from 150 to 250 μA, delivered at a duty factor between 75 and 90%, while the fission detectors subtended a solid angle of 70 mΩ. This large solid angle was obtained by operating the solid state detectors within 10 cm of the target. It should be clear that operation of these detectors in such close proximity to the electron beam would have been impossible had the beam had any significant halo or positional instability.

A subset of the data obtained from this experiment is shown in figure 1. The SCR beam time required to complete the work was about 150 hours. Because of a combination of duty factor limitations, beam halo, and positional instabilities, the experiment could not have been done in less than 10,000 hours at any other facility in the world.

The second experiment is the free electron laser [2]. In this experiment a spatially periodic transverse magnetic field allows a relativistic electron beam to exchange energy with a photon beam. When properly adjusted there is a net transfer of energy from the electrons to the photons. We use this gain mechanism to power an optical oscillator [3] which we refer to as a free electron laser.

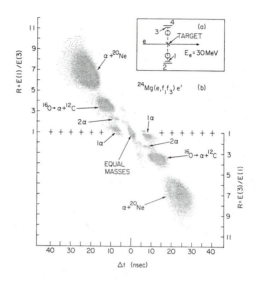

FIGURE 1. (a) The coincidence-detection geometry for the ^{24}Mg fission experiment. (b) A density plot of the ratio of the energy of the light fragment to that of the heavy fragment as a function of the time difference between their arrival in opposite counters. The predominant peaks are indicated. Those labeled "2α" are from ^{24}Mg $\rightarrow^{16}$O$+^{8}$Be where both alphas from ^{8}Be are detected, while the "1α" peaks result from the detection of only one of these alphas.

For fundamental reasons having to do with the electron-photon interaction, the emittance of the electron beam must be very small. (The beam emittance can be defined as the product of the beam's diameter at a minimum and its angular divergence far from the minimum.) For our experiment, with a 5 meter interaction length and a 3.2 cm period, the beam emittance is limited to 0.05π mm mr. This implies that the electron beam diameter can remain less than about 0.7 mm throughout the entire interaction region.

The same reasoning which places limits on the emittance of the beam also places limits on its energy width, energy stability and positional stability. In our case the energy must remain fixed within a few parts in 10^4 and the beam position at a minimum must remain fixed to within about 10% of the minimum size on a time scale of an hour or so.

Finally, since the gain at low power is directly proportional to the instantaneous current density, it is clear that this experiment belongs to the class in which instantaneous peak current is of major importance, as contrasted with the fission experiment, where average current and high duty cycle are critical. For the success of this experiment we took advantage of the microscopic time structure of the beam inherent in any linear accelerator to provide peak currents which were approximately 20,000 times the average current! Electron bunches in the SCA normally are present for about 1.8° out of each 360° per rf cycle (1 cycle ≃ 770 ps). This automatically provides a peak to average current ratio of 200. (This time structure does not affect the fission experiment due to the short time scale involved.) The peak to average ratio is further enhanced by a high speed pulsed gun [7] which was designed so that electron bunches are present only during each 110th rf cycle. The resulting peak currents are of the ampere level.

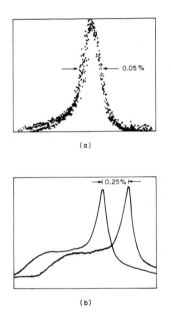

(a)

(b)

FIGURE 2. (a) Energy resolved beam of the SCA operating in the intense pulsed mode (see text) at 44 MeV and 1.2 A peak current. The energy width is 0.05% (FWHM). The energy stability of the beam is indicated by the fact that the figure is a superposition of four separate spectra, taken over an interval of 40 minutes. (b) Energy spectrum of above beam after passing through free electron laser. The two spectra have been displaced by 0.25% in energy. The centroid of the distribution has shifted about 0.1% low, representing the increase in energy of the photon beam.

Data taken in March 1981 are shown in figure 2. For all cases the SCA is delivering a 44 MeV beam with peak currents of 1.2 A. The energy stability of the beam is indicated by figure 2a which shows the energy dispersed beam. The energy width is 0.05% (FWHM), and the figure is composed of four superimposed traces, taken over a period of forty minutes. Figure 2b again shows the energy dispersed beam, but after it has passed through the laser. As the two traces are shifted by 0.25%, it is obvious that the energy exchange with the photon beam was substantial. The centroid of the distribution has shifted some 0.1% low, representing the increase in energy of the optical beam. It is interesting to note in passing that although the net energy shift of the beam was downward, a significant number of electrons gained energy. The upper limit of the energy gain was about 500 keV, and is probably the highest electron energy gain yet produced by a laser.

The combination of beam parameters required for operation of the free electron laser can be satisfied only by the SCA. No other accelerator can satisfy all of the requirements simultaneously.

The last experiment is the study of the giant dipole resonance in ^{12}C through $(e, e'p)$ and $(e, e'\alpha)$ coincidence experiments [4]. In these experiments the electron is detected in a spectrometer of 3.6 mΩ acceptance and the proton (or alpha) is detected in a solid state detector telescope of 40 mΩ (or 60 mΩ) acceptance. As in the fission experiment, the solid state detectors are located within 10 cm of the target. Some results from the $(e, e'p)$ experiment are shown in figure 3.

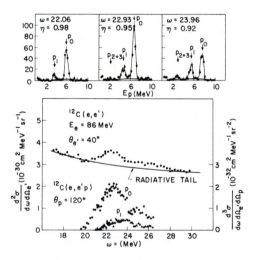

FIGURE 3. Top: Three coincidence proton spectra from the experiment of [4]. The spectra are for various values of ω (energy transfer) with $\Delta\omega = 150$ keV. The parameter η is the relative efficiency of the relevant electron channel. Accidental coincidence backgrounds are shown as solid lines. Bottom: The ^{12}C(e, e') and ^{12}C$(e, e'p)$ cross sections measured in the same experiment.

The first two experiments relied upon the properties of the linac beam, but this experiment required beam energies greater than those available at reasonable duty cycle from the linac alone. Thus, the recirculation system was used to multiply the energy of the linear beam by a factor of 2 or 3, and the giant resonance experiments were done with beams of 80 to 120 MeV at about 75% duty factor. The average beam current delivered at 80 MeV was typically 25–35 μA (limited by pileup in the solid state detectors), and at 120 MeV was typically 15–20 μA (limited by regenerative beam breakup). As in the previous experiment, beam quality and stability were important to the success of the giant resonance experiments. At 80 MeV the measured beam emittance was less than 0.01π mm mr, and the beam halo was less than 2×10^{-3} outside a radius of 1.5 mm with a beam spot on target of 0.7 mm FWHM. The energy width of the beam was 13 keV.

Using an array of proton (or alpha) detectors, it is possible to reduce the data collection time for a typical giant resonance experiment to approximately 600 hours. This beam time is longer than desired and is fixed by the small acceptance of the electron spectrometer used. Nonetheless, this experiment cannot be mounted at any other existing facility in the world.

3. DESCRIPTION OF
THE SUPERCONDUCTING RECYCLOTRON

In this section we briefly describe the superconducting linear accelerator and its associated energy multiplying beam recirculating system. In addition, we outline some of the advantages realized through the use of superconductivity and superfluidity.

In any linear accelerator, particles obtain energy from high frequency fields inside microwave cavities. Currents are induced in the cavity walls by these fields, and cause a power loss which is proportional to the electrical resistivity of the surface. In a conventional linear accelerator these losses can be staggering. For instance, at SLAC (Stanford Linear Accelerator Center) the power required just to make up for these wall losses is several *billion* watts.

It is, of course, absurd to consider operating a machine which consumes that quantity of power on a continuous basis. Simply to prevent the cavities from melting, the rf fields are present for only one to ten microseconds every 1000. Thus the beam duty cycle is limited to 0.1 to 1%.

We have already noted the disadvantage this brings to coincidence experiments, in which an attempt is made to separate out particular events from a large background by coincident detection of two particles. A further disadvantage of short pulse length operation is the difficulty of making an electron beam that is homogeneous in energy for high resolution experiments. The difficulty arises from the fact that fluctuations which occur during the pulse are so rapid that it is very difficult to do anything to compensate for their effect.

The contribution of superconductivity is that it drastically reduces the power losses required to sustain the electrical fields in the cavities, thus allowing continuous operation of the accelerator. At the same time that this improves the duty cycle, the long pulses allow easy application of reasonably standard feedback techniques to stabilize the beam energy and position. Finally, it is worthwhile to design the entire system so that the energy width and emittance of the beam are improved (compared with conventional machines) to be compatible with the energy and position stability.

In order to attain the improvements which superconductivity would appear to make possible, it soon becomes apparent that the superconducting cavities need to be operated in an environment of extreme thermal and mechanical stability. For once nature seems to be on our side, as she has provided superfluidity as a low temperature companion to superconductivity. The ability of superfluid liquid helium to carry heat with very little temperature drop, coupled with its large heat capacity, makes it an ideal fluid in which to immerse an accelerator. It cools the metal accelerator parts to the required temperatures, and simultaneously provides the necessary stable environment.

It is true that we cannot get all of these benefits without paying some penalties. The necessity of having to operate in the superfluid cryogenic environment is clearly an inconvenience, and mechanical inefficiencies coupled with thermodynamic realities conspire to require that we pay about 1000 watts of room temperature power for every 1 watt removed from the superconducting accelerator. On the other hand, when we realize that the surface losses of the superconducting cavities are only one part in 10^5 of those of the normal cavities, it is clear that we are still ahead in total power consumption by a factor of 100 or so. Finally, even if the power costs were comparable in the two machines, the beam improvements made available by the low temperatures would warrant the development of the technology.

Since the details of the linac have been published elsewhere [5,6], only a brief description will be given here. Electron bunches are carefully formed in a room temperature system and injected into the first superconducting section, a one meter long structure called the capture section. The entering electrons have an average kinetic energy of 100 keV and a spread of 2 keV. They occupy about 10° of an rf cycle. When they leave the capture section they have an average energy of about 2.5 MeV and they occupy a phase space area of about 15 keV by 1.5°. They then encounter a 3 meter long structure referred to as the pre-accelerator where they are accelerated to about 5 MeV. The phase space shape is unchanged.

Further acceleration is accomplished by a series of modular structures each 6 meters long. The average field gradient which can be maintained at

TABLE 1. TYPICAL BEAMS DELIVERED BY THE SUPERCONDUCTING LINAC FOR EXPERIMENTAL PHYSICS.

Energy (MeV)	62	40	30	25
Duty factor	5%	90%	90%	90%
Peak current (μA)	15	220	220	220
Average current (μA)	0.9	200	200	200
Resolution, $\Delta E/E$ (FWHM)	0.016%	0.025%	0.03%	0.035%
Emittance (π mm · mr)	0.02	0.025	0.03	0.04

a 75% duty cycle is about 1.75 MeV/meter. The amplitudes and phases of the fields in each superconducting cavity are electronically regulated to 0.01% and 0.1°, respectively.

Some representative beams produced by the linac for use in experimental physics are listed in table 1 along with some of the more important descriptive parameters.

The recirculation system [8] shown schematically in figure 4a is used to allow the beam to pass through the linac several times, thus effectively multiplying the length of the accelerator. With this system, beam energies from two to five times the linac energy can be produced.

Recirculation of the beam is based on a multichannel magnet. The principle of this magnet is illustrated in figure 4b, where it can be seen that

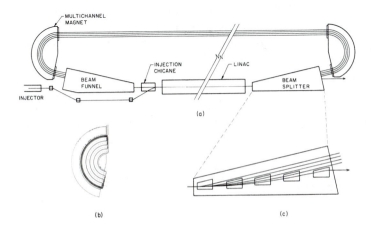

FIGURE 4. Schematic of the superconducting recyclotron.

there are four channels and that the windings of the coil are arranged in four groups. All four groups produce flux in the outer channels, three groups produce flux in the next, *etc.* If, as in the figure, the groups are in series, the fields will be in the ratios 1:2:3:4, and, neglecting the different radii of the channels, beams with these momentum ratios will be bent properly by the magnet. In practice, the coils can be independently powered, so the momentum ratios are not fixed as above.

In this design, the magnetic field in any channel is proportional to the beam energy in that channel. The field is small in the low energy channels, and large in the high energy channels. Thus magnet weight and power requirements are substantially less than for a uniform field magnet as would be required by a simple race-track microtron, particularly at energies above a few hundred MeV.

The arrangement of the magnets which split the overlapping beams from the linacs into four parallel channels is shown in figure 4c. Constant orbit separation is achieved by allowing the *nth* orbit beam to pass through n equally spaced equal strength bending magnets. This arrangement allows the beam to be extracted at any orbit by reversing the field in the appropriate magnet. The beams are recombined for injection into the linac by a system of magnets which is a mirror image of those in figure 4c.

In order to compensate for the real geometry of the system, for the velocity variations of the electrons with energy, for thermal expansion, *etc.*, and in general to provide fine adjustments of the accelerating phase of each pass, the path length of each orbit is adjustable. This is achieved by means of horizontal steering coils placed at the entrance and exit of each multichannel magnet. Each pair of coils is located at conjugate points of unit magnification so that the deviation introduced by the first coil is exactly canceled by the second, while the path length through the magnet is variable. Beam focusing is provided on each orbit by three quadrupole doublets, located in the beam splitter, in the beam funnel, and in the center of the return path. This arrangement provides a wide range of possible beam optical conditions.

The recirculation hardware has been designed for five accelerations of the beam (passes through the linac). The system has been tested [9] with four passes, and it operates routinely with two and three passes. Typical beams used for nuclear physics experiments are described in table 2.

The properties of superconducting materials and superfluid helium have been exploited to construct an electron accelerator which provides a beam of outstanding quality. The unmatched combination of beam emittance, energy resolution, beam current, duty factor and stability has been used to carry out various experiments which would have been impossible at any other existing facility.

TABLE 2. BEAMS DELIVERED BY THE SUPERCONDUCTING RECYCLOTRON.

Energy (MeV)	80	117	175	230
Passes through linac	2	3	3	4
Duty factor	75%	75%	20%	15%
Peak current (μA)	50	15	40	8.5
Average current (μA)	40	12	8	1.3
Transmission	100%	100%	100%	~85%
Resolution, $\Delta E/E$ (FWHM)	0.02%	0.015%	0.012%	0.029%
Emittance (π mm mr)	<0.01	<0.01	—	—

Acknowledgments

This work was supported by National Science Foundation grant PHY-79-05286-02.

References

[1] A. M. Sandorfi, J. R. Calarco, R. E. Rand and H. A. Schwettman, *Phys. Rev. Lett.* **45**, 1615 (1980).

[2] *Free Electron Generators of Coherent Radiation*, edited by S. F. Jacobs, H. S. Pilloff, M. Sargent, III, M. O. Scully and R. Spitzer. This is Vol. 7 of the *Physics of Quantum Electronics* series (Addison-Wesley, Reading, MA, 1980).

[3] D. A. G. Deacon, L. R. Elias, J. M. J. Madey, G. J. Ramian, H. A. Schwettman and T. I. Smith, *Phys. Rev. Lett.* **38**, 892 (1977).

[4] J. R. Calarco, in *Proceedings of the 1980 RCNP International Symposium on Highly Excited States in Nuclear Reactions*, edited by Ikegami and Muraoka (Osaka University, Osaka, Japan, 1980), p. 543; J. R. Calarco, J. Arruda-Neto, K. Griffioen, S. S. Hanna, D. H. H. Hoffmann, M. S. McAshan, R. E. Rand, A. M. Sandorfi, H. A. Schwettman, T. I. Smith, J. P. Turneaure, K. Wienhard and M. R. Yearian, "Decay of the ^{12}C Giant E1 Resonance from ^{12}C$(e, e'p)^{11}$B Coincidence Measurements," High Energy Physics Laboratory Report 881 (Stanford Unversity, Stanford, CA, 1980).

[5] J. R. Calarco, M. S. McAshan, H. A. Schwettman, T. I. Smith, J. P. Turneaure and M. R. Yearian, *IEEE Trans. Nucl. Sci.* **NS-24**, 1091 (1977).

[6] M. S. McAshan, K. Mittag, H. A. Schwettman, L. R. Suelzle and J. P. Turneaure, *Appl. Phys. Lett.* **22**, 605 (1973).

[7] J. M. J. Madey, G. J. Ramian and T. I. Smith, *IEEE Trans. Nucl. Sci.* **NS-27**, 999 (1980).

[8] R. E. Rand and T. I. Smith, in *Proceedings of Conference on Future Possibilities for Electron Accelerators*, edited by J. S. McCarthy and R. R. Whitney (University of Virginia, Charlottesville, 1979), Paper X.

[9] C. M. Lyneis, M. S. McAshan, R. E. Rand, H. A. Schwettman, T. I. Smith and J. P. Turneaure, *IEEE Trans. Nucl. Sci.* **NS-28**, 3445 (1981).

Large Solid Angle Superconducting Pion Channel for the Radiotherapeutic Treatment of Cancer

Peter Fessenden, Malcolm A. Bagshaw, Douglas P. Boyd, David A. Pistenmaa, H. Alan Schwettman and Carl F. von Essen

Conventional modern treatment of cancer using ionizing radiation had its beginning approximately 25 years ago with the advent of megavoltage sources of photons. Prior to that time, only photon energies up to about 300 keV were routinely available, and these attenuated so rapidly in tissue that delivery of a tumoricidal absorbed energy deposition, or dose, to most regions of the body was impossible. In spite of the wide availability of penetrating radiation sources today, there still are about 60,000 cancer deaths annually, in the United States alone, attributable to local failure after treatment with ionizing radiation [1].

It has been recognized for some time [2,3] that beams of negative pi mesons, or pions, may be particularly advantageous for use in cancer therapy. Unlike photons, which attenuate nearly exponentially as they penetrate matter, negative pions are unique (figure 1). A fast pion with a therapeutically useful range will penetrate matter losing energy *via* atomic collisions in a manner similar to any fast, charged particle until it has nearly

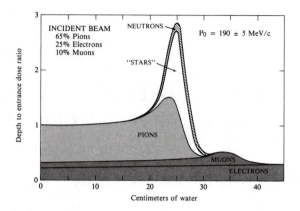

FIGURE 1. Localized energy deposition due to the negative pion Bragg peak enhanced by nuclear disintegration. (Figure, courtesy of S. B. Curtis and M. R. Raju.) The muons and electrons are contaminates focused along with the pions because they have the same momentum.

come to rest. This slowing down process produces the familiar Bragg peak, somewhat tempered in the pion case because of enhanced scattering of the relatively light pion. The unique phenomenon is the subsequent capture of the negative pion in atomic-like orbits forming a pionic atom. Because the pion is 278 times more massive than an electron, the lower energy levels of this pion-nucleus system result in a strong overlap of the pion and nuclear wave functions. This nuclear capture is not stable, and nuclear disintegration, releasing approximately 100 MeV of kinetic energy, results. Most of this energy consists of "stars" (so called because of the star-like appearance of the multipronged event in a nuclear emulsion detector) and neutrons, as shown in figure 1. The energy is deposited relatively locally, and produces an enhanced Bragg peak. In addition, a significant fraction of this enhanced energy deposit is due to heavy charged particles, and is densely ionizing. Since there is very strong evidence [4] that densely ionizing radiation is more biologically effective than sparsely ionizing radiation, the Bragg peak region is effectively even more enhanced than the physical depth-dose curve of figure 1.

In the late 1960's, William Fairbank and others, in a collaboration between the Stanford Physics and Radiology departments, conceived the idea of a large solid angle multibeam pion channel for cancer radiotherapy [5]. These ideas culminated in the construction of the Stanford medical pion generator in 1975 [6,7]. The entire device is shown schematically in figure 2. The system is, in effect, a 60-beam magnetic spectrometer, which is designed around two 60-coil superconducting toroidal magnets. The use of state-of-the-art superconducting technology allowed strong magnetic

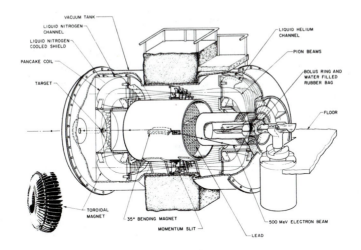

FIGURE 2. The Stanford Medical Pion Generator (SMPG) completed in 1975 to serve as a prototype of multibeam pion radiotherapy.

fields with a minimum of magnet hardware, thus providing for both a large, one-steradian solid angle and a short, six-meter flight path. The result is efficient collection and focusing of the pions while minimizing pion decay in flight. A further improvement in dose delivery is inherent in the design as shown in figure 4; the 60 beams converge nearly radially toward the axis, giving an added geometrical advantage with respect to dose localization.

The Stanford medical pion generator, constructed on site at the Stanford High Energy Physics Laboratory, served as a prototype to explore the advantages of simultaneous multiport pion radiotherapy, and preclinical testing and evaluation studies were completed in 1979. Figure 3 shows

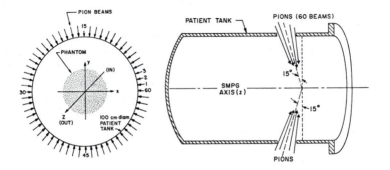

FIGURE 3. Geometry of the pion beams entering the patient tank.

the geometrical configuration of the pion beams and patient tank area, including a cylindrical "phantom" which facilitated much of the preclinical physics, dosimetry and radiobiology experiments. When a small ionization chamber, constructed from tissue-like materials, is scanned through the phantom, relative dose profiles such as those of figure 4 result. The X, Y

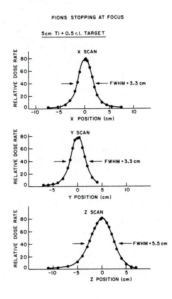

FIGURE 4. Three orthogonal energy deposition, or dose, profiles with all 60 beams stopping near the center of a cylindrical water phantom.

and Z axes correspond to the schematic of figure 3, and illustrate the good three-dimensional localization of dose at depth. These data resulted from the superconducting toroid current being such that all pions had a range in water nearly equal to the radius of the phantom. When the Stanford medical pion generator is operated with pions of less range, the 60-beam configuration leads to a "smoke ring," or annulus, of high dose. Figure 5a is the result of a film exposure showing this two dimensionally. Figure 5b is a set of one-dimensional ionization chamber scans, illustrating the result of progressively larger annular rings as the toroid current is reduced, with the accompanying reduction of magnetic field and pion range.

An extensive program of assessing the quality of the radiation dose was undertaken. This was done by exploring the cell-killing ability for various configurations of the pions, in addition to doing proportional counter studies, investigating on an event-by-event basis the spatial density of energy deposition. Although the cell killing in the so-called plateau region of the

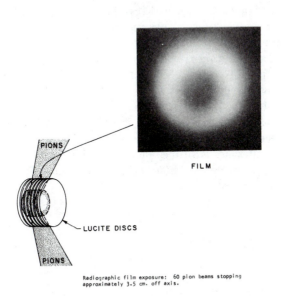

FIGURE 5a. Two-dimensional "smoke ring" resulting from exposure of a film with pions of insufficient range to reach the axis. The film exposure resulted from 60 pion beams stopping approximately 3.5 cm off axis.

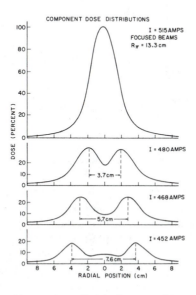

FIGURE 5b. Dose profiles in the cylindrical water phantom for pions of different ranges. Double peaking occurs when the toroid current is less than 515 amperes, resulting from pions stopping short of the phantom axis.

pion beams (corresponding to the left of figure 1, well before the enhanced Bragg peak) is very similar to that for conventional radiation therapy using x-rays, the cell killing near the region where the pions are stopping and nuclear disintegration is occurring, is significantly increased. This is illustrated in figure 6, where the logarithm of mammalian cell survival is plotted against dose for irradiation with "focused" pions (bottom curve) or x-rays (top curve). In this particular configuration of cell irradiation on the phantom axis with the pions all stopping near the axis, the relative biological effectiveness (RBE) of the pions for 50% survival is about 2.8. In other words, it takes 2.8 times more radiation for the same cell killing using conventional radiation than it does using radiation associated with the stopped pion region.

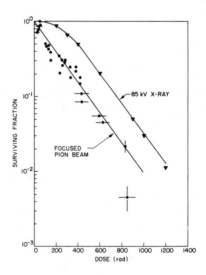

FIGURE 6. Mammalian cell survival for different doses of pions and conventional radiation (x-rays), showing the greater lethality of pions. (Figure, courtesy of G. C. Li.)

The average rate of energy deposition in tissue with conventional x-ray or electron radiotherapy is approximately a few keV/micron. The average deposition rate was mapped for various configurations of pion beams in the Stanford medical pion generator, and found to vary from a few keV/micron in the pion plateau, or entrance region, to about 60 keV/micron in the pion stopping region. Some of these data are included in the summary diagram of figure 7, which is a scattergram relating RBE_{50} for mammalian cell survival to the average deposition rate of beams of the Stanford medical pion generator and heavy ions used for radiotherapy elsewhere [8]. The

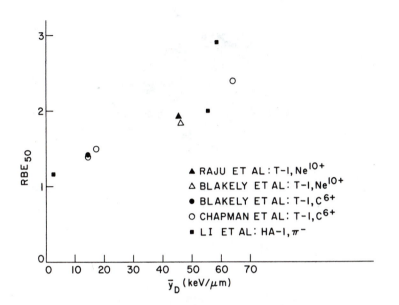

FIGURE 7. Experimental data showing the relative biological effectiveness (RBE) of various radiations as a function of $\bar{y}_D$, which is a quantity closely related to energy deposition per unit distance. The square point at the left is for plateau pions near the surface. Square points near the upper right result from measurements near the peaks of dose distributions for two different configurations of the 60 beams of the SMPG. (Figure, courtesy of G. Luxton.)

abscissa, $\bar{y}_D$, is the dose average of the "lineal" energy, which is the energy deposited in a small spherical gas proportional counter, divided by the counter diameter. The proportional counter is constructed of tissue-like materials, and has a muscle tissue equivalent diameter of 2 microns. RBE and energy deposition phenomena are too complicated to expect a simple correlation to exist. However, these data, collectively, strongly imply that biological advantages are associated with beam configurations presenting increased energy deposition rates.

The ability to localize dose mainly to the region of interest is illustrated in the hypothetical treatment in figure 8. This calculation for irradiation of a pancreatic tumor was done by adding together a number of smoke ring distributions with successively larger diameters (lower pion range). For each ring, there was a given weighting (time of irradiation) and a certain configuration of beams on. (Each beam has its own set of slits that can be varied between fully opened and fully closed.) A more illustrative example of the dose tailoring that can be done with multiple converging pion beams is shown in figure 9. In this case, all beams are on and have a range sufficient to cause all 60 stopping regions to overlap, or form a focus.

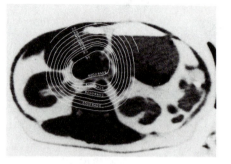

FIGURE 8. One plane of a hypothetical treatment for a patient with carcinoma of the pancreas, planned by using the properties of pion beams from the SMPG. (Figure, courtesy of C. H. Yuen.)

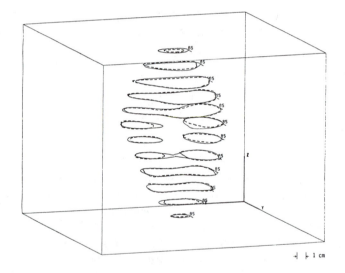

FIGURE 9. Multiple planes of a hypothetical treatment of a bladder cancer and lymph nodes at risk, planned by using the properties of the PIOTRON, which is the 60-beam channel treating cancer patients at the Swiss Institute for Nuclear Research. The dotted lines are the desired positions of the 85% dose level, while the solid lines are those achieved in the calculated plan. (Figure, courtesy of E. Pedroni.)

The high dose focus, or spot, is then raster scanned *via* translating the patient point-by-point through a three-dimensional grid. By varying the irradiation (essentially the time of irradiation) at each point of the grid, the dose delivery is accurately tailored to the target region (in this case, the bladder and accompanying regions of lymph nodes at risk).

Figure 9 is the result of treatment planning with the 60-beam pion radiotherapy facility at the Swiss Institute for Nuclear Research (SIN) in Villigen, Switzerland. This machine, called the PIOTRON, is a fully operational clinical facility [9], which was completed in 1979 and is currently treating cancer patients [10]. It is one of the most sophisticated cancer radiation units in the world. However, it will likely be five years to a decade before clinical trials with this state-of-the-art device can demonstrate definitively the anticipated improvements in cancer treatment.

References

[1] H. D. Suit and M. Goitein, *Europ. J. Cancer* **10**, 217 (1974).

[2] P. H. Fowler and D. H. Perkins, *Nature* **189**, 524 (1961).

[3] C. Richman, H. Aceto, M. R. Raju *et al.*, *Am. J. Roentgen* **96**, 777 (1966).

[4] G. W. Barendsen, in *Current Topics in Radiation Research*, edited by M. Ebert and A. Howard (North-Holland Publishing Co., Amsterdam, 1968), Vol. IV, p. 295.

[5] H. S. Kaplan, H. A. Schwettman, W. M. Fairbank, D. P. Boyd and M. A. Bagshaw, *Radiology* **108**, 159 (1973).

[6] D. P. Boyd, H. A. Schwettman and J. Simpson, *Nucl. Instr. and Meth.* **111**, 315 (1973).

[7] D. A. Pistenmaa, P. Fessenden, D. A. Boyd, G. Luxton, R. C. Taber and M. A. Bagshaw, *Radiology* **122**, 527 (1977).

[8] G. Luxton, P. Fessenden and H. D. Zeman, *Radiat. Res.* **85**, 238 (1981).

[9] G. Vescey, I. Horvath and J. Zellweger, in *Proceedings of the VI International Conference on Magnet Technology* (Slovak Academy of Science, Bratislava, 1977), p. 361.

[10] C. F. von Essen, H. Blattmann, J. F. Crawford, P. Fessenden, E. Pedroni, C. Perret, M. Salzmann, K. Shortt and E. Walder, *Int. J. Radiat. Oncol. Biol. Phys.* **8**, 1499 (1983).

Development of the Superconducting Cavity Oscillator

J. P. Turneaure and S. R. Stein

1. INTRODUCTION

Extremely high frequency precision and even higher frequency stability have been achieved with radio frequency oscillators (or clocks). For example, the frequency of a cesium beam atomic clock can be achieved with a precision of a few times $10^{-14}(\delta\nu/\nu)$ [1], and the superconducting cavity oscillator (discussed in this paper) has achieved a frequency stability of $2 \times 10^{-16}(\delta\nu/\nu)$ for sampling times on the order of 100 sec. Because of this high frequency precision and stability, oscillators play an important role in physics experiments. For example, a gravitational red-shift experiment employing hydrogen masers, one on the earth and one in a Scout D rocket sent to a peak altitude of 10^4 km, verified the predicted red shift to a precision of 70 parts per million [2]. This red-shift experiment required hydrogen masers which had a frequency stability on the order of $10^{-14}(\delta\nu/\nu)$.

In this paper the development of the superconducting cavity oscillator (SCO) and its application to two fundamental physics experiments are recounted. The two experiments are an experiment to set an upper limit on the time drift of the fine structure constant, and a null gravitational red-shift experiment to test the metricity of gravitational theory. This work on the SCO was possible at Stanford because of the development of high-Q superconducting cavities for the superconducting accelerator (see paper by

H. A. Schwettman entitled "RF Superconductivity and its Applications" in this volume). For a more general article on superconducting cavity oscillators and their applications see [3].

2. DEVELOPMENT OF THE SUPERCONDUCTING CAVITY OSCILLATOR

An rf oscillator which can achieve high frequency stability requires several features. First, it requires a resonator, which is the frequency-determining element of the oscillator. For the SCO the resonator is a superconducting microwave cavity, which is a volume under vacuum enclosed by a superconductor. The cavity is operated in one of its electromagnetic modes which must have an extremely stable resonant frequency ν_0. This mode must have a very high quality factor Q. This Q is related to the cavity bandwidth $\Delta\nu_b$ by the equation

$$Q = \nu_0/\Delta\nu_b \quad .$$
(1)

Second, an appropriate oscillator circuit is needed which transfers the high intrinsic frequency stability of the cavity to that of the oscillator's useful power output.

2.1 High-Q Superconducting Cavities

The work at Stanford to achieve high-Q superconducting cavities began in about 1961. The goal of this work was to produce superconducting cavities that could be used to construct a high duty factor electron accelerator. The initial work concentrated on the electroplating of a thin layer (25 μm) of the superconductor lead on a copper cavity structure. Both simple cavities and more complex accelerator cavities were investigated. The highest Q, 3.7×10^{10}, was achieved for a simple circular cylindrical cavity operated in the TE$_{013}$ mode [4]. During this same period of time there was also an interest in exploring superconducting niobium cavities. The study of niobium cavities began in earnest in the late 1960's. This study eventually led to a Q of 10^{11} for a superconducting niobium cavity operated in the TM$_{010}$ mode [5]. This type of niobium cavity, although not optimized for use as the frequency-determining element of an oscillator, was used for the SCO development since all of the needed technologies to make these high-Q niobium cavities were available.

2.2 Frequency Stability of Superconducting Cavity

The approximate resonant frequency for a particular cavity mode is found by solving Maxwell's equations and assuming that the cavity boundaries are perfect reflectors. The cross section of the niobium cavity used for the

SCO is shown in figure 1. This cavity is operated in the TM_{010} mode. If one makes the simple assumption that this cavity is formed by a cylinder of radius R with two flat ends, the resonant frequency of the TM_{010} mode can be easily calculated and the result is

$$\nu_0 = 2.405 \ c/R \quad , \tag{2}$$

where c is the velocity of light. As can be seen in figure 1, the cavity shape is more complicated than assumed; however, the resonant frequency can still be expressed as

$$\nu_0 = b \ c/\ell \quad , \tag{3}$$

where ℓ is a length which characterizes the size of the cavity and b is a constant which depends on the cavity shape and mode. The niobium cavities used for the SCO's all have had resonant frequencies of about 8.6 GHz. Equation (3) does not quite give the correct expression for the resonant frequency since the superconductor is not a perfect reflector. The largest correction to the frequency given in equation (3) is due to the rf skin depth which for frequencies up to 10 GHz is nearly equal to the dc superconducting penetration depth.

To understand the stability of the cavity resonant frequency, the sources of resonant frequency variation must be understood. Three important sources of resonant frequency variation, which are discussed below, are those due to the rf field level in the cavity, the temperature of the niobium cavity structure, and the deformation of the cavity structure by its own weight in gravity. There are, of course, many other sources which produce resonant frequency variations (see [3]). These include quantum fluctuations, electromagnetic and mechanical thermal noise, mechanical vibration, the vacuum condition within the cavity, external pressure on the niobium cavity structure, charged particles in the cavity vacuum caused by radiation, and structural changes within the niobium walls due to stress relaxation, creep, and radiation damage.

2.2.1 Field Dependence

The frequency of a cavity is perturbed by the stored electromagnetic energy. This perturbation is due to two effects. First, the electromagnetic field applies a pressure to the surface of the superconductor which is proportional to the square of the field. This pressure stresses the cavity so its resonant frequency is lowered. Second, the superconducting rf skin depth is perturbed by the magnetic field [6]. The resulting increase in skin depth is proportional to the square of the magnetic field and also lowers the resonant frequency. Since both of these effects are proportional to the square of the electromagnetic field, the resonant frequency shift is directly proportional to the cavity stored energy. For the niobium cavity with massive

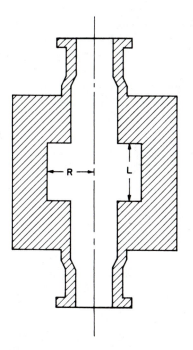

FIGURE 1. Cross-sectional view of TM_{010}-mode cavity for SCO. For this cavity $R = 1.40$ cm and $L = 1.65$ cm.

walls shown in figure 1, the two effects are estimated to be about equal. Measurements give a fractional frequency shift for this cavity of

$$\Delta\nu/\nu_0 = -1.65 \times 10^{-6} \, U \quad , \tag{4}$$

where U is the stored energy in joules.

If the cavity stored energy varies, its frequency will also vary. The stored energy varies primarily because of variations in the magnitude of the rf power source driving the cavity. Assuming that the Q is constant, the stored energy is directly proportional to the applied rf power. Thus the resonant frequency will vary as the magnitude of the power varies. It would seem that the solution would be to decrease the stored energy to a very low level so the effect of power variations would be negligible. However, the rf power (hence, the stored energy) is required to overcome noise in the SCO circuit. Thus, these competing requirements yield an optimum level of power applied to the superconducting cavity.

2.2.2 Temperature Dependence

The cavity resonant frequency is dependent on temperature due to a number of effects: the temperature dependence of the superconducting surface reactance, and the thermal expansion due to both the lattice and the electrons. These effects act to decrease the resonant frequency with increasing temperature. SCO's using niobium cavities were employed by C. M. Lyneis and J. P. Turneaure to measure the temperature dependence of the resonant frequency of niobium cavities, since this information is useful in the study of the superconducting surface reactance and of the lattice thermal expansion at low temperature [7]. Figure 2 is a graph of the temperature

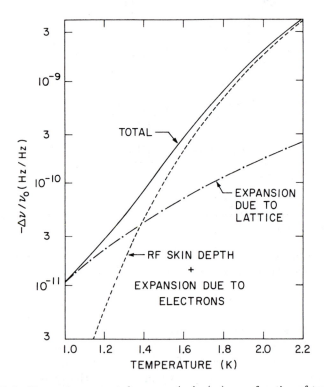

FIGURE 2. Change in resonant frequency $(-\Delta\nu/\nu_0)$ as a function of temperature. Shown in the figure are the total change in the resonant frequency and also the contributions to the total due to the lattice thermal expansion, proportional to T^4, and due to the superconducting rf skin depth and the thermal expansion of the electron system, approximately proportional to $\exp(-\Delta/kT)$. Δ is the superconducting energy gap, k is the Boltzmann constant, and T is the temperature.

dependence of resonant frequency $(\Delta\nu/\nu_0)$ as a function of temperature [8]. Since the resonant frequency will vary as the temperature of the niobium cavity varies, it is essential that the superconducting cavity be temperature controlled.

2.2.3 Gravity

Acceleration on a cavity can produce strain in the niobium cavity structure and thus perturb the resonant frequency. The largest source of acceleration in an earth laboratory is the acceleration due to gravity. As is observed in figure 5 (presented in a subsequent section of this paper), the niobium cavity is supported from its top end and thus the cavity is loaded by its own weight. The effect of the weight is to bend the two end plates and to place the cylindrical section under tension. The frequency shift due to this strain is positive and is estimated to be about $4 \times 10^{-9}(\Delta \nu / \nu_0)$ for an acceleration of 1 g. This frequency shift makes the resonant frequency sensitive to both tilt and variation of g. With the cavity oriented in the vertical direction as shown in figure 5, the resonant frequency variation with tilt is small since by symmetry it is at the top of a parabolic dependence. The principal source for variations of the local acceleration of gravity are earth tides, which are periodic deformations of the earth principally due to the moon. At Stanford the peak-to-peak variation in g is about 2×10^{-7} g. Thus, the earth tides can cause frequency variations on the order of $10^{-15}(\delta \nu / \nu_0)$. The frequency variations due to earth tides are not observed in a frequency comparison among SCO's which use identical cavities in the same laboratory, since each of their cavities undergoes the same resonant frequency variation.

2.3   An Oscillator Circuit

The superconducting cavity was used as the frequency determining element in an oscillator circuit. Any number of circuits could have been employed; however, it was desired to use a circuit that required a minimum of development effort. One of the authors (Turneaure) invented a circuit which is a modification of the Pound cavity stabilization circuit which utilizes an intermediate modulation frequency [9]. He presented this idea at a seminar only to find that the same circuit had already been employed by W. J. Trela and W. M. Fairbank to make measurements on superfluid helium flow [10], although using a higher modulation frequency of 30 MHz. At that time the other author of this paper (Stein) was working on an extension of those helium flow measurements. Since we had a mutual interest in this oscillator circuit, we began a joint effort on the SCO development. Our improvement of this oscillator circuit was later used at Stanford by C. A. Waters, H. A. Fairbank, J. M. Lockhart and K. W. Rigby to make measurements of the low temperature transition in the microwave surface impedance of copper. (See their paper entitled "Low Temperature Transition in the Microwave Surface Impedance of Copper" in this volume.)

A schematic of the SCO circuit is shown in figure 3. There are two important features of this circuit. The first, which kept the development effort to a minimum, is that all of the active components are standard

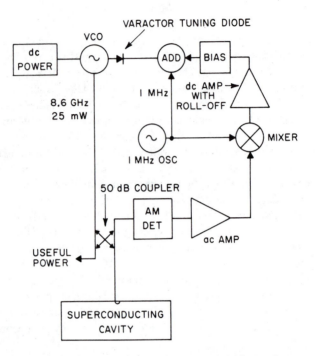

FIGURE 3. Schematic diagram of the one-port, passive SCO circuit.

commercial devices and are at room temperature. The second feature is that only a single transmission line is required to connect the room temperature microwave electronics to the 1.2 K niobium cavity, and the SCO circuit is insensitive to the length of this line, as is described below. The primary oscillator in the SCO circuit is a voltage controlled oscillator (VCO) which is a Gunn oscillator whose frequency can be controlled by a voltage applied to a varactor diode. The SCO circuit locks the VCO to the superconducting cavity resonant frequency by means of a frequency stabilization loop.

The frequency stabilization loop works as follows. The VCO is frequency modulated at 1 MHz, producing a pure phase modulation (PM) output consisting of a carrier, 1 MHz sidebands and, to a smaller extent, sidebands at harmonics of 1 MHz. The typical output of a Gunn oscillator is on the order of 25 mW and most of this power is available as useful power output. A small portion (typically a few hundred nanowatts) of the VCO power output is extracted by a coupler and is directed toward the superconducting cavity. It is important to note that the bandwidth of an 8.6 GHz cavity with a Q of 10^{10} or greater is less than 1 Hz. If the rf carrier is close to the resonance, the phase and amplitude of the reflected carrier will depend on the difference between the carrier and cavity frequencies. However, the phase modulation sidebands, which are far from the resonance, are reflected with almost no influence from the cavity resonance. When the reflected carrier and modulation sidebands are combined, the result is an admixture of phase and amplitude modulation. The magnitude and phase of the amplitude modulation (AM) depend upon the difference between the carrier and cavity frequencies. The amplitude modulation is detected by the AM detector. Figure 4 shows the output of the AM detector as a function of the carrier frequency relative to the cavity resonant frequency. As shown in figure 4, the AM output is linear near resonance and is exactly zero at resonance. This output, after ac amplification, synchronous detection at 1 MHz, dc amplification and appropriate filtering, is used as a control signal to stabilize the VCO. A very important feature of this circuit is that the carrier and phase modulation sidebands utilize the same transmission line between the VCO and the AM detector. This fact makes the frequency stabilization relatively insensitive to either transmission line length or its variations.

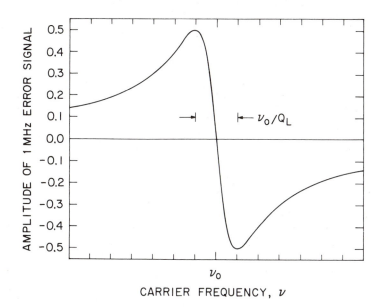

FIGURE 4. Output of AM detector as a function of VCO carrier frequency in the vicinity of the cavity resonant frequency.

This SCO circuit does have limitations. For example, the frequency modulation of the VCO may produce an output with a small amount of

AM, the directional coupler may have a dispersive characteristic which converts the PM partially into AM, and the AM modulation detector itself may convert some of the PM modulation to an AM output. This unwanted AM produces an AM bias which is added to the error signal as shown in figure 4. If this AM bias varies, the output frequency of the SCO will also vary. The effect of AM bias on the SCO output frequency can be decreased by utilizing a cavity with a larger Q since the slope of the error signal increases with increasing Q.

2.4   Low Temperature Configuration of SCO

Figure 5 illustrates the configuration of the SCO components at low temperature. The TM_{010}-mode niobium cavity after final preparation is connected at its upper end to a coaxial feedthrough with a ceramic vacuum seal. The lower end of the cavity is connected to a flange with a copper pump-out port. Both ends are sealed with indium wire seals. The copper pump-out port is then connected to an ultra high vacuum pumping system, pumped out, and baked under vacuum, typically at 100 °C. After bakeout the cavity reaches a pressure below 10^{-9} torr at which time the copper pump-out port is pinched off leaving the cavity under a self-contained vacuum. Figure 6 is a photograph of an evacuated cavity assembly.

Also illustrated in figure 5 are other important low temperature components. The coaxial feedthrough, which is also a coax-to-waveguide transition, is mounted on a copper waveguide so the cavity axis is vertical. About 2 cm before the waveguide passes through the wall of the vacuum can the waveguide material is changed to low thermal conductivity stainless steel. The stainless steel waveguide continues upward to room temperature for connection to the 50 dB coupler at room temperature. The cavity and other components are surrounded with a vacuum can (only a portion is shown) for two purposes: (1) to protect the cavity from helium-bath pressure variations, and (2) to allow high stability temperature regulation of the cavity. The vacuum can is surrounded with a 1.1 K liquid helium bath.

The remaining components inside the vacuum can are part of the cavity temperature regulation system. They include a germanium resistance thermometer mounted on the copper waveguide for sensing the cavity temperature, a copper thermal heat leak mounted between the copper waveguide and vacuum can wall for conducting heat to the helium bath, and a resistance heater for increasing the cavity temperature above the helium bath temperature. With appropriate room temperature electronics, the thermometer and heater are used to regulate the temperature of the cavity to about 1 μK.

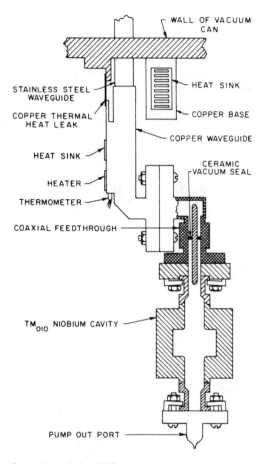

FIGURE 5. Configuration of the SCO low temperature components. These components are surrounded by a vacuum can which is immersed in liquid helium.

FIGURE 6. Photograph of an evacuated TM_{010}-mode niobium cavity mounted to coaxial feedthrough and pinched-off pump-out port.

2.5   Frequency Stability of SCO

The SCO was developed over a number of years during which frequency stability data were taken. The latest data, which are presented here, were taken in preparation for and during a null gravitational red-shift experiment [11] in 1978. This experiment involved an ensemble of three SCO's and two hydrogen masers. The three SCO's were designated SCO#1 through SCO#3 and had unloaded Q's of 3.0×10^{10}, 3.8×10^{10} and 9.0×10^{10}, respectively. Of these three, SCO#3 generally produced the highest frequency stability and lowest frequency drift, probably because of its higher Q.

TABLE 1. CHARACTERISTICS OF SCO #3 SUPERCONDUCTING CAVITY AT 1.2 K.

Resonant frequency	8.618 GHz
Unloaded Q	9.0×10^{10}
Loaded Q	2.0×10^{10}
Coefficient of the temperature dependence of $\Delta v/v_0$	-1.4×10^{-10} K^{-1}
Coefficient of the stored energy dependence of $\Delta v/v_0$	-1.65×10^{-6} J^{-1}
$\Delta v/v_0$ due to acceleration of gravity	$\sim 4 \times 10^{-9}$

The principal operating characteristics of SCO#3 are summarized in two tables. Table 1 lists the characteristics of the superconducting cavity for SCO#3 at 1.2 K. Table 2 lists the typical operating conditions which are

TABLE 2. TYPICAL OPERATING CONDITIONS OF SCO #3.

Power incident on cavity	3.0×10^{-7} W
1 MHz modulation index	1.0
Cavity stored energy	1.0×10^{-7} J
Maximum cavity surface field	2.3 G
$\Delta v/v_0$ due to stored energy	-3.4×10^{-13}

imposed by the room temperature SCO circuit. As shown in the second table, the 3.0×10^{-7} W applied incident power yields a maximum cavity surface magnetic field of 2.3 G and a $\Delta\nu/\nu_0$ due to the stored energy of -3.4×10^{-13}.

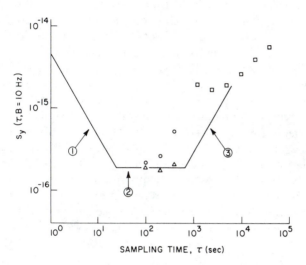

FIGURE 7. Frequency stability of an SCO as a function of sampling time. See the text for an explanation of the data shown in the figure.

The estimated frequency stability of SCO#3 is presented in figure 7. This figure contains different types of data and an approximate curve which is an estimate of the frequency stability of SCO#3. The frequency stability of SCO#3 is measured by comparing its frequency with that of other oscillators, which in this work are the other SCO's and hydrogen masers.

First, SCO#3 is compared with the other two SCO's. Since the comparison with SCO#2 yields the highest frequency stability, the frequency stabilities from this comparison are given in the figure, and they are represented by $\bigcirc$ and $\triangle$. Since this comparison involves similar types of oscillators, the frequency fluctuation variances are divided by two so the data in the figure refer to a single SCO. The data represented by $\bigcirc$ are the square root of the standard two-sample Allan variance. The Allan variance is a measure of the frequency fluctuations and is a function of the sampling time τ and measurement bandwidth B. As shown in the figure, the square root of the Allan variance has a minimum value of 2.1×10^{-16} for $\tau = 100$ sec and increases to 5.2×10^{-16} for $\tau = 400$ sec. This increase with τ is thought to be the result of frequency drift. Since comparison of the SCO's with the hydrogen masers indicates that SCO#3 drifts much less than SCO#2, a three-sample Hadamard variance which removes linear drift is calculated from the same data. The square root of the Hadamard variance, represented by $\triangle$, is nearly flat at 2.0×10^{-16} for τ between 100 and 400 sec.

Second, SCO#3 is compared with the hydrogen masers. The square root of the Allan variances, calculated from these data and represented by $\square$, are also given in figure 7. Since this comparison involves dissimilar types of oscillators, the Allan variance is not divided by two so the data in the figure refer to the sum of the variances of both oscillators. For the shortest sampling time of 1200 sec, the frequency stability is 1.3×10^{-15}, which is probably dominated by the stability of the hydrogen maser. However, for the longest sampling time of 3.8×10^4 sec, the frequency stability is 5.9×10^{-15}, which is dominated by the stability of the SCO.

The estimated frequency stability of SCO#3 is approximated by three straight lines in figure 7. Line 1 lies in a region which is dominated by mechanical vibration of the superconducting cavity structure and the resulting modulation of the resonant frequency. In this region the frequency stability is proportional to τ^{-1} and its magnitude is dependent on the measurement bandwidth. Line 2 lies in a transitional region between lines 1 and 3. Line 3 lies in a region where the frequency stability is dominated by frequency drift and is approximately proportional to τ.

3. APPLICATION TO FUNDAMENTAL PHYSICS EXPERIMENTS

An SCO can be applied to fundamental physics experiments in a variety of ways. It can, for example, be used as an *oscillator* or *clock* for very long baseline radio interferometry [12] and as a very sensitive *transducer* to detect energy absorbed in a gravity wave antenna [13]. Perhaps the most direct applications of an SCO to fundamental physics experiments, however, are those that explore the physical significance of the SCO frequency ν_0 (also the cavity resonant frequency).

Let us express ν_0 in terms of the fundamental constants. According to equation (3) ν_0 depends on the cavity shape, the cavity size ℓ, and the velocity of light. If it is assumed that the cavity is formed from a single crystal, ℓ can then be expressed as the lattice spacing of the crystal times the number of lattice spacings in the direction of ℓ. Since the lattice spacing is, to first order in the fundamental constants, equal to some number times the Bohr radius a_0, the resonant frequency can be expressed approximately as

$$\nu_0 = b' \, c/a_0 \, , \quad \text{where } b' = \left(\frac{a_0}{\ell} \right) b \quad . \tag{5}$$

a_0 can be replaced by its expansion in terms of the fundamental constants to yield

$$\nu_0 = b'' \, mce^2/h^2 \quad , \tag{6}$$

where b'' is a dimensionless number independent of the fundamental constants, m and e are the mass and charge of an electron, and h is Planck's constant. This dependence of ν_0 on the fundamental constants is the starting point for the discussion of the following two experiments.

3.1 Secular Drift of the Fine Structure Constant

It has been proposed that the fine structure constant α may vary with time [14]. This possibility has been investigated using both astronomical [15] and geophysical [16] data. This possible time variation is characterized by R_α which is defined as

$$R_\alpha = \frac{1}{\alpha} \frac{d\alpha}{dt} \quad . \tag{7}$$

The lowest upper limit on this quantity has been estimated from geophysical data and is $|R_\alpha| \leq 10^{-17}$ year^{-1} for a measurement time of about 10^9 years. Limits placed on $|R_\alpha|$ from astronomical data also involve measurement times on the order of 10^9 years.

An upper limit may be placed on $|R_\alpha|$ using SCO's. This can be done by comparing the frequency of an SCO with that of a cesium frequency standard or a hydrogen maser. Both the cesium frequency standard and the hydrogen maser are based on hyperfine transitions which have a dependence on the fundamental constants different than that for the SCO frequency. The ratio of the hyperfine transition frequency ν_{HF} to the SCO frequency ν_{SCO} is

$$\frac{\nu_{HF}}{\nu_{SCO}} = \text{constant} \cdot g \left(\frac{m}{M} \right) \alpha^3 \quad , \tag{8}$$

where g is the nuclear g-factor and M is the nuclear mass for either the hydrogen or cesium nucleus. If it is assumed that g and the ratio m/M are independent of the fundamental constants to first order, then

$$|R_\alpha| = \frac{1}{3} \left[\frac{d}{dt} \left(\frac{\nu_{HF}}{\nu_{SCO}} \right) \right] \bigg/ \left(\frac{\nu_{HF}}{\nu_{SCO}} \right) \quad . \tag{9}$$

An experiment to measure the possible time variation of α was made in 1975 [17]. In this experiment, three essentially independent SCO's were compared with the cesium frequency standard ensemble at Hewlett-Packard in Santa Clara, California, for a measurement time of 12 days. The data from this experiment yielded an $|R_\alpha|$ which has a 68% probability of being 4.1×10^{-12} year^{-1} or less. This result, although about 4×10^5 times greater than the result based on geophysical data, has been determined with a measurement time which is 10^{-10} of the geophysical time.

3.2   A Null Gravitational Red-Shift Experiment

One consequence of Einstein's equivalence principle is that the gravitational red shift is universal and independent of the nature of the type clock being studied, and for a weak gravitational field is given by

$$Z = \frac{\Delta \nu}{\nu} = \frac{\Delta U}{c^2} \quad , \tag{10}$$

where ν is the clock frequency and U is the Newtonian gravitational potential. Thus a comparison of the red shifts of two types of clocks based on different physical principles provides a test of Einstein's equivalence principle. Such a test [11] was performed at Stanford in 1978 using an ensemble of three SCO's built at Stanford and two hydrogen masers built at the Smithsonian Astrophysical Observatory. The frequencies of these two devices are based on different physical principles, as discussed above. For the SCO the frequency depends on the velocity of light divided by the cavity size ℓ, and for the hydrogen maser the frequency depends on a hyperfine transition.

The experiment was carried out by taking advantage of the red shift that occurs in an earth laboratory due to the time variation of distance from the laboratory to the sun. There are two terms. One is the result of the daily rotation of the earth about its axis, and the second is the result of the annual motion of the earth in a slightly eccentric orbit about the sun. The experiment was carried out over a short time of ten days and close in time to the maximum rate of change of the distance between the earth and the sun. These conditions allow the annual variation to be approximated by a linear term in the time. Thus for this experiment the laboratory variation of the dimensionless reduced solar gravitational potential u is approximated by

$$u = \frac{U}{c^2} = -3.2 \times 10^{-13} \cos[2\pi(t - t_0)] + 2.8 \times 10^{-12}(t - t_0) \quad , \tag{11}$$

where t is the time measured in solar days and t_0 is noon April 4, 1978.

The experiment was interpreted in the following way. The red shift of each type of clock is written as

$$Z_{SCO} = (1 + \alpha_{SCO})u \quad \text{and} \quad Z_{HF} = (1 + \alpha_{HF})u \quad , \tag{12}$$

where α_{SCO} and α_{HF} are the deviation of the red shifts from the accepted values. The experiment described above yields a measure of

$$Z_{SCO} - Z_{HF} = (\alpha_{SCO} - \alpha_{HF})u \quad . \tag{13}$$

The net result of the experiment is an upper limit

$$|\alpha_{SCO} - \alpha_{HF}| \leq 1.7 \times 10^{-2} \quad . \tag{14}$$

This result has also been interpreted by Will [18] in terms of the $TH\epsilon\mu$ formalism developed by Lightman and Lee [19].

4. CONCLUSION

A superconducting cavity oscillator utilizing a high-Q superconducting niobium cavity for its frequency determining element has achieved very high frequency stability: $2 \times 10^{-16}(\delta\nu/\nu)$ for a sampling time of 100 sec. The SCO frequency stability has been limited by practical considerations rather than by any fundamental ones. To achieve better frequency stability, improvements in several aspects of the SCO need to be made. These aspects involve (1) the dependence of the cavity resonant frequency on electromagnetic field, temperature and acceleration of gravity, and (2) the ability of the SCO circuit to maintain the oscillator at the center of the cavity resonance. Substantial improvement can be made in all of these aspects. For example, the sensitivity of the cavity resonant frequency to the acceleration of gravity can be reduced by supporting the superconducting cavity in a symmetrical manner so that the resulting strain averaged over the cavity yields a resonant frequency shift near zero. Also it should be possible to regulate the temperature of the superconducting cavity to a level much better than 1 μK. (See paper by J. A. Lipa entitled "High Resolution Heat Capacity Measurements Near the Lambda Point" in which he reports resolving temperature differences of less than 1 nK.) With improvements in all of these aspects, it should be possible to achieve a short-term frequency stability which extends into the 10^{-17} or $10^{-18}(\delta\nu/\nu)$ range and also to achieve much lower long-term frequency drift. Alternatives to the SCO discussed above have been suggested by Braginsky [20] and others [21,22]. These SCO alternatives, which utilize sapphire loaded superconducting cavities, may also lead to improved frequency stability. SCO frequency stability reaching into the $10^{-18}(\delta\nu/\nu)$ range appears to be possible and should provide new opportunites for application of these oscillators to fundamental physics experiments.

Acknowledgments

Work described in this paper was supported in part by the U.S. Office of Naval Research and the U.S. National Aeronautics and Space Administration.

References

[1]   D. J. Wineland, Metrologia 13, 121 (1977).

[2]   R. F. C. Vessot, M. W. Levine, E. M. Mattison, E. L. Blomberg, T. E. Hoffman, G. U. Nystrom, B. F. Farrel, R. Decher, P. B. Eby, C. R. Baugher, J. W. Watts, D. L. Teuber and F. D. Wills, Phys. Rev. Lett. 45, 2081 (1980).

[3]   S. R. Stein and J. P. Turneaure, in *Future Trends in Superconductive Electronics*, edited by H. C. Wolfe (American Institute of Physics, New York, 1978), Conference Proceedings No. 44, p. 192.

[4]   J. M. Pierce, *J. Appl. Phys.* **44**, 1342 (1973).

[5]   J. P. Turneaure and Nguyen Tuong Viet, *Appl. Phys. Lett.* **16**, 333 (1970).

[6]   J. Halbritter, in *Externer Bericht* (Kernforschungszentrum, Karlsruhe, 1968), 3/68-8.

[7]   C. M. Lyneis, Ph.D. Thesis, Stanford University, 1974.

[8]   C. M. Lyneis and J. P. Turneaure, unpublished.

[9]   R. V. Pound, *Rev. Sci. Instr.* **7**, 490 (1946).

[10]   W. J. Trela and W. M. Fairbank, *Phys. Rev. Lett.* **9**, 822 (1967).

[11]   J. P. Turneaure, C. M. Will, B. F. Farrell, E. M. Mattison and R. F. C. Vessot, *Phys. Rev. D* **27**, 1705 (1983).

[12]   For a review of frequency stability requirements see W. K. Klemperer, *Proc. IEEE* **60**, 602 (1972).

[13]   For example, see article by D. G. Blair, in *Gravitational Radiation*, edited by N. Deruelle and T. Piran (North-Holland, Amsterdam, 1983), p. 339.

[14]   For a review of this subject see F. J. Dyson, in *Aspects of Quantum Theory*, edited by A. Salam and E. P. Wigner (Cambridge University Press, Cambridge, 1972), p. 216.

[15]   A. M. Wolfe, R. L. Brown and M. S. Roberts, *Phys. Rev. Lett.* **37**, 179 (1976).

[16]   A. I. Shlyakhter, *Nature* **264**, 340 (1976).

[17]   J. P. Turneaure and S. R. Stein, in *Atomic Masses and Fundamental Constants 5*, edited by J. H. Sanders and A. H. Wapstra (Plenum, New York, 1976), p. 636.

[18]   C. M. Will, *Theory and Experiment in Gravitational Physics* (Cambridge University Press, Cambridge, 1981), p. 62.

[19]   A. P. Lightman and D. L. Lee, *Phys. Rev. D* **8**, 364 (1973).

[20]   V. B. Braginsky and V. I. Panov, *IEEE Trans. Magnetics* **MAG-15**, 30 (1979).

[21]   D. M. Strayer, G. J. Dick and E. Tward, *IEEE Trans. Magnetics* **MAG-19**, 512 (1983).

[22]   D. G. Blair and I. Evans, *J. Phys. D.* **15**, 1651 (1982).

The Development of
the Free Electron Laser

John M. J. Madey

1. INTRODUCTION

Some years ago an effort was begun in the High Energy Physics Laboratory (HEPL) at Stanford to demonstrate a free electron laser (FEL), that is, a laser based upon the radiation emitted by a beam of electrons in transitions between free and unbound initial and final states of motion.

With the enthusiastic support of William Fairbank, the effort was successful, and has led to a fairly extensive national effort to further develop and exploit this new class of laser. The occasion of the Near Zero conference and the publishing of the *Near Zero* book give us an opportunity to review the physics of the FEL, the history and accomplishments of the effort at HEPL, and the prospects for the further development and application of these devices.

2. BASIC PHYSICS

While the special properties of laser light—coherence, low angular divergence and high power—appear extraordinary in comparison with the properties of the light generated by ordinary thermal or electrically excited sources, the mechanism responsible for laser action, stimulated emission, is a pervasive physical effect which has long been known to play a critical role even in such ordinary phenomena as black body radiation. It has also long been clear that the attainment of laser action is not contingent upon the use of any particular or generic radiative transitions. Although early lasers made extensive use of transitions between the bound states of individual atoms and molecules, any transition which resulted in the emission of a photon could in principle be exploited to secure laser operation. The Jello™ laser [1] demonstrated by A. L. Schawlow several years ago has been cited as evidence that anything will lase if it is pumped hard enough, and extensive commercial use is now made of the excimer laser in which the transition responsible for radiation occurs between a bound excited state and the unbound ground state of a rare gas-halide molecular complex.

Prior to the development of lasers and masers, essentially all generators of coherent electromagnetic radiation were based on the properties and interactions of electron beams in the presence of time-varying or static electric potentials or magnetic fields. This technology was, in fact, astonishingly varied and successful.

The formalisms used to develop these prior electron beam devices were, in general, quite distinct from the principles invoked to describe the first atomic and molecular lasers. Most of these devices could be explained using either the theory of space-charge limited current flow (negative grid amplifiers and oscillators), transit time bunching (the klystron), or Pierce's generalized traveling wave analysis (traveling wave tubes) [2]. Systems designed employing these concepts continue to dominate applications involving the generation of high power radiation at wavelengths from about 1 meter to 1 millimeter. However, although numerous efforts had been made to develop electron beam oscillators and amplifiers capable of operating beyond 300 GHz in the days before the laser, success proved elusive due to the actual or perceived technical requirements for short wavelength operation.

Given both the differences in formalism and the astonishing success of the laser in generating power at optical wavelengths, there was a general perception in the years immediately following the first laser demonstrations that laser technology and the prior electron beam technology were completely distinct. It was therefore interesting to ask if perhaps something had been missed: whether, in particular, there was some aspect of an electron's interaction with light that had been overlooked in the prior years of effort, and whether such an effect, if it existed, might be exploited to achieve a short wavelength capability comparable to that attained with the new atomic and molecular laser sources.

It was, of course, well known that free electrons could participate in interactions in which light quanta were emitted. Well-established examples of such interactions include bremsstrahlung and Cerenkov radiation. Bremsstrahlung, in which light is emitted by an electron due to its acceleration by an external electric or magnetic field, seemed a particularly attractive possibility since the spectrum of the emitted radiation was known to extend through the optical region. Given the presence of stimulated emission there was at least the hope of laser operation in the visible spectrum and beyond.

At the time I became interested in the problem, a number of possibilities had been examined, including the classic example of the bremsstrahlung radiation emitted by electrons scattered by the coulomb potential of the ion cores in a crystal lattice [3]. Stimulated emission was found to be possible in this case, although the absorption caused by the acceleration of the electrons by the light wave as they passed the ion cores was found to have a comparable magnitude and spectral dependence. The net available amplification, the difference between the amplification due to stimulated emission and the attenuation due to absorption, was therefore small.

To secure a larger net amplification, it seemed necessary to utilize a radiation mechanism in which the net probability or frequency dependence of the absorption process differed more substantially from the probability or frequency dependence of stimulated emission. A natural alternative to consider for this purpose was Compton scattering, in which electrons scatter the photons present in an initial electromagnetic pump wave. Compton scattering can be considered a form of bremsstrahlung in which the electrons are periodically accelerated by the initial electromagnetic wave. In Compton scattering the recoil associated with emission shifts the peak of the emission spectrum to a longer wavelength than the peak of the absorption spectrum. If the widths of the emission and absorption spectra, as determined by the electrons' velocity spread and the spectral content of the initial electromagnetic wave, are sufficiently narrow, the spectral separation of the emission and absorption processes can be exploited to generate significant and useful levels of amplification.

The stimulated Compton scattering mechanism was first discussed by Kapitza and Dirac [4], and its application to secure laser operation was first proposed by Pantell, Soncini and Puthoff [5]. Given the bremsstrahlung analogy, it was natural to suggest the replacement of the initial electromagnetic wave assumed in Compton scattering by a spatially periodic transverse magnetic field, a configuration first suggested by Motz [6] and Phillips [7]. As noted by Weizsäcker and Williams [8], the acceleration, and hence the radiation spectrum produced by such a static magnetic field, is indistinguishable from the acceleration produced by a real electromagnetic wave for relativistic electrons. However, it is in general easier to produce strong static fields than real electromagnetic waves; since the stimulated emission transition rate is proportional to the energy density of the initial field, significantly higher gains can be achieved through the use of the static field configuration.

This chain of reasoning led to the configuration shown schematically in figure 1. An electron beam, generated by an external electron gun or accelerator (not shown), is directed through a periodic transverse magnetic field with period λ_q. It was predicted that a light beam sent through the magnetic field with the electrons would be amplified if the wavelength of the light were set to couple preferentially to the peak of the spectrum for stimulated emission.

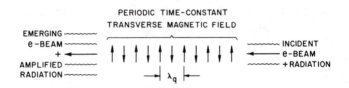

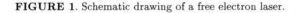

FIGURE 1. Schematic drawing of a free electron laser.

For a given initial electron energy, the wavelength corresponding to the peak of the spontaneous and stimulated emission lineshapes is given approximately by:

$$\lambda_{stim} \approx \lambda_q(1 + \alpha^2 B^2)/2\gamma^2 \quad ,$$

where

$$\gamma^2 = 1/\left[1 - \left(\frac{v}{c}\right)^2\right] \quad ,$$

$$\alpha^2 B^2 = \left[\lambda_q eB/2\pi mc^2\right]^2 \quad ,$$

$B = $ magnitude of periodic magnetic field (gauss) ,

$v = $ electron velocity (cm/sec) ,

$m = $ electron mass (grams) ,

$e = $ electron charge (statcoulombs) .

From time reversal, it is evident that the peak of the absorption lineshape must then be given by

$$\lambda_{absorption} \approx \left(1 - \frac{2h\nu}{\gamma mc^2}\right)\lambda_{stim} \quad ,$$

where $\nu = c/\lambda$. The effect of the electron's recoil during emission and absorption is to reduce the wavelength for absorption by the factor $2h\nu/\gamma mc^2$. Although this is a small shift, of the order of 1×10^{-7} for a typical infrared or visible FEL, it is sufficient to provide useful net amplification on the long wavelength side of the spontaneous lineshape. The shift is also sufficiently small to permit representation of the gain function by the derivative of the spontaneous emission lineshape [9].

Although the search for a free electron laser was begun on quite theoretical and abstract grounds, the available gain, tuning range, and power output predicted even for existing electron beam sources were high enough to inspire serious practical interest in the device. Although relativistic electron energies would be required, operation at wavelengths from 100 μ to 100 Å appeared possible given electron energies in the range 10–1000 MeV. Net gains in excess of 10% per pass also appeared possible if electron currents of the order of 0.1–10 A could be obtained. A saturated laser power output proportional to the spontaneous emission linewidth was predicted, and it was suggested that higher power outputs could be obtained by reducing the undulator period or magnetic field to match the change in energy of the decelerating electrons.

3. THE FIRST EXPERIMENTS

Bill Fairbank became increasingly interested in the free electron laser problem as the theoretical possibilities unfolded, and strongly encouraged me to write up the analysis as part of my Ph.D. dissertation. Bill's encouragement was absolutely critical; to appreciate this it is only necessary to think back to one's days as a graduate student and the uncertainty one felt in committing time to a new and speculative venture.

Of course, the publication of this work led to another question. What could be done to experimentally verify the physics of the FEL, and could we actually demonstrate such a device? This was a time, in the early 70's, of many new ideas in the low temperature group, including the detection of gravitational radiation, measurement of the gravitational mass of electrons and positrons, sensitive measurements of magnetic susceptibility, the search for fractional charge, the gyro relativity experiment, the ^{3}He electric dipole experiment, the magnetic measurement of heart muscle activity and blood flow, and the use of negative pions transported and focussed by a large superconducting spectrometer for radiotherapy. In this context, the proposal to pursue research on a free electron laser did not seem so crazy.

In particular, there was a project underway at Stanford, the superconducting accelerator (SCA), which was to prove critical to the demonstration of the FEL. The idea of the SCA had grown from Bill's research on superconducting microwave cavities at Yale, in which he demonstrated the possibility of Q's greatly in excess of those attainable with copper cavities. Given a material with a sufficiently high critical field, Bill pointed out that these high Q's would make it possible to operate a microwave electron linac in a highly stable, cw configuration, with a gradient comparable to that attained in conventional pulsed linacs. With strong contributions from Alan Schwettman, Todd Smith, Michael McAshan, John Turneaure, Larry Suelzle and Perry Wilson, this concept had materialized by the early 70's in the form of a unique and advanced accelerator system employing super-fluid helium cooled niobium cavities driven by a phase-locked, amplitude-stabilized high power rf system.

The key advantage offered by the SCA for FEL experiments was its small energy spread. Because the FEL gain varies inversely as the square of the spontaneous linewidth, it is highly desirable to reduce the incident energy spread to a value significantly below the homogeneous spontaneous linewidth. Since the linewidth scales as 1/(number of magnet periods), and the experimental conditions appeared to favor a long wiggler, it was necessary in the first experiment to reduce the fractional energy spread to the order of 0.1% or less. While this was a very stringent constraint, the SCA had been designed to provide an energy spread lower by a factor of two.

Bill strongly advocated the FEL as a possible experiment to be run on the new accelerator, and was also instrumental in convincing the Air Force Office of Scientific Research to initiate a substantial research program at Stanford to construct the necessary hardware and instrumentation, and to support the subsequent research program.

The experimental effort to demonstrate FEL operation began in 1972 and continued to 1984. Although the work was funded primarily by AFOSR, additional funds were contributed in the later years of the program by the Office of Naval Research and the Department of Energy. The research yielded the first confirmation of the FEL gain equation and the estimates for saturated power output, and also the first demonstration of an infrared FEL oscillator. More subtle data were also obtained, particularly on start-up effects, including the evolution of the optical and electron momentum spectra, and on the dependence of power on cavity length.

The experiment and its principal results are summarized in figures 2–5. The first version of the experiment was set up to measure the magnitude of the FEL gain and its dependence on electron energy and optical power density. A high power pulsed CO_2 laser was sent through the periodic magnetic field with a 24 MeV electron beam from the SCA, and a fast detector was used to measure the amplitude modulation imposed on the laser beam by the bunched electron beam from the linac (figure 2) [10]. The magnitude of the measured gain matched the theoretical prediction to within 10%, and the gain lineshape matched the derivative of the lineshape for spontaneous emission (figure 3). The system could be driven quite hard by operating the CO_2 laser at full power; with the 160 period wiggler used in the experiment, it was observed that up to 0.2% of the electrons' energy could be converted to light without loss of gain.

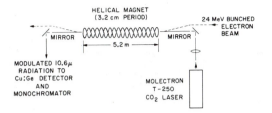

FIGURE 2. The 1976 FEL amplifier experiment. A 24 MeV beam from the superconducting accelerator was used to amplify a 10.6-micron high power CO_2 laser beam.

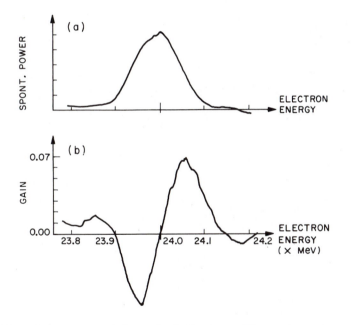

FIGURE 3. Spontaneous emission and gain lineshapes. The spontaneous power radiated by the electron beam in the amplifier experiment could be measured by observing the detector output with the CO_2 laser beam turned off (upper panel). When the CO_2 laser was turned on, much higher power was radiated by the electrons due to stimulated emission. The measured gain lineshape was proportional to the derivative of the spontaneous power spectrum as indicated by theory (lower panel).

Following the amplifier experiments, the system was converted to an oscillator by installing a pair of spherical mirrors around the wiggler magnet to provide feedback (figure 4). Operation of the oscillator was first achieved in January 1977, at a wavelength of 3.3 microns [11]. Although the amplifier experiment had decisively confirmed the basic theory for

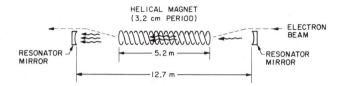

FIGURE 4. The 1977 oscillator experiments. In the oscillator, the electron energy was raised to 43 MeV, the CO_2 laser driver was removed, and feedback was provided by means of a confocal optical resonator.

the FEL, the oscillator experiment had a more substantial practical impact. For the general laser community, it was the first demonstration of an FEL in a "useful" configuration, while for those of us concerned more immediately with FEL physics it disclosed more subtle aspects of operation which had not been revealed in the amplifier experiment [12]. Subsequent operation of the oscillator yielded detailed time-resolved measurements of the output optical and momentum spectra (figure 5), and also the dependence of power output, gain, and spectral content on cavity length. The shift to lower momentum in the electron spectra as the laser turns on is very obvious in these data, as is the shift to longer optical wavelength.

These experiments would not have been possible without Bill's support, the use of the SCA, and the enthusiastic efforts of the FEL, SCA, and HEPL research staffs. The excellence of the High Energy Physics Laboratory and its staff set a standard in these experiments which subsequent research efforts have had to work very hard to equal.

4. PRESENT AND POSSIBLE FUTURE DEVELOPMENTS

The success of the amplifier and oscillator experiments opened the door to a broader national program in FEL physics and technology. Measured by the level of effort, the greater part of this program has been directed toward the enhancement of output power and efficiency. In particular, our early theoretical conjectures concerning the use of tapered, variable period wigglers to improve power output have been confirmed in an impressive series of experiments [13] performed by Edighoffer and his colleagues at TRW using the EG&G linac and the Stanford SCA, by Brau and his colleagues at Los Alamos National Laboratory, by Slater and his colleagues using the Boeing linac, and by Prosnitz and his colleagues at Lawrence Livermore Laboratory. Present plans call for the development of operational high power FEL oscillators at both Boeing and Livermore.

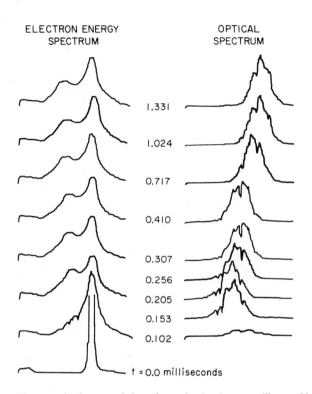

FIGURE 5. Time resolved spectral data from the 3 micron oscillator. Very detailed data were obtained on the evolution of the electron momentum and optical spectra in the final experiments with the 3 micron oscillator. The left panel shows the qualitative evolution of the electron momentum spectrum from a nearly monoenergetic initial distribution to the characteristic double-humped spectrum obtained from constant period FEL's at saturation. The right panel shows the initial chirp and subsequent slow drift of the optical spectrum.

Significant further theoretical developments were also encouraged by these experiments, including the development of a classical theory for the FEL which parallels the earlier traveling wave tube theory [14]. Considered in the context of questions which motivated the original research, this was a fascinating development, for it showed that the theoretical seeds of the free electron laser were actually contained within prior electron device technology.

Of perhaps more general interest to the scientific community has been the effort to realize the FEL's potential for operation at wavelengths, power outputs, or pulse lengths inaccessible to other laser sources. The second operational FEL oscillator was constructed at Orsay, France, in 1983 and used the 200 MeV beam circulating in an electron storage ring to achieve operation in the visible at 6300 Å [15]. It is expected that future efforts in the development of storage ring-based FEL's will yield oscillators operating at wavelengths well below 1000 Å in the extreme ultraviolet.

The long-wavelength regime between 1 mm and 10 μ may also benefit from the availability of FEL sources. Pioneering efforts in the development of sub-mm and far-infrared FEL's have been initiated by Elias at UC Santa Barbara and by Shaw and Patel at AT&T Bell Laboratories. The ability to generate high peak powers and short pulse lengths in this wavelength range should greatly enhance research capabilities in the electronic and magnetic properties of solids by facilitating the excitation of these systems on a scale comparable to their resonant response or damped relaxation times.

High peak power, picosecond pulse length, near-infrared and visible FEL sources may also prove of interest, even given the existing range of conventional laser sources at these wavelengths. Megawatt-level peak powers on a picosecond time scale are not readily available from conventional laser sources, particularly in the infrared. The availability of flexible, high peak power FEL's in this wavelength range could significantly enhance research capabilities in the study of molecular or electronic relaxation, or more general nonlinear effects in which it is necessary to efficiently excite some short-lived transient state of a system. Other possible applications include biomedicine and surgery in which the tunability and power output of the FEL could significantly enhance the ability to optimize transmission or absorption in tissues and structures characterized by varying spectral absorptance and scatter.

Since FEL's require a substantial investment in accelerator hardware for operation, the future utility of these devices will be determined by both the extent of the existing or planned accelerator facilities which might be available for use, and the extent to which the possible special applications of the FEL may generate support for new accelerator facilities. It is certainly plausible at this point to expect that FEL's will become more generally available to the research community through use of the existing linear accelerator and storage ring facilities and that an additional number of small dedicated systems will be built.

5. CONCLUSION

The development of the FEL incorporated a number of unique elements, from the basic physics of the concept to the special role of the HEPL low temperature facilities and the prospective applications of the technology. These elements contribute to an account which is interesting it its own right. But the picture becomes truly impressive when this story is viewed as part of the greater range of research activities in Bill's low temperature group at Stanford.

References

[1] T. W. Hänsch, M. Pernier and A. L. Schawlow, *IEEE J. Quantum Elec.* **QE-7**, 45 (1971).

[2] J. R. Pierce, *Traveling-Wave Tubes* (Van Nostrand, New York, 1950).

[3] D. Marcuse, *Bell System Tech. J.* **41**, 1557 (1962); and **42**, 415 (1963).

[4] P. L. Kapitza and P. A. M. Dirac, *Proc. Camb. Phil. Soc.* **29**, 297 (1933).

[5] R. H. Pantell, G. Soncini and H. E. Puthoff, *IEEE J. Quantum Elec.* **QE-4**, 905 (1968).

[6] H. Motz, *J. Appl. Phys.* **22**, 527 (1951); and **24**, 826 (1953).

[7] R. M. Phillips, *IRE Trans. Elec. Devices* **ED-7**, 231 (1960).

[8] See W. Heitler, *The Quantum Theory of Radiation* (Clarendon, Oxford, England, 1960), p. 414.

[9] J. M. J. Madey, *J. Appl. Phys.* **42**, 1906 (1971).

[10] L. R. Elias, W. M. Fairbank, J. M. J. Madey, H. A. Schwettman and T. I. Smith, *Phys. Rev. Lett.* **36**, 717 (1976). Special credit is also due M. McAshan for his contributions to the design, construction and operation of the cryogenic systems for the SCA and FEL.

[11] D. A. G. Deacon, L. R. Elias, J. M. J. Madey, G. J. Ramian, H. A. Schwettman and T. I. Smith, *Phys. Rev. Lett.* **38**, 892 (1977).

[12] S. Benson, D. A. G. Deacon, J. N. Eckstein, J. M. J. Madey, K. E. Robinson, T. I. Smith and R. Taber, *J. de Phys. Colloq. C1* **44**, 353 (1983).

[13] These and other recent experiments are described in the annual proceedings of the International Free Electron Laser Conference, published as Volumes 5, 7, 8 and 9 of *The Physics of Quantum Electronics* (Addison-Wesley, Reading, Mass.).

[14] W. B. Colson, *Phys. Lett.* **59A**, 187 (1976); F. A. Hopf, P. Meystre and M. O. Scully, *Opt. Communications* **16**, 413 (1976).

[15] M. Billardon, P. Elleaume, J. M. Ortega, C. Bazin, M. Bergher, M. Velghe, Y. Petroff, D. A. G. Deacon, K. E. Robinson and J. M. J. Madey, *Phys. Rev. Lett.* **51**, 1652 (1983).

The Development of
the Superconducting Susceptometer
and Its Biophysical Applications

Edmund P. Day and John S. Philo

1. INTRODUCTION

William Fairbank was intrigued by Fritz London's conjecture [1] that the macroscopic order of large biomolecules might have the same cause as the macroscopic order of superfluids and superconductors. It was natural, then, for Fairbank to follow up the discovery of flux quantization [2,3] with attempts to apply this discovery to the study of biomolecules. The outcome of Fairbank's interest in biophysics has been a versatile new instrument capable of studying the weak magnetic effects found in biomolecules. London's conjecture has not been verified. However, the instrument that grew from it has begun to play an increasingly important role in biophysics research.

The development of a superconducting susceptometer for biophysical applications was begun at Stanford in 1963 by John Pierce [4] and pursued by Edmund Day [5] and John Philo [6]. The original concepts developed at Stanford for an ultrasensitive magnetometer [4] were superceded by the prediction and discovery of the Josephson effect in 1962 [7,8]. This led to the development elsewhere of the superconducting quantum interference device (SQUID) [9–12]. It is this magnetometer which is used in all current versions of the superconducting susceptometer. But for this change, today's superconducting susceptometer is identical to that conceived by William Fairbank in the early 1960's immediately following the discovery of flux quantization. In particular, all versions of this instrument developed at Stanford or elsewhere depend upon flux quantization for a stable magnetic field at the sample. All depend upon flux quantization in a closed superconducting loop for the transfer of information from the sample to the SQUID. Even the SQUID, which relies upon the Josephson effect, depends upon flux quantization in a superconducting cylinder for its sensitivity. The superconducting susceptometer is a perfect example of an instrument made possible by the discovery of flux quantization in a superconducting ring.

The value of the superconducting susceptometer lies in its high sensitivity, its large dynamic range, its fast response, and its versatility. This instrument can make precise absolute or relative or true difference static susceptibility measurements on very small samples in high or low magnetic field. It can detect either nuclear or electronic magnetic resonance. It can be used to follow photo-induced susceptibility transients. The capability of making one or more of these measurements simultaneously or in succession within one instrument on a small sample is invaluable for biophysical applications. This versatility of the superconducting susceptometer when applied to biophysical measurements has been our primary motivation. From the start the goal has been to push the sensitivity and speed of the instrument as far as possible. This has not been done solely to produce the world's most precise static susceptibility data, although this has been achieved. Rather, the aim has been to reach a goal still before us of an instrument unique in its capability for multiple measurements.

In this chapter we will describe the superconducting susceptometer, sketch its development at Stanford, and discuss results indicating its range of application to biophysics. Two examples of biophysics research made possible by the development of the superconducting susceptometer will be mentioned. These are the hemoglobin kinetics studies by Philo of the University of Connecticut and the saturation magnetization studies of nitrogenase by Day at the University of Minnesota.

2. THE SUPERCONDUCTING SUSCEPTOMETER

The concept of a superconducting susceptometer is indicated in figure 1. The sample to be studied is placed within a superconducting detection coil located at the center of a persistent mode superconducting magnet. The magnet produces a large, stable magnetic field at the sample. The detection coil surrounding the sample is part of a continuous, closed superconducting circuit which has a second coil at the magnetometer. Because this two-coil circuit is equivalent to a single closed superconducting ring, the magnetic

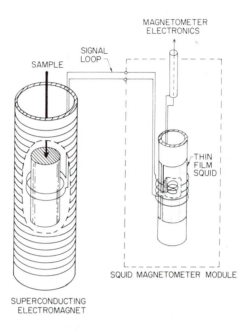

FIGURE 1. Conceptual drawing of a superconducting susceptometer. The solenoid, signal loop and thin film are superconducting.

flux threading this circuit is quantized and conserved. Flux quantization within this circuit will result in a transfer to the magnetometer of information about magnetization changes at the sample.

Magnetization changes at the sample coil can be induced in a variety of ways. Static susceptibility is measured by inserting a sample into this coil while recording the change in SQUID output. Magnetic resonance is detected with the sample in place by inducing changes in nuclear magnetization with applied radio frequency radiation or by inducing changes in electronic magnetization with applied microwave radiation. In either case the SQUID records the resulting magnetization changes. Finally, a light pulse can be used to induce a transient susceptibility change at the sample which is monitored with the SQUID.

The concept of the superconducting susceptometer as pictured schematically in figure 1 is deceptively simple. To implement this concept our focus has been the sample region. The superconducting susceptometer is limited not by the need for a faster or more sensitive or more stable SQUID magnetometer but by the requirements of increased bandwidth, lower noise, and greater field and temperature stability at the sample coil. The difficulty in developing this instrument has been to limit the magnetization and field noise introduced by the apparatus itself at this pickup coil. This is an especially challenging problem in an instrument designed to study room temperature samples. A further challenge in the study of biophysical samples is to develop sample preparation and handling techniques as well as data collection and analysis procedures to avoid magnetic noise generated by the sample holder, background water, and unavoidable paramagnetic impurities in a metalloprotein sample.

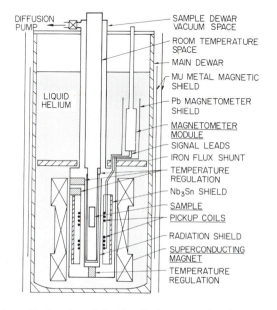

FIGURE 2. Schematic drawing of the Stanford superconducting susceptometer. The time response of this instrument is limited to milliseconds by eddy currents in the normal metal coil former and dewar walls between the sample and the superconducting pickup coils.

A schematic cross section of the Stanford superconducting susceptometer for biophysical applications is given in figure 2. This instrument represents fifteen man-years of development work by John Pierce, Edmund Day and John Philo as graduate students of William Fairbank. It is the first superconducting susceptometer designed to study room temperature samples and the first to function at high fields in the type II region of superconductivity. Its sensitivity is two orders of magnitude beyond that of commercial superconducting susceptometers available up to the end of 1983. Omitted from the drawing is the sample handling component of the susceptometer. The insertion mechanism for precision static susceptibility measurements [6,13–15], the radio frequency coil for nuclear magnetic resonance [16], and the light pipe for photo-triggered susceptibility transients [6,13,17] are detailed elsewhere.

An inside-out dewar arrangement designed by Pierce allows samples at or above room temperature to be inserted into the 4.2 K pickup coils. Magnetization changes are detected by a SQUID magnetometer. In the original instrument a dc SQUID adapted by Day from designs worked out by others [18–20] was used. Commercial SQUID's are now available. The stable magnetic field at the sample is produced by the combination of the persistent mode superconducting magnet and enclosed differential superconducting shield. Magnetization changes are transferred from the sample to the magnetometer by the closed superconducting pickup coil circuit. Field noise and vibration noise are minimized by the homogeneous magnet, matched opposing pickup coil pair, and rigid differential shield-coil unit designed by Day [5]. He found that the limiting noise in this susceptometer arises from the nuclear paramagnetism of the copper coil former [5]. Millidegree temperature drifts of this copper former of the pickup coil cause magnetization drifts five orders of magnitude larger than the noise level of the SQUID detector. Philo controlled this noise by regulating the temperature of both the coil former and the enclosed radiation shield. This improved the instrument's sensitivity by two orders of magnitude [6,13]. More details of this instrument and its development can be found in the Ph.D. theses of Pierce [4], Day [5] and Philo [6] and in an article by Philo and Fairbank [13]. We now change our focus from the instrument itself to its biophysical applications.

3. TIME-RESOLVED MAGNETIC SUSCEPTIBILITY: HEMOGLOBIN KINETICS

One of the unique aspects of the susceptibility program initiated by Bill Fairbank has been the exploitation of the time resolution of the SQUID to study rapid changes in the magnetic properties of a sample. Living systems are dynamic. Our understanding of the structure and function of biomolecules requires study of their kinetics and the properties of short-lived reaction intermediates. The Time REsolved Magnetic Susceptibility (TREMS) technique offers a new tool for the study of the short-time behavior of biomolecules. It is a tool whose potential is still being explored.

The focus of TREMS applications has thus far been the hemoglobin molecule. Hemoglobin is by far the most studied protein, and is the archetype of a molecular machine. The property of hemoglobin that is so interesting (and physiologically important!) is its use of positive feedback to make a molecular "switch." Each hemoglobin molecule contains four binding sites for O_2, which are separated by about 30 Å. As each site binds O_2, positive feedback within the molecule alters the remaining vacant sites to increase their affinity for O_2. The affinity for binding the fourth O_2 is about 250 times greater than for the first. This positive feedback makes the molecule act as a switch, with the states of zero or four bound O_2's highly favored. This enables it to acquire a full "load" of four oxygens in the lungs and switch to zero bound in the tissues. Such feedback interactions (positive or negative) between different sites on the same macromolecule are an important control mechanism which occurs widely in living systems.

The iron in deoxyhemoglobin is in a high-spin, $S = 2$, paramagnetic state, while in oxyhemoglobin it is in a low-spin, $S = 0$, diamagnetic state. This difference in paramagnetism forms the basis for using TREMS to follow the kinetics of O_2 binding. Moreover, there may be a close relationship between the changing spin states and the mechanism of cooperativity between the binding sites. The iron in hemoglobin is located in the center of a large, planar aromatic ring compound called porphyrin. X-ray crystallographic studies of hemoglobin have shown that when the iron is in a high-spin state ($S = 2$ for ferrous iron, $S = 5/2$ for ferric iron) it moves out of the plane of the porphyrin ring, whereas for low-spin states ($S = 0$ or $1/2$) it is in or very near the porphyrin plane. A simplified picture to explain why a high-spin iron moves out of the plane is that its ionic radius is too large to fit in the central hole of the porphyrin. Thus the high-spin iron in deoxyhemoglobin must move about 0.6 Å toward the porphyrin plane in order to bind O_2. It is this motion which is the key to the mechanism for site-site interactions postulated by Perutz [21–23]. The iron is linked to a section of the protein *via* its bond to a histidine residue, and therefore when it moves it drags a large segment of protein along with it. This motion of the protein segment may, in turn, trigger changes at the other binding sites. Moreover, this motion may provide the means by which the protein alters the O_2 binding properties; since movement of the iron must occur for O_2 to be bound, forces generated by the protein on the iron which aid or oppose this motion can alter the energetics of O_2 binding. Thus, the changes in position of the iron are the key to the present best model for cooperativity; oxygen binding at one site moves the protein to trigger changes at the other sites, which make it easier for the iron at these sites to move into the porphyrin plane, thus raising the probability that these vacant sites will bind O_2. Since these changes in iron position are intimately related to changes in spin state, magnetic susceptibility measurements can provide important tests of whether this model of Perutz is correct.

What time resolution is needed to study the magnetic changes in hemoglobin? The binding of O_2 occurs on a time scale of milliseconds while the transfer of information between binding sites occurs on a time scale of hundreds of microseconds. The time resolution of a commercial rf SQUID is

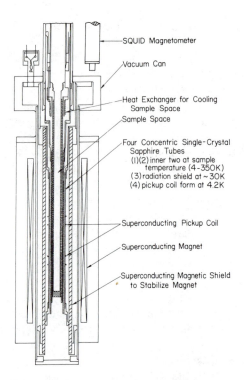

FIGURE 3. Scaled cross-sectional drawing of the sample region of the Connecticut superconducting susceptometer. The time response of this instrument is limited by the SQUID magnetometer.

about one microsecond. For a susceptometer, however, the time resolution is usually limited to milliseconds or longer by eddy currents in the metal walls of the sample dewar. Susceptometer design involves an unavoidable trade-off between the use of metals to improve thermal stability to achieve equilibrium sensitivity and the removal of metals to achieve time resolution. The thin metal walls of the Stanford instrument permitted millisecond time resolution. This was sufficient for pilot studies by Philo of CO binding to hemoglobin. The binding of CO is exactly analogous to that of O_2 including cooperativity except that CO can be easily driven off the hemoglobin by a pulse of light. The first TREMS experiments showed that the magnetic technique can give CO binding rates in excellent agreement with those obtained by optical techniques [3, 16]. Moreover, Philo was able to transiently generate the high affinity form of hemoglobin and show that under some conditions its magnetic moment differs from that of the low affinity form.

Philo has now built (fall 1983) a new susceptometer at Connecticut which is optimized for kinetics experiments (see figure 3). By using single crystal sapphire to construct the sample dewar, all normal metals are eliminated and the response time is limited by that of the SQUID. This increased time resolution has been achieved at the expense of a loss of a factor of ten in equilibrium sensitivity compared with the Stanford instrument. Figure 4 shows data from the first kinetics experiment with the Connecticut instrument taken at a time resolution of about 10 microseconds. A pulse of light from a dye laser drives off all the CO, leaving the iron in a paramagnetic state. As the CO rebinds (returning the system to equilibrium), this paramagnetism disappears. Prior to the flash, the protein is in the high affinity, rapid-binding state. After the flash, it begins to switch to the low affinity, slow-binding state. However, at the high CO concentration in this experiment about 50% of the CO recombines before the switch to the low affinity state can occur, *i.e.*, our time scale is now fast enough to "see" the rate at which the affinity changes occur. Thus the TREMS studies now underway should provide a crucial test of the Perutz model.

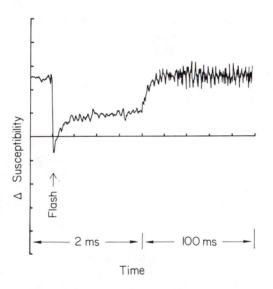

Time

FIGURE 4. Changes in magnetic susceptibility following photolysis of a 100 μM carboxyhemoglobin solution under 1 atm of CO at 20 °C (see text). The light flash arrives at 400 microseconds. The time scale changes by a factor of 50 at 2 milliseconds. The trace is a sum of ten transients. The ringing beginning at approximately 20 milliseconds in this first experiment is caused by the sound of the laser when it is fired.

A new class of experiments has been created by the high field and low temperature capabilities of the commercially available superconducting susceptometer built by SHE Corporation [24]. This instrument was developed from the Stanford design with extended field (up to 5 T) and temperature (down to 2 K) capability. This makes it possible to study the saturation magnetization behavior of metalloenzymes, an area of research being developed by Day at the University of Minnesota.

4. SATURATION MAGNETIZATION STUDIES OF A COMPLEX ENZYME: NITROGENASE

Magnetic changes lie at the heart of metabolism. In photosynthesis, in nitrogen fixation, and in respiration, electrons move rapidly in a controlled fashion through a sequence of trap sites (redox centers). Each redox center can exist in two or more states (oxidation states) defined by the number of electrons at the redox site. Typical redox centers consist of one or more of a small number of transition metal ions in an organic matrix provided by the protein. Measuring the magnetic properties of the redox centers of these metalloproteins in each of their redox states is a necessary first step in understanding the metalloprotein's function. The magnetic signature of these redox changes is usually the best, and often the only, means of tracing the electron's path through the sequence of trap sites.

The symmetry at the transition metal ion site within a protein matrix rarely approaches that of an isolated metal ion. Consequently orbital angular momentum is quenched and spin-only magnetic behavior is observed. Within this framework a specified oxidation state of a redox center can still vary from high spin to low spin depending on the details (number and nature of nearest neighbors) of the local environment. For this reason, determining the spin and zero field splitting parameters of each redox state at each redox center yields informaion on the symmetry, coordination number and nearest neighbors of the transition metal ion.

The zero field splittings of the magnetic sites of a metalloprotein are typically on the order of ± 1 to ± 10 cm^{-1}. This corresponds to temperatures of 1.4 K to 14 K and to fields of 1.1 to 11 T (for an S = 1/2, g = 2 site). Therefore, magnetization saturation studies from 2 to 20 K in fields up to 5 T can be used to measure zero field splittings and the spin of the redox centers of metalloproteins. These studies are difficult because even concentrated metalloprotein samples are diamagnetic, even at low temperatures. The saturable magnetization to be studied at 2 K is typically on the order of a few percent of the sample's diamagnetism. Routine measurement of such signals requires measurement of a sample's magnetization to parts per ten thousand. This requires not only the sensitivity of the superconducting susceptometer, but also new methods of sample preparation and characterization as well as proper techniques for data collection and analysis.

In the nitrogenase data to be presented here, a matched control was measured over the same field and temperature range and subtracted. This was done to take out the effects not only of the diamagnetism of the helium gas, delrin sample holder, and frozen water, but also of the nuclear paramagnetism of the protons in the delrin and water. Each of these effects changes with temperature on a scale comparable with the saturation magnetization under study. The nuclear paramagnetism of the protons must be dealt with carefully, due to their varied and sometimes very slow ($\sim$ hours) relaxation rates.

Fundamental to these studies is the successive measuremment of the metalloenzyme sample using electron spin resonance (ESR) and Mössbauer spectroscopy prior to study with the superconducting susceptometer. In this way the redox state of the susceptibility sample is determined, as is the level of impurity paramagnetism. Proper interpretation of the sample's saturation magnetization is difficult to impossible without these additional data.

If hemoglobin represents a relatively simple protein, nitrogenase represents the other extreme. Nitrogenase is functionally complex. This enzyme catalyzes the reduction of gaseous atmospheric nitrogen to ammonia in a free energy consuming (MgATP requiring) process involving eight electrons:

$$N_2 + 8H^+ + 8e^- \rightarrow 2NH_3 + H_2 \quad . \tag{1}$$

Nitrogenase is structurally complex. This enzyme has two components of approximate molecular weight 60,000 and 200,000. Saturation magnetization studies of this complex enzyme by E. Day in collaboration with E. Münck of the University of Minnesota and with W. H. Orme-Johnson of MIT have just begun (fall 1983). The data described here are preliminary and intended to indicate the nature of this new area of metalloenzyme research rather than to describe final results.

The larger of the two components of nitrogenase, called the MoFe protein, contains two molybdenum atoms and 30 iron atoms [25]. Mössbauer studies combined with ESR [26–29] have established that these metal ions are grouped into two M centers each containing six iron atoms and, presumably, one molybdenum atom, four P clusters each containing four iron atoms, and two iron atoms behaving spectroscopically identically (the S site iron). Figure 5 gives a schematic representation of these results. Three redox states of the MoFe protein have been studied to date. As isolated under reducing and strictly anaerobic conditions, the MoFe protein is in its native state. This isolated protein can be reversibly oxidized with thionine. It is quickly and irreversibly damaged by exposure to oxygen. Full reversible oxidation with thionine involves the removal of six electrons. A third redox state of the MoFe protein can be reached only in the presence of the smaller nitrogenase component (the Fe protein), a suitable source of

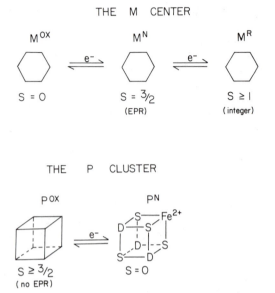

FIGURE 5. Schematic diagram of the redox centers of the MoFe protein of nitrogenase. Each of two M centers contains 6 Fe and each of four P clusters contains 4 Fe. No specific geometry is implied (adapted from [28]).

electrons (reductant), and excess free energy (MgATP). Under these conditions nitrogenase will fix nitrogen to ammonia, and the M centers have one more electron than in the native state.

The first saturation magnetization study of the MoFe protein is aimed at determining the spin of the S site iron atoms in the native MoFe protein. In the native state the two M^N centers each give an ESR signal characterized as having spin $S_M = 3/2$ with zero field splitting parameters $D_M = 5.5 \pm 1.5$ cm^{-1} and $(E/D)_M = 0.055$. Mössbauer spectroscopy has established that the native state of the P^N clusters is diamagnetic ($S_P = 0$) and that the S site iron atoms are ESR silent with integer spin ($S_S = 0, 1, 2, \ldots$).

Preliminary saturation magnetization data for native MoFe protein are given in figure 6. The molar magnetizations at four fixed temperatures (open square 15 K; closed circle 5 K; open circle 3.4 K; closed square 1.9 K) are plotted in SI units against $\mu_B H/kT$. The three labeled solid lines below the data curves are theoretical curves for the magnetization of the two M centers of native MoFe protein. These curves were calculated assuming $S_M = 3/2$, $D_M = 6$ cm^{-1}, and $(E/D)_M = 0.055$. The differences between these calculated curves and the data are due either to impurities or to the S site iron. We will proceed by studying this sample after it has been treated to remove more of the impurity iron. We will also study a

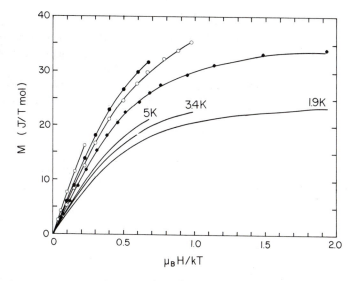

FIGURE 6. Saturation magnetization of the MoFe protein of nitrogenase in its native state. Saturation was observed at each of four fixed temperatures (open square 15 K; closed circle 5 K; open circle 3.4 K; closed square 1.9 K) by increasing the field to 5 T. The solid lines are the theoretical isotherms at the indicated temperatures for the two M centers ($S_M = 3/2$, $D_M = 6$ cm^{-1}, and $(E/D)_M = 0.055$).

second preparation of the native MoFe protein. The preliminary data of figure 6 rule out a spin of $S_S = 2$ as a possibility for the S site iron.

Saturation magnetization studies of nitrogenase represent a particularly challenging application of the superconducting susceptometer. The preparative techniques of biochemistry must be improved to limit the iron impurities in the sample. The successive measurement of the Mössbauer and ESR spectra of the susceptibility sample has proven essential. The groundwork has been laid for study of the magnetic properties of ESR silent states in metalloproteins. Day's study of the saturation magnetization of nitrogenase using the superconducting susceptometer in combination with Mössbauer and ESR is an example of a general new technique in biophysics. A remaining goal is to use information derived from low temperature saturation magnetization studies to design room temperature kinetics studies of complex enzymes such as nitrogenase.

Addendum

Substantial progress has been made in SQUID saturation magnetization studies of metalloproteins since this paper was written. The preliminary data of figure 6 have now been improved by deuterating the buffer and using a boron nitride sample holder in place of a delrin sample holder.

These procedures eliminate the effects of the very slow relaxation of spin $I = 1/2$ nuclei from the magnetization data [30,31]. Isofield data rather than isothermal data are now collected to avoid the field dependent effects of ferromagnetic impurities in the holder [31]. Magnetization studies have determined that the S site of the native MoFe protein is diamagnetic [31]. Significant results have been obtained for the Fe protein of nitrogenase [32].

References

[1] F. London, *Superfluids* (Dover, New York, 1961), Vol. 1, p. 8.

[2] B. S. Deaver, Jr., and W. M. Fairbank, *Phys. Rev. Lett.* **7**, 43 (1961).

[3] R. Doll and M. Näbauer, *Phys. Rev. Lett.* **7**, 51 (1961).

[4] J. M. Pierce, Ph.D. Thesis, Stanford University, 1967.

[5] E. P. Day, Ph.D. Thesis, Stanford University, 1972.

[6] J. S. Philo, Ph.D. Thesis, Stanford University, 1977.

[7] B. D. Josephson, *Phys. Lett.* **1**, 251 (1962).

[8] P. W. Anderson and J. M. Rowell, *Phys. Rev. Lett.* **10**, 230 (1963).

[9] R. C. Jaklevic, J. Lambe, A. H. Silver and J. E. Mercereau, *Phys. Rev. Lett.* **12**, 159 (1964).

[10] J. E. Zimmerman and A. H. Silver, *Phys. Lett.* **10**, 47 (1964).

[11] W. W. Webb, *IEEE Trans. Magn.* **8**, 51 (1972).

[12] R. P. Giffard, R. A. Webb and J. C. Wheatley, *J. Low Temp. Phys.* **6**, 533 (1972).

[13] J. S. Philo and W. M. Fairbank, *Rev. Sci. Instr.* **48**, 1529 (1977).

[14] J. S. Philo and W. M. Fairbank, *J. Chem. Phys.* **72**, 4429 (1980).

[15] R. K. Kunze, Jr., and E. P. Day, *J. Chem. Phys.* **72**, 5809 (1980).

[16] E. P. Day, *Phys. Rev. Lett.* **29**, 540 (1972).

[17] J. S. Philo, *Proc. Natl. Acad. Sci. USA* **74**, 2620 (1977).

[18] M. R. Beasley, Ph.D. Thesis, Cornell University, 1968.

[19] A. F. Hebard, Ph.D. Thesis, Stanford University, 1970.

[20] B. Cabrera, Ph.D. Thesis, Stanford University, 1975.

[21] M. F. Perutz, *Nature* **228**, 726 (1970).

[22] M. F. Perutz, *Sci. Am.* **239**, 92 (1978).

[23] M. F. Perutz, *Ann. Rev. Biochem.* **48**, 327 (1979).

[24] SHE Corporation, San Diego, CA.

[25]   M. J. Nelson, P. A. Lindahl and W. H. Orme-Johnson, *Adv. Inorg. Biochem.* **4**, 1 (1982).

[26]   E. Münck and R. Zimmermann, in *Mössbauer Effect Methodology*, edited by I. J. Gruvermann and C. W. Seidel (Plenum, New York, 1976), Vol. 10, p. 119.

[27]   R. Zimmermann, E. Münck, W. J. Brill, V. K. Shah, M. T. Henzl, J. Rawlings and W. H. Orme-Johnson, *Biochim. Biophys. Acta* **537**, 185 (1978).

[28]   B. H. Huynh, M. T. Henzl, J. A. Christner, R. Zimmermann, W. H. Orme-Johnson and E. Münck, *Biochim. Biophys. Acta* **623**, 124 (1980).

[29]   B. H. Huynh, E. Münck and W. H. Orme-Johnson, in *The Biological Chemistry of Iron*, edited by H. B. Dumford, D. Dolphin, K. N. Raymond and L. Sicker (Reidel, Boston, 1981), p. 241.

[30]   A. Roy, E. P. Day and D. M. Ginsberg, *Bull. Am. Phys. Soc.* **29**, 368A (1984).

[31]   E. P. Day, T. A. Kent, P. A. Lindahl, E. Münck, W. H. Orme-Johnson, H. Roder and A. Roy, (submitted to *Biophys. J.*).

[32]   P. A. Lindahl, E. P. Day, T. A. Kent, W. H. Orme-Johnson and E. Münck, *J. Biol. Chem.* **260**, 11160 (1985).

Superconducting Magnetometry for Biomagnetic Measurements

John P. Wikswo, Jr., and Mark C. Leifer

1. HISTORICAL INTRODUCTION

The Stanford biomagnetic research effort was conceived in the late 1960's when William Fairbank and John Madey discussed the possibility of using superconducting shielding and superconducting magnetometers being developed at Stanford to detect the magnetic field of the human brain. The advantages provided by the use of low temperatures were obvious. In the early 1960's, Baule and McFee [1] used a pair of million-turn pickup coils to detect the 100 pT magnetic field associated with the electrical activity of the human heart. The signal was termed the magnetocardiogram (MCG). The sensitivity of their measurements, made in a pasture on the outskirts of Syracuse, New York, was limited not by ambient noise but by the Johnson noise of their room temperature detector coils. With the advent of the SQUID magnetometer, measurements of the cardiac magnetic field were no longer limited by instrument sensitivity but by external magnetic noise [2]. Two approaches were taken to overcome this noise: the first was to use extensive magnetic shielding [2] to reduce ambient noise levels and the second was to use gradiometers or differential magnetometers that are insensitive to distant noise sources [3].

Both noise reduction techniques were used for biomagnetic measurements which John Wikswo undertook for his Ph.D. research at Stanford. The first step was the design and construction of a 13 meter deep, 2 meter diameter molypermalloy shield for a low magnetic field facility already under construction for use by several experiments. Inside this shield (figure 1) the geomagnetic field was reduced to 35 nT and the 60 Hz fields to less than 1 nT.

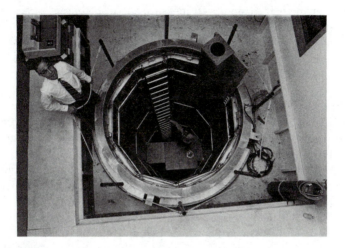

FIGURE 1. The 25 foot deep magnetic shield constructed at Stanford in 1973 for the MSPG and MCG measurements. At the left is W. M. Fairbank. John Wikswo is standing on a platform 6 feet above the bottom of the shield.

At this time, interest shifted from neuromagnetic measurements to an exciting new cardiovascular measurement. In 1971, Fairbank, James Opfer and Wikswo began exploring the various ways that magnetic susceptibility measurements, which had proven quite valuable for the study of biological molecules, could be applied to the study of human cardiovascular function *in vivo*. They realized that the displacement of blood by the pumping action of the heart and the replacement of this volume with lower density lung tissue would result in periodic changes in the diamagnetic susceptibility distribution within the torso. Application of an external magnetic field and measurement of the time-dependent susceptibility-induced changes in this field would provide a measurement of cardiac mechanical activity. While the mass redistribution is significant, on the order of hundreds of milliliters moving several centimeters back and forth every second, the susceptibility changes are small since both blood and cardiac muscle exhibit a diamagnetic susceptibility nearly that of water. In order to make magnetic measurements of cardiac volume changes it would be necessary to detect magnetic field changes eight orders of magnitude smaller than the applied magnetic field. With conventional susceptometers, a small sample is placed near the center of a strong magnet and the field changes are measured by a closely coupled magnetometer; the susceptometer for the cardiac measurements would have human beings for the samples and would pose new problems in sensitivity and noise rejection.

Bauman and Harris in Cleveland had encountered these problems a few years earlier in their attempts to measure the iron content of the liver [4]. They had successfully used a susceptometer having copper coils wound on an iron core with a mouse-sized air gap, but they concluded that it was unlikely that this type of susceptometer could be used at the human scale. They stopped work on their instrument, but continued their research in iron storage diseases in the liver.

At Stanford, cryogenic technology was found to be the key to successful whole-body magnetic susceptibility measurements. A superconducting magnetometer could measure the small signals which would be generated by even the largest magnet system. The required field stability could most easily be obtained with a persistent current magnet, and the requirement for mechanical stability of the magnetometer in the large applied field could be met by rigidly mounting the magnetometer and the magnet in the same dewar. Rejection of external magnetic noise would be provided by the use of a SQUID magnetometer whose pickup coil was configured as a long baseline (15 cm) gradiometer, and by the molypermalloy shield shown in figure 1. By 1973 Wikswo had constructed a superconducting magnetic susceptometer along these principles. It contained a persistent-current magnet that produced 10^{-2} T at the heart [5]. With the magnet energized, the SQUID gradiometer detected field changes associated with cardiac mechanical activity. The technique was eventually named magnetic susceptibility plethysmography (MSPG) in deference to accepted medical terminology. The gradiometer could also be used without the applied field to detect the MCG. Thus one instrument was capable of making magnetic measurements of both the mechanical and electrical activity of the heart.

Since the SQUID magnetometer, the magnetic shield and the data acquisition system developed for the MSPG could also be used to study the MCG, the project at Stanford was extended to include research on both signals. By the early 1970's, other researchers using SQUID magnetometers had begun to map the magnetic field produced by cardiac electrical activity [6]. Baule and McFee had described the theoretical differences between the spatial sensitivity of the electrocardiogram (ECG) and the MCG [7], and several people suggested possible arrangements of cardiac current sources that could be of interest [7-9]. Cohen and Kaufman demonstrated that the ability of a SQUID to measure the magnetic field from steady currents could be used to study the low frequency currents that flow in the heart during myocardial infarction [10]. Plonsey [11] pointed out that the electric and magnetic field equations derived by Geselowitz [12] to describe the ECG and MCG would allow the MCG to contain new, independent information about cardiac electrical activity not obtainable electrically. The practical question now facing magnetocardiography was whether new, useful information did in fact exist, and whether the spatial sensitivity differences described by Baule and McFee could have diagnostic importance in clinical cardiology.

The MCG research program at Stanford took a unique approach toward answering this question. While most of the previous MCG research had been based on measurements of only one component of the cardiac magnetic field, the MCG work at Stanford was directed toward understanding the vector nature of the MCG and its relationship to the vector electrocardiogram. The three components of the vector MCG were measured by rotating a tilted-coil gradiometer developed by Opfer at Develco, Inc. Mathematical techniques were devised by Wikswo and Jaakko Malmivuo to determine the magnetic dipole that best described the vector MCG. Finally, ongoing patient studies were begun to evaluate the clinical effectiveness of the MCG in diagnosing heart disease.

Biomagnetic research has continued for over a decade at Stanford as a joint project of the physics department and the cardiology division and has provided a clear understanding of the MSPG and an improved description of the MCG. The whole-body susceptometer introduced at Stanford has evolved through the efforts of Farrell, Harris and others at Case Western Reserve University into a clinical instrument for measuring liver iron content. Finally, work by Wikswo at Vanderbilt to understand the information content of the MCG and other biomagnetic measurements has led to the first measurements of the magnetic field of an isolated nerve and the development of miniature, high sensitivity magnetometers with room temperature pickup coils, thereby returning full circle to the early discussions between Madey, Wikswo and Fairbank, and to the general type of magnetometers first built by Baule and McFee. With this background, we can now proceed to discuss in detail the application of superconducting magnetometry to biomagnetic measurements.

2. BIOMAGNETIC SUSCEPTOMETRY

The origin of the MSPG signal can be explained by examining the behavior of a time-dependent distribution of magnetic material in an applied steady magnetic field. For small applied fields body tissue and blood are magnetically linear, isotropic and usually diamagnetic. Thus a uniform external magnetic field $\mathbf{H}_0$ induces a magnetization $\mathbf{M}(\mathbf{r}, t)$ throughout the body given by

$$\mathbf{M}(\mathbf{r}, t) = \chi(\mathbf{r}, t)\mathbf{H}_0 \qquad (1)$$

where $\chi(\mathbf{r}, t)$, the dimensionless volume susceptibility, has a magnitude of approximately -9.5×10^{-6} in SI units. If only the time-varying portion of the resultant field outside the body is examined, the field strength during the cardiac cycle is given by [13]

$$\mathbf{B}(\mathbf{r}, t) = \frac{\mu_0}{4\pi} \int_v \chi(\mathbf{r}, t) \left[\frac{3\mathbf{H}_0 \cdot (\mathbf{r} - \mathbf{r}')}{|\mathbf{r} - \mathbf{r}'|^5}(\mathbf{r} - \mathbf{r}') - \frac{\mathbf{H}_0}{|\mathbf{r} - \mathbf{r}'|^3} \right] d^3 r' \quad . \quad (2)$$

The integration need extend only over regions having time-varying susceptibility. For a fixed $\mathbf{r}$, $\mathbf{B}(\mathbf{r}, t)$ is the MSPG signal detected by a magnetometer at $\mathbf{r}$; integration by equation (2) for a specific susceptibility distribution constitutes a solution of the forward MSPG problem. The maximum change in $\chi(\mathbf{r}, t)$ is directly related to the net change in cardiac volume between systole (heart contracted) and diastole (heart relaxed)

$$\Delta\chi(\mathbf{r}) = \chi(\mathbf{r}, t_s) - \chi(\mathbf{r}, t_d) \quad (3)$$

where t_d and t_s are the times of diastole and systole, respectively. $\Delta\chi(\mathbf{r})$ is positive for some regions and negative for others, corresponding to a net decrease or increase of diamagnetic material at the point $\mathbf{r}$. If only the time-varying part of the field is desired, equation (2) can be used to give the MSPG signal amplitude $\Delta\mathbf{B}(\mathbf{r})$ in terms of $\Delta\chi(\mathbf{r}')$. These equations can be used with simple cardiovascular models to predict the MSPG waveform and to estimate various other contributions to the signal [14].

To extract clinically useful information from this measurement one must solve the inverse problem, that is, obtain a description of cardiac mechanical activity from MSPG signals. Typical of all such inverse problems, there is no unique solution. Instead, parameters describing a physiologically realistic model are adjusted to fit the data, and the values of these parameters yield information about cardiac performance. The ease with which this can be done is determined in part by the complexity of the model. Fortunately, even simple physiological models give reasonably good fits to recorded MSPG data [13].

The susceptometer used in the original MSPG measurements, shown schematically in figure 2, had a superconducting magnet with a highly nonuniform 10^{-2} T magnetic field [5,14]. Recent studies of the MSPG have placed both the patient and a simple SQUID differential magnetometer, such as is used to measure the MCG, within a large set of Helmholtz coils. While the uniformity of the field simplifies the mathematical analysis of the signals, motion of the magnetometer relative to the coils is a severe problem, and the applied fields are generally smaller than those obtainable with superconducting magnets. This in turn produces an MSPG signal that may be comparable in magnitude to the MCG. The distortion of the MSPG signal by the MCG can be removed by subtracting two sequential MSPG

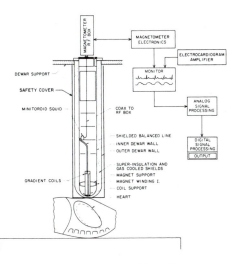

FIGURE 2. A schematic representation of the original MSPG susceptometer positioned over a subject's chest.

recordings made with oppositely applied fields so that the field-independent MCG signal is canceled. Boismier *et al.* [15] used a SQUID gradiometer inside a 1.3 m diameter Helmholtz set and examined MSPG signals in fields of 1.7×10^{-4} T. They showed rough agreement between measured data and a simple hemispherical model. Katila *et al.* [16] used a similar apparatus with 2 m coils to perform detailed mapping over the chests of five subjects. One of their typical MSPG traces, recorded directly over the ventricles of the heart, is reproduced in figure 3. The authors discussed at length the correlation between features of the MSPG and physiological events in the heart and great vessels. Using a spherical blood-mass model they were able to measure total cardiac volumes to within 20–30% of that obtained echocardiographically. This work represents the most accurate magnetic determination of cardiac volume changes made so far.

Since the MSPG signals are eight orders of magnitude smaller than the applied field, and the MCG is approximately seven orders of magnitude smaller than the magnetic field of the earth, there should be an observable susceptibility-related distortion of MCG signals recorded in the earth's magnetic field. Figure 4 shows an MCG recorded from a subject in the geomagnetic field, with arrows indicating the contribution of cardiac susceptibility changes to the signal. To avoid this distortion, measurements of the MCG should be made in ambient fields on the order of 5% of that of the earth. If care is taken to minimize power supply ripple and vibrational noise, the same coils used to apply the magnetizing fields needed for MSPG measurements could be used to null the geomagnetic field. Thus the MCG

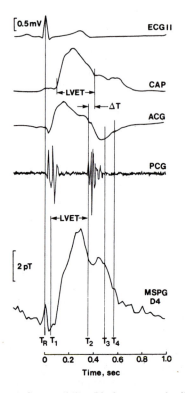

FIGURE 3. The Magnetic Susceptibility Plethysmography (MSPG) signal (lower trace) recorded directly over the ventricles in a 3.4×10^{-4} T uniform magnetic field by the Helsinki Biomagnetism Group. The MCG has been canceled by subtraction. Other physiological traces are the electrocardiogram (ECG), carotid artery pulse (ACG), and phonocardiogram (PCG). The interval LVET is the left ventricular ejection time. From Katila *et al.* [16].

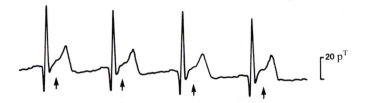

FIGURE 4. The magnetocardiogram (MCG) of a normal male subject recorded at Stanford in the geomagnetic field. The arrows point to the distortion caused by the MSPG signal.

and MSPG have essentially identical instrumentation requirements. Application of the MSPG technique to problems in clinical cardiology is awaiting a clear demonstration of the capabilities of the technique. Measurements of field changes perpendicular to the applied field are needed since models suggest that these changes contain information about the motion of the heart. More accurate models are needed for both the forward and the inverse problems so that the capabilities and limitations of the technique can be more clearly assessed.

Superconducting susceptometry has been successfully applied to the clinical detection of a class of liver diseases known as iron overload. As mentioned above, most body tissue is diamagnetic, with a small negative susceptibility. Iron is stored in the liver in the form of ferritin and hemosiderin, which are paramagnetic with large positive susceptibilities. In cases of pathologic iron overload, sufficient ferritin and hemosiderin are stored to increase dramatically the liver tissue susceptibility so that the entire liver appears paramagnetic. Harris *et al.* [17] were able to detect iron overload by simply scanning a SQUID gradiometer laterally across the torso while in the geomagnetic field. Figure 5 shows scans from this study, where a clear

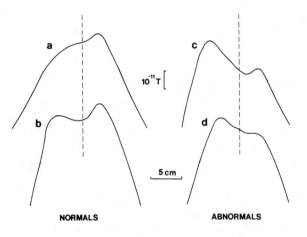

FIGURE 5. Torso scans of the magnetic susceptibility of two normal subjects (left) and two subjects with iron overload (right). Dotted lines represent torso midline; the liver is to the right in the traces. From the Case biomagnetism group (Harris *et al.* [17]).

depression of the signal occurs over the paramagnetic livers (right sides of the torsos) of the abnormal subjects. These investigators have recently constructed a superconducting susceptometer with a persistent current magnet [18,19]. The pickup coils of the gradiometer and the nonuniform field produced by the superconducting magnet are designed to limit the sensitivity of the instrument to a small region encompassing the liver. A novel water displacement technique allows accurate measurements of liver susceptibility relative to that of water by simply moving the susceptometer up to and away from the subject. The group is extending the technique to the more difficult problems of diagnosing iron deficiency, where the susceptibility of the liver tissue differs little from that of normal tissue, and of detecting iron overload in the heart.

An alternative approach to susceptometry utilizes low frequency alternating magnetic fields. This technique suffers from the problems of low field strength and vibrational noise discussed above and is unsuitable for dynamic measurements such as bloodflow. However, the ac technique offers noise reduction for static liver measurements by operating the SQUID above its 1/f noise regime, and by use of phase sensitive detection. Bastuscheck et al. [20] were able to record high quality liver scans in a noisy laboratory using an ac susceptometer operating at 23 Hz.

Regardless of which instrument design eventually proves most practical, biomagnetic susceptometry has already established itself as a viable noninvasive diagnostic tool. Considerations of cost, ease of signal interpretation, and information content relative to other diagnostic techniques will determine whether biomagnetic susceptometry becomes a widely accepted clinical tool.

3. MAGNETOCARDIOGRAPHY

In order to explain what we have learned about the MCG-ECG relationship, we must first describe the theory that governs the electric and magnetic fields associated with cardiac electrical activity. The analysis and calculation of the electric and magnetic fields produced by biological systems is involved because the voltage sources and currents are distributed throughout a conducting medium. In particular, the current in the law of Biot and Savart must include not only the current in the passive conductor, but also the current in the region that contains the bioelectric sources. In a very simple model of the heart, the electrically active region can be approximated by a cup-shaped wavefront that propagates outward at approximately 1 m/sec, as shown in figure 6a. This wavefront serves as the battery that drives the current in the passive conductor. Because bioelectric sources act more often like high impedance current sources than like low impedance voltage sources, we introduce an "impressed current density" $\mathbf{J}^i(\mathbf{r})$. The total current density at any point is the sum of the impressed current density (if the point is within an active region) and the passive current density that obeys Ohm's law

$$\mathbf{J}(\mathbf{r}) = -\sigma(\mathbf{r})\nabla V(\mathbf{r}) + \mathbf{J}^i(\mathbf{r}) \quad . \tag{4}$$

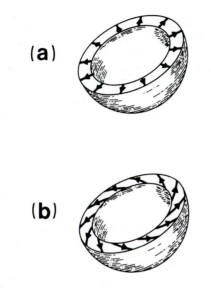

FIGURE 6. Two hypothetical models of the impressed current density in the cardiac activation wavefront that produces the ECG and the MCG. In (a), the impressed currents are perpendicular to the wavefront, in (b) they have a tangential component. The two sources would produce identical ECG signals, but differing MCG signals [22].

The impressed current density is a current dipole density, which is conjugate to the permanent polarization in electrets and the permanent magnetization in ferromagnets.

A pair of equations describe the electric potential and the magnetic field produced by impressed current sources in a finite, piecewise homogeneous conductor [12]

$$V(\mathbf{r}) = -\frac{1}{4\pi\sigma(\mathbf{r})} \int_v \frac{\nabla' \cdot \mathbf{J}^i(\mathbf{r}')}{|\mathbf{r}-\mathbf{r}'|} d^3r'$$
$$- \frac{1}{4\pi\sigma(\mathbf{r})} \sum_j \int_{S_j} (\sigma' - \sigma'') V(\mathbf{r}') \nabla' \frac{1}{|\mathbf{r}-\mathbf{r}'|} \cdot d\mathbf{S}'_j \tag{5}$$

$$\mathbf{B}(\mathbf{r}) = \frac{\mu_0}{4\pi} \int_v \frac{\nabla' \times \mathbf{J}^i(\mathbf{r}')}{|\mathbf{r}-\mathbf{r}'|} d^3r'$$
$$+ \frac{\mu_0}{4\pi} \sum_j \int_{S_j} (\sigma' - \sigma'') V(\mathbf{r}') \nabla' \frac{1}{|\mathbf{r}-\mathbf{r}'|} \times d\mathbf{S}'_j \tag{6}$$

where the summation is over all surfaces separating regions of differing conductivity. Equation (5) is difficult to solve analytically because the potential of interest appears not only on the left of the equal sign but also within the second integral on the right. In each equation, the first integral is over the impressed current, or the bioelectric sources, while the second integral accounts for the effects of inhomogeneities in the conducting medium. Given a description of the impressed currents, we can use equations (5) and (6) to calculate the ECG and the MCG, *i.e.*, we can solve the forward problem.

Note from these equations that the electric potential is determined by the divergence of the impressed currents, while the magnetic field is determined by their curl. There is an interesting, and somewhat controversial, observation that can be made about the impressed current distribution. If $\mathbf{J}^i(\mathbf{r})$ is an arbitrary vector field, it can be represented by a Helmholtz decomposition

$$\mathbf{J}^i(\mathbf{r}) = \mathbf{J}^i_F(\mathbf{r}) + \mathbf{J}^i_V(\mathbf{r}) \tag{7}$$

where

$$\nabla \times \mathbf{J}^i_F(\mathbf{r}) = 0 \quad \text{and} \quad \nabla \cdot \mathbf{J}^i_V(\mathbf{r}) = 0 \quad . \tag{8}$$

$\mathbf{J}^i_F$ has no curl and is called a flow field, while $\mathbf{J}^i_V$ has no divergence and is called a vortex field. In an infinite homogeneous conductor, represented by the first terms in equations (5) and (6), the electric potential is determined solely by $\mathbf{J}^i_F$ and the magnetic field is determined only by $\mathbf{J}^i_V$; if $\mathbf{J}^i_F$ and $\mathbf{J}^i_V$ are independent, then V and $\mathbf{B}$ will likewise be independent. In a conductor with boundaries or internal inhomogeneities, V is still determined solely by $\mathbf{J}^i_F$, but $\mathbf{B}$ now receives contributions from both $\mathbf{J}^i_V$ and $\mathbf{J}^i_F$ through the appearance of V in the second integral of equation (6). The magnetic field may still contain independent information, but this now depends upon the relative contributions of the first and second integrals in equation (6).

The relationship between the ECG and MCG signals that would be produced by the simple electrical source in figure 6a can be determined from equations (5) and (6). If we make the appropriate assumptions, this source reduces to a uniform electric double layer, so that the electric potential at any point in a homogeneous medium is proportional to the solid angle subtended at that point by the rim of the double layer. We have shown [21] that the magnetic field is given by a particular line integral over the rim of the double layer. Since the rim determines both signals, one can argue that the two signals contain the same information, but that they have differing spatial sensitivities. Thus this source is an example of one in which physiological constraints cause the flow and vortex sources to be directly related.

If we relax the assumption that the impressed currents are perpendicular to the cardiac activation wavefront, current sources such as the one in figure 6b can be considered [22]. This type of source might result from the spiral geometry of cardiac muscle fibers within the heart. Since this source has the same divergence as that in figure 6a, the two sources will produce identical electrical fields. The second source differs from the first only by a divergence-free vortex component of $\mathbf{J}^i$ parallel to the wavefront surface. This component will result in the two sources having differing curls and thus differing magnetic fields, so that only magnetic measurements could distinguish between such sources if they indeed exist in the heart.

A simple test for independent information in the MCG has been conducted at Stanford [21,23,24]. Sufficient simultaneous ECG and MCG data were recorded to allow determination of the components of the electric dipole $\mathbf{p}$ and the magnetic dipole $\mathbf{m}$ that describe the ECG and MCG. If the ECG and MCG were produced by the source in figure 6a in a spherical conductor, then one would expect that

$$\mathbf{m} = \frac{1}{2}\mathbf{r} \times \mathbf{p} \tag{9}$$

where $\mathbf{r}$ is the location of the center of the rim of the wavefront. This equation requires $\mathbf{m}$ and $\mathbf{p}$ to be perpendicular, as is seen to be approximately the case in the vector displays in figure 7. In a group of 20 normal subjects [24], the mean angle between the peak ECG and MCG dipole moments was found to be $77° \pm 21°$. This relationship has been confirmed by studies in France and Finland [25,26]. In subjects with abnormal cardiac activation, as shown in figure 7b, the relationship between the signals becomes more complicated. While the observed departures from the expected relationship could be due to sources like figure 6b, it is more likely that they result from the nondipolarity of complex sources based on figure 6a and from the irregular shape and internal inhomogeneities of the human body.

The ideal test of the relationship between the ECG and the MCG would be to record both signals and to attempt to predict one from the other. While it is possible to obtain beautiful MCG and ECG data, unfortunately the actual source is so much more complicated than our simple uniform double layer source that the modeling required for such a test is almost impossible. Therefore any attempts to identify independent information must first be performed on simpler systems. For this reason the biomagnetic work at Vanderbilt has been directed toward understanding the relationship between the electric and magnetic fields produced by single nerve axons, which can be readily modeled as a one-dimensional system. The electrical behavior of nerve axons can be described analytically to a high degree of accuracy, which allows detailed predictions of the magnetic field from electric data. Thus this system is ideal for examining the relationship between bioelectric and biomagnetic fields.

The principal features of the electric and magnetic fields surrounding an active nerve axon are diagrammed in figure 8a. The thin lines show one pair of a set of azimuthally symmetric current loops outside a nerve in a conducting medium [27]. The bands represent the magnetic field encircling the

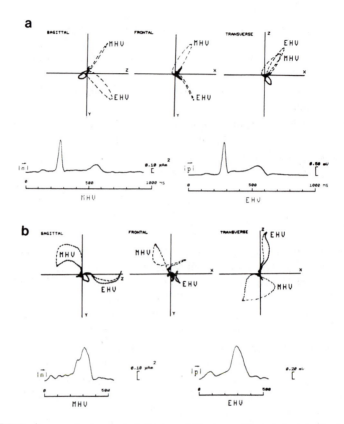

FIGURE 7. A computer-generated display of the magnetic and electric dipole models for the ECG and MCG, termed the magnetic heart vector (MHV) and electric heart vector (EHV), obtained from two subjects in a 1977 Stanford study [23]. The loop displays are projections of the trajectory of the tip of the MHV and EHV in three coordinate planes. The vector magnitudes are also shown. In the upper section, the data are for a representative normal subject and demonstrate the near perpendicularity of the two dipole moments at the time of their peak magnitude. The data in the lower section are from a subject with abnormal cardiac activation. The peak **m** and **p** vectors are not perpendicular.

nerve, and the arrows on the nerve axis are equivalent dipole current sources associated with the propagating action potential. The magnetic field at a distance r from the nerve is given by Ampère's law. Close to the surface of the nerve the current enclosed by a circular path around the nerve equals the current flowing internal to the membrane, and the magnetic field falls off linearly with distance from the nerve. As the distance from the nerve increases, an increasing fraction of the magnetic field from the internal current is cancelled by that from the return current in the external medium and the field falls off more steeply with distance, eventually as $1/r^3$.

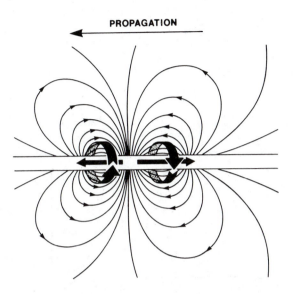

FIGURE 8a. A schematic representation of the electric and magnetic fields of a single nerve axon. The thin lines represent one pair of a set of azimuthally symmetric current loops outside a nerve in a conducting medium. The bands represent the magnetic field encircling the nerve, and the arrows on the nerve axis are equivalent dipole current sources associated with the propagating action potential [27].

We have developed SQUID magnetometers with room temperature pickup coils that can detect the magnetic field from isolated frog sciatic nerves [27–29], which are bundles of 2000 axons. The nerve is immersed in Ringer's solution and is threaded through the ferrite core of a toroidal pickup coil so that electrical current within the nerve will thread the toroid but the return current external to the nerve will not. The time-dependent 100 pT magnetic field produced by the cellular action current induces current in the pickup coil that is detected by the SQUID with signal-to-noise ratios of 10:1 in a 2 kHz bandwidth. The cryogenic environment required for the operation of the SQUID is provided by a small dip-probe to house the SQUID sensor and to allow its operation in a liquid helium storage dewar. We have recently developed a room temperature semiconductor amplifier that has sufficiently low noise $(0.4 \, \text{nV}/\text{Hz}^{1/2})$ and low input impedance $(1 \, \Omega)$ to replace the SQUID in measuring the currents induced in the toroidal coil by the neuromagnetic field [30]. We achieved the necessary amplifier performance by using 20 parallel input stages, each of which uses a precisely matched low-noise transistor pair. This amplifier was used to make the first measurements of the magnetic field of a single nerve axon [31], shown in figure 8b.

In order to obtain insight into the magnitude and interpretation of the magnetic field of nerves, we have developed a model that uses solutions of

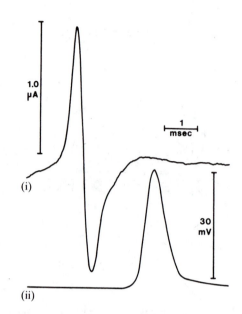

FIGURE 8b. The first measurement of the magnetic field of a single axon [31]. (i) A one-hundred sweep average of the output of the semiconductor current to voltage converter when the pickup coil was placed around an isolated medial axon from a 2 kg lobster. The peak is produced by the forward-directed axial depolarization current as it passes through the toroid, and the negative deflection is produced by the repolarization current. The initial slope of the signal is the response of the amplifier to the stimulus artifact. This can be avoided by inductive artifact cancellation. The slight overshoot after depolarization is the result of the low frequency cutoff of the toroids which can be corrected by frequency compensation. (ii) The transmembrane action potential recorded with a glass microelectrode 1 cm past the toroid. The delay between the signals results from the finite impulse propagation velocity.

Laplace's equation to obtain the magnetic field from the currents inside and outside the nerve and within the nerve membrane [32]. This calculation proves that extracellular measurements of the magnetic field can be used to determine intracellular currents. We have also examined a one-dimensional cable model, known as the core-conductor model, which demonstrates that extracellular magnetic measurements can be used to obtain the transmembrane action potential, the resistivity of the cytoplasm within the axon, and the time dependence of membrane conductivity without the use of microelectrodes.

These calculations have enabled us to make three very straightforward observations regarding the usefulness of magnetic measurements at the cellular level.

- Magnetic measurements provide a direct measurement of bioelectric currents without assumptions regarding resistivity.

- If magnetic and electric measurements are used to provide independent measurements of current and voltage, respectively, then the resistivity can be calculated directly.
- If practical limitations make the measurement of voltages by use of electrodes inconvenient or impossible, magnetic measurements of current and a knowledge of the resistivities can provide the desired voltage information.

Note that the first statement reflects Ampère's law, while the second is a form of Ohm's law! The first two observations might provide significant insight in those biophysical situations where resistivities are both important and difficult to characterize. One example is the complex electrical connections between cardiac cells that govern the spread of activation through the heart; another is the electrical connections between developing cells. The third observation may prove significant in the use of magnetometers to monitor the functional integrity of an entire *in vivo* human nerve bundle during surgery without damaging or puncturing it once it has been exposed, or in long term monitoring of nerves using implanted toroids. In addition to the situations covered by these observations, the development of more sophisticated instrumentation and experiments may allow identification of systems of cells which behave like the current sources in figure 6b. If these sources are to be found, they must be found magnetically since they are electrically silent.

4. CONCLUSION

The biomagnetic research at Stanford in the 1970's led to the development of superconducting susceptometers that are being used for noninvasive measurement of cardiac mechanical activity and liver function. The work on the vector magnetocardiogram has improved our understanding of the relationship between the electric and magnetic fields of the heart and has led to development of a new technique for studying current flow in isolated cells. William Fairbank provided the original stimulation and continuing enthusiasm for this work, and the authors wish to express their sincere gratitude.

Acknowledgments

A large number of people worked with W. M. Fairbank and the authors on the biomagnetic measurements at Stanford, most notably Drs. Donald Harrison, William Barry and Jerry Griffin of the Cardiology Division, Drs. James Opfer and Robin Giffard of the Physics Department, and Jaakko Malmivuo from Finland. The recent work at Vanderbilt has involved Dr. John Barach of the Department of Physics and Astronomy and Dr. John Freeman of the Department of Anatomy. The biomagnetic research effort at Stanford was originally supported by grants from the RANN program of the National Science Foundation and the Air Force Office of Scientific Research. It is presently supported by grants from the National Institutes of Health and the NASA Biomedical Applications Transfer Program. The biomagnetic research at Vanderbilt is supported in part by an Alfred P. Sloan fellowship and by Office of Naval Research contracts N00014-80-C-0883 and N00014-82-K-0107.

References

[1]   G. Baule and R. McFee, *Am. Heart J.* **55**, 95 (1963).

[2]   D. Cohen, E. A. Edelsack and J. E. Zimmerman, *Appl. Phys. Lett.* **16**, 278 (1970).

[3]   J. E. Zimmerman and N. V. Frederick, *Appl. Phys. Lett.* **19**, 16 (1971).

[4]   J. H. Bauman and J. W. Harris, *J. Lab. Clin. Med.* **70**, 246 (1967).

[5]   J. P. Wikswo, Jr., J. E. Opfer and W. M. Fairbank, *AIP Conf. Proc.* **18**, 1335 (1974).

[6]   D. Cohen and D. McCaughan, *Am. J. Cardiol.* **29**, 678 (1972).

[7]   G. Baule and R. McFee, *Am. Heart J.* **79**, 223 (1970).

[8]   D. Cohen, *Phys. Today* **28**, 34 (1975).

[9]   E. Lepeshkin, *Adv. Cardiol.* **10**, 319 (Karger, Basel, 1974).

[10]   D. Cohen and A. L. Kaufman, *Circ. Res.* **36**, 414 (1975).

[11]   R. Plonsey, *IEEE Trans. Biomed. Eng.* **BME-19**, 239 (1972).

[12]   D. B. Geselowitz, *IEEE Trans. Mag.* **MAG-6**, 346 (1970).

[13]   J. P. Wikswo, Jr., *Medical Physics* **7**, 297 (1980).

[14]   J. P. Wikswo, Jr., J. E. Opfer and W. M. Fairbank, *Medical Physics* **7**, 307 (1980).

[15]   G. Boismier, L. Hlatky and N. Tepley, *J. Appl. Phys.* **50**, 2490 (1979).

[16]   T. Katila, R. Maniewski, T. Tuomisto, T. Varpula and P. Siltanen, *IEEE Trans. Biomed. Eng.* **BME-29**, 16 (1982).

[17]   J. W. Harris, D. E. Farrell, M. J. Messer, J. H. Tripp, G. M. Brittenham, E. H. Danish and W. A. Muir, *Clin. Res.* **504A**, 26 (1978).

[18]   D. E. Farrell, J. H. Tripp, P. E. Zanzucchi, J. W. Harris, G. M. Brittenham and W. A. Muir, *IEEE Trans. Mag.* **MAG-16**, 818 (1980).

[19]   D. E. Farrell, J. H. Tripp, P. E. Zanzucchi, J. W. Harris, G. M. Brittenham and W. A. Muir, in *Biomagnetism*, edited by S. N. Erne, H.-D. Hahlbohm and H. Lubbig (de Gruyter, Berlin, 1981), p. 507.

[20]   C. M. Bastuscheck and S. J. Williamson, *J. Appl. Phys.* **52**, 2581 (1981).

[21]   J. P. Wikswo, Jr., J. A. V. Malmivuo, W. H. Barry, M. C. Leifer and W. M. Fairbank, in *Cardiovascular Physics*, edited by D. N. Ghista, E. Van Vollenhoven and W. Yang (Karger, Basel, 1979), p. 1.

[22]   J. P. Wikswo, Jr., and J. P. Barach, *J. Theoretical Biol.* **95**, 721 (1982).

[23]   W. H. Barry, D. C. Harrison, W. M. Fairbank, K. Lehrman, J. A. V. Malmivuo and J. P. Wikswo, Jr., *Science* **198**, 1159 (1977).

[24]   M. C. Leifer, J. C. Griffin, E. J. Iufer, J. P. Wikswo, Jr., W. M. Fairbank and D. C. Harrison, in *Biomagnetism*, edited by S. N. Erne, H.-D. Hahlbohm and H. Lubbig (de Gruyter, Berlin, 1981), p. 123.

[25]   B. Denis, J. Machecourt, C. Favier and P. Martin-Noel, *Ann. de Cardiologie et D'Angéiologie* **27**, 81 (1978).

[26]   P. K. Karp, T. E. Katila, M. Saarinen, P. Siltanen and T. T. Varpula, *Circ. Res.* **47**, 117 (1981).

[27]   J. P. Barach, J. A. Freeman and J. P. Wikswo, Jr., *J. Appl. Phys.* **53**, 4532 (1980).

[28]   J. P. Wikswo, Jr., J. P. Barach and J. A. Freeman, *Science* **208**, 53 (1980).

[29]   J. P. Wikswo, Jr., *Rev. Sci. Inst.* **53**, 1846 (1982).

[30]   J. P. Wikswo, Jr., P. C. Samson and R. P. Giffard, *IEEE Trans. Biomed. Eng.* **BME-30**, 215 (1983).

[31]   J. P. Wikswo, Jr., J. O. Palmer and J. P. Barach, *Bull. APS* **26**, 1223 (1981).

[32]   K. R. Swinney and J. P. Wikswo, Jr., *Biophys. J.* **32**, 719 (1980).

Digital Electronics Near Zero Kelvin

J. N. Hollenhorst

1. INTRODUCTION

An exciting new chapter is now being written in the application of super-conductivity. This is the development of a new digital integrated circuit technology based on Josephson junctions [1–6]. It now appears possible that Josephson circuits will be applied within the next decade to the construction of miniature computers with an order of magnitude better performance than possible with conventional electronic devices.

This is certainly a fitting topic for a book in honor of William Fairbank. The technology combines many of Bill's loves such as superconductivity, low temperatures and flux quantization. More than this, however, it is the kind of subject Bill enjoys. A back-of-the-envelope calculation shows performance advantages of many orders of magnitude. The approach is unconventional. It looks difficult. Let's try it!

The tremendous proliferation of digital computers over the past twenty years has led some to suggest that we are in the throes of a second industrial revolution. This development is due in large part to the steady progress made over the years in electronic technology. Most of us are familiar with "Moore's Law" which states that every year the number of components on a silicon integrated circuit doubles [7]. This progress, beginning in the late 1950's, continues unabated today with the number of components approaching one million. The most dramatic impact of this has been the miniaturization, cost reduction and performance increases for small computers. At the other end of the cost scale, however, the progress has been less dramatic. Until recently the performance of high speed "supercomputers" had been improving by about a factor of 2 every 2 years; this trend is now leveling off [8]. In fact, most of the recent advances in supercomputer performance have resulted from increasingly sophisticated computer architectures rather than from improved integrated circuits. The fundamental difficulty is the relative resistance of large, high speed electronic systems to miniaturization. Seymour Cray, the designer of many supercomputers [9], has identified the two biggest problems as the "thickness of the mat" and "getting rid of the heat" [10]. To explain these problems we must first understand the importance of miniaturization.

In any computer, the rate at which sequential operations can be performed is limited both by how fast signals propagate through individual logic gates and by how fast the signals propagate from point to point in the computer. If great care is taken in engineering, the signals will propagate at velocities close to the speed of light. This means that in a computer with a cycle time of 10 ns the longest total wiring length must be somewhat shorter than 3 meters and thus the computer itself must be considerably smaller than 3 meters in its maximum dimension. Indeed, if one contemplates a Josephson supercomputer with a cycle time of about 1 ns, it would require a package less than 10 cm on a side and occupying a volume of less than one liter!

The "thickness of the mat" refers to the physical size of the interconnection wiring in a supercomputer. The wiring complexity of such a computer is enormous and difficult to miniaturize. As wires are reduced in size they become more resistive and consequently less able to pass the increasingly high speed signals. In the IBM 3081 computer each circuit board is fabricated with 33 layers of wiring in order to accommodate all the high speed connections. The availability of superconductivity greatly alleviates this difficulty by allowing one to make nearly ideal microstrip transmission lines with dimensions 100 to 1000 times smaller than printed circuit wiring.

The most difficult problem is "getting rid of the heat" from a smaller computer package with hundreds of thousands of logic gates and many millions of memory bits. High speed logic circuitry is always obtained at the expense of increased power consumption per function. An important figure of merit is the switching energy of a logic gate, defined as the product of propagation delay and power consumption. In a given technology it is usually possible to trade off power dissipation for speed, but the switching energy stays roughly constant. The Cray 1 computer consumes approximately 100 kilowatts of regulated power and is cooled by circulating freon through pipes connected to each circuit board.

If the Cray 1 could be scaled down to a 1 ns computer with circuitry with the same switching energy, one would need to remove 1 megawatt of power from a 1 liter package! We will see that Josephson circuits can be made with power dissipation 10^5 times smaller than the circuits in the Cray 1. A 1 ns Josephson computer would dissipate only about 10 watts, an easy cooling job for liquid helium.

2. THE VIRTUES OF LOW TEMPERATURE

As readers of this volume are well aware, Bill Fairbank has always been an exponent of the virtues of low temperature. Before discussing the specific benefits of superconductivity in digital electronics, let us consider some fundamental reasons why we can expect improved electronic devices at low temperature. We will start with a fundamental but not very practical limit on digital circuits.

The term "information" has become almost a cliché in the computer field. In the study of communication theory developed by Shannon and others [11], information is given a precise definition. Its basic unit is the binary digit (bit), and the function of all digital systems is the processing of information. A familiar result of Shannon's concerns the inevitable energy cost of commmunication [12]. In the presence of thermal noise with temperature T the cost of transmitting one bit of information is always greater than $kT \ln 2$, where k is Boltzmann's constant. A similar result applies to information processing. When two binary signals are processed by a logic gate to produce a single binary output, information is irretrievably lost. The input signals cannot be reconstructed by examining the output. From the irreversibility of the basic logic operation Landauer, Keyes and others [13,14] have argued that each elementary logic operation is accompanied by the dissipation of an energy of at least $kT \ln 2$. While it is in principle possible to build a reversible computer which gets around this limitation, all practical digital circuits employ irreversible binary logic gates.

Present day circuits have switching energies which are many orders of magnitude larger than kT. It is interesting, though, that from very general thermodynamic arguments one finds a fundamental limit on switching energy that is about two orders of magnitude smaller at liquid helium temperature than at room temperature. In fact, some proposed Josephson logic gates come very close to the thermodynamic limit [15,16].

A more practical limitation is imposed by the need for nonlinearity in digital systems. Nonlinearity is required in order to avoid the degradation of logic signals as they propagate through a series of logic gates. Keyes [14] has argued that thermal noise will wash out any nonlinearity which occurs on a voltage scale smaller than kT/e, where e is the electron charge. This relationship is obeyed for semiconductor circuits where, for example, the nonlinearity of an ideal p-n junction is given by $I = I_0(\exp(eV/kT) - 1)$. Digital circuits must therefore operate at voltages large compared to kT/e.

In both Josephson technology and semiconductor technology the dominant contribution to switching energy is due to the charging of parasitic capacitances. This energy depends quadratically on the voltage since the energy stored in a capacitor is $\frac{1}{2}CV^2$. If we could scale the voltages down by a factor of 100 by going to lower temperature, we could achieve an improvement of 10^4 in switching energy. We will see that the typical voltages in Josephson circuits are about 1000 times smaller than in conventional semiconductor circuits. This is the principle reason for the near zero switching energy of Josephson circuits.

It is not uncommon for the bulk of the power supplied to a high speed integrated circuit (IC) to be consumed by the low impedance drivers needed to communicate signals off-chip. By the time signals get off-chip, the wires must be treated as transmission lines which require a power V^2/Z_0, where Z_0 is the characteristic impedance. It is difficult to make Z_0 bigger than about 100 ohms. Again we see the quadratic dependence on voltage. In Josephson technology, the signals are fast enough that even on-chip wiring must be treated as transmission lines. As we will see, however, the reduced voltages allow us to use matched transmission lines throughout without excessive power dissipation.

Let us conclude this section by listing a few ancillary advantages of low temperature operation. As conventional circuits are scaled to smaller sizes, one soon finds that the interconnect resistance becomes unacceptable. Low temperature operation allows much lower resistance for elemental conductors like copper and aluminum and, of course, zero resistance for superconductors. Another worry with IC scaling is the failure of conductors due to electromigration. This leads to the use of very thick conductors which are difficult to fabricate reliably. Since the problem gets worse as the third power of the linewidth [17], this is a very serious problem for very large scale integration (VLSI). Electromigration is a thermally activated process and is essentially nonexistent at 4 degrees kelvin. In fact, most other failure modes of integrated circuits such as corrosion and diffusion are thermally activated and thus virtually eliminated at low temperatures. This gives us reason to hope that a low temperature computer might be much more reliable than a conventional computer.

3. SUPERCONDUCTING ELECTRONICS

At temperatures below about 15 K, useful superconducting materials become available. Recently a set of superconducting components has been developed which in principle would allow a complete computer to be built. The active element in these components is the Josephson tunnel junction.

Many of the advantages of low temperature operation are not specific to Josephson technology. Indeed, it is possible to operate semiconductor devices at low temperatures and obtain improved performance. However, Josephson devices still have an advantage of about three orders of magnitude in switching energy over the best semiconductor devices ever reported, including both 0.25 μm MOSFET's [18] and the new high electron mobility GaAs FET's operated at 77 K [19].

3.1 Superconducting Microstrip Transmission Lines

Josephson circuits are connected to one another by superconducting wires which offer zero electrical resistance to dc currents. If a narrow superconducting strip is placed over a wide superconducting ground plane with an insulator between them, the resulting structure is a superconducting microstrip transmission line. If the dimensions are properly controlled, the line will have a well-defined characteristic impedance and will propagate high frequency signals with very little dispersion and dissipation.

Kautz [20] has analyzed the propagation characteristics of both normal and superconducting microstrips; some of his results are summarized in figure 1. In figure 1b the evolution of a one-picosecond pulse (full width at half maximum) is shown as it propagates down three types of microstrip line. All the lines are assumed to have a dielectric thickness of 2000 Å and a width much larger than this. The choice of 2000 Å is a reasonable estimate of what is required to make high speed, low crosstalk lines with linewidths comparable to those of semiconductor devices currently in production (~ 2 μm). For a copper line at room temperature, a line length of 1 mm appears unsuitable due to the attenuation and distortion of the pulse. Copper at 4 K is considerably better due to its lower resistance and might be usable to lengths of 1 mm. Niobium is easily usable to 1 cm lengths. While pulse widths of 1 ps are not actually required, these plots are indicative of the superior performance of 4 K microstrips and superconducting microstrips in particular. In figure 1a we see that a 100 ps pulse is unchanged after propagating 10 cm in a niobium line while being seriously degraded in a copper line even at low temperature.

3.2 The Josephson Tunnel Junction

The discovery of flux quantization by Deaver and Fairbank [21] and Doll and Näbauer [22] provided the first confirmation of the macroscopic quantum coherence of superconductors. A short while later Brian Josephson [23] predicted an effect which is a direct consequence of this coherence. Previously, Ivar Giaever [24] had observed the energy gap in superconductors by looking at the flow of current through a thin insulating barrier between two superconductors. The current flow in such a junction is dominated by the

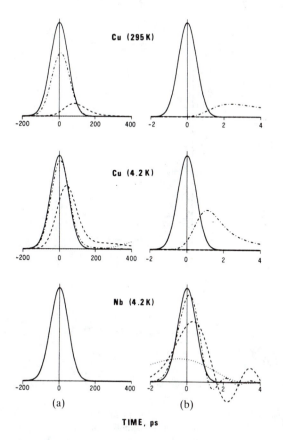

FIGURE 1. Propagation of pulses down a microstrip transmission line. (a) shows propagation of a 100 ps pulse (full width at half maximum). (b) shows a 1 ps pulse. Results are shown for nonsuperconducting copper lines at both 295 K and 4.2 K and for superconducting niobium at 4.2 K. The pulse shape is shown after the following propagation distances: zero (solid curve), 1 mm (dot-dash), 1 cm (dash) and 10 cm (dot). These curves are taken from Kautz [20].

quantum mechanical tunneling of electrons[1] through the potential barrier imposed by the insulating layer. The tunneling current depends on the density of states in the superconducting electrodes, which leads to a strong nonlinearity in the I-V characteristic. Figure 2a shows the current-voltage characteristics of this "Giaever tunneling" current. The most prominent feature is a pronounced jump in current at a voltage $V_g = 2\Delta/e$, where Δ is the energy gap in the superconducting electrodes.

---

[1] Actually, we are referring here to the tunneling of quasiparticles but for simplicity we call them electrons.

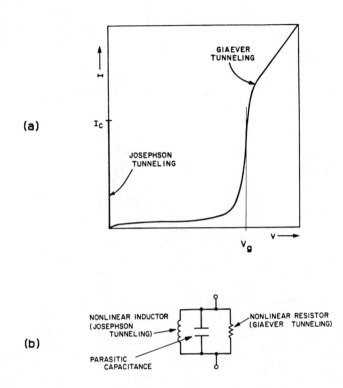

(a)

(b)

FIGURE 2. (a) *I-V* characteristic of a Josephson tunnel junction. The curve shows a superconducting branch due to Josephson tunneling and a normal branch due to Giaever tunneling. (b) Equivalent circuit of a Josephson junction. The Giaever tunneling is represented by a nonlinear resistor, the Josephson tunneling by a nonlinear inductor, and the displacement current in the tunneling barrier by a parasitic capacitance.

Josephson was curious about what effect the quantum coherence would have on this tunneling current. He discovered that in addition to the tunneling of electrons one should also observe currents of similar magnitude due to tunneling of Cooper pairs. He found that there is a discontinuity in the macroscopic quantum phase as one crosses the tunnel junction, and the magnitude of the pair current is related to the phase jump θ by the equation:

$$I = I_c \sin \theta \quad , \tag{1}$$

where the critical current I_c is a characteristic of the junction. He further showed that the voltage across the junction is proportional to the rate of change of the junction phase:

$$V = \frac{\Phi_0}{2\pi} \frac{d\theta}{dt} \quad . \tag{2}$$

Here $\Phi_0 = \frac{h}{2e}$ is the quantum of magnetic flux, so dear to the heart of Bill Fairbank. Equations (1) and (2) are the celebrated Josephson equations. We can eliminate the phase θ by writing:

$$V = L\frac{dI}{dt} \ ,$$

$$\text{with } L = \frac{\Phi_0}{2\pi I_c \, \cos\left[\frac{2\pi}{\Phi_0}\int V \, dt + \theta_0\right]} \ . \tag{3}$$

Thus the Josephson effect is equivalent to a nonlinear inductance. The inductance L depends on the time integral of the voltage across the junction. The constant θ_0 is the only remaining vestige of macroscopic quantization in equation (3); it is constrained by the condition of flux quantization.

An equivalent circuit of a Josephson junction is shown in figure 2b [25,26]. This equivalent circuit is called the resistively shunted junction model (RSJ) and seems to predict very well the behavior of today's digital Josephson circuits. The capacitance in the model arises from the very closely spaced electrodes in a tunnel junction. At sufficiently high frequencies and voltages the RSJ model breaks down and the behavior of Josephson junctions is considerably more complicated.

The Josephson effect produces a new piece in the I-V characteristic in figure 2a. Since the Josephson inductance has zero series resistance, it can support the flow of current at zero voltage just as a bulk superconductor can. It is clear from equation (1) that a junction can support a supercurrent only as great as the critical current I_c. This gives rise to the vertical branch in the I-V curve in figure 2a. If we set V equal to a constant nonzero value in equation (2), it is clear that θ must increase linearly. It follows from equation (1) that the supercurrent oscillates sinusoidally at a frequency ν proportional to the voltage, $\nu = V/\Phi_0$. The constant of proportionality is given by $\Phi_0^{-1} = 483$ GHz/mV. Even at the small voltages typical of Josephson circuits ($V \sim 1$ mV), the Josephson oscillation is near 10^{12} Hz. At these frequencies the Josephson current is almost completely shunted by the junction capacitance, and for most purposes we need only consider the Giaever current at finite voltages.

4. DIGITAL JOSEPHSON CIRCUITS

4.1 Fabrication

Digital Josephson circuits are fabricated in a manner very similar to the fabrication of semiconductor integrated circuits [27]. Superconductors, insulators and resistors are deposited on a suitable substrate and patterned using state-of-the-art photolithographic techniques. Tunnel junctions are formed by growing a thin native oxide barrier on the surface of one superconducting film and then depositing another superconducting film on top. The most critical step in fabrication is the formation of the junction since the critical current depends exponentially on the thickness of the oxide barrier (~ 30 Å) [28]. A change of several angstroms in the barrier thickness can lead to an order of magnitude change in the critical current. The favored superconducting materials at present are lead alloys, but there is a trend toward refractory materials such a niobium and niobium nitride.

4.2 Tunnel Junction as Digital Switch

Consider what happens if we try to put current into a Josephson junction which is loaded with a resistance R_L as in figure 3a. Referring to the I-V

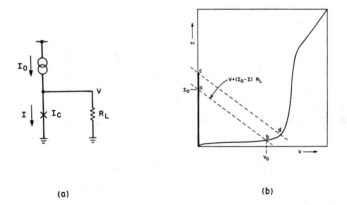

(a)   (b)

FIGURE 3. Biasing a Josephson junction as a current diverting switch. (a) The circuit diagram shows a junction represented by an X and biased by a current source I_0. The resistor R_L is the load resistance. (b) The I-V characteristic shows the stable operating points which must lie on both the load line (dashed line) and the junction characteristic (solid line). For a bias current I_0 the device is stable at either point "a" or "b."

characteristic in figure 3b, it is clear that with a current bias I_0 the voltage across the resistor must be somewhere on the load line $V = (I_0 - I)R_L$. Since the voltage must also lie on the I-V curve of the junction, there are two possible operating points "a" and "b" where the load line intersects the I-V curve. At point "a" the device is biased on the superconducting branch and has zero resistance. At point "b" it is biased on the nonlinear resistive branch and has a high resistance. The device is hysteretic since the past history of the bias determines which bias point is realized at any given time. If we bring up the current from zero, the junction will be in the zero voltage state until the critical current is reached at point "c." The device then switches to the finite voltage state at point "d" and will follow the Giaever tunneling characteristic until the bias current is reduced to zero.

The high and low impedance states "a" and "b" allow one to use a Josephson junction much like a switch which is either opened or closed. The switch opens when the critical current is exceeded and closes again when the current is reduced to zero. When the switch opens, the output voltage responds with a time constant $R_L C$, where C is the junction capacitance. In digital applications the state "a" may be used to represent a binary zero. In this state zero current flows in the load resistor. In state "b" almost all of the current flows in the load and this represents a binary one.

To design a logic gate we need to find a way to cause a junction to switch states when the proper input signals are applied. The most straightforward approach is to arrange things so that the inputs supply additional current to the junction causing the critical current to be exceeded. Another approach is to make a device whose critical current can be depressed below the bias current I_0 by applying the input signals. We will discuss an example of each of these approaches. Both yield latching gates in that the output will not change state when the inputs are removed. The gate must be reset by removing the bias current completely. While this latching logic complicates the design of digital circuits, it is not a serious shortcoming and, in fact, has certain advantages.

Before discussing a specific gate design, let us examine the performance of a tunnel junction as a switch. Some typical values for the circuit parameters in figure 3 are $I_c = 100$ μA, $V_0 = 1$ mV and $R_L = 10$ Ω. These figures are appropriate for devices made with a minimum feature size of about 5 μm, significantly larger than semiconductor units now in production. The voltage scale is set by the energy gap voltage $V_g = \frac{2\Delta}{e}$. In the BCS theory $V_g = 3.52$ $\frac{kT_c}{e}$, where T_c is the critical temperature of the superconducting electrodes. Since the operating temperature is typically close to T_c and the voltage close to V_g, Josephson devices work very close to the kT/e limit discussed above. For lead $T_c \sim 7.2$ K and $V_g \sim 2.7$ mV. When the junction is in the voltage state, the load dissipates a power $(100\ \mu\text{A})(1\text{ mV}) = 100$ nW. In a real circuit perhaps four times this much power would be dissipated in the supply bus for a total power per gate of about $P = 0.5$ μW. The switching speed of a gate is usually limited by the charging time of the junction capacitance. A typical capacitance value is about 1 pF, giving a risetime of about $(10\ \Omega)(1\text{ pF}) = 10$ ps. In an actual gate two junctions may have to be charged, giving a propagation delay $\tau = 20$ ps. The switching energy is $E = P\tau = 10^{-17}$ joules. This switching energy is about seven orders of magnitude smaller than that of a typical ECL (emitter coupled logic) gate in the Cray 1 computer. It is the combination of very high switching speed and very low power dissipation which makes Josephson circuitry attractive.

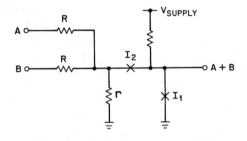

FIGURE 4a. A two-input JAWS gate. The circuit has two inputs A and B and an output A+B which is the logical OR of the inputs.

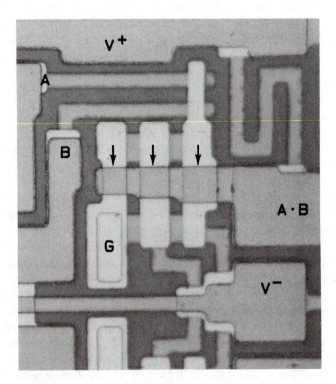

FIGURE 4b. Photomicrograph of a two-input JAWS AND gate which is slightly more complex than the OR gate shown above. The smallest lines in this photo are 4 μm wide.

4.3   JAWS Gate

Figure 4a shows a simple Josephson OR gate which switches by direct current injection [29,30]. In keeping with a long-standing tradition in the Josephson field, this gate is given the aquatic name JAWS (Josephson Atto-Weber Switch). If no inputs are applied, the supply current is shorted to ground by junction I_1 and the output is at zero volts. If either input A or B is turned on, a current will flow through I_2 and I_1, causing I_1 to switch to its finite voltage, high impedance state. This will cause most of the supply current to divert through I_2 and through the small resistor $r \ll R$ to ground. Now I_2 will switch, isolating the input from the output, and the supply current will be diverted to the output circuit where it may be fed to one or more downstream gates. Figure 4b is a photograph of a JAWS gate made in our laboratory using a lead alloy based technology.

4.4   SQUID Gate

Figure 5a depicts another type of Josephson OR gate called a supercon-ducting quantum interference device (SQUID) [31–34]. If two junctions are placed in parallel, the resulting circuit behaves like a Josephson junc-tion with twice the critical current. If some extra inductance is included in the circuit, we may couple a magnetic field into it with a transformer. In this case the current flowing in the parallel circuit will divide between the two junctions in a complicated way determined by the magnetic field, the circuit inductances and the nonlinear Josephson inductances. The net result is that the device can be designed so that a magnetic field will either increase or decrease the critical current of the circuit. A two-input OR gate can be made as in figure 5a by building an input transformer with two primaries and biasing in such a way that the device switches when either input is applied. Figure 5b shows the symbol we use to represent this gate, while figure 5c shows a photograph of an actual device. The two-input transformer is made by simply passing two microstrip lines over a wider microstrip line. The junctions may be seen as the two circular regions in this photomicrograph.

4.5   Terminated Lines

Figure 6 shows how the output of a SQUID gate is connected to the in-puts of several downstream gates. When the first gate switches, its supply currrent is diverted into the fanout line and flows through the input trans-formers of gates G1, G2 and G3. The inputs of the downstream gates are connected in series. Finally, the output line is terminated in the load re-sistor R_L. A great advantage of the Josephson technology is realized by setting the load resistor R_L equal to the characteristic impedance Z_0 of the

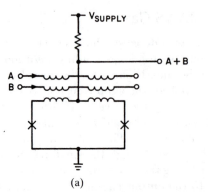

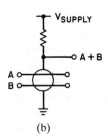

(a)

(b)

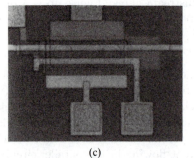

(c)

FIGURE 5. A two-input SQUID gate. (a) The circuit diagram shows a superconducting loop containing two junctions and a transformer. The transformer has two primaries A and B which are the inputs,to the gate. The output A+B is the logical OR of the inputs. (b) This schematic of the gate in 5a is used in figure 6. (c) The photomicrograph shows a gate as in 5a; the supply resistor is not shown. The transformer primaries consist of two microstrip transmission lines which pass over a wider line which is the transformer secondary. The two circular regions are the Josephson junctions which are formed where windows in a thick insulating layer allow two superconducting layers to contact each other separated only by the thin tunneling barrier.

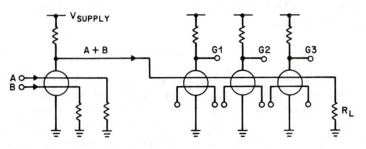

FIGURE 6. Connection of a SQUID logic gate to downstream gates. The output A+B of the first gate is connected in series to the inputs of gates G1, G2 and G3 and finally terminated in the matched load R_L.

fanout line. The fanout line is a superconducting microstrip transmission line, as discussed earlier. The output signal travels from the first gate to the terminating resistor R_L at nearly the speed of light. If the line were not terminated properly, a reflection would occur and propagate back to the gate where another reflection would occur. Under these circumstances, it would take many times the round trip travel time for the signal to settle to its steady state value. In semiconductor circuits, it is generally not possible to terminate circuits in this manner. The most important reason is that it consumes too much power. As mentioned previously, the characteristic impedance cannot easily be made greater than about 100 ohms, and the terminating resistor consumes a power V^2/Z_0. Since V^2 is typically about 10^6 times as great in semiconductor circuits and Z_0 approximately the same, we see that Josephson circuits have a clear advantage.

4.6 Josephson Memory Cell

One of the remarkable features of superconductivity is the infinite lifetime of a current circulating in a superconducting loop. This phenomenon may be used to store information as illustrated in figure 7. Imagine, as in the

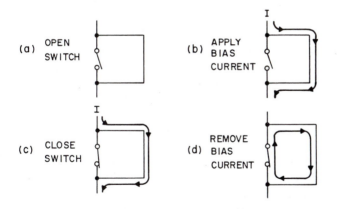

FIGURE 7. A simple superconducting memory cell. A persistent current is written into the loop by opening the switch (a), applying a bias current (b), closing the switch (c) and removing the bias current (d). This persistent current may be used to represent a logical "1." A "0" may be written by opening the switch and then closing it again with no bias applied.

figure, that we can construct a superconducting loop with a switch in it. If the switch is opened and a bias current applied, the current will flow on the right hand side of the loop as in figure 7b. If the switch is now closed, the current will continue to flow on the right as in figure 7c. If the bias current is removed, a circulating current will be trapped in the loop as in figure 7d. This persistent current may be used to represent a binary one. The absence of a persistent current would represent a binary zero. A zero is easily written by opening and then closing the switch with no bias applied. This is a nonvolatile memory in that it consumes zero power when the data are not being read or written. The information will be lost, of course, if the liquid helium boils away.

A practical memory cell may be made by replacing the switch in figure 7 with a SQUID write gate [35]. If a two-input AND gate is used, the data may be written by a coincidence of two control lines. This allows the storage cells to be placed into a rectangular matrix and accessed by means of horizontal and vertical control lines. A means of sensing the presence of current in a storage cell must also be provided. This may be accomplished by adding a SQUID read gate to each cell. This gate would perform a three-input AND function, switching only if the proper horizontal and vertical lines are selected and only if a persistent current is trapped in the loop. Figure 8 shows several cells from a memory array of this sort fabricated in our laboratory. The write gate is a three-junction SQUID, and read gate a two-junction SQUID.

5. A JOSEPHSON COMPUTER

A great deal of the pioneering work on Josephson junctions was performed at Bell Laboratories, beginning with the first observation of the Josephson effect by Anderson and Rowell [36]. Much of the subsequent technological development of Josephson digital electronics was performed at IBM Corporation. In this section I would like to describe a hypothetical Josephson computer as envisioned by the IBM group. Additional details may be found in the original papers [1,37]. All the basic elements of this computer have been demonstrated at IBM and at other laboratories in the United States, Japan and Europe. What remains to be shown is that these components can be produced reliably and economically and integrated into a complete functioning computer system.

The hypothetical computer is modeled after one of IBM's large mainframes, the IBM 3033. This computer is representative of high-end general purpose mainframes and not the highly vectorized scientific supercomputers such as the Cray 1 or CDC Cyber 205. Table 1 gives a comparison of some of the major performance features of the 3033 and the Josephson computer. The 3033 design was mapped circuit by circuit into IBM's Josephson and packaging technologies. The worst case propagation paths were then analyzed and used to estimate performance. This approach will tend to underestimate the performance of a real Josephson computer since the design is optimized for the characteristics of the 3033 circuit technology.

FIGURE 8. Photomicrograph of a Josephson memory array. Each cell employs a three-junction SQUID as a write gate and a two-junction SQUID as a sense gate. The write gate can be seen in the photograph as the device with three circular Josephson junctions. The storage loop is the rectangular conductor pattern which is interrupted on one side by the write gate and which passes directly over the sense gate on the other side.

A major accomplishment of IBM's has been the design of a miniature cryogenic packaging technology [38,39]. In their system, Josephson IC's are mounted on silicon circuit boards which in turn are mated to silicon microsocket boards. The microconnectors consist of 75 μm diameter pins spaced on 300 μm centers; these plug into mercury filled cavities in a silicon plugboard. Miniature connectors are necessary to prevent problems due to the parasitic inductance of the contacts. The connector density is more than 70 times greater than that available with a more conventional center to center spacing of 0.1 inches. All packaging components are made of silicon to prevent problems with differential thermal contraction and because of the extensive knowledge of silicon etching processes.

The hypothetical computer fits into a package measuring 15 cm × 30 cm × 20 cm (a volume of 9 liters). About 98% of this volume is occupied by the main memory of 512 million bits. This is eight times the amount of

TABLE 1. PERFORMANCE COMPARISON OF IBM 3033 AND
HYPOTHETICAL JOSEPHSON COMPUTER.

	IBM 3033	Josephson
Number of logic circuits	200,000	200,000
Cycle time	57 ns	4 ns
Cache memory capacity (bits)	512 K	512 K
Cache memory access time	57 ns	0.5 ns
Main memory capacity (bits)	64 M	512 M
Main memory access time	285 ns	15 ns
Computation rate*	4.9 MIPS	70 MIPS
Regulated power	50 kW	12.5 W
Total power	190 kW	15 kW
Critical path length	850 cm	54 cm
Total system volume	16 m³	1.2 m³

* Measured in millions of instructions per second (MIPS).

memory in the 3033. It is estimated that this amount of additional memory would be required because of the wider gap in access time between the main memory and mass storage disk memory.

The Josephson computer can be run with a 4 ns cycle time with 60% of the delay due to the propagation distance. Since the cycle time is a factor of about 15 higher in the Josephson computer, its performance should be about 70 MIPS. The total regulated power consumption is estimated to be about 12.5 watts compared to 50 kW for the 3033. Again, the bulk of this dissipation is in the eight times larger memory. A closed cycle cryocooler to provide this cooling power would consume 15 kW. This should be compared to the total power consumption of 190 kW for a 3033, including power for cooling and airconditioning. The total volume of the Josephson computer including cryostat, cryocooler and compressor is 1.2 m³ compared to 16 m³ for the 3033.

In summary, a Josephson computer would have fifteen times the throughput and eight times the memory, and would consume thirteen times less total power and 4000 times less regulated power while occupying thirteen times less volume than an IBM 3033.

6.   CONCLUSION

The application of superconductivity to digital electronics has resulted in devices with impressive performance. Gate delays below 20 ps are routinely achieved with conservative lithographic design rules [40]. Power dissipation is measured in microwatts per gate. No transistor circuits have come close to duplicating this combination of characteristics. The technology seems most suited to the needs of very large electronic systems whose miniaturization is limited by problems of heat removal and wiring complexity. The technology has progressed to the point where all components necessary for a large computer have been demonstrated. Chips as complex as a 12×8-bit parallel multiplier have been successfully fabricated [41]. It is not yet clear whether Josephson circuits can be manufactured and tested cost-effectively. Some of the problems which need further work are nonuniformity and nonreproducibility of junction critical currents, junction failure during temperature cycling, and trapping of magnetic flux in superconducting devices. Further work is also needed toward economical testing of circuits.

It is my belief that in 1992 when we are celebrating Bill Fairbank's 75th birthday few will doubt the advantages of low temperature operation of large computers. It is my hope that the first Josephson computers will be entering the market place at that time. Perhaps our biggest worry then will be occasional errors due to magnetic monopoles passing through the computer!

References

[1]   W. Anacker, *IEEE Spectrum* **16**, 26 (1979).

[2]   J. Matisoo, *Sci. Am.* **242**, 50 (1980).

[3]   W. Anacker, *IBM J. Res. Dev.* **24**, 107 (1980).

[4]   J. Matisoo, *IBM J. Res. Dev.* **24**, 113 (1980).

[5]   T. R. Gheewala, *IEEE Trans. Electron Devices* **ED-27**, 1857 (1980).

[6]   H. H. Zappe, *IEEE Trans. Electron Devices* **ED-27**, 1870 (1980).

[7]   G. Moore, *IEEE Spectrum* **16**, 30 (1979).

[8]   R. Sugarman, *IEEE Spectrum* **17**, 28 (1980).

[9]   Seymour Cray is the designer of the CDC 6600, CDC 7700, CDC 8600, Cray 1 and Cray 2 computers.

[10]   I. E. Sutherland and C. A. Mead, *Sci. Am.* **237**, 210 (1977).

[11]   C. E. Shannon, *Bell Syst. Tech. J.* **27**, 379 (1948).

[12]   A. D. Wyner, *Proc. IEEE* **69**, 239 (1981).

[13] R. W. Landauer, *IBM J. Res. Dev.* **5**, 183 (1961).

[14] R. W. Keyes, *Proc. IEEE* **69**, 267 (1981).

[15] K. K. Likharev, *IEEE Trans. Magn.* **MAG-13**, 245 (1977).

[16] S. Shin, T. Moriya, F. Shiota and K. Hora, *Jap. J. Appl. Phys.* **10**, 2011 (1979).

[17] C. A. Mead and L. Conway, *Introduction to VLSI Systems* (Addison-Wesley, 1980), p. 52.

[18] D. L. Fraser, Jr., H. J. Boll, R. J. Bayruns, N. C. Wittwer and E. N. Fuls, Seventh European Solid State Circuits Conference, Freiburg, Germany (1981).

[19] C. Cohen, *Electronics* **54**, 73 (1981).

[20] R. L. Kautz, *NBS J. Res.* **84**, 247 (1979).

[21] B. S. Deaver, Jr., and W. M. Fairbank, *Phys. Rev. Lett.* **7**, 43 (1961).

[22] R. Doll and M. Näbauer, *Phys. Rev. Lett.* **7**, 51 (1961).

[23] B. D. Josephson, *Phys. Lett.* **1**, 251 (1962).

[24] I. Giaever, *Phys. Rev. Lett.* **5**, 147 (1960).

[25] W. C. Stewart, *Appl. Phys. Lett.* **12**, 277 (1968).

[26] D. E. McCumber, *J. Appl. Phys.* **39**, 3113 (1968).

[27] J. H. Greiner, C. J. Kirchner, S. P. Klepner, S. K. Lahiri, A. J. Warnecke, S. Basavaiah, E. T. Yew, J. M. Baker, P. R. Brosious, H.-C. W. Huang, M. Murakami and I. Ames, *IBM J. Res. Dev.* **24**, 195 (1980).

[28] S. Basavaiah, J. M. Eldridge and J. Matisoo, *J. Appl. Phys.* **45**, 457 (1974).

[29] T. A. Fulton, S. S. Pei and L. N. Dunkleberger, *Appl. Phys. Lett.* **34**, 709 (1979).

[30] T. A. Fulton, S. S. Pei and L. N. Dunkleberger, *IEEE Trans. Magn.* **MAG-15**, 1876 (1979).

[31] R. C. Jaklevic, J. Lambe, J. E. Mercereau and A. H. Silver, *Phys. Rev.* **140**, A1628 (1965).

[32] T. A. Fulton, L. N. Dunkleberger and R. C. Dynes, *Phys. Rev.* **B6**, 855 (1972).

[33] D. L. Stuehm and C. W. Wilmsen, *Appl. Phys. Lett.* **20**, 456 (1972).

[34] H. H. Zappe, *Appl. Phys. Lett.* **27**, 432 (1975).

[35] S. M. Faris, W. H. Henkels, E. A. Valsamakis and H. H. Zappe, *IBM J. Res. Dev.* **24**, 143 (1980).

[36] P. W. Anderson and J. M. Rowell, *Phys. Rev. Lett.* **10**, 230 (1963). U.S. Patent 3,281,609 filed on 1/17/64 and issued on 10/25/66.

[37] M. J. Marcus, *Research Report RC-8010*, (IBM Thomas J. Watson Research Center, Yorktown Heights, New York, 1979).

[38] A. V. Brown, *IBM J. Res. Dev.* **24**, 167 (1980).

[39] M. B. Ketchen, B. J. van der Hoeven, J. Matisoo, J. H. Greiner, D. J. Herrell, R. H. Wang, R. W. Guernsey, C. J. Anderson, P. C. Arnett, S. Berman, H. R. Bickford, A. A. Bright, P. Geldermans, T. R. Gheewala, K. R. Grebe, H. C. Jones, M. Klein, S. P. Klepner, P. A. Moskowitz, M. Natan, S. Puroshothaman, J. Sokolowski, J. W. Stasiak, D. P. Walkman, A. J. Warnecke, C. T. Wu and T. Yogi, *IEEE Electron Device Letters* **EDL-2**, 262 (1981).

[40] S. S. Pei, *Appl. Phys. Lett.* **40**, 739 (1982).

[41] T. A. Fulton and L. N. Dunkleberger, *Bell System Tech. J.* **61**, 931 (1982).

CHAPTER V

Fundamental Particles

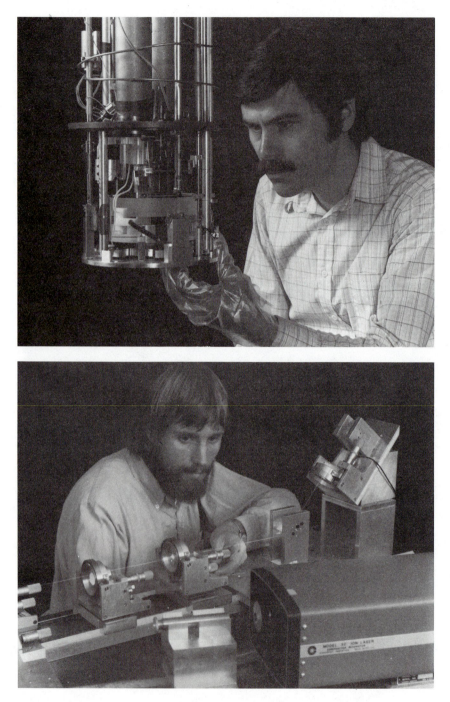

(*Top*) James Phillips adjusting the fractional charge apparatus. (*Bottom*) William Fairbank, Jr., aligning the single atom detection apparatus.

Fundamental Particles: Introduction

W. Peter Trower

Upon meeting William Fairbank you notice his intense interest in what he is doing. Soon you feel a strong pull as he tries to connect these interests with what you are doing. This Fairbank characteristic is responsible for fundamental particles being an important part of the life work of this low temperature physicist. Now high energy physics has contact with super-conductivity, as Fermilab's Tevatron accelerator attests. But Fairbank has not only provided particle physics devices, he has attacked basic issues in the field.

The search for the elementary structure of matter pauses temporarily at what appears to be indivisible entities. Earth, chemical elements, atoms, nuclei and nucleons were regarded as irreducible, only to be revealed as composite. At each step the constituent mass and extent diminished in magnitude, relentlessly approaching near zero, the recurrent theme of Fairbank's work.

The current myth has not only nucleons made up of constituents (three colored quarks of a presumed six flavors) but also mesons (two same-colored quarks, antiparticles of each other). This idea has utility and therefore acceptance; the existence of every known particle is explained [1] and their interactions are understood. However, this beautiful model states that quarks cannot be isolated, a prohibition that flies under the rubric of quark confinement. Now "cannot" is a red flag to Fairbank; it motivates him to thought, conversation and action, all with ornery skepticism.

This chapter's initial contributions describe experiments which focus on the question of quark confinement. First, S. J. Freedman and J. Napolitano describe their fruitless search which is exemplary of the many conducted at accelerators and with cosmic rays [1]. Then present and former students of Fairbank review their decade of evidence for fractional charge resident on some levitated superconducting niobium balls. Their experiments are uncorroborated by other levitation experiments [2], but the techniques and the nature of the balls are quite different. A datum discarded by Millikan during his demonstration of charge quantization is intriguing [3]. Most encouraging would be a blind test [4] yielding results consistent with those discussed in this chapter [5].

The Fairbank obsession with getting near zero has been visited upon his biological and intellectual progeny alike, as seen in the next paper. Here son William, although steering a determinedly independent course, exhibits the paternal influence. He discusses a method he pioneered [6] with which he studies a near zero population of atoms to detect a few which might be quarked.

Robert Wagoner's contribution follows, in which he discusses the mechanism whereby free fractional charges might be produced in the early universe and still be present in stable matter. Once again Fairbank's laboratory work attains cosmological significance.

A Fairbank student, Blas Cabrera, developed an ingenious technique to produce near zero magnetic fields in large volumes, which he describes in paper III.5. That technique allowed the important problems described in the next two papers to be addressed.

Magnetic monopoles, by their absence, capture the attention of physics students first encountering Maxwell's formulation of electromagnetism. That monopoles might be theoretically *desirable* was established in Dirac's elegant conjecture [7]. Indeed, monopoles might be *indispensable* if spontaneously broken non-Abelian gauge theories unify the electroweak and strong interactions [8]. Cabrera, M. Taber and S. Felch review their discovery of a candidate monopole event and discuss their plans to experimentally pursue this lead. In the intervening time, neither Cabrera and coworkers nor some two dozen other groups throughout the world have seen another such candidate. The original event survives unimpugned but, with the passage of time, it is clear that "one" becomes increasingly near zero.

The demonstration of an electric dipole moment, be it in nucleus or particle, would be direct evidence for the violation of time reversal invariance, once a sacrosanct physical principle. Fairbank's early interest in this problem has resulted in the construction of a ^{3}He gyroscope, which Taber and Cabrera describe in the final paper of this chapter.

Despite all it contains, this chapter on fundamental particles does not provide a complete picture of Fairbank's contribution to particle physics.

The positron-electron free fall experiment, discussed in paper VII.2 by James Lockhart and F. C. Witteborn, bears on the essential distinction between particle and antiparticle. Fairbank's early involvement with liquid helium bubble chamber development [9] is not recounted here. Stanford's superconducting particle accelerators, the latest of which is described in paper IV.2 by H. A. Schwettman, have long held Fairbank's interest.

For someone of talent and energy whose exclusive attention was directed to particle physics, the foregoing would constitute an impressive career at his sixty-fifth birthday celebration. But that this field is only one of many in which Fairbank has worked provocatively and productively demonstrates the extraordinary talents of the man.

References

[1] C. G. Wohl, R. N. Cahn, A. Rittenberg, T. G. Trippe, G. P. Yost, F. C. Porter, J. J. Hernandez, L. Montanet, R. E. Hendrick, R. L. Crawford, M. Roos, N. A. Törnqvist, G. Höhler, M. Aguilar-Benitez, T. Shimada, M. J. Losty, G. P. Gopal, Ch. Walck, R. E. Shrock, R. Frosch, L. D. Roper, W. P. Trower and B. Armstrong, [Particle Data Group], *Rev. Mod. Phys.* **56**, S1 (1984).

[2] M. Marinelli and G. Morpurgo, *Phys. Rep.* **85**, 161 (1982); and D. Liebowitz, M. Binder and K. O. H. Ziock, *Phys. Rev. Lett.* **50**, 1640 (1983).

[3] R. A. Millikan, *Phil. Mag.* **19**, 209 (1910).

[4] L. W. Alvarez, private communication.

[5] W. M. Fairbank and J. D. Phillips, in *Inner Space/Outer Space*, edited by E. W. Kolb, M. S. Turner, D. Lindley, K. Olive and D. Seckel (U. Chicago Press, Chicago, 1985), p. 563.

[6] W. M. Fairbank, Jr., T. W. Hänsch and A. L. Schawlow, *J. Opt. Soc. Am.* **65**, 199 (1975).

[7] P. A. M. Dirac, *Proc. Roy. Soc. London A* **133**, 60 (1931).

[8] G. 't Hooft, *Nucl. Phys. B* **79**, 276 (1974) and **105**, 538 (1976); and A. M. Polyakov, *Sov. Phys.-JETP* **20**, 194 (1974).

[9] W. M. Fairbank, E. M. Harth, M. E. Blevins and G. G. Slaughter, *Phys. Rev.* **100**, 971 (1955).

Limits on Fractionally Charged Particle Production from Cosmic Rays and Accelerators: How Near Zero?

S. J. Freedman and J. Napolitano

1. INTRODUCTION

Understanding the implications of the evidence for fractionally charged matter provided by the experiments of Fairbank, Hebard, LaRue and Phillips [1] represents a challenge to workers in many fields of physics. One tempting explanation is that free quarks have been discovered. However current theoretical arguments support the idea that quarks are permanently confined within hadrons [2]. Thus we are prevented from ever observing their one-third integer charge. Moreover, numerous experimental attempts to produce quarks at accelerators or to observe them in cosmic rays have produced no convincing evidence that they exist free.

How significant are these failures to discover free quark production? This question is hard to answer. No available theory of free quarks lends itself to reliable calculation, so it is not possible to make a strong case that the null experiments are incompatible with the positive signal detected by Fairbank *et al.*

It is important to improve experimental sensitivities whenever possible, but it is even more important to investigate untested mechanisms for quark production. This last approach concerns us here. We discuss recent experiments that have attempted to find quarks that may have escaped detection in previous searches.

We make no attempt at completeness. An exhaustive review of quark searches up to 1977 has been published by Jones [3], and other reviews exist [4]. After a brief introduction to the most common experimental method, we discuss recent developments in the hunt for fractionally charged particles in cosmic rays and accelerators.

2. EXPERIMENTAL METHOD

Nearly all particle physics quark searches utilize measurements of energy loss to infer the electric charge of particles. The basic physics is summarized by the Bethe-Bloch formula [5] which relates the average ionization energy loss $(dE/dx)_{av}$ to the particle's electric charge (Q) and velocity $\beta(\beta = v/c)$:

$$\left(\frac{dE}{dx}\right)_{av} = \frac{Q^2}{\beta^2} f(\beta) \quad . \tag{1}$$

Here $f(\beta)$ is a slowly varying function which depends upon the material in which the energy is deposited. At high energies $(dE/dx)_{av}$ becomes insensitive to energy over a large range of energies and has a minimum value. The energy loss again rises at the highest energies ("relativistic rise"). The extent of this rise depends on the details of the materials. Most experiments have relied on the "minimum ionization" region where energy loss is essentially directly a measure of charge. Of course, the sign of the charge is not determined.

The shape of the energy loss spectrum reflects the statistics of the energy loss process [6] and the statistics of the detector (*e.g.*, photon-electron statistics in a plastic scintillator). Figure 1 shows a typical energy loss spectrum for a thin (2.54 cm thick) plastic scintillator using normal single particle cosmic rays.

The typical quark search experiment consists of many dE/dx counters (plastic scintillators, proportional chambers, *etc.*), and some include provisions for measuring particle velocities or momenta.

3. COSMIC RAY QUARK SEARCHES

Cosmic ray searches can be divided into two categories: (a) "single particle" searches in which charge is measured when one particle passes through the apparatus, and (b) "air shower" searches which attempt to

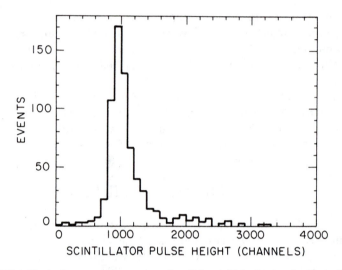

FIGURE 1. Typical energy loss spectrum in a 2.5 cm thick plastic scintillator for single particle cosmic rays. The data are taken from [15].

detect quarks within the core of an air shower, or as massive, slow particles delayed with respect to the air shower front. Positive results have been occasionally reported but none have stood up against subsequent scrutiny.

In general, cosmic ray searches place an upper limit on the flux of cosmic ray quarks with $Q = 1/3$ or $Q = 2/3$ at the level of about 10^{-10} (cm^2 sr sec)$^{-1}$ for both single particle and air shower experiments. The significance of the limit depends on several properties of the hypothetical quark. In general, quarks must be penetrating, relativistic, and spatially separated from other cosmic ray particles to be detected.

If quarks are produced only in extremely violent collisions, they might appear in high energy air showers. The well-known experiment of McCusker *et al.* [7] yielded a positive result, but the conclusion was soon criticized [8] and is not generally accepted. Another idea [9] is to separate quarks from the rest of the air shower by using the atmosphere as an extended absorber. This suggests that single particle searches performed at large zenith angles may prove fruitful. Until recently, only three such experiments were reported. Two [9,10] had rather low sensitivity; the third [11] was restricted only to particles with charge 1/3 in the minimum ionization region.

Yock [12] pointed out that quarks with low velocity had not been thoroughly examined, except as delayed particles following air shower fronts [3]. Using a range telescope with time of flight aimed vertically, Yock found three fractional charge candidates. However, continued running yielded no candidates when events with spurious spark chamber hits were ignored [13,14].

Recently, an investigation of fractionally charged cosmic rays [15] has been carried out using a highly redundant detector originally designed for e^+e^- annihilation physics at the PEP colliding beams facility at Stanford Linear Accelerator Center. The detector was sensitive over a large range of velocities and was oriented toward large zenith angles. The detector consisted of two symmetric arms, each containing 12 scintillation counter hodoscopes, 4 multiwire proportional chambers, and a layer of lucite Cerenkov counters. The scintillation counters measured dE/dx and single particle tracks were reconstructed with the proportional chambers. Five of the scintillation counter layers provided time of flight information and thus measurements of particle velocities. The acceptance was 4×10^3 cm^2 sr for zenith angles between 45° and 90°. No evidence for fractional charge was observed and the corresponding limit (90% confidence limit) on the quark flux was $\sim 3 \times 10^{-10}$ cm^{-2} sr^{-1} sec^{-1} for velocities between 0.10 c and c. (It is assumed that low velocity quarks were able to penetrate through the apparatus, and therefore were massive [15].)

Bjorken and McLerran [16] and McCusker [4] have investigated the possibility that the "Centauro" events may be due to quarks tightly bound to nuclei, forming a new state of matter. Centauro events are characterized by very high primary energies (hundreds of TeV) and large hadron multiplicities (50–100) but few photons, implying few π_0's. One explanation is that they are due to the breakup of large nuclei which, because they are bound to quarks, have anomalously long path lengths in the atmosphere. The conclusions are, of course, very speculative. Events of this type have not appeared so far in recent studies of high energy $p\bar{p}$ collisions at CERN (Geneva, Switzerland) in which the center of mass energy is 540 GeV [17]. This leaves open the possibility that massive nuclei bound to quarks are responsible in cosmic rays.

4. ACCELERATOR SEARCHES

Quarks have been sought in many different experiments in most of the world's accelerators. We discuss here some of the more recent experiments in previously untested areas.

By far the majority of accelerator quark searches used hadron beams on hadron targets. In this category, experiments attempt to discover quarks using the highest available center of mass energies. Results are often expressed as limits on the number of quarks relative to the number of unit charged particles. The resulting limits are in the range 10^{-9} to 10^{-11} [3].

The experiments of Cutts *et al.* [18] and Bussiere *et al.* [19] examine the production of massive particles with charge $Q = 2/3$ in proton-nucleus collisions at beam energies of 400 GeV and 240 GeV, respectively. Neither experiment saw any evidence for quarks. They set production limits of about 10^{-11} of that of ordinary particles. Antreasyan et al. [20] pointed out that quark production at high P_T (momentum perpendicular to the beam axis) had not been investigated. They conducted an experiment to search in this region. For $P_T = 4.10$ GeV/c for $Q = 2/3$ and $P_T = 2.05$ GeV/c for $Q = 1/3$ they found no evidence for quarks at the level of 10^{-4} to 10^{-5}. Finally, two experiments have searched for quarks at the highest center of mass energy available in proton colliding beams. Basile et al. [21] searched at the CERN ISR (Intersecting Storage Rings) pp collider with 62.2 GeV center of mass energy. One candidate was observed consistent with a particle having $Q = 1/3$ and $\beta = 1$, but the event was discounted (see [21]). This represents about 10^{-9} of the unit charged particle flux. Very recently, the CERN UA2 collaboration [22] looked for quarks in $p\bar{p}$ collisions using the present highest available center of mass energy of 540 GeV. Only preliminary results from a short period of data taking have been announced. There is no evidence for free quarks.

Before the operation of high energy e^+e^- colliding beam accelerators (PEP and PETRA), massive quark production in electromagnetic interactions had not been tested. Now several experiments have studied this intriguing possibility. Results are generally presented as ratios of cross sections relative to $\mu^+\mu^-$ pair production, separated into exclusive (two oppositely charged quarks produced alone) and inclusive (quarks produced along with ordinary particles) production. Two general purpose detectors, the Mark II at SPEAR [23] and JADE at PETRA [24] have published results. Neither reports evidence for quarks. The cross section limits are on the order of 10^{-4} of the μ-pair cross section from the Mark II for quark masses between 1 and 3 GeV/c^2. Since the beam energy is higher at PETRA and the $\mu^+\mu^-$ cross section is smaller, JADE reports higher limits (10^{-2} to 10^{-3}) for quarks with masses less than 12 GeV/c^2. The Mark II results apply to quarks with $Q = 2/3$ and (for inclusive production) only for the negative charge state. JADE provides limits for $Q = 2/3$ and $Q = 1/3$ for inclusive production but only $Q = 2/3$ for exclusive production (see figure 2).

The PEP free quark search experiment described earlier was specifically designed to search for quarks with $Q = 1/3$ and $Q = 2/3$ in both inclusive and exclusive reactions. In addition, the time of flight system allows sensitivity up to high masses for exclusive production (about 14 GeV/c^2), even though the center of mass energy was about the same, $E_{cm} \sim 29$ GeV, as at PETRA. Results have been reported for both exclusive [25] and inclusive [26] reactions. No evidence for fractional charge was observed by either mechanism, with cross section limits of about 10^{-2} times the μ-pair cross section. The mass dependence of these limits is displayed in figure 2

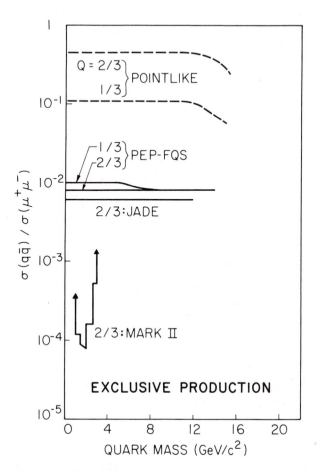

FIGURE 2. Limits (90% confidence limit) on quark production in e$^+$e$^-$ annihilation. See text for details.

where the limits from this experiment are compared to the results from the Mark II and JADE. A naive model of (unconfined) point-like spin-1/2 fractionally charged quarks would suggest cross sections of 1/9 and 4/9 of the μ-pair cross section for $Q = 1/3$ and $Q = 2/3$, respectively. Such particles are clearly ruled out for masses less than 14 GeV/c^2.

Until recently, neutrino production of free quarks had been untested. A 1978 reanalysis of data from previous bubble chamber experiments [27] yielded a quark production cross section less than 4×10^{-40} cm^2. (Typical neutrino cross sections are on the order of 10^{-38} cm^2.) A dedicated experiment at CERN [28] searched for quarks using both avalanche chambers and scintillators to measure ionization. No signal corresponding to free quarks was reported. The authors concluded that quarks are produced at

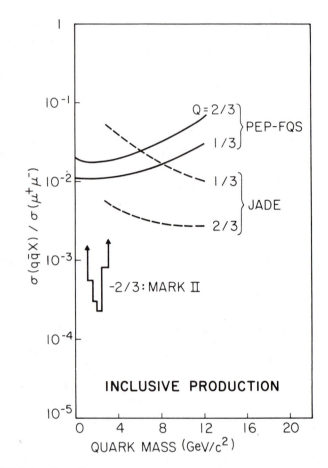

FIGURE 2 (*continued*).

a rate less than about 10^{-5} per neutrino interaction. These experiments are continuing.

5. THEORIES OF FREE QUARKS

Before concluding, we briefly mention developments in the theory of free quarks. Spurred by Fairbank's positive signal, some theorists have sought reconciliation between confinement (implied by nonobservation of quarks in cosmic rays and at accelerators) and an apparent abundance of about 10^{-20} quarks per nucleon in niobium.

These theories generally assume that (a) quantum chromodynamics (QCD) correctly describes the strong interactions, and (b) QCD confines quarks when expressed in its simplest form. The first attempt to modify QCD to allow free quarks was by DeRujula, Giles and Jaffe [29]. They invoked spontaneous symmetry breaking of the SU(3) group of QCD. They found that free quarks would be produced but with very small cross sections, and probably with anomalously large cross sections for interacting with matter, which makes them doubly difficult to see at accelerators or in cosmic rays. Slansky, Goldman and Shaw [30] took a similar approach with gauge symmetry broken down from SU(3) to SO(3). In their model, quarks are still confined, but fractionally charged di-quarks are possible. As in the DeRujula-Giles-Jaffe scheme, quark production in cosmic rays and at accelerators is suppressed, although di-quark interactions in matter should be about as strong as normal hadrons.

Because of the indications, experimental and theoretical, that quarks are not produced at "low" energies, an abundance in matter of 10^{-20} quarks per nucleon is sometimes attributed to relics left over from the big bang. This problem has most recently been investigated by Wagoner and Steigman [31], and by Kolb, Steigman and Turner [32]. Both investigations found that such an abundance is not inconsistent with known aspects of cosmology and that relatively low mass quarks are possible.

A different approach is discussed by Wagoner, Schmitt and Zerwas [33] who postulate the existence of electrically neutral colored objects. These may be bound to fractionally charged quarks to produce colorless fractionally charged particles. Color confinement is maintained. They obtain a very small cross section for e^+e^- production, but unless the quark mass is greater than about 10 GeV/c^2, quarks should be produced relatively copiously in hadron reactions. In addition, one would expect [34], even for extremely large masses, an abundance in matter much larger than that observed by Fairbank.

6. CONCLUSIONS

Table 1 lists the experiments we have discussed, with the limits obtained. Recent experiments have gone further to exclude quark production by various untested mechanisms. However, these results do not necessarily conflict with the results of Fairbank et al. In addition, as pointed out by Orear [35], if free quarks have high interaction cross sections they would be stopped before being detected in nearly every experimental apparatus used so far. The PEP free quark search experiment is preparing to publish the first results on fractionally charged particles with up to 100 times normal hadronic cross section [36]. Though no evidence is seen, we believe this is an important consideration for designing future experiments. The door is open for further experimental investigation. The challenge is great but the results may be well worth the effort.

TABLE 1. RECENT QUARK SEARCH EXPERIMENTS.

Author	Reference	Type	Limit/Signal (90% C.L.)	Comments
Yock (1978)	[12]	Low velocity cosmic rays; vertical	Flux $\cong 3 \times 10^{-9}$ cm^{-2}sr^{-1} sec^{-1}	Three candidates found. None, however, was reproduced in subsequent efforts [13], [14]
Marini et al. (1982)	[15]	Large zenith angle cosmic rays	Flux $< 3 \times 10^{-10}$ cm^{-2}sr^{-1} sec^{-1}	Wide range of velocities examined
Cutts et al. (1978)	[18]	400 GeV p-Be collisions	Flux $< 10^{-11}$ of charged particles	$Q \geq 2/3$ only; mass must be at least several GeV
Bussiere et al. (1980)	[19]	240 GeV p-Be and p-Al collisions	Flux $< 10^{-11}$ of charged particles	$Q \geq 2/3$ only; mass dependence of limits is discussed
Andreasyan et al. (1977)	[20]	400 GeV p-Cu collisions at high transverse momentum	Flux $< 10^{-4} - 10^{-5}$ of charged particles	Momentum transverse to beam must be at least 2 GeV/c ($Q = 1/3$) or 4 GeV/c($Q = 2/3$)
Basile et al. (1977)	[21]	pp collisions at 62.2 GeV in center-of-mass	Flux $< 10^{-9}$ of charged particles	One candidate found consistent with $Q = 1/3$
Weiss et al. (1981)	[23]	e^+e^- collisions at 4–7 GeV in center-of-mass	$\sigma_q < 10^{-4} \, \sigma_{(\mu^+\mu^-)}$	$Q = \pm 2/3$ only for exclusive $Q = -2/3$ only for inclusive
Bartel et al. (1980)	[24]	e^+e^- collisions at 31 GeV in center-of-mass	$\sigma_q < 10^{-3} - 10^{-2} \, \sigma_{(\mu^+\mu^-)}$	$Q = 2/3$ only for exclusive $Q = 1/3$ and 2/3 for inclusive
Marini et al. (1982)	[25]	e^+e^- collisions at 29 GeV in center-of-mass	$\sigma_q < 10^{-2} \, \sigma_{(\mu^+\mu^-)}$	$Q = 1/3$ and 2/3 in exclusive reactions
Ross et al. (1982)	[26]	e^+e^- collisions at 29 GeV in center-of-mass	$\sigma_q < 10^{-2} \, \sigma_{(\mu^+\mu^-)}$	$Q = 1/3$ and 2/3 in inclusive reactions
Morrison (1978)	[27]	Neutrino production in bubble chambers	Number of quarks $< 10^{-2}$ per interaction	Reexamination of bubble chamber data
Basile et al. (1980)	[28]	Neutrino production in CERN SPS wide-band beam	Number of quarks $< 10^{-5}$ per neutrino interaction	Continuing experiments with similar apparatus

References

[1] G. S. LaRue, W. M. Fairbank and A. F. Hebard, *Phys. Rev. Lett.* **38**, 1011 (1977); G. S. LaRue, W. M. Fairbank and J. D. Phillips, *Phys. Rev. Lett.* **42**, 142 (1979); G. S. LaRue, J. D. Phillips and W. M. Fairbank, *Phys. Rev. Lett.* **46**, 967 (1981); see also A. F. Hebard, G. S. LaRue, J. D. Phillips and C. R. Fisel in this volume.

[2] For a review, see M. Bander, *Phys. Rep.* **75**, 205 (1981).

[3] L. W. Jones, *Rev. Mod. Phys.* **49**, 717 (1977).

[4] G. Barbiellini *et al.*, DESY 80-042 (1980); O. W. Greenberg, *Ann. Rev. Nucl. Sci.* **28**, 327 (1978); L. Lyons, Oxford University Report OUNP 80-38 (1980); G. Sussino, Frascati Report LNF-81/47-P (1981); and C. B. A. Mc-Cusker, Sydney Report 81-0411 (1981).

[5] See J. M. Paul, *Nucl. Instrum. Methods* **96**, 51 (1971).

[6] L. Landau, *J. Phys.* **8**, 201 (1944); P. V. Vavilov, *Sov. Phys.–JETP* **5**, 749 (1957).

[7] I. Cairns *et al.*, *Phys. Rev.* **D18**, 1394 (1969); C. B. A. McCusker and I. Cairns, *Phys. Rev. Lett.* **23**, 658 (1969).

[8] For a review see F. A. Ashton, in *Cosmic Rays at Ground Level*, edited by A. W. Wolfendale (Institute of Physics, London, 1973).

[9] R. B. Hicks, R. W. Flint and S. Standil, *II Nuovo Cim.* **14A**, 1, 65 (1973).

[10] P. Franzini and S. Shulman, *Phys. Rev. Lett.* **21**, 1013 (1968).

[11] T. Kifune *et al.*, *J. Phys. Soc. Japan* **36**, 629 (1974).

[12] P. C. M. Yock, *Phys. Rev.* **D18**, 641 (1978).

[13] P. C. M. Yock, *Phys. Rev.* **D22**, 1 (1980); *Phys. Rev.* **D23**, 1207 (1981).

[14] P. C. M. Yock, in *XXI International Conference on High Energy Physics, Paris* (1982). This experiment did produce interesting evidence for a new $Q = 1$ particle of mass ~ 4.2 GeV/c^2 consistent with a similar observation in [13]. This particle was not detected in another cosmic ray seach (see [15]) but the inconsistency of the experiments may be due to differences in the detectors. In particular, Yock's detector was aimed vertically while the detector in [15] looked at zenith angles between 45° and 90°.

[15] A. Marini *et al.*, *Phys. Rev.* **D26**, 1777 (1982); and J. Napolitano *et al.*, *Phys. Rev.* **D25**, 2837 (1982).

[16] J. D. Bjorken and L. D. McLerran, SLAC-PUB-2308 (1979).

[17]   A. Odian, private communication.

[18]   D. Cutts et al., Phys. Rev. Lett. **41**, 363 (1978).

[19]   A. Bussiere et al., Nucl. Phys. **B174**, 1 (1980).

[20]   D. Antreasyan et al., Phys. Rev. Lett. **39**, 9, 513 (1977).

[21]   M. Basile et al., II Nuovo Cim. **40A**, 41 (1977).

[22]   L. diLella, presented at the American Physical Society spring meeting, Washington, D.C., 1982.

[23]   J. M. Weiss et al., Phys. Lett. **101B**, 6, 439 (1981).

[24]   W. Bartel et al., Zeit. Phys. **C6**, 295 (1980); J. Burger, in Proceedings of the 1981 International Symposium on Lepton-Photon Interactions, (Bonn, 1981), p. 115.

[25]   A. Marini et al., Phys. Rev. Lett. **48**, 1649 (1982).

[26]   M. Ross et al., Phys. Lett. **B118**, 199 (1982).

[27]   D. R. O. Morrison, in Proceedings of the XIX-International Conference on High Energy Physics, Tokyo (1978).

[28]   M. Basile et al., II Nuovo Cim. **A45**, 281 (1978); Lett. Nuovo Cim. **29**, 8, 251 (1980).

[29]   A. DeRujula, R. C. Giles and R. L. Jaffe, Phys. Rev. **D22**, 1, 227 (1980); and A. DeRujula, R. C. Giles and R. L. Jaffe, Phys. Rev. **D17**, 1, 285 (1978).

[30]   R. Slansky, T. Goldman and G. L. Shaw, Phys. Rev. Lett. **47**, 13, 887 (1981).

[31]   R. V. Wagoner and G. Steigman, Phys. Rev. **D20**, 4, 825 (1979).

[32]   E. W. Kolb, G. Steigman and M. S. Turner, Phys. Rev. Lett. **47**, 19, 1357 (1981).

[33]   R. Wagoner, I. Schmitt and P. M. Zerwas, Phys. Rev. **D27**, 1696 (1983).

[34]   R. Wagoner, private communication.

[35]   J. Orear, Phys. Rev. **D18**, 9, 3504 (1978).

[36]   W. Guryn et al., Phys. Lett. **B139**, 313 (1984).

Search for Fractional Charge

A. F. Hebard, G. S. LaRue,
J. D. Phillips and C. R. Fisel

1. INTRODUCTION

One of the outstanding aspects of William Fairbank's character is his ability to combine unique physical intuition with unbounded enthusiasm in the pursuit of difficult objectives. These characteristics have been an important component of the glue which has held together the *fractional charge search* experiment from its inception in the early sixties to its present day status. The above authors, listed in chronological order with respect to their involvement in this effort, feel privileged to have been offered the challenge and opportunity of working as Ph.D. students on a project with such profound implications about the physical world around us. Prompted by this occasion of Bill Fairbank's sixty-fifth birthday celebration, it is our pleasure to have an opportunity to describe the experiment, present some recent new data and reveal our present thinking about new directions in this research.

In 1964 M. Gell-Mann [1] and, independently, G. Zweig [2] made an insightful suggestion that all baryons and mesons might themselves be made up of fractionally charged entities. Gell-Mann gave these fundamental building blocks of matter the fanciful name "quarks" [1] and the complex modern theory of these particles (Quantum Chromodynamics— or QCD) is peppered with even more fanciful descriptions such as "up," "down," "strange," "charmed," "bottom," "color," "flavor," "gluon," *etc.*

The novel and exciting aspect of the theory, however, is that the quarks carry an electric charge equal to 1/3 or 2/3 that of an electron. Various combinations of these quarks by pairs (mesons) or by triplets (baryons) give rise to theoretically predicted and experimentally observed structure in hadronic matter. However, it is commonly believed that quarks are bound forever by a potential which increases with separation, and therefore that they cannot exist as fractionally charged entities in the free, unbound state. A search for free fractional charge in matter is a stringent test of this theory. The beauty of fractional charges, from an experimental point of view, is that their fractional charge makes them uniquely detectable and that once free, as long as charge is conserved, a fractionally charged particle cannot lose its fractional charge without finding and joining with a partner, or partners, also fractionally charged. Thus if *any fractionally charged quarks or leptons* escaped the big bang in an unconfined state they must still be present on the matter in the universe.

The experimental techniques which have been used in the search for free quarks [3] can be summarized in three categories: accelerator production experiments, cosmic ray experiments, and content-of-matter experiments. Our experiment belongs to this last category and assumes that free quarks are already distributed at some low density in matter. The free quarks may be attached to nuclei, thereby forming "quarked" atoms with fractional charge, or may in fact diffuse through matter as a hydrogen-like impurity. In 1964 it seemed to us that the most promising way of detecting the presence of quarked atoms in a content-of-matter search was to directly measure the fractional part of the charge on a macroscopic chunk of material. R. A. Millikan pioneered these concepts in his famous oil-drop experiments in which he determined that the charge on drops with mass of approximately 10^{-11} g appeared as integral multiples of the electron charge e. In order to increase the probability of finding the sparsely distributed quarked atoms, we decided to look for fractional charge on superconducting niobium spheres having a mass approximately 10^7 times that of a typical oil drop. An appealing aspect of the experiment was that the sphere, approximately 1/4 mm in diameter, could be recovered and remeasured to check the reproducibility of the results.

2.   APPARATUS

The experimental configuration is shown schematically in figure 1. A superconducting niobium sphere, approximately 1/4 mm in diameter, is supported against gravity in a vacuum at 4.2 K on the magnetic field produced by underlying coaxial superconducting coils. The suspension system is extremely stable because zero resistance supercurrents flow both on the surface of the sphere and in the support coils. An alternating electric field

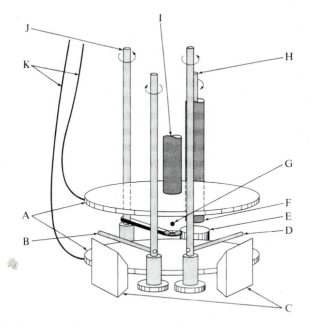

FIGURE 1. Schematic diagram of low temperature measurement region (not to scale). [A] Capacitor plates, [B] Electron source, [C] Mirror mounts, [D] Positron source, [E,F,H] Carousel containing eight niobium balls, [G] Levitated niobium ball, [I] Magnetometer input coil, [J] Ball arm, [K] High voltage leads.

E_A is applied by capacitor plates above and below the ball. The potential applied to the plates is externally generated with the proper phase and at exactly the same frequency as the vertical mechanical oscillations of the sphere. This electric field provides a vertical force on the sphere which is directly proportional to its charge. Accordingly, the steady-state resonant amplitude of motion or the rate of amplitude change (since the damping is low) is a measure of the charge. A sensitive SQUID magnetometer located immediately above the top capacitor plate is used to detect the ball's motion.

The initial charge on a levitated sphere is typically on the order of $\pm 10^5$ e and therefore must be neutralized to within a few unit charges of zero by exposure to electron or positron sources (see figure 1). If the charge cannot be brought to zero then there is a residual charge on the sphere. It is important to emphasize that this experiment does not measure the *absolute charge* on the electron as did Millikan's experiment, but rather the *residual charge* calibrated with respect to unit charge changes. The charge is determined modulo one and it is therefore impossible to tell the difference between $+1/3$ and $-2/3$ of an electron charge.

3. THEORY AND RESULTS

In order to obtain the residual charge from the ball's motion, we must take account of background forces. The alternating force on the ball is

$$F_A = (q_r + ne)E_A + \mathbf{P}_A \cdot \nabla E_F + \mathbf{P}_F \cdot \nabla E_A + F_M + F_Q \qquad (1)$$

where q_r is the residual charge, n is the integer charge number, R is the radius of the ball, E_A is the z component of the applied electric field $\mathbf{E}_A$, $\mathbf{P}_F$ is the permanent electric dipole of the ball, $\mathbf{P}_A$ is the induced electric dipole of the ball, F_M is the magnetic force, F_Q is the quadrupole force, and E_F is the z component of the fixed electric field arising from contact potential variations on the plates near the ball. The measured residual force F_A^r is defined as the value of F_A for $n = 0$ in (1). It can be measured to within $0.01\ eE_A$ with the technique described above.

We have shown [4–7] by measuring $\mathbf{P}_F$ and $\partial E_A/\partial z$ that all known electromagnetic multipole forces are negligible at z_0 (where $\partial E_A/\partial z = 0$) except for the force due to dc electric field gradients from the plates. Eliminating these negligible forces, (1) becomes

$$F_A^r = q_r E_A + R^3 E_A \left(\partial E_F/\partial z\right) + P_{Fz}\left(\partial E_A/\partial z\right) \quad . \qquad (2)$$

We measure F_A^r as a function of each ball's position. q_r is independent of position. The second term in (2) represents the interaction of the ball's alternating electric dipole moment $R^3 E_A$ with the dc electric field gradient $\partial E_F/\partial z$ due to contact potential variations on the plates and is a function of position.

The third term in (2) represents the interaction of the ball's dipole moment P_{Fz} with the alternating electric field gradient $\partial E_A/\partial z$ between the plates. The gradient $\partial E_A/\partial z$ is primarily due to the images of the ball's dipole moment in the plates, and is a calculable function of position. We measure $\partial E_A/\partial z$ independently of the ball's charge by inducing a large dc dipole on the ball with a dc ("battery") potential difference between the plates. The measurements are plotted in figure 2. Both the battery measurements and the third term in (2) are proportional to $\partial E_A/\partial z$ and thus have the same position dependence. We extract the proportionality constant $P_{Fz}/\left(2R^3 E_{Batt}\right)$ by a least squares fit with a separate value of P_{Fz} for each levitation. If P_{Fz} is 1×10^{-7} esu cm (our largest dipole has been 2×10^{-7} esu cm), $P_{Fz}\left(\partial E_A/\partial z\right)$ is 1.5% of the battery curve at all positions z, and both vanish at $z = z_0$.

Now we have

$$(F_A^r)_{z_0} = q_r E_A + R^3 E_A \left(\frac{\partial E_F}{\partial z}\right) \quad . \qquad (3)$$

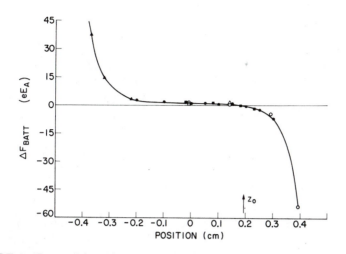

FIGURE 2. "Battery" force ΔF_{Batt} as a function of ball position. ΔF_{Batt} is proportional to the alternating electric field gradient $\partial E_A/\partial z$; we measure it by superimposing a dc ("battery") electric field on the ac field between the plates used to measure residual charge. The strong position dependence of these measurements (which are reproducible for all levitations) is used to determine accurately the relative positions of balls in different levitations. A dc electric dipole moment due to contact potential variations on the ball gives a force during residual charge measurements of 1.5% of that plotted here for the maximum vertical dipole usually observed, $P_{Fz} = 1.0 \times 10^{-7}$ esu cm. In more recent measurements with optically flat capacitor plates, the position z_0 at which $\partial E_A/\partial z = 0$, has been within 0.04 cm of the center of the plates.

From this point on, the analysis falls into three cases listed below, depending on whether $\partial E_F/\partial z$ is zero, and on whether balls of different radii are measured. Of course, $\partial E_F/\partial z$ must remain constant for this analysis to be valid.

Case A—The patch effect field gradient $\partial E_F/\partial z$ is zero. Figure 3 shows measurements [5] of F_A^r on three different balls. Since $\partial E_F/\partial z$ is zero to within experimental error, $q_r = F_A^r/E_A$ at z_0. These measurements show residual charges near zero and $1/3\ e$.

Case B—The patch effect field gradient $\partial E_F/\partial z$ is not zero and all balls measured have the same radius R. With a constant $R^3 E_A\,(\partial E_F/\partial z)$ at z_0, a difference in F_A^r/E_A for these balls is a difference in q_r. Data taken since our most recent publication are shown in figure 4. The differences between F_A^r/E_A of each curve with respect to the average of the four middle curves are $+0.000\ e$, $-0.003\ e$, $-0.002\ e$, $-0.256\ e$, $+0.259\ e$, $+0.004\ e$. The statistical errors for all are $\pm 0.005\ e$. There is a magnetic background force which is proportional to the angle between the magnetic levitation field and g. In the run shown here, begun in September 1980, this force could have contributed a systematic error of up to $\pm 0.07\ e$. In the runs before that one

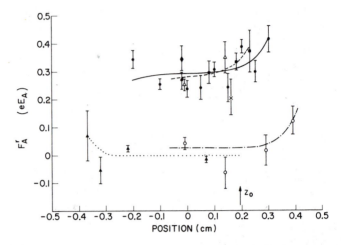

FIGURE 3. The position dependence of the residual force F_A^r for two runs early in 1977. The patch effect electric field gradient is 0 (case A in text). The curves are least squares fits to $F_A^r = q_r E_A + P_{Fz} \, (\partial E_A / \partial z)$. The constant vertical difference between these curves, except for the upturns at the ends due to a dc electric dipole moment on the ball, indicates residual charges for the two upper curves of $+0.313 \pm .019 \, e$ and $+0.344 \pm .010 \, e$.

(data above the dashed line in figure 6), the angle between the magnetic field and g was checked and adjusted so that the magnetic background force would be less than $0.01 \, eE_A$. This angle has also been checked and adjusted in all later runs. Other systematic errors, summarized in reference [6], are $\lesssim 0.03 \, e$. The ball in the fourth levitation of this set is 8% smaller than the other balls. This gives rise to an additional $\pm 0.03 \, e$ uncertainty in its charge, depending on the value of $\partial E_F / \partial z$.

The lower curve in figure 4 shows

$$R^3 E_A \, (\partial E_F / \partial z) = F_A^r - q_r E_A - P_z \, (\partial E_A / \partial z) \tag{4}$$

for each measurement. Since these points all fall on the same curve, we conclude that $\partial E_F / \partial z$ remained constant. The constancy of $\partial E_F / \partial z$ is crucial to subtracting out this background force.

Case C—The patch effect field gradient $\partial E_F / \partial z$ at z_0 is not zero and a number of balls of different radius are measured. Figure 5a shows previously published data on five balls of different radius. If all balls have $q_r = 0$ then $(F_A^r)_{z_0} \, (R_6 / R_i)^3 = R_6^3 E_A \, (\partial E_F / \partial z)_{z_0}$, where R_i is the radius of ball i. This quantity should be the same for every levitation. The two curves shown differ by a constant. Figure 5b shows these data adjusted for differences in radius, residual force and dipole moment. They all fit the same curve, implying that $\partial E_F / \partial z$ remained constant during the measurement.

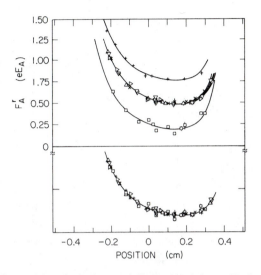

FIGURE 4. The position dependence of the residual force F_A^r for a run begun in September 1980. The patch effect electric field gradient $\partial E_F / \partial z$ is not zero, and balls of the same radius were measured (case B in the text). The curves in the upper half are least squares fits to $F_A^r = q_r E_A + P_{Fz}(\partial E_A / \partial z) + R^3 E_A(\partial E_F / \partial z)$, where a function representing $R^3 E_A\,(\partial E_F / \partial z)$ is chosen to interpolate between the points. The constant differences between these curves indicate residual charge differences for the upper and lower curves of $+0.259\ e$ and $-0.256 \pm .005\ e$ (statistical) $\pm .07\ e$ (systematic). The absolute values of the residual charges cannot be determined. In the lower curve, the data are offset by residual charge and corrected for dc electric dipole moment, showing that $\partial E_F / \partial z$ remained the same for these levitations.

Figure 5c shows $(F_A^r)_{z_0}$ plotted against R_i^3. According to (3), the lines have slope $E_A\,(\partial E_F / \partial z)_{z_0}$ and intercepts which are the residual charges. Note that each of the points fall on one of these lines. Thus these data are consistent with four measurements of zero residual charge and one of $-(1/3)\ e$ residual charge.

4. CONCLUSION

We have made 45 independent measurements on 16 balls. The results are shown in figure 6 in chronological order from the bottom to the top. The last six measurements, shown in figure 4, have been made since our most recent publication [7] and are not included in figure 6 because only residual charge differences were measured. The statistical errors represent 1 standard deviation. Out of 30 repeat measurements, we have observed 12 residual charge changes, in each case of $\pm(1/3)\ e$. These changes have occurred only when a ball was brought into contact with other surfaces. We have made 14 independent measurements on ball 6 over a 3 1/2 year

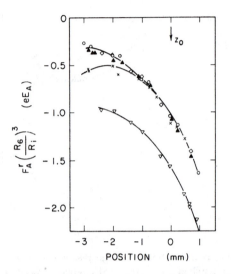

FIGURE 5a. The position dependence of the volume-normalized residual force $F_A^r (R_6/R_i)^3$ for a run begun in January 1980. The patch effect electric field gradient $\partial E_F/\partial z$ is not zero, and balls of differing radius were measured (case C in text). The curves are least squares fits to $F_A^r (R_6/R_i)^3 = q_r E_A (R_6/R_i)^3 + P_{Fz}(\partial E_A/\partial z)(R_6/R_i)^3 + R_6^3 E_A(\partial E_F/\partial z)$ (plotted in units of e), where a function to interpolate $\partial E_F/\partial z$ is chosen as in figure 4.

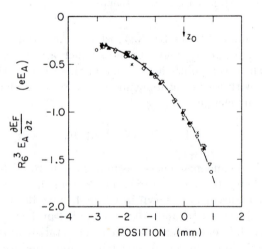

FIGURE 5b. The data of figure 5a offset by the residual charge term $q_r E_A$, and corrected for permanent electric dipole P_{Fz}. Since the points all fall on the same curve, we conclude that the patch effect electric field gradient $\partial E_F/\partial z$ remained the same during all five levitations.

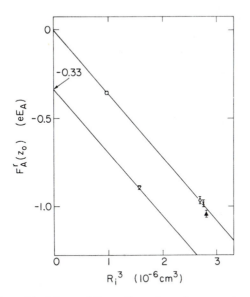

FIGURE 5c. The residual force of figure 5a evaluated at z_0, where no correction is necessary for a permanent electric dipole moment on the ball, plotted against R_i^3. The lines have slope $\partial E_F / \partial z$ and intercept q_r, the residual charge. Ball 13, with $R_i^3 = 1.6 \times 10^{-6}$ cm^3, has a residual charge of $-0.324 \pm .014$ e.

period, every measurement consistent with 0 e or $\pm(1/3)$ e, and have observed 9 changes of $(1/3)$ e. For the 27 measurements since reference [6], the ball stayed at low temperatures during any one cooldown. All of the balls measured since reference [5] were heat-treated on a tungsten plate at $1800°$ C for 17 hr in a vacuum of 10^{-9} torr.

Our apparatus measures the residual force on the ball to 0.01 eE_A. We have shown by measurement of the relevant parameters that all known electromagnetic multipole forces are negligible except for that force due to patch effect fields and have shown that we can take them into account and so obtain the true residual charge. In order for a measurement to be valid, the patch effect field must not change with time. In all cases where this was true, the value of the residual charge was 0 e or $\pm(1/3)$ e.

We are improving the experiment to further study the background forces and to gather data more rapidly on the abundance of fractional charge. In order to assay other substances, we are constructing a room temperature ferromagnetic levitation experiment [8]. This apparatus will allow us to measure iron spheres. Other substances can be levitated inside split hollow iron spheres. These spheres, of about 300 microns diameter with 50 micron walls, will be made for us by the laser fusion target group at Lawrence Livermore Laboratory.

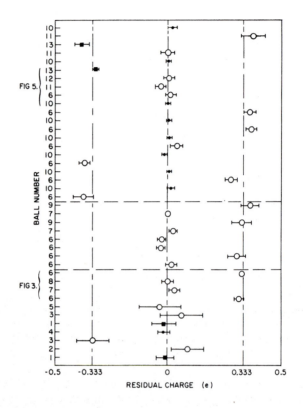

FIGURE 6. Histogram of all data taken up to May 1980. Residual charge differences (not shown here) from the September 1980 run are +0.001 e, −0.001 e, −0.002 e, −0.258 e, +0.259 e, +0.002 e; ±.005 e (statistical) ±.07 e (systematic).

We plan to extend the room temperature apparatus by converting it into a mass spectrometer in which solids are evaporated by a high power laser and deposited on the levitated iron balls. The mass spectrometer we are proposing is unique in that it has an unambiguous fractional charge detector. Knowledge of the charge to mass ratio will allow the large scale separation of fractionally charged particles.

Addendum

Since the writing of this paper, we have discovered experimentally and theoretically a new magnetic background force which could have been as large as $1/3\ eE_A$ in all of our runs [9,10]. The effect averages to zero if the ball spins about a vertical axis, appears as a drift during the measurements if the ball spins very slowly, and contributes a random apparent fractional charge if the ball does not spin. We have spun a ball with a rotating magnetic field even though the balls are superconducting and experience no eddy current torque. We are now perfecting techniques for spinning every ball we measure, and for verifying that they are spinning. We are proceeding with new runs in which the effect will be eliminated by spinning the ball by tilting the charge-measuring electric field or applying a rotating magnetic field, or by reshaping the levitation field.

The new effect would contribute a random apparent charge, whereas our published data (figure 6) peak sharply near zero and $\pm 1/3$ e, and are self-consistent. We have recently analyzed the conditions under which the charge-measuring electric field would have spun the ball during charge measurements. This analysis shows that the ball was probably spinning when the early data were taken, averaging the magnetic effect to zero, so that the early data may well represent true fractional charge.

Nonetheless, we claimed that we had measured the fractional charges of these spheres only after we had taken into account all known background forces *and* obtained statistical evidence in the form of a peaked residual charge distribution, as shown in figure 6. With the discovery of the new magnetic background effect, we must await the results of our new experiments before we can claim positively the presence or absence of fractional charge on these niobium samples.

Acknowledgments

This work was supported in part by the National Science Foundation.

References

[1]   M. Gell-Mann, *Phys. Lett.* **8**, 214 (1964).

[2]   G. Zweig, CERN Reports No. 8182/TH401 and No. 8419/TH412 (1964).

[3]   L. W. Jones, *Rev. Mod. Phys.* **40**, 717 (1977).

[4]   G. S. LaRue, Ph.D. Thesis, Stanford University, 1978 (unpublished).

[5]   G. S. LaRue, W. M. Fairbank and A. F. Hebard, *Phys. Rev. Lett.* **38**, 1011 (1977).

[6]   G. S. LaRue, W. M. Fairbank and J. D. Phillips, *Phys. Rev. Lett.* **42**, 142, 1019(E) (1979).

[7]   G. S. LaRue, W. M. Fairbank and J. D. Phillips, *Phys. Rev. Lett.* **46**, 967 (1981).

[8]   C. R. Fisel, Ph.D. Thesis, Stanford University, 1986 (unpublished).

[9]   J. D. Phillips, Ph.D. Thesis, Stanford University, 1983 (unpublished).

[10]   W. M. Fairbank and J. D. Phillips, in *Inner Space/Outer Space*, edited by E. W. Kolb *et al.* (University of Chicago Press, Chicago, 1986), p. 563.

Single Atom Detection with Lasers: Applications to Quark and Superheavy Atom Searches

William M. Fairbank, Jr.

It is probably quite apparent from the title of this paper that William Fairbank, Sr., has had a significant influence on my research. This is indeed very true. Although he did not push me in any way toward a career in physics, I am sure that his obvious love of science affected my career decisions. The experiments which I will describe really originated in the many discussions which my father and I had during my early graduate years at Stanford. Actually, I had tried to avoid working with Dad when I came to Stanford because I wanted to establish my own identity. I chose instead to pursue a career in lasers with Arthur Schawlow. This worked fine for the first half of my Ph.D. program, but every time I saw Dad he kept telling me about Art Hebard's exciting new results on the quark experiment. Many of you know how this can go. He will come up to you with a new idea for an experiment. As the discussion proceeds, he gets more and more enthusiastic. You tell him, "Well, Dad, I've got these other things to do ...," but he just keeps going on. Before you know it you're a collaborator of his, whether you want to be or not.

We talked quite a bit about how lasers could be used to learn more about the fractional charges in Hebard's niobium balls. We discussed, for example, using a high power laser as the vaporization source for a mass spectrometer. Since the sample could then be completely vaporized by the laser, there would not be any questions about whether or not the quarked atoms evaporated, as there were in some previous quark searches. Schawlow and Ted Hänsch were eventually drawn into this discussion. They suggested an even better idea: The quarked atoms should have a characteristic spectrum which could be detected sensitively by laser-excited resonance fluorescence. Perhaps even a single quarked atom could be seen. I decided to pursue these ideas for my Ph.D. thesis and have been working on them since that time. In this paper I will first review briefly the different methods of laser single atom detection and some of the work I have done. Then I will discuss our current quark and superheavy atom searches, which are just getting under way.

In the last ten years there have been considerable advances in the field of single atom detection with lasers. The two basic methods which are now in use are illustrated in figures 1 and 2. Both techniques are extremely selective because a signal is generated only when a dye laser is tuned exactly to the resonant wavelength of a ground state transition in the atom of interest.

PHOTON BURST DETECTOR

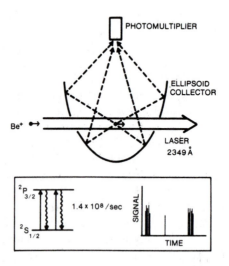

FIGURE 1. Single atom detection by laser resonance fluorescence. When the laser is tuned to resonance a burst of spontaneous emission is given off by a single atom as it passes through the laser beam. An ellipsoidal mirror focuses the light on a photomultiplier tube.

I have been working mainly on the resonance fluorescence method (figure 1). In this technique a low power, continuous dye laser is used. When an atom in the laser beam absorbs a photon, it goes to the excited state. It will decay 10^{-7} to 10^{-8} seconds later, spontaneously emitting a photon in a random direction. This process can be repeated up to 10^7 to 10^8 times per second. Thus one can, in principle, get a large burst of photons from a single atom as it passes through a resonant laser beam. One of the most efficient ways to collect the fluorescence is with an ellipsoidal mirror. If the atom intersects the laser beam near one focus of the ellipse, the emitted light will be concentrated at the other focus, where the photomultiplier is placed. In our experiments we are able to collect 57% of the emitted photons from a single atom. With a good photomultiplier of quantum efficiency 20–30%, 1/6 to 1/10 of the fluorescence photons can be recorded as counts.

The photon burst method can probably be used today to detect up to half of the atoms in the periodic table and 15% of the ions in the periodic table, but in some cases you have to work quite hard to get the needed wavelength. Note, however, that this is with true single atom sensitivity; when the atom is there, you record its presence by the photon burst generated. When the atom is absent you do not get a significant number of photons detected.

The second method, Resonance Ionization Spectroscopy (RIS), has been developed mainly by Sam Hurst and his colleagues at Oak Ridge National Laboratory (figure 2). In this case a high power pulsed laser excites an atom to the first excited state and then immediately ionizes it before it can return to the initial state by emitting a photon. Then the ion or the electron that is created by the ionization process is detected. Depending on the atom and the selectivity desired, several lasers or two-photon transitions may be used. Due to the many nonlinear processes now available for generation of ultraviolet wavelengths by high power lasers, the RIS method is capable of detecting every atom in the periodic table except helium and neon.

Since these two techniques are both extremely sensitive and quite general, a variety of applications is possible. Sam Hurst has compiled a list of more than one hundred areas of science where laser single atom detection could have an impact [1]. Some of these, such as the quark and superheavy atom searches which I will describe later, involve the forefront of basic physics. Others are more practical. Most of the experiments I have done to date fall into the latter class. The measurements of low sodium vapor densities that Hänsch, Schawlow and I did at Stanford in 1974 were the first experiments which demonstrated that detection of atoms by laser resonance fluorescence can be extremely sensitive [2,3]. We were able to measure sodium densities as low as 100 atoms/cm^3 (figure 3). This was seven orders of magnitude lower than other methods had achieved. One might say that we were on a "frontier of near zero" in these experiments in the sense that there was

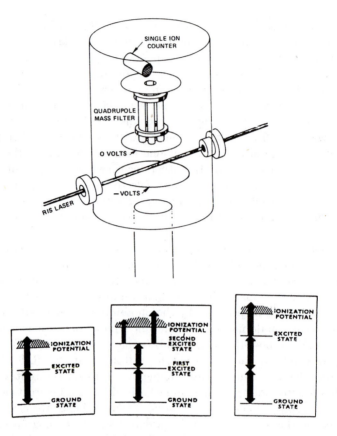

FIGURE 2. Single atom detection by resonance ionization spectroscopy. Resonant pulsed lasers drive an atom to an excited state and then ionize it. The positive ion is usually detected, sometimes through a mass spectrometer, as is shown here. Three possible RIS schemes are illustrated.

only about one atom on the average in our detection volume at the lowest density.

George Greenlees' group at the University of Minnesota was the first to propose and demonstrate true single atom detection by the photon burst method in 1977 [4]. They have detected single sodium and barium atoms in an atomic beam in vacuum. In the last five years at Colorado State University, we have been trying to monitor the motion of single atoms in buffer gases using this method [5]. Our apparatus is shown in figure 4. We seed sodium atoms into a helium flow and record the temporal behavior of their photon burst with a real-time digital correlator. A sample correlation function of a single sodium atom diffusing in 200 torr of helium is shown in figure 5. The vertical intercept indicates that about 40 photocounts

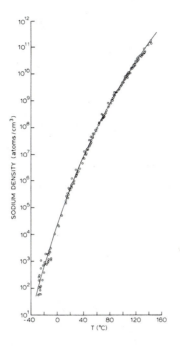

FIGURE 3. Sodium vapor pressure measurements made with the resonance fluorescence method. The lowest density measured by other methods is 10^9 atoms/cm^3.

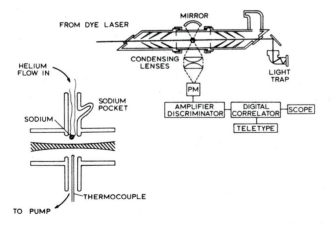

FIGURE 4. Apparatus for fluorescence correlation spectroscopy. In the side view at the bottom left, sodium atoms are seeded into a helium flow and cross the laser beam near the focus of the ellipse. The diaphragms in the sidearms (top view) help keep stray light scattered at the windows from reaching the photomultiplier (PM). Autocorrelation functions of the photon burst from single or multiple atoms are recorded in the real-time digital correlator. This apparatus has been used to make diffusion and flow velocity measurements. In the velocity experiments, two parallel laser beams are often used.

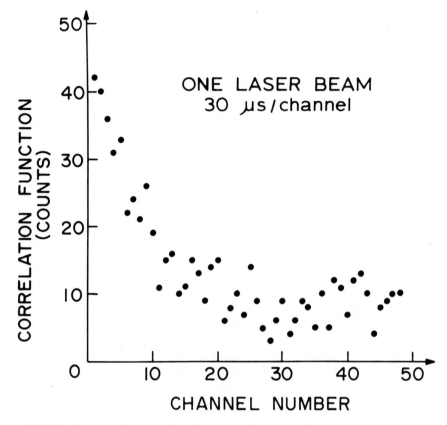

FIGURE 5. A sample correlation function of the photon burst from a single sodium atom diffusing through the laser beam in helium at 200 torr.

were detected in the burst from this particular atom. The decay of the correlation function is a measure of the time the atom spent in the laser beam, about 300 μsec.

Our ability to record the time duration of the burst from an atom allows us to do several interesting experiments which are otherwise hard to do. Some examples are listed in table 1. The diffusion coefficient of sodium atoms in gases is normally difficult to measure because sodium is so reactive. With our method we determine an average diffusion time by integrating over many correlation functions such as that shown in figure 5. From a measurement of the laser beam diameter, a diffusion coefficient can be extracted. We have done such experiments with sodium in helium, neon

TABLE 1. APPLICATIONS OF FLUORESCENCE CORRELATION SPECTROSCOPY.

1. Diffusion coefficients of reactive atoms:
   D for Na in He, Ne, Ar measured

2. Diffusion of excited atoms:
   D* for Na* in He measured; D*/D = 1.2 ± .2

3. Measurement of fast chemical reaction rates—ground-state and excited-state atoms:
   No measurements yet

4. Measurement of fast gas flows, e.g., wind tunnels:
   Mach 0.03 measured with single atoms
   Mach 0.25 measured with 5000 atoms
   Mach 1.0 expected to be measurable with single atoms

and argon [6,7]. A related application which is unique to this method is the possibility of determining the diffusion coefficients for excited atoms. As the intensity of the laser beam is increased, the atom will spend a higher and higher fraction of its time in the excited state while it is diffusing. Hence, the observed diffusion coefficient should be more and more characteristic of the resonant excited state, rather than the ground state. We have found in this way that the diffusion coefficient of the excited state in sodium is actually larger than the ground state [6,7]. This was surprising to us at first, because the larger excited atom should diffuse slower. Nevertheless, estimates based on measured helium-sodium molecular potentials do agree with the experimental results. One should also be able to measure fast chemical reaction rates of ground state and excited state atoms by this method. But we have not yet had the time to do any experiments of this type.

We have been funded primarily to investigate this technique as a tool for measuring the speed of fast flows in wind tunnels. Currently the best noninvasive method of velocity measurement is laser Doppler velocimetry. It uses light scattering from dust particles. There are questions about its accuracy at sonic speeds, however, because the dust particles tend to lag behind the flow. An atom would be a more ideal probe. It should show no lag. So far we have measured 10 m/sec flows with single sodium atoms and 90 m/sec flows (Mach 0.25) with a one-second integration [6,7]. We project that a new detector will allow us to measure Mach 1 laboratory flows using single atoms. No tests have yet been made in real wind tunnels.

Because of the many quark experiments which have failed to produce evidence for free fractional charges, it has been difficult until recently to obtain funding for an optical quark search [8]. The positive evidence which

George LaRue and Jim Phillips have provided in the last few years has changed the picture considerably [9]. It is now important not only to confirm the existence of free fractional charge, but also to isolate and identify the fractionally charged particle, to find out to what nuclei the particles bind, and to concentrate these particles, if possible. Our single atom detection apparatus may be able to do some of these things.

Many people have speculated privately on what the fractional charges observed on the niobium balls might be. Unfortunately these experiments provide essentially no information on this subject, although Schiffer believes that $+1/3\ e$ is favored statistically over $-1/3\ e$ [10]. If the quarks are positive, e.g., $+1/3\ e$, $+2/3\ e$, or $+4/3\ e$, they may bind a single electron to become a hydrogen-like quark-atom, which I call $HQ^{-2/3}$, $HQ^{-1/3}$, or $HQ^{+1/3}$. About half of the metals, generally those toward the left of the periodic table, absorb hydrogen exothermically. They may also collect hydrogen quark-atoms.

Using data for normal hydrogen in metals, I have roughly estimated in table 2 the solution energy for quarks in metals [11]. The reactions which have been calculated are

$$(Q^{+1/3})_{\text{gas}} \rightarrow (Q^{+1/3})_{\text{metal}} \quad , \tag{1}$$

$$(\text{H} \cdot HQ^{-1/3})_{\text{gas}} \rightarrow (\text{H}_2)_{\text{gas}} + (Q^{+2/3} + e^-)_{\text{metal}} \quad , \tag{2}$$

$$\frac{1}{2}(\text{H}_2)_{\text{gas}} \rightarrow (\text{H}^+ + e^-)_{\text{metal}} \quad , \tag{3}$$

$$(\text{H} \cdot HQ^{+1/3})_{\text{gas}} \rightarrow (\text{H}_2)_{\text{gas}} + (Q^{+4/3} + e^-)_{\text{metal}} \quad . \tag{4}$$

The $+1/3\ e$ quarks are taken to be unbound to electrons or atoms in the gas because their binding is relatively small. This assumption does not change the calculation significantly. The other two quarks are assumed to have an electron in the gas phase and to form hydrogen-like molecules with H. Of course, there are many other possibilities. Note that in table 2 the $+1/3\ e$ and $+2/3\ e$ quarks are predicted to dissolve exothermically in all metals, while solution of $+4/3\ e$ quarks is endothermic in all cases except Sr. It is interesting that niobium is predicted to be among the strongest binders of $+1/3\ e$ and $+2/3\ e$ quarks, while tungsten is relatively poor for both. Perhaps the niobium balls which were heat treated on W received their fractional charges from the W substrate during the heat treatment. Thus we are planning to look for the spectrum of $HQ^{-2/3}$ and $HQ^{-1/3}$ in niobium and other metals with various heat treatments.

If the free quarks are negative, they may be found bound electrostatically to any nucleus. Since our techniques are quite general, we have great freedom in choosing which quarked atom to look for. In general, the $-1/3\ e$

TABLE 2. ENERGIES OF SOLUTION, ΔH IN eV, FOR QUARKED ATOMS ($Q_{1/3}$) AND MOLECULES ($HQ_{2/3}$, H_2, AND $HQ_{4/3}$) IN METALS. A negative value indicates that location in the metal is energetically preferred over the gas. Note that this is true for $+1/3$ and $+2/3$ quarks, but not for $+4/3$ quarks. Rows 1–4 represent reactions (1)–(4), or nuclear charges $+1/3$, $+2/3$, $+1$, and $+4/3$, respectively.

K	Ca	Sc	Ti	V	Cr	Mn	Fe	Co	Ni	Cu	Zn	Ga	Al
−1.54	−1.53	−1.47	−1.33	−1.32	−1.20	−1.29	−1.21	−1.16	−1.16	−1.19			−1.22
−1.33	−1.86	−2.26	−2.55	−2.44	−2.18	−2.16	−2.39	−2.53	−2.72	−2.29			−2.01
−0.31	−0.78	−0.90	−0.50	−0.31	0.54	0.09	0.28	0.37	0.17	0.47			0.60
0.81	0.41	0.69	2.04	2.37	4.03	2.91	3.69	4.11	3.91	4.00			3.94

Rb	Sr	Y	Zr	Nb	Mo	Tc	Ru	Rh	Pd	Ag	Cd	In
−1.55	−1.58	−1.51	−1.37	−1.32	−1.20		−1.17	−1.18	−1.18	−1.17		
−1.24	−1.78	−2.03	−2.40	−2.47	−2.28		−2.29	−2.56	−2.94	−2.18		
−0.28	−0.95	−0.88	−0.52	−0.67	0.41		0.56	0.28	−0.11	0.71		
0.75	−0.11	0.41	1.80	2.33	3.87		4.22	3.93	3.54	4.42		

Cs	Ba	Ce	Hf	Ta	W	Re	Os	Ir	Pt	Au	Hg	Tl
−1.55	−1.56	−1.51	−1.37	−1.32	−1.18			−1.07		−1.13		
−1.22	−1.82	−1.84	−2.25	−2.46	−2.15			−2.61		−2.70		
−0.26	−0.90	−0.70	−0.38	−0.35	0.67			0.76		0.37		
0.76	0.06	0.59	1.92	2.29	4.30			5.17		4.35		

quark-atoms are preferred over the $+1/3\ e$ quark-atoms because their resonant wavelengths are longer and easier to obtain. We further narrow the choice by considering the accuracy to which the quark-atom spectra can be predicted. It would be nice to have 1 Å accuracy so that the laser could be set to the predicted wavelength with a 1 Å bandwidth, and no scanning would be required. Accurate calculations from first principles are tedious except for the one-electron atoms. Hence, fits to isoelectronic sequences are usually used. Table 3 presents a summary of the types of

TABLE 3. ACCURACY OF ISOELECTRONIC SEQUENCE CALCULATIONS OF QUARK-ATOM TRANSITIONS. The uncertainties in the given wavelengths are hard to estimate and should be taken only as a rough guide to the accuracy of the calculations.

Quark-Atom Type	Wavelength Uncertainty	Examples
Noble gases	~1 Å	$ArQ^{-1/3}\ 1p_0 - 2p_8\ 2484.4 \pm 0.9$ Å
		$NeQ^{-1/3}\ 1p_0 - 2p_8\ 1878.9 \pm 0.5$ Å
Halogens	~1 Å	$ClQ^{-1/3}\ 2P^{\circ}_{3/2} - 2P_{3/2}\ 1813.1 \pm 1.5$ Å
		$ClQ^{-1/3}\ 2P^{\circ}_{3/2} - 2D^{\circ}_{3/2}\ 3222.5 \pm 1.3$ Å
Alkali $(+1/3e)$	~10 Å	$NaQ^{+1/3}\ 3^2S_{1/2} - 3^2P_{1/2}\ 4277 \pm 15$ Å
Alkali $(-1/3e)$	~100 Å	$NaQ^{-1/3}\ 3^2S_{1/2} - 3^2P_{1/2}\ 9700 \pm 300$ Å
		$NaQ^{-1/3}\ 3^2S_{1/2} - 4^2P_{1/2}\ 6000 \pm 100$ Å
Alkaline earth	~10 Å	
Transition metals	Large	
Rare earths	Large	
Niobium	Very large	
Hydrogen	Small except for quark mass uncertainty	$HQ^{-2/3}\ 1s\text{-}2p\ 10932.2$ Å $(m = \infty)$ $HQ^{-1/3}\ 1s\text{-}2p\ 2733.0$ Å $(m = \infty)$

results which can be expected from isoelectronic sequence fits. Note that reasonably accurate predictions can be obtained with this method for atoms on the left and right sides of the periodic table, except for the $-1/3\ e$ alkali atoms. Results are poorer in the middle of the table where sequences are shorter, and many level crossings occur. Niobium is, unfortunately,

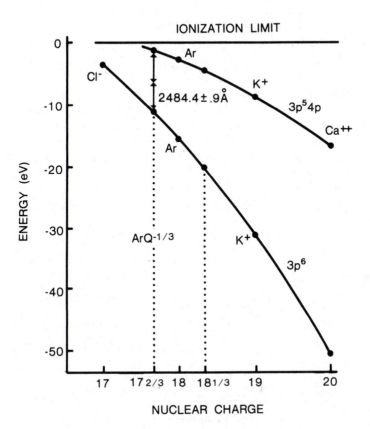

FIGURE 6. Energies of the ground state and one 4p excited state in the argon iso-electronic sequence. The energy levels of argon quark-atoms can be found by locating the points on the curves at $Z = 17\frac{2}{3}$ and $Z = 18\frac{1}{3}$. The two-photon resonance in the $ArQ^{-1/3}$ atom is found to be at $2484.4 \pm .9$ Å$(1p_0 \rightarrow 2p_8)$.

one of the worst cases because data are not available for the spectrum of the third member of its isoelectronic sequence, Tc^{++}. An example of an isoelectronic sequence fit, the argon series, is given in figure 6. In this case a two-photon resonance is predicted at $2484.4 \pm .9$ Å in the $ArQ^{-1/3}$ atom $(Z = 17\frac{2}{3})$. One of the main reasons this result can be so precise is that the ground state energy of Cl^- is known so accurately. Thus an interpolation rather than an extrapolation is being done. In comparison, the $-1/3\ e$ alkali quark-atoms require extrapolation and are much more uncertain.

The quark-atoms from table 3 which have the most accurate predicted spectra, noble gases, alkalis and halogens, are fortunately also the quark-atoms which have the simplest geochemistry. Thus some predictions can be made as to where they should concentrate in nature. I expect to find most of these quark-atoms in the ocean [11]. Thus our second set of experiments will involve looking for these quark-atoms in seawater samples.

My current quark experiments are being done in collaboration with Sam Hurst at a company in Oak Ridge, Atom Sciences, Inc., which Sam co-founded. Atom Sciences has designed an advanced single atom detection apparatus called SIRIS (Sputter Initiated Resonance Ionization Spectroscopy) which will be used for commercial ultrasensitive analysis of solids. With a few modifications it is nearly ideal for optical quark and superheavy atom searches.

SCHEMATIC DIAGRAM OF SIRIS APPARATUS

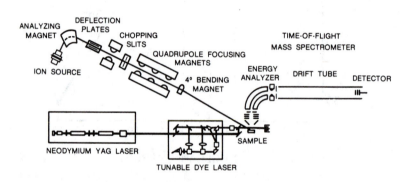

FIGURE 7. The SIRIS apparatus for quark-atom detection. A pulsed ion beam sputters the sample. Negative quarked atoms $(Q^{-1/3})$ are accelerated toward the region where the laser beam will pass. Positive ions are returned to the sample. The laser resonantly ionizes only the quarked atoms. The quark-ion $(Q^{+2/3})$ is accelerated and passed through an electrostatic analyzer. A time-of-flight mass spectrometer allows the observation of all masses simultaneously. This apparatus will also be used for a super-heavy atom search. In those experiments a photon burst detector will eventually be placed in front of the charged particle detector.

Our setup for the quark experiment is illustrated in figure 7. A pulsed ion source is used to vaporize the sample by sputtering. This process is relatively independent of binding, so we do not have to worry about whether or not the quarks stick to the sample, as we do with thermal vaporization. At the end of the current pulse, a resonant laser ionizes selected quark-atoms from charge $-1/3$ e to charge $+2/3$ e. With careful selection of potentials in the apparatus, the ionized quark-atoms can be separated from a possible small nonresonant background by an electrostatic analyzer. This idea is not mine; it is borrowed from several groups who are doing quark searches with accelerator mass spectrometers [12]. Integer-charged possibilities which do not get through the analyzer are $0 \rightarrow +1$, $-1 \rightarrow 0$, and $-1 \rightarrow +1$; the only integer combination which simulates a fractional charge is $-1 \rightarrow +2$, which is very unlikely in our case.

Since the mass of a free quark is unknown, we will use a time-of-flight mass spectrometer to detect the quark-atoms rather than a magnetic sector. In that way all masses can be recorded in each laser shot. We expect that the background will be very low in our experiments, perhaps zero, especially at the heavier masses. Note that a $+2/3$ e quark-ion acts in the mass spectrometer as if it had $1\frac{1}{2}$ times its mass. The expected signal rate based on the average quark concentration in Dad's experiments, 0.6 quarks per large ball, is on the order of one detected quark-atom per 100 hours of operation. This is fairly low, but it may be higher than the background. Note, for example, that detector dark current is negligible because of the low duty cycle of our machine. Note also that no enrichment procedure has been utilized. With enrichment or a more optimum choice of sample (Pd or Zr?), a quite reasonable detection rate might be achieved. We are anxiously awaiting the arrival of the SIRIS machine so that real experimental values for the background levels can be determined.

In the seawater quark searches, some processing must be done before the sample can be placed in the SIRIS apparatus at high vacuum. We plan to use a simple two-step enrichment procedure which is illustrated in figure 8. In step 1 the water would be evaporated slowly in a modest vacuum. The quarked atoms will probably be left bound to the sludge at the bottom of the container by chemical and image charge forces. But just to be sure, a small positive potential will be applied to the container in order to attract any evaporating negative quark-atoms back to the surface. After the sample is dried, we will vaporize it slowly with a continuous ion beam. We will collect the negatively charged particles and bend them in a magnetic field. A new sample plate then will collect only those ions with mass-to-charge ratio m/q greater than 100 amu/e. Note that an $ArQ^{-1/3}$ atom would have $m/q = 120$ or greater, depending on the mass of the free quark. Most of the vaporized atoms will be neutral or positive, and most of the negative ions will be light. Only a small fraction of the normal atoms should hit the collector plate. Based on data from sputtering of salts, we estimate that an enrichment of greater than 10^6 can be obtained in two cycles of this procedure. There are, of course, some uncertain aspects of this enrichment procedure due to the lack of information on the chemical properties of quarked atoms. This is a problem in any quark enrichment process. We have tried to make our procedure as simple as possible in order to minimize the uncertainty. The final clean enriched sample will be placed in the SIRIS apparatus and analyzed for noble gas and halogen $-1/3$ e quark-atoms. Some of the calculated wavelengths are given in table 3.

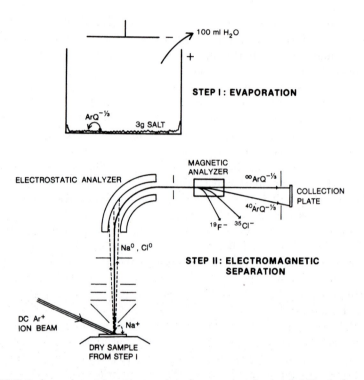

FIGURE 8. A quark enrichment procedure for seawater. In step I 100 ml of seawater is evaporated in a modest vacuum, leaving 3 g of salts on the bottom of the container. A potential is applied to make sure that any evaporating $-1/3\,e$ quark-atoms are returned to the sample. In step II the argon quark-atoms are separated from most of the normal atoms electromagnetically. Sample atoms are vaporized in vacuum by sputtering with a continuous ion beam. Negative ions are collected and separated according to m/q by a magnetic field. Argon quark-atoms will have $m/q = 120$ to ∞. Major negative ion components such as Cl^- will miss the enriched sample plate. A procedure similar to step II can also be used to enrich samples for the superheavy atom search.

Another fundamental problem which requires very high sensitivity and selectivity is the possible existence in the universe of stable elementary particles with integer charge and mass greater than the proton. A few such particles may exist around us as a relic of the big bang. Typical concentrations predicted by cosmological theories are 10^{-10} [13]. Presumably these superheavy particles are bound to nuclei and/or electrons and should appear to be superheavy isotopes of normal elements.

A number of predictions for such particles have been made in the past decade, the most interesting of which are those which arise from unified field theories. The unification of electromagnetic and weak forces is now generally accepted by physicists. Recent theories which add the strong force, the so-called grand unified theories, are of several types and have little experimental verification to date. Nevertheless, they do make some rather striking predictions: the decay of the proton, a matter/antimatter asymmetry, the existence of a unification particle at 10^{15} GeV, possibly a magnetic monopole, *etc.* Two versions due to Weinberg [14] and Susskind [15] are especially interesting to me because they predict a new type of force called technicolor and a new type of quark called techniquark. The lowest mass combination of techniquarks should be a stable technibaryon (lifetime $\sim 10^{76}$ years) with a mass of 1000–3000 amu [15]. Some unified theories which incorporate gravity have also been constructed. They predict a new symmetry called R parity. All known particles are R-even. The lowest mass R-odd particles should be stable [16] and have mass in the range 10^2–10^7 amu [17].

To date there have been several experimental searches for superheavy hydrogen and oxygen atoms using mass spectrometers and cyclotrons [18], but the highest mass which has been probed is 1100 amu. This is just the beginning of the technibaryon range. We hope in our optical experiments to look for superheavy atoms which are cosmologically more favorable [19] and to probe masses well above 1000 amu.

The apparatus which we will use is similar to that of the quark experiment (figure 7). Of course the special techniques for selecting fractional charges will not work in this case. The needed high selectivity against normal isotopes will be obtained in two ways: (a) a large isotope shift, especially for low-Z elements, and (b) a very late arrival in the time-of-flight mass spectrometer. Note in figure 9 that we can tune our laser to excite and ionize all the interesting superheavy lithium isotopes without exciting the normal atoms. If two resonance steps are used, the selectivity will be even greater.

One of the main problems we will have in our initial experiments is the difficulty of detecting the slow moving superheavy ions. Experiments with normal ions at low velocity have indicated a threshold for secondary electron emission at $v = 5.5 \times 10^4$ m/sec [20]. In other words, a 40 kV ion with mass greater than 2550 amu will probably produce no response in a conventional charged particle detector. Thus our first experiments may be limited to the 100–2500 amu range. We are planning to eventually place a photon burst detector at the end of the time-of-flight tube. It is interesting that the photon burst method is complementary to the conventional ion detector; it works better for heavier ions which move more slowly and generate a larger burst. Perhaps even a 10^7 amu ion could then be detected. Two of the most interesting superheavy ions, Be and Tc, should work well in the photon burst method. Some others, such as Li, may be detectable if they are first neutralized in a charge transfer cell.

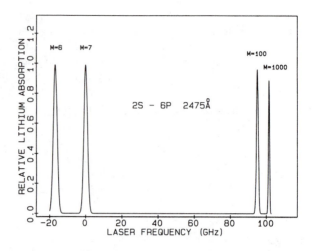

FIGURE 9. Predicted absorption lineshapes of sputtered lithium isotopes for the 2s-6p resonance at 2475 Å. Doppler broadening, fine structure and isotope shifts have been included. Note that the superheavy atoms are well separated from the normal isotopes. A laser with a 10 GHz bandwidth covering the range from 93 to 103 GHz will be able to ionize all the superheavy isotopes without ionizing the normal isotopes.

The previous hydrogen experiments, which had large enhancement factors due to electrolytic separation of heavy water, had a sensitivity of about 10^{-28} up to 1100 amu. We hope in our experiments to be able to probe higher masses at a sensitivity corresponding to big bang superheavy particle concentrations of 10^{-20} to 10^{-26} [21]. There are some possibilities for much lower limits if large scale chemistry is used.

I would like to conclude by saying that with laser techniques we can get to one "frontier of near zero"—near zero numbers of atoms in a volume. My fervent hope is that in these last two experiments which I have described we will have *near zero* rather than *absolutely zero* particles of interest in our samples.

Addendum

Since the Near Zero conference when this material was presented, experimental searches for fractional charges and superheavy atoms have been performed. Initial results are reported in references [11], [22] and [23].

Acknowledgments

This research was supported in part by the National Science Foundation and the Research Corporation.

References

[1] Some of these applications are discussed in G. S. Hurst, M. G. Payne, R. D. Willis, B. E. Lehmann and S. D. Kramer, "Resonance Ionization Spectroscopy: Counting Noble Gas Atoms" in *Laser Spectroscopy V*, edited by A. R. W. McKellar, T. Oka and B. P. Stoicheff (Springer-Verlag, Berlin, 1981), p. 59.

[2] W. M. Fairbank, Jr., T. W. Hänsch and A. L. Schawlow, *J. Opt. Soc. Am.* **65**, 199 (1975).

[3] W. M. Fairbank, Jr., Ph.D. Thesis, Stanford University, 1974.

[4] G. W. Greenlees, D. L. Clark, S. L. Kaufman, D. A. Lewis, J. F. Tonn and J. H. Broadhurst, *Opt. Commun.* **23**, 236 (1977).

[5] C. Y. She and W. M. Fairbank, Jr., *Opt. Lett.* **2**, 30 (1978); C. L. Pan, J. V. Prodan, W. M. Fairbank, Jr., and C. Y. She, *Opt. Lett.* **5**, 459 (1980).

[6] W. M. Fairbank, Jr., C. Y. She and J. V. Prodan, *Proc. Soc. Photo-Opt. Instr. Eng.* **286**, 94 (1981); C. Y. She and W. M. Fairbank, Jr., *ibid.* **426**, 49 (1983); J. V. Prodan, C. Y. She and W. M. Fairbank, Jr., *Opt. Commun.* **43**, 215 (1982).

[7] J. V. Prodan, Ph.D. Thesis, Colorado State University, 1981.

[8] For a review of experimental searches for free quarks, see L. W. Jones, *Rev. Mod. Phys.* **49**, 717 (1977); *Phys. Today* **26** (5), 30 (1973); and Y. S. Kim, *Contemp. Phys.* **14**, 289 (1973).

[9] G. S. LaRue, W. M. Fairbank and A. F. Hebard, *Phys. Rev. Lett.* **38**, 1011 (1977); G. S. LaRue, W. M. Fairbank and J. D. Phillips, *ibid.* **42**, 142 (1979); G. S. LaRue, J. D. Phillips and W. M. Fairbank, *ibid.* **46**, 967 (1981).

[10] J. P. Schiffer, *Phys. Rev. Lett.* **48**, 213 (1982).

[11] W. M. Fairbank, Jr., G. S. Hurst, J. E. Parks and C. Paice, in *Resonance Ionization Spectroscopy 1984*, edited by G. S. Hurst and M. G. Payne (Institute of Physics, Bristol, 1984), p. 287.

[12] K. H. Chang, A. E. Litherland, L. R. Kilins, W. E. Kieser and E. L. Hallin, "A Mass-Independent Search for Fractionally Charged Particles," and D. Elmore, T. Gentile, H. Kagan and S. L. Olson, "Fractional Charge Spectroscopy with a MP Tandem Accelerator" at the Quark Searchers Conference, San Francisco State University, June 5–8, 1981.

[13] C. B. Dover, T. K. Gaisser and G. Steigman, *Phys. Rev. Lett.* **42**, 1117 (1979); S. Wolfram, *Phys. Lett.* **82B**, 65 (1979).

[14] S. Weinberg, *Phys. Rev. D* **13**, 974 (1976); *ibid.* **19**, 1277 (1979).

[15]   L. Susskind, *Phys. Rev. D* **20**, 2619 (1979); E. Farhi and L. Susskind, *ibid.* **20**, 3404 (1979).

[16]   S. Weinberg, *Phys. Rev. Lett.* **48**, 1303 (1982).

[17]   H. Pagels and J. R. Primack, *Phys. Rev. Lett.* **48**, 223 (1982).

[18]   P. F. Smith and J. R. J. Bennett, *Nucl. Phys. B* **149**, 525 (1979); P. F. Smith, J. R. J. Bennett, G. J. Homer, J. D. Lewin, H. E. Walford and W. A. Smith, "A Search for Heavy Charge +1 Particles in Enriched D_2O," Quark Searches Conference, San Francisco State Unviersity, June 5–8, 1981; R. Middleton, R. W. Zurmuhle, J. Klein and R. V. Kellarits, *Phys. Rev. Lett.* **43**, 429 (1979); R. A. Muller, L. W. Alvarez, W. R. Holley and E. J. Stephenson, *Science* **196**, 521 (1977); T. Alvager and R. Naumann, *Phys. Lett. B* **24**, 647 (1967); G. Kukavadze, L. Memelova, L. Suvorov, *Sov. Phys.–JETP* **22**, 272 (1966).

[19]   R. N. Cahn and S. L. Glashow, *Science* **213**, 607 (1981).

[20]   L. A. Dietz and J. C. Sheffield, *J. Appl. Phys.* **46**, 4361 (1975); B. L. Schran *et al.*, *Physica* **32**, 749 (1966).

[21]   Cosmological enhancement factors discussed in [19] are included in these calculations.

[22]   W. M. Fairbank, Jr., W. F. Perger, E. Riis, G. S. Hurst and J. E. Parks, in *Laser Spectroscopy VII*, edited by T. W. Hänsch and Y. R. Shen (Springer-Verlag, Berlin, 1985), p. 53.

[23]   W. M. Fairbank, Jr., E. Riis, R. D. LaBelle, J. E. Parks, M. T. Spaar and G. S. Hurst, in *Resonance Ionization Spectroscopy 1986*, edited by G. S. Hurst (Institute of Physics, Bristol, 1986).

Production of Fractional Charge

Robert V. Wagoner

Following evidence from their earlier experiments, members of the Stanford group have obtained evidence that 1/3 integer charges exist on superconducting spheres of niobium, with an abundance $\gtrsim 10^{-20}$ fractional charges per nucleon [1]. Although no detection of fractional charges has been claimed in other experiments at similar or smaller abundance levels, their interpretation is subject to severe chemical effects [2]. As we shall see, any fractionally charged particles that are discovered in ordinary matter would most likely be relics of the early universe, providing a new probe of both cosmology and particle physics [3]. It is typical of William Fairbank that he recognized the importance of such experiments, and pioneered the method of detection (magnetic levitation) that is still the cleanest.

Various schemes for the production of fractionally charged particles have been proposed. Two models involving broken quantum chromodynamics (QCD) have been explored, in which the gluon acquires (at an energy either greater than [4] or less than [5] $\Lambda \sim 200$ MeV) an effective mass within its experimentally allowed range $\mu \lesssim 60$ MeV. Within unbroken QCD, two models have also been proposed. One introduces fractionally charged leptons, while the other introduces quarks with nonstandard electric charges. Both of these possibilities can be manifested within certain types of grand unified theories [6]. Here we shall briefly consider a specific model for the final possibility, but also mention general conclusions within any model employing unbroken QCD. Full details are provided by Wagoner, Schmitt and Zerwas [7].

We shall assume the existence of Q quarks, which are triplets under $SU(3)_{\text{color}}$ but have no electroweak charges. They are thus counterparts to the leptons. To obtain the minimum abundance which survives from the early universe, we assume that Q and $\bar{Q}$ are equally abundant.

In the era preceding the quark–hadron transition, the Q (and $\bar{Q}$) quarks remained in statistical equilibrium with the other particles until their rate of annihilation into gluons became less than the rate at which their (equilibrium) abundance was changing due to the expansion of the universe. If their mass $M_Q \gtrsim 10$ GeV, this "freeze-out" occurred before the quark–hadron transition. In fact, an analysis of their production in hadron–hadron (figure 1a) and $e^+ - e^-$ (figure 1b) collisions gives this mass limit.

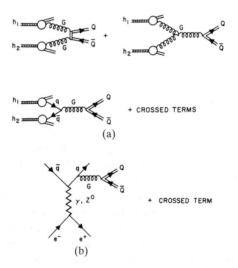

FIGURE 1. Diagrams governing the production of Q-$\bar{Q}$ pairs in (a) hadron–hadron and (b) e^+-e^- collisions, *via* ordinary quarks (q) and gluons (G).

During the quark–hadron transition ($kT \sim \Lambda$), all the surviving Q and $\bar{Q}$ quarks formed color-singlet states (mesons and baryons) with the much more abundant ordinary quarks. The flavors likely to be stable are $(\bar{Q}u)$, (Qud), and their antiparticles. These fractionally charged hadrons should have typical hadronic cross sections. Annihilations of these fractionally charged hadrons occurred for a brief period after they were formed, resulting in a final abundance relative to photons (today) of

$$Q/\gamma = \bar{Q}/\gamma \simeq 2 \times 10^{-20} \quad . \tag{1}$$

Their abundance relative to nucleons is then $Q/N \gtrsim 3 \times 10^{-11}$. In order that their present mass density not exceed the inferred upper limit on the total universal mass density, $M_Q \lesssim 10^{12}$ GeV.

Although their annihilations ceased when $kT \sim 25$ MeV, the fractionally charged hadrons interacted with the nucleons present and those nuclei produced at $kT \sim 0.1$ MeV. It appears likely that the binding energy of fractionally charged hadrons to the nucleus is comparable in magnitude with the typical nuclear potential energy. (As J. Martoff has pointed out, Λ–nucleus interactions may give us relevant information.) The absolute value of the charge of most fractionally charged hadrons should be no greater than $2e/3$ during nucleosynthesis.

We thus expect that the nuclear evolution of fractionally charged hadrons will be similar to that of neutrons, or of the uncharged heavy baryons studied by Dicus and Teplitz [8]. If so, then most fractionally charged hadrons should emerge from primordial nucleosynthesis bound to a ^{4}He nucleus, with a relative abundance

$$\frac{(F\ ^4\mathrm{He})}{^4\mathrm{He}} = \frac{(\bar{F}\ ^4\mathrm{He})}{^4\mathrm{He}} \simeq \left(\frac{p}{^4\mathrm{He}}\right)\left(\frac{Q}{p}\right) \simeq 10\left(\frac{Q}{N}\right) \gtrsim 3 \times 10^{-10} \quad , \qquad (2)$$

where the symbols F and $\bar{F}$ mean the number densities of fractionally charged hadrons (quark and antiquark, respectively) associated with the atoms denoted by the succeeding symbol. The relative abundance of other fractionally charged nuclei should be roughly similar to those of the ordinary nuclei produced in the early universe, except for the possible presence of $F\ ^5$He and $F\ ^8$Be and a suppression of $(F\ \mathrm{H})/\mathrm{H}$.

With virtually all fractionally charged particles emerging from the early universe in the form of fractionally charged nuclei, the resulting Coulomb barriers will prevent subsequent annihilation at stellar interior temperatures $T \lesssim 10^7$ K. We might also expect that (on the average) stellar nucleosynthesis would produce abundances of fractionally charged nuclei per ordinary nucleus of the same order of magnitude as the fraction per nucleon in He, hence $F\ ^AZ/^AZ \gtrsim 10^{-10}A$. ($A$ is the mass number and Z is the charge.)

Another way to produce fractionally charged hadrons is through the interaction of cosmic rays with the earth, moon, $etc.$ Using the known cosmic ray flux of protons, the hadronic production cross section, and the relevant mixing depth of the surface gives $Q/N \lesssim 6 \times 10^{-27}$ for the earth and $Q/N \lesssim 6 \times 10^{-22}$ for the moon.

It appears that this model involving electroweak neutral quarks produces the smallest primordial cosmic abundance of fractional charge of all models incorporating QCD. The first reason is that any model containing unequal numbers of particles and antiparticles must produce a higher final abundance than the symmetric case. The second reason is the fact that the abundance of fractional charge which survives annihilation in any symmetric model is inversely proportional to its annihilation cross section. Since the annihilation cross section of fractionally charged hadrons (whether they contain charged or uncharged Q quarks) appears to be an upper bound, their surviving abundance should represent a lower bound. For instance, fractionally charged leptons survive with an abundance $\gtrsim 10^{-3}$ per nucleon [9]. In figure 2 we summarize the predictions for the production of fractionally charged particles within QCD, and indicate the abundance limits from the two types of magnetic levitation experiments.

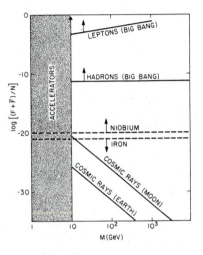

FIGURE 2. Predicted (solid) and observed (dashed) abundances of fractionally charged particles (F) plus antiparticles $(\bar{F})$ per nucleon (N) plotted against their masses (M). Production in the early universe and by cosmic rays impinging on the earth and moon is included. The lower limit observed from the levitation of niobium spheres and the upper limit observed from the levitation of iron (steel) spheres are indicated.

Thus we conclude that a (nonzero) cosmic abundance of fractionally charged particles per nucleon smaller than $\sim 10^{-11}$ (outside stars) would be incompatible with the validity of (confining) QCD and the standard big bang model.

The critical question which remains is the actual abundance of fractional charge. It is clear that the chemistry of fractionally charged atoms and molecules must be better understood before any conclusions can be drawn. For instance, fractionally charged atoms can combine into neutral molecules during the evolution of the collapsing nebula which produced the solar system [10]. However, from the interstellar abundance of grains $[N(grain) \sim 4 \times 10^{-13} N(\mathrm{H})]$, one can calculate that the time required for any fractionally charged particle (atom or molecule) to collide with and stick to a grain relative to the time required to combine with a particle of the opposite fractional charge is given by

$$\frac{t(grain)}{t(neutralized)} \sim 10^5 \frac{N(\bar{F})}{N(\text{H})} \tag{3}$$

Thus it appears likely that most fractionally charged particles would have existed on separate grains when the earth formed.

At present, the best example of unaffected matter is probably the atmospheres of stars. The presence of spectral lines due to fractionally charged atoms at or somewhat above our predicted minimum level of abundance can certainly not be ruled out.

Acknowledgments

This work was supported in part by the U.S. National Science Foundation (grant PHY 81-18387) and by the U. S. National Aeronautics and Space Administration (grant NAGW-299).

References

[1] G. S. LaRue, J. D. Phillips and W. M. Fairbank, *Phys. Rev. Lett.* **46**, 967 (1981); W. M. Fairbank and J. D. Phillips, in *Inner Space/Outer Space*, edited by E. W. Kolb, *et al.* (University of Chicago Press, Chicago, 1986), p. 563.

[2] K. S. Lackner and G. Zweig, *Phys. Rev. D* **28**, 1671 (1983).

[3] R. V. Wagoner, in *Physical Cosmology*, Les Houches, Session XXXII, edited by R. Balian, J. Audouze and D. N. Schramm (North-Holland, Amsterdam, 1980), p. 395.

[4] R. V. Wagoner, in *Essays in Nuclear Astrophysics*, edited by C. A. Barnes, D. D. Clayton and D. N. Schramm (Cambridge University Press, Cambridge, 1982), p 495.

[5] E. W. Kolb, G. Steigman and M. S. Turner, *Phys. Rev. Lett.* **47**, 1357 (1981).

[6] L. F. Li and F. Wilczek, *Phys. Lett.* **107B**, 64 (1981); H. Goldberg, T. W. Kephart and M. T. Vaughn, *Phys. Rev. Lett.* **47**, 1429 (1981).

[7] R. V. Wagoner, I. Schmitt and P. M. Zerwas, *Phys. Rev. D* **27**, 1696 (1983).

[8] D. A. Dicus and V. L. Teplitz, *Phys. Rev. Lett.* **44**, 218 (1980).

[9] H. Goldberg, *Phys. Rev. Lett.* **48**, 1518 (1982).

[10] L. J. Schaad, B. A. Hess, Jr., J. P. Wikswo, Jr., and W. M. Fairbank, *Phys. Rev. A* **23**, 1600 (1981).

A Superconductive Detector to Search for Cosmic Ray Magnetic Monopoles

B. Cabrera, M. Taber and S. B. Felch

1. INTRODUCTION

On February 14, 1982, a prototype superconductive detector designed to search for a cosmic ray flux of magnetic monopoles observed a single candidate event. Six weeks later, the preliminary results from the initial 151 days of operation were presented for the first time at the Near Zero conference honoring William Fairbank, where they generated much interest. This velocity and mass independent search for any flux of monopoles is made by continuously monitoring the current in a 20 cm^2 superconducting loop. Only the one candidate event has occurred during five runs now totaling 382 days. The data set an upper limit of 2.4×10^{-10} cm^{-2} sec^{-1} sr^{-1} on the flux of all magnetically charged particles passing through the earth's surface. Even though it has not been possible to rule out a spurious cause, the possible existence of superheavy magnetically charged particles passing through the earth's surface has produced enormous interest within the scientific community. If the entire local missing mass is made up of 10^{16} GeV/c^2 monopoles traveling with typical galactic velocities of 300 km/sec, then a flux of 4×10^{-10} cm^{-2} sec^{-1} sr^{-1} would result. In this paper we summarize the theory for detection of monopoles using superconductors, describe the operation of the prototype detector and discuss our plans for future larger detectors.

2.   THE MAGNETIC MONOPOLE HUNT

Serious interest in magnetic monopole searches began in 1931, when P. A. M. Dirac [1] proposed the existence of magnetically charged particles to explain the observed quantization of electric charge. He showed that only integer multiples of a fundamental magnetic charge g (called the Dirac charge) are consistent with quantum mechanics. Many years of experimental searches using conventional ionization detectors produced no convincing candidates.

During the 1960's Luis Alvarez and his coworkers at Berkeley developed induction detectors [2] and soon realized that the current change in a superconducting loop provided the best resolution. Samples, with a total weight of 30 km, were passed through their superconductive detector, with each sample making about 1000 passes per measurement. Many samples contained material from exotic locations such as the moon, the north and south poles and ocean bottom where enhanced densities might be expected. No candidates were found.

In the early 1970's, one of us (Blas Cabrera) became interested in magnetic monopoles when Bill Fairbank and Art Hebard, and later George LaRue and Jim Phillips, began to suspect that fractional electric charges exist in matter [3]. An earlier theoretical paper written by Julian Schwinger had suggested that quarks might carry magnetic charge, accounting for their large binding energies without the need of introducing a new force. A sensitive detector was built based on superconductive technologies developed within Bill Fairbank's group. These include ultra low magnetic field shielding (see article entitled "Near Zero Magnetic Fields with Superconducting Shields" in this volume) and improved SQUID magnetometry. A sensitivity more than 5000 times greater than the early Alvarez detectors allowed sufficient resolution to detect a Dirac charge making a single pass through the superconducting ring. A modest number of samples, including niobium spheres from the fractional electric charge experiments, were tested [4]. Again no candidates were found.

Over the last decade, work on unification theories has unexpectedly yielded strong renewed interest in monopoles. In 1974, 't Hooft and independently Polyakov [5] showed that in true unification theories (those based on simple or semi-simple compact groups) magnetically charged particles are necessarily present. These include the standard SU(5) grand unification model. The modern theory predicts the same long range field and thus the same charge g as the Dirac solution; now, however, the near field is also specified leading to a calculable mass. The standard SU(5) model predicts a monopole mass of 10^{16} GeV/c^2, horrendously heavier than had been considered in previous searches.

Such supermassive magnetically charged particles would possess qualitatively different properties from those assumed in earlier searches [6]. These include necessarily nonrelativistic velocities from which follow weak ionization and extreme penetration through matter (passing through the earth with little loss of kinetic energy). Thus such particles may very well have escaped detection in earlier searches.

A flux of cosmic ray particles possessing magnetic charge can be detected by monitoring the current in a superconducting loop. Whereas multipass techniques were sufficient for static matter searches, here single pass detection is essential. Such superconductive detectors directly measure the magnetic charge independent of particle velocity, mass, electric charge and magnetic dipole moment. In addition, the detector response is based on simple and fundamental theoretical arguments which are extremely convincing. Before describing the prototype superconductive detector, we present a simple derivation of the Dirac theory and discuss its similarities to flux quantization in superconductors.

3. THE DIRAC MONOPOLE THEORY

Dirac asked whether the existence of magnetically charged particles could be made consistent with quantum mechanics [1]. He considered a single electron in the field of a magnetic charge and found that for the electron wavefunction to remain single valued a quantization condition must exist between the elementary electric and magnetic charges given by

$$eg = (1/2)\hbar c \quad . \tag{1}$$

Thus, if magnetic charges existed they would explain the experimentally observed quantization of all electric charges. No other theoretical explanation for this quantization exists.

We can understand the Dirac quantization using an argument very similar to the derivation for flux quantization in superconductors (see equation (8) in paper entitled "Rotating Superconductors and Fundamental Physical Constants" in this volume). Here we consider the Schrödinger current density for a single electron in the field of a magnetic charge

$$\mathbf{j} = (\hbar e/2im)(\psi^*\boldsymbol{\nabla}\psi - \psi\boldsymbol{\nabla}\psi^*) - (e^2/mc)\psi^*\psi\mathbf{A} \quad . \tag{2}$$

Then for any path Γ not through the monopole

$$mc/e^2 \oint_\Gamma \mathbf{j} \cdot d\mathbf{l}/\psi^*\psi = n(hc/e) - \int_{S_\Gamma} \mathbf{B} \cdot d\mathbf{S} \quad . \tag{3}$$

Flux quantization is not present because atomic distances are much smaller than the London penetration depth, and thus j is not negligible. In fact, the term on the left side of equation (3) is typically 10^5 times larger than the magnetic flux term on the right side. However, if we take two surfaces S_Γ and S_Γ' each bounded by the path Γ and with the two together enclosing the monopole, then the current density term of equation (3) is identical for both. Subtracting the two equations we obtain

$$k(hc/e) = \oint_{S_\Gamma - S_\Gamma'} \mathbf{B} \cdot d\mathbf{S} = 4\pi g \quad , \qquad (4)$$

where $k = n - n'$ is also an integer and the closed surface integral equals the total flux emanating from the pole. Thus, equation (1) is obtained for the smallest nontrivial solution $k = 1$.

The Dirac condition quantizes but does not symmetrize electric and magnetic charges because the elementary magnetic charge is predicted to be much stronger than the elementary electric charge. From equation (1), g given by

$$g = e(1/2\alpha) \qquad (5)$$

is $137/2$ times greater than the elementary electric charge e. Therefore, two magnetic charges a given distance apart feel a force which is $(137/2)^2$ greater than that between two electric charges the same distance apart. The coupling constant, $g^2/\hbar c = \alpha(g/e)^2 = 34$, would thus be even stronger than for the nuclear force.

Note also that from equation (4) the flux emanating from a Dirac charge is

$$4\pi g = hc/e = 2\phi_0 \quad , \qquad (6)$$

exactly twice the flux quantum of superconductivity! This result should not be surprising, for Dirac imposed the same single valued condition on a single electron wavefunction in the field of a monopole as was used in deriving flux quantization from the coherent many-body Cooper pair order parameter. The Cooper pairs possess twice the electron charge, accounting for the factor of two in equation (6). Thus the observation of flux quantization in a superconducting loop demonstrates sufficient resolution for detecting Dirac magnetic charges passing through it.

4.   MONOPOLE COUPLING TO A SUPERCONDUCTING RING

As a Dirac magnetic charge passes through a superconducting loop, all of the flux from the monopole couples to the loop. Thus a supercurrent is induced in the loop which corresponds to a flux change of exactly $2\,\phi_0$.

For example, a magnetic charge g moving at velocity v along the axis of a superconducting ring of radius b produces an induced supercurrent given by

$$I(t) = (\phi_0/L) \left(1 + vt/(v^2 t^2 + b^2)^{1/2} \right) \quad , \tag{7}$$

where L is the self inductance of the loop and the particle is in the plane of the loop at $t = 0$. As t increases from $-\infty$ to $+\infty$ the total flux change is exactly $2 \; \phi_0$.

For an arbitrary trajectory which passes through the ring, the number of flux quanta threading the ring will change by two; whereas, for a trajectory that does not pass through, the flux will remain unchanged. In all cases the supercurrent change occurs over a characteristic time b/v. Present SQUID electronics cannot resolve expected rise times of 10–100 nanosec for particle velocities between 10^{-3} and 10^{-4} c, and only the resulting net dc shift can be detected for such a particle which passes through the loop.

A superconductive system based on these properties is sensitive only to magnetic charges. The passage of any presently known particle possessing electric charge or magnetic dipole moment would cause very small transient signals but no dc shifts. Thus, a cosmic ray flux of magnetically charged particles such as the predicted supermassive GUT monopoles can be detected by monitoring the current in a superconducting loop.

5. THE PROTOTYPE COSMIC RAY MONOPOLE DETECTOR

To test the feasibility of such devices, an existing instrument, originally built as an ultrasensitive absolute magnetometer (see figure 4 in the article entitled "Near Zero Magnetic Fields with Superconducting Shields" in this volume), was used in a prototype search for supermassive monopoles [7]. This instrument has been operated as a monopole detector for a total of 382 days. It consists of a four turn 5 cm diameter loop made of 0.005 cm diameter niobium wire. As shown in figure 1, the coil is positioned with its axis vertical and is connected to the superconducting input coil of a SQUID. The passage of a single Dirac charge through the loop would result in an 8 ϕ_0 change in the flux through the superconducting circuit, comprised of the detection loop and the SQUID input coil (2 ϕ_0 couple to each of the four turns). The SQUID and the loop are mounted inside a superconducting shield in an ambient ultra low magnetic field of 5×10^{-8} gauss (see the article entitled "Near Zero Magnetic Fields with Superconducting Shields" in this volume).

The voltage output of the SQUID electronics is directly proportional to the supercurrent in the detection loop, and is continuously recorded onto a strip chart recorder through a 0.1 Hz low pass filter. Several intervals

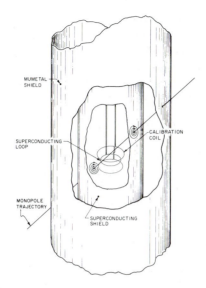

FIGURE 1. Schematic of prototype detector used in monopole search.

throughout a continuous one-month time period are shown in figure 2a, where no adjustment of the dc level has been made. Typical disturbances caused by daily liquid nitrogen and weekly liquid helium transfers are evident. A single large event was recorded (figure 2b). It is consistent in magnitude with the passage of a single Dirac charge within an uncertainty of ±5%. It is the largest event of any kind in the record.

In addition to the strip chart recording, during the last 231 days of operation the data were also continuously sampled with a computer at

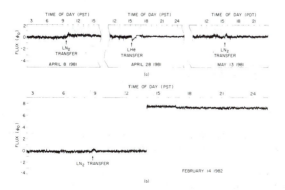

FIGURE 2. Data record showing (a) typical stability and (b) the candidate monopole event.

20 points per second. These data were digitally filtered to an equivalent bandwidth of 0.1 Hz and permanently stored. A sensitive accelerometer (better than a milli-g resolution) was mounted on the instrument, and its output was also recorded. No other large events were seen.

The induced current change expected from the passage of a Dirac charge through the loop can be written as

$$\Delta I = (8\phi_0/L)\,(\eta - \zeta(A_\ell/A_s)) \tag{8}$$

where $A_\ell/A_s = 1/16$ is the ratio of the loop to shield areas, $\eta = \pm 1$ for a trajectory intersecting the loop and zero otherwise, and ζ is a form factor with value between -1 and $+1$, which depends on the trajectory impact parameter and inclination angles. The extremum values for ζ (± 1)

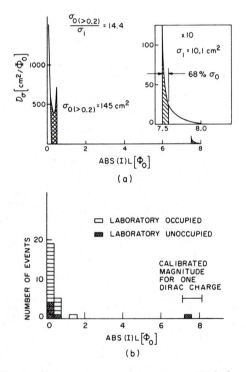

FIGURE 3. (a) Density of sensing area *versus* event magnitude for prototype detector. (b) Histogram of all spontaneous event magnitudes.

are obtained for trajectories that are parallel or antiparallel to the shield axis and within the shield cross section; ζ is zero for trajectories which are perpendicular to the shield axis. The density of sensing area *versus* the event magnitude has been calculated and is shown in figure 3a for an isotropic distribution of Dirac magnetic charges passing through the detector. The sensing area density is proportional to the probability of an event with a given induced current.

An analysis of the first 151 days of operation has found 27 spontaneous events exceeding a noise threshold of 0.2 ϕ_0 which remain after excluding known disturbances such as transfers of liquid helium and nitrogen. The magnitude distribution of these is plotted in figure 3b. An event is defined as an offset with stable levels for at least one hour before and after. Only six events were recorded during the 70% of the run time that the laboratory was unoccupied. The analysis for the remaining 231 days of operation includes a similar distribution of small events and no large events.

The most likely sources for spurious signals are associated with the mechanical sensitivity of the apparatus. Mechanically induced offsets have been intentionally generated. These could be caused by shifts of the four-turn loop wire geometry, which would produce inductance changes and inversely proportional current changes (the magnetic flux must remain constant). Alternatively, trapped supercurrrent vortices in or near the SQUID could move from mechanically induced local heating. In any case, raps with a screwdriver handle against the detector assembly cause offsets.

Although we have not been able to produce any events which appear as clean as the candidate event, magnitudes close to the candidate event magnitude have been generated. If we further assume that mechanical stresses frozen into the detector upon initial cooling from room temperature could spontaneously release, it would be possible to imagine a large spurious signal. The difficulty with easily accepting this scenario as a plausible explanation for the candidate event is that the data from this detector have been accumulated over five separate runs, each commencing with a cooldown of the apparatus from room temperature, and no other such signals were seen. Nevertheless, we feel that *it is not possible to rule out a spurious cause for the February candidate event.*

6. LARGER DETECTORS ARE NEEDED

Since observing the candidate event our primary effort has gone into the design and construction of a detector with larger sensing area and with sufficient redundancy to definitively rule out all spurious causes for any new events. This apparatus consists of three mutually orthogonal loops each twice the diameter of the prototype loop. The apparatus, shown in figure 4, has seven times the loop area averaged over solid angle and an additional factor of seven in "near miss" area. This "near miss" sensing area corresponds to events from trajectories which pass through the superconducting shield surrounding the loops but miss the loops. The magnetic field change from the doubly quantized supercurrent vortices, which appear in the shield walls at the entrance and exit points along the trajectory, couples to the loops, producing smaller but detectable induced current changes.

FIGURE 4. Photograph of new three-loop detector.

Yet larger detectors in the future cannot use larger loops because the induced current from the 2 ϕ_0 flux change decreases as the self inductance and loop area increase. To overcome this limitation a design is being developed where the surface of a large cylindrical shell of superconductor will be periodically scanned for newly appearing double quantized vortices (see figure 5). Only a magnetically charged particle penetrating such a detector (usually twice) would leave this magnetic signature, much as an electrically charged particle would leave a track in a photographic emulsion. As long as the cylinder remained superconducting, strong pinning forces would prevent vortex motion. The trapped flux pattern would be periodically recorded with a small scanning coil coupled to a SQUID. Simulated scans over a 10 cm × 10 cm area are shown in figure 6. These include typical SQUID noise levels. Sensing areas greater than 1 m^2 can be achieved.

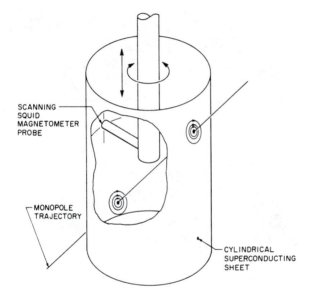

FIGURE 5. Proposed scanning detectors.

7. CONCLUSIONS

Superconductive devices are the only known monopole detectors which are sensitive to arbitrarily low particle velocities, and their response is exactly calculable using simple fundamental concepts. Our prototype detector has seen one uncorroborated event which might signify the existence of supermassive magnetically charged particles passing through the earth. Several generations of larger detectors are being built which will either find very convincing events or relegate the Valentine's Day event to the category of improbable nonreproducible results.

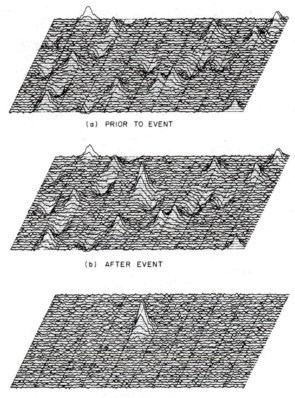

(a) PRIOR TO EVENT

(b) AFTER EVENT

(c) BACKGROUND SUBTRACTION

FIGURE 6. Simulated scans showing a Dirac charge event.

Acknowledgments

It is appropriate that the preliminary results from the prototype detector were presented for the first time at the Near Zero conference honoring Bill Fairbank. He has taught us to look for experiments which span the traditional divisions of physics, in this case applying the unique advantages of superconductivity to particle physics.

The equipment and techniques used in this research were funded in part by the National Bureau of Standards, the National Aeronautics and Space Administration and the National Science Foundation.

References

[1]   P. A. M. Dirac, *Proc. Roy. Soc.* **A133**, 60 (1931); *Phys. Rev.* **74**, 817 (1948).

[2]   P. Eberhard, D. Ross, L. Alvarez and R. Watt, *Phys. Rev.* **D4**, 3260 (1971).

[3]   G. S. LaRue, W. M. Fairbank and A. F. Hebard, *Phys. Rev. Lett.* **38**, 1011 (1977); G. S. LaRue, W. M. Fairbank and J. D. Phillips, *Phys. Rev. Lett.* **42**, 142, 1019(E) (1979); G. S. LaRue, J. D. Phillips and W. M. Fairbank, *Phys. Rev. Lett.* **46**, 967 (1981).

[4]   B. Cabrera, *Proceedings of LT14* **4**, 270 (1975); B. Cabrera, *AIP Conference Series* **44**, 73 (1978).

[5]   G. 't Hooft, *Nucl. Phys.* **79B**, 276 (1974) and *Nucl. Phys.* **105B**, 538 (1976); A. M. Polyakov, *JETP Lett.* **20**, 194 (1974).

[6]   For a detailed review of recent monopole theory and experiment see *Magnetic Monopoles*, edited by R. A. Carrigan, Jr., and W. P. Trower (Plenum, New York, 1983).

[7]   B. Cabrera, *Phys. Rev. Lett.* **48**, 1378 (1982); see also papers by B. Cabrera in reference [6] and *Phys. Rev. Lett.* **51**, 1933 (1983).

The ^{3}He Nuclear Gyroscope and a Precision Electric Dipole Moment Measurement

M. Taber and B. Cabrera

1. INTRODUCTION

In 1962 William M. Fairbank and William O. Hamilton proposed a high resolution experiment to search for an electric dipole moment on the ^{3}He nucleus. The existence of an EDM on any particle or system of particles in a stationary quantum state would be direct evidence for violation of time reversal (T) invariance and, less importantly, violation of parity (P). The T operation is equivalent to observing the dynamics of a system on a movie film running backwards. If the reverse sequence is physically allowed then the system is invariant under the T operator. A magnetic dipole precessing in a magnetic field satisfies T invariance since the direction of the currents producing the magnetic field as well as those producing the dipole moment reverse. The result is a precession in the opposite direction which is consistent with the time reversed sequence. However, for an EDM precessing in an electric field, only stationary electric charges are present and the opposite precession direction observed in the reverse time sequence is at odds with the unchanged physical precession.

Once T invariance, along with parity (P) and charge conjugation (C) invariances, were thought to be independent fundamental physical symmetries. The combined operation of CPT was known to be equivalent to relativistic Lorentz invariance and even today is not expected to be violated. However, with the discovery of P nonconservation in ^{60}Co beta decay in 1957 and then of CP and T nonconservation in the decay of K^0 in 1964, it is clear that T violating processes do exist in nature and that elementary EDM's can also exist.

In the early 1950's, before any experimental evidence had been found or theoretical interest had been developed, Smith, Purcell and Ramsey questioned the validity of P invariance and looked for an EDM on the neutron. More than thirty years of work on ever more precise neutron experiments have lowered the limits of the neutron EDM by many orders of magnitude (now $< 1.6 \times 10^{-24}$ e cm at 90% confidence level), but none has been found [1].

EDM experiments play a crucial role as a test for theoretical models which have attributed T violation to electromagnetic, strong, weak and superweak interactions, frequently in second order or higher. Many models have been excluded by experiment, but the basic mechanism for T violation remains obscure.

Although the electrical neutrality of the neutron has made it a natural candidate for EDM experiments, the importance of T violation and the very limited experimental evidence for its existence (only in K^0 decay) have led to searches for EDM's in other particles or systems of particles. These include heavy atoms such as Cs and Xe as well as molecular TlF. At Stanford we have pursued the use of ^{3}He not for ease of electric field application, but rather because of the simple structure of the ^{3}He atom and its suitability in a cryogenic environment where a number of powerful techniques can be brought to bear.

The Fairbank and Hamilton proposal to develop a ^{3}He nuclear gyroscope was based on an analysis by Leonard Schiff. At first glance the application of an electric field on the ^{3}He nucleus seems impossible since in the steady state an electrically charged particle would be at a field null. However, Schiff showed that the additional magnetic interaction of the magnetic dipole moment of the ^{3}He nucleus with the distorted electron cloud would result in a finite first order electric field at the nucleus, since now the sum of two forces is zero in equilibrium [2]. An electric field $\sim 10^{-7}$ times that of the applied field is seen at the nucleus. Although a shielding factor of this magnitude is a significant liability when compared to neutrons in an unattenuated electric field, there are several compensating features.

For a spherical sample cell containing a dilute mixture of spin-polarized ^{3}He in ^{4}He, relaxation times approaching one week are expected and have been experimentally verified. Ultra low magnetic field shielding (see paper entitled "Near Zero Magnetic Fields with Superconducting Shields" in this volume) and SQUID magnetometry allow full utilization of these long relaxation times for EDM measurements. The neutron lifetime precludes observation times longer than 10^3 sec, and even the most sensitive recent experiments are statistically limited by the number of confined neutrons in each measurement, typically under 1000. Because of the longer relaxation times and the much larger number of polarized atoms (about 10^{16}), measurements of the ^{3}He EDM can reach comparable sensitivities to that of the neutron measurements.

Early work by R. H. Romer at Duke University on ^{3}He relaxation [3] and by King Walters at Texas Instruments and later at Rice University on optical pumping polarization techniques [4] laid the experimental groundwork. At Stanford, Isaac Bass began work on the room temperature optical pumping techniques in the late 1960's, and in 1969 Bass and the authors made the first unsuccessful attempts to observe the polarization signal with a SQUID. By the mid 1970's we had completed the construction of a dedicated apparatus for making relaxation measurements. It included an ultra low field superconducting shield. Measurements completed in 1978 demonstrated relaxation times in excess of 140 hours using a hydrogen wall coating [5]. Since that time we have completed the design of a precision quartz ^{3}He nuclear gyroscope. Once having demonstrated precise angular resolution of the magnetization vector, high voltage electrodes will be added for application of electric fields and EDM measurements.

This paper begins with a brief summary of the nuclear gyroscope theory, followed by a description of the relaxation time measurements and of the design of a nuclear gyroscope for use in future high precision EDM measurements on the ^{3}He nucleus.

2. THEORY OF ^{3}HE NUCLEAR GYROSCOPE

The quantum mechanical equations of motion for the expectation value of a nuclear magnetic dipole moment $\boldsymbol{\mu}$ are

$$d\langle\boldsymbol{\mu}\rangle/dt = \gamma\langle\boldsymbol{\mu}\rangle \times \mathbf{B} \tag{1}$$

for the nucleus placed in a magnetic field $\mathbf{B}$, where $\boldsymbol{\mu} = \gamma\hbar\mathbf{I}$, the gyromagnetic ratio γ times the nuclear spin vector operator $\hbar\mathbf{I}$. This equation is formally identical to the classical equation of motion obeyed by any freely spinning body possessing a magnetic dipole moment aligned along its angular momentum vector. For spin 1/2 particles (such as the ^{3}He nucleus), there are no higher multipole moments than dipole. Thus equation (1) represents the total interaction with an external magnetic field.

If a large number of noninteracting ^{3}He atoms are placed in a sample volume and are subjected to a homogeneous magnetic field $\mathbf{B} = \mathbf{B}_0$, then the motion of the total angular momentum can be treated classically, provided the components of the total angular momentum are large compared to $\hbar$. Thus we may replace the expectation values of the individual spin vectors in equation (1) by the classical variable $\mathbf{M}$, whence

$$d\mathbf{M}/dt = \gamma(\mathbf{M} \times \mathbf{B}) \quad , \qquad (2)$$

where $\mathbf{M}$ is the sample magnetization.

We can understand equation (2) by transforming it to an arbitrary rotating frame using the kinematic relation

$$(d/dt)_{\text{inertial frame}} = (d/dt)_{\text{rot frame}} + \boldsymbol{\omega} \times \quad , \qquad (3)$$

where $\boldsymbol{\omega}$ is the angular velocity of the rotating frame. Thus

$$(d\mathbf{M}/dt)_{\text{rot frame}} = \mathbf{M} \times (\gamma\mathbf{B}_0 + \boldsymbol{\omega}) \quad . \qquad (4)$$

Now if $\boldsymbol{\omega} = -\gamma\mathbf{B}_0$, then $(d\mathbf{M}/dt)_{\text{rot frame}} = 0$. Thus the application of a uniform magnetic field has the effect of causing $\mathbf{M}$ to rotate (precess) with angular velocity $\boldsymbol{\omega}_0 = -\gamma\mathbf{B}_0$. The application of a magnetic field to a spin is equivalent to a rotation where the proportionality constant, γ, the gyromagnetic ratio, depends on the internal structure of the particle or system of particles. For ^{3}He, $\gamma = 2.038 \times 10^4$ rad sec^{-1} gauss^{-1}.

The simplicity of the equation of motion leads to the possibility of using the angular momentum of nuclear spins such as ^{3}He for making a precise gyroscope. In general the following conditions must be satisfied:

(1)   The spins must be polarized, *i.e.*, preferentially aligned so that there is an easily detectable net magnetization. We use an optical polarization technique which provides polarization levels many orders of magnitude larger than those attainable with equilibrium thermal polarization in a magnetic field. This process is equivalent to spinning up a mechanical rotor.

(2)   The nuclear relaxation time, equivalent to the spin-down time of a mechanical gyro, must be usably long. Nuclear relaxation arises from spin-spin interactions modulated by relative interatomic motions and from spin-field interactions caused by diffusion through magnetic field gradients, which must be kept to a minimum.

(3)   The magnetic field environment must be intrinsically very stable. We use superconducting shelding to provide both a low absolute magnetic field and high stability. Inside of this environment a uniform magnetic field, $\mathbf{B}_0$, is applied with precision coils and a persistent supercurrent. The component of $\mathbf{M}$ transverse to $\mathbf{B}_0$ will precess at the Larmor frequency. A rotation about $\mathbf{B}_0$ will result in a phase shift of the Larmor frequency which can be precisely measured. Rotations transverse to $\mathbf{B}_0$ will produce a much smaller reponse, provided that they are slow compared to ω_0. Thus, if rotations about all three orthogonal directions are to be detected, three independent systems with orthogonal axes should be used.

(4)   The final requirement for a nuclear gyro is to detect the precessing magnetization and its frequency or phase. We use a SQUID magnetometer which provides very high sensitivity and a minimum of back reaction on the nuclear spin system.

Our goal is to develop a ^{3}He gyro suitable for an EDM experiment. The effect of an electric field parallel to $\mathbf{B}_0$ interacting with a hypothetical EDM is to produce a small frequency shift $\Delta\omega$ in the Larmor frequency ω_0. The change in B_0 that would correspond to $2\Delta\omega$ (the frequency shift from reversing the electric field) is given by

$$\Delta B_0 = 2fEDe/\gamma\hbar \quad , \tag{5}$$

where $f \sim 10^{-7}$ is the Schiff screening factor, E is the applied electric field, and De is the EDM. For $E = 100$ kV/cm and $D = 10^{-24}$ cm, the present level of measurements on the neutron, we obtain $\Delta B_0 = 10^{-15}$ gauss! With the best dc SQUID's available today, it would take several hours of integration to detect such a change. The small size of ΔB_0 indicates how important it is to achieve the utmost in magnetic field stability.

3.   PROGRESS TOWARDS A CRYOGENIC ^{3}HE NUCLEAR GYRO

3.1   Relaxation Time Studies of ^{3}He-^{4}He Mixtures

From the outset, we knew that it was important to achieve relaxation times of days. Liquid ^{3}He had been measured to have a relaxation of 300–500 sec and was known to be dominated by intrinsic spin-spin interactions. Early work by Romer [3] had shown that a relaxation time of 2 hrs could be achieved using a dilute 1.7% ^{3}He in liquid ^{4}He. If this result was dominated by ^{3}He-^{3}He interactions rather than wall induced relaxation, then further improvements could be achieved by further dilution. A density concentration below 0.1% should then yield relaxation times in excess of a day.

We designed an apparatus (figure 1) to test the feasibility of such long relaxation times and to approximate the operation of a ^{3}He gyro. Measurements were made inside an 8 inch diameter ultra low field shield with an ambient magnetic field of 2 μG (see paper entitled "Near Zero Magnetic Fields with Superconducting Shields" in this volume). The sample cell was a 1 cm diameter blown Pyrex bulb connected to a 0.5 mm Pyrex capillary. Helmholtz field coils and an rf-biased SQUID with superconducting pickup loop coupled to the sample bulb completed the cryogenic apparatus. The room temperature portions of the apparatus consisted of gas handling and purification equipment and an optical pumping setup capable of polarizing, to 5–10%, 300 cm^3 of pure ^{3}He at 1 torr pressure [5].

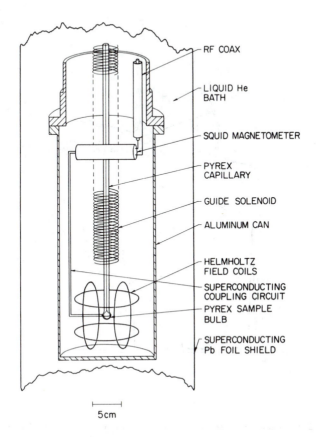

RF COAX

LIQUID He
BATH

SQUID MAGNETOMETER

PYREX
CAPILLARY

GUIDE SOLENOID

ALUMINUM CAN

HELMHOLTZ
FIELD COILS

SUPERCONDUCTING
COUPLING CIRCUIT

PYREX SAMPLE
BULB

SUPERCONDUCTING
Pb FOIL SHIELD

5cm

FIGURE 1. Schematic of cryogenic portion of apparatus for measuring polarized ^{3}He relaxation times.

Once a sample of ^{3}He was polarized, highly purified ^{4}He was admitted to the optical pumping cell and the mixture was then allowed to flow down the 2 m long capillary to the cryogenic sample cell. A solenoid around the capillary was used to insure that the ^{3}He nuclei were not subjected to any abrupt changes in field on their path to the cell. With this precaution we found that a mixture of 0.07% ^{3}He in ^{4}He retained 40–50% of its initial polarization after filling the sample cell. The resulting sample magnetization was sufficient to yield a signal-to-noise ratio of 1000 in a 3 Hz bandwidth.

With ample signal, we carried out a series of relaxation time studies primarily at 0.07% concentration. The longitudinal relaxation time T_1 is defined as the exponential time constant of the magnetization aligned with the applied field $\mathbf{B}_0$. Although at the lowest magnetic fields we found T_1 to be very short and dominated by ferromagnetic contamination in the pickup

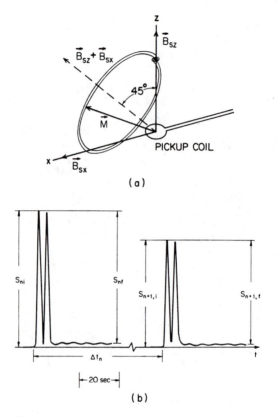

FIGURE 2. (a) Diagram of the transitory precession technique used to measure the magnitude of **M**. (b) Typical data record of two sequential measurements of **M**. The time between measurements, Δt_n, ranged to values greater than a day for the longest T_1's.

loop coil form, by applying a much larger field with the Helmholtz pair, we were able to determine the value that T_1 would have in the absence of the contamination, and obtained 40 hrs (see figures 2a and 2b). When a thin solid H_2 wall coating was used in the sample cell, this T_1 rose to 140 hrs (over 5 days!), very near to the theoretical estimate of 160 hrs from the ^3He-^3He interactions alone. The results show that the relaxation in the bare Pyrex bulb is in fact dominated by the wall relaxation and a T_1 of days is feasible particularly using the H_2 coating.

However, it is T_2, the exponential time constant for the precessing component of the magnetization, that is important for gyro and EDM applications. In general $T_1 \geq T_2$, with T_2 approaching T_1 when the diffusion of ^3He across the sample cell is sufficiently rapid and the magnetic field gradient is sufficiently small that the magnetization remains uniform throughout the cell. It is also important that the sample be highly spherical since any nonellipsoidal magnetized sample will not produce a uniform internal field. In our original apparatus this was not the case and T_2 was limited to 30 min in a gas mixture with small magnetization.

The problems associated with sample asphericity and ferromagnetic contaminants are not particularly troublesome since they are well understood and easily eliminated by proper design and choice of materials. Of somewhat greater concern is the reproducibility of the effectiveness of the H_2 wall coating. As each warm gas sample is introduced into the cryogenic cell, the H_2 may vaporize and recondense. This process has an unknown effect on the H_2 distribution and the ortho-para (spins aligned/spins antialigned) ratio. However, both theoretical analysis and experimental results indicate good prospects for the development of an ultrastable ^{3}He gyro and a ^{3}He EDM experiment competitive with the neutron measurements, provided a carefully designed precision apparatus is built.

3.2   New Precision ^{3}He Gyro Apparatus

We have designed and built the components of an ultrastable ^{3}He gyro. Figure 3a shows a schematic of the apparatus, which is made of ground fused silica (quartz). The inner surface of the outer housing is accurately

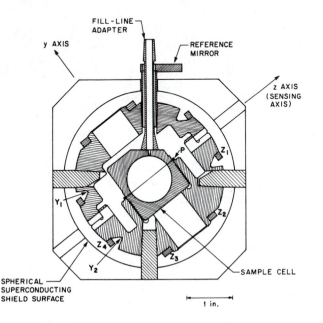

FIGURE 3a. Schematic diagram of design for ultrastable ^{3}He gyroscope.

FIGURE 3b. Components of the fused silica gyroscope housing.

spherical and concentric with the sample cell and serves as a substrate for a superconducting shield. The sample cell is substantially larger than in the previous apparatus (3.8 cm *versus* 1 cm) to minimize asphericity and surface effects. The two halves of the sample cell are to be optically contacted for the final assembly. The primary field coils are oriented such that B_0 can be aligned with the earth's axis of rotation so that small tilts will not adversely affect gyro stability. The quartz components of this new apparatus are shown in figure 3b. We are eager to complete the assembly and test the stability of this new gyro. Later we intend to include electric field electrodes. The space for these has also been designed into the quartz structure.

Acknowledgments

We would like to thank Bill Fairbank for many early ideas on the ^{3}He gyro and EDM experiments and for his encouragement and perpetual enthusiasm. Bill Hamilton and Isaac Bass also contributed much to the earlier phases of the work. Most recently this work has been supported by the U. S. Air Force Office of Scientific Research under contract no. F49620-78-C-0088.

References

[1] See, for example, W. B. Dress, P. D. Miller, J. M. Pendelbury, P. Perrin and N. F. Ramsey, *Phys. Rev. D* **15**, 9 (1977).

[2] L. I. Schiff, *Phys. Rev.* **132**, 2194 (1963).

[3] R. H. Romer, *Phys. Rev.* **115**, 1415 (1959).

[4] W. A. Fitzsimmons, L. L. Tankersley and G. K. Walters, *Phys. Rev.* **179**, 156 (1969).

[5] M. A. Taber, Ph.D. Thesis, Stanford University, 1978 (unpublished), and *J. de Physique* **39**, C6–192 (1978).

CHAPTER VI

Gravitation and Astrophysics

(*Top*) Leonard Isaac Schiff (1915–1971). (*Middle*) Daniel DeBra, William Fairbank, Francis Everitt and Robert H. Cannon, Jr., examining a scale model of the GP-B orbiting gyroscope experiment , 1980. (*Bottom*) William Fairbank, Michael McAshan and Peter Michelson discussing plans for cooling the 4800 kg gravitational wave detector to ultralow temperature.

Gravitation and Astrophysics: Introduction

P. F. Michelson and C. W. F. Everitt

While still at Duke, William Fairbank became interested in the experimental study of the gravitational interaction. As early as 1957 he was considering methods of measuring the force of gravity on antimatter. Soon after Fairbank arrived at Stanford, L. I. Schiff, Fairbank and R. Cannon proposed a method of using an orbiting gyroscope to perform a new test of Einstein's general theory. Shortly after the discovery of the Crab pulsar, Fairbank and W. O. Hamilton began to consider the advantages of cooling a Weber-type gravity wave detector to extremely low temperatures. A few years later, P. W. Worden, Jr., and C. W. F. Everitt began the development of a cryogenic experiment to test the equivalence principle.

All of these measurements require techniques of extreme sensitivity, low noise and great mechnical and electrical stability. From the beginning of these investigations, Fairbank's philosophy recognized that two types of advantages emerge from cooling to very low temperatures: one based on macroscopic quantum effects, namely superconductivity and superfluidity, and the other based on the general reduction of kT thermal noise, thermal expansion, thermal emf, creep, *etc.* Most of the papers in this chapter are concerned with the application of this philosophy to fundamental experiments in gravitation.

Kip S. Thorne discusses the theoretical foundations and astrophysical significance of the Stanford gyroscope experiment. An overview of the experiment and its historical origins are presented in the paper by Everitt. The important technical details of the experiment are discussed in a series of papers by J. A. Lipa, G. M. Keiser, J. T. Anderson, J. P. Turneaure, E. A. Cornell, P. D. Levine, R. A. Van Patten, J. V. Breakwell and D. B. De-Bra. A related paper by Remo Ruffini summarizes a variety of investigations in cosmology and astrophysics, many of which also involve gravitational effects of rotation.

The subject of cryogenic gravity wave detectors is discussed in three separate papers by contributors from the three institutions that began a collaboration more than a decade ago. P. F. Michelson and colleagues describe the Stanford University detector and efforts there to build a quantum-limited detector. W. O. Hamilton (Louisiana State University) discusses the fundamental aspects of gravity wave detection using resonant-mass detectors and compares different approaches to solving the electromechanical transduction problem. Edoardo Amaldi describes the efforts at the University of Rome to develop a quantum-limited detector capable of seeing predicted events due to gravitational collapses occurring in the Virgo cluster.

In an area related to gravity wave detection, Ho Jung Paik describes precision gravity experiments that use superconducting accelerometers. The accelerometer he describes is an extension of the transducer technology developed for the Stanford gravity wave detector. A different type of accelerometer with a different application is described in P. W. Worden's paper on orbiting and ground-based equivalence principle experiments.

J. M. Goodkind describes the measurement of small gravity changes on the surface of the earth using a superconducting gravimeter that he and his colleagues have developed. V. S. Tuman discusses the possibility of using a similar instrument to detect small changes in the eigenvibrations of the earth. The earth eigenmodes may be excited by low frequency gravitational radiation.

In a somewhat different area, Robert Hofstadter discusses the development of a high energy gamma ray astronomy experiment that will soon be placed in orbit by the space shuttle. Fundamental techniques in high energy physics developed by Hofstadter and his colleagues have made possible the construction of this observatory. F. C. Witteborn, J. H. Miller and M. W. Werner describe the low temperature optics being developed for an experiment at the opposite end of the electromagnetic spectrum, the Space Infrared Telescope Facility. These two orbiting astrophysical facilities will observe the universe in regions of the electromagnetic spectrum that are inaccessible to earth-based observations and therefore require the "near zero" atmospheric pressure of space.

Gravitomagnetism, Jets in Quasars, and the Stanford Gyroscope Experiment

Kip S. Thorne

William Fairbank is tremendously lucky—or is he clairvoyant? He initiated the Stanford gyroscope experiment in 1961 as a search for the dragging of inertial frames by the earth's rotation—an effect so small as to be interesting in principle, but not in practice. However, today, twenty years later, just when technology development for the gyroscope experiment has reached completion, the theoretical framework for the experiment is changing. Physicists now see the experiment as a search for the gravitational analog of a magnetic field; and astrophysicists now invoke this "gravitomagnetic field" as a power source and alignment force for recently discovered jets squirting out of quasars and galactic nuclei. Suddenly the gyroscope experiment is a crucial test of the mechanism of the most violent explosions in our universe.

In this paper[1] I shall not discuss the gyroscope experiment itself; for that see Everitt [1]. Rather, I shall describe the new astrophysical motivations

---

[1]Some of this paper is adapted, with changes, from §3 of the author's chapter in *Quantum Optics, Experimental Gravitation, and Measurement Theory*, edited by P. Meystre and M. O. Scully (Plenum Press, New York, 1983).

for it; and while doing so I shall introduce you to some unusual but powerful viewpoints about general relativistic gravity: (i) the split of the spacetime metric $g_{\alpha\beta}$ and its associated forces into a "gravitoelectric field" **g**,
a "gravitomagnetic field" **H**, and a space curvature (*not spacetime* curvature) with metric g_{jk}; and (ii) the "membrane paradigm" for black holes,
with its membrane-like event horizon endowed with electric charge, electric
current, electric resistance, and an electric battery.

1. THE SPLIT OF SPACETIME INTO SPACE PLUS TIME

When spacetime is highly dynamical, for example around two colliding
black holes, there is no natural, preferred way to split spacetime up into
space plus time. This fact has driven relativists, beginning with Einstein,
to describe gravity in terms of a unified, four-dimensional spacetime with
dynamically evolving four-dimensional curvature.

On the other hand, astrophysicists and experimental physicists usually
deal with situations where spacetime is stationary or nearly stationary
rather than dynamical, for example the spacetime around the earth or
around a quiescent black hole. In such cases stationarity dictates a preferred way to slice spacetime up into three-dimensional space plus one-
dimensional time. Although such "3 + 1 splits" are not treated in standard textbooks on general relativity, they are used widely by professional
relativists—for example, in numerical solutions of the Einstein field
equations (*e.g.*, Smarr [2]), in the quantization of general relativity (*e.g.*,
Wheeler [3]), in astrophysical studies of black holes (*e.g.*, Macdonald and
Thorne [4]), and in analyses of laboratory experiments to test general relativity (*e.g.*, Braginsky, Caves and Thorne [5]). There is no approximation
inherent in such a $3+1$ split; it is merely a rewrite of full general relativity
in a new mathematical language.

The 3 + 1 split regards three-dimensional space as curved rather than
Euclidean; its metric g_{jk} (in an appropriate coordinate system) is just the
spatial part of the spacetime metric $g_{\alpha\beta}$. In this curved 3-space reside two
gravitational potentials: a "gravitoelectric" scalar potential Φ, which is
essentially the time-time part g_{00} of the spacetime metric; and a "gravitomagnetic" vector potential γ, which is essentially the time-space part g_{0j}
of the spacetime metric. The decomposition of $g_{\alpha\beta}$ into g_{jk}, Φ, and γ is
analogous to the decomposition of the electromagnetic four-vector poten-
tial A_{α} into an electrical scalar potential $\psi = -A_0$ and a magnetic vector
potential $\mathbf{A} = A_j$.

The analogy between gravity and electromagnetism is especially remarkable in the case of systems with weak gravity and low velocities ($v \ll c$),

such as the earth and a gyroscope orbiting it. For such systems a crucial role is played by the "gravitoelectric field" **g** and the "gravitomagnetic field" **H**, which are constructed from Φ and γ in a manner familiar from electromagnetism:

$$\mathbf{g} = -\boldsymbol{\nabla}\Phi, \quad \mathbf{H} = \boldsymbol{\nabla} \times \boldsymbol{\gamma}, \quad \Phi = -\frac{1}{2}(g_{00} + 1)c^2, \quad \gamma_j = g_{0j} \quad . \quad (1)$$

(Here c is the speed of light.) It turns out that **g** is just the Newtonian gravitational acceleration, and **H** is a force field of which Newton was unaware because in the dynamics of the solar system its effects are $\sim 10^{12}$ times smaller than those of **g**.

For weak gravity, low velocity systems the general relativistic field equations for **g** and **H** become almost identical to Maxwell's equations, and the geodesic equation of motion for an uncharged particle is identical to the Lorentz force law (see *e.g.*, Forward [6]; Braginsky, Caves and Thorne [5]):

$$\boldsymbol{\nabla} \cdot \mathbf{g} \;=\; \overset{\downarrow}{-}\, 4\pi G\rho, \quad \boldsymbol{\nabla} \times \mathbf{g} = \overset{\downarrow}{0} \;,$$

$$\boldsymbol{\nabla} \cdot \mathbf{H} \;=\; 0, \quad \boldsymbol{\nabla} \times \mathbf{H} = \overset{\downarrow}{4}\, [\overset{\downarrow}{-}\, 4\pi G\rho \mathbf{v}/c + (1/c)(\partial \mathbf{g}/\partial t)] \;\; ; \tag{2}$$

$$d\mathbf{v}/dt = \mathbf{g} + (\mathbf{v}/c) \times \mathbf{H} \;\; . \tag{3}$$

Note that the only differences from Maxwell's equations are (i) minus signs in the source terms (vertical arrows), which cause gravity to be attractive rather than repulsive; (ii) a factor 4 in the strength of **H**, presumably due to gravity being associated with a spin-2 field rather than spin-1; (iii) the replacement of charge density by mass density ρ times Newton's gravitation constant G; (iv) the replacement of charge current by $G\rho\mathbf{v}$ where **v** is the velocity of the mass ρ; and (v) the absence of $-(1/c)(\partial\mathbf{H}/\partial t)$ in the $\boldsymbol{\nabla} \times \mathbf{g}$ equation. In fact, $-(1/c)(\partial\mathbf{H}/\partial t)$ is absent because I have limited myself to terms which are first order in the v/c of the gravitating mass; including second order terms restores the $-(1/c)(\partial\mathbf{H}/\partial t)$ but also introduces non-Maxwell-like terms elsewhere in the field equations; see Braginsky, Caves and Thorne [5] for details and for other approximations underlying equations (2) and (3).

2. THE EXTERIOR OF
A ROTATING SPHERICAL BODY

From our electrodynamical experience we can infer immediately that any rotating spherical body (*e.g.*, the sun or the earth) will be surrounded by a radial gravitoelectric (Newtonian) field **g** and a dipolar gravitomagnetic field **H**:

$$\mathbf{g} = -\frac{GM}{r^2}\mathbf{e}_r \ , \quad \mathbf{H} = \frac{2G}{c}\left[\frac{\mathbf{S} - 3(\mathbf{S} \cdot \mathbf{e}_r)\mathbf{e}_r}{r^3}\right] \tag{4}$$

The gravitoelectric monopole moment is the body's mass M; the gravito-magnetic dipole moment is its spin angular momentum **S**.

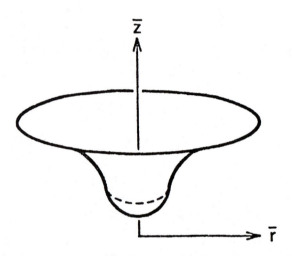

FIGURE 1. Embedding diagram for the equatorial plane of a gravitating spherical body. The curved space outside the body's surface (dashed circle) is described by equations (5) and (6). For a mathematical description of the bowl-like interior see, *e.g.*, [7], pp. 612–615.

The curvature of the space around the spherical body is constant in time and can be described either by the weak-field limit of Schwarzschild's spatial metric

$$ds^2 = (1 + r_g/r)dr^2 + r^2(d\theta^2 + \sin^2\theta d\phi^2) \ , \quad r_g \equiv 2GM/c^2 \ll r \quad , \tag{5}$$

or, pictorially, as follows. As we move from one equatorial circle surrounding the body out to another, the circle's circumference increases less rapidly than Euclid would demand: $d(\text{circumference})/d(\text{radius}) < 2\pi$. If we imagine extracting the equatorial plane from the curved space of the body and embedding it with unchanged geometry in a flat Euclidean 3-space with coordinates $\bar{r}, \bar{z}, \bar{\phi}$, then we obtain the paraboloidal surface (figure 1)

$$\bar{z} = 2\sqrt{r_g(\bar{r} - r_g)} \ , \quad ds^2 = d\bar{z}^2 + d\bar{r}^2 + \bar{r}^2 d\bar{\phi}^2 \quad ; \tag{6}$$

cf. pp. 612–615 of Misner, Thorne and Wheeler [7].

3. RELATIVISTIC PRECESSIONS OF GYROSCOPES

Consider a gyroscope with spin angular momentum **s** in orbit around a rotating earth (the Stanford gyroscope experiment). The three aspects of the earth's gravity [**g** field, **H** field, and space curvature; equations (4), (5), (6)] will each produce a precession of the gyroscope relative to the distant stars.

The interaction of the gyroscope's spin **s** with the earth's gravitomagnetic field **H** is analogous to the interaction of a magnetic dipole $\boldsymbol{\mu}$ with a magnetic field **B**. Just as a torque $\boldsymbol{\mu} \times$ **B** acts in the magnetic case, so a torque $(1/2)$**s** $\times$ **H**$/c$ acts in the gravitational case. [Equations (1), (2), (4) dictate that $\boldsymbol{\mu} \to (1/2)$**s**$/c$, **B** $\to$ **H**.] The gyroscope's angular momentum is changed by this torque:

$$\frac{d\mathbf{s}}{dt} = \frac{1}{2c}\mathbf{s} \times \mathbf{H} \; ; \qquad \begin{array}{l} \text{\textbf{s} precesses with the "gravitomagnetic"} \\ \text{angular velocity } \boldsymbol{\Omega}_{GM} = -\mathbf{H}/2c \end{array} \qquad (7)$$

Note that $\boldsymbol{\Omega}_{GM}$ is independent of the structure of the gyroscope; this is a manifestation of the principle of equivalence, and it permits one to regard the precession as a "dragging of inertial frames" by the rotation of the earth. This gravitomagnetic precession is often called the "Lense-Thirring" precession, since Lense and Thirring were the first to discover it in the equations of general relativity. For a gyroscope in the 500 km high polar orbit of the Stanford experiment, by averaging $\boldsymbol{\Omega}_{GM}$ over the orbit and using equation (4) for **H**, we obtain

$$\Omega_{GM} = \frac{G}{2c^2}\frac{S}{r^3} \simeq 0.05 \; \frac{\text{arc-seconds}}{\text{year}} \quad , \qquad (8a)$$

a precession 50 times greater than the design sensitivity of the Stanford experiment.

The interaction of the gyroscope with the earth's gravitoelectric field **g** is analogous to the interaction of a classical spinning electron with the Coulomb electric field **E** of an atomic nucleus. Just as motion of the electron through **E** induces in the electron's rest frame a magnetic field $\mathbf{B}_{induced} = -(\mathbf{v}/c) \times \mathbf{E}$ and a torque $\boldsymbol{\mu} \times \mathbf{B}_{induced}$ and a resulting precession of the electron spin ("atomic spin-orbit coupling"), similarly motion of the gyroscope through **g** produces in the gyroscope's rest frame a gravitomagnetic field $\mathbf{H}_{induced} = -(\mathbf{v}/c) \times \mathbf{g}$ and a torque $((1/2)\mathbf{s}/c) \times \mathbf{H}_{induced}$ and a resulting "spin-orbit" precession of the gyroscope with

$$\Omega_{SO} = -\frac{1}{2c}H_{induced} = \left[\frac{r_g}{2r}\right]^{5/2} \frac{n}{r_g/c} \simeq 2.3 \; \frac{\text{arc-seconds}}{\text{year}} \quad . \qquad (8b)$$

Here $r_g = 2GM/c^2 = 0.89$ cm is the earth's "gravitational radius," **n** is the unit normal to the orbital plane, and the orbit is assumed to be 500 kilometers high ($r = 6371 + 500$ km).

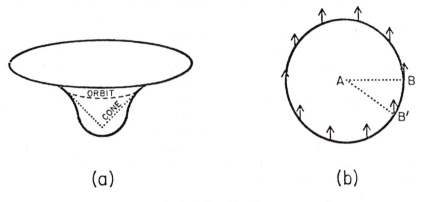

(a)   (b)

FIGURE 2. Precession of a gyroscope induced by the curvature of space.

In the hydrogen atom there is also a "Thomas precession" which results from the fact that the product of two "velocity boosts" is a combined "boost and rotation." In our gravitational problem the Thomas precession is absent because the gyroscope is presumed to be in a free fall orbit—*i.e.*, it is not accelerated relative to local inertial frames; there are no "boosts." On the other hand, the gravitational field has an aspect, the curvature of space, with no electromagnetic analog; and that space curvature produces an unfamiliar type of precession. In figure 2a we see an embedding diagram (*cf.* figure 1) for the curved space of the earth. The dashed line is the circular orbit of the gyroscope. The gyroscope can only feel that part of space which is in the immediate neighborhood of its orbit. This allows the pedagogical simplification of replacing the paraboloidal embedding surface of the real curved space by a cone (dotted line) which is tangent to the paraboloid at the gyroscope's orbit. Such a cone can be constructed by drawing a circle on a flat sheet of paper (figure 2b), cutting the pie slice $(B-A-B')$ out of it, and joining the edges $A-B$ and $A-B'$ together. As the gyroscope orbits around the cone it always keeps its spin in the same fixed direction on the flat-sheet-of-paper geometry of the cone's surface (arrows in figure 2b; no local precession). However, as one easily sees by cutting out the pie slice and pasting the cone together, there will be a net precession of the spin when the gyroscope returns to its starting point $B = B'$. From the shape of the embedding surface [equation (6)] and the cone construction one easily finds that the net precession angle after a single Keplerian orbital period $(\pi r_g/c)(2r/r_g)^{3/2}$ corresponds to a precession angular velocity

$$\Omega_{SC} = 2\left[\frac{r_g}{2r}\right]^{5/2}\frac{\mathbf{n}}{r_g/c} = 2\Omega_{SO} \quad . \tag{8c}$$

Note that this "space-curvature precession" has twice the angular velocity of the spin-orbit precession. The two together bear the name "geodetic precession":

$$\Omega_{geo} = \Omega_{SO} + \Omega_{SC} = 3 \left[\frac{r_g}{2r}\right]^{5/2} \frac{\mathbf{n}}{r_g/c} \simeq 6.9 \frac{\text{arc-seconds}}{\text{year}} \ . \tag{8d}$$

The reader may find it enlightening to compare the above derivations of the relativistic precession formulae with the standard derivation in [7], §40.7.

4. THE MEMBRANE PARADIGM FOR BLACK HOLES

As a prelude to my discussion of gravitomagnetic effects in quasars and galactic nuclei, I shall describe the "membrane paradigm" for black holes [18].

"Paradigm" is a word used by the historian of science Thomas Kuhn [8] to describe the body of problem solving techniques, mental pictures and mathematical formalisms used by a specific community of scientists in doing research on a specific set of topics. "Magnetohydrodynamics," with its magnetic field lines like stretchy rubber bands frozen into a plasma, is one paradigm. The "quasilinear theory" of weak plasma turbulence, with its elementary-particle-like plasmons, is another quite different paradigm.

Each paradigm has its own realm of validity and its own regime of computational and conceptual power. Occasionally there occurs a scientific revolution in which an old paradigm (e.g., the "old quantum theory" of pre-19[1424]) is replaced by a new paradigm (e.g., the wave mechanics of Schrödinger). On other occasions two very different paradigms exist side by side with a finite domain of overlap—each mathematically equivalent to the other in the domain of overlap, but each having a very different mathematical formalism and set of mental pictures. (Example: magnetohydrodynamics and quasilinear theory with an overlap domain which includes Alfvén waves.)

The theory of black holes was dominated before the mid-1960's by a "frozen-star paradigm," which made extensive use of "Schwarzschild coordinates" with their infinite gravitational redshift at the horizon, and which emphasized mental pictures of collapsing stars that become "frozen" at the horizon because of the redshift. (See, e.g., Zel'dovich and Novikov [9].) A scientific revolution in the 1960's replaced this frozen-star paradigm by the "black hole paradigm," which makes extensive use of "Kruskal coordinates," "Eddington-Finkelstein coordinates," "Penrose diagrams," and mental pictures of stars that collapse quickly through the horizon and into a singularity. (See, *e.g.*, [7].) Since the mid-1970's a third paradigm has been taking hold. First codified by Damour [10], this "membrane paradigm" (also called "bubble paradigm") treats the horizon of a black hole as a two-dimensional membrane in three-dimensional space—a membrane endowed with electric charge, electric current, electric resistivity, an electric battery, viscosity, surface pressure, temperature, entropy and gravitoelectric and gravitomagnetic fields. The membrane paradigm is mathematically equivalent to the black hole paradigm everywhere outside the horizon—*i.e.*, everywhere of relevance for astrophysics. But the membrane paradigm loses its validity inside the horizon. An observer who falls through the horizon discovers, for example, that the horizon is not really endowed with electric charge and current; it merely looked that way from the outside. The membrane paradigm is still under development; for recent work which married it to the 3 + 1 split of spacetime, see Macdonald and Thorne [4], and Thorne and Macdonald [11]; for a detailed presentation of the paradigm, see [18].

Figure 3 depicts several electromagnetic aspects of the membrane paradigm. At the horizon (membrane) of a black hole the normal component of the electric field $\mathbf{E}_\perp$ is terminated by surface electric charges (charge density σ_H; Gauss's law), and the tangential component of the magnetic field $\mathbf{B}_H$ is terminated by surface currents (current density $\mathbf{J}_H$; Ampère's law). Charge is conserved at the horizon: any volume currents $\mathbf{j}$ flowing into and out of the horizon produce time changes of σ_H and/or divergences of $\mathbf{J}_H$. The tangential component of the electric field, $\mathbf{E}_H$, is *not* terminated at the horizon; rather, it extends into the horizon where it drives the surface currents in accordance with Ohm's law:

$$\mathbf{E}_H = R_H \mathbf{J}_H \ , \quad R_H = 4\pi/c = 377 \text{ ohms} \quad . \tag{9}$$

The normal component of the magnetic field $\mathbf{B}_\perp$ is also not terminated at the horizon; rather, it extends into the horizon where it interacts with the *gravito*magnetic potential γ of equation (1) to produce electric potential drops (battery effect) in the horizon:

$$\int \mathbf{E}_H \cdot d\boldsymbol{\ell} = \Delta V = \int \gamma \times \mathbf{B}_\perp \cdot d\boldsymbol{\ell} \quad . \tag{10}$$

This battery will be crucial in the discussion of power sources for quasars, below. (For further details on figure 3 and on equations (9) and (10), including several subtle but crucial issues in the definitions of $\mathbf{E}_H$ and $\mathbf{B}_H$, see Macdonald and Thorne [4] or see [18].)

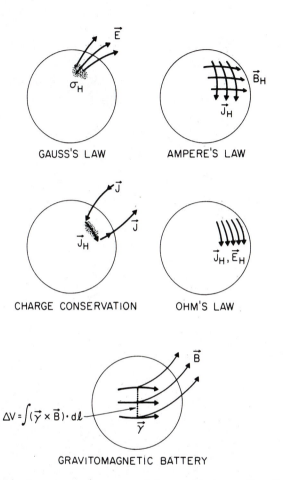

FIGURE 3. Some electromagnetic aspects of the membrane paradigm for black holes.

5. GRAVITOMAGNETISM AND RELATIVISTIC PRECESSION

It is now clear that quasars and other strong extragalactic radio sources are fed power ($P \gtrsim 10^{44}$ erg/sec) by jets of gas and magnetic field, and that each jet is generated by a compact supermassive object ($M \gtrsim 10^7$ solar masses) in the nucleus of a galaxy; *cf.* Begelman, Blandford and Rees [12]. A supermassive black hole is the prime candidate for the compact object.

In some radio sources (*e.g.*, NGC 6251, figure 4), the compact object must hold the jet direction constant for times as long as 10 million years. The only way a black hole can do this is by the gyroscopic

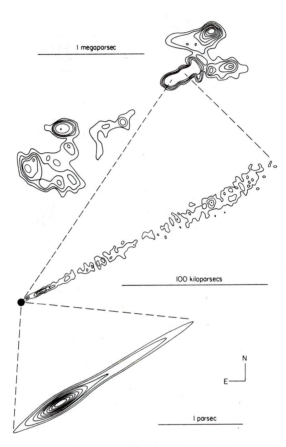

FIGURE 4. Radio maps of the jets in the galaxy NGC 6251; from Readhead, Cohen and Blandford [17].

action of its spin; and the only way it can communicate the direction of its spin to the jet is *via* its gravitomagnetic field **H**. In fact, **H** should produce a gravitomagnetic precession of any accretion disk encircling the hole, and that precession together with the disk's viscosity should drive the inner region of the disk into the hole's equatorial plane (Bardeen and Petterson [13]; figure 5). The resulting configuration has only two preferred directions along which to send jets: the north and south poles of the hole. This mechanism of jet alignment is widely believed by astrophysicists.

Some observed jets precess with precession periods $\gtrsim 10^4$ years. Astrophysicists attribute this to a precession of the central hole's spin. The most promising way that the spin can be made to precess is by orbital motion of the hole (now acting as a "test gyroscope") around a companion hole

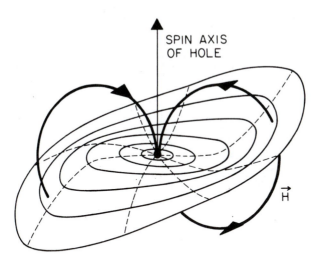

FIGURE 5. The Bardeen-Petterson effect. An accretion disk is driven into the equatorial plane of a black hole.

or other massive object (which acts as a source of **g** and **H** fields and of space curvature). The resulting geodetic precession will be faster than the gravitomagnetic precession, and will have a period

$$\frac{2\pi}{\Omega_{geo}} \sim 10^4 \text{ years } \left[\frac{\text{distance between holes}}{0.01 \text{ parsecs}}\right]^{5/2}$$

$$\left[\frac{\text{mass}}{10^8 \text{ solar masses}}\right]^{-3/2} \tag{11}$$

(Begelman, Blandford and Rees [14]).

There are several plausible models for generation of the jets. One of the most attractive (Blandford and Znajek [15]; Macdonald and Thorne [4]) relies on the rotational energy stored in the hole's gravitomagnetic field as the power source, and relies on the horizon's gravitomagnetic battery, equation (10), as that power source's agent.

Consider a rotating black hole surrounded by a magnetized accretion disk (figure 6). As the disk's plasma accretes, it drags magnetic field lines with itself, depositing them on the horizon. Although the **B** fields in the disk may be very chaotic, when deposited on the horizon they quickly slide around ("imperfect magnetohydrodynamics"; "finite conductivity of the horizon"; time scale of sliding $\sim r_g/c$) until they become very orderly. Their ultimate, nearly uniform configuration, as depicted in figure 6, is that which minimizes the horizon's ohmic dissipation (Macdonald and Thorne [4]). These quasiuniform **B**-field lines are held on the hole by Maxwell

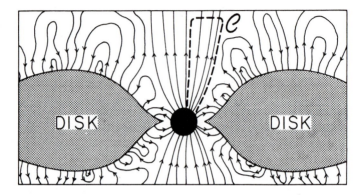

FIGURE 6. Electromagnetic extraction of rotational energy of a black hole ("Blandford-Znajek process").

pressure from the surrounding chaotic field lines, which in turn are anchored in the disk by currents. If the disk were suddenly removed, the field would slide off the horizon, convert itself into radiation, and fly away in a time $\sim r_g/c$.

Near the horizon the magnetic field **B** will be so strong that currents cannot flow across it; but far from the horizon, where **B** is weaker, they can. Consequently, the **B** field acts like a dc transmission line. The horizon's gravitomagnetic battery, equation (10), drives currents around closed loops, such as curve $\mathcal{C}$ of figure 6: up the **B** field from the horizon to a weak-**B** region, across the **B** field there, and then back down the **B** field to the horizon and through the horizon's battery to the starting point. These currents transmit power in the form of Poynting flux from the horizon to the weak-**B** region, where it is deposited into charged particles and accelerates them to ultrarelativistic energies. Both the horizon with its gravitomagnetic battery, and the weak-**B** "acceleration region" with its particles soaking up power, possess total electric resistances of order 30 ohms (Znajek [16]; Damour [10]; Macdonald and Thorne [4]). Thus, the battery and its load are impedance matched, and this "Blandford-Znajek" [15] process has the optimum possible efficiency for depositing the hole's gravitomagnetic, rotational energy into ultra-relativistic charged particles—particles that might well generate the jets observed in quasars and galactic nuclei. Numbers for typical jet models are

$$\begin{bmatrix} \text{Battery} \\ \text{voltage} \end{bmatrix} \sim \tfrac{1}{4} \left[\frac{S}{S_{max}} \right] B_\perp r_g$$

$$\sim [10^{20} \text{ volts}] \left[\frac{S}{S_{max}} \right] \left[\frac{B_\perp}{10^4 \text{ gauss}} \right] \left[\frac{M}{10^9 \, M_\odot} \right] \quad ,$$

(12a)

$$\begin{bmatrix} \text{Power} \\ \text{output} \end{bmatrix} \sim \begin{bmatrix} \frac{(\text{Voltage})^2}{4\mathcal{R}_H} \end{bmatrix}$$
$$\sim \begin{bmatrix} 10^{45} \, \frac{\text{erg}}{\text{sec}} \end{bmatrix} \begin{bmatrix} \frac{S}{S_{max}} \end{bmatrix}^2 \begin{bmatrix} \frac{B_\perp}{10^4 \, \text{gauss}} \end{bmatrix}^2 \begin{bmatrix} \frac{M}{10^9 \, M_\odot} \end{bmatrix}^2 \, , \tag{12b}$$

where S/S_{max} is the black hole's angular momentum ("gravitomagnetic dipole moment") in units of the maximum possible angular momentum GM^2/c of a hole, $M/10^9 M_\odot$ is the hole's mass in units of 10^9 solar masses, $B_\perp$ is the strength of the ordered magnetic field threading the hole, and $\mathcal{R}_H = (\text{length/circumference}) \times R_H \sim 30$ ohms is the total electrical resistance between the horizon's polar regions and its equatorial regions.

A large number of astrophysicists (not including me) are working on detailed models for the creation and acceleration of jets by the Poynting flux that emerges along a hole's **B**-field lines, and models for the collimation of the jets by magnetic field tension, by the walls of a funnel-shaped accretion disk, and by nozzles formed in surrounding gas. For a detailed review, see Begelman, Blandford and Rees [12].

6. CONCLUSION

It is truly remarkable that the gravitomagnetic potential γ and field **H**, which are so weak in the solar system that humans have never seen them, are predicted to be so strong near black holes that they are crucial elements in astrophysical model building. As an astrophysicist, I eagerly await the culmination of Bill Fairbank's dreams and planning, and of the decades of brilliant and meticulous technology development by Francis Everitt and others in the Fairbank-Everitt group—a culmination in the human race's first personal encounter with gravitomagnetism and relativistic gyroscope precession.

Acknowledgments

This work was supported in part by the National Science Foundation (AST82-14126 and PHY77-27084).

References

[1]   C. W. F. Everitt, in *Experimental Gravitation: Proceedings of Course 56 of the International School of Physics "Enrico Fermi,"* edited by B. Bertotti (Academic Press, New York, 1974), p. 331.

[2]   L. Smarr, *Sources of Gravitational Radiation,* (Cambridge University Press, Cambridge, 1979).

[3]   J. A. Wheeler, in *Battelle Recontres: 1967 Lectures in Mathematics and Physics*, edited by C. DeWitt and J. A. Wheeler (Benjamin, New York, 1968), p. 242.

[4]   D. Macdonald and K. S. Thorne, *Mon. Not. Roy. Astron. Soc.* **198**, 345 (1982).

[5]   V. B. Braginsky, C. M. Caves and K. S. Thorne, *Phys. Rev. D* **15**, 2047 (1977); esp. §III.B.

[6]   R. L. Forward, *Proc. IRE* **49**, 892 (1961).

[7]   C. W. Misner, K. S. Thorne and J. A. Wheeler, *Gravitation* (W. H. Freeman and Co., San Francisco, 1973).

[8]   T. Kuhn, *The Structure of Scientific Revolutions* (University of Chicago Press, Chicago, 1962); see especially pages 174-181 of the second edition, 1970.

[9]   Ya. B. Zel'dovich and I. D. Novikov, *Relativistic Astrophysics* (University of Chicago Press, Chicago, 1967), Vol. 1.

[10]   T. Damour, *Phys. Rev. D* **18**, 3598 (1978).

[11]   K. S. Thorne and D. Macdonald, *Mon. Not. Roy. Astron. Soc.* **198**, 339 (1982).

[12]   M. C. Begelman, R. D. Blandford and M. J. Rees, *Rev. Mod. Phys.* **56**, 255 (1984).

[13]   J. M. Bardeen and J. A. Petterson, *Astrophys. J. Letters* **195**, 65 (1975).

[14]   M. C. Begelman, R. D. Blandford and M. J. Rees, *Nature* **287**, 307 (1980).

[15]   R. D. Blandford and R. L. Znajek, *Mon. Not. Roy. Astron. Soc.* **179**, 433 (1977).

[16]   R. L. Znajek, *Mon. Not. Roy. Astron. Soc.* **185**, 833 (1978).

[17]   A. C. S. Readhead, M. H. Cohen and R. D. Blandford, *Nature* **272**, 131 (1978).

[18]   *Black Holes: The Membrane Paradigm*, edited by K. S. Thorne, R. H. Price and D. M. Macdonald (Yale University Press, New Haven, 1986).

The Stanford
Relativity Gyroscope Experiment
(A): History and Overview

C. W. F. Everitt

1. BACKGROUND

Kip Thorne in the preceding paper has expounded the reasons for seeking to measure the relativistic precessions of gyroscopes in earth orbit. This and the next six papers describe the experiment we are developing at Stanford under NASA support, usually referred to by its NASA denomination Gravity Probe B. Nowhere in all the work inspired by William Fairbank is the "near zero" principle so broadly exemplified as here.

The idea of testing general relativity by means of gyroscopes was separately discussed by Schouten [1], Fokker [2] and Eddington [3] soon after Einstein had advanced the theory in 1915. De Sitter [4] in 1916 had calculated that the earth-moon system would undergo a relativistic rotation in the plane of the ecliptic of about 19 m arc-sec/yr due to its motion around the sun. Fokker proposed searching for a corresponding precession of the earth's axis, while Eddington observed that "if the earth's rotation could be accurately measured by Foucault's pendulum or by gyrostatic experiments, the result would differ from the rotation relative to the fixed stars by this amount." This prediction, which Eddington credited to Schouten, omits the effects of the earth's rotation on the gyroscope, investigated many years later by L. I. Schiff [5].

Eddington's discussion stimulated Blackett [6] in the 1930's to examine the prospect for building a laboratory gyroscope to measure the 19 m arc-sec/yr precession. He concluded that with then-existing technology the task was hopeless. There the matter rested until 1959, when, two years after the launch of Sputnik, and following also upon the improvements in gyroscope technology since World War II, Schiff and G. E. Pugh [7] independently proposed to test Einstein's theory by observing the precessions with respect to a distant star of one or more gyroscopes placed in an earth orbiting satellite. According to Schiff's calculations such a gyroscope may be expected to undergo relativistic precessions given by

$$\Omega = \frac{3}{2}\frac{GM}{c^2R^3}(\mathbf{R} \times \mathbf{v}) + \frac{GI}{c^2R^3}\left[\frac{3\mathbf{R}}{R^2}(\boldsymbol{\omega} \cdot \mathbf{R}) - \boldsymbol{\omega}\right] \quad , \tag{1}$$

where $\mathbf{R}$ and $\mathbf{v}$ are the instantaneous position and velocity of the gyroscope, and M, I and $\boldsymbol{\omega}$ are the mass, moment of inertia and angular velocity of the rotating central body, in this instance the earth.

The first term in equation (1) is the *geodetic* precession Ω_G resulting from the motion of the gyroscope through the curved space-time around the earth, the counterpart of de Sitter's effect from motion about the sun. In a 650 km near-circular orbit around an ideal spherical earth, it amounts to 6.6 arc-sec/yr in the plane of the orbit. The second term in the equation is the *motional* (now sometimes called *gravitomagnetic*) precession due to the rotation of the central body. The integrated value in a 650 km polar orbit is 42 m arc-sec/yr in the plane of the earth's equator. The precessions are measured with respect to the line of sight to a suitable guide star and must, as Schiff pointed out, be corrected for the aberration in the apparent position of the star. As will be explained in section 5.2, three other effects of general relativity appear in the data, of which the largest is the 19.0 m arc-sec/yr solar geodetic precession.

The experiment, as now conceived, places four gyroscopes and a reference telescope, all at liquid helium temperature, in a polar orbiting statellite. The duration of the mission is between one and two years, and the aim is to measure the two principal relativity effects, plus other effects to be discussed in sections 5.2 and 5.6, to rather better than 1 m arc-sec/yr. Thus the geodetic precession will be determined to about 1 part in 10^4 (the most precise test yet attempted of any effect of general relativity) and the motional precession to between 1 and 2 percent. Figure 1 illustrates the orientation of the gyroscopes. The spin vectors of all four lie approximately in the plane of the orbit and along the line of sight to the guide star, Rigel, with one pair spinning clockwise and the other counterclockwise. This configuration makes the two effects Ω_G and Ω_M predicted by Schiff appear at

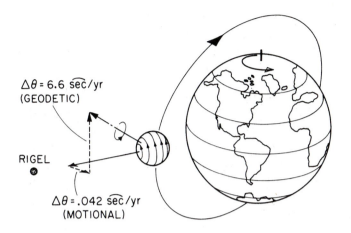

$\Delta\theta = 6.6 \ \widehat{\text{sec}}/\text{yr}$
(GEODETIC)

RIGEL

$\Delta\theta = .042 \ \widehat{\text{sec}}/\text{yr}$
(MOTIONAL)

FIGURE 1. Relativity effects as seen in gyroscopes with spin vector lying parallel to line of sight to star.

right angles in the output of each gyroscope, giving fourfold redundancy in the measurement of each effect. The spacecraft rolls slowly with 10 minute period about the line of sight to the star. The technique for reducing the data is indicated in section 4.2 and explained in detail in section 5.

The program is being developed, under NASA support, jointly by the Stanford Physics and Aero-Astro Departments with Lockheed Missiles and Space Corporation as aerospace subcontractor. The first formal contact with NASA was a two-page letter from Fairbank and Schiff to Dr. Abe Silberstein, dated January 27, 1961, describing an instrument to be mounted on the Orbiting Astronomical Observatory that would measure the geodetic precession to a few percent. This early proposal, short as it was, embodied several ideas that were to prove crucial in the experiment as we now conceive it.

Section 2 below describes the origin of the idea in the work of Schiff and Pugh. Section 3 briefly summarizes the history of the program. Sections 4 through 8 describe the experiment, further details being given in succeeding papers.

2. ORIGIN OF THE IDEA

Historians of science have often descanted on the mysterious phenomenon of simultaneous discovery. Pugh and Schiff illustrate the theme. The two men came upon the idea of the orbiting gyroscope experiment within a few weeks by different paths, knowing nothing about each other's work, and each contributed significantly to the scheme of the experiment as it is now conceived. Schiff's investigation, begun in November 1959 after an earlier broad study of experimental tests of general relativity, first appeared in an article submitted to *Physical Review Letters* on February 11, 1960. Pugh's took the form of a proposal, dated 12 November 1959, in the unlikely sounding Weapons System Evaluation Group Research Memorandum Number 11, from the Pentagon. The two men learned about each other toward the end of February 1960, and exchanged manuscripts and pleasingly cordial letters shortly thereafter. For clarity I take Schiff's contribution first despite Pugh's priority. In retrospect Pugh must be seen as unlucky in not having chosen a more traditional vehicle for publication. His work, though lacking the theoretical depth of Schiff's, was of exceptional quality and deserves wider recognition than it has had.

2.1 Schiff's Investigation

The path by which Schiff arrived at the idea of the gyroscope experiment was oddly circuitous. His interest in general relativity went back a long way, as far as 1939, when under Oppenheimer's influence he wrote a paper on the application of Mach's principle to rotating electric charges. Considerations about Mach's principle were, as we shall see, crucial to Schiff's work on the gyroscope experiment; appropriately in 1964 he contributed a paper "On the Observational Basis of Mach's Principle" to the issue of *Reviews of Modern Physics* in honor of Oppenheimer's sixtieth birthday.

Another of Schiff's interests was in the equivalence principle, which he applied in 1958 in a very ingenious way to a topic that happened also to attract Bill Fairbank's interest at the same time, the question whether there is a gravitational repulsion between particles and antiparticles which might account for the separation of matter and antimatter in the universe. By considering the gravitational status of virtual particles, Schiff deduced that positrons like electrons should fall downwards. This unexpected insight, right or wrong, seems to have led Schiff to look for other novel applications of the equivalence principle, and these came to a focus in an article "On Experimental Tests of the General Theory of Relativity" [8] submitted to the *American Journal of Physics* on October 6, 1959. There Schiff demonstrated just how tenuous the evidence for Einstein's theory was.

Einstein in 1915 had identified three observational tests of general relativity. the gravitational redshift, the deflection of starlight by the sun, and the precession of the perihelion of the planet Mercury. Forty-four years later, in 1959, these remained the only feasible observations, regarded by most people as "crucial tests" of Einstein's theory. Schiff set out to "examine to what extent the full formalism of general relativity is called upon in the calculation of these three effects, and to what extent they may be correctly inferred from weaker assumptions that are well established by other experimental evidence" [8]. The redshift certainly does not require the full formalism. Einstein in 1911, before general relativity existed, had computed it on the basis of special relativity and the equivalence principle, results which are established by the Eötvös experiment and observations with high energy particles. In formulating his critique in this way Schiff, while right on the main point, was tacitly ignoring the conceptual extension involved in applying equivalence to calculate the redshift, an omission which led R. H. Dicke, who refereed Schiff's paper for the *American Journal of Physics*, to write a forceful accompanying paper urging the importance of directly testing Einstein's redshift formula. The exchange between Schiff and Dicke bore fruit later in Dicke's work on the experimenal basis of gravitational theories and in the "Schiff conjecture" about the relationship between weak and strong equivalence [9].

What of the deflection of starlight? Conventional wisdom had it that this does require the full formalism of general relativity, because when Einstein, in the same paper of 1911, had applied special relativity and the equivalence principle to calculate a deflection, he obtained a formula identical with the one given by a classical ballistic theory of light, but only half of what he was to get four years later from general relativity. It was just this doubling of the classical (and the special relativistic) deflection that Eddington made such play of in 1919 when he claimed that the solar eclipse observations by Dyson, Davidson and himself supplied a decisive proof of general relativity.

Schiff thought otherwise. Applying a line of reasoning previously sketched out in a different context by W. Lenz [10], he claimed an oversight in Einstein's 1911 calculation. In addition to the deflection from time dilation, worked out by Einstein, there is a second special relativistic effect from the "FitzGerald contraction" of space radially in toward the sun. This term (analogous to the space-curvature contribution to gyroscope precession discussed by Kip Thorne) Schiff found to be equal in magnitude to the Einstein term and in the same direction. The sum of the two just equals the observed deflection.

Schiff's argument on light deflection has been sharply criticized, a point to which I shall return in section 2.3. Meanwhile for Schiff himself in 1959, two of the three "crucial tests" of general relativity failed of being crucial. The third, the precession of the perihelion of Mercury, was a real test, first because it could not be calculated without introducing an equation of motion (the geodetic equation), and second because in applying that equation changes in clock rate of order $(GM/c^2r)^2$ must be taken into account, and these "cannot be found by the methods of this paper " [11]. Schiff's prognosis was gloomy. Improved redshift or light deflection experiments offered little hope of anything new, and

> By the same token, it will be extremely difficult to design a terrestrial or *satellite* [my italics] experiment that really tests general relativity,

and does not merely supply corroborative evidence for the equivalence principle and special relativity. To accomplish this it will be necessary either to use particles of finite rest mass so that the geodesic equation may be confirmed beyond the Newtonian approximation, or to verify the extremely small time or distance changes of order $(GM/c^2 r)^2$. For the latter the required accuracy of a clock is somewhat better than one part in 10^{18}.

Within three months Schiff had conceived in the gyroscope experiment not one but two new tests of general relativity, to be done in a satellite, both checking the equations of motion beyond the Newtonian approximation, and one (the measurement of the motional precession) checking a wholly new aspect of the theory.

Evidence for the development of Schiff's ideas comes from 68 pages of handwritten notes, dated 12/20/59 with the parenthetic comment "work started about a month earlier," plus the recollections of Mrs. Schiff, Bill Fairbank, Bill Little and Bob Cannon. The notes are in fair copy, Schiff's habit being to destroy his first rough calculations, but internal evidence suggests that only the first 11 or so pages were about work done before December 20. This reconstruction differs from the recollections of Fairbank and Cannon in that it puts the critical events after the start of Christmas vacation, but is, I believe, necessitated by the evidence of the notes.

In the December 1959 issue of *Physics Today* there appeared an advertisement for the Jet Propulsion Laboratory [12] trumpeting "important developments at JPL" through an account of "The Cryogenic Gyro. A fundamentally new type of gyroscope with the possibility of exceptionally low drift rates," accompanied by an artist's rendition of a spinning metal sphere afloat on the distorted lines of magnetic force above a horizontal loop of wire. Schiff's notes start with a description of such a gyroscope as he envisioned it, followed by 11 pages about two scientific experiments that might be done with it: a test of Mach's principle and a relativistic clock experiment. Marginally inserted next to the JPL reference is "See also GE article in *Wall St Journal* 1/21/60," an account of the parallel program "Project Spin" at General Electric [13]. The insert clearly followed the bulk of Schiff's work; Bill Fairbank's recollection, on the other hand, is that even before the JPL advertisement he and Schiff had talked about an article from MIT on uses of magnetically levitated superconductors. Knowing how good Bill's memory is for details of this kind, I have little doubt that such an article will turn up; however, with *Physics Today* coming out in the middle of the preceding month as it then did, Schiff's statement "work begun about a month earlier" is consistent with the JPL advertisement's being the trigger for his thinking.

Neither of the experiments analyzed in the first 11 pages of Schiff's notes is the gyroscope experiment as we now know it. His test of Mach's principle had not yet gone beyond Oppenheimer. Relativistic predictions did not enter; the idea was to check agreement between the local and general frames by comparing the direction of a gyroscope or rather "several gyroscopes pointing in different directions, and started at different times of the day and the year" with the line of sight to some suitable heavenly object. As Schiff put it,

> This experiment is much in the spirit of the Eötvös experiment, a negative result is expected, and if established with precision and reliability is useful. A positive result, definitely established, would be the scientific event of the century! (Non-visible matter in the universe may be determining.) [14]

The gyroscopes were to be ground-based at the earth's equator; the aim was for "an accuracy comparable with the best astronomical observations, something of the order of $0.01''$ to $1''$ of arc"; and there was a long and technically intriguing analysis of disturbances acting on the gyroscope and of two methods of reading out the direction of spin, one due to Bill Little, the other to Schiff himself.

If Schiff did not then know that such a gyroscope would undergo a relativistic precession, what led him to the idea? The answer lies in the second experiment where the gyroscope is treated as a clock. Schiff, of course, was hoping for a clock good enough to measure the second order redshift. A spinning superconductor, being free of eddy-current disturbances and temperature-dependent changes in dimension, holds promise of being an exceptionally good clock. Schiff's notes and Bill Fairbank's recollections trace the course of this idea as applied in measuring frequency differences between clocks at the top and bottom of Hoover tower on the Stanford campus (300 ft), between a balloon at 100,000 ft and the ground, and finally (p. 9 of notes): "For a satellite in an eccentric orbit (as proposed by Zacharias), Δh might be 600 miles or 10^8 cm, an effect of fractional amount 10^{-10}."

At page 10 Schiff lists six points to be investigated. The first is "Does the gyro act as a clock in the relativistic sense?" The others are technical questions on clock performance. The remaining 58 pages of notes never take him beyond point one.

If a gyroscope is a clock in the relativistic sense, its spin rate will change with changes in the gravitational potential $\phi = gh/c^2$. Applying a technique from the paper on gravitational properties of antimatter [15], Schiff examined the effects of ϕ on the Hamiltonians of oscillator clocks and gyroscopes. For an oscillator, the redshift formula $\omega = \omega_0(1 + \phi)$ came out easily; for a rotator the same holds provided the angular momentum J does not change from motion through the field. Schiff added "as expected" to the sentence stating this result; after which, heavily lined out in characteristic

Schiff fashion, comes (p. 11, my italics): *"Now J is not changed as the rotator is moved in the field since no torques are exerted. Motion (polar vector) cannot produce torque (axial vector)."*

This, I conjecture, is the point Schiff had reached by December 20. Abruptly the next paragraph reverses direction with "We must now see how J changes with the motion in the field," followed by reference to a paper of 1951 "Spinning test-particles in general relativity" by E. Corinaldesi and A. Papapetrou [16]. Schiff at this point evidently suspected that a rotational clock would not obey the Einstein redshift formula, and indeed one of his first calculations (p. 14) gave an expression for the spin rate of a gyroscope with axis normal to the orbit plane of a satellite, equivalent to $\omega = \omega_0(1 + 2\phi)$, twice the Einstein value. Long before reading Schiff's notes I recall Bill Fairbank's telling me how Schiff, while working on the clock problem, telephoned him to say that there was a mistake in Papapetrou's paper. The issue, as Schiff makes clear in his final published result [17], is the choice of supplementary conditions on the equation of motion. In the notes, Schiff reached the proper answer after 20 pages. Changes in gravitational potential do indeed alter the gyro spin speed in accordance with the standard redshift formula. Motion along Newtonian equipotentials leaves the spin speed unaffected but causes a precession: the geodetic precession of the first term in equation (1) above.

When and why did Schiff start thinking about a space experiment? From one side he was preconditioned to it through thinking about the orbital clock experiment "as proposed by Zacharias," as well by his own earlier reflections on satellite experiments. From another side he must in all probability have seen, through analyzing the mass unbalance torques on a ground-based gyroscope (pp. 3–8 of the notes), that gyro performance would be enormously improved by operating in zero-g. But what really put him into space was—*mathematical convenience!* As Schiff noted (pp. 11–12) the Corinaldesi-Papapetrou equations "only apply to free motion in the Schwarzchild field & it seems very difficult to work out a constrained motion." Hence (new paragraph) "We first consider the motion of the gyroscope in a free satellite."

The next 30 pages of notes, with one short exception, all relate to the satellite experiment. The exception "suggested by Fairbank" (p. 18) is a preliminary calculation for a ground-based gyroscope yielding a precession that is "of the order of the Mach effect & must therefore be considered there as a correction, along with the Thirring-Lense effect," together (at this stage) with a 12 hour periodic variation in clock rate. All this "assumes the C-P equations are valid for constrained motion." Later (p. 40) Schiff derived proper constrained equations, eliminated the periodic effect, and showed that the precession for an earth-based gyroscope, allowing for the reduction in translational velocity and presence of a Thomas precession term is about 0.4 arc-sec/yr instead of 6.6 arc-sec/yr.

After this elaborate investigation Schiff quickly worked out the seemingly more difficult motional effect. The metric was known from the work of de Sitter (1916) [18] and Lense and Thirring (1918) [19]. Lense and Thirring had deduced that a moon orbiting a rotating planet undergoes a relativistic advance of its ascending node. Schiff's formula for gyroscope precession (the second term in equation (1)), was more complex, giving a precession which, unlike the Lense-Thirring drag, depends on orbit inclination and reverses sign in equatorial orbits. Schiff concluded the notes with investigations of sundry other effects, the aberration of starlight, gravity gradient disturbances on the gyroscope, and the relativistic effects of the sun, the moon and the galaxy. The article in *Physical Review Letters* was published on March 1, 1960, followed later in the year by the definitive paper in the *Proceedings of the National Academy*.

Just as Pugh and Schiff lighted on the experiment almost together, so in the fall of 1959 came an unforeseen confluence of interests at Stanford. Schiff, through writing his *American Journal of Physics* article, had fixed on the need for new tests of general relativity. Bill Fairbank, new to Stanford from Duke, was focusing on experiments that would go beyond traditional low temperature phenomena into wider applications. No topic could be more apt than the superconducting gyroscope. Meanwhile Bill had met Bob Cannon, also new to Stanford (from MIT), who had been thinking about the large improvements that could be gained in the performance of air bearing or electrically suspended gyroscopes by operating them in space. Cannon has often told the story of Fairbank introducing him to Schiff at the Stanford swimming pool, and of three "dripping men" deciding there and then to pursue the experiment, with Cannon saying to his wife that evening, "I have met a man who needs a gyroscope even better than the ones we have been talking about." Bill Fairbank's fertile imagination and powers of intellectual catalysis drew this diverse group together and started the adventure we are now on.

2.2   Pugh's Proposal

After Schiff's elaborately roundabout, though scientifically profound, route to the gyroscope experiment, Pugh's approach seems almost embarrassingly simple. In 1958 H. Yilmaz [20] had advanced as an alternative to general relativity a new generally covariant theory of gravitation, which yielded identical predictions to Einstein's for the three "crucial tests," though it has since been shown to be nonviable on other grounds. Originally Yilmaz did not discuss frame-dragging, but at the January 1959 New York meeting of the American Physical Society he gave a 10 minute talk on experimental consequences of generally covariant scalar, vector and tensor theories of gravitation. All would yield identical results for the three "crucial tests" but a possible experiment to distinguish the nature of the interaction would be "to launch an artificial satellite whose plane contains the axis of rotation of the earth" [21]. According to Yilmaz, a vector or tensor interaction would cause a Lense-Thirring drag of about 0.56 arc-sec/yr, a scalar interaction would cause no drag [22].

Pugh's "Proposal for a Satellite Test of the Coriolis Prediction of General Relativity" started where Yilmaz had left off. Yilmaz's suggestion, argued Pugh,

> does not appear feasible as a definitive test, because of the extremely large magnitude of perturbations that are not accurately known, such as the quadrupole and higher order moments of the earth's gravitational field. However the spin of a satellite is much less influenced by such effects and may provide a feasible technique for the experiment. [23]

Pugh's initial idea, then, was to place a large spinning body in orbit around the earth and compare its direction of spin with the line of sight to a guide star. The reference telescope would be mounted on the spinning body, aligned closely enough with its maximum axis of inertia to keep the star image within the field of view. The output would consist in observing gradual changes in diameter of the circles traced by the star image in the focal plane of the telescope. To accommodate a 10 inch aperture telescope and still provide sufficient gryoscopic action, Pugh proposed using a 1 m diameter satellite weighing about half a ton. Variants on Pugh's scheme were afterwards studied by groups led by Howard Knoebel at the University of Illinois [24] and David Frisch at MIT [25].

Pugh assumed that the satellite axis would precess through the same angle as the orbit-plane, and since the nodal drag calculated for him by Yilmaz was 0.36 arc-sec/yr (twice the correct value) the precession he gave was eight times what it should have been. From Peter Bergmann, Pugh learned of the de Sitter-Fokker precession, for which he gave, with acknowledgement to F. Pirani, the correct value of 6.6 arc-sec/yr. With these figures as goals for measurement, Pugh proceeded to analyze nonrelativistic disturbances on the satellite's spin axis from atmospheric drag, gravity gradients and magnetic effects.

Atmospheric drag was the Achilles' heel of the experiment. There seemed no way of making the disturbances from it small enough with a bare satellite. Pugh reacted with the brilliant suggestion that the primary satellite should be

> encased in a larger hollow sphere or "tender" satellite having the same center of mass. The tender satellite could be equipped with

light beams or other sensing mechanisms to monitor the position of the primary satellite *without* exerting significant forces or torques on the primary satellite. The use of *external* vernier rocket jets would allow repositioning of the tender with respect to the primary satellite so that the two would not collide during the course of the experiment. [26]

Thus was born the concept of the drag-free satellite. Pugh at once pointed out almost all the applications of drag-free technology that have since been pursued by others: aeronomy, geodesy, gravitational modeling of the earth, navigational satellites, the correction of "classical perturbations in orbital-type relativity" (*i.e.*, perihelion tests), even (though neither Paul Worden nor I knew of this when we began) a test of the equivalence principle.

It was Bergmann who put Schiff and Pugh in touch with each other. Pugh in a letter to Schiff (March 2, 1960) noted regretfully that "I must multiply my estimated effect by .25 for a polar orbit and −.5 for an equatorial orbit," and observed that "You could hardly pick a worse location than Stanford to measure the Lense-Thirring effect." Schiff replied (March 14), "Your double satellite experiment and my gyro within a satellite are really the same thing. I would certainly like to see this done, although it is possible that the earth-based gyro will not be as difficult at this stage of satellite development." Concerning their views of theory, Schiff continued,

We differ slightly in motivation in that you regard an experiment that would distinguish between Einstein's and other theories of gravitation as being of primary importance, whereas I would regard an experiment that tests any of the theories, Einstein's in particular, beyond the equivalence principle as being of primary importance. Hence you tend to stress the Lense-Thirring effect, whereas I tend to regard it and the non-rotating earth effect as of equal importance. [27]

2.3   Significance of the Effects

Even today, 25 years after Schiff, general relativity lacks a secure experimental foundation. Einstein advanced a theory of great conceptual elegance, radically different from Newtonian theory, with few testable consequences. One result over a long period was a mad proliferation of rival theories all claiming to account for the three "crucial tests."

A partial answer came in the late 1960's through the formulation of the PPN (parametrized post-Newtonian) framework for classifying gravitational theories and comparing their experimental consequences. The idea goes back to a discussion in Eddington's *Mathematical Theory of Relativity* [28] revived by Schiff in 1960 [29]. In the hands of Nordvedt and Will, PPN analysis has eliminated a large number of theories previously thought to be viable, either by disclosing internal inconsistencies or by showing that the theories lead to peculiar effects not present in general relativity or nature.

Will [30] concludes that of some fifty to eighty theories of gravitation once held, only about half a dozen survive, all of which yield identical or nearly identical results with general relativity for solar system tests.

Invaluable as this demolition work is, it does not establish Einstein. Ultimately perhaps the most interesting consequence of the PPN investigation is the discovery of how many potentially occurring effects vanish in general relativity. Einstein's is a minimalist theory. Apart from some rather marginal deductions about gravitational radiation in the Taylor-Hulse binary pulsar, the one truly new positive discovery since 1960 has been the Shapiro time delay effect [31], the relativistic increase in transit times of radar signals reflected from bodies (planets or spacecraft) as they pass behind the sun. Time delay is closely related to light deflection. Measurements of it, light deflection and redshift in recent years all support the Einstein predictions within experimental limits (2 parts in 10^4 for redshift in the Vessot-Levine experiment, 1 part in 10^2 for light deflection in VLBI measurements on radio stars in the ecliptic plane, and 1 part in 10^3 for Viking Mars lander time delay measurements). Each result has been analyzed with unusual care, but some reserve is appropriate since each depends on elaborate data modeling. Also there is the peculiar status of the precession of the perihelion of Mercury, which has oscillated between agreement and disagreement with the Einstein prediction with changing data on the sun's oblateness. The latest results suggest a 1% discrepancy with general relativity [32].

If Schiff's 1959 argument were correct that redshift and light deflection (and by the same token radar time delay) only test special relativity and the equivalence principle, this would be a meager crop. Actually the argument has been disputed, and is now usually thought to have been disproved, though Schiff himself continued to grant it heuristic value. To follow the issues, certain distinctions have to be made.

First, in the Schwarzschild solution of Einstein's field equations around a static spherically symmetric body, the departure from flat space-time, when written in standard rather than isotropic coordinates, does indeed involve a modification of the time coordinate and a modification of the radial component of the space coordinates. The modification to time is just the Einstein clock shift; the modification to space is such that the radial distance from the center to a sphere of surface area A, as measured by observers at rest around the central body, is somewhat greater than the square root of $A/4\pi$. In stating that Einstein had omitted space curvature from his 1911 argument, we are on safe ground.

Schiff (and several other writers) sought to repair that omission. Their line of reasoning is roughly as follows [33]. Assume that in weak fields, such as exist around the earth and the sun, the gravitational potential is given to first approximation by the Newtonian formula GM/r, so that a reference frame R falling inwards freely from infinity will at distance r have velocity $v = \sqrt{2GM/r}$ with respect to the central body. According to the equivalence principle, the frame R is inertial so the laws of special relativity apply to it. Suppose an observer on R possesses clocks and measuring rods which he compares with clocks and rods attached to a static spherically symmetric coordinate frame at distance r from the origin. The time dilation and space contraction of special relativity will affect the comparisons. If the inertial observer measures the interval between ticks of the clock in the static frame S as δt, its proper time in S is actually the shorter interval $\delta t\sqrt{1 - v^2/c^2}$. If he measures the radial distance interval as δr, its proper value in S is $\delta r/\sqrt{1 - v^2/c^2}$. Comparisons of transverse distances remain unaffected. If then one can use these intervals δr and δt as differentials for the radial and time coordinates in the stationary coordinate system S, one arrives at the curved three-dimensional space and modified time of the Schwarzchild metric (and hence at the general relativistic light deflection) without invoking the elaborate machinery of general relativity. The seductiveness of the argument is enhanced when one learns that it can indeed be made rigorous for the falling clock, where of course it agrees with Einstein's 1911 result and with other simple arguments that yield a gravitational redshift without invoking general relativity.

An obvious objection, first advanced by Schild [34], is that in the PPN framework the value for the light deflection (and likewise for the time delay) is proportional to $(1 + \gamma)GM/c^2r$, where the 1 represents the contribution from the Einstein clock shift and the γ the contribution from space curvature, equal to 1 in general relativity but not in all metric theories of gravity. In the Brans-Dicke scalar-tensor theory, for example, with Dicke's original value of 6 for the scalar parameter ω, γ is 0.92. The Brans-Dicke theory is certainly consistent with special relativity and the equivalence principle, yet matter in it produces less curvature of space than in Einstein's theory. It does so because of the admixture in Brans-Dicke of a scalar potential along with Einstein's tensor potential. Tucked away in Schiff's argument seems to be a hidden assumption that gravitation depends solely on a tensor potential.

But Schiff's argument can be objected to on other grounds. A certain intellectual legerdemain appears in the transfer from contraction of a falling body to Schwarzchild curvature. In 1968 Sacks and Ball [35], and independently Rindler [36], exposed various fallacies in Schiff's and other simple derivations of space curvature. Rindler indeed claimed to have found a decisive counterexample to all such derivations. My own impression is that Rindler proved too much and that one version put forward by Tangherlini [37] in 1962 might with proper modification be made rigorous. However, pursuit beyond a certain point is futile. The true value of such arguments is pedagogical. If they remove some of the mystique of incomprehensibility from general relativity and give some hint as to what aspect of the theory is under test, they have served their turn.

In treating the gyroscope experiment, Schiff remarked that calculation of *either* the motional or the geodetic effect involves the larger structure of general relativity since each requires an equation of motion beyond the Newtonian approximation. Here Schiff followed the thought of his *American Journal of Physics* article where the precession of the perihelion of Mercury could not be calculated "by an extension of the methods of this paper [since] we require in addition an equation of motion for a particle of finite rest mass, to replace the argument used above that the speed of light measured by a [falling clock] B is c" [8]. This check of the equation of motion was what made Schiff tend to regard the nonrotating earth effect as equal in importance with the motional or Lense-Thirring effect.

Most physicists, including Schiff himself in other contexts, have viewed the motional effect as the more important because the dragging of the inertial frame by rotating matter is an aspect of general relativity different in kind from anything seen in earlier tests. The difference may be put in various ways: measuring off-diagonal terms in the metric, searching (as in Kip Thorne's elegant discussion) for a gravitomagnetic field, determining a particular combination of PPN parameters. The most searching discussion is C. N. Yang's. After pointing out that general relativity "though profoundly beautiful, is likely to be amended," and that the amendment is likely to involve a new, beautiful and symmetrical geometrical concept, he asks,

> What is this new geometrical symmetry? We do not know ... However, many of us believe that whatever this new geometrical symmetry will be, it is likely to entangle with spin and rotation, which are related to a deep geometrical concept called torsion. But, no one has figured out what precise new concept related to rotation is the relevant one. From the viewpoint of gauge theory, I had pointed out [*Phys. Rev. Lett.* **33**, 445 (1974)] that the natural amalgamation of Einstein's theory with gauge theory is to involve the derivatives of R_{ik}, hence to involve spin.

> That the amendment of Einstein's theory may not disturb the usual tests is easy to imagine, since the usual tests do not relate to spin. The proposed Stanford experiment is especially intesting since it *focusses on the spin*. I would not be surprised at all if it gives a result in disagreement with Einstein's theory. [38]

What then of the geodetic effect, and equations of motion? Certainly the calculations involve, as Schiff said, terms in an equation of motion beyond the Newtonian approximation, as may be seen from Schiff's paper on the gyroscope and more simply from de Sitter's investigation of the earth-moon system, where the precession comes from a relativistic perturbation on the

Newtonian equations for lunar motion [39]. Two statements of clarification
are needed, however. First, in general relativity the equations of motion are
not independent postulates. They follow with almost sinister inevitability
either from an extension of a variational principle known to apply in special
relativity or, more unexpectedly and unlike any other theory in physics,
by direct deduction from the field equations. Schiff's easy separation of
different aspects of the theory does not quite work. Second, the rotation
of direction is not restricted to a gyroscope. Eddington made the point
when he said that the de Sitter effect "does not have exclusive reference to
the moon: in fact the elements of the moon's orbit do not appear in [the
final equation]. It represents a property of the space around the earth—a
precession of the inertial frame in this region relative to the general inertial
frame of the sidereal system." [40]

The weak-field result can be understood by referring to Kip Thorne's discussion. There the geodetic precession comes in two parts: two-thirds from
lateral translation of the gyroscope through the curved three-dimensional
space around the earth, one-third from spin-orbit coupling in the gravitoelectric field. In this treatment the experimentally observed decrease in
clock rate with distance from the center of the earth is replaced by an
artificial universal time. Sticking to the empirical (or at least notionally
empirical) one may continue to attribute two-thirds of the geodetic precession to space-curvature, as defined earlier, while reinterpreting the other
one-third as an effect of the lateral motion of the gyroscope through the
radial gradient in the time dimension of the Schwarzchild metric. For theories within the PPN framework the total precession is $(1 + 2\gamma)GM/2c^2r$,
with γ being the measure of space curvature. A measure of the geodetic
precession effects two things: (a) assuming a precision of 1 part in 10^4 on
Ω_G, it fixes γ to a part in 7000, rather more than a factor of ten better
than the best radar ranging determination; (b) it tests to a part in 3000 the
precession from lateral motion of the gyroscope through a time gradient.
An interesting investigation not yet attempted would be to see whether the
time gradient part of the gyro precession can be derived from the equivalence principle by an argument analogous to Einstein's argument for light
deflection. Schiff would have said no. The significance of the geodetic
measurement of course depends on whether the result agrees or disagrees
with general relativity. Barring conpensating deviations, agreement would
confirm both the time and the space aspects of the theory. Disagreement
would on first presumption be attributed to γ's being different from 1, but
only because one is thinking within a PPN framework.

Having studied weak field effects, it is instructive to see what happens
to a gyroscope orbiting a black hole. Consider a nonrotating black hole
of Schwarzchild radius r_0. The extreme of light deflection and also of
radar time delay occurs when the light or radar pulse goes into orbit; this

occurs at radius $3r_0$. For a gyroscope, the parallel question is at what distance does the spin axis become locked in through geodetic precession radially towards the center of the black hole (or tangential to a circular orbit around it)? Since the lowest stable captive orbit for massive objects is $6r_0$, one may conjecture that the locking radius also is $6r_0$, a result which if true neatly illustrates the difference between experiments with gyroscopes and experiments with electromagnetic signals. Mark Jacobs and I are investigating this problem. For rotating black holes, the story is more complicated. With an equatorial orbit the Lense-Thirring-Wilkins [41] drag on the orbiting body alters the radius of the lowest stable orbit, decreasing it for corotation and increasing it for counterrotation, and the motional precession of the gyroscope has the opposite sense from the Wilkins effect. For counterrotating orbits, locking becomes impossible. For corotating orbits, its occurrence would depend on whether the reverse effect of the motional precession is outweighed by the enhanced geodetic precession for the closer-in orbit. These and the more complicated effects in inclined orbits offer an intriguing field of study.

3. EARLY DAYS AND THE CONCEPT OF THE LONDON MOMENT READOUT

I have described in the introduction to this volume my own first encounter with Bill Fairbank and the gyroscope experiment. Upon arriving at Stanford in 1962 I found that, even though NASA funding had yet to commence, significant work was already going on in both physics and aero-astro departments. Morris Bol, with Bill, had demonstrated the principle of a new kind of gyroscope readout based on the Mössbauer effect and was beginning his doctoral research to detect the London moment in a spinning superconductor; Roger Bourke and Benjamin Lange, with Bob Cannon, were respectively studying the dynamics of a magnetically supported superconducting rotor [42] and the design and performance of the drag-free satellite [43].

As originally conceived, the experiment used a spherical superconducting gyroscope, magnetically levitated, with a Mössbauer readout. This readout, a typically ingenious Fairbank idea, was based on having a small ^{57}Fe source on the gyro rotor, and a detector mounted on a corotating cylinder interposed between the gyroscope and the reference telescope. Any misalignment between the axes of the gyroscope and the cylinder would result in a periodic linear displacement between the source and detector, and the instantaneous velocity associated with this displacement could be measured by the Mössbauer effect. A separate measurement would then be made of the orientation of the cylinder with respect to the telescope.

Relativity data would come from differencing three signals: telescope-to-star, cylinder-to-telescope, gyroscope-to-cylinder.

One concern with a magnetically suspended superconducting gyroscope is the effect of the London moment in the spinning superconductor. Fritz London, in his book on *Superconductivity* [44], extending an earlier investigation of Becker, Sauter and Heller [45], had shown that a spinning superconductor develops a magnetic moment proportional to spin speed aligned with the instantaneous axis of rotation. A simple calculation reveals that the torque from the magnetic support field may easily cause gyro drift several orders of magnitude larger than the 10 m arc-sec/yr limit aimed at by Fairbank and Schiff. It was partly from this concern as well as from interest in the phenomenon itself, that Bill and Morris Bol decided in 1961 to set about measuring the magnitude and properties of the London moment. Similar experiments were started independently about the same time by Hildebrandt [46] and by King, Hendricks and Rorschach [47], but Bol and Fairbank's [48] work was especially interesting because in addition to detecting the London moment in a superconductor spun up below its transition temperature, they demonstrated that a solid strain-free superconductor, spun in the normal state, would generate a London moment spontaneously on cooling through the transition. This effect, the analog of the Meissner field exclusion effect, is one of the many intriguing consequences of superconductivity's being an equilibrium state.

A cardinal principle of physical experimentation is that one should try to convert obstacles into advantages. A few months after my arrival at Stanford I struck upon two possible ways of exploiting this bothersome London moment. One was to apply the magnetic torque on the gyroscope in aligning the spin axis with the reference direction. The other (very tentative) was that since the London moment is tied to the direction of spin, it might supply the basis for a gyro readout. My original notion here was to adapt one of the Blackett astatic magnetometers we had used in paleomagnetic measurements at Imperial College. This, as now appears, would have been problematical. The Blackett magnetometer [49], though a brilliantly conceived instrument deserving of far more acclaim than it has received, would not have fitted well. The delicate mechanical suspension would have made it hard to adapt to space operations. The reaction torque on the gyroscope from the detecting magnets would have caused difficulty. The sensitivity, which was at best 5×10^{-11} G/cm in a 0.03 Hz bandwidth, would have yielded a readout resolution of 1 arc-sec in 100 sec of time, marginal for an experiment at 10 m arc-sec/yr and unacceptable for the 1 m arc-sec/yr we now seek.

Bill Fairbank and I discussed these ideas, but did nothing with them. Then a month later Bill came rushing up to me bursting with excitement about the London moment readout. Apparently he had forgotten our

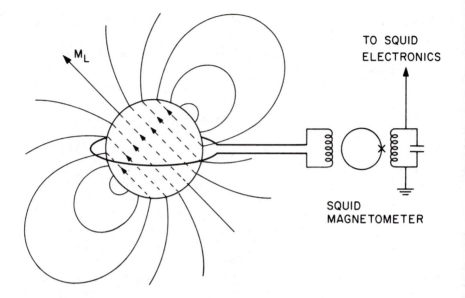

FIGURE 2. London moment readout of the gyro spin axis.

earlier conversations and had come upon the same idea independently. Be that as it may, he brought a new and crucial factor into the discussion: the modulated inductance magnetometer that he, Bascom Deaver and John Pierce [50] were just then beginning to work on. If this could reach the hoped-for precision, it would be the key to a successful gyro readout. We quickly saw that the right design was to have a readout loop centered on the gyroscope, and wrapped closely around it in order to make use of the self-shielding action of the superconducting rotor in rejecting changes in the external field from the readout (figure 2). Later we abandoned the modulated inductance magnetometer for the more subtle and sophisticated SQUID (Superconducting QUantum Interference Device), whose invention by others still lay ahead, but that was a detail, albeit an important one. The London moment readout unlocked the riddle, first by allowing us to eliminate the awkward intermediate rotating cylinder between the telescope and gyroscope, and second by making us realize that with a magnetic readout one should substitute for the magnetic suspension an electrical suspension system of the kind invented by the late Arnold Nordsieck [51] of the University of Illinois and developed commercially by Honeywell and Rockwell.

At this point a conversation with Howard Knoebel, Nordsieck's successor at Illinois, is worth recalling. Knoebel visited Stanford shortly after I arrived and before we had thought of the London moment readout. Rather cryptically even to myself, I expressed concern about gyro readout, and he immediately responded, "Well, you are worrying about the right problem; the readout is the difficult part." Correctly so. Paradoxical as it may sound, development of a gyroscope with drift rate at the milliarc-sec/yr level is not the hard problem. The real difficulties are two: reading out the direction of spin and spinning up the gyroscope.

The London moment readout has four advantages: (a) since the London moment is tied to the gyroscope's instantaneous axis of spin rather than its body axes, the readout can be applied to an ideally round, ideally homogeneous rotor; (b) it has adequate angular resolution; (c) it is sensitive only in second order to the centering of the gyro rotor in the readout loop; (d) the reaction torque of the readout current on the London moment is negligible.

More detailed information on all four points is given in the paper by Anderson (paper (C) of this series).

3.1 Building the Research Group

NASA funding commenced in March 1964 retroactive to November 1963, with initally a supplement from the U.S. Air Force. The original proposal was "To Develop a Zero-G Drag-Free Satellite and Perform a Gyro Test of General Relativity in a Satellite," with Cannon and Fairbank as co-principal investigators. The place of "drag-free satellite" in the title was significant. From the beginning, our intention was to develop the gyroscope experiment jointly between the Stanford Physics and Aero-Astro Departments, but Bob Cannon, unlike Bill and myself, foresaw that this would be a long process. Perhaps even Cannon did not realize just how long it would be, but in any event he wisely set for the Aero-Astro Department an independent goal of gaining early flight experience through development of the drag-free satellite. Cannon's hope was to apply the drag-free satellite in aeronomy and geodesy, as Pugh had suggested. As it turned out, the first application was to the U.S. Navy's TRIAD Transit Navigation Satellite, build by Johns Hopkins Applied Physics Laboratory and launched in July 1972, which carried the DISCOS (DISturbance COmpensation System) drag-free controller developed by the Stanford Guidance and Control Group under Daniel B. DeBra [52].

In parallel with the main program Lange [53] explored a variant on Pugh's idea of making the gyroscope itself a drag-free proof mass. He proposed using a carefully mass-balanced silicon rotor a few cm in diameter, for which the direction of spin would be read out from an optical flat accurately positioned on the pole of the rotor's maximum axis of inertia. The telescope was to be mounted on the outer satellite, not on the spinning proof mass as in Pugh's proposal. Although this "unsupported gyroscope" was never reduced to practice, it led to important research by several graduate students, not least among them Bradford Parkinson, who has now returned to Stanford in another role as program manager for the relativity gyroscope experiment.

Once NASA-Air Force funding was available, the aero-astro side of the enterprise rapidly expanded. Dan DeBra, James Mathiesen and Richard Van Patten all joined the professional staff in 1964 (at first principally to work on the drag-free satellite), and several graduate students followed. Within physics we did not expand so fast, but a close working relationship established with DeBra and Van Patten from 1965 onwards led to many of the basic ideas for the flight mission, especially in attitude and translational control of the spacecraft, in work on the reference telescope, and in the conception of a "science data instrumentation system" to subtract and process the gyroscope and telescope signals [54].

Meanwhile, in cooperation with Honeywell Incorporated we started the long and arduous process of developing the gyroscope, both in its mechanical design and in adapting the electrical suspension system to our application. Daniel Bracken, then a physics graduate student, played a critical role through his ingenious work on the gas spin up system [55] (see paper (B) of this series). Another important event in 1965 was the start of our long and happy collaboration with Donald Davidson (then of Davidson Optronics Incorporated, later of Optical Instrument Design Co.), who built the fused quartz telescope for the experiment [56] and originated many of the ideas then and later for the design and manufacture of quartz gyro housings and magnetic shielding assemblies. Somewhat later, from 1968 on, Wilhelm Angele of NASA Marshall Center also contributed to gyroscope development, especially in his work on methods of manufacture for extremely spherical gyro rotors [57].

By 1968 we had developed many of the basic concepts for the experiment, but had not begun to have a working system. At this point John Lipa arrived from Australia on a CSIRO fellowship. He at once took charge of dewar and gyroscope development, which he led with brilliant insight and determination full time from 1969 to 1979 and part time thereafter. He was soon joined by John Anderson on gyro readout, Jack Gilderoy, Jr., on mechanical fabrication, and John Nikirk, who until his untimely death in 1975 held responsibility with Dick Van Patten for electronics development.

It would be invidious to try to apportion credit for what has been supremely a team effort, requiring tenacious application from many different people. In addition to those already mentioned, three physics graduate students deserve special recognition: Peter Selzer for developing the porous plug device for controlling the flow of liquid helium from a dewar under zero-g conditions [58], Daniel Wilkins for analyzing the correction to Schiff's geodetic formula arising from the earth's oblateness [59], and Blas Cabrera for developing ultralow magnetic field shielding techniques [60] which later he, as a member of the research staff, applied directly to the gyroscope experiment. Other research staff members who worked on the program up to 1981 were Robert Clappier, Frank van Kann, Graham Siddall, Brian Leslie, G. M. Keiser, Stephen Cheung and John Turneaure. Among aero-astro graduate students, work of especial note was done by John Bull, David Klinger, Richard Vassar, J.-H. Chen and Thierry Duhamel. In all, 23 graduate students have obtained doctoral degrees on the program, either in physics or in aeronautics and astronautics, another 11 have worked part time, and 15 undergraduates have made significant contributions, including several honors theses. One undergraduate, Charles Marcus, was selected as a 1984 national finalist for the American Physical Society's Apker award [61].

Recent developments in the group are discussed in the next subsection.

3.2   The Path to a Flight Program

The gyroscope experiment has the curious distinction of having had the longest running single continuous grant ever awarded by NASA, from November 1963 to July 1977. The closeout of this grant in 1977 [62], besides being a legal necessity, had symbolic significance in marking the end of the exploratory phase of the program.

NASA had begun examining the feasibility of a flight experiment in the early 1970's. In 1971 Ball Brothers Research Corporation completed a "Mission Definition Study" which contained a first look at the spacecraft layout and a program plan. The plan advanced there, with some prescience, was for a three-flight program with (a) a dewar test flight, (b) an engineering test flight of the gyroscopes, (c) the science mission. Our current plan has the dewar test flight already completed through the successful flight in 1982 of the IRAS (Infra-Red Astronomy Satellite) dewar, whose design was largely based on the one worked out for the gyroscope experiment by Ball Brothers Research Corporation and Stanford in 1971. The approach to engineering and science flights will be described in a moment.

To start a flight program in 1971 would have been premature, but by the late 1970's the situation had changed. In 1978 the Space Sciences Board appointed an *ad hoc* Gravitational Physics Committee under the chairmanship of I. I. Shapiro to formulate "A Strategy for Gravitational Physics in the 1980's"; its report, published in 1981 [63], put the gyroscope experiment as the number one priority, the only dedicated flight mission recommended by the board. In 1980 NASA conducted a major review of technological readiness under the chairmanship of Jeffrey Rosendhal and concluded that "the remarkable technical accomplishments of the dedicated Stanford experiment team give us confidence that, when they are combined with a strong engineering team in a flight development program, this difficult experiment can be done" [64].

With these and other endorsements, the stage was set for planning the flight program. NASA had already completed a Phase A study in-house at Marshall Center in 1980. In 1982 a much more extensive Phase B study was completed [65], also in-house, but this for a variety of technical reasons led to a somewhat large spacecraft (weight 5300 lb, power 576 W) and a too expensive mission. Accordingly, in 1983 we undertook an extensive restructuring of the program, which cut the weight to 2800 lb and the power to 143 W without sacrificing any of the essential science goals, and lowered the anticipated cost to about $130 M [66].

One concern throughout the Phase B study has been how to keep the risk of the program, especially the cost risk, within reasonable bounds. In 1983, after much thought, we settled on a plan that would separate the development costs of the high technology instrument from the "marching army" costs of spacecraft construction by proceeding in two phases, with the dewar and instrument being built first and tested in a 7 day engineering flight on shuttle in 1989, and then brought back for minor refurbishment and integration with the spacecraft for the science mission, to be launched in 1991. The first phase is called STORE (Shuttle Test Of the Relativity Experiment).

This plan was endorsed by the NASA administrator, James M. Beggs, in March 1984. In November 1984 Stanford selected Lockheed Missiles and Space Corporation as its aerospace subcontractor on STORE, with Stanford providing the central gyro package and Lockheed the dewar/probe and electronics packages. John Turneaure heads the Stanford hardware development group.

4. GENERAL APPROACH TO EXPERIMENT

4.1 A Near Perfect Gyroscope

In concept, the GP-B relativity gyroscope experiment is simplicity itself: an earth-orbiting spacecraft containing one or more precise gyroscopes referenced to a precise telescope pointing at a stable guide star. The difficulties lie only in the precisions needed. Doing a 1 m arc-sec/yr experiment calls for a gyroscope with an absolute drift rate of about 10^{-18} rad/sec, some nine orders of magnitude less than the absolute drift rates of very good inertial navigation gyroscopes, and six orders of magnitude less than the compensated drift rates obtained in the best such gyroscopes by modeling out predictable errors. More precise readouts are needed for both the gyroscopes and the telescope than in the best conventional instruments. Not surprisingly has the development of the experiment taken so long.

It might seem absurd, in view of the great effort and ingenuity applied to inertial navigation instruments over the past thirty years, that a university research team, however well supported by NASA and industry, should aim to produce a gyroscope with performance a million times better than the best hitherto available. But the task we are attempting is different. Inertial navigation gyroscopes have to operate in submarines, aircraft and missiles under very high g loads. They must be small, light, and—in commercial applications, at least—cheap. While cost is indeed a concern to us, sizing of the total package is less so, and, most important, the experiment operates not at levels of 1 g to 30 g as in submarines or missiles but in the nearly weightless environment of space. Since the limitations on certain inertial navigation gyroscopes, especially those using a suspended spinning sphere as the reference element, come principally from support torques, it is plain that the performance of such gyroscopes should greatly improve in space. To see the possibilities for improvement, it is only necessary to examine the simplest support-dependent torque: the mass unbalance torque Γ_u. Consider a spinning sphere that is perfectly round but not quite homogeneous. Let the mass of the sphere be M and the distance between its center of geometry and center of mass along the spin axis be δr. If the gyroscope is supported about its center of geometry and subjected to a transverse acceleration f, then $\Gamma_u = Mf\delta r$ and the resulting drift rate $\Omega_u = \Gamma_u/I\omega_s$ is

$$\Omega_u = \frac{5}{2}\frac{f}{v_s}\frac{\delta r}{r} \quad , \tag{2}$$

where r is the radius and v_s the peripheral velocity of the sphere (say, 2000 cm/sec). To do an experiment on earth, $\delta r/r$ would have to be 4×10^{-17}. In space with a 10^{-10} g acceleration, it need be only 4×10^{-7}.

Arguments of this type led us to the concept of a gyroscope in the form of a very round, very homogeneous fused quartz sphere, coated with superconductor, operating at cryogenic temperatures in the nearly zero-g environment of space. The gyroscope is weakly suspended by electrical fields in an evacuated spherical cavity, spun up initially by gas jets, and has for its angular readout the London moment readout scheme described in section 1.2. Details are given in paper (B) of this series by J. A. Lipa and G. M. Keiser. The marriage of cryogenic techniques with space techniques also solves, as we shall see, various problems in the operation of the reference telescope and the spacecraft control systems.

4.2   Experiment Configuration

Figure 3 is a view of the instrument, the main structural element of the spacecraft. The dewar has an annular helium well with capacity 1580 L, supported from 12 "passive orbital disconnect struts" (PODS) within a shell of 69 inch diameter and 106 inch overall length. A neck tube joins the helium well to the outer shell of the dewar so as to form a continuous

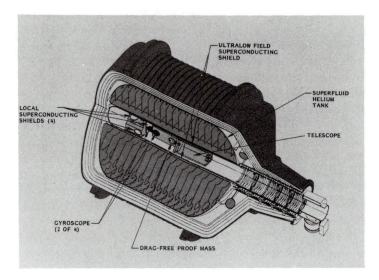

FIGURE 3. General view of GP-B instrument and dewar.

enclosed cavity of 10 inch inner diameter into which the instrument fits. The dewar is superinsulated and has four vapor-cooled radiation shields arranged to provide cooling also for the neck tube. The dewar is designed to operate at 1.8 K and hold helium for approximately two years, the proposed duration of the GP-B mission. Boiloff of the helium is controlled by the porous plug device to which we have already referred. The escaping helium gas is vented to space through proportional thrusters, also developed at Stanford, which provide control authority for pointing the spacecraft, as well as for a translational control system referenced to the drag-free proof mass.

The instrument comprises a quartz block structure, including the reference telescope, four gyroscopes and drag-free proof mass, enclosed in an evacuated cylindrical chamber 10 inches in diameter and 96 inches long, with its own insulating neck tube, all forming an independent assembly that can be inserted or removed as a unit in the inner cavity of the dewar. The gyro-telescope structure is held together in molecular adhesion by "optical contacting" for maximum mechanical stability.

The four gyroscopes are in a straight line on the instrument axis with their spin axes aligned parallel to the line of sight to the guide star, Rigel, two rotating clockwise and two counterclockwise. This configuration, as mentioned earlier, puts the spin vectors parallel to the line of sight causing the two precession effects, Ω_G and Ω_M, to appear simultaneously in each gyroscope. The resulting total precession is $\sqrt{\Omega_G^2 + \Omega_M^2}$, with a phase angle

ϕ with respect to the plane formed by the earth's axis and the line of sight given by $\tan\phi = \Omega_M/\Omega_G$. Since the spacecraft rolls about the line of sight with an angular velocity ω of 2.5×10^{-4} rad/sec (10 minute period), the gyro readout will record a sinusoidal signal of amplitude $\sqrt{\Omega_G^2 + \Omega_M^2}$ and frequency ω, whose phase can be determined separately by means of a "star blipper," attached to the spacecraft, which picks up signals each revolution from one or more suitably bright stars situated at points on the celestial sphere nearly 90° away from the guide star. The roll greatly reduces certain drift torques, eliminates the effect of 1/f noise in the SQUID, and eliminates errors from null drifts in the gyro and telescope readout.

Operation in space reduces the support torques on the gyroscopes, but does not make them zero. With a satellite of typical area/mass ratio in a 650 km orbit, the average residual acceleration on the spacecraft (and hence on the gyroscopes) from air drag, solar radiation pressure, *etc.*, is a few times 10^{-8} g. Low as this is, it is not low enough, so a drag-free proof mass is placed near the center of mass of the spacecraft. The proof mass is located in a cavity in the quartz block as shown in figure 3. The control system provides signals to the same proportional thrusters that are used for pointing the spacecraft. The level of translational control needed is about 10^{-10} g, a factor of 20 less stringent than that already demonstrated by the DISCOS controller for TRIAD.

4.3   Magnetic Shielding

Crucial to the success of the London moment readout of the gyroscope are two constraints on magnetic fields. (a) The gyroscope must operate in a very *low* magnetic field (less than 10^{-7} G) to keep trapped flux in the rotor at an acceptable level. (b) The gyroscope must operate in a very *stable* field (effective variations less than 2×10^{-13} G) to prevent changes in the external field from disturbing the readout.

Trapped flux in the rotor appears in the readout as an ac signal at the 170 Hz spin speed superimposed on the essentially dc signal from the London moment. In itself this ac signal is not deleterious; indeed, it can be a useful aid as a diagnostic for gyro torques and a calibrating signal for the gyro scale factor. If, however, the amplitude exceeds the linear range of the readout amplifiers, rectification offsets will occur. Hence the 10^{-7} G requirement.

The requirement is met by making use of one of the expanded balloon ultralow magnetic field shields conceived by W. O. Hamilton [67], developed by B. Cabrera, and described by Cabrera elsewhere in this volume. The shield, located between the gyro probe and the inner well of the flight dewar as shown in figure 3, is 70 inches long and 10 inches in diameter and closed at the lower end. It is made of 2.5 mil thick lead foil. To protect the superconductor at the open end from going normal in the earth's field, a mu-metal shield (not shown) is incorporated in the dewar.

Since the lead bag has an open end, the earth's magnetic field enters it and may affect the gyroscopes. In fact, the combination of mu-metal shield and lead bag attenuates the transverse component of the field by about seven orders of magnitude, leaving a need for nearly six orders of magnitude additional attenuation to reduce the field variations as the satellite orbits the earth to the 2×10^{-13} G requirement. The extra attenuation is achieved by (a) surrounding each gyroscope with a transverse cylindrical superconducting shield (figure 3), which provides nearly four orders of magnitude of attenuation through a combination of direct shielding and symmetry, and (b) exploiting the self-shielding effect of a gyro readout loop tightly coupled to the superconducting rotor such that the external field penetrates only the narrow annular gap between the ball and the loop.

The inner shields also eliminate cross talk between the gyroscope readouts. For further details see paper (C) of this series.

4.4   Telescope and Data Instrumentation System

The star-tracking telescope has to give a precise reference with resolution, linearity and null stability adequate to avoid errors of a milliarc-sec during the lifetime of the experiment. A scheme is also needed to subtract and process the gyroscope and telescope signals into a form suitable for storage and transmission to ground. The designs of the telescope and data instrumentation systems are bound up with the design of the attitude control system of the spacecraft. The proposed telescope is linear to 0.3 m arc-sec over a range of ±60 m arc-sec. Provided the telescope and gyro readouts are scaled correctly, the requirement is to point the telescope within this range of the apparent star position. Pointing to 60 m arc-sec is feasible with a two-loop control system. The proportional thrusters point the spacecraft to rather better than an arc-sec; within the instrument there are cryogenic actuators which tilt the gyro-telescope structure with respect to the dewar, and keep the telescope pointed within the desired 60 m arc-sec range.

The telescope is illustrated in figure 4. It is a folded Schmidt-Cassegrain system of 150 inch focal length and 5.6 inch aperture, held together entirely by optical contacting. The physical length is 13 inches. The addition of the tertiary mirror to the conventional Cassegrain design puts the focal plane at the front of the telescope. This is done primarily for structural convenience; but it also gives an opportunity to improve baffling of stray light. Image division occurs in the light-box contacted to the corrector plate. The light first passes through a beam-splitter near the focal plane to give two star-images, one for each readout axis. Each image then falls on the sharp

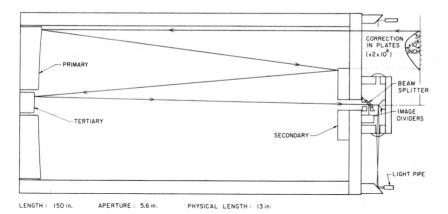

LENGTH : 150 in.   APERTURE : 5.6 in.   PHYSICAL LENGTH : 13 in.

FIGURE 4. Design layout of cryogenic star-tracking telescope.

edge of a roof prism, where it is again subdivided and passed through light-pipes to a chopper and photodetector at ambient temperature.

The ultimate limit to sensitivity of the telescope is set by photon noise from the starlight. The noise equivalent angle θ_t for each readout axis is calculated from the fluctuation in intensity of the light falling on the photocell during each signal period. For a star of magnitude M and color temperature $\bar{\lambda}$, the result is approximately

$$\theta_t = 2 \times 10^{-6} \frac{\delta}{D} \frac{\sqrt{2.51^{-M}\Delta\nu}}{\bar{\lambda}\epsilon\eta} \quad , \tag{3}$$

where δ is the image diameter, D the telescope aperture, ϵ the effective light loss in each channel, η the quantum efficiency of the detector, and $\Delta\omega$ the bandwidth. For diffraction limited optics, δ may be taken as 2.44 λ/D. A telescope of 5.6 inch aperture, having $\epsilon\eta$ about 0.001 (which is very conservative), resolves the direction of a first magnitude star to 10 m arcsec in 0.1 seconds of time. Rigel is of 0.16 magnitude.

Figure 5 shows the principle of the science data instrumentation system. The gyro and telescope signals are subtracted and summed with the final signal in the precision summing amplifier Σ_1 and then filtered in the integrating data loop represented by the heavy lines on the diagram. The output of Σ_1 is an amplitude modulated suppressed carrier alternating current signal; it is processed in a sampling demodulator and filter to give a direct current output with extremely low offset, and then integrated by means of an 18 bit up-down binary counter, which contains the readout signal for storage and telemetry. The integrating loop is closed by an 18 bit digital-to-analog converter summed into Σ_1. Call the gyro ouput G, the telescope output T, and the signal in the up-down counter R. The summing amplifier yields the function $(T - G + R)$, which is maintained at

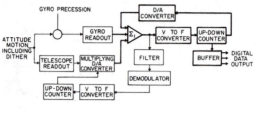

RELATIVITY DATA INSTRUMENTATION SYSTEM

FIGURE 5. Instrumentation system for processing the GP-B science data.

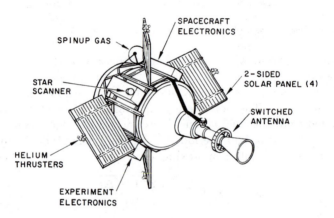

FIGURE 6. GP-B spacecraft, with telescope sun shield and other support equipment.

TABLE 1. PRELIMINARY WEIGHT BREAKDOWN OF SPACECRAFT (Kg).

Structure	273
Dewar (including helium)	743
Instrument	155
Control systems	25
Command and data management system	77
Electrical power	273
Thermal control	16
Total	1562

null, making the final signal R equal to $(G - T)$, the quantity of interest in the experiment.

The scale factors of the gyro and telescope readouts have to be nearly matched; otherwise, a pointing error of 50 m arc-sec might cause a null shift in R indistinguishable from the relativity signal. In the flight experiment, the match should be to about 2%. It is achieved by introducing a low frequency dithering motion in the pointing control to make the gyro-telescope package swing back and forth across the line of sight to the star with an amplitude of about 30 m arc-sec at about 1 minute period. If the scale factors of the two readouts are not equal, a signal appears at the output of Σ_1, where it is synchronously detected and used to drive an automatic gain control on the telescope output by the second loop shown in figure 6.

4.5   Spacecraft

Figure 6 shows the layout of the spacecraft. The extension at the front end is the sunshield for the telescope. The solar arrays provide an initial power of 230 W and a final power after one year's lifetime of 200 W. The overall diameter of the spacecraft (excluding solar panels) is 8 feet, the length including the sunshield is 14 feet 8 inches. It is designed to stand upright on a single shuttle pallet mounting and so occupy one-fifth of the shuttle bay. Table 1 shows the weight breakdown. The power requirement is about 150 W.

5.   REDUCTION OF THE DATA

5.1   Explanation

A gyroscope moving in an ideal polar orbit experiences two principal relativity effects Ω_G and Ω_M, both, as we have seen, causing linearly increasing changes in direction of the spin vector, with the geodetic precession Ω_G lying in the plane of the orbit and the motional precession Ω_M lying in the plane of the celestial equator. These, being at right angles, are easily separated. In nonpolar orbits the terms become mixed in a way that seems troublesome but in fact is not.

The succeeding sections cover:

(1)   procedures for separating the two Schiff terms in any orbit

(2)   the inclusion of three smaller relativistic effects in addition to the Schiff terms

(3)   a method of calibrating the scale factor of the experiment absolutely through starlight aberration signals

(4)   the combination of (1), (2), (3) in a Kalman filter covariance analysis of the data

(5)   an application of the experiment to improve the measurement of the distance to Rigel and potentially to improve our knowledge of the nearby distance scale for the universe

(6)   the effect of uncertainties in the proper motion of the guide star.

For convenience we take (2) first.

5.2   Five Relativity Effects

An experiment at the milliarc-sec/yr level will detect five distinct relativity effects, Ω_G and Ω_M from a spherical earth, and three additional terms:

- the de Sitter-Fokker solar geodetic precession

- a correction to the Schiff terrestrial geodetic precession due to the earth's oblateness

- the deflection of the light from the guide star by the gravitational field of the sun.

For the starlight deflection term one is, as it were, turning the experiment around and using the gyroscope as a reference for the telescope.

The first of these additional effects was drawn to my attention by Blackett in 1962 in the same letter [6] in which he recounted his reflections on the experiment in the 1930's. Schiff when thinking of a 10 m arc-sec/yr experiment had dismissed it as negligible [5]. Published discussions of its importance in the gyroscope experiment were given simultaneously by Barker and O'Connell [68] and by Wilkins [59]. The second effect was investigated independently by Wilkins [69] and O'Connell [70]; in 1978 J. V. Breakwell [71] gave the elegant treatment published for the first time in paper (F) of this series. The third was first pointed out by O'Connell and Surmelian [72], while a method of extracting it from the data *via* the Kalman filter covariance was derived by Duhamel [73]. Table 2 lists the five relativity effects, with numerical values for Ω_G, Ω_M and the oblateness correction to Ω_G computed for a satellite in a 650 km polar orbit. The solar geodetic and starlight deflection effects are independent of the orbit. Each of the three additional terms can be treated satisfactorily in data reduction, though each has a different logical status. The explicit formulae relating the oblateness and solar geodetic effects to the Schiff geodetic and motional coefficients being determined by the experiment are given below in section 5.4.

o   *de Sitter-Fokker effect*: This effect, which lies in the plane of the ecliptic, being identical in character with the terrestrial geodetic precession, is computed from the same formula with the mass and distance of the sun

TABLE 2. THE FIVE RELATIVITY EFFECTS.

Schiff geodetic effect Ω_G	6600 m arc-sec/yr
Schiff motional effect Ω_M	42.0 m arc-sec/yr
DeSitter-Fokker solar geodetic effect Ω_{DS}	19.0 m arc-sec/yr
Oblateness correction to Ω_G	-7 m arc-sec/yr
Starlight deflection (with Rigel as guide star)	$+14.4$ m arc-sec maximum

substituted for the mass and distance of the earth. The logical relation is a bootstrap one in which a combination of the terrestrial geodetic precession and a component of the solar geodetic precession is determined experimentally, and then used to calibrate the remaining component of the solar geodetic precession. See changes given below in equations (6) and (7).

o *Oblateness correction*: This is computed from Breakwell's formula, based on a generalized application of Schiff's geodetic expression $\Omega_G = 3(\mathbf{g} \times \mathbf{v})/2c^2$ derived from equation (1). Logically the calculation rests on the same assumption as the one underlying the computation of the de Sitter-Fokker effect, but the relationship of the oblateness correction to Ω_G is a more direct one, since the effect comes from the same source and (for a polar orbit) lies in the same plane.

o *Starlight deflection*: This causes a deflection away from the sun in two axes, making the apparent position of the star describe the curve shown in figure 7 with a maximum deflection of 14.4 m arc-sec on June 10. Since the time signature of the effect is different from any of the gyroscope precession terms, it can be separated from them in the data reduction. The precision depends on the choice of SQUID magnetometer in the gyro readout. Duhamel [74] has shown that a conventional SHE Incorporated 19 MHz SQUID yields a 4% measurement of the starlight deflection coefficient and a Clarke double-junction SQUID a 1.4% measurement.

5.3 Aberration of Starlight and Scale Factor Calibration

A critical issue in the experiment is the calibration of the scale factor of the gyroscope. We have seen (section 4.4) how the telescope scale factor

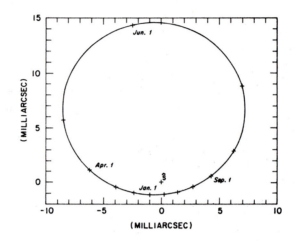

FIGURE 7. Starlight deflection measurement.

is forced to agree with that of the gyroscope by means of the dithering technique, but no process has been described that will insure that a given voltage out of the gyro readout circuit will correspond to a given angular displacement of the spin axis. Nor, in the laboratory, is it easy to find such a method of calibration.

In space, however, by remarkable good luck, nature has supplied a ready-made yardstick. Superimposed on the relativity terms to be measured, and completely distinguishable from these terms, are other signals of known amplitude and phase from the aberration of starlight. As the earth orbits the sun, there is a ±20.408 arc-sec variation in the apparent position of the guide star due to the motion of the telescope across the line of sight; as the satellite orbits the earth, there is a corresponding ±5.5 arc-sec variation in the orbit plane. The annual aberration is known from JPL ephemerides data to 0.07 m arc-sec or 3 parts in 10^6, and the orbital aberration from tracking data to about the same precision. These signals appear in the subtracted gyro-telescope output of the experiment and establish an absolute scale factor for it.

The role of aberration in scale factor calibration was first pointed out in 1968 [75]. Occasionally these signals have been regarded as an obstacle to a successful relativity experiment; in reality, they provide a further illustration of the principle enunciated earlier that an obstacle seen rightly may turn into an advantage.

The standard expression for the angular deflection from aberration is $(v \sin \theta)/c$, where $v \sin \theta$ is the velocity of the telescope across the line of sight and c is the velocity of light. An intriguing point made independently by P. Stumpff [76] and T. Duhamel [77] is that in computing the annual aberration at the milliarc-sec level, it is necessary to include a special relativistic correction $-(v^2 \sin 2\theta)/4c^2$ in the calculation.

5.4   Inclined Orbits and the Kalman Filter Covariance Analysis

So far I have only discussed effects in an ideal polar orbit. In an inclined orbit a component of the geodetic precession becomes superimposed on the motional precesson, and if the coinclination angle $(90° - i)$ exceeds a certain value, there is the further complication that certain Newtonian torques on the gyroscope resulting from the gradient in the earth's gravitational field cause drift rates in excess of 1 m arc-sec/yr. These gravity gradient disturbances are of two kinds: a *direct* term from the interaction of the earth's monopole field with the quadrupole mass moment of the spinning gyro rotor, and *indirect* terms from the effects of gravity gradient accelerations that act on the gyroscopes because they are not at the center of mass of the spacecraft. The latter produce gyro drifts through the mass unbalance torque of equation (2) and the various suspension torques discussed in section 6 of paper (B) in this series of papers.

There is yet another complication. In a nonpolar orbit the earth's oblateness makes the right ascension of the ascending node of the orbit regress at a rate

$$\omega_n = -\frac{3}{2} J_2 \left[\frac{R_e}{a}\right]^2 \bar{\omega}_o \cos i \quad , \tag{4}$$

where R_e and J_2 are the mean radius and oblateness coefficient of the earth, and a, $\bar{\omega}_o$ and i are the semimajor axis, mean motion and inclination of the orbit. At 650 km altitude $\omega_n = -7.03 \cos i$ deg/day. In the discussion which follows, "near polar" refers to an orbit for which ω_n is a few degrees or less per year, and "nonpolar" to one for which ω_n is many degrees per year.

A first impression would be that all these complications make the doing of an experiment in anything other than an ideal polar orbit hopelessly complicated. Once again, however, an apparent obstacle (the regression of the orbit plane) turns out to be an advantage [78]. The regression modulates the effects in such a way that the relativity effects can be distinguished uniquely from gravity gradient terms.

A full discussion is given elsewhere [79,80]. Here it is sufficient to write down what happens in a near polar orbit and state the results for other orbits qualitatively. The results are expressed in terms of the time variation $\dot{\mathbf{n}}_s$ of the unit vector $\mathbf{n}_s$ along the direction of spin of the gyroscope (related to the precession vector $\boldsymbol{\Omega}$ by the usual formula $\dot{\mathbf{n}}_s = \boldsymbol{\Omega} \times \mathbf{n}_s$) and coefficients A_G, A_M and A_g corresponding to the numerical values from geodetic, motional and gravity gradient precessions of the gyroscope, A_G and A_M

being the quantities we want to determine from the experiment and compare with Schiff's predictions. The experimentally significant quantities are the rates of change of $\mathbf{n}_s$ in the north-south and east-west planes and the second derivative of $\mathbf{n}_s$ in the north-south plane. Let us write these as $(\dot{\mathbf{n}}_s^{ns})_{\text{MEAS}}$, $(\dot{\mathbf{n}}_s^{ew})_{\text{MEAS}}$ and $(\ddot{\mathbf{n}}_s^{ns})_{\text{MEAS}}$, it being understood that each term is averaged over many orbits. If we work out the expected gyroscope precessions in a near polar orbit *including the effects of the oblateness and solar geodetic terms* (but assuming that the starlight deflection term has been removed from the data by virtue of its different time signature), we find, after stripping the equations of all terms smaller than a milliarc-sec, that

$$A_g = \frac{(\ddot{n}_s^{ns})_{\text{MEAS}}}{\omega_n} , \tag{5}$$

$$A_G = (\dot{n}_s^{ns})_{\text{MEAS}} \left[1 - q \sin I + \frac{9}{8} J_2 \left(\frac{R_e}{\bar{r}} \right)^2 \right] - (\ddot{n}_s^{ns})_{\text{MEAS}} \frac{\phi_0}{\omega_n} , \tag{6}$$

$$A_M = (\dot{n}_s^{ew})_{\text{MEAS}} \cos \delta + (\dot{n}_s^{ns})_{\text{MEAS}} (i' - q \cos I) . \tag{7}$$

Here J_2 and R_e are as before the oblateness coefficient and mean radius of the earth, δ is the declination angle of the guide star from the celestial equator, $\bar{r}$ is the mean radius of the satellite orbit (see paper (F) below), i' is the coinclination of the orbit plane ($i' = 90° - i$), ϕ_0 is the initial misalignment angle between the orbit plane and the line of sight to the star, I is the inclination angle of the earth's axis to the ecliptic plane, and q is given by

$$q = \frac{M_s}{M_e} \left[\frac{\bar{r}}{\bar{r}_s} \right]^{5/2} , \tag{8}$$

where M_e and M_s are the masses of the earth and the sun and $\bar{r}_s$ is the mean distance from the earth to the sun.

Equations (5), (6) and (7) are the heart of the data reduction process in a near polar orbit. Three points need to be made, illustrated numerically for a 650 km orbit with 0.1 deg coinclination i' and 0.1 deg initial misalignment ϕ_0. The nodal regression rate for an i' of 0.1 deg is 4.5 deg/yr.

First, and most important, the determination *via* equation (7) of the smaller, scientifically more significant of the quantities, the motional coefficient A_M, is notably free from pollution. Neither the gravity gradient precession terms A_g nor the oblateness correction Ω_Q affect it. Nature is on our side. The corrections for the terrestrial and solar geodetic precessions Ω_G and Ω_{DS} *are* appreciable, being respectively 11.5 and 17.7 m arc-sec/yr, but both are known with extreme precision from the 1 part in 10^4 measurement of the geodetic coefficient A_G effected *via* equation (6). They can be removed from the data with great confidence.

The one significant complication in measuring A_M is the uncertainty in proper motion of the guide star. See section 5.5.

Second, the numerical value of the gravity gradient coefficient A_g (lumped together from all sources) is known on other grounds to be of order 100 m arc-sec/yr. Cross multiplying equation (5) to get $\ddot{n}_s^{ns} = A_g \omega_n$, we find that in a 650 km orbit with coinclination 0.1 deg, the cumulative precession from this term is about 4 m arc-sec after one year and 16 m arc-sec after two years, all in the north-south plane. Thus the effect is small, and distinguishable from the relativity terms in data reduction through its quadratic form.

Third, although the last term of equation (6) shows a contribution to the north-south precession from the initial misalignment of the orbit plane with the line of sight to the star, with an A_s of 100 m arc-sec/yr and an i' of 0.1 deg, this term contributes only 0.17 m arc-sec/yr and can therefore be dropped from the equation in most instances, further simplifying data reduction. The corrections to A_G from $q \sin I$ and $J_2(R_e/\bar{r})^2$ yield drift rates of 6.8 and 7.0 m arc-sec/yr, respectively. Since the quantities q, I, J_2, R_e and $\bar{r}$ are each known to a part in 10^6, the computational margin is enormous.

Thus the approach to data reduction in a near polar orbit is simple and direct: A_G is determined from equation (6) and A_M from equation (7), and the only significant contribution of the gravity gradient effects is a small quadratic drift in the north-south plane which is readily separable from A_G and has no effect on A_M. Indeed, if the determination of A_M in a near polar orbit were the only goal, we could probably relax some of the manufacturing constraints on the gyroscope stated in paper (B) below. We choose, however, to retain them.

In nonpolar orbits the separation of terms is more complex but still unambiguous. Before giving results we must briefly consider the Kalman filter covariance analysis developed by John Breakwell, Richard Vassar and Thierry Duhamel [79,81]. The Kalman filter may be regarded as an extension of the Gaussian least squares method of data fitting, which takes into account parameter variations, for example variations in the roll rate of the satellite and the scale factor of the gyroscope. It provides the technique for utilizing the starlight aberration signals to calibrate the gyro scale factor, and also for separating the relativity signals from the gravity gradient signals in the regressing orbit.

The relativistic precessions comprise terms that either are linear in time or depend on the first harmonic of the nodal regression rate ω_n; the gravity gradient precessions comprise in addition terms depending on higher harmonics of ω_n. The direct gravity gradient torque, the mass unbalance torque and the odd harmonic suspension torque all yield precessions of the same signature containing terms in $2\omega_n$; the even harmonic suspension torques yield precessions containing terms in $2\omega_n$, $3\omega_n$ and $4\omega_n$. The covariance analysis finds coefficients for these effects from the higher frequency components and uses them to compute uncorrupted relativity signals, a linear east-west term of rate

$$\bar{n}_s^{ew} = \left[A_G \cos i - \frac{1}{2} A_M (1 + 3 \cos 2i) \right] \cos \delta \quad , \tag{9}$$

and two periodic terms, one in the north-south direction of amplitude

$$\Delta n_s^{ns} = \frac{1}{\omega_n} (A_G - 3 A_M \cos i) \sin i \quad , \tag{10}$$

the other in the east-west direction of amplitude $\Delta n_s^{ew} = \Delta n_s^{ns} \sin \delta$. Since the nodal regression rate ω_n varies as $\cos i$, the amplitudes of Δn_s^{ew} and Δn_s^{ns} both depend on inclination angle as $(A_G \tan i - 3 A_M \sin i)$. Equations (9) and (10) may be solved to determine A_G and A_M.

Use of a nonpolar orbit has two intriguing consequences. First, it partially eliminates proper motion errors. Since the proper motion of the guide star will be sensibly uniform, the sinusoidal components of the relativistic precessions will not be affected by it and the coefficient $(A_G - 3 A_M \cos i)$ can be determined absolutely. Second, surprisingly, certain nonpolar orbits yield a more precise measurement than a polar orbit. Assume, as is indeed the case, that the precision is limited by noise in the gyro readout. In a polar orbit the line of sight to Rigel is occulted by the earth for nearly half of each orbit and no data can be taken then. Now take a slowly precessing orbit chosen so that at the beginning and end of the year the ascending node lies in a plane 90° away from the line of sight. At these times, which, being at the extremities of the measurement curve, are the most critical, the star is continuously visible, and nearly twice as much data is available for fixing the shape of the curve. Figure 8 illustrates the variation of resolution with inclination found for simulated gyro readout noise by Richard Vassar [82]. The curve has two local minima for orbits with inclination angles of 78 deg and 86.25 deg, corresponding to nodal regressions of 577 deg and 182 deg. The expected error in the 86.25 deg orbit is slightly lower than in an ideal polar orbit.

Another outcome of the covariance analysis is the evolution of the measurement error with time. Naively one would expect a measurement limited by gyro readout noise to improve with time as $t^{-3/2}$ [83], since the relativity signals (in polar orbit) increase linearly with t and the SQUID noise averages as $t^{-1/2}$. Reality is more subtle. The Kalman filter makes use of the readout data for three distinct purposes: (a) to measure the gyro angle, (b) to compare the relativity signals with the starlight aberration signals, (c) to aid in processing the satellite roll phase information. All three contribute to the overall noise of the measurement. Figure 9, also due to Vassar

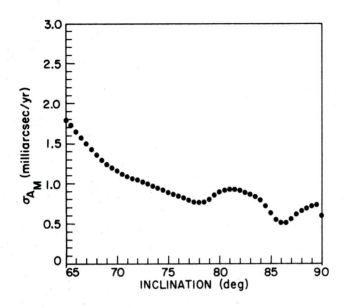

FIGURE 8. Expected error in measuring the motional precession *versus* inclination, assuming optimum launch month and known proper motion ($i = 65 - 90$ deg).

[84], illustrates the time development of the uncertainties σ_{A_M} and σ_{A_G} in the determinations of the motional and geodetic precessions, obtained for a polar orbiting mission with a September launch date, assuming an intrinsic readout noise of 1 m arc-sec in 70 hours, as discussed by Anderson in paper (C) of this series. Superimposed on the expected smooth improvement in resolution is a six-month periodicity consequent upon a frequency doubling of the annual aberration signal used in scale factor calibration.

Breakwell and his students have extended the Kalman filter covariance analysis to investigate many different aspects of the data reduction process. One already discussed is the computation of the relativistic deflection of starlight from the experiment. Others include studies of the effects of the polhode motion of the gyro rotor [85], methods of handling interruptions of data, including ones resulting in temporary malfunction of the gyroscope [86], and a method of improving the measurement resolution by making use of the trapped flux in the rotor to aid in scale factor calibration [87].

In conclusion, I remark that intriguing as nonpolar orbits are, it is wise, in the first instance anyway, to stick with the near polar orbit. Doing so minimizes the burden of data reduction and hence the possibility of error in the reduction process. Particularly important is the fact that in a near polar orbit gravity gradient torques have no influence on the determination of the motional precession of the gyroscope.

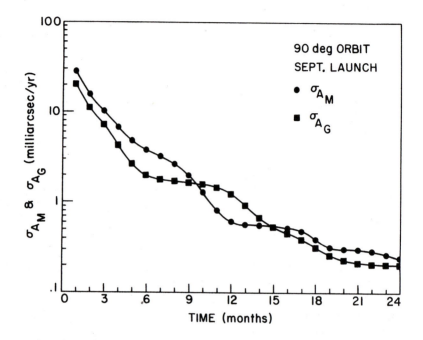

FIGURE 9. Time history of σ_{A_M} and σ_{A_G} for a 90 deg orbit, September launch.

5.5 The Proper Motion Error

The largest known source of error in the experiment is, as already remarked, the uncertainty in proper motion of the guide star. A current best estimate for the uncertainties in the absolute proper motion of Rigel is 0.9 m arc-sec/yr in declination and 1.7 m arc-sec/yr in right ascension [88]. The uncertainty in right ascension contributes a 4% error to the determination of the motional coefficient A_M.

Ultimately the proper motion error is of no great concern because knowledge of proper motions will improve with time and can be applied retroactively to improve the result. Data from the European Space Agency's HIPPARCOS astrometric satellite and from the Hubble Space Telescope may have appreciably reduced the uncertainty by the time the experiment flies. Beyond that there are (in addition to the partial solution for an inclined orbit described in the preceding subsection) two possibilities. One, suggested by R. H. Dicke [89], is to design a special earth-based instrument to measure the motion of Rigel with respect to a local background field of distant stars rather than to solve the general astrometric problem. The other, suggested in 1965 by I. I. Shapiro [90], is to compare results from two gyroscope experiments flown at different orbit altitudes.

5.6 Ranging to Rigel, and a Potential Method for Improving the Nearby Distance Scale of the Universe

The apparent position of Rigel varies not only through aberration but also through parallax. Being about 250 parsecs from the solar system, its annual parallactic variation is about 4 m arc-sec, and this variation, being 90 deg out of phase with the annual aberration, produces a phase shift ϵ ($\sim 2 \times 10^{-4}$ rad) in the measured aberration signal. Currently the distance to Rigel, inferred from a distance scale based on its brightness and star type, is known to about 25 percent. Vassar and Duhamel [91] have shown that the distance to Rigel can be fixed to 3% in a gyroscope experiment with readout based on the 19 MHz SHE SQUID, and to 0.7% in one with a readout based on the Clarke double junction SQUID.

If the method were extended to measure ranges to the group of Cepheid variables within our galaxy, the experiment could help significantly in improving the overall distance scale for the universe.

6. PROSPECTIVE: SIX CONCEPTS FOR IN-FLIGHT CALIBRATION OF THE EXPERIMENT

A measurement as difficult and important as the one we are undertaking demands, whatever its outcome, exceptionally severe scrutiny. Only a modest knowledge of the history of physics is needed to make one aware how easily results in agreement with a cherished theory are accepted and those in disagreement explained away. However great our confidence in the error analysis, any spirit of idle complacency about it would be an abrogation of scientific responsibility. A rigorous plan of in-flight check and countercheck is a *sine qua non* of the gyroscope experiment.

Systematic consideration of in-flight calibration followed upon a discussion initiated by Rainer Weiss of MIT and David Wilkinson of Princeton during the visit of the NASA Technology Review Committee to Stanford in August 1980. At that meeting two countervailing principles were brought out. One, emphasized by Weiss and Wilkinson, is the physicist's principle: "Vary everything you can." To assure that the instrument is working as planned, the gyro operating conditions (pressure, support voltage, spin speed and so forth) should be varied widely during the course of the experiment. The other, emphasized by Israel Tabeck, former project manager of the Viking mission at NASA Langley Center, is the engineer's principle, "If it ain't broke, don't fix it." The following discussion, which is a long way from being final, outlines an in-flight calibration plan in terms of six concepts: *redundancy, variation, enhancement, separation, continuity* and *absolute relationships.*

The flight instrument has four gyroscopes, each of which is designed to measure all five relativity effects. If all the gyroscopes agree to 1 m arc-sec or better throughout the year, one has gained *redundancy* of the most valuable kind, because it is a redundancy combined with *variation.* The four gyroscopes are independent not just in being separate but in the deeper sense of having different characteristics and operating conditions. Two spin clockwise and two counterclockwise (which makes the direct, but not the indirect, gravity gradient torques on them yield drifts of opposite sign); each rotor has a different shape and mass distribution (which makes the support and mass unbalance torques on each slightly different); each gyroscope is at a different location with respect to the center of mass of the spacecraft (which makes the indirect gravity gradient effects on each one different). If with all these variations the four gyroscopes yield identical relativity signals, as they ought to, our confidence in the result is much higher than it would be if the gyroscopes were simple replicas of one another.

Two other possible redundancies deserve consideration. One which we wished for but have now abandoned would be for each gyroscope to have two orthogonal readout loops. J. M. Lockhart [92] has shown that the extra loops give data of only marginal usefulness and are best removed on grounds of simplicity. The second redundancy, still under study, hinges on whether the four precise readout loops for the gyroscopes are coplanar or whether two are referred to the x-axis readout of the telescope and two to the y-axis readout. The latter arrangement, though a complication in design, has operational advantages and also has the potential for allowing cross-checks on certain kinds of telescope error. If adopted, it is an exemplification also of the principle of *separation* to be discussed below.

Our next concept for in-flight calibration is *enhancement.* A cardinal principle of physical experimentation, to my knowledge first systematically applied by Henry Cavendish [93] during his torsion balance measurement of the gravitational constant G in 1798, is that if one is uncertain how large some disturbance on an apparatus is, one should deliberately increase it and see how big it is. The gyroscope experiment depends preeminently on making a great number of effects "near zero." A corollary is that at some period in the mission one should invert the strategy and make the effects large.

To serve as a useful diagnostic, *enhancement* requires *separation.* It is no good making all the disturbances large at the same time. Examples of productive enhancements follow.

○ *Drag-free bias*: Control to 10^{-10} g appears to be essential in reducing the mass unbalance and primary *odd* harmonic suspension torques on the gyroscope to an acceptable level while gathering relativity data. Now these torques are linear in the applied acceleration. Apply a bias of say 10^{-7} g and their effects will be enhanced three orders of magnitude, making the gyroscope precess in 8 hours through an angle equal to its total annual precession from these sources under normal operating conditions.

o *Suspension level*: In normal operating conditions the preload acceleration on the gyroscope needs to be about 10^{-7} g to minimize the primary *even* harmonic suspension torques (see papers (B) and (E)). Raise the preload to say 10^{-4} g and the effect from the first term in equation (4) of paper (B) will be enhanced three orders of magnitude, so again in 8 hours the gyroscope will precess through an angle equal to its total annual precession from this source under normal operating conditions.

o *Spacecraft roll*: Roll of the spacecraft appears to be essential in reducing the gas torques to an acceptable level. It also contributes in various degrees of significance to removing other sources of gyro drift, for example, the preload terms in the even harmonic primary suspension torques, the miscentering torques, the magnetic torque due to action of residual trapped flux in the gyro shield on the London moment, and the electric torque due to the interaction of a charge on the rotor with corresponding induced charges on the electrodes. For each, the averaging factor is a few parts in 10^4. (For another class of torques, roll averaging yields only a factor of two improvement.) Stop the roll and again in 8 hours the gyro precessions from these sources will equal or exceed their total annual values in a rolling spacecraft. To further distinguish the effects, one holds the spacecraft in a fixed orientation and varies other parameters such as preload, magnetic field and charge on the rotor. Note that since spacecraft roll is also utilized to get rid of gyro readout errors, the diagnostic program must include periods during which the spacecraft stays fixed in inertial space to let the precessions occur, and periods during which it is rolled to allow them to be measured. The offsets between measurements in the presence and absence of roll determine the effectiveness of roll in reducing readout errors.

Other potential *enhancements* include raising the gas pressure and lowering the spin speed of the gyroscope. Here one must pause to reflect. Some people have found the notion of varying the spin speed particularly appealing, but this of all the enhancements is the hardest, the one most likely to end in catastrophe. Acknowledging the force of Israel Tabeck's warning, it would seem best to divide the mission into three phases:

(1) initialization (one to two weeks),

(2) gathering of relativity data (a year or more),

(3) post-experiment calibration tests (about two months).

Tests based on varying the preload and drag-free level and on stopping the roll would be done in phase (3), a test at reduced spin speed in phase (1).

An attractive approach is to utilize trapped flux in the gyro rotor to study gyroscope performance before spin up. Consider a torque Γ_i independent of spin speed. For the spinning gyroscope, $\Omega_i = \Gamma_i/I\omega_s$; for the nonspinning one, in time t the rotor will slew through an angle $\theta_i = \Gamma_i t^2/2I$. If Ω_0 is the desired performance, the maximum allowable slew angle is

$$(\theta_i)_{max} < \frac{1}{2}\Omega_0\omega_s t^2 \quad , \tag{11}$$

which for an Ω_0 of 0.3 m arc-sec/yr (5×10^{-17} rad/sec) corresponds to a rotation of 38 arc-sec in a day. This is easily resolved. Since the trapped field at 10^{-7} G is three orders of magnitude less than the London moment field for a gyroscope spinning at 170 Hz, the resolution is found by substituting 1 arc-sec for 1 m arc-sec in the figures for gyro readout, $i.e.$, 1 arc-sec in 70 hours with an SHE 19 MHz SQUID and in 16 hours with a Clarke double-junction SQUID. A day is more than enough. In reality the gyroscope will have some initial slow spin established during levitation, but that is an advantage rather than a disadvantage.

Evidently a program of enhancement tests can be worked out for the nonspinning gyroscope.

What of tests during phase (2) of the experiment when relativity data is being gathered? Here we apply our fifth and sixth calibration concepts: *continuity* and *absolute relationships*.

Continuity may be thought of as a modest violation of the "near zero" principle. The use made of orbital aberration signals in calibrating the gyro scale factor is one example. Having such a precisely known signal with periodicity 98 min acting continually throughout the year is an elegant diagnostic. For usefulness in "continuity" an effect needs to be repetitive, distinguishable from other terms, large enough to be detected but still sufficiently "near zero" not to upset the experiment. Take four examples: (a) cyclic gravity gradient accelerations, (b) the quadratic component of gyro drift, (c) residual trapped flux in the gyro rotor, (d) the action of the earth's magnetic field on the gyro readout.

○ *Cyclic gravity gradient accelerations*: The gravity gradient acceleration acting on a gyroscope ℓ cm from the center of mass of the spacecraft has a component of amplitude $1.7 \times 10^{-9} \ell g$ at twice orbital period lying in the orbit plane. (See equation (4) of paper (B).) This acceleration will appear in the output of the gyro suspension system; its continuity throughout the year is a check of the suspension system's performance and hence of the gyro performance.

o *Quadratic component of gyro drift*: As was explained in section 3.4, there is a small quadratic component of precession in the north-south plane arising from the changes caused in various torque terms by the changing gravity gradient in a regressing near polar orbit. Continuity of this effect throughout the year in proper relationship to the orbit plane checks the gyro performance and fixes the magnitude of the coefficient A_g for the sum of mass unbalance, direct gravity gradient and odd-harmonic suspension torques. It does more. The quadratic precession is the result of applying to the gyroscope a linearly increasing acceleration of known magnitude at right angles to the orbit plane. For a gyroscope 30 cm from the spacecraft center of mass, in an orbit regressing 4 deg in a year, the total change in acceleration over the year is 3.4×10^{-9} g, or 34 times the drag-free limit on the spacecraft. Once A_g has been found from the quadratic term, it can be combined with the drag-free limit to set an upper bound on the uncertainty introduced into the experiment through the action of the residual drag-free acceleration on mass unbalance, direct gravity gradient and odd-harmonic suspenion torques.

o *Trapped flux in the gyro rotor*: The use of trapped flux in assessing gyro performance prior to spin up has already been discussed. It also aids during the mission. The trapped flux signals provide a repeating pattern at the spin frequency, modulated by the polhoding of the body. Details are given elsewhere; three points are significant. (i) The carrier frequency gives the spin speed and spin down rate of the gyro rotor, (the spin down rate yielding incidentally the best measure of the gas pressure in the cavity). (ii) Measurement of the amplitude of the trapped signals is, as was mentioned in section 3.4, a useful aid in calibrating the scale factor of the gyro readout, specifically in reducing effects of short-term fluctuations [87]. (iii) Continuity in the polhode pattern (there should be no detectable variation throughout the year) gives insight into certain kinds of gyro disturbance [94].

o *Action of the earth's magnetic field*: To prevent signals at the 1 m arc-sec level getting into the gyro readout from the earth's magnetic field, the field has to be attenuated by some thirteen orders of magnitude. With less attenuation, a signal at twice orbital period, modulated at 24 hour period because of the earth's rotation, will appear in the output, but higher order terms will be negligible, being smaller and more strongly attenuated. Two points can be made. First, a small disturbance of this kind will not impair the relativity data because the doubly periodic signature is different from that of any of the relativity terms; one may if one chooses relax the shielding requirement slightly. Second, by keeping watch on signals doubly periodic with the orbit, one can check the integrity of the magnetic shielding continually throughout the mission.

Tact is needed in balancing "near zero" and "continuity." The right initial approach is a strenuous emphasis on "near zero." Our original rhetoric was first make each disturbing effect absolutely zero, then average it in every conceivable way possible; then, finally, the experiment may just become feasible. But "continuity" skillfully applied has its own potency. The secret is to have effects that are known with great assurance and sharply distinguished from each other. The use of the quadratic component of gyro drift to determine A_g is a good example. The applied acceleration, being fixed by geometrical considerations, is known very exactly. Other cases may require more thought. Thus the magnetic shielding effect seems unambiguous in that it is doubly periodic with the orbit; but suppose the doubly periodic gravity gradient acceleration disturbed the gyro readout in some way. Would this cause an ambiguity, and if so would the ambiguity matter? Probably not, because any gravity gradient effect should be 90° out of phase with the direct magnetic effect; but these are the kinds of issues that need to be thought through.

Lastly, *absolute relationships*. Two examples will suffice: (a) the absolute relationship between the planes in which the aberration signals occur and the planes in which the relativity signals are expected to occur; and (b) the absolute relationship between the magnitudes of the relativistic starlight deflection and the relativistic gyro precessions.

The orientation of the rolling spacecraft is established once every 10 min by the combination of rate-integrating roll-reference gyroscopes and a star blipper (see section 2.2). Appearing in the output are the orbital and annual aberration signals, each of which has a known roll phase tied respectively to the plane of the orbit and the plane of the ecliptic. A suggestion by T. M. Spencer [95], confirmed by R. Vassar and J. V. Breakwell, is that the aberration signals may themselves provide a roll reference, possibly even to the exclusion of the star blipper. In fact, Vassar has shown [96] that an experiment done thus without the blipper is degraded only by about a factor of two. It seems best, however, to keep the star blipper, not just to recover the factor of two, but because doing so provides an end-around check of the total process used in separating the two relativity terms. Admittedly, the deepest problem of the experiment, the error from the uncertainty in Rigel's proper motion, remains, but it is nice to bridge over the other uncertainties.

Similarly for the relativistic deflection of starlight. A properly working experiment measures starlight deflection to about 1 percent. But of course the deflection coefficient is known at least that well. Our best programmatic is to treat starlight deflection as a relativistic calibrating signal, whose relation to the relativistic gyro precessions adds a further end-around check on the experiment.

Proper motion is the one shaft that finds a real chink in our armor. Only by accepting the risks of an inclined orbit can we gain any in-flight calibration of that uncertainty, and then only a partial one. The need for certainty on this point is a pressing one.

The in-flight calibration process is made up of many interlocking pieces whose whole, if properly put together, is greater than the sum of its parts. The six concepts advanced here, *redundancy, variation, enhancement, separation, continuity* and *absolute relationships*, are guides in the difficult process of establishing a safe, systematic and searching plan for the three program phases: initialization, gathering of relativity data, and post-experiment testing.

7. RETROSPECTIVE I:
TEN FUNDAMENTAL REQUIREMENTS

To perform an experiment that will determine the relativistic drifts of a gyroscope to a precision of 1 m arc-sec/yr with respect to the inertial frame, ten separate requirements have to be met. An enumeration of these will usefully aid in retrospective summary:

(1) a gyroscope with drift rate less than 1 m arc-sec/yr,

(2) a gyro readout system that is linear and stable, and has a resolution better than 1 m arc-sec over a range of about ±100 arc-sec,

(3) a telescope readout that has a resolution better than 1 m arc-sec sufficiently stable and linear over a range of about ±60 m arc-sec,

(4) a gyro-telescope structure that is mechanically and optically stable in inertial space to better than a milliarc-sec/yr,

(5) a pointing system that keeps the telescope aligned with the reference star to within the telescope's linear range,

(6) a science-data instrumentation system capable of (a) subtracting the gyro and telescope signals from each other to a precision better than 1 m arc-sec over a total range of ±100 arc-sec, (b) ensuring that the scale factors of the gyro and telescope readouts are matched with the required accuracy over the range of the telescope readout, so that a subtraction made when the system is not pointing at the star remains an honest one,

(7) a means of eliminating effects of electronic, magnetic or optical bias drifts in the gyroscope and telescope readouts and in the science-data instrumentation system,

(8) a calibration of the combined scale factor of the gyro and science-data instrumentation systems, so that one knows that a particular digital count corresponds to a particular number of arc-sec, despite the long-term and random short-term variations in these scale factors,

(9)   a means of separating the geodetic and motional precessions observed by each gyroscope,

(10)   adequate knowledge of the absolute proper motion of the guide star.

If any of the foregoing ten requirements is not met, the experiment will fail, although inadequate present knowledge of (10) could be corrected for by retroactive extrapolation of later improved data.

In the gyroscope experiment, (1) is achieved by space operation combined with high vacuum and low magnetic field techniques and spacecraft roll, (2) by the London moment readout with an appropriate feedback loop, (3) by the cryogenic optically contacted quartz telescope with roof prism image dividers, manufactured in a particular manner, (4) by operating at low temperatures in the vacuum of space, (5) by a precision pointing control system with proportional thrusters, (6) part (a) by an 18 bit integrating data loop, and part (b) by an automatic gain control loop which forces the telescope scale factor to match the gyroscope scale factor through injecting a "dither" signal into the pointing system and looking synchronously in the output of the summing amplifier of the data integration loop for a mismatch signal which can be used for reference in automatic gain control, (7) by rolling the spacecraft about the line of sight to the star with a 10 min roll period, which chops the signal and eliminates any drifts with frequency longer than that for the roll (care is needed to scrutinize possible rectification effects at 10 min, due, for example, to temperature dependent null drifts in the electronics systems), (8) by making use of (a) the annual aberration signals to provide absolute long-term scale factor calibration, (b) the orbital aberration signals to provide both absolute short-term and absolute long-term scale factor calibration, (c) trapped flux signals which provide the best relative short-term scale factor calibrations, (9) from spacecraft roll through measurement of roll phase with the external gyroscopes and star-blipper. Requirement (10) is at present the weakest link in the chain because no one is sure how far to trust the astrometers.

With the present uncertainty in the absolute proper motion of Rigel ascension, the resultant uncertainty in determining the 44 m arc-sec/yr motional precession is 3.8%, and this at present is the dominant error in the experiment.

8.   RETROSPECTIVE II:
THREE INTRINSIC AND
SEVEN EXTRINSIC "NEAR ZEROS"

I remarked earlier that the gyroscope experiment preeminently illustrates the "near zero" principle. If anything can be regarded as "near zero" it is the 10^{-11} deg/hr absolute drift which is required of this gyroscope.

The 10^{-11} deg/hr limit on drift rate is an *intrinsic* "near zero," a requirement the gyroscope has to meet, whatever the details of its design. Two other intrinsic "near zeros" follow from the fundamental one, related to the observation (section 1.2) that the two real difficulties in the experiment are (a) reading the direction of spin and (b) spinning the gyroscope to begin with. Take spin up. To spin a gyroscope one must apply a torque Γ_s parallel to the axis of spin. Assuming that Γ_s is constant and that there are no drag torques, one finds that the total angular momentum $I\omega_s$ of the gyroscope is just equal to $\Gamma_s t_s$, where t_s is the spin time. After spin up, Γ_s is reduced, hopefully to zero. Suppose there remains a small component of torque Γ_r at right angles to the spin axis, then since $\Omega_r = \Gamma_r/I\omega_s$, we get the following requirement on torque switching:

$$\frac{\Gamma_r}{\Gamma_s} < \Omega_0 t_s \quad , \tag{12}$$

which for a gyroscope with an Ω_0 of 5×10^{-17} rad/sec and a spin time of 2000 sec means Γ_r/Γ_s has to be less than 10^{-13} ... truly a "near zero" requirement, but one that is indeed achieved by the gas spin up system in a rolling spacecraft.

The third intrinsic "near zero" applies to the reaction torque on the gyroscope from the readout system. The readout current acts back on the London moment causing a torque; there is also a differential damping torque from the trapped flux in the rotor. Both are negligible. In a rolling spacecraft the readout current torque for a gyroscope whose spin axis is misaligned with the readout plane by an angle α is given approximately by

$$\Omega_{rr} = \frac{15}{16\pi} \left(\frac{mc}{e}\right) \frac{\omega_s}{\rho r L K} \sin 2\alpha \quad , \tag{13}$$

where ω_s and (mc/e) are, as before, the gyro spin rate and the mass-to-charge ratio for the electron (in electromagnetic units), ρ and r are the density and radius of the ball, L is the inductance of the readout loop (in electromagnetic units), and K is the loop gain of the feedback servo. For small angles, equation (13) reduces to

$$\Omega_{rr} = 3 \times 10^{-15} \left(\frac{\omega_s}{\rho r L K}\right) \alpha = 2 \times 10^{-15} \frac{\alpha}{K} \quad , \tag{14}$$

so for a gyroscope misaligned by 20 arc-sec, the drift rate from this source is $1.2 \times 10^{-3}/K$ m arc-sec. Since the servo gain is 10^3 or 10^4, the resultant effect is very near zero indeed.

The three intrinsic "near zeros" lead in turn to a different category of "near zeros" which may be called *extrinsic*. These are the particular design constraints that have to be met to reach the desired gyro performance; there are seven of them as listed in table 3. All are attainable. Note the complementary character of the constraints on electric and magnetic torques. The gyroscope has a nonzero magnetic dipole moment (the London moment) and must therefore operate in near zero magnetic field, but has nonzero electric fields around it (the suspension field) and must therefore have near zero electric dipole moment.

Wisdom for the gyroscope experiment rests on these seven pillars.

TABLE 3. SEVEN EXTRINSIC "NEAR ZEROS."

	Requirement	Reason
Gyro rotor		
Inhomogeneity	3×10^{-7}	Mass unbalance and gravity gradient torques
Out of roundness	5×10^{-7}	Suspension torques
Electric dipole moment	10^{-10} e.s.u.	Electric torque
Environment		
Temperature	1.8 K	Superconductivity Mechanical stability
Acceleration	$10^{-10}\,g$	Suspension and mass unbalance torques
Magnetic field	$10^{-7}\,G$	Readout magnetic torques
Gas pressure	10^{-10} torr	Gas torques

References

[1]   Eddington cites J. A. Schouten, *Proc. Kon. Akad. Weten. Amsterdam* **21**, 533 (1918). I have not yet personally verified this reference; however, the paper by Schouten in [39] strongly suggests that Eddington was right.

[2]   A. D. Fokker, *Proc. Kon. Akad. Weten. Amsterdam* **23**, 729 (1920).

[3]   A. S. Eddington, *Mathematical Theory of Relativity* (Cambridge, 1924), p. 99.

[4]   W. de Sitter, *Mon. Not. Roy. Astron. Soc.* **77**, 172 (1916).

[5]   L. I. Schiff, *Proc. Nat. Acad. Sci.* **46**, 871 (1960); also *Phys. Rev. Lett.* **4**, 215 (1960).

[6]   P. M. S. Blackett, Letter of 20 November 1962 to C. W. F. Everitt.

[7]   G. E. Pugh, WSEG Research Memorandum Number 11 (Weapons Systems Evaluation Group, The Pentagon, Washington 25 D.C., November 12, 1959).

[8]   L. I. Schiff, *Am. J. Phys.* **28**, 340 (1960).

[9]   R. H. Dicke, in *Relativity, Groups and Topology*, edited by C. DeWitt and B. DeWitt (Gordon & Breach, New York, 1964), p. 165. For the Schiff conjecture, see footnote in [8] p. 343 and numerous later discussions, *e.g.*, C. M. Will, *Theory and Experiment in Gravitational Physics* (Cambridge, 1981), pp. 38, 50–53.

[10]   In A. Sommerfeld, *Lectures on Theoretical Physics Vol. III, Electrodynamics* (Academic Press, New York, 1964), p. 313. The planned publication referred to there never appeared.

[11]   Reference [8], p. 43.

[12]   *Physics Today* **12**, 29 (Dec. 1959).

[13]   *Wall Street Journal*, Thursday, January 21, 1960, p. 26.

[14]   Schiff's notes, p. 1. I am grateful to Frances Schiff for making the notes available to me.

[15]   L. I. Schiff, *Proc. Nat. Acad. Sci.* **45**, 69 (1960).

[16]   E. Corinaldesi and A. Papapetrou, *Proc. Roy. Soc.* **A209**, 259 (1951)). This was preceded by a separate paper by Papapetrou alone, to which Schiff refers in the margin of his notes and in his published paper.

[17]   Reference [5], p. 876.

[18]   W. de Sitter, *Mon. Not. Roy. Astron. Soc.* **76**, 727 (1916).

[19]   J. Lense and H. Thirring, *Phys. Zeits.* **19**, 156 (1918).

[20]   H. Yilmaz, *Phys. Rev.* **111**, 1417 (1958).

[21]   H. Yilmaz, *Bull. Amer. Phys. Soc.* **4**, 65 (1959), paper Y2.

[22]   Thus reference [21]; but Pugh [7], p. 5, says that Yilmaz in his talk exhibited several alternative formulations of general relativity which "do not predict the Lense-Thirring effect or predict it in lesser degree."

[23]   Reference [7], p. 5.

[24]   D. H. Cooper, G. R. Karr, J. L. Myers and D. Skaperdas, Report R-378 (Coordinated Science Laboratory, University of Illinois, Urbana-Champaign, 1968), Final Report NASA Research Grant NSG 443; James

L. Myers, Ph.D. Thesis, University of Illinois, Urbana-Champaign (1968), Report R-396 (Coordinated Science Laboratory, University of Illinois, Urbana-Champaign 1968).

[25] D. H. Frisch and J. F. Kasper, Jr., *J. Appl. Phys.* **40**, 3376 (1969); D. I. Shalloway and D. H. Frisch, *Astrophys. and Space Sci.* **10**, 106 (1971).

[26] Reference [7], p. 12.

[27] Mrs. Schiff supplied copies of the Pugh-Schiff letters.

[28] Reference [3], p. 105.

[29] Reference [5], pp. 871, 872.

[30] C. M. Will, *Theory and Experiment in Gravitational Physics* (Cambridge, 1981), gives an up-to-date review.

[31] I. I. Shapiro, *Phys. Rev. Lett.* **13**, 789 (1964).

[32] H. A. Hill, *Int'l. J. Theoret. Phys.* **23**, 683 (1984).

[33] The version given here is that of K. E. Eriksson and S. Yngström, *Phys. Lett.* **5**, 119 (1963). One of the more persuasive treatments will be found in R. Adler, M. Bazin and M. Schiffer, *The General Theory of Relativity* (McGraw-Hill, New York, 1965), p. 194.

[34] A. Schild, *Am. J. Phys.* **28**, 778 (1960).

[35] W. M. Sacks and J. A. Ball, *Am. J. Phys.* **36**, 240 (1968).

[36] W. Rindler, *Am. J. Phys.* **36**, 540 (1968); *ibid.* **37**, 72 (1969).

[37] F. R. Tangherlini, *Nuovo Cim.* **25**, 1081 (1962).

[38] Letter of December 5, 1983 to James M. Beggs, NASA administrator.

[39] Reference [4], p. 167, equation (87) and subsequent equations. In other words, the gyroscope obeys Fermi-Walker transport. The point seems to have been first made by J. A. Schouten, *Proc. Kon. Akad. Weten. Amsterdam* **12**, 1 (1918), to whom perhaps belongs credit for "Fermi-Walker" transport.

[40] Reference [3], *ad. loc.*

[41] D. C. Wilkins, *Phys. Rev.* **D5**, 814 (1972).

[42] R. D. Bourke, Ph.D. Thesis, Stanford University, SUDAAR Report 189 (1964).

[43] B. O. Lange, *Am. Inst. Aero. Astro. J.* **2**, 1590 (1964); Ph.D. Thesis, Stanford University, SUDAAR Report 194 (1964).

[44] F. London, *Superfluids Vol. 1: Macroscopic Theory of Superconductivity* (Wiley, New York, 1953), p. 83.

[45] R. Becker, F. Zauter and G. Heller, *Zeit. Phys.* **85**, 772 (1933).

[46] A. F. Hildebrandt, *Phys. Rev. Lett.* **12**, 190 (1964).

[47] J. B. Hendricks, A. King, Jr., and H. E. Rorschach, *Proceedings of the IX International Conference on Low Temperature Physics* (Plenum, New York, 1965), p. 466.

[48] M. Bol and W. M. Fairbank, *ibid.* p. 471.

[49] P. M. S. Blackett, *Phil. Trans. Roy. Soc.* **245**, 309 (1952).

[50] J. E. Opfer, *Rev. de Phys. Appl.* **5**, 37 (1970).

[51] H. W. Knoebel, *Control Engineering*, 70 (February 1964).

[52] The Staffs of the Space Dept. of the Johns Hopkins University Applied Physics Laboratory, and the Guidance and Control Laboratory of Stanford, *J. Spacecraft and Rockets* **11**, 637 (1974).

[53] B. O. Lange, "The Unsupported Gyroscope," paper presented at the Unconventional Inertial Sensors Symposium, New York (November, 1964).

[54] Contemporary accounts of this work may be found in the fifth, eighth and ninth "Semi-Annual Reports on the Physics Portion of a Program to Develop a Zero-G Drag-Free Satellite and Perform a Gyro Test of General Relativity in a Satellite" (Stanford University, 1966, 1968).

[55] T. D. Bracken and C. W. F. Everitt, *Adv. Cry. Eng.* **13**, 168 (1968).

[56] W. M. Fairbank and C. W. F. Everitt, "Final Report on Contract NAS8-25705, To Build and Test a Precision Star-Tracking Telescope" (1972).

[57] W. Angele, *Prec. Eng.* **2**, 119 (1980). See also C. W. F. Everitt, J. A. Lipa and G. J. Siddall, *ibid.* **1**, 5 (1979); J. A. Lipa and G. J. Siddall, *ibid.* **2**, 123 (1980); J. A. Lipa and J. Bourg, *ibid.* **5**, 101 (1983).

[58] P. M. Selzer, W. M. Fairbank and C. W. F. Everitt, *Adv. Cry. Eng.* **16**, 277 (1960).

[59] D. C. Wilkins, *Ann. Phys. (New York)* **61**, 277 (1970); reference [62], appendix A.

[60] B. Cabrera, "Near Zero Magnetic Fields with Superconducting Shields," in this volume.

[61] C. M. Marcus, *Rev. Sci. Instrum.* **55**, 1475 (1984).

[62] "Final Report on NASA Grant 05-020-019 to Perform a Gyro Test of General Relativity in a Satellite and Develop Associated Control Technology," edited by C. W. F. Everitt (HEPL, Stanford University, 1977). Contributors: J. T. Anderson, B. Cabrera, R. R. Clappier, D. B. DeBra, J. A. Lipa, G. J. Siddall, F. J. van Kann and R. A. Van Patten.

[63]   Committee on Gravitational Physics, Space Science Board, *Strategy for Space Research in Gravitational Physics in the 1980's* (National Academy of Sciences, Washington, D.C., 1981).

[64]   "An Assessment of the Technological Status of the Stanford Gyrorelativity Experiment," chaired by J. Rosendhal (September 1980).

[65]   "Gravity Probe B Phase B Final Report," George C. Marshall Space Flight Center, Alabama (February 1983).

[66]   "Account of the Restructuring of the GP-B Relativity Gyroscope Program," edited by C. W. F. Everitt (HEPL, Stanford University, 1983).

[67]   B. Cabrera and W. O. Hamilton, in *Science and Technology of Superconductivity*, edited by W. D. Gregory, W. N. Mathews and E. A. Edelsack (Plenum, New York, 1973), Vol. 2, p. 587.

[68]   B. M. Barker and R. F. O'Connell, *Phys. Rev. Lett.* **25**, 1511 (1970).

[69]   Reference [59]. Wilkins' first document on this topic is a Stanford University memorandum dated November 8, 1968.

[70]   R. F. O'Connell, *Astrophys. Space Sci.* **4**, 199 (1969); B. M. Barker and R. F. O'Connell, *Phys. Rev. D* **6**, 956 (1972).

[71]   First reported in section M of [62].

[72]   R. F. O'Connell and G. L. Surmelian, *Phys. Rev. D* **4**, 286 (1971).

[73]   T. Duhamel, Ph.D. Thesis, Stanford University, SUDAAR Report 540 (1984), p. 31.

[74]   *Ibid.*, p. 35.

[75]   C. W. F. Everitt, W. M. Fairbank and L. I. Schiff, in *The Significance of Space Research for Fundamental Physics*, ESRO SP-52, edited by A. F. Moore and V. Hardy (ESRO, Paris, 1971), p. 33.

[76]   P. Stumpff, *Astron. Astrophys.* **84**, 257 (1980).

[77]   T. Duhamel [73], appendix A.

[78]   As pointed out in C. W. F. Everitt, *Proceedings of the Conference on Experimental Tests of Gravitation Theories*, JPL Tech. Memo. 33-499, edited by R. W. Davies (JPL, Pasadena, 1971), pp. 77, 80.

[79]   R. Vassar, J. V. Breakwell, C. W. F. Everitt and R. A. Van Patten, *J. Spacecraft* **19**, 66 (1982).

[80]   C. W. F. Everitt, to be submitted to *Phys. Rev. D*.

[81]   Reference [73], p. 15.

[82]   Reference [79] and R. Vassar, Ph.D. Thesis, Stanford University, 1982, SUDAAR Report 531 (1982), p. 92.

[83]   As was stated in [78], p. 78.

[84]   Reference [79]; also [82], p. 101.

[85]   Reference [73], p. 49.

[86]   *Ibid.*, p. 37.

[87]   *Ibid.*, p. 77.

[88]   J. T. Anderson and C. W. F. Everitt, "Limits on the Measurement of Proper Motion and the Implications for the Relativity Gyroscope Experiment," memorandum, HEPL, Stanford University (1979).

[89]   R. H. Dicke, private communication, *ca.* 1980.

[90]   I. I. Shapiro, private communication, *ca.* 1965.

[91]   Reference [79], and T. Duhamel, private communication.

[92]   J. M. Lockhart, private communication.

[93]   H. Cavendish, *Phil. Trans. Roy. Soc.* **83**, 470 (1798). Reprinted in the 1809 abridgment of the *Transactions*, edited by C. Hutton, G. Shaw and R. Pearson (Royal Society, London, 1809), Vol. 18, p. 389.

[94]   See C. W. F. Everitt, "Commentary on a paper by Barker and O'Connell on 'Effects of the Gyroscope's Quadrupole Moment on the Relativity Gyroscope Experiment'," memorandum, HEPL, Stanford University (1975), p. 23.

[95]   T. M. Spencer, in "Mission Definition Study of Stanford Relativity Satellite," Ball Brothers Research Corporation Report F71-07 (1971).

[96]   R. Vassar, private communication.

The Stanford
Relativity Gyroscope Experiment
(B): Gyroscope Development

J. A. Lipa and G. M. Keiser

1. INTRODUCTION

The central element of the Stanford relativity experiment is a cryogenic gyroscope with a performance in space many orders of magnitude superior to that of conventional earth-bound devices. In the early phases of the program, a great deal of thought was given to the problem of achieving the performance goal, and an entirely new gyro concept was developed by William Fairbank and Francis Everitt, based on the application of cryogenic techniques to electrostatic gyroscopes. In paper VI.3(A) Francis Everitt has given a history and overview of the experiment. In this paper we describe the construction of the gyroscope, outline its development into an operational laboratory device, and present some of the results obtained with it. We also give a brief account of the torque analysis which demonstrates that it is feasible to make a gyroscope with the desired performance.

2. GENERAL DESCRIPTION

The gyroscope is an adaptation of the electrically suspended gyroscope (ESG) invented by A. Nordsieck in 1952 [1] and subsequently developed in variant forms by Honeywell and Rockwell. The Stanford version of the gyroscope has five main differences from conventional ESG's: (i) an operating temperature of around 2 K, (ii) a gas spin up system, (iii) a superconducting readout system, (iv) ideally equal moments of inertia, and (v) more stringent manufacturing tolerances for the rotor. These differences stem from a number of interwoven requirements placed on the experiment which must be met if the underlying scientific goal is to be achieved. For example, the readout system makes use of the London moment (described in the previous paper) not only because it works so well with a perfectly spherical superconducting rotor, but also because it gives an extremely linear readout, which is essential for calibrating the experiment, and because it is relatively immune to rotor mis-centering errors.

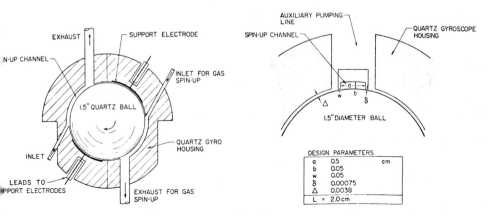

FIGURE 1. Cross-section of the gyroscope with details of spin up channel.

The gyroscope is illustrated in figure 1. It consists of a 1.5 inch diameter ball of extremely homogeneous fused quartz, ground and polished to a sphericity of better than 1 μin, and coated with a thin film of superconducting niobium. The ball is suspended within a spherical cavity by voltages applied to circular electrodes sputtered onto the inside wall of the housing. The suspension system uses ac support voltages and sensing signals as described below. On earth a field of about 700 V rms/mil must be applied to support the rotor. In space, with the gyroscope nearly in free fall, the support field is lowered to about 0.2 V/mil. The rotor is spun to its operating speed of about 170 Hz by means of a gas jet system shown in the figure. After spin up, the pressure is reduced to about 10^{-10} torr using a technique described in a following paper by Turneaure, Cornell, Levine and Lipa, and the ball is allowed to coast freely. Its spin down time constant at 10^{-10} torr is about 3000 years. As the rotor is spun up, it develops a magnetic dipole moment, called the London moment, which is aligned parallel to its spin axis, and is maintained by the superconducting properties of the rotor coating. To obtain readout information, a superconducting pickup loop is placed around the rotor and connected to a SQUID magnetometer. Any change in the pointing angle of the dipole relative to the loop normal gives a signal in the magnetometer.

A theoretical analysis of gyro performance, described briefly below, has established that a gyroscope with drift rate significantly less than 1 milliarcsec/year can be built with existing technology provided that a number of severe constraints on the gyroscope and its environment are met. The principal manufacturing constraints chosen for the gyroscope after making various tradeoffs are given in table 1. The roundness requirements on the rotor and housing come from an analysis of gyro suspension torques, plus the torque due to residual electric charge on the rotor. The requirements on homogeneity and coating uniformity come from an analysis of torques of two kinds: (i) the action of the residual acceleration of the spacecraft on a ball whose center of mass and center of geometry do not coincide (mass-unbalance torque), and (ii) the action of the gradient in the earth's gravitational field on the quadrupole mass-moment of the ball (gravity gradient torque). The two effects yield drift rates which set comparable limits on rotor homogeneity.

TABLE 1. MANUFACTURING TOLERANCES FOR THE RELATIVITY GYROSCOPE.

Sphericity of rotor	$<0.8\ \mu$in
Variation in thickness of rotor coating	$<0.3\ \mu$in
Rotor homogeneity	$\sim 3 \times 10^{-7}$
Sphericity of housing cavity	$<20\ \mu$in

3. GYRO ROTORS

There are four main steps needed to fabricate a rotor suitable for the gyro experiment. First, material of acceptable homogeneity must be selected. At present this is done using interferometric techniques. Second, an accurate sphere must be cut. Next, it must be measured to a precision approaching 0.1 μin in roundness, and finally it must be coated with a robust, thin superconducting film. These steps are outlined briefly below.

Theoretical analysis and empirical evidence [2] establish a relationship between variations in the density ρ of transparent materials and variations in their refractive index n of the approximate form $\Delta\rho/\rho \sim 2.3\Delta n/n$. Hence, if one can measure variations in the refractive index of the fused quartz from which the rotor is to be made to about 1 part in 10^7, one can determine the density variations to the 3 parts in 10^7 requirement of table 1.

In 1979 J. Bates and M. Player at the University of Aberdeen commenced a program to develop an improved measuring instrument, initially for checking cubes but with the possibility of being adapted to make measurements on the finished rotors. The system measures optical path length using a laser interferometer. To avoid errors from surface irregularities, the cube is immersed in a cell containing a fluid mixture with a refractive index very nearly equal to that of the quartz. One beam of the interferometer is sent through the cell at a fixed location, and the cube is translated from side to side on a carriage immersed in the fluid. The total mechanical path length is a constant, being the sum of the distances through the cell windows, the fluid and the cube. The index-matching technique eliminates optical path length changes due to sample thickness changes, leaving only the contribution from the internal density variations. The second beam does not intersect the cube and is reflected from a fixed mirror to provide a reference path length. Figure 2 shows measurements obtained by Player

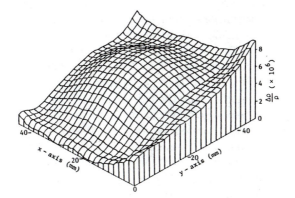

FIGURE 2. Refractive index variations across the face of a 2 inch quartz cube.

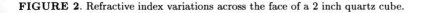

and Dunbar [3] of the integral refractive index variations parallel to one axis of a quartz cube. Similar maps were made for the other two axes. For this particular cube, the range of index variations across the face corresponded to density variation $\Delta\rho/\rho$ of 8×10^{-6}. To obtain full information on the density distribution within a cube, a tomographic method must be used, which is still under development [4]. In our case we are primarily interested in setting bounds on the low frequency density variations, so the integral path length data is acceptable. Material of much higher quality than that used for the preliminary measurements described here can be obtained. By the use of well-controlled precision anneal cycles, density variations as low as a few parts in 10^7 over rotor-sized areas have been reported [5].

Once the material is selected, it is rough ground to a spherical shape and then lapped and polished to the right diameter and sphericity. Since it is more convenient to fit the rotor to the gyro housing than the housing to the rotor, the sphere is brought to size to within about 20 μin. The diameter is controlled by monitoring the polishing rate. After many experiments, a lapping machine was designed by W. Angele [6], capable of routinely producing rotors with maximum peak-to-valley departures from sphericity of about 1 μin. The machine has four laps, arranged tetrahedrally, and is enclosed in a lucite box for cleanliness and temperature stability. The laps are all driven at the same speed, but their directions of rotation are reversed in a programmed sequence. An important step in achieving the final accuracy has been to develop an alignment procedure that makes the axes of rotation of the four laps coincide at the center of the ball to high precision.

To verify the sphericity of a rotor, a complete three-dimensional map of the surface of the sphere with an accuracy of about 0.1 μin is needed. Machines capable of measuring roundness to this precision in one plane have been available for a number of years. With the Talynova-73 computer-aided roundness measuring system we use, the workpiece is held fixed and the measurement is made with a rotating stylus transducer mounted on a precision spindle. To generate the data for complete maps of spherical rotors, we constructed a fixture which allows the ball to be turned through a succession of known angular increments about a horizontal axis. A series of great circle measurements on the ball is digitally processed to give either latitude-longitude maps or contour maps of the surface [7]. Figure 3 shows contour maps of a ball rotated in steps of 90° about the vertical axis. This rotor has a peak-to-valley maximum variation of about 2.4 μin. The information of most use in calculating gyro torques is the numerical value of each coefficient in a spherical harmonic expansion of the rotor shape. This too can be computed from the Talynova measurements [8].

The quartz rotor is coated with niobium to provide readout and suspension capability. If the coating is nonuniform, it will make the rotor

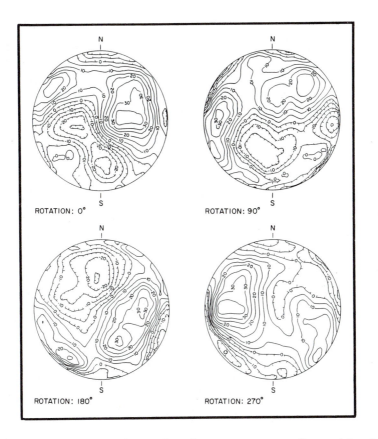

FIGURE 3. Contour maps of the surface of a quartz gyro rotor. Contour intervals are 5 nanometers, with tie-marks indicating areas below the mean surface level.

out-of-round, and because the density of niobium differs from that of quartz, it could also disturb its mass balance. A simple calculation shows that the requirement on mass balance sets a limit of 0.3 μin on the maximum peak-to-valley thickness variation that can be allowed. This in turn sets a limit on the thickness of the coating, depending on how uniformly it can be applied. To obtain a highly robust coating, the technique of sputter deposition is used. In this method a somewhat diffuse stream of niobium atoms is transferred from a target to the rotor. To obtain a uniform coating, the rotor must be rolled during the deposition to present all elements equally to the target. A number of techniques to do this were developed, the most successful to date using a microprocessor-controlled two-axis roller system. With such a device, a uniformity of about 2% has been achieved [9].

In early laboratory tests of the gyroscope, we encountered difficulty in levitating rotors with thin coatings owing to catastrophic electrical breakdown at high voltage. The difficulty was partly a matter of cleanliness and partly a matter of adhesion; dirt in the housing and sharp edges around a puncture in the film (whether on the rotor or the electrode) would both cause breakdown. The problem was avoided by using coatings 100 μin thick. Such a coating would have to be uniform to 0.3% in both thickness and density to meet the requirements of the flight experiment. More recently we have been experimenting with the levitation of rotors with coatings in the 5–20 μin range. With the use of careful procedures aimed at reducing particle contamination we have now been successful in obtaining reliable long term operation with the thin films. To achieve this it was also necessary to improve the deposition technique substantially, and we were assisted by J. Siebert of Ball Aerospace in this area.

4.   ROTOR SPIN UP AND HOUSING DESIGN

The gas jet system used to spin the rotor is shown schematically in figure 1. The rotor must be spun up while it is in the superconducting state in order to develop the London moment. The use of a cryogenic spin up system poses problems because of the very low viscosity of the helium gas, and its relatively short mean free path. These two factors dictate that the side walls of the spin up channel shown in the lower portion of the figure must approach to within 300 μin of the rotor. Without this narrow clearance it would be impossible to hold the pressure in the electrode area low enough to avoid both electrical breakdown and excessive drag while applying enough torque within the channel to reach the operating speed. On the other hand, the rotor-to-electrode gap must be large enough to reduce the torques on the gyro due to electrode imperfections to a tolerable level. With a 20 μin sphericity requirement on the electrodes, a gap of 1.5 mils to the electrodes is needed to meet the flight performance goal. Thus the raised ridges around a spin up channel are an essential feature of the design in figure 1. This detail of the design has a major influence on the techniques used to fabricate the housing.

By far the best method for making a spherical surface within a housing formed from two hemispherical shells is "tumble-lapping." In this procedure, first developed by Honeywell, two roughed out hemispheres are pinned together in their final configuration with a weighted lap and grinding compound in the cavity, and then shaken about two or more axes on a special vibration machine. Sphericities of 5 μin are achieved, allowing a 10–15 μin margin for the uniformity of the sputtered electrodes. Because tumble lapping is intrinsically a full sphere cutting procedure, it is necessary to form the raised ridges after the electrode surfaces are completed. Only two approaches are available for making the raised ridges after tumble lapping: deposition of metal or quartz on the ridges, or insertion of a

separate quartz piece containing the spin up channels. So far, the deposition technique has been used successfully only on housings made from
alumina ceramic. Severe difficulties were encountered when quartz housings were used, due primarily to the brittle nature of the material. For
laboratory testing a ceramic housing is acceptable, but questions exist concerning its shape at low temperatures. Thus, for the flight gyro, quartz
and the insert approach are used.

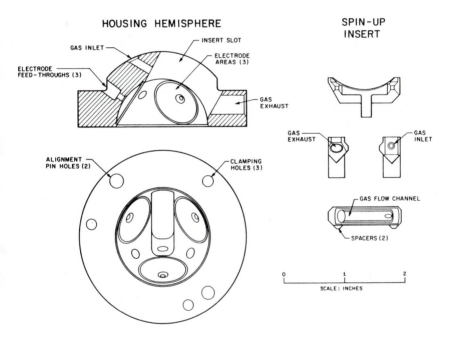

FIGURE 4. Design of quartz housing with inserts.

Figure 4 shows the design of a quartz housing with inserts. In this design
each spin up channel is fabricated as a single unit, independent of the gyro
housing, and glued laterally to the walls of a slot cut through the housing.
The most critical problem is cementing the pieces together in such a way
that they are aligned within a 50 μin tolerance and can withstand repeated
cycling to low temperatures. Quartz wedges (not shown) are used to hold
the insert away from the wall of the slot opposite the insert spacers, and
the correct elevation of the ridges above the electrode surface is established
by cementing the inserts into place with the housing temporarily assembled
and containing a wedged-in tooling ball of radius 300 μin greater than the
radius of the gyro rotor. This housing meets all the requirements for a
flight gyroscope.

During the course of gyro testing it proved important to have accurate centering adjustments on each axis of the electrical suspension system. The ability to move the ball precisely to any position in the housing suggests a variant method of spin up. Instead of passing gas through two balanced channels with raised lands, one makes use of a single channel without raised lands and moves the ball towards that channel, temporarily reducing the rotor-housing gap Δ to the value δ required for spin up. Although the use of only one channel halves the area of gas in contact with the ball, the ultimate spin speed can remain the same because the drag in the housing can be reduced by a similar factor. Single channel spin up allows a simpler housing design. Also, it allows one to maintain a larger rotor-electrode gap during normal gyro operation. Small raised pads in the housing prevent the ball from sitting directly on the electrodes.

Optimization of a single channel system follows similar lines to that of a dual channel system [10]. One difference is that there is an upper limit on the channel length. Since the rotor and housing are of different radii, the rotor-land gap for the translated rotor is not uniform. Simple geometric considerations show that if the half-angle subtended by the channel at the center of the rotor exceeds 30°, gas leakage at the ends of the channel will become excessive. A redesign for a significantly shorter channel could allow an increase of up to 20% in electrode area, which is good. With a shorter channel, the momentum transfer efficiency is reduced, so even though the gas pressure (and hence the mass throughput) is higher, the gyroscope takes longer to spin up. With a 5% efficiency, the mass of helium needed is about 250 g per gyro. Although the gas pressure is higher than with dual channel spin up, the reduction in channel area makes the lateral acceleration exerted by the gas on the ball acceptable, certainly less than 0.2 g. We have demonstrated single channel spin up in existing housings, and are designing a new housing with a configuration close to the optimum.

5. ROTOR SUSPENSION

The suspension system used in ground-based testing of the gyroscope was designed by the late J. R. Nikirk, following principles applied in the Honeywell gyroscope. Suspension is by 20 kHz voltages applied to the three pairs of electrodes, in the form of a constant or "preload" component and a variable or "control" component. The control voltage is added to one electrode and subtracted from the opposite electrode to generate a net force along the given axis. This approach linearizes the system; the force, even though it depends on the square of the voltage, is directly proportional to the control component. The rotor is maintained at zero voltage by making the system three-phase, each phase being associated with one of the three orthogonal support axes. The position of the rotor in the housing is measured by applying 1 MHz sensing signals to the same electrodes as are used for support. Opposed pairs of electrodes form two arms of a capacitance bridge; the unbalanced output from the bridge serves as the control signal for adjusting the 20 kHz support voltages. The control bandwidth is 4000 rad/sec. The voltage needed to support the ball on earth with a 1.5 mil rotor-electrode gap is about 1 kV rms. With a sensing voltage of 3 V, the centering accuracy is better than 1 μin.

Paper VI.3(E) by R. Van Patten describes the extension of this concept to a multilevel suspension system for use in the actual flight experiment.

6.   LABORATORY TESTING

The first cryogenic gyro test facility was completed in 1973 and used for general gyro testing, spin up demonstration, London moment observations and simple gyro dynamics studies. A schematic view of the apparatus is shown in figure 5. The dewar was equipped with two layers of conventional magnetic shielding to achieve a residual field of about 10^{-5} gauss after

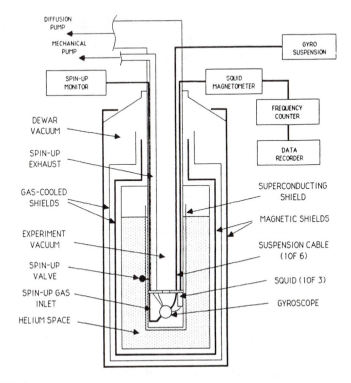

FIGURE 5. Schematic view of first gyro test facility.

careful adjustment of the remanent magnetization, and with a superconducting shield to stabilize the field after cool down. Spin up was performed with the dual channel system described above and readout with a three-axis SQUID magnetometer system fabricated by R. Clappier and J. Anderson. Because of the relatively high level of background field, it was difficult to use the London moment as a readout of the spin axis direction, and observations [11] were confined to measurements of its amplitude as a function of spin speed. On the other hand, the trapped flux from the background field could independently be used to observe not only the motion of the spin axis as a function of time but also the motion of the body axes relative to the spin axis, the polhoding motion.

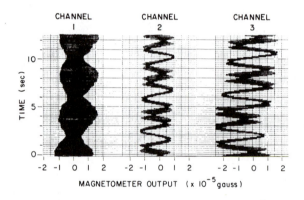

FIGURE 6. Trapped flux signals from a gyro rotor undergoing precession.

An example of the type of output obtained from the trapped flux signal is shown in figure 6. The left hand trace shows the output observed with a pickup loop in the horizontal plane, with time running upward, and the other two traces show signals from the two orthogonal loops with their planes vertical. The high frequency signal is due to the rotation of the trapped flux around the spin axis at the spin speed. In the center and R.H. traces, this signal is modulated by an intermediate frequency which is 90° phase shifted between the loops. This corresponds to a precession of the spin axis. Since this signal is not present in the L.H. trace, we can conclude that the motion is confined to precession about the vertical axis, similar to a simple top. The L.H. trace shows a strong modulation at an even lower frequency, and a similar effect can be seen on the other traces. This signal is in phase on all three channels and corresponds to the polhode motion of the spin axis relative to the rotor body axes. Thus one can see that the trapped flux signal is a simple and powerful tool for studying the dynamical behavior of the gyroscope. However it is not easy to turn this signal into a precision readout suitable for the relativity experiment. For this we need the full London moment readout.

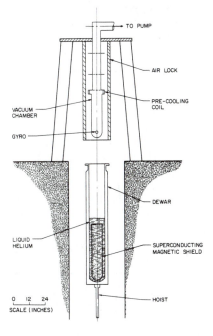

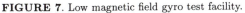

FIGURE 7. Low magnetic field gyro test facility.

In 1975 we decided to build a new apparatus in which the gyroscopes would be placed in an ultralow magnetic field shield made by the techniques described by B. Cabrera elsewhere in the volume. This apparatus [12] is shown in figure 7. The gyroscope is mounted on a support structure hanging inside an experiment chamber that can be exhausted to a pressure of a few times 10^{-7} torr. The top plate of the chamber is attached to a rigid frame which stands on a concrete isolation pad in order to provide a very stable mount for the experiment. A helium dewar, kept permanently cold, contains the ultralow magnetic field shield. The dewar can be raised and lowered on a hydraulic piston and used to cool the apparatus to cryogenic temperatures. An airlock prevents solid air from condensing into the dewar while it is being lifted into the operating position.

Two types of rotors have been studied in the low field test facility, a quartz rotor and a hollow beryllium rotor. This latter rotor is of the Honeywell type with a 10% difference in moments of inertia, but coated with

a superconducting film to allow readout. With the large moment of inertia ratio, the polhode frequency is about 10% of the spin speed, and thus
can easily be placed far above the precession frequency. Simple filtering
techniques can then be used to eliminate the ac trapped flux signal from
the magnetometer outputs, suppressing both the spin speed signal and the
polhode signal. In figure 8 we show the resultant signal, of which a few

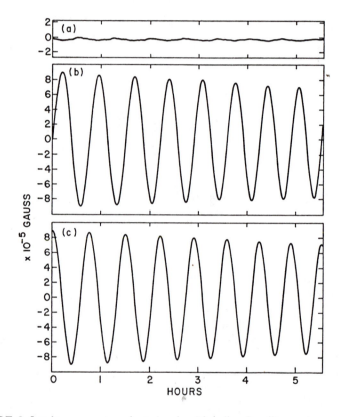

FIGURE 8. London moment readout signals with hollow beryllium rotor.

percent is due to dc trapped flux and the remainder due to the London moment. The upper trace is the signal from a loop with its plane horizontal
and the other two from crossed loops with their planes vertical. Again the
precession signal is primarily due to spin axis motion about the local vertical. With this system we have obtained an angular resolution equivalent
at a 200 Hz spin speed to about 1 arc-sec in a 10 sec integration time, in
agreement with that expected from magnetometer noise considerations. In

most of the tests the gyro rotor has had a large mass unbalance, principally because with the thick rotor coatings it is not easy to get a well-balanced ball, but also because it is convenient in developmental work to take advantage of the rapid precession of the gyroscope about the vertical axis caused by the mass-unbalance torque.

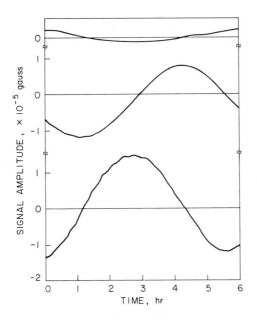

FIGURE 9. London moment readout signals with quartz rotor.

In figure 9 we show the readout signals obtained with a quartz rotor. Here the polhode frequency is of necessity very low and it cannot readily be filtered from the data. The single sine wave is due to precession and the "ripple" to polhoding. The London moment, again comprising the bulk of the signal, does not give rise to the polhoding signal; it is due to modulation of the residual trapped flux. This record shows significantly more noise than that in figure 8. This is due primarily to the use of a five times heavier rotor and the related increase in the suspension system output voltages. Extensive shielding of the magnetometer input circuit is necessary to reduce pick up from the suspension system to an acceptable level. Of course for the flight experiment this source of interference will be negligible, due to the low support forces applied in the close-to-free fall environment of the spacecraft. The upper trace contains a noticeable component of the basic precession sine wave. This is probably due to the effect of the earth's rotation on the precession.

7. TORQUE ANALYSIS

Gyro torques may be divided into two classes: (i) support-dependent, (ii) support-independent. The support-dependent torques include the mass-unbalance torque and the torques from the suspension system acting on the out-of-roundness of the gyro rotor. The support-independent torques include the various effects of gravity gradients, electric charge on the rotor, magnetic fields, residual gas, cosmic rays, photon bombardment, all of which are negligible in an earth-based gyroscope and therefore need careful study to make sure that they can be held below the 0.3 milliarc-sec/year design goal of the experiment.

Equation (1) in the previous paper shows that the mass-unbalance torque is proportional to the residual acceleration transverse to the gyroscope spin axis multiplied by $\delta r/r$, the ratio of the displacement between center of geometry and center of mass to the radius of the gyro rotor. Assuming a uniform density gradient from one pole of the spinning ball to the other, $\Delta r/r$ may be replaced by $3/8$ $(\delta\rho/\rho)$, where $\delta\rho$ is the difference between the densities at the two poles. If, to obtain a gyroscope with a drift rate below 0.3 milliarc-sec/year, we stipulate that no individual torque term shall contribute an error greater than 0.1 milliarc-sec/year, then we find that the requirement on density is $\delta\rho/\rho < 3 \times 10^{-7}$, as given in table 1.

Recently [13] analysis has greatly improved and simplified the evaluation of support-dependent torques by expanding the shape of the rotor in spherical harmonics, and calculating the torque from the differential with respect to angle of the energy stored in the support field. The calculation is effected with the aid of the D-matrices used in nuclear physics to rotate a set of spherical harmonics from one system of coordinates to another. The expression for the primary torques depending only on rotor shape is

$$\Gamma_n = -\frac{\partial}{\partial\eta} \sum_{\ell p} C'_{\ell p} B_{\ell o} Y_{\ell p}(\beta, \alpha) \quad , \tag{1}$$

where η is the rotation angle, the $Y_{\ell p}$ are spherical harmonics, the $B_{\ell o}$ are coefficients in the expansion of the rotor out-of-roundness in the rotor coordinate frame, and the $C'_{\ell p}$ are coefficients which express the demarcation of the electrode boundaries in terms of a spherical harmonic expansion in the housing coordinate frame, separately summed over the squares of the voltages applied to the six electrodes. The only integration needed to reduce equation (1) to numerical form is the one to find the $C'_{\ell p}$, and this integration may be done by hand from a table of Legendre polynomials.

When the results from the primary suspension torques are translated into expressions giving the drift rate Ω of the gyroscope in terms of applied acceleration, they reduce to two forms, one for odd and one for even harmonics of the rotor shape. If the suspension system is assumed to have a fixed preload and the gyro spin vector is assumed to point (as it will very nearly do in space) along a line in the y-z plane lying exactly midway between two electrodes whose axes are labeled y and z, the resulting expressions may be written

$$\Omega_p^{\text{odd}} = \frac{f}{v_s} \sum_{\ell \text{ odd}} B_{\ell o} A_\ell \tag{2}$$

$$\Omega_p^{\text{even}} = \left[\zeta h + \frac{f_z^2 - f_y^2}{h} \right] \sum_{\ell \text{ even}} B_{\ell o} A_\ell' . \tag{3}$$

In these equations v_s is the peripheral velocity of the rotor, A_ℓ, A_ℓ' are numerical coefficients, h is the "preload acceleration," *i.e.*, the acceleration that has to be applied parallel to an electrode axis to drive the voltage on the opposite electrode to zero, f_z and f_y are components of acceleration parallel to the z and y electrode axes, and $\zeta = (h_z - h_y)/h$ is the preload compensation factor, a numerical coefficient expressing the extent to which the preloads in different axes are matched. The equation for odd harmonics is identical in form to the equation for the mass-unbalance torque; the equation for even harmonics involves the preload acceleration also.

To reduce equation (2) to numerical predictions, one needs to know the structure of the acceleration $\mathbf{f}$ acting on the gyroscope, which is made up of four terms: $\mathbf{f} = \mathbf{f}_d + \mathbf{f}_c + \mathbf{f}_g + \mathbf{f}_\nu$, where $\mathbf{f}_d$ is the limit on the dragfree controller, $\mathbf{f}_c$ is the centrifugal acceleration due to spacecraft roll, $\mathbf{f}_g$ is the gravity gradient acceleration which arises from the gyroscope's not being at the center of mass of the spacecraft, and $\mathbf{f}_\nu$ is the acceleration produced by noise in the pointing servo. For odd harmonic torques and the mass-unbalance torque, one takes averages of the accelerations; for even hormonic torques, one has to take into account the possibility of rectification from the squared terms f_z^2 and f_y^2 in equation (3).

The centrifugal acceleration $\mathbf{f}_c$ is well averaged by roll in its effect on both odd and even harmonics; however, it has been shown that the slight misalignment of the gyro spin vector with the spacecraft roll axis gives rise to a small torque from the preload term h in equation (3), in consequence of which for a preload of 10^{-6} g the average position of the gyro spin vector over the year has to be held to within about 10 arc-sec of the roll axis. For a 10^{-7} g preload, the alignment can be correspondingly relaxed.

The noise acceleration $\mathbf{f}_\nu$ is surprisingly large, about 3×10^{-6} g at the 25 rad/sec bandwidth of the inner loop of the pointing servo. However, most of the noise comes in above the 4 rad/sec bandwidth of the suspension system and has no effect on the gyroscope. Also, the drift rate from the residual effect comes only through the misalignment of the gyro spin vector. The magnitude is at most 10^{-2} milliarc-sec/year.

For the gravity gradient term, we consider a gyroscope displaced from the center of mass of the spacecraft through a distance ℓ along the roll axis. In a satellite moving in a near-circular orbit, and having its roll axis misaligned with the orbiting plane by an angle α, the acceleration $\mathbf{f}_g$ is

$$\mathbf{f}_g = \frac{g'\ell}{2R} \left[(1 + 3 \cos 2\omega_o^t)\mathbf{n}_\ell + 3 \sin 2\omega_o^t \mathbf{n}_p + \alpha \mathbf{n}_o \right] , \tag{4}$$

where R is the radius of the orbit, g' is the earth's acceleration at orbit altitude, ω_o^t is the mean motion, and $\mathbf{n}_\ell$, $\mathbf{n}_o$ and $\mathbf{n}_p$ are unit vectors in directions respectively parallel to ℓ, along the orbit normal, and in the orbit plane perpendicular to both ℓ and the orbit normal. For an ℓ of 20 cm, the quantity $g'\ell/2R$ is about 1.5×10^{-8} g. Thus in the orbit plane there is a secular acceleration parallel to ℓ (and therefore nearly parallel to the gyro spin vector) of magnitude 1.5×10^{-8} g, and a cyclic term rotating at twice the orbital rate, of amplitude 4.5×10^{-8} g. Perpendicular to the orbit plane there is a secular term of magnitude 1.5×10^{-8} αg, which with an α of $2°$ would yield 10^{-9} g. The secular term is in the direction $\mathbf{n}_\ell$, and though large in comparison with the 10^{-10} g residual acceleration $\mathbf{f}_d$ from the drag free controller, has negligible effect on the gyroscope since it is nearly parallel to the gyro spin vector. The cyclic acceleration term averages over the orbit for odd harmonics in the shape of the rotor, and averages over the combination of roll and orbit for the even harmonics. Only the secular term perpendicular to the orbit plane is significant. It yields a torque proportional to α which causes a precession of the gyroscope in the orbit plane, $i.e.$, in the same plane as the geodetic precession. Since in a regressing orbit α changes linearly with time, the resultant drift rate will be quadratic in time, so that even if α is large enough for the term to introduce a significant error in the measurement of the geodetic precession, the error can be removed in data reduction.

Summing up the argument, we find that the only terms in $\mathbf{f}$ that are significant are the component of gravity gradient acceleration $(g'\ell\alpha/2R)\mathbf{n}_o$ along the orbit normal and the component of residual acceleration $(\mathbf{f}_d)_i$ from imperfect dragfree performance left in inertial space perpendicular to the roll axis after allowing for any averaging from the roll. The gradient term is known exactly once the orbit is known. For the residual drag we make the conservative assumption that $(\mathbf{f}_d)_i$ is 10^{-10} g, $i.e.$, that there is no significant benefit from roll. Once the accelerations are known, upper limits on the drift rates of the gyroscope due to mass-unbalance and primary suspension torques may be calculated from the known limits on homogeneity and out-of-roundness, together with the known preload and the average misalignment of the gyro spin vector. The conclusion of the analysis is that drifts of the gyroscope due to mass-unbalance and suspension torques can both separately be held below 0.1 milliarc-sec/year.

TABLE 2. SUMMARY OF GYROSCOPE DRIFT ERRORS.

Torque	Drift Rate (milliarc-sec/year)
Mass unbalance	<0.1
Suspension:	
Primary	<0.1
Secondary	<0.03
Gravity gradient	<0.05
Magnetic:	
Direct	$<10^{-5}$
Differential damping	$<10^{-8}$
Electric:	
Charge on ball	<0.001
Patch effect	<0.01
Gas:	
Differential damping	<0.1
Brownian motion	$<6 \times 10^{-5}$
Photon bombardment	$\sim 10^{-6}$
Cosmic rays:	
Primary (iron nuclei)	$<10^{-3}$
South Atlantic anomaly	$\sim 4 \times 10^{-4}$
Showers from sun	$\leq 4 \times 10^{-4}$
Worst-case-sum	<0.38
Root-mean-square sum	<0.18

The support-independent torques have been treated exhaustively elsewhere [14]. Table 2 gives upper limits on the principal drift terms for a gyroscope in a satellite moving in an orbit with coinclination 10 arc-min.

References

[1]   As quoted by H. W. Knoebel, *Control Engineering* **11**, 70 (1964). Earlier reports are not widely available.

[2]   G. J. Siddall, private communication.

[3]   M. Player and G. Dunbar, private communication.

[4]   L. V. de Sa, Thesis, University of Aberdeen (1981).

[5]   I. M. Siddiqui and R. W. Smith, *Optica Acta* **25**, 737 (1978).

[6]   W. Angele, *Prec. Engr.* **2**, 119 (1980).

[7]   J. A. Lipa and J. Bourg, *Prec. Engr.* **5**, 101 (1983).

[8]   J. A. Lipa and G. J. Siddall, *Prec. Engr.* **2**, 123 (1980).

[9]   P. Peters, private communication.

[10]   T. D. Bracken and C. W. F. Everitt, *Adv. Cry. Eng.*, **13**, 168 (1968), and G. R. Karr, J. B. Hendricks and J. A. Lipa, *Physica* **107B**, 21 (1981).

[11]   J. A. Lipa, J. R. Nikirk, J. T. Anderson and R. R. Clappier, *LT-14 Proceedings* (North-Holland, Amsterdam, 1975), Vol. 4, p. 250.

[12]   B. Cabrera and F. J. van Kann, *Acta Astronautica* **5**, 125 (1978).

[13]   G. M. Keiser, to be published.

[14]   "Report on a Program to Develop a Gyro Test of General Relativity in a Satellite and Associated Control Technology," edited by C. W. F. Everitt (HEPL, Stanford University, 1980), p. 560. Contributors: J. T. Anderson, J. V. Breakwell, B. Cabrera, R. R. Clappier, D. B. DeBra, P. Eby (NASA Marshall Center), J. J. Gilderoy, Jr. , G. M. Keiser, B. C. Leslie, J. A. Lipa, G. J. Siddall, F. J. van Kann, R. A. Van Patten and R. Vassar.

The Stanford Relativity Gyroscope Experiment (C): London Moment Readout of the Gyroscope

John T. Anderson

1. INTRODUCTION

It has always been an objective of the relativity gyroscope experiment that the gyroscope should be a near perfect uniform solid sphere. It should be featureless and balanced. Under these conditions, it appears that the number of methods to determine the direction of spin of the gyroscope is limited. The conventional approaches used for free precession gyroscopes employ either unequal moments of inertia in the gyroscope, patterning on the surface, or a flat area on a pole of the gyroscope that is tracked by optical means. Objections to these approaches are based on both unacceptable drift and inadequate sensitivity.

The relativistic precessions of the gyroscope are small, 6.6 and 0.042 seconds of arc respectively for the geodetic and motional precessions. To make a meaningful measurement of the motional effect requires that the drift and other errors in the gyroscope be much smaller. Gyroscope errors below 0.001 seconds of arc are reasonable both in terms of the scientific value of the measurement and from the practical point of what we believe can be achieved. This goal cannot be reached without an unrelenting search for, and understanding of the sources of, Newtonian torques on the gyroscope. It was in the course of calculating the torque from the magnetic moment of the superconducting niobium coating on the gyroscope that C. W. F. Everitt hit upon the idea of using the magnetic moment as the basis of the angular readout of the gyroscope. This magnetic moment, the London moment [1], appears along the axis of rotation in any spinning superconducting object. The corresponding magnetic field B_L inside the superconductor is given by

$$B_L = (2m/e)2\pi f_r \quad , \tag{1}$$

where m and e are the mass and charge respectively of the electron, and f_r the rotation rate of the gyroscope in hertz. Interactions of the London moment with magnetic fields or magnetic or diamagnetic materials can lead to torques on the gyroscope. With proper design of magnetic shielding around the gyroscope, such torques can be made insignificant while leaving the London moment as a measureable indicator of gyroscope angle.

In simplest concept, the London moment readout [2] consists of a magnetometer capable of determining the direction of the magnetic moment. In practice, conventional magnetometers have inadequate sensitivity to perform the function. However, a combination of the Josephson effect based SQUID magnetometer [3], superconducting signal pickup loops, and an experiment geometry that requires a relatively limited angular readout range provides a gyroscope readout system having the sensitivity to reach the desired experiment accuracy of 0.001 seconds of arc in its one-year lifetime.

2. THE LONDON MOMENT READOUT

The London moment readout is shown schematically in figure 1. The superconducting readout loop lies in a plane containing the roll axis of the satellite. The roll axis is chosen to point in the direction of the guide star Rigel, selected according to specific criteria, which serves as the inertial reference against which to measure the motion of the gyroscope. The apparent position of the guide star is dependent upon stellar aberration, relativistic bending of starlight near the sun, stellar parallax, and proper motion, but, with the exception of the last, these are either directly calculable or deducible from the data as desirable measurements. Proper motion must be obtained by independent means and at present amounts to an error of approximately 0.002 seconds of arc [4]. The gyroscope spin axis will initially be aligned to within 10 seconds of arc of the nominal line of sight to Rigel, and the maximum departure from the nominal position, primarily from stellar aberration, will be under 50 seconds of arc. Relativity data

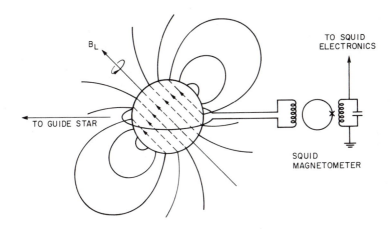

FIGURE 1. Basic configuration of the London moment readout.

are obtained from the difference between the line of sight to Rigel and the gyroscope spin axis as measured by the gyroscope readout system.

Conservation of magnetic flux in a closed superconducting path requires that the flux in the superconducting circuit consisting of the pickup loop and the SQUID magnetometer input coil remain a constant, independent of the amount of magnetic flux threading the pickup loop from the London moment of the gyroscope. Flux conservation is maintained in the presence of an applied field by a flux of the opposite sense generated by a shielding current in the readout circuit. If it is assumed that the gyroscope initially lies in the plane of the pickup loop and that there is no trapped flux in the readout circuit, then the current I_p that flows in the readout circuit owing to a deflection θ of the gyroscope spin axis is

$$I_p = \frac{\phi}{L_{tot}} \sin\theta \quad , \tag{2}$$

where

$$\phi = \pi r_0^2 B_L \sum_i^N (r_0/r_{li}) \quad , \tag{3}$$

and where r_0 is the radius of the gyroscope, r_{li} are the radii of each of the pickup loop turns, L_{tot} is the total inductance of the readout circuit, and N is the number of turns on the pickup loop. The ratio of radii is a geometric factor that comes about because the pickup loop is necessarily larger than the rotor radius and does not intercept all of the flux originating in the rotor. For the small angles under consideration, it is sufficient to replace $\sin\theta$ by θ. Although equation (2) contains the scaling of I to θ, in practice a direct calibration of the SQUID output to θ can be more accurately obtained from both daily and annual stellar aberrations of 5 to 20 seconds of arc respectively. The presumed constancy of the magnetic signal from an almost inevitable small amount of trapped flux spinning with the gyroscope can be used to provide short-term gain calibrations [5].

In its earliest concept, the London moment readout was to have a single-turn pickup loop. It was held that while the flux increased as N and the inductance as N^2, the coil energy, which is proportional to ϕ^2/L, did not depend upon N. In practical terms, however, the increase in inductance is useful. A single-turn pickup loop has ~ 0.2 μH inductance, while typical SQUID's under consideration have ~ 2 μH. The multi-turn pickup loop provides efficient energy transfer to the SQUID without the signal losses associated with a matching transformer. Further, by the proper spacing of turns, one can actually achieve a sensitivity increase. Spacing the turns a modest 0.02 r_0 significantly decreases the mutual inductance between turns so that for a small number of turns the inductance increases more slowly than N^2 without any significant loss of intercepted flux. Less of the field energy in the effective volume enclosed by the pickup loop is lost to energy stored in the mutual inductance between turns.

It is useful to have a perspective on the relevant numbers. The ratio of the maximum angle to be sensed to the system resolution is

$$\text{dynamic range} \quad = 100''/0.001''$$

$$= 10^5 \quad .$$

This is not an extraordinarily high level of precision although it does require considerable care. Had the experiment geometry not been arranged to give readout angles close to null, the dynamic range would have been $\sim 2 \times 10^8$. However, in terms of sensitivity, the actual magnetic field corresponding to 0.001 seconds of arc is extremely small. The full London moment at a gyroscope rotation rate of 170 Hz is 1.2×10^{-4} gauss. To sense the field change for a gyroscope precession of 0.001 seconds of arc requires a field sensitivity of 7×10^{-13} gauss!

If a wide dynamic range were needed, then the quantization of magnetic flux could be applied to the readout to extend its range of precision [6]. Flux quantization is central to the operation of the SQUID sensor. In the absence of linearizing feedback from the SQUID electronics, the SQUID output would be periodic with each quantum of flux generated within the SQUID by the input current. By operating the SQUID and electronics in a linear mode but resetting the electronics to zero each time the SQUID contains one full flux quantum, the readout current can be measured in precise, countable units and fractions thereof over an extended range. The stability is determined by the mechanical and electrical stability of the SQUID at liquid helium temperatures, which are excellent. The scaling of the readout for the gyroscope experiment would be $\sim$ 1300 seconds of arc per flux quantum, or $\sim$ 240 flux quanta for one quadrant.

Clearly the most significant issue to secure the success of the London moment readout is obtaining adequate magnetometer sensitivity. Additionally, one must be concerned with magnetic torques generated by the readout. As the environment of the gyroscope is one filled with electromagnetic noise from the electrostatic suspension system, the magnetometer must in some way be protected from this source of interference. We shall see that these issues are interrelated.

3. SENSITIVITY

To do a meaningful measurement of a quantity implies the ability to obtain a statistically valid number on a time scale no longer than the allowable duration of the measurement based upon other factors. These factors may be either technical or practical, such as the lifetime of a transient event like a supernova or the lifetime of the experimenter. In the case of the gyroscope experiment, the lifetime is set by the holdtime for the liquid helium in the dewar, which is a minimum of one year. It is a design objective that data and noise accumulated from one year of operation give a sensitivity of 0.001 seconds of arc. The dominant source of noise is that from the gyroscope readout system, but, as will be shown below, it is sufficiently low to reach the experiment objectives.

If the satellite is in a nonregressing orbit, the relativistic gyroscope precession signal is a slowly building ramp. Spacecraft roll was initially introduced to reduce torques on the gyroscope, but serves also to sinusoidally modulate the relativity signal. This shifts the spectral content of the relativity signal from frequencies on the order of once per year to once per satellite revolution, which currently is 10 minutes. The importance of this can be seen from the SQUID noise curves shown in figure 2 [7,8,9,10]. SQUID's, like all electronic measuring instruments, have what is known as 1/f noise. The noise power rises inversely with frequency below some corner frequency. The benefit of shifting the readout signal frequency spectrum to higher frequencies is that the SQUID noise is lower. The corner frequency in many SQUID systems is on the order of 1 hertz. It would be desirable to modulate the relativity signal at greater than 1 hertz, but practical satellite balance and control authority limitations preclude roll frequencies greater than about 0.017 Hz (10 minute roll period).

On any signal measured, the amount of noise associated with the measurement increases with the bandwidth of the measuring instrument. For "white noise," in which the noise power per unit bandwidth is constant, the amplitude of the noise increases as the square root of the bandwidth of the measurement. For other noise distributions, one must integrate the noise

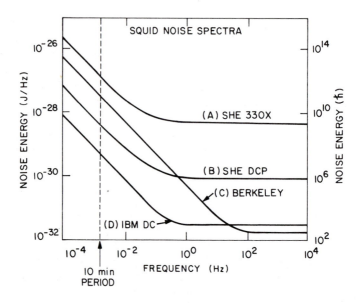

FIGURE 2. Simplified power spectra for several SQUID's. With the exception of A, these are dc SQUID's. The sources of the data for curves A through D are, respectively, references [7] through [10]. The simplification that has been used is the elimination of the statistical roughness of the data by the approximate fitting of a model consisting of a 1/f section and a white noise section.

density over the measurement bandwidth. The usual strategy for making a measurement minimally corrupted by noise is to reduce the measurement bandwidth. For dc measurements, this is usually accomplished by putting the signal through a low-pass filter. In a simple measurement such as checking the voltage of a battery, there is an implied low frequency limit to the measurement which is on the order of the reciprocal of the time duration of the measurement. If one measures the battery voltage for 1 second, the data contain little information on how the battery voltage varies over a 1000 second period. It makes little sense to filter a 1 second measurement through a 0.001 Hz filter; the measurement itself becomes just high frequency noise that is removed by the filter. Thus all measurements are in fact made through a bandpass filter. The least noise corruption of the measurement comes from the minimum bandpass that can be used consistent with the frequency content of the signal.

Direct current measurements and some very low frequency measurements are affairs of the mind. Because of the very long time scales involved, one is often inclined to think of them differently from events occurring at frequencies of approximately 1 Hz or above. They are not different. In the case of the gyroscope experiment, the roll frequency of the satellite, and hence the primary frequency contained in the readout signal, is 0.0017 Hz, but the same signal-processing rules and techniques applicable at 1 kHz or 1 MHz are applicable there as well. Therefore the noise on the readout signal can be reduced by passing it through a bandpass filter centered on the satellite roll frequency with as narrow a passband as possible. The minimum bandwidth is the reciprocal of the observation time, which in this case is the one year hold-time of the dewar. Conceptually, it is meaningful to calculate the frequency or observation time to reach a given resolution, 0.001 seconds of arc being the obvious choice here.

In most SQUID applications, the SQUID input coil can be treated as a lossless inductance. As such, it draws no signal power but it does store signal energy. The input of a dissipative device draws signal power, and the noise of such a device is referred to the input in units of power/unit bandwidth. Correspondingly, when the input is an energy, the noise is in units of energy/unit bandwidth, hence the units in figure 2. The energy in the SQUID input coil is

$$ E = \frac{1}{2} \left(\frac{L_i I_p^2}{2} \right) \quad , \tag{4} $$

where L_i is the SQUID input inductance. The additional factor of one-half comes from the fact that the rolling of the satellite produces a sine wave from a dc signal. The bandwidth B to detect this signal is found simply from

$$ B = \frac{E}{e_n} \quad , \tag{5} $$

where e_n is the energy noise density at the frequency of interest.

In figure 3 is shown a realization of the readout circuit with typical circuit values. The addition of the filter resistor is of practical rather than fundamental importance. It is needed to remove high frequency magnetic noise which otherwise would impair the operation of the SQUID system. Signals for which the inductive reactance of the readout circuit is greater than the resistance of the filter resistor are shunted by the resistor. The source of the noise is suspension system currents at 20 kHz, 1 MHz, and their harmonics in and around the gyroscope. Laboratory versions of the readout have used a transformer and "damping cylinder" to inductively couple the pickup loop to the SQUID. The damping cylinder is a resistive cylinder interposed between the two windings of the transformer. The induced currents flowing in the cylinder at high frequencies shield the transformer winding attached to the SQUID, but at low frequencies they are damped by the resistance of the cylinder, which allows the dc and low frequency signals to pass. Improvements to the suspension electronics will eliminate the need for this relatively inefficient filtering method.

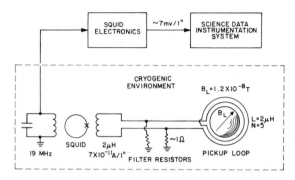

FIGURE 3. Typical circuit values for a practical implementation of the London moment readout.

Using the values given in figure 3 and equations (1) through (4), the energy in the SQUID input coil for a gyroscope angle of 0.001 seconds of arc is 2.5×10^{-33} J. From curve A of figure 2, the SQUID energy noise density at 0.0017 Hz is 1.6×10^{-27} J/Hz; since the bandwidth will be small, this can be treated as a constant over the bandwidth of interest. The required bandwidth is 1.6×10^{-6} Hz. The reciprocal of B gives the observation time. In fact, since we presume to know the phase of the relativity signal (the satellite roll phase), the noise in the out-of-phase component can be rejected so that the observation time is reduced by one-half. The approximate observation time is thus 89 hours.

The choice of curve A of figure 2 for the noise data as a baseline is guided by the considerable operational experience that has accumulated for that SQUID. One prefers to choose mature components for a space mission where possible. However, the quality of the experiment, system performance, and perhaps experiment content are enhanced by having a SQUID sensor of considerably lower noise. As shown in figure 2, lower noise SQUID's are available. With the best of these in the $1/f$ noise region, the IBM SQUID (curve D of figure 2), the observation time to reach 0.001 seconds of arc is 12 seconds! At present, most of the advanced SQUID's are either more complex to operate, suitable only for specialized applications, or not generally available, but we anticipate the eventual incorporation of advanced SQUID technology into the gyroscope readout.

That one can actually measure such a small signal is shown in figure 4 [11,12]. The prominent peak is the simulation of a gyroscope signal obtained by producing in a SQUID the same magnetic flux as would be produced by the current in a single-turn readout loop for a 0.020 seconds of arc gyroscope angle and a satellite roll rate of 0.017 Hz (1 minute period). The amplitude of the noise at frequencies away from the roll frequency is ~ 0.0007 seconds of arc rms. Data were taken for 140 hours and analyzed

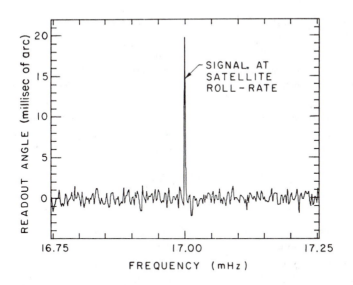

FIGURE 4. Noise spectrum of the SQUID output for a signal input corresponding to that from the readout loop for a 0.020 seconds of arc simulated gyroscope signal. The amplitude of the noise, 0.0007 seconds of arc rms, at frequencies away from that of the signal, is a measure of the sensitivity at the signal frequency.

to recover the signal. The noise level and observation time are in agreement with equation (5) and the observed noise spectrum of the SQUID used in this test.

Other methods have been suggested to improve the sensitivity of a magnetic gyroscope readout. Everitt has examined the use of a spinning ferromagnetic material [13], which produces a magnetic moment known as the Barnett moment. Although the field could be relatively large, the presence of the ferromagnetic material in the pickup loop would act to attract stray magnetic fields into the loop, thereby degrading the signal-to-noise ratio. The London moment, while small, is generated in a diamagnetic material which provides shielding of the pickup loop. Hendricks has considered a low-inductance pickup loop consisting of a relatively wide band close to the gyro rotor that includes the SQUID [14]. There are technical problems in integrating the SQUID with the gyro, and possibly in null stability. One approach that appears promising is the use of trapped flux in the gyro rotor. In such a case, the flux is now locked to the rotor surface rather than to the spin axis, so the operation is somewhat different from that of the London moment readout. The primary advantage is that by choosing to observe the rotating component of the trapped field, the readout can take advantage of the much lower SQUID noise at the gyroscope rotation speed of 170 Hz instead of the SQUID noise at the satellite rotation speed of 0.0017 Hz. Perhaps a larger field could be used than that provided by the London moment, but other considerations may limit its size. Assuming the use of one of the advanced SQUID's, observation times of minutes instead of tens of hours would be needed to reach 0.001 seconds of arc. At the present level of understanding, the ac trapped flux readout requires complex signal processing. However, the possibility of an exceptional improvement in sensitivity demands that it be studied further. Some operational experience will be gained through its use in the gimbaled gyroscope test apparatus now being built.

4. TORQUES INDUCED ON THE GYROSCOPE BY THE READOUT

A measurement, whether quantum limited or macroscopic, alters the state of the system being measured. In the former the Heisenberg uncertainty principle offers a guide to the ultimate accuracy of the measurement, while in the latter such pedestrian effects as the current drawn by a voltmeter will influence the measurement. In the case of the gyroscope, the perturbation of interest is the torque the readout causes. The source of the torque is the interaction between the London moment of the gyroscope and the magnetic moment of the readout loop from any current flowing in it. Given a maximum drift rate of 0.001 seconds of arc/year, the maximum torque from all sources must be less than $\Gamma_{max} = 1.5 \times 10^{-18}$ N m. For purposes of illustration, we will allocate all of this to the readout. The torque Γ_l on the gyroscope from a current I_l in the readout loop is given by

$$\Gamma_l = \pi r_0^2 I_l B_L \sum_i^N \frac{r_0}{r_{li}} \cos \theta \quad . \tag{6}$$

If I_l is an ac current, proper attention must be paid to the phase. Since the satellite is rolling, a trapped current in the readout loop produces an alternating torque on the gyroscope, and no net drift. The readout current itself does, however, produce a drift. While the current through the SQUID alternates polarity as the satellite rolls, the magnetic field generated by the current is always in opposition to that of the gyroscope. For gyroscope angles of less than 1600 seconds of arc, the torque produced by the readout current will be less than Γ_{max}, a factor of 32 greater than the angular range required by the experiment. As shown in figure 3, the readout will use current feedback to the readout input circuit, reducing the readout circuit current and thereby reducing the torque on the gyroscope from this source by the loop gain of the feedback system. It can be shown that the readout reaction torque is the angular derivative of the energy in the readout circuit. Therefore any method of increasing readout sensitivity that relies upon increasing the signal in the readout loop also increases the readout reaction torque. However, the use of current feedback will, for most situations, reduce the torque to acceptable or negligible values. Those methods that rely upon improving the sensor noise do not generally influence the torque.

Trapped flux in the gyroscope rotor provides another source of torque. The flux rotating with the rotor within the pickup loop acts an an ac generator. Whereas the dissipation of the readout circuit at dc is zero, it is not generally so above dc. The filter resistor shown in figure 3 provides the largest source of dissipation, but most SQUID's have resistive shunts built into them either for L/R filtering or to prevent Josephson junction hysteresis. Torques from this source fall into a broad category known as differential damping torques, which act selectively on different components of the gyroscope spin. In this case, there is no torque generated by the trapped flux rotating around the component of spin lying perpendicular to the plane of the pickup loop.

For small angles, the drift from ac trapped flux is given by:

$$\Omega = \frac{15}{64} \frac{\left(B_{ptf} \sum_i^N r_0/r_{li}\right)^2 F(f_r, f_0)}{\rho r_0^5 L_{tot}} \theta \quad , \tag{7}$$

where B_{ptf} is the average perpendicular component of the trapped flux, and ρ is the rotor density. $F(f_r, f_0)$ contains the frequency dependence, and is given by

$$F(f_r, f_0) = \frac{1}{f_r} \frac{(f_r/f_0)}{1 + (f_r/f_0)^2} \quad . \tag{8}$$

If the trapped field in the rotor differs significantly from that of a dipole, the additional harmonic content of the trapped flux signal increases the drift by an amount that is found by summing the contributions from each of the harmonics. For readout loop configurations under consideration, trapped fields ~ 50 times the London moment field would cause unacceptable drift, but this is many orders of magnitude larger than any that could be tolerated on other grounds. Current feedback reduces the drift by the square of the feedback gain, so that this source of torque is also negligible.

In practice, none of the sources of torque from the readout appear to present any difficulty.

Acknowledgments

The author thanks his colleagues Blas Cabrera, Robert Clappier, James Lockhart (who now carries the responsibility for the readout system) and Richard Van Patten at Stanford, and Palmer Peters at the NASA Marshall Space Flight Center, for their contributions to this work.

References

[1]   F. London, *Superfluids* (Dover, New York, 1964), p. 83.

[2]   J. T. Anderson and R. R. Clappier, in *Future Trends in Superconductive Electronics*, edited by B. S. Deaver, Jr., C. M. Falco, J. H. Harris and S. A. Wolf (American Institute of Physics, New York, 1978), p. 155.

[3]   R. P. Giffard, R. A. Webb and J. C. Wheatley, *J. Low Temp. Phys.* **6**, 533 (1971).

[4]   J. T. Anderson and C. W. F. Everitt, "Limits on the Measurement of Proper Motion and the Implications for the Relativity Gyroscope Experiment" (unpublished report, W. W. Hansen Laboratories of Physics, Stanford University, 1979).

[5]   T. G. Duhamel, Stanford University Department of Aeronautics and Astronautics Report 540 (Stanford University, 1984), p. 77.

[6]   J. T. Anderson and C. W. F. Everitt, *IEEE Trans. Magn.* **MAG-13**, 377 (1977).

[7]   SHE Corporation, data sheet for the model 330X rf SQUID system (San Diego).

[8]   SHE Corporation, data sheet for the model DCP dc SQUID system (San Diego).

[9]   J. Clarke, private communication.

[10]   C. D. Tesche *et al.*, in *Proceedings of the 17th International Conference on Low Temperature Physics*, edited by U. Eckern, A. Schmid, W. Weber and H. Wühl (North-Holland, Amsterdam, 1984), p. 263.

[11]   B. Cabrera and J. T. Anderson, in *Future Trends in Superconductive Electronics*, edited by B. S. Deaver, Jr., C. M. Falco, J. H. Harris and S. A. Wolf (American Institute of Physics, New York, 1978), p. 161.

[12]   J. T. Anderson and B. Cabrera, *J. Physique* **C-6**, 1210 (1978). Figure 4 is the amplitude of the component of the noise spectrum of the data in phase with the simulated readout signal, whereas the corresponding figure in this reference, figure 2, is the logarithm of the power spectrum.

[13]   "Final Report on NASA Grant 05-020-019 to Perform a Gyro Test of General Relativity in a Satellite and Develop Associated Control Technology," edited by C. W. F. Everitt (HEPL, Stanford University, 1977), p. 70.

[14]   J. B. Hendricks, *IEEE Trans. Magn.* **MAG-11**, 712 (1975).

The Stanford
Relativity Gyroscope Experiment
(D): Ultrahigh Vacuum Techniques
for the Experiment

J. P. Turneaure, E. A. Cornell,
P. D. Levine and J. A. Lipa

1. INTRODUCTION

The gyro relativity experiment described in the preceding papers of this chapter makes use of a rotor spinning in a vacuum at a pressure of 10^{-10} torr or less to measure the very small precessions predicted by Einstein's general theory of relativity. The method to spin up the rotor, which uses a helium gas jet [1,2], leads to adsorption of helium on the gyroscope rotor and housing surfaces. This adsorbed helium yields a pressure above the required 10^{-10} torr during part of the gyro relativity flight experiment. This excessive helium gas pressure may lead to spurious gyro torques [3]. To keep the resulting spurious precession due to helium pressure below an acceptable level (about 0.3 m arc-sec) for most of the flight experiment duration, it is important that the time required for the helium pressure to reach 10^{-10} torr be short compared to the flight duration. It is possible to calculate this time if one has a knowledge of the adsorption properties of helium films for materials internal to the vacuum system in which the gyroscopes are placed, along with knowledge of the temperature of these materials and the pumping speed for helium.

In this article, the description of an apparatus for measuring the adsorption properties of helium films on a substrate and the initial experimental results for a copper substrate are presented. Also, a low temperature bakeout procedure, which has been developed to quickly achieve the required pressure in the gyroscope housing while still maintaining the other required conditions of the experiment, is described; and the results of a bakeout cycle which approximately simulates this procedure are presented. The apparatus, which has been developed, can be extended to investigate other materials, such as fused quartz, which are to be used in the construction of the gyro relativity experiment.

2. EXPERIMENTAL METHODS

The principal purpose of this work is to obtain helium pressure data for substrates of interest as a function of temperature and amount of adsorbed helium. For this reason the apparatus was designed to observe the properties of adsorbed helium films by measurement of the helium pressure in the volume rather than by measurement of the adsorption film specific heat. Although the direct measurements are of the pressure in the volume adjacent to the film, data from measurements can be used to infer some film properties since the chemical potential of the gas and the film are equal under the conditions of this experiment.

A schematic of the apparatus used in this work is shown in figure 1. The principal parts of the apparatus are the copper substrate; a helium supply coupled to the vacuum system by a variable leak; pressure measuring equipment including an ion gauge and a quadrupole mass analyzer; components for measurement and control of the substrate temperature; ion and turbo pumps; various vacuum lines and valves to connect the copper substrate to the pressure measuring equipment, helium supply and pumps; and a helium dewar.

The vacuum system is constructed using ultrahigh vacuum (UHV) techniques. The turbo pump is used for roughing the vacuum system and for pumping when the vacuum system is at higher pressures or at moderate partial pressures of helium. The ion pump, which is of the triode type and is capable of pumping helium, is used when the vacuum system is at lower pressures. Prior to an experimental run, the room temperature portion of the vacuum system is baked out at about $200°$ C and the low temperature portion at about $100°$ C. After bakeout the total pressure measured by the ion gauge is about 10^{-9} torr; however, after the low temperature portion is cooled down to 4.2 K, the total pressure drops to about 10^{-10} torr.

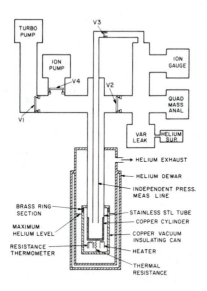

FIGURE 1. Schematic of apparatus to measure adsorption properties of helium.

The copper substrate is largely made of OFHC copper which, after machining and cleaning, was fired in dry hydrogen at about 1000° C to braze it to the stainless steel pumping line. The total surface area of the copper substrate is 210 cm². Nine percent of the copper substrate area is chromium-copper alloy (1% Cr).

Metered quantities of helium are inserted into the vacuum system by utilizing a constant volume of 45.9 cm³, a capacitance pressure gauge to measure the helium pressure in the constant volume, and a variable leak to allow a portion of the helium to be transferred to the vacuum system. The constant volume and variable leak are baked out to remove contamination. Pure helium gas is then inserted into the constant volume after it has passed through a liquid helium cryotrap to remove impurities. The pressure in the constant volume is typically brought to 1 torr. The error in the amount of helium inserted into the vacuum system is about 1% plus 2×10^{15} atoms.

The helium pressure in the vacuum system is measured by two means: an ion gauge and a quadrupole mass analyzer. The ion gauge is calibrated by inserting a measured amount of helium into the vacuum system and noting the ion gauge reading. The helium pressure for this reading is determined from the ratio of the vacuum system volume to the constant volume and from the pressure change in the constant volume. This calibration is accurate to a few percent, and it has proved to be stable. At lower pressures where the total pressure is not dominated by helium, a quadrupole mass analyzer is used to measure the partial pressure of helium. The quadrupole mass analyzer is calibrated at intermediate pressures by transferring the ion gauge calibration to the quadrupole mass analyzer. This pressure measuring instrumentation allows the pressure to be observed in the range of 10^{-12} to 10^{-4} torr. The pressures, which are measured at room temperature usually through the 12 mm diameter line with valves V_2 closed and V_3 open, are corrected for the thermomolecular effect.

As shown in figure 1, the copper substrate is located in a vacuum-insulating can so the substrate can be maintained at a temperature different from that of the liquid helium. The temperature of the substrate is measured with a calibrated germanium resistance thermometer which has an error of a few millikelvin for temperatures in the range of 4 to 10 K. The temperature, which is determined by the liquid helium temperature, the thermal resistance and the power applied to the heater, is regulated by a servo loop using the resistance thermometer as a sensing element and the heater as a control element.

The copper substrate forms a volume which is connected to room temperature by a 41 mm diameter stainless steel line for pumping, and by a smaller 12 mm diameter stainless steel line for pressure measurement. Helium adsorption on the surface of these stainless steel lines can produce a large uncertainty in the helium surface coverage for the copper substrate. To reduce this uncertainty, the two stainless steel lines are kept at a higher temperature than the copper substrate. The larger stainless steel line is kept at a higher temperature with a combination of the brass ring section (see figure 1), heaters and insulation; its temperature is monitored with resistance thermometers. For this system to function correctly, it is necessarry that the maximum helium level be kept below the brass ring section. The smaller stainless steel tube is warmer since it is located in the vacuum system and is connected only at the room temperature end.

3. ADSORPTION DATA FOR COPPER

The measurements shown in figure 2 were made as follows. Both valves V_1 and V_4 were closed, which left the vacuum system without pumping. First, the amount of helium corresponding to curve A in figure 2 was inserted into the probe vacuum system (0.327 atoms/nm^2). The copper substrate was then raised in temperature to the highest temperature shown for the curve, 9.5 K for curve A. At this temperature the pressure and related adsorption coverage came to equilibrium in a few seconds. Measurements at lower temperature were then made by stepping the temperature downward through the rest of the points on the curve. The next curve to the right was then made by inserting an additional known amount of helium and then repeating the above measurement process.

The curves shown in figure 2 have a range of coverage from about 0.03 to 1.5 of a monolayer. Some care needs to be exercised in using the pressure range shown in the figure. At the highest pressures, a large fraction of the helium gas is found in the volume of the vacuum system. For each curve, the pressure at which 10% of the helium is found in the volume has been estimated. These estimates are indicated by the dashed curve in the figure. Above and to the left of this dashed line more than 10% of the helium is in the vacuum system volume. At the lowest pressures shown in the figure, there is an experimental difficulty. The pressure is beginning to approach a residual value, as seen by the curvature, rather than continuing an exponential decrease with $1/T$. This is thought, but not proved, to be due to the way that gas is inserted into the vacuum system. As the apparatus is currently used, the helium gas is inserted down the independent pressure measuring line. Although the helium gas is rather pure, there may be contamination which condenses on the lower and cold end of this line. This contamination may provide sites on which the helium can be adsorbed. At higher pressures helium is adsorbed at these sites, and at lower pressures the desorption of helium looks like a virtual leak in the pressure measuring line. Evidence indicates that this explanation may be correct since the residual pressure does slowly decrease with time.

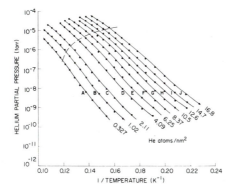

FIGURE 2. Helium pressure as a function of inverse temperature and surface coverage for a copper substrate.

The data shown in figure 2 are used to determine the parameters of an appropriate expression which characterizes the adsorption properties. This expression can be used together with known pumping and temperature conditions to calculate the pressure in a copper structure as a function of time. The following equation, which assumes the ideal condition that the binding energy for helium, ϵ_0, is uniform over the substrate, is the starting point for such an expression:

$$P = (2\pi m/h^2)^{3/2}(kT)^{5/2}[x/(1-x)] \exp(-\epsilon_0/kT) \quad , \tag{1}$$

where the factor in brackets is the Langmuir isotherm factor which is dependent on x, the fraction of monolayer coverage [4]. Since the binding
energy is not expected to be uniform over the copper substrate and the
coverage exceeds one monolayer, equation (1) is modified to the following equation by allowing the Langmuir isotherm factor and ϵ_0 to become
arbitrary functions of x:

$$P = (2\pi m/h^2)^{3/2}(kT)^{5/2}a(x) \exp(-\epsilon(x)/kT) \quad . \tag{2}$$

Equation (2) is fit to the set of data below the dashed curve in figure 2.
Before fitting, the residual pressures are removed from the data. Figure 3
is a plot of the resulting $a(x)$ and $\epsilon(x)/k$ as a function of surface coverage.

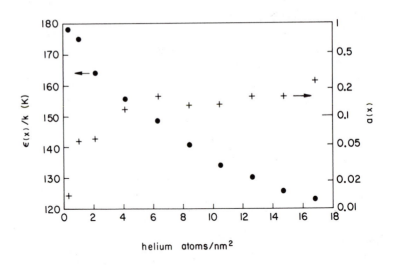

helium atoms/nm²

FIGURE 3. $\epsilon(x)/k$ (represented by •) and $a(x)$ (represented by +) as a function of
surface coverage for a copper substrate.

As shown in the figure, $\epsilon(x)/k$ varies from 178 K at low coverage to 123 K
at high coverage (x about 1.5). This dependence on coverage is usual for
inhomogeneous substrates since the binding energy associated with defects
on the substrate is typically larger than that for the rest of the substrate
[5]. Further evidences of inhomogeneity of the substrate are the absence
of any marked behavior in the data at the expected monolayer completion
(about 11 atoms/nm²) and the large departure of $a(x)$ from the Langmuir
isotherm factor.

constituent: the "inos." The introduction of these particles is important since they determine the sizes, masses and times of formation of galaxies and clusters of galaxies. In this new scenario [50], a first generation of stars and globular clusters is formed at $z \sim 10^3$, just after the decoupling between matter and radiation, but galaxies are formed much later at $z \sim 10$ by the subsequent formation of "ino" halos. If this scenario is correct it should naturally lead to the formation and morphology of galaxies, to the understanding, from few basic principles, of quasars, and to the identification of their formation era.

In all of the above considerations, little is said about the initial perturbation spectrum governing the formation of galaxies. Yakov Borisovich Zel'dovich [51] has postulated a given perturbation spectrum to be inserted in the Lifshitz formalism. My feeling now is that this approach also should be reexamined, and that a physical explanation of the process of fragmentation should give, from first principles, the spectral distribution of the perturbations. This should be done even at the price of abandoning the analyticity of the functions describing Lifshitz perturbations in the fragmentation era [52].

Time will tell if one of the many new techniques introduced into physics by Bill Fairbank will make possible the direct observation of these "inos" with earthbound experiments.

Before ending, I would like to recall on a more personal note the warm and intense friendship Jane and Bill Fairbank have and had with my mother, my sister and myself. These precious contacts through the years have enriched our lives.

References

[1] R. Ruffini and S. Bonazzola, *Phys. Rev.* **187**, 767 (1969).

[2] Returning to Princeton, I worked on this subject with J. A. Wheeler. At the ESRO colloquium in Interlaken (Switzerland) in September 1969, we presented considerations on the cross section of gravitational wave antennas, on the intensity and distribution of gravitational wave sources, and on neutron stars. These results, further expanded, appear in [5]. See also R. Ruffini and J. A. Wheeler, *Acc. Naz. Quaderno* **157** (1971).

[3] See R. Penrose, *Rivista Nuovo Cim.* **1**, 272 (1969).

[4] These results were finally published in [5]. Mention of them was made in D. Christodoulou, *Phys. Rev. Lett.* **22**, 1596 (1970); see, as well, R. Floyd and R. Penrose, *Nature* **229**, 177 (1971).

[5] R. Ruffini and J. A. Wheeler, in ESRO Report SP52 (Paris, 1971). Reproduced in M. Rees, R. Ruffini and J. A. Wheeler, *Black Holes,*

Gravitational Waves and Cosmology (Gordon and Breach, New York, London, 1974).

[6]   D. Christodoulou and R. Ruffini, *Phys. Rev. D* **4**, 3552 (1971).

[7]   D. Wilkins, *Ann. Phys.* **61**, 277 (1970).

[8]   D. Wilkins, *Phys. Rev. D* **5**, 814 (1972).

[9]   M. Johnston and R. Ruffini, *Phys. Rev. D* **10**, 2324 (1974).

[10]   R. Hanni and R. Ruffini, *Phys. Rev. D* **8**, 3259 (1973), and *Nuovo Cim. Lett.* **15**, 189 (1976).

[11]   R. Ruffini, in *Black Holes*, edited by B. and C. DeWitt (Gordon and Breach, New York, London, Paris, 1973), p. 499.

[12]   See, *e.g.*, E. Amaldi, P. Bonifazi, P. Bordoni, G. Castellano, C. Cosmelli, V. Ferrari, S. Frasca, M.-K. Fujimoto, F. Fuligni, U. Giovanardi, V. Iafolla, I. Modena, G. V. Pallottino, B. Pavan, G. Pizzella, R. Rapagnani, F. Ricci, S. Ugazio and G. Vannaroni, in *Proceedings of the Second Marcel Grossmann Meeting on General Relativity, Part B*, edited by R. Ruffini (North-Holland, Amsterdam, 1982), p. 1211.

[13]   See, *e.g.*, W. Hamilton, D. Darling, J. Kadlec and D. DeWitt, in *Proceedings of the Second Marcel Grossmann Meeting on General Relativity, Part B*, edited by R. Ruffini (North-Holland, Amsterdam, 1982), p. 1175.

[14]   T. Regge and J. A. Wheeler, *Phys. Rev.* **108**, 1063 (1957).

[15]   F. Zerilli, *Phys. Rev. D* **2**, 2141 (1970), and *J. Math. Phys.* **11**, 2203 (1970).

[16]   M. Davis and R. Ruffini, *Nuovo Cim. Lett.* **2**, 1165 (1972); M. Davis, R. Ruffini, W. H. Press and R. H. Price, *Phys. Rev. Lett.* **27**, 1466 (1971); R. Ruffini, *Phys. Rev. D* **7**, 972 (1973).

[17]   M. Davis, R. Ruffini, J. Tiomno and F. Zerilli, *Phys. Rev. Lett.* **28**, 1352 (1972).

[18]   M. Johnston, R. Ruffini and F. Zerilli, *Phys. Rev. Lett.* **31**, 1317 (1973).

[19]   M. Johnston, R. Ruffini and F. Zerilli, *Phys. Rev.* **49B**, 185 (1974).

[20]   M. Davis, R. Ruffini and J. Tiomno, *Phys. Rev. D* **5**, 2932 (1972); W. Haxton and R. Ruffini, *Ann. Phys.* **95**, 1 (1975).

[21]   R. Ruffini, in *Physics and Astrophysics of Neutron Stars and Black Holes*, edited by R. Giacconi and R. Ruffini (North-Holland, Amsterdam, 1978), p. 287.

[22]   See, *e.g.*, *Sources of Gravitational Radiation*, edited by L. Smarr (Cambridge University Press, Cambridge, 1979), and references therein.

[23] T. Piran, *Phys. Rev. Lett.* **41**, 1085 (1978); T. Piran, in *Proceedings of the Second Marcel Grossmann Meeting on General Relativity, Part A,* edited by R. Ruffini (North-Holland, Amsterdam, 1982), p. 679.

[24] R. F. Stark and T. Piran, in *Proceedings of the Fourth Marcel Grossmann Meeting on General Relativity,* edited by R. Ruffini (North-Holland, Amsterdam, 1986).

[25] See, *e.g.,* R. Ruffini, *Bull. Am. Phys. Soc.* **18**, 109 (1973). See also R. Ruffini and J. A. Wheeler, *Phys. Today* **24**, 30 (1971); R. Leach and R. Ruffini, *Ap. J. Lett.* **180**, L15 (1973); C. Rhoades and R. Ruffini, *Phys. Rev. Lett.* **32**, 324 (1974).

[26] In 1974 W. Fairbank invited me to spend a quarter teaching at Stanford. I remember always with pleasure that period and some of the students and auditors, among whom were Wick Haxton, Blas Cabrera, Kyle Baker and Mark Peterson.

[27] Some simple and basic ideas on the magnetosphere of a collapsed object were considered in R. Ruffini and A. Treves, *Astrophys. Lett.* **13**, 109 (1973); see as well paragraph 2.6 in [11]; and see J. R. Wilson, in *Proceedings of the First Marcel Grossmann Meeting on General Relativity,* edited by R. Ruffini (North-Holland, Amsterdam, 1976), p. 393.

[28] R. Ruffini and J. Wilson, *Phys. Rev. D* **12**, 2959 (1975).

[29] R. Ruffini, *Ann. N.Y. Acad. Sci.* **262**, 95 (1975).

[30] J. Wilson, *Ann. N.Y. Acad. Sci.* **262**, 123 (1975).

[31] N. Deruelle and R. Ruffini, *Phys. Rev. Lett.* **52B**, 437 (1974), and **57B**, 248 (1975).

[32] T. Damour and R. Ruffini, *Phys. Rev. D* **14**, 332 (1976); T. Damour, N. Deruelle and R. Ruffini, *Nuovo Cim. Lett.* **15**, 252 (1976); T. Damour and R. Ruffini, *Phys. Rev. Lett.* **35**, 463 (1975).

[33] T. Damour, R. S. Hanni, R. Ruffini and J. Wilson, *Phys. Rev. D* **17**, 1518 (1978).

[34] These results have been summarized in [21], translated into a variety of languages, and more recently expressed, as well, in 3+1 formalism.

[35] E. M. Lifshitz, *J. Phys. U.S.S.R.* **10**, 116 (1946).

[36] See, *e.g.,* G. Gamow, *Phys. Rev.* **74**, 505 (1948).

[37] See, *e.g.,* S. M. Faber and J. S. Gallagher, *Ann. Rev. of Astron. and Astrophys.* **17**, 135 (1979).

[38] See, *e.g.,* V. C. Rubin, W. K. Ford, N. Thormard, D. Bernstein, *Ap. J.* **261**, 439 (1982).

[39] See, *e.g.*, D. Fabricant, M. Lecar and P. Gorenstein, *Ap. J.* **241**, 552 (1980), and D. Fabricant and P. Gorenstein, *Ap. J.* **267**, 535 (1983).

[40] See, *e.g.*, R. Fabbri, R. T. Jantzen and R. Ruffini, *Astron. and Astrophys.* **114**, 219 (1982).

[41] M. Baldeschi, G. Gelmini and R. Ruffini, *Phys. Lett.* **B97**, 388 (1983).

[42] G. Ingrosso and R. Ruffini, in *Proceedings of the Fourth Marcel Grossmann Meeting on General Relativity*, edited by R. Ruffini (North-Holland, Amsterdam, 1986).

[43] M. Merafina and R. Ruffini, *ibid.*

[44] A. Malagoli and R. Ruffini, *Astron. and Astrophys.* **157**, 292 (1986).

[45] M. Arbolino and R. Ruffini, submitted to *Astron. and Astrophys.*, 1986.

[46] S. Filippi, I. D. Novikov and R. Ruffini, *Nuovo Cim. Lett.* **39**, 165 (1984).

[47] R. Ruffini, D. J. Song and L. Stella, *Astron. and Astrophys.* **125**, 265 (1983).

[48] R. Ruffini and D. J. Song, *Astron. and Astrophys.*, in press, 1986.

[49] M. Baldeschi, R. Ruffini and D. J. Song, in preparation.

[50] R. Fabbri and R. Ruffini, *Astrohys. and Space Sci.* **82**, 249 (1982).

[51] See, *e.g.*, Ya. B. Zel'dovich and I. D. Novikov, *Stroienie i Evolutsia Vselennoj* (Nauka, Moscow, 1975).

[52] R. Ruffini, D. J. Song and S. Taraglio, submitted to *Astron. and Astrophys.*, 1986.

Near Zero:
Toward a Quantum-Limited Resonant-Mass
Gravitational Radiation Detector

P. F. Michelson, M. Bassan,
S. Boughn, R. P. Giffard,
J. N. Hollenhorst, E. R. Mapoles,
M. S. McAshan, B. E. Moskowitz,
H. J. Paik and R. C. Taber

1. INTRODUCTION

The gravitational radiation detection program at Stanford University was begun in 1969 as a cooperative effort with Louisiana State University (see paper by W. O. Hamilton entitled "Near Zero Force, Force Gradient and Temperature: Cryogenic Bar Detectors of Gravitational Radiation" in this volume) and with the University of Rome (see paper by E. Amaldi entitled "Large Mass Detector of Gravitational Waves" in this volume). The earliest ideas about cryogenic gravitational radiation detectors resulted from discussions that W. M. Fairbank had with Hamilton while Hamilton was at Stanford in the mid-1960's. In the early stages of development, the project was greatly aided by the participation of James Opfer while he was an assistant professor in the low temperature physics group at Stanford.

From the start of these projects there have been two objectives:

(1)   to directly establish the existence of gravitational radiation, predicted early in this century by Albert Einstein [1], and

(2)   to use gravitational radiation as a tool for astronomical observation.

Efforts to detect gravitational radiation have focused on the strongest signals expected from extraterrestrial sources. A significant flux of gravitational wave energy is radiated only by events of astrophysical scale because of the weakness of the gravitational interaction and the principle of equivalence. The latter implies that the lowest order mass multipole that can radiate gravitational waves is the quadrupole. A mass quadrupole is also necessary for detection. This property of gravitational radiation should be compared with the more familiar situation in electrodynamics where dipole radiation is possible.

Any source of gravitational radiation must have a time-dependent mass distribution that is not spherically symmetric. The gravitational wave luminosity L of a source with a time-varying mass quadrupole Q_{ij} is given by

$$L \simeq \sum_{ij} \left(G/5c^5 \right) \langle (d^3 Q_{ij}/dt^3)^2 \rangle \tag{1}$$

where

$$Q_{ij} = \int \rho \left(x_i x_j - r^2/3 \right) d^3x \quad . \tag{2}$$

This formula can be easily derived in the limit of weak gravitational field and slow motion of the source [2]. The more general validity of the quadrupole formula has been widely discussed [3].

Let us consider the flux predicted by equation (1) for a possible laboratory source consisting of a rapidly rotating rod of mass M and length ℓ shown in figure 1. The luminosity calculated from equation (1) is

$$L = \frac{8}{9} \left(\frac{G}{5c^5} \right) M^2 \ell^4 \omega^6 \quad . \tag{3}$$

If we take $M = 5$ tons, $\ell = 3$ meters and $\omega = 10^4$ rad/sec, then $L \sim 10^{-13}$ erg/sec. The rotation rate assumed is near the upper limit possible without exceeding the elastic limit of the bar thus causing the experiment to self-destruct. Even for this extreme case, the laboratory source produces a gravitational wave luminosity that is practically undetectable. To see why, let us now consider the problem of detecting gravitational radiation.

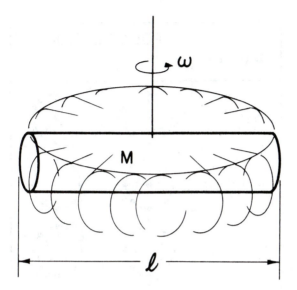

FIGURE 1. A laboratory generator of gravitational radiation. Even if the rod is rotated at a rate near the upper limit possible without exceeding the rod's elastic limit, the gravity wave signal is too weak to detect.

2.   INTERACTION OF GRAVITATIONAL WAVES WITH A DETECTOR

Because of the equivalence principle a single point mass cannot be used as a detector of gravitational radiation; at least two masses separated by a non-zero distance ℓ are required. A gravitational wave interacting with this system will cause a time-dependent change in the distance between the two test masses, $\Delta\ell$. Just like electromagnetic radiation, the gravitational radiation predicted by general relativity is transverse and has two independent polarizations. However, gravitational radiation is spin-two in character, unlike the electromagnetic case. It is convenient to characterize the two independent polarizations of the wave by dimensionless amplitudes h_+ and $h_\times$. The energy flux of the gravitational wave is given by [2]

$$\Phi = \left(c^3/16\pi G\right) \langle h_\times^2 + h_+^2 \rangle \quad . \tag{4}$$

Figure 2 illustrates the effect of a gravitational wave of each polarization on a set of test masses placed in a circle of radius R. In general the motion will be a superposition of both polarizations. For simplicity it is assumed that $R \ll \lambda$, where λ is the wavelength of the radiation. Notice that the interaction of the test masses with the wave induces a time-dependent mass quadrupole moment relative to the center of mass of the configuration. This distortion can be thought of as being produced by a time-dependent tidal

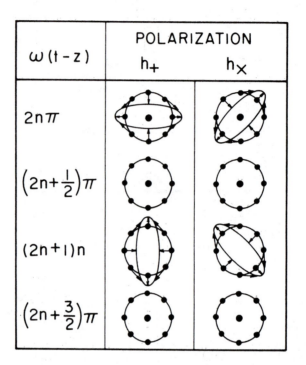

$\omega(t-z)$	POLARIZATION	
	h_+	$h_\times$
$2n\pi$		
$\left(2n+\dfrac{1}{2}\right)\pi$		
$(2n+1)n$		
$\left(2n+\dfrac{3}{2}\right)\pi$		

FIGURE 2. The effect of polarized gravitational radiation on a set of test masses in a circle of radius R. The gravitational wave induces a time-dependent mass quadrupole moment relative to the center of mass of the configuration.

force acting on the mass distribution. As long as the dimensions of the antenna are small compared to the wavelength of the radiation, the tidal force on a mass element Δm at position $\mathbf{r}$ relative to the center of mass has components given by

$$F_i = -\sum_j \Delta m R_{i0j0}(t)r_j \quad , \tag{5}$$

where $R_{i0j0}(t)$ are components of the Riemann curvature tensor. The response of a detector of arbitrary shape can be found by solving the equation of motion of each mass element under the influence of the gravitational tidal force and any elastic and dissipative forces that are present. For the case of two free point masses initially at rest and separated by distance $\ell \ll \lambda$, a gravitational wave of amplitude $h(t)$, incident with optimum polarization and direction, will change the separation by

$$\frac{\Delta\ell(t)}{\ell} = h(t) \quad . \tag{6}$$

For a monochromatic source with luminosity L, distance r from the detector, and frequency ω_g, the amplitude of the dimensionless strain at the detector is obtained from equation (4) as

$$h^2 \simeq \left(\frac{16\pi G}{c^3}\right) \left(\frac{L}{4\pi r^2}\right) \left(\frac{1}{\omega_g^2}\right) \tag{7}$$

after averaging over source polarization and direction. For the laboratory source considered earlier this gives $h \sim 10^{-37}$ for $r = \lambda$. Clearly, detecting a laboratory source of gravitational radiation would require the measurement of very small displacements. Such measurements do not appear feasible even with optimistic projected improvements in present technology.

Radiation emitted during the gravitational collapse of a star offers much better prospects for direct detection. For example, a gravity wave pulse of 1 msec duration corresponding to the radiation of the energy equivalent of 10% of a single solar mass at the center of our galaxy will produce a flux at the earth of about 2×10^7 J m^{-2} sec^{-1}. This corresponds to a dimensionless strain amplitude h of about 10^{-17}. Of course the cataclysmic collapse of a star occurs only once in its lifetime and the rate of occurrence of such events in the galaxy is very uncertain, probably about one every ten to thirty years [4]. For observations at a reasonable rate, we would need to be able to detect such events in all galaxies within about 6×10^{23} m, the distance characterizing the M101 cluster containing several hundred galaxies. These events might result in about one pulse per month with a flux of $\lesssim 50$ J m^{-2} sec^{-1} at the earth. Thus strain sensitivities to impulsive events of 10^{-20} to 10^{-21} are probably necessary. At the present time it appears possible to build detectors with this sensitivity. Already in 1981 we operated a detector at Stanford University with a dimensionless strain sensitivity at the noise level, for impulsive events, of 10^{-18} [5]. Some of the important characteristics of this detector are described in a later section of this paper.

The possibility of directly detecting continuous astrophysical sources of gravitational radiation also exists. In this case a long integration time can be used to improve the strain sensitivity of the detector relative to that for burst signals. In either case the sensitivity is inevitably limited by noise of thermal origin or noise associated with the detector's amplifier readout.

The type of gravitational wave detector first developed by Joseph Weber [6] and subsequently used by other experimental groups, including the Stanford group, consists of an extended resonant-mass object, usually in the form of a right cylinder, with a fundamental longitudinal elastic resonance f_o around 1 kHz, where the signal spectral energy density is expected to be largest. The bar, or "antenna," is equipped with a strain or motion detector that monitors the dynamic strain induced in the fundamental mode by interaction with the gravitational wave. As a gravitational wave passes by perpendicular to the axis of the antenna, the length of the bar will be modulated. Because of the forces responsible for the elasticity of the bar, the gravitational wave does work and thus imparts energy to the odd-order longitudinal modes. For a burst of short duration, the excitation will appear as a sudden ringing of the bar.

Because the size of the mechanical resonator is determined by the velocity of sound in the material used, which is much less than the velocity of gravitational waves (equal to the velocity of light), the resonant-mass antenna is always much smaller than the radiation wavelength. This large mismatch accounts, in part, for the relative insensitivity of gravitational radiation detectors when compared with radio antennas. Long baseline laser interferometers have been proposed as an alternative means of detection which avoids this limitation [7,8]. Although excellent progress is being made in the development of such optical readout, free-mass detectors, these have not yet reached a competitive level of sensitivity.

3.   SENSITIVITY OF A RESONANT-MASS DETECTOR

The sensitivity of any gravitational radiation detector depends not only on the noise characteristics of the detector but also on the spectral character of the signals. We can broadly classify the expected signals into three categories: burst signals from stellar collapse and collision, periodic signals from stellar oscillations or binary systems, and stochastic background radiation.

A resonant-mass gravitational radiation detector is shown schematically in figure 3. The mechanical oscillations of the antenna are transformed into an electrical signal by a motion transducer and then amplified by an electrical amplifier. Because of noise in the detector the output is filtered to optimize the signal-to-noise ratio. The principal sources of noise which determine the sensitivity to burst sources are antenna and transducer thermal noise, transducer electrical noise, and amplifier noise. The requirements on the transducer-amplifier combination are much less difficult to meet in detecting periodic sources. In this case the antenna thermal noise is the limiting noise source.

Given the form of the signal, the signal-to-noise ratio is maximized by means of an optimum linear filter [9] whose frequency response is given by

$$K(\omega) = M^*(\omega)/S_n(\omega) \qquad (8)$$

where $M^*(\omega)$ is the conjugate of the Fourier transform of the expected signal and $S_n(\omega)$ is the noise power spectral density. The signal-to-noise ratio at the output of the optimum filter is

$$\rho_o = \frac{1}{2\pi} \int_{-\infty}^{\infty} |M(\omega)|^2/S_n(\omega)d\omega \quad . \qquad (9)$$

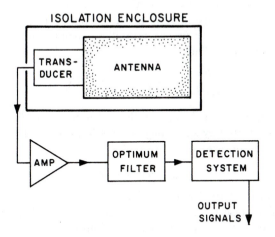

FIGURE 3. Basic components of a resonant-mass gravitational radiation detector. The transducer converts the mechanical oscillations of the antenna into an electrical signal that is amplified. Because each of these components including the electronic amplifier contributes noise, the output is filtered to optimize the signal-to-noise ratio.

For burst detection, as long as the optimum filter bandwidth is small compared to the expected burst bandwidth, the form of the filter is unique for a particular detector and independent of the detailed time dependence of the burst. In practice the output of the detector is digitized and the filtering is implemented with a computer.

4. DETECTOR NOISE SOURCES

The optimum signal-to-noise ratio given by equation (9) depends in general on all the parameters of the system. By studying how ρ_o varies with respect to the adjustable parameters, it is possible to optimize the detector's sensitivity.

The linear amplifier can be characterized by a series voltage noise source with spectral density $S_V(\omega)$, and a shunt current noise source with spectral density $S_I(\omega)$, followed by a noise-free amplifier. We will assume that this amplifier is a current amplifier with zero input impedance. In the Stanford detector the amplifier is a SQUID magnetometer which can be represented by an ideal current amplifier in series with an inductor [10].

The noise-match impedance of the amplifier is defined as

$$Z_n = \{S_V(\omega)/S_I(\omega)\}^{1/2} \quad .$$

(10)

If the input to the amplifier is terminated in this impedance, then the voltage and current noise contributions to the amplifier output are equal.

The noise temperature of the amplifier T_{amp} is defined as the temperature which a resistor with resistance Z_n at the amplifier input would have if it contributed a noise power at the amplifier output equal to that of the amplifier noise sources. Thus

$$\frac{\hbar\omega}{\exp\left(\hbar\omega/kT_{amp}\right) - 1} = \{S_V(\omega)S_I(\omega)\}^{1/2} \quad . \tag{11}$$

The quantum limit for a linear amplifier, $(S_V S_I)^{1/2} = \hbar\omega$, occurs when the amplifier is limited by spontaneous emission noise at the input. In this case the limiting noise temperature is [11]

$$T_{amp} = \hbar\omega/k \ln(2) \quad , \tag{12}$$

which is equal to 7×10^{-8} K at 1 kHz.

The amplifier noise temperature and the noise-match impedance completely characterize the amplifier in calculations of the signal-to-noise ratio.

The thermal noise of the antenna can be characterized by the spectral density of the Nyquist force noise applied to the antenna, given by

$$S_f(\omega) = 2kTM_a\omega_a/Q_a \quad , \tag{13}$$

where T, M_a, ω_a and Q_a are the antenna's temperature, effective mass, resonant frequency and mechanical quality factor, respectively. The thermal noise sources associated with the transducer can be described in a similar way. Notice that $S_f(\omega)$ is a white noise source (*i.e.*, independent of frequency). An impulsive gravitational wave signal applied to the antenna also has a white spectrum. Thus the signal-to-noise ratio is independent of frequency if the limiting noise source is the antenna thermal noise. A bandwidth restriction arises only when one considers the problem of matching the amplifier to the antenna.

The noise power spectral density at the detector output, $S_n(\omega)$ in equation (9), is made up of contributions from all the noise sources discussed above. If the detector is described by linear equations of motion and the noise sources are uncorrelated, then $S_n(\omega)$ is given by

$$S_n(\omega) = \sum_j S_j(\omega)|T_j(\omega)|^2 \tag{14}$$

where $T_j(\omega)$ is the transfer function for the jth noise source $S_j(\omega)$.

5. THE STANFORD DETECTOR

The resonant-mass detector which we have constructed at Stanford University illustrates many of the points we have discussed above. The detector,

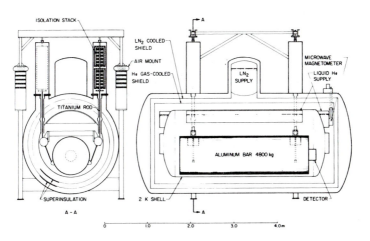

FIGURE 4. Schematic diagram of the low temperature, 4800 kg resonant-mass detector constructed at Stanford University. In its present form the detector consists of a cylindrical aluminum antenna coupled to a SQUID amplifier.

shown in figure 4, has been operated in a mode to detect impulsive gravitational wave events. In its present form the detector consists of a 4800 kg cylindrical aluminum antenna with a fundamental longitudinal resonance at 840 Hz. A resonant superconducting transducer [12] converts the mechanical motion of one end of the antenna into an electrical signal which is amplified by a SQUID amplifier. The amplifier output is fed to a digital filter which closely approximates the characteristics of the optimum filter required to maximize the sensitivity to short-duration gravitational wave signals. The output of the filter with a known time delay resulting from signal processing is recorded on magnetic tape, together with timing information derived from WWVB.

It is convenient to characterize the detector sensitivity in terms of T_d, the effective noise temperature for the detection of short pulses, and σ, the integrated cross section of the antenna [13]. The condition for detection of short gravitational wave pulses with energy spectral density $\tilde{F}(\omega)$ at 50% efficiency is given approximately by

$$\tilde{F}(\omega_a) \geq \frac{kT_d}{\sigma} \ln\left(1/\tau_d R\right) \quad , \tag{15}$$

where τ_d is the correlation time of the filtered detector output and R is the accidentals rate. For the detector we have operated at Stanford University the appropriate value of σ for spin-two tensor gravitational waves is 4×10^{-25} m^2 Hz. Practical considerations prevent this quantity from being much larger for a single antenna.

Another quantity of interest is the dimensionless wave amplitude h_{burst} for a pulse signal of duration τ and characteristic frequency $1/(2\pi\tau)$. Using equation (4), the detection condition given by equation (15) can also be written as

$$h_{burst}^2 \geq \frac{4G}{c^3} \frac{kT_d}{\sigma} \ln(1/\tau_d R) \quad . \tag{16}$$

The detector noise temperature can be calculated using equation (9) as

$$kT_d = \left(\frac{M_a}{2\pi} \int_{-\infty}^{\infty} \frac{d\omega}{S_n(\omega)} \right)^{-1} \tag{17}$$

where $S_n(\omega)$ is the output noise power spectrum referred to the antenna input. At sufficiently low temperature the limiting value of this expression is $2kT_{amp}$ [14].

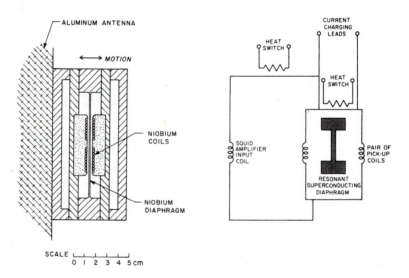

FIGURE 5. Superconducting motion transducer.

The motion transducer shown schematically in figure 5 makes use of the low mechanical and electrical losses in superconducting materials. Oscillations of the antenna are coupled to the fundamental eigenmode of a superconducting diaphragm, which modulates the inductance of current-carrying superconducting pickup coils, causing an ac voltage proportional to the velocity of the diaphragm to appear at the output terminals. The output signal from the transducer is fed to the input coil of the SQUID.

Because of the magnetic fields produced by the stored currents, displacement of the diaphragm not only causes an induced proportional current to flow in the SQUID coil but also produces an electromagnetic force tending to return the diaphragm to its equilibrium position. This additional restoring force has the effect of raising the resonant frequency of the transducer. The use of a large magnetic field (limited in practice by the H_{c1} field of the diaphragm material) results in a large change in the resonant frequency.

The transducer is usually operated in a resonant mode when mounted on the gravitational wave antenna. In this mode, the diaphragm is made so that the optimum field and coil spacing tune its resonant frequency exactly to the bar fundamental. The bar and the diaphragm act as a pair of coupled oscillators, and the resonances split into two closely spaced eigenfrequencies seperated by $(m_t/M_a)^{1/2} \omega_a$ where m_t is the effective mass of the diaphragm. It can be shown that a given amplitude of bar motion is associated with an amplitude of diaphragm motion which is larger by a factor of $(M_a/m_t)^{1/2}$. The present transformation ratio is about 300, resulting in a useful bandwidth of about 1 Hz. In this mode of operation the diaphragm resonance provides a suitable noise-match between the impedance of the antenna and the much higher impedance characterizing the electromechanical transduction process.

The mechanical to magnetic coupling of the transducer can be characterized by

$$\beta = K_m / \left(2\omega_t^2 m_t\right) = 1 - \omega_o^2/\omega_t^2 \qquad (18)$$

where K_m is the spring constant due to the electromagnetic coupling to the SQUID amplifier input circuit, ω_o is the transducer resonant frequency with no current stored, and ω_t is the tuned frequency with current stored. In the usual resonant mode of detector operation $\omega_t = \omega_a$.

Because of the close coupling between the antenna and the transducer, a high value of Q_t, the transducer Q, is important. It has been found that as the transducer is tuned, Q_t decreases approximately as

$$\frac{1}{Q_t} = \frac{1}{Q_m} + \frac{\beta}{Q_e} \quad , \qquad (19)$$

where Q_m is the mechanical Q-value with no current stored. This relationship indicates that a damping mechanism exists which is equivalent to a fixed loss in the electrical circuit that can be characterized by an electrical Q-value of Q_e. At present we find $Q_e = 3 \times 10^4$. There is experimental evidence that substantially higher Q-values can be obtained.

Figure 6 shows the measured output noise power spectrum of the present Stanford detector, along with the relative contributions from the amplifier noise sources and the detector's thermal noise sources. The solid curve in figure 7 is the signal-to-noise ratio per unit bandwidth for an impulsive signal. The area under this curve is inversely proportional to the detector noise temperature T_d. Notice that the signal-to-noise ratio peaks between

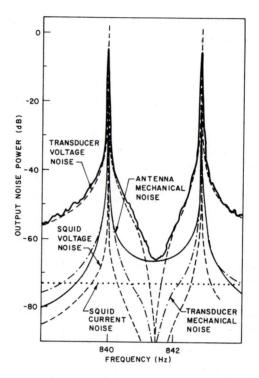

FIGURE 6. The measured output noise power spectrum of the Stanford detector (solid line). The relative contributions from the amplifier current noise, the transducer electrical noise, the transducer thermal noise and the antenna thermal noise are also shown.

the normal modes in this particular case even though the signal is largest at the normal mode frequencies. The frequency dependence of the signal-to-noise ratio in general is a function of detector parameters such as transducer mass and Q, for example. The dotted curve in figure 7 is the signal-to-noise ratio for a detector with a transducer mass 4.6 times heavier than that in the present system. The other detector parameters have remained fixed.

The detector noise temperature T_d predicted for the conditions of figure 6 is 6.4 mK [15]. The lowest noise temperature actually achieved with the 4800 kg detector was 10 mK. The discrepancy is believed to be due in part to the use of a filter which is not optimal.

6. OBSERVATIONS WITH THE DETECTOR

Since the detector noise sources discussed in section 4 are expected to be Gaussian distributed, the linearly-filtered detector output in the absence

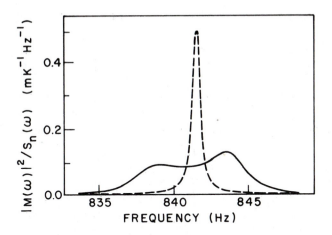

FIGURE 7. The signal-to-noise ratio per unit bandwidth for an impulsive signal. The solid curve is computed for the noise spectrum of figure 6. The dotted curve is the signal-to-noise ratio for a detector with a transducer mass 4.6 times heavier than that in the present system.

of signals or any non-Gaussian distributed disturbances is also expected to be Gaussian distributed. Thus the expected distribution of the squared envelope of the filtered detector output $\epsilon_f(t)$, in the absence of signals, is an exponential distribution given by

$$P(\epsilon_f) = (1/kT_d) \exp(-\epsilon_f/kT_d) \quad . \tag{20}$$

If a signal of strength ϵ_s is known to be present (for example, a calibration pulse) the expected output distribution function is a Rician distribution given by

$$P(\epsilon_f \mid \epsilon_s) = (1/kT_d) \exp[-(\epsilon_s + \epsilon_f)/kT_d] I_o\left(\sqrt{\epsilon_s\epsilon_f}/kT_d\right) \tag{21}$$

where I_o is the modified Bessel function of zero order. In the limit $\epsilon_s \gg kT_d$, this distribution function is a Gaussian distribution with mean value ϵ_s and width kT_d. Note that once the detector is suitably calibrated the noise temperature T_d can be obtained either from the average mean square of the filtered output or from the slope of the distribution as shown by equation (20).

The 4800 kg detector described above was operated for more than one year with a noise temperature typically near 20 mK [5]. Figure 8 shows a histogram of the squared envelope of the filtered detector output for a typical 24 hour period. The results are well described by equation (19).

During the entire period of observations in 1981, the distribution of the data below 150 mK fits equation (19) very well. At higher energies there were significantly more events than predicted by equation (19), indicating that the detector was affected by one or more nonstationary or non-Gaussian processes. Apart from gravitational radiation, the detector can be excited by a number of internal and external disturbances including occasional spontaneous acoustic emission. Some of these events can easily be eliminated from consideration as possible gravitational wave events because they are coincident with a significant excitation of the second longitudinal mode of the antenna, which because of symmetry does not couple to gravitational waves.

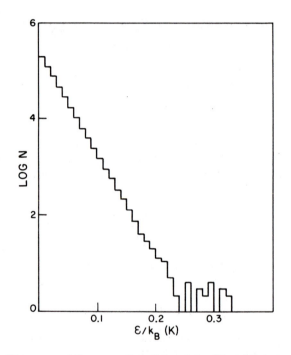

FIGURE 8. Histogram of the squared envelope of the filtered detector output for a 24 hour period.

Without coincidence operation with another detector of similar sensitivity, it is impossible to draw positive conclusions about the remainder of the data. However, using the event distribution remaining after application of the second longitudinal mode veto, we were able to establish a new upper limit on the distribution of gravitational wave pulses reaching the earth [5]. Figure 9 shows a smoothed estimate at the 3σ confidence level based on the measured event rate. Also shown is the improvement that would be expected if two detectors with this sensitivity were operated in coincidence.

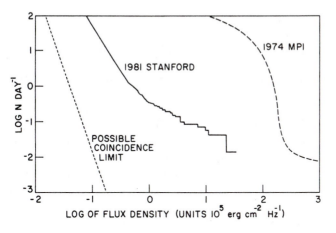

FIGURE 9. Upper limit on the distribution of 1 kHz gravitational wave pulses reaching the earth. Also shown is the previous upper limit established by a coincidence experiment carried out by the Max Planck Institute.

7. TOWARD THE LINEAR AMPLIFIER QUANTUM LIMIT

Even if gravitational wave signals are detected at the present sensitivity, the continuing development of still more sensitive detectors is essential. These will be required not only for detecting more distant and thus more frequent events, but also for extracting source properties from the observed waveforms. The development of wide-band detectors will be important for the study of the received waveforms. This will yield information about the coherent bulk motions of the matter generating the radiation. In this regard, gravitational waves are a unique tool for astronomical observation because they can "reveal features of their sources which one could never learn by electromagnetic, cosmic ray or neutrino studies." [3]

At first it might appear that a resonant-bar detector is particularly unsuited for use as a wide-band detector. While it is true that the maximum signal amplitude is achieved on resonance, the signal-to-noise ratio is independent of frequency if the limiting noise source is the Brownian motion noise of the antenna. A bandwidth restriction arises when one considers matching the amplifier to the antenna. Of course the amplifier can be most easily matched near the antenna's resonant frequency. It appears that with the technology used in the present Stanford detector the SQUID amplifier can be noise matched over a bandwidth approaching $0.15 f_o$.

Using linear transducer-amplifier readout schemes, the energy sensitivity of resonant-mass detectors can be improved by more than four orders of magnitude before being limited by the uncertainty principle. In this limit the detector noise temperature is given by equations (12) and (16) as

$$T_d = \frac{2\hbar\omega_a}{k \ln(2)} \tag{22}$$

which is equal to 1.4×10^{-7} K for an antenna resonant at 1 kHz. Ignoring the logarithmic statistical factor, the minimum detectable value of h_{burst} given by equation (16) is about 5×10^{-21} for the Stanford 4800 kg antenna. This sensitivity is sufficient to detect the radiation of less than 10^{-7} of a solar mass into gravitational waves of 1 kHz bandwidth at the galactic center. More sensitivity and bandwidth can be achieved with an array of detectors. For example, with 25 detectors, each with a suitably chosen resonant frequency, the minimum detectable value of h_{burst} is about 10^{-21} in a bandwidth approaching 3 kHz.

The possibility of sensitivity beyond the linear amplifier quantum limit was recognized several years ago by Braginsky [16]. There is now mounting interest in applying unconventional quantum detection strategy to avoid the linear detection limit (see, for example, references [17,18]). However, since none of the techniques proposed for more subtle detection schemes mitigate any of the problems which limit the existing technology, it seems that the use of the largest antenna coupled to a linear detector with the lowest attainable noise level is the best strategy available at present. It offers the exciting possibility of seeing predicted gravitational events.

It should be kept in mind that while astrophysicists are putting much effort into estimates of gravitational radiation strengths and rates, these estimates remain very uncertain due to the lack of firm knowledge about the nature of the universe. Since the kind of information carried by gravitational radiation is almost orthogonal to that carried by electromagnetic waves, it is doubtful that this uncertainty will be reduced until the actual detection and measurement of gravity waves.

In the last few decades astronomers have opened a succession of new windows onto the universe. Each new window has dramatically improved our understanding of the cosmos. The operation of gravitational wave antenna systems capable of detecting gravity waves offers us the prospect of opening still another window. The possibility exists that a fortunate act of nature might at any moment reward attentive experimentalists with proof that gravitational radiation can be directly detected.

Addendum

On February 23, 1987, neutrinos from a supernova (SN 1987A) in the Large Magellanic Cloud (LMC) were detected [19,20], thereby opening another window for astronomy. Unfortunately with respect to gravitational waves, no second-generation resonant-mass or laser-interferometric gravity wave detectors were operating at the time the neutrino bursts were detected. Since the only gravitational wave detectors operating then were two room temperature bar detectors, limits on the observed flux due to prompt emission of a burst of gravitational waves from SN 1987A will necessarily be set by these less sensitive detectors.

At the time of publication of this volume, second-generation resonant-mass detectors have been operated with rms dimensionless strain sensitivities of 10^{-18}. Within about 2 years the sensitivity of these detectors is expected to approach 10^{-20}. The largest plausible burst signal predicted from a supernova at the distance of the LMC is $h_{burst} = 6 \times 10^{-19}$. Clearly the operation of the next generation of detectors at the predicted sensitivity would provide interesting data from such a collapse. It will be important not only to do coincidence observations with two or more gravity wave detectors but also to search for coincident events between neutrino burst detectors and gravity wave detectors.

Finally, we note that even though SN 1987A is the first supernova observed in the neighborhood of our galaxy since the one observed optically by Kepler in 1604, a calculation of the expected rate of stellar deaths in the galaxy [21], using a detailed model of the distribution of stars in the disk and standard values for stellar evolutionary lifetimes, gives a collapse rate of about 1 stellar collapse every 8 years. A study of the distribution of radio and x-ray neutron star pulsars and their lifetimes [22] suggests that the stellar collapse rate is 1 collapse every 1 to 5 years. Obviously most of these collapses are not seen optically. This may be because many of the collapses either occur in dense molecular clouds from which electromagnetic radiation cannot escape or occur without generating a large electromagnetic signal. The ability to detect neutrinos from these collapses has signaled that we are on the verge of opening a new era of gravitational wave astronomy with the next generation of resonant-mass detectors.

Acknowledgments

This work was supported by the National Science Foundation. Throughout the past ten years the Stanford gravity wave detection project has benefited enormously from the technical assistance of Norm Rebok.

References

[1]   A. Einstein, *Konig. Preuss Akad. der Wissenschaften, Sitzungs-berichte*, Erster Halbband, 688 (1916).

[2]   L. D. Landau and E. M. Lifshitz, *The Classical Theory of Fields*, (Pergamon Press, Oxford, 1975), 4th edition, p. 354.

[3]   K. S. Thorne, *Rev. Mod. Phys.* **52**, 285 (1980).

[4]   R. D. Davis, in *Proceedings of an Informal Conference on Neutrino Physics and Astrophysics* (F. I. Acad. Sci. USSR, Moscow, 1978), Vol. 2, p. 99.

[5]   S. P. Boughn, W. M. Fairbank, R. P. Giffard, J. N. Hollenhorst, E. R. Mapoles, M. S. McAshan, P. F. Michelson, H. J. Paik and R. C. Taber, *Ap. J. Lett.* **261**, L19 (1982).

[6]   J. Weber, *Phys. Rev.* **117**, 306 (1960).

[7]   R. Weiss, *MIT Quarterly Progress Report, Research Laboratory of Electronics* **105**, 54 (1972).

[8]   R. W. P. Drever, in *Gravitational Radiation, Les Houches Summer School, June 1982*, edited by N. Deruelle and T. Piran (North-Holland, Amsterdam, 1983), p. 321.

[9]   L. A. Wainstein and V. D. Zubakov, *Extraction of Signals from Noise* (Prentice-Hall International, London, 1962).

[10]   J. N. Hollenhorst and R. P. Giffard, *J. Appl. Phys.* **51**, 1719 (1980).

[11]   H. Heffner, *Proc. IRE* **50**, 1604 (1962).

[12]   H. J. Paik, *J. Appl. Phys.* **47**, 1168 (1976).

[13]   C. W. Misner, K. S. Thorne and J. A. Wheeler, *Gravitation* (W. H. Freeman Co., San Francisco, 1973), 1st edition.

[14]   R. P. Giffard, *Phys. Rev. D* **14**, 2478 (1976).

[15]   P. F. Michelson and R. C. Taber, *J. Appl. Phys.* **52**, 4313 (1981).

[16]   V. B. Braginsky, in *Topics in Theoretical and Experimental Astrophysics*, edited by V. de Sabbata and J. Weber (Plenum, London, 1977).

[17]   C. M. Caves, K. S. Thorne, R. W. P. Drever, V. D. Sandberg and M. Zimmermann, *Rev. Mod. Phys.* **52**, 341 (1980).

[18]   K. S. Thorne, R. W. P. Drever, C. M. Caves, M. Zimmermann and V. D. Sandberg, *Phys. Rev. Lett.* **40**, 667 (1978).

[19]   K. Hirata *et al.*, *Phys. Rev. Lett.* **58**, 1490 (1987).

[20]   R. M. Bionta *et al.*, *Phys. Rev. Lett.* **58**, 1494 (1987).

[21]   J. N. Bahcall and T. Piran, *Ap. J. Lett.* **267**, L77 (1983).

[22]   D. G. Blair and W. Candy, *Mon. Not. Roy. Astron. Soc.* **212**, 219 (1985).

Near Zero Force, Force Gradient and Temperature: Cryogenic Bar Detectors of Gravitational Radiation

A Primer for Nonspecialists

William O. Hamilton

1. INTRODUCTION

I believe that the manner in which William Fairbank came into the field of gravitational radiation research should be held up as an example to students of how physics should be done. It does not make any difference if you are an expert in a field; if the research looks interesting and if you see a way to make a contribution, then you should go ahead and try. In the case at hand, Bill and I had talked about gravitation experiments late at night when I was working on my thesis for George Pake and Arthur Schawlow. We were in no way experts in gravitation. My thesis work involved measuring specific heats of powdered organic free radicals, and the equilibrium times in the experiment were hours. There was no one around in the old physics corner after 3:00 a.m. except Bill and me, me flipping a switch every half hour or so for a second or two and waiting for the temperature of the sample to stabilize, and Bill keeping me company with exciting stories about the experiments yet to be done in low temperature physics. We had cooked up a scheme to measure the speed of propagation of gravitational attractions by looking at the phase of a gravitationally driven pendulum and were a little discouraged when the phase shift we had to measure was 10^{-18} radian. However, between 3 and 5 in the morning all sorts of possible ways to measure such things come to mind. Our main interest centered about a much easier experiment: to measure the electric dipole moment of ^{3}He to a precision greater by a factor of 10^{15} than previously done. After I got my degree I started an NSF postdoc with Bill with the idea of doing that experiment.

Our factor of 10^{15} changed to 10^6 as others discovered things we had missed. The experiment was to see if ^{3}He spins precess in an electric field and zero magnetic field. Isaac Bass and Mike Taber developed the optical pumping apparatus to get the polarized ^{3}He while Blas Cabrera developed the necessary technique to make zero magnetic field. In the meantime, however, we began to hear about Joe Weber's experiments at Maryland [1] and the old interest in gravitation reasserted itself. Bill and I visited Weber at Maryland in 1967 and saw what he was doing. It was clear that we could build on his work and use low temperatures to gain greater sensitivity. With no more knowledge than that gained between 3 and 5 a.m. on countless mornings in the old physics corner, a visit to Maryland, and what had rubbed off from people like Leonard Schiff who knew something about gravitation, we proceded to design an experiment which we thought should be the ultimate low temperature gravity wave experiment. Why was Bill interested? Because the physics looked interesting. Mr. Chang, a longtime friend of the physics department, once told me about an old Chinese curse: *"May you live in interesting times."* I inferred that interesting was a synonym for difficult. We had no idea what an interesting experiment this would turn out to be. This paper is intended to be an aid to those who know about as much about gravitational waves as we did when we started. I hope that they will appreciate how physics and engineering come together in an experiment and how designing for the ultimate experiment at the beginning may in the long run lead to the most interesting results.

2. WHAT ARE GRAVITATIONAL WAVES?

Gravitational waves are a result of energy being transmitted through the gravitational field, in the same way as electromagnetic waves are the manifestation of energy being transmitted through the electromagnetic field [2,3,4,5]. We all know the usual picture from a freshman physics book: an electric field points first up, then down, with the associated magnetic field at right angles to the electric field. We can think of that electric field as representing a force being transmitted from a distant source to a charge.

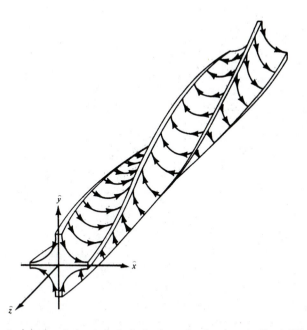

FIGURE 1. A right-hand circularly polarized gravitational wave (from [2]).

The wave impinges on the charge causing it to accelerate. A gravitational wave may be thought of in the same way, except that the wave will not accelerate an isolated point mass. It does not represent a propagating force but instead it is a propagating force gradient or tidal force. Figure 1, from Misner, Thorne and Wheeler, represents a gravitational wave stopped in mid-squeeze as it were. The lines show the direction of the strain induced in a solid body as the wave impinges on it. At the origin a circular ring would be distorted into an ellipse with its major axis in the x direction.

Thorne [6] points out that mathematically one can say that the gravitational wave does accelerate mass in a direction transverse to the wave's k vector, but since the acceleration is proportional to the ratio of gravitational mass to inertial mass the local frame itself is accelerated and hence the acceleration is unobservable locally. All we can observe is the difference in acceleration between one point and another, and thus we interpret the wave's effects as a propagating strain or tidal force.

In any case it is enough for the experimentalist to realize that a gravitational wave will induce strain in any material object and that the appropriate instrumentation to develop is that to measure dynamic strains. To that end Weber developed the bar antenna, where by monitoring the amplitude of the bar's fundamental longitudinal mode of vibration he measured

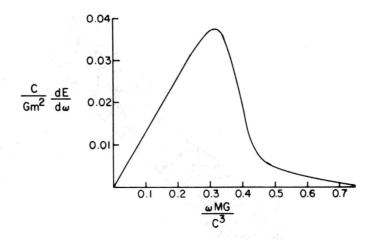

FIGURE 2. The energy spectrum generated by a point mass of mass m falling into a black hole of mass M (from Davis, Ruffini, Press and Price).

the strain induced in the bar by a variety of force gradients including, he hoped, gravitational waves [7,8]. In Weber's original gravity wave antenna the strain is converted to electrical signals by an array of piezo-electric crystals glued to the antenna. The cryogenic bar antennae are second generation Weber bars where, in many cases, the piezo-electric crystals have been replaced by other instrumentation more appropriate to the cryogenic environment. To see the need for cryogenics and exotic instrumentation we must ask ourselves about the size of strain that we might expect a gravitational wave to induce in a material antenna.

This question has been asked by a number of authors. One of the most interesting results is that obtained by Davis, Ruffini, Press and Price [9]. They considered a point mass falling into a black hole and asked how much gravitational radiation would emerge from such an encounter and what the frequency would be? The energy spectrum they obtained is shown in figure 2. The units have been added to show the physical quantities that could be measured.

If we allow the point mass to be one solar mass and the black hole to be 10 solar masses, we find that the frequency at which the spectrum is a maximum will be about 2 kHz. This then is the frequency of the strain which we must measure. If we assume that the energy from the collision is radiated in all directions and also assume that such a cataclysmic event occurs at the center of our galaxy where the density of stars is greatest, we calculate that the energy flux corresponds to a strain of 10^{-17} or so induced in material bodies on earth. Thus we have identified our problem: we must detect strains of 10^{-17} at frequencies of 1 kHz or below [10,11].

3.   DESIGNING THE EXPERIMENT

At the time we began the gravity wave effort most of the details I have given about gravity waves were known only to Weber or to theorists. We knew what strain Weber felt he was able to see in his antenna. Bill Fairbank very quickly wrote down the equipartition theorem for the antenna's fundamental longitudinal mode,

$$\frac{1}{2}m\omega_0^2\langle x^2\rangle = \frac{1}{2}\,kT \quad,$$ (1)

and determined for Weber's room temperature antenna that

$$\langle x^2\rangle^{1/2} \simeq 1.8 \times 10^{-14}\,\mathrm{cm} \quad,$$ (2)

whereupon he invoked Fairbank's rule: *"Any experiment is better if it is done at low temperature."* While we believe that subsequent events have verified the rule in this instance they have also verified Hamilton's corollary: *"Any experiment will be harder if it is done at low temperature."*

Once it is decided to do a cryogenic experiment, however, it is not any harder to do it right. One of the first questions we were asked when we were first thinking about starting the gravity wave research was, *"Why don't you just cool to liquid nitrogen temperature?"* We decided that it involved nearly as much effort to build a dewar to go to 77 K as to go to 4.2 K. As a matter of fact we then asked how much incremental effort was involved to build an experiment to run at the lowest temperature we knew how to obtain, 0.002 K. It involved very little extra effort in the dewar construction so we decided to start to build with that temperature in mind, and a potential improvement in temperature of 150000 a possibility.

The next question we had to confront was the size of the antenna. What is a reasonable size antenna to cool to 0.002 K? Weber's antenna was 3 feet in diameter and 5 feet long. We decided to plan for one twice as long and of the same diameter. There is good reason for selecting this size antenna. A gravity wave of the type described previously, propagating as a pure strain, could not excite the second harmonic of an antenna. Thus since the second harmonic of a 10 foot bar would be close to the fundamental frequency of a 5 foot bar we could investigate whether the events Weber was seeing were truly due to a wave which appeared as a propagating strain. The dewar that resulted from these considerations is shown in figure 3. It was designed by Stener Kleve and Bruce Pipes at Louisiana State University working with Bill Montgomery at NASA's Mississippi Test Facility and was built at MTF under the supervision of Tom Bernat [12]. Many design ideas were adapted from the Stanford superconducting accelerator project. The arrangement for this construction and for the procurement of the aluminum was greatly facilitated by Joe Reynolds of LSU.

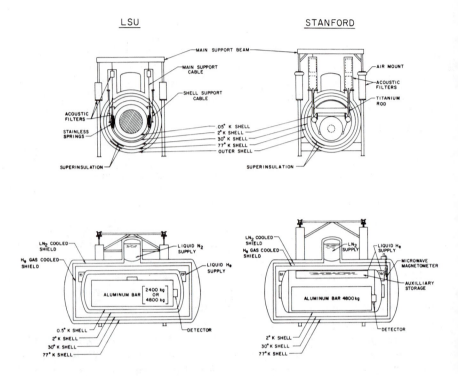

FIGURE 3. Schematic of the dewar. Modifications have been made by both LSU and Stanford. The identical dewars were constructed by Mississippi Test Facility and LSU.

FIGURE 4. Cryogenic gravity wave antenna located at Louisiana State University. The antenna is a 2400 kg solid aluminum cylinder.

4. EXPERIMENTAL DETAILS

Over the past 10 years other groups have begun to work on cryogenic bar detectors. One of the earliest was the Rome group. LSU, Rome and Stanford started a collaboration to build three as nearly identical antennae as possible and to try to look for coincidences between the individual independent groups. Other groups are at Rochester, Maryland, Perth, Tokyo, Beijing and Guangzhou. Room temperature detectors were built at Glasgow, Bell Labs, the Max Planck Institute and Moscow. Laser detectors are under construction at Cal Tech, Glasgow, MIT and the Max Planck Institute.

All of the groups realized that the size of the strain that could theoretically be detected was actually much smaller than the estimate Bill made using the equipartition theorem. Wang and Uhlenbeck worked out, in the 1940's, the statistical mechanics of a harmonic oscillator. The results are physically very reasonable [13].

We all know that the arguments that lead to the equipartition theorem also lead to Boltzmann statistics. The probability of measuring an energy E in a system at temperature T is

$$P(E) \propto e^{-E/kT} = \exp -\frac{M\omega_0^2 x^2}{2kT} \quad . \tag{3}$$

This is a Gaussian probability density which we all have seen treated in numerous math courses and in sophomore modern physics. We recognize the variance or width of the Gaussian as

$$\sigma^2 = \frac{kT}{M\omega_0^2} \quad . \tag{4}$$

However this probability is not what the experimentalist measures. The experimentalist measures the amplitude of excitation of the antenna as a function of time. If he has digital capability he samples the amplitude with a characteristic sampling time τ_s. The probability that the experimentalist measures is a conditional one. What is the probability of measuring an amplitude x at time τ_s, given that the amplitude measured earlier at time zero was x_0? This is what Wang and Uhlenbeck calculated and their result, somewhat simplified, was

$$P(x, \tau_s \mid x_0, 0) \propto \exp\left[-\frac{M\omega_0^2 (x - \bar{x})^2}{2kT(1 - e^{-\tau_s/\tau^*})}\right] \quad , \tag{5}$$

where $\bar{x} = x_0 e^{-(\tau_s/\tau^*)}$ and τ^* is the natural decay time of the damped oscillator, $\tau^* = 2Q/\omega_0$. Note that $\bar{x}$ represents the classical decay of the oscillator from its previously measured amplitude x_0.

This is again a Gaussian probability but now centered not on $x = 0$ but on $x = \bar{x}$. The probability has a time dependent variance

$$\sigma^2 = \frac{kT(1 - e^{-\tau_s/\tau^*})}{M\omega_0^2} \quad . \tag{6}$$

This result makes perfect physical sense. If we measure an amplitude of oscillation x_0 at $t = 0$ we do not expect a totally random result on the next measurement. The energy in the oscillator is dissipated only in the characteristic relaxation time τ^* and hence we expect our successive measurements to be close to the preceding one but reduced by that energy lost in the damping.

The variance represents the result of the fluctuating random forces which drive the Brownian motion. Since a harmonic oscillator is a tuned system, its response to noise is also restricted to a narrow range of frequencies. The time dependence of the variance reflects the narrow-band nature of the noise. Physically it expresses the fact that, while the noise is proportional to temperature, the fluctuating forces cannot appreciably affect the uncertainty in successive measurements of x if the sampling time τ_s is short compared to the relaxation time τ^*. If we assume that this is the case and expand the exponential, keeping only the linear term, we find

$$\sigma^2 = \frac{kT\tau_s}{\tau^* M\omega_0^2} \tag{7}$$

and comparing with the variance for the usual Boltzmann statistics we can identify an effective temperature

$$T_{eff} = \frac{T\tau_s}{\tau^*} = T\frac{\omega_0\tau_s}{2Q} \quad . \tag{8}$$

This effective temperature reflects how large the uncertainty in successive measurements is. This effect of the narrow-band noise explains how some room temperature detectors have operated with noise temperatures (T_{eff}) of 4 K when the ambient temperature was closer to 290 K. It also explains why many investigators have searched for new materials with low losses (high-Q) so that the noise temperature can be reduced.

This analysis also indicates the beginning of the physical ideas associated with "optimal filters." By subtracting the signal received at $t = \tau_s$ from that received at $t = 0$, we expect that the difference should contain only the result of forces or force gradients applied between the two sampling times. Hence most filtering algorithms consist largely of a subtraction of two successive events [14–17].

We note also that the effective temperature can be reduced to zero if τ_s is small enough and τ^* long enough. All experimental physicists realize that this is a nonphysical situation. Something has been left out.

5. TRANSDUCERS AND AMPLIFIERS

What has been ignored is the true physical nature of the device we use to convert the mechanical energy of the antenna to an electrical signal (the transducer) and the amplifier that converts that electrical signal to a useful level.

If we attempt to shorten τ_s in order to reduce the effective temperature due to Brownian motion we must, by necessity, use amplifiers and transducers of a wide bandwidth to reproduce those signals that change in a short time. The noise associated with these devices will then interfere with the signals we wish to measure. To see how that interference occurs we shall look at the general case and investigate two specific examples, the Stanford transducer and the LSU transducer.

While it has not yet been demonstrated, it is generally accepted that it is theoretically possible to construct a noiseless transducer. By that I mean a device that converts the mechanical acceleration and velocity of the antenna into electrical voltage and current without adding additional noise.

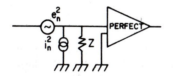

FIGURE 5. The representation of an amplifier required by quantum mechanics.

Weber [18] and Heffner [19] demonstrated that quantum mechanics requires that any amplifier exhibit noise, and established the lower limits on that noise. Engineering practice is to represent that noise by two noise generators as shown in figure 5. The action of these noise generators is crucial to the understanding of the behavior of the gravity wave antenna. We can look separately at the current and voltage generators and their effect on the transducer and antenna. The following amplifier is assumed to be perfect, providing noiseless gain.

For most of the transducers that have been used, the effect of the current generator is to drive a current through the electrical impedance that the antenna offers to the transducer. Since the mechanical analogue of the current in these transducers is the velocity of the end of the antenna, or the time rate of change of the strain, the effect of the amplifier noise current generator is to increase the velocity of the random excitation of the antenna. Since the antenna offers a low impedance at resonance, the effect is concentrated at the antenna resonant frequency. In fact, this current acts in exactly the same manner as the fluctuating forces which drive the Brownian motion and so its net effect is to cause an increase in the temperature of the mechanical mode to which it is coupled. The apparent effect of the amplifier current generator is to add to the antenna narrow-band noise.

The voltage generator on the other hand adds to the voltage presented to the amplifier input terminals. While the sharply tuned impedance of the antenna will have some effect on the results, the voltage noise will tend to be broad-band. As a result, while it will contribute to the apparent temperature of the mechanical mode in question, it will have its largest effect when the Brownian motion signal is filtered to extract the magnitude of the change in the antenna state which occurred because of forces which acted between successive samples. The broad-band voltage noise has its largest effect on the noise temperature of the antenna.

The general rule used in optimizing the transducer and amplifier is to design so that the contribution to the system noise from the amplifier noise current generator and noise voltage generator are equal. This match, known as noise matching, generally means that the amplifier is not impedance matched to its source. Instead the transducer is designed so that it presents a resistive load to the amplifier,

$$R_n = \frac{\langle e_n \rangle}{\langle i_n \rangle} \tag{9}$$

where $\langle e_n \rangle$ and $\langle i_n \rangle$ are the rms values of the voltage and current noise generators. These problems are discussed in detail by Motchenbacher and Fitchen [20] and to some extent by Giffard [21].

It is crucial to understanding the rationale of the experiments to realize the effect of the amplifier on the measurement. We can reduce the Brownian motion of the antenna to a nearly negligible level by cooling the antenna and selecting an antenna with a high enough Q. The fact that quantum mechanics dictates that any amplifier must have noise, and the experimental fact that any noisy amplifier may be represented by a voltage and current noise generator and hence can actually drive the antenna as well as add noise to the signal, makes the design of these experiments somewhat an art. The search for better transducers is controlled by these considerations.

We discuss two transducers to give a general feeling as to what is involved in converting small mechanical strains or displacements into electrical signals. The first transducer is that developed by Ho Jung Paik in his thesis and used in the Stanford experiment. It is a mechanically tuned accelerometer and hence has a much better impedance match to the displacement of the bar [22].

Figure 6a shows the Stanford accelerometer. A trapped persistent current applies a magnetic field to both sides of a niobium diaphragm. The

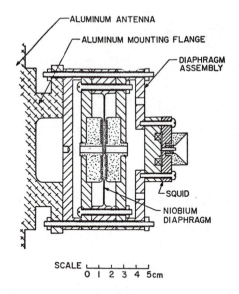

FIGURE 6a. The Stanford superconducting resonant accelerometer.

lowest membrane mode of the diaphragm is made to coincide with the antenna resonant frequency. The accelerometer mounts on the end of the bar and thus the diaphragm is excited by the bar motion. Movement of the diaphragm pushes flux from either of the two coils that conduct the persistent current and the resulting voltage drives a current through the third inductance. The resultant magnetic field is detected by a SQUID magnetometer and amplified by the associated SQUID electronics.

Figure 6b shows the LSU accelerometer [23]. This utilizes two superconducting cavities, sharply re-entrant to make their resonant frequency very sensitive to relative spacing of the center post and end plate. The cavities are separated by a thin niobium diaphragm to which the cavity center posts are attached. The accelerometer mounts on the end of the antenna. In contrast to Paik's accelerometer the mechanical resonant frequency of the diaphragm is chosen to lie below the antenna frequency so that as the end of the bar moves, the center posts tend to hold a fixed position in inertial space, and the resonant frequencies of the cavities are modulated, one increasing as its gap opens while the other decreases. The cavities are excited at their common electrical frequency (600 MHz) from the same oscillator and the phase shift of the signals transmitted through the cavities is measured by passing them through a double balanced mixer acting as a phase detector. The resultant signal is then amplified either by a conventional UHF amplifier or by a yet-to-be-tested Josephson junction parametric amplifier [24].

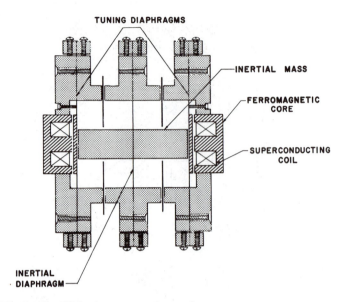

FIGURE 6b. The LSU microwave cavity accelerometer.

A number of other transducers, including piezo-electric crystals, are described by Amaldi and Pizzella [10].

I have described transducers and amplifiers at some length because the operation of these determines the noise level of the experiment and the algorithm that will be used to extract the data. We can make a naive estimate of the energy sensitivity of a gravity wave detector by saying that the energy deposited in the bar must be greater than the energy change due to Brownian motion plus the apparent energy due to amplifier noise:

$$E_s \; > \; kT(1 - e^{-\tau_s/\tau^*}) + kT_n \left(\beta \omega \tau_s + \tfrac{1}{\beta \omega \tau_s} \right)$$

$$\simeq \; \tfrac{kT\tau_s}{\tau^*} + kT_n \left(\beta \omega \tau_s + \tfrac{1}{\beta \omega \tau_s} \right) \quad .$$

(10)

The first term is the Brownian motion uncertainty that was described earlier. The second term is the amplifier noise, referenced to the input of the transducer. The transducer coupling constant is β and the two parts of the amplifier noise contribution are due to the current and voltage noise, respectively. T_n is the noise temperature of the amplifier. The narrow-band character of the current noise is apparent in that it is of the same form as the Brownian motion term. The coupling constant enters directly into the current noise contribution while the inverse of the coupling constant multiplies the voltage noise. This reflects the fact that the current noise directly puts energy into the antenna, and hence is directly proportional to β, while the voltage noise is additive and must be referred to the antenna, hence its effect is inversely proportional to β. In the limit that τ^* is long and T small so that the Brownian motion becomes negligible, the sensitivity of the antenna is completely controlled by T_n and the minimum detectable energy becomes

$$E_s > 2kT_n \quad \text{when} \quad \tau_s = \frac{1}{\beta\omega} \quad . \tag{11}$$

Weber first showed that quantum mechanics limits T_n. The amplifier noise temperature has a lower limit

$$T_n > \frac{\hbar\omega}{k\ln 2} \quad , \tag{12}$$

and thus we have a fundamental limit to the size of the signal we may meaningfully extract from our amplifier.

Braginsky, Thorne, Caves and others [25–28] have carried these ideas further and have asked what, if any, are the ultimate limits to measurement of small strains. Can one do better? To establish the noise temperature Weber and Heffner have used the usual uncertainty relation of the form

$$\Delta x \Delta p > 1/2 \, |\langle [x,p] \rangle| \tag{13}$$

and have assumed the minimum uncertainty wave packet has equal uncertainty in the two noncommuting variables. Braginsky, Thorne and Caves have asked if one can structure a measurement process so that it is physically impossible to measure one of the noncommuting variables. If so, then since the uncertainty in the variable not measured is very large, one should be able to measure the other to arbitrary precision.

For the case of gravity wave antennae one must investigate the quantum properties of a harmonic oscillator. Most conventional measurements of the properties of a harmonic oscillator consist of what are called amplitude and phase measurements; they are structured so that the displacement of the oscillator from its equilibrium position and its phase with respect to a perfect external oscillator can be determined simultaneously. As a result, amplitude and phase measurements obey the usual uncertainty relationships. Caves showed directly that the usual output of a lock-in detector is an amplitude and phase measurement, and by introducing the complex amplitude showed that the outputs of a two-phase lock-in detector are conjugate variables, *i.e.*,

$$\Delta x_1 \Delta x_2 > \hbar/2 \quad , \tag{14}$$

where x_1 and x_2 are the in-phase and quadrature voltages, respectively.

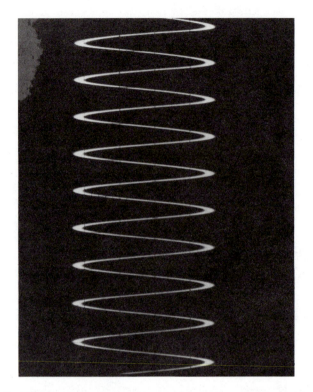

FIGURE 7a. The signal output from the monitor cavity when the diaphragm moves with an amplitude of 10^{-11} meters. There is no field in the measuring cavity.

The important question to ask, however, is how to structure a measurement so that physically the experimenter can only extract one of the components of the complex amplitude, either x_1 or x_2. Thorne and Caves, and independently Oelfke [29], showed that if an oscillator's position is measured by an electromagnetic wave of the form

$$E(t) = A \cos \omega_m t \cos \omega_{rf} t \quad , \tag{15}$$

where ω_m is the frequency of the mechanical oscillator, the electromagnetic wave can provide information only about the component of the displacement which is in phase with the external oscillator at frequency ω_m. This form of modulation has been called "two stick" modulation by Braginsky because the spectrum consists of two distinct lines at $\omega_{rf} \pm \omega_m$.

Two stick modulation is the form of electromagnetic radiation which is needed to avoid perturbing an oscillator and hence introducing uncertainty into the measurement of the desired component of the complex amplitude. It is also the form of modulation that is required to prevent the current or

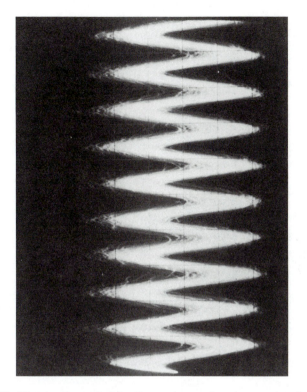

FIGURE 7b. The signal output from the monitor cavity when a noisy electric field is in the measuring cavity during an amplitude and phase measurement.

voltage noise of the amplifier from exciting the bar in the undesired component. Oelfke, Spetz and Mann [30] have demonstrated the behavior of a harmonic oscillator when its amplitude is measured by an electromagnetic wave and a noisy amplifier. The pictures of the data are shown in figures 7a, 7b and 7c, and the apparatus used to generate the data is shown in figures 8a and 8b. The experiment will be explained in the next section.

6. SOMETHING NEW

The apparatus used by Oelfke, Spetz and Mann is a modification of the LSU accelerometer described earlier. The central niobium diaphragm is the mechanical harmonic oscillator. The position of the diaphragm can be monitored with conventional amplitude and phase measurement by exciting one cavity with very low level rf power (-50 dbm) and detecting the transmitted rf power with a double balanced mixer so as to produce a voltage proportional to the diaphragm displacement. At this level of excitation

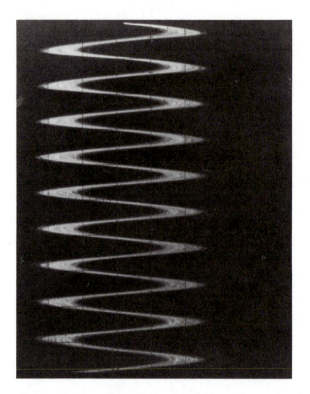

FIGURE 7c. The signal output from the monitor cavity when a noisy electric field is present in the measuring cavity during a back-action-evading measurement.

the force applied to the diaphragm by the electromagnetic field is much less than the fluctuating forces which maintain thermal equilibrium, and as a result it is negligible. The second cavity is driven at rf powers of -15 to -30 dbm in order to produce easily observable forces on the diaphragm. The two cavities are completely independent electromagnetically, one being resonant at 595 MHz and the other at 605 MHz.

If one makes a noisy amplitude and phase measurement of the diaphragm the resultant diaphragm displacement is shown in figure 7b. Figure 7a shows the displacement without the electromagnetic field in the second cavity. The diaphragm is being driven to an amplitude of 10^{-11} m by a mechanical driver. The effect of the electromagnetic field on the displacement signal can be seen. The setup for this experiment is shown in figure 8a.

If two stick modulation is used as shown in figure 8b then the effect on the diaphragm is shown in figure 7c. Note that all the noise is put into one component, that one in phase with the diaphragm. Contrast figure 7c with figure 7b. There is no phase noise when two stick modulation is used.

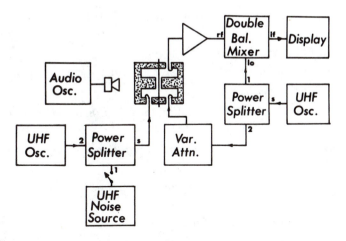

FIGURE 8a. Apparatus set up to obtain the amplitude and phase measurement of figure 7b.

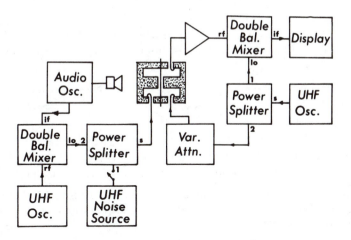

FIGURE 8b. Apparatus set up to obtain a back-action-evading measurement.

I want to emphasize that this experiment is not a quantum-non-demolition experiment. Oelfke, Spetz and Mann observed a back-action-evading state but at an amplitude much above that where quantum mechanics becomes the limiting factor. This is, however, a demonstration that a quantum-nondemolition measurement corresponds to a back-action-evading measurement. And it demonstrates that, as we attempt to limit the effects of amplifiers, we are also making the correct moves to surpass the standard quantum limit.

References

[1]   J. Weber, *Phys. Rev.* **117**, 305 (1960).

[2]   C. W. Misner, K. S. Thorne and J. A. Wheeler, *Gravitation* (W. H. Freeman, San Francisco, 1973).

[3]   R. Adler, M. Bazin and M. Schiffer, *Introduction to General Relativity*, (McGraw Hill, New York, 1975), 2nd Ed.

[4]   J. Weber, *General Relativity and Gravitational Waves* (Wiley (Interscience), New York, 1961).

[5]   H. C. Ohanian, *Gravitation and Spacetime* (Norton, New York, 1976).

[6]   K. S. Thorne, *Rev. Mod. Phys.* **52**, 285 (1980).

[7]   J. Weber, *Phys. Rev. Lett.* **18**, 498 (1967).

[8]   J. Weber, *Phys. Rev. Lett.* **20**, 1307 (1968).

[9]   M. Davis, R. Ruffini, W. H. Press and R. H. Price, *Phys. Rev. Lett.* **27**, 1466 (1971).

[10]   E. Amaldi and G. Pizzella, in *Relativity, Quanta and Cosmology*, edited by Pantaleo and deFinis (Johnson Reprint Corp., New York, 1979).

[11]   D. H. Douglass and V. B. Braginsky, in *General Relativity*, edited by S. W. Hawking and W. Israel (Cambridge University Press, Cambridge, 1979).

[12]   W. O. Hamilton, P. B. Pipes, S. Kleve, T. P. Bernat, D. G. Blair, D. H. Darling, D. DeWitt, M. S. McAshan, R. Taber, S. Boughn, W. Fairbank, W. P. Montgomery and W. C. Oelfke, *Cryogenics* **107** (1982).

[13]   M. C. Wang and G. E. Uhlenbeck, *Rev. Mod. Phys.* **17**, 323 (1945); also in *Selected Papers on Noise and Stochastic Processes*, edited by N. Wax (Dover, New York, 1954).

[14]   P. Bonifazi, P. Ferrari, S. Frasca, C. Pallottino and G. Pizzella, *Il Nuovo Cimento* **1C**, 465 (1978).

[15]   J. B. Thomas, *Introduction to Statistical Communication Theory* (Wiley, New York, 1969).

[16]   A. D. Whalen, *Detection of Signals in Noise* (Academic Press, New York, 1971).

[17]   P. F. Michelson and R. C. Taber, *J. Appl. Phys.* **52**, 4313 (1981).

[18]   J. Weber, *Rev. Mod. Phys.* **31**, 681 (1959).

[19]   H. Heffner, *Proc. IRE* **50**, 1604 (1962).

[20] C. D. Motchenbacher and F. C. Fitchen, *Low Noise Electronic Design* (Wiley (Interscience), New York, 1973).

[21] R. P. Giffard, *Phys. Rev.* **14D**, 2478 (1976).

[22] H. J. Paik, *J. Appl. Phys.* **47**, 1168 (1976).

[23] W. C. Oelfke and W. O. Hamilton, *Acta Astronautica* **5**, 87 (1978).

[24] J. Kadlec and W. O. Hamilton, in *SQUID '80*, edited by Hahlbohm and Lubbig (de Gruyter, Berlin, 1980).

[25] V. B. Braginsky, Y. I. Vorontsov and K. S. Thorne, *Science* **209**, 547 (1980).

[26] C. M. Caves, K. S. Thorne, R. W. P. Drever, V. D. Sandberg and M. Zimmermann, *Rev. Mod. Phys.* **52**, 341 (1980).

[27] K. S. Thorne, C. M. Caves, V. D. Sandberg and M. Zimmermann, in *Sources of Gravitational Radiation*, edited by L. Smarr (Cambridge University Press, Cambridge, 1979).

[28] J. N. Hollenhorst, *Phys. Rev.* **15D**, 1669 (1979).

[29] W. C. Oelfke, in *Quantum Optics, Experimental Gravitation and Measurement Theory*, edited by P. Meystre and M. O. Scully (Plenum Press, New York, 1983).

[30] G. W. Spetz, A. G. Mann, W. O. Hamilton and W. C. Oelfke, *Phys. Lett.* **104A**, 335 (1984).

Large Mass Detector of Gravitational Waves

Edoardo Amaldi

The name of William M. Fairbank was for a long time well known to me from his work "near zero," but I had the pleasure of meeting him personally only at the end of the 1968-69 winter when he was in Rome for a few weeks, visiting my colleague and friend Giorgio Careri, who at the time was working on the quantum vorticity induced by ions in liquid helium.

Talking with Careri, Fairbank mentioned his interest in the detection of gravitational waves by using resonant detectors, as Joseph Weber had done in his pioneering work, but with bars at very low temperatures where not only their Brownian motion is much lower but also a few low temperature techniques can be advantageously employed.

Giorgio Careri knew of my years-long interest in starting an experimental activity on gravitational radiation at the University of Rome. Therefore he arranged a very agreeable encounter between Fairbank and me, which was the first of a number of conversations. These brought us to the agreement of starting, as soon as possible, a collaboration, the main scope of which was the detection of gravitational radiation of extraterrestrial origin in co-incidence between three stations located at Stanford University, Louisiana State University and the University of Rome.

On that occasion I was impressed by the determination of Professor Fairbank when he stated that the final goal of our collaboration had to be the construction at each of the three stations of at least a resonant antenna of about 5000 kg operated at a temperature of about 3 mK.

Today a few laboratories, and in particular the Fairbank laboratory, are not too far from reaching these conditions. In light of our present understanding of the various facets of the problem, this appears to be about the best that can be done within the limits of today's technology for succeeding in observing bursts of gravitational waves due to supernovae as far as those in the Virgo cluster.

In order to summarize the arguments involved in the choice of the antenna parameters, let us first recall that the minimum detectable spectral energy density at the resonant frequency ν_R can be cast in the form [1]

$$f(\nu_R) = \frac{kT_{\text{eff}}}{\Sigma}\psi \ \ (\text{J/m}^2 \ \text{Hz}) \tag{1}$$

where ψ is a statistical factor that I will leave out of our discussion, k the Boltzmann constant,

$$\Sigma = \frac{8}{\pi}\frac{G}{c}\left(\frac{v_s}{c}\right)^2 M \ \ (\text{m}^2 \ \text{Hz}) \tag{2}$$

the cross section of the bar [2], and T_{eff} the *effective temperature* which depends on many parameters of the antenna, the associated electronics and the filter used in the analysis of the data.

The product kT_{eff} represents the total noise of the system. From (1) and (2) we see that a high sensitivity (*i.e.*, a low value of $f(\nu_R)$) can be obtained by choosing the mass M of the bar as large as possible.

Giffard showed in 1976 [3] that the important quantity T_{eff}, under appropriate conditions and with a proper data analysis, can take its minimum value

$$(T_{\text{eff}})_{min} = 2T_n \quad , \tag{3}$$

where T_n is the noise temperature of the amplifier connected to the transducer that detects the vibrations of the bar:

$$T_n = \frac{\sqrt{V_n^2 I_n^2}}{k} \quad , \tag{4}$$

where V_n^2 and I_n^2 are the mean square values of the voltage and current (bilateral) spectral density (V^2/Hz, A^2/Hz).

In the frame of a straightforward application of standard quantum considerations [4], the ultimate limit to the noise temperature T_n of an amplifier is given by

$$kT_n = \frac{\hbar\omega_R}{\ln 2} \tag{5a}$$

which, for $\omega_R = 2\pi \times 900$, gives

$$T_n = 0.62 \times 10^{-7}\text{K} \quad . \tag{5b}$$

Today we believe that, by employing quantum nondemolition methods [5], the lower limit (5) can be surpassed. I will not, however, take into account this recent approach, which probably will require some time before becoming of practical use. Thus, I will proceed in the discussion of the minimum detectable spectral energy density under the assumption of the strict validity of conditions (3) and (5), *i.e.*, I will remain in the framework of the standard phase-amplitude approach, which most probably represents what will still actually be done a few years from now.

A detailed analysis of the problem, carried out by Pizzella [6] and generalized by Pizzella and Pallottino [7], leads to the following expression for T_{eff}:

$$T_{eff} = NT_n \sqrt{1 + \frac{1}{\lambda_0^2}} \sqrt{1 + \frac{2T\lambda_0}{\beta QT_n}} \quad , \tag{6}$$

where

$$\lambda_0 = \frac{V_n}{I_n} \frac{1}{|Z_{22}|} \tag{7}$$

is what we call the *electric matching parameter*, and Z_{22} the electrical impedance of the impedance matrix (Z-matrix) of the two-port model of the electromechanical transducer [3]:

$$f(t) = Z_{11}\dot{x}(t) + Z_{12}i(t) \quad , \tag{8}$$

$$v(t) = Z_{21}\dot{x}(t) + Z_{22}i(t) \quad . \tag{9}$$

In the expression (6) also appears the usual parameter

$$\beta = \frac{\text{electric energy in the transducer}}{\text{total energy in the antenna}} \tag{10}$$

and a numerical factor N, the value of which depends on the algorithm adopted for the data analysis. For example, $N = 2.42$ for the difference algorithm; $N = 2$ for the optimal Wiener-Kolmogorov filter [7,8]. In the latter case, the best sensitivity, corresponding to the Giffard condition (3),

$$f(\nu_R) = \frac{2kT_n}{\Sigma} \quad , \tag{11}$$

is obtained by imposing the following "matching conditions":

$$\frac{\beta QT_n}{2T} \gg \lambda_0 \quad , \tag{12a}$$

$$\lambda_0 \gg 1 \quad . \tag{12b}$$

These relations, however, cannot be easily fulfilled. If the matching conditions (12) are replaced with the much less demanding conditions

$$\frac{\beta Q T_n}{2T} = \lambda_0 = 1 \quad , \tag{13}$$

we lose a factor 2, which at present does not appear too bad.

The difficulty in fulfilling condition (12a) is clearly connected with the difficulty of obtaining large values of Q and β and low values of T. Notice that nonresonant piezoelectric as well as capacitive transducers, in practice, always have values of β of the order of 10^{-5}, and values about 10 times greater in the more favorable case of inductive transducers.

If one adopts a resonant transducer (two-mode system) [9] in which an oscillator of frequency very close to that of the bar and mass $m \ll M$ is coupled to the bar, a value of β not far from 1 can be obtained. In this case, however, the sampling time Δt should be larger than the beating period

$$\Delta t > T_b = \frac{2\pi}{\omega_b} \quad . \tag{14}$$

This relation, combined with the optimal value of the sampling time

$$\Delta t_{opt} = \frac{2\lambda_0}{\beta \omega_R} \tag{15}$$

gives a restriction on the β value:

$$\beta < \sqrt{\mu \frac{1 + \lambda_0^2}{4}} \sim 3.3 \times 10^{-2} \quad , \tag{16}$$

with $\mu = m/M \sim 2 \times 10^{-3}$.

The use of a three-mode system as proposed by Richard [10] has the advantage of reducing T_b and therefore of releasing condition (16). For $\mu \sim 10^{-2}$ one can envisage using values of β of the order of 0.1.

In order to detect, under condition (13) and (5b), a spectral density of

$$6 \times 10^{-6} \text{ J/m}^2 \text{ Hz} \quad , \tag{17}$$

corresponding to a collapse taking place in the Virgo cluster with 5% of $6 \ M_\odot$ converted into gravitational waves [2], we need

$$\Sigma \geq 5.70 \times 10^{-25} \text{ m}^2 \text{ Hz} \quad ,$$

i.e.,

$$M \geq 3600 \text{ kg} \quad .$$

For example, with the value (16) of β, the following "solutions" can be adopted:

$$T = 0.1 \text{ K} \qquad Q = 10^8 \qquad \lambda_0 \sim 1 \quad , \qquad (18a)$$

or

$$T = 3 \text{ mK} \qquad Q = 3 \times 10^6 \qquad \lambda_0 = 1 \quad . \qquad (18b)$$

The conditions (18b) are just those originally proposed by Fairbank, showing once more his extraordinary intuition.

The conditions that some of the experimentalists set as a practical goal to be reached in the next few years are, however, closer to (18a).

If the value (17) used here is too optimistic, as suggested by some authors, then the use of quantum nondemolition methods will become a necessity, and the detection of supernovae in the Virgo cluster will become possible only by means of resonant detectors of a successive generation.

References

[1]   E. Amaldi and G. Pizzella, in *Relativity, Quanta and Cosmology*, edited by F. De Finis (Johnson Reprint Corporation, New York, 1979), Vol. 1, p. 9.

[2]   M. Rees, R. Ruffini and J. A. Wheeler, *Black Holes, Gravitational Waves and Cosmology* (Gordon and Breach, New York, 1974).

[3]   R. P. Giffard, *Phys. Rev. D* **14**, 2478 (1976).

[4]   J. Weber, *Phys. Rev.* **94**, 215 (1954); H. Heffner, *Proc. IRE* **50**, 1604 (1962).

[5]   K. S. Thorne, R. W. P. Drever, C. M. Caves, M. Zimmermann and V. D. Sandberg, *Phys. Rev. Lett.* **40**, 667 (1978). For a recent review of the whole subject see C. M. Caves, in *Quantum Optics, Experimental Gravitation, and Measurements Theory*, edited by P. Meystre and M. O. Scully (Plenum Press, in press).

[6]   G. Pizzella, *Nuovo Cimento* **2C**, 209 (1979).

[7]   G. Pizzella and G. V. Pallottino, *Nuovo Cimento* **4C**, 237 (1981).

[8]   P. Bonifazi, V. Ferrari, S. Frasca, G. V. Pallottino and G. Pizzella, *Nuovo Cimento* **1C**, 465 (1978).

[9]   H. J. Paik, *J. Appl. Phys.* **47**, 1168 (1976).

[10]   J. P. Richard and C. Cosmelli, in *Proceedings of 9th International Conference on General Relativity and Gravitation*, Jena, July, 1980.

Precision Gravity Experiments Using Superconducting Accelerometers

Ho Jung Paik

1. INTRODUCTION

Precision gravity experiments require availability of "near zero" motion detectors. The reason for this is clear. Gravity is the weakest of all known forces in nature, and, invariably, post-Newtonian or relativistic corrections are looked for in modern precision experiments on gravity. For example, in order to detect gravitational waves of extraterrestrial origin, which is a general relativistic phenomenon, it is expected [1] that the motion of a massive mechanical antenna needs to be determined to an accuracy of 10^{-16} to 10^{-20} cm. Fortunately, the properties of superconductors and recent advances in SQUID (Superconducting Quantum Interference Device) technology permit construction of superconducting motion detectors which could satisfy these requirements.

In this paper, we describe the principle and development history of a superconducting accelerometer (section 2) and superconducting gravity gradiometers (section 3), both of which were initiated at Stanford University under the direction of William Fairbank. We then discuss two precision gravity experiments in which such superconducting motion detectors play essential parts: a null test of the inverse square law of gravitation (section 4), and gravity mapping of the earth and other planets (section 5). We report recent progress and current status of these experiments. The application of the superconducting accelerometer in a gravitational wave detector is described elsewhere in this chapter.

2. A SUPERCONDUCTING ACCELEROMETER

Development of a sensitive superconducting accelerometer was suggested to me by Bill Fairbank in 1970 as a Ph.D. thesis project. The need and urgency of such development existed because of the ongoing project of setting up a coincidence network of 3 mK, 5 ton gravitational wave detectors at Stanford, Baton Rouge and Rome "in a few years." The original idea of Fairbank was to levitate a superconducting inertial mass on a superconducting magnet, as has been carried out by Worden [2] for an improved Eötvös experiment, and to use the principle of Opfer's vibrating plane magnetometer [3] for motion detection. It became clear, however, that the emerging technology of rf SQUID's [4] was more promising. Therefore, a design was born in which modulation of persistent current, as suggested by Fairbank, was combined with a readout by an rf SQUID.

Soon the magnetic levitation of the proof mass gave way to a semi-rigid mechanical suspension. Two factors contributed to this choice. The first was the discovery by the author [5] of the principle of a "resonant transducer," which eliminated the need for an extremely soft suspension of the proof mass. The second was a discovery by Ceperley [6] of a very high mechanical quality factor ($Q > 10^5$) of niobium (Nb) at liquid helium temperatures in a niobium helical cavity. This pointed to the possibility of constructing an extremely high-Q coupled oscillator system between the aluminum gravity wave antenna and a niobium proof mass. Incidental benefits coming from the new design were ruggedness and simplicity of construction.

Figure 1 is a schematic diagram of a superconducting accelerometer. The device is composed of a circular Nb diaphragm (proof mass), superconducting coils with a persistent current (electromechanical transducer), and an rf SQUID (amplifier). A persistent current I_0 is stored in the superconducting loop formed by the two sensing coils L_1 and L_2. The superconducting input coil, L_3, of the SQUID is connected in parallel with L_1 and L_2. Since the magnetic field produced by I_0 cannot penetrate the Nb diaphragm due to the "Meissner effect," the moving diaphragm acts as a piston pumping magnetic flux in and out of L_1 and L_2. The "flux quantization" in a superconducting loop containing L_3 then causes some magnetic field to appear in L_3, which is detected by the SQUID. This circuit has been analyzed fully and optimized for a maximum signal-to-noise ratio for various applications [7,8]. Figure 2 is a photograph of the components of a superconducting accelerometer constructed at Stanford.

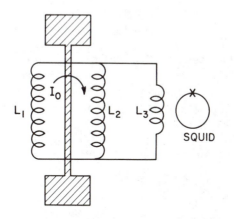

FIGURE 1. A schematic diagram of a superconducting accelerometer.

FIGURE 2. A photograph of the components of a superconducting accelerometer. The niobium diaphragm is shown at the center. Coil forms carrying superconducting "pancake" coils and Nb covers for the assembly are shown on the two sides of the diaphragm. A mirror surface of the diaphragm with a large crystal structure was obtained by a combination of heat treatment and deep electropolishing. As a result of these treatments, mechanical Q's in excess of 10^7 have been measured in these Nb diaphragms [7–9]. The high-Q nature of the device combined with the remarkable noise performance of a SQUID magnetometer makes the superconducting accelerometer an extremely sensitive instrument.

For a nonresonant application of the device, the acceleration sensitivity improves as the suspension frequency f_0 is lowered [8,10]. Therefore, a magnetically levitated proof mass, as originally suggested by Fairbank, would be ideal. When a commercial rf SQUID [11] is employed for sensing magnetic field, a superconducting accelerometer with a proof mass of $m = 400$ g and $f_0 = 30$ Hz has an instrument noise level [8] of 10^{-12} $g_{\rm E}$ Hz$^{-1/2}$, where $g_{\rm E} = 9.8$ m sec^{-2}. If the resonance frequency of the suspension could be lowered to 0.1 Hz, the sensitivity should improve to a level of 10^{-16} $g_{\rm E}$ Hz$^{-1/2}$ between the $1/f$ noise corner frequency of the SQUID, 0.01 Hz, and the resonance frequency, 0.1 Hz. Further improvements in sensitivity may be possible using a quieter dc SQUID.

3. SUPERCONDUCTING GRAVITY GRADIOMETERS

As a further technical spinoff from the gravitational wave experiment, two types of superconducting gravity gradiometers have been conceived and developed at Stanford as a joint effort between the physics and aeronautics/astronautics departments starting in 1975. The scientific basis for such development was obvious. Stability of persistent currents and material properties at cryogenic temperatures as well as the enormous sensitivity available in a superconducting device suggested a possibility of constructing a reliable, compact, rugged, sensitive gravity gradiometer by combining a pair of superconducting accelerometers in a differencing mode. The motivation for such work, however, was not clear from the physics point of view initially, although Dan DeBra, of the aeronautics/astronautics department, realized tremendous potential applications of such a device in inertial and gravity survey and navigation. When Long [12] announced in early 1976 a discovery of an inverse square law failure in gravitation, Fairbank quickly realized the usefulness of a sensitive gravity gradiometer in an improved test of Newton's law. Null tests of the gravitational inverse square law, which has since been the driving force behind the development of superconducting gravity gradiometers at Stanford and at the University of Maryland, will be discussed in section 4.

Two versions of superconducting gravity gradiometers are shown schematically in figure 3. The first (a) is a "displacement-differencing" gravity gradiometer. A single sensing coil L_1 sandwiched between the two superconducting proof masses, M_1 and M_2, forms a superconducting loop with a SQUID input coil L_2. When a persistent current I_0 is stored in this loop, a *displacement* difference caused by a gravity gradient $g_2 - g_1$ modulates L_1 and causes a time-varying current to flow through L_2. In the second version (b), two superconducting accelerometers discussed in section 2 share a common SQUID. When two persistent currents I_1^0 and I_2^0 are stored in the two superconducting sensing loops with opposite polarity,

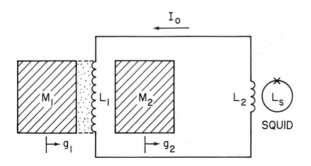

FIGURE 3a. Schematic diagram of superconducting displacement-differencing gravity gradiometer.

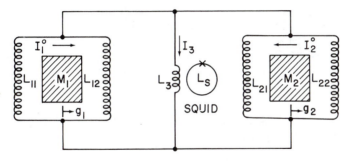

FIGURE 3b. Schematic diagram of superconducting current-differencing gravity gradiometer.

and the ratio I_2^0/I_1^0 is properly adjusted, signal *currents* proportional to g_1 and g_2 are differenced by the SQUID input coil. Thus it is called a "current-differencing" gravity gradiometer. In either case, the gravity signal is differenced in the superconducting circuit *before detection*. This is an important feature for a sensitive differential instrument working in a large common-mode background.

Prototype models of both types of superconducting gravity gradiometers were developed and tested in the years 1976–1978 [12]. Detailed analyses and test results of the current-differencing [13] and the displacement-differencing [14] gravity gradiometers are available.

Based on this early work, more advanced systems have been developed since 1978. A more sophisticated model of the displacement-differencing device has been constructed and tested at Stanford [14], whereas a new program at Maryland starting in 1979 concentrated on an improved version of the current-differencing gradiometer [15]. The Maryland program involves development of a three-axis in-line component gravity gradiometer which is a repetition of a single-axis gravity gradiometer in three orthogonal directions. The gravity gradient is a second derivative of a scalar potential with respect to spatial coordinates

$$\Gamma_{ij} = \partial^2 \phi / \partial x_i \partial x_j \quad , \tag{1}$$

which is a symmetric tensor by construction. A full tensor instrument can be constructed by extending the technique of a single-axis superconducting gravity gradiometer [16].

The Maryland instrument employs proof masses of $m = 400$ g and $f_0 = 25$ Hz and a baseline of $\ell = 15$ cm. The expected instrument noise level is approximately 0.1 E $Hz^{-1/2}$, where 1 E (Eötvös unit) = 10^{-9} sec^{-2}. A single-axis portion of this instrument has been completed and is undergoing tests. In a preliminary test, we observed calibrated noise levels of 2 E $Hz^{-1/2}$ below 1 Hz and 0.2 E $Hz^{-1/2}$ between 15 and 20 Hz [17]. The excess noise at the low frequency end comes from an insufficient balance of the ground vibration. We expect that the instrument-limited noise level will be realized at all frequencies when the system is cooled in a mu-metal shield. The potential sensitivity of the present Maryland gradiometer is better than 10^{-2} E $Hz^{-1/2}$ when it is combined with an available low-noise dc SQUID.

4. NULL TEST OF THE INVERSE SQUARE LAW OF GRAVITATION

In a precision test of a fundamental law, a null experment is always desirable and aesthetic since the quantity looked for is usually a near zero or exactly zero quantity. In the outset of this program in 1976, we considered various possibilities of implementing a gravitational analog of the Maxwell-Cavendish null experiment for Coulomb's law. It was immediately understood that, although the basic laws and the objectives of the experiments are the same, there is one marked difference between electrostatics and gravitation. It is the immobility of gravitational "charges" which, unlike electric charges, cannot distribute themselves to guarantee a null field inside an arbitrary closed shell. Therefore, either a perfect spherical shell or a perfect infinitely long cylindrical shell would be needed for a true null experiment. Because of technical difficulties involved in constructing sources of these geometries, a truncated cylindrical shell was chosen as a practical compromise. However, the need to correct for Newtonian contributions from missing masses was quite dissatisfying and seemed to mar the beauty of a null experiment.

During a conference trip in March 1978, relief came. I suddenly realized that one can obtain a *source-independent* quantity by adding gravity gradients in three orthogonal directions. Moreover, this is a *null* quantity in vacuum. The sum of orthogonal gradients is the trace of the gravity gradient tensor given by equation (1), which is the Laplacian of the gravitational potential. Therefore, the new experimental principle is embedded in the well-known Poisson's equation

$$\nabla^2 \phi = \sum_i \Gamma_{ii} = 4\pi G\rho \quad , \tag{2}$$

which is a differential equivalent of the inverse square force law. At last, a truly source-geometry-independent, null experiment was obtained [18]. I was so excited that I could not sleep at all that night. Early in the morning, I had to "break into" Bill Fairbank's hotel room, which happened to be next door to mine, to discuss the new idea. It was in a way returning in kind for many "break-ins" that Bill had made to my lab in early morning hours to discuss his new ideas.

A cylindrical shell experiment has been developed by Mapoles [14] at Stanford. At Maryland, we started preparation for a Laplacian experiment for which a three-axis gravity gradiometer had to be developed. For a dynamic source of gravity, a rotating dumbbell [19] and a swinging pendulum [20] have been analyzed. For a laboratory experiment, a periodic signal from a dynamic source is measured with a stationary Laplacian detector and is averaged over many cycles to improve the signal-to-noise ratio. The source-independent nature of the new scheme permits a precision test of the law on a geological scale (between 100 m and ~ 10 km), where experimental data are most scarce. An ocean tide could be used to generate a time-varying gravity signal, or a stationary geological object such as a mountain or the ocean could be used while the detector would be moved periodically [19,20].

Figure 4 summarizes existing data in the gravitational inverse square law and shows potential sensitivities of the Laplacian experiments in three different settings: a laboratory, a geological scale and an earth orbit. A deviation from the inverse square law of the type

$$\phi(r) = -\frac{GM}{r} \left(1 + \alpha e^{-\mu r} \right) \tag{3}$$

has been assumed for these plots. For detailed references of previous experiments and sensitivity computations for the Laplacian experiments, see reference [20].

At the University of Maryland, a single-axis portion of the three-axis gravity gradiometer has been completed and is undergoing tests. The instrument is fully operational and exhibits a noise spectrum as discussed in section 3. Figure 5 is a photograph of the single-axis superconducting gravity gradiometer in a so-called "umbrella" orientation. The sensitive

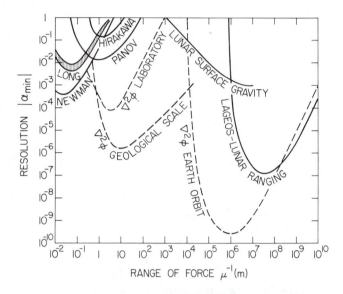

FIGURE 4. Resolution of various experiments for the inverse square law of gravitation.

FIGURE 5. A photograph of a single-axis gravity gradiometer in an "umbrella" orientation.

axis of the gradiometer is tilted in such a direction that a $\pm120°$ rotation of the device around the vertical axis brings the x-axis into the y- and z-axes. By rotating the sensitive axis discretely in this manner and summing the outputs for the three measurements, we have obtained a preliminary result of the null experiment. A lead pendulum weighing 1600 kg was used to generate a time-varying gravity signal at an average source–detector distance of 2.3 m. Before an appropriate control is made in metrology, an upper limit of 0.1 is obtained for any departure from the inverse square law at 2.3 m. A more detailed account of this experiment has been published [21].

We are planning to build a dynamic source of larger mass (10 tons) and extend the source–detector distance to more than 10 m. After completion of the laboratory experiment, we hope to carry out the much desired geological-scale experiments. Advantages of the new approach will then be truly manifest.

5. OTHER PRECISION GRAVITY EXPERIMENTS

The development of a three-axis gravity gradiometer is funded by NASA with a goal to fly an advanced three-axis model with a sensitivity of 10^{-4} E Hz$^{-1/2}$ in a low altitude earth orbit to obtain a precision gravity map of global coverage. The resulting precision gravity data of the earth will make an immeasurable contribution to geophysicists in understanding geodynamics and also aid inertial survey and navigation engineers in improving accuracy. Such a mission will also provide an excellent opportunity to improve the inverse square law data at the distance scale of 10^2 to 10^4 km, as shown in figure 4. The satellite could be launched initially into an elliptical orbit for this experiment [20].

The equivalence principle experiment discussed by Worden elsewhere in this chapter uses similar techniques.

A gravity gradiometer is by itself a gravitational wave detector. So a gigantic version of a superconducting gravity gradiometer could be used as a "tunable" detector for a monochromatic gravitational wave signal coming from pulsars. Such a detector has been proposed [22] and may be constructed in the future. The extremely small strain $h \lesssim 10^{-26}$ expected from pulsars will come into the range of detectability of a superconducting detector if a $Q \lesssim 10^9$ is achieved at pulsar frequencies in a circuit carrying a large persistent current [22]. Such a Q requires three orders of magnitude improvement from the state of the art but is well within a theoretical limit.

In this paper, we attempted to describe some of the fundamental gravity experiments that had been initiated under the direction of William Fairbank and to explain how immense technical challenges had been tackled with his ideas and influences. A fundamental experiment, searching for gravitational waves, created a need for improvement of the state of the art for acceleration measurement. The resulting device finds many practical applications and opens up opportunities to perform other fundamental experiments. This is an example of how one experiment has led to opening up many new branches of technology and experimental science. At the heart of this success is Fairbank's style of attacking fundamental problems using elegant detection strategies in which physical parameters are often required to have "near zero" values.

References

[1] See, for example, W. H. Press and K. S. Thorne, *Ann. Rev. Astron. and Astrophys.* **10**, 335 (1972).

[2] P. W. Worden, Jr., Ph.D. Thesis, Stanford University (1976), unpublished.

[3] J. E. Opfer, *Rev. Phys. Appl.* **5**, 37 (1970).

[4] An rf SQUID is a Josephson junction magnetometer with extremely low noise performance. For its principle of operation, see R. P. Giffard, R. A. Webb and J. C. Wheatley, *J. Low Temp. Phys.* **6**, 533 (1972).

[5] H. J. Paik, in *Experimental Gravitation*, edited by B. Bertotti (Academic Press, New York, 1974), p. 515. Paper presented at the International School of Physics "Enrico Fermi," Varenna, Italy (1972).

[6] P. H. Ceperley, Ph.D. Thesis, Stanford University (1971). HEPL Report 655.

[7] H. J. Paik, Ph.D. Thesis, Stanford University (1974). HEPL Report 743.

[8] H. J. Paik, *J. Appl. Phys.* **47**, 1168 (1976).

[9] E. R. Mapoles, private communications.

[10] H. J. Paik, *J. Astronaut. Sci.* **29**, 1 (1981).

[11] System 330X, SHE Corporation, San Diego, California.

[12] H. J. Paik, E. R. Mapoles and K. Y. Wang, in *Proceedings of Conference on Future Trends in Superconductive Electronics*, edited by B. S. Deaver, Jr., C. M. Falco, J. H. Harris and S. A. Wolf (Charlottesville, Virginia, 1978), p. 166.

[13] K. Y. Wang, Ph.D. Thesis, Stanford University (1979), unpublished.

[14] E. R. Mapoles, Ph.D. Thesis, Stanford University (1981), unpublished.

[15] H. A. Chan, Ph.D. Thesis, University of Maryland (1982), unpublished.

[16]   H. J. Paik, *J. Astronaut. Sci.* **29**, 1 (1981).

[17]   M. V. Moody, H. A. Chan and H. J. Paik, *IEEE Trans. Mag.* **Mag-19**, 461 (1983).

[18]   H. J. Paik, *Phys. Rev.* **D19**, 2320 (1979).

[19]   H. J. Paik and H. A. Chan, in *Proceedings of the Second Marcel Grossmann Meeting on General Relativity*, edited by R. Ruffini (North-Holland, Amsterdam, 1982), p. 1013.

[20]   H. A. Chan and H. J. Paik, *Phys. Rev. Lett.* **49**, 1745 (1982).

[21]   H. A. Chan, M. V. Moody and H. J. Paik, in *Proceedings of the Third Marcel Grossmann Meeting on General Relativity*, edited by Hu Ning (Science Press and North-Holland, 1983).

[22]   R. V. Wagoner, C. M. Will and H. J. Paik, *Phys. Rev.* **D19**, 2325 (1979).

Almost Exactly Zero:
The Equivalence Principle

Paul W. Worden, Jr.

1. INTRODUCTION

For several years I have been developing an equivalence principle experiment at Stanford in association with Francis Everitt. The purpose of this experiment is to put the smallest possible upper limit on any violation of the equivalence of gravitational and inertial mass, making use of cryogenic technology in an earth-orbiting satellite. The use of cryogenics and of the zero-g environment of space to perform an experiment which seeks to detect a minute difference in acceleration between two masses strongly exemplifies William Fairbank's near zero principle, and my work on the experiment has benefited greatly from his advice and encouragement.

1.1 The Equivalence Principle

The equivalence principle is really two principles, which are known as the strong (or Einstein) equivalence principle and the weak equivalence principle (or uniqueness of free fall). It is with the latter that this experiment is concerned. The principle says that all objects in the same gravitational field fall with the same acceleration, independent of their composition and internal structure. Einstein generalized the weak equivalence principle to say that all of the nongravitational laws of nature (*i.e.*, the laws of special relativity) are the same in any local, freely falling frame of reference. This, then, is the strong equivalence principle, which is the founding postulate of general relativity. The weak equivalence principle follows naturally from Einstein's strong principle; conversely, a violation of the weak principle is also a violation of the strong principle. Both ideas are based on observation.

The strong equivalence principle is what allows gravity to be treated as the geometry of space-time. If two otherwise dissimilar test bodies are falling together and stay at rest with respect to each other, they are following essentially the same path through space-time; and since the freely falling bodies are by definition acted on only by gravity, the identity of paths allows one to treat the motion as being force-free in a curved space-time, which would be impossible if the bodies fell at different rates. This is what Einstein did. Since the strong equivalence principle embodies the weak principle, any violation of the weak principle would undermine the foundation of Einstein's theory of general relativity.

A conjecture by the late L. I. Schiff, now proved for several cases, is that

> Any complete self-consistent theory of gravitation, embodying special relativity and the weak equivalence principle, is a metric theory [1].

Many of the theories competing with general relativity are metric theories—theories which depend on the use of a background geometry—but some are not. Improved measurements of the equivalence principle can strengthen the case for or against the metric basis of the theory.

Predictions of possible violations of the equivalence principle start from various points of view. Mathison in 1937 [2] showed that even in general relativity a spinning body experiences a small deviation in motion from a geodesic path. The effect is a periodic acceleration perpendicular to the plane of the orbit. For rapidly spinning rings it might be as large as a few parts in 10^{16} of the gravitational acceleration in earth orbit.

More interesting was the suggestion by Lee and Yang in 1955 of a possible connection between the violation of equivalence and baryon number conservation [3]. Much the most straightforward way of accounting for exact baryon number conservation is by a massless vector boson coupled to the baryon number. Lee and Yang's argument was that, in analogy with electrodynamics, local symmetry of a Yang-Mills field leads to baryon number conservation and the existence of a Coulomb-type repulsion between baryons. If we write Q_b for the "baryon charge," we have the total force between two neutral massive bodies

$$F = -G\frac{M_1 M_2}{R^2} + Q_b^2 \frac{A_1 A_2}{R^2} \quad , \tag{1}$$

where M_1, M_2 are the gravitational masses and A_1, A_2 the mass numbers for the two bodies. Since the ratio M/A differs for different elements, the Lee-Yang argument leads to a violation of equivalence at some level. Unfortunately we do not have at present any independent means of estimating a likely magnitude for Q_b. At the current limit of knowledge of the equivalence principle, Q_b^2/Gm_p^2, where m_p is the mass of the proton, is less than 6×10^{-8}. There are reasons for thinking that the measurement could be pushed to a precision of a part in 10^{11} in a cryogenic earth-based experiment using solid hydrogen and copper as test masses, and to a part in 10^{16} in an orbital experiment.

Interest in the Lee-Yang hypothesis has revived recently through the failure of experiments to detect the decay of the proton predicted by the SU_5 model. Other arguments for possible isolation of equivalence have been based on hypotheses that there may be something anomalous about the gravitational equivalence of weak interaction energy. Whatever one may think of these various arguments, the experimentalist has a motive of his own to pursue the measurement. His interest is not simply to prove or disprove theories but to find out how the real world behaves. As William Fairbank's career has abundantly verified, it is by looking in unexplored places, or in known places with greater acuity, that the unexpected is discovered. Part of the justification of this experiment is the mountain climber's "because it is there." But it is nice that this particular mountain happened also to be the foundation stone on which general relativity stands.

1.2   Conceptual Development and Historical Background

The story of Galileo and the leaning tower of Pisa has made it common knowledge that two objects of different weight dropped simultaneously will strike the ground at the same time. (The prevailing theory in 1600 predicted otherwise.) Galileo, however, was not the first to perform the experiment, if in fact he did [4]. In the fifth century A.D. Ioannes Philiponus, a scholar in Byzantium, recorded, and possibly performed, the same experiment:

> For if you let fall from the same height two weights of which one is many times as heavy as the other, you will see that the ... difference in time is a very small one [5].

Galileo's major contribution was not the experiment of dropping weights but rather the concept of *inertia*, the tendency of matter to remain in its state of motion. A century later Isaac Newton formulated his theory of gravity, from which we get the concept of mass as a source of gravity. This kind of mass is now called *active* gravitational mass, to distinguish it from *passive* gravitational mass, which is what responds to gravity and gives rise to weight. Newton himself may have had some confusion between the three ideas. Inertia, active and passive gravitational mass are regarded as being equivalent quantities, that is, they all measure exactly the same thing, which is *quantity of matter.* There is no such equivalence for forces other than gravity. Thus in electromagnetism, the amount of charge (the source of electric field corresponding to active mass) is not proportional to the inertia, and likewise neither are the "charges" associated with the weak and strong forces.

Both for measuring violations of the equivalence principle and for comparing the sensitivities of experiments on it, we need a consistent measure of departures from perfect equivalence. Such a measure is the Eötvös ratio η, which is the difference in acceleration of two falling test bodies, divided by their average acceleration. With this definition, the value does not depend on the particular gravitational field used in the experiment. A mathematically equivalent (but conceptually more difficult) definition of η is the difference in the ratio of passive gravitational mass to inertia of the two bodies, divided by the average ratio. Mathematically this is expressed as

$$\eta = \frac{2(X_A - X_B)}{(X_A + X_B)} = \frac{2\left[(M_p/M_i)_A - (M_p/M_i)_B\right]}{\left[(M_p/M_i)_A + (M_p/M_i)_B\right]} \quad , \tag{2}$$

where X denotes the accelerations, and (M_p/M_i) the ratios of passive to inertial mass, of body A and body B, respectively. The first definition of η emphasizes the equality of the accelerations of the test bodies; the second emphasizes the proportionality between the mass (*i.e.*, inertia) and its weight, which in a given gravity field is measured by the passive mass. The closer η is to zero, the more accurately the proportionality holds. Both conceptions of the equivalence principle are important historically.

Questions of the reality of Galileo's experiment aside, he was aware that a cannon ball reaches the ground ahead of a musket ball by about 10 cm in a fall of 100 meters [6]. Assuming the balls were of different materials (stone or iron for the cannon ball, lead for the musket ball), this constitutes a test of equivalence . The result is $\eta < 10^{-3}$, that is, Galileo could not have detected a violation of the equivalence principle smaller than this. Galileo realized that the difference in rate of fall that he did observe was due to the resistance of the air, and so had an intuitive concept of the equivalence principle.

Newton was the first to experimentally investigate the equivalence principle as such. In section VI, cor. vii of the *Principia* he states:

> And hence appears a method both of comparing bodies one with another, as to the quantity of matter in each, and of comparing the weights of the same body in different places, to know the variation of its gravity. And by experiments made with the greatest accuracy, I have always found the quantity of matter in bodies to be proportional to their weight.

Newton's experiments used pendulums, "very accurately made," to determine the proportionality between mass and weight. Unfortunately, he could only demonstrate $\eta < 10^{-3}$, no better than Galileo's observation. In pendulum experiments the measurement of η depends on a small difference between large and separately measured quantities (*i.e.*, the lengths of the pendulums) rather than on η directly. A quite small inaccuracy in the measurements may be very large compared to the difference, and completely overwhelm any small effect. Also it is quite difficult to make a really accurate pendulum.

Further free fall and pendulum experiments were performed up to this century, as well as measurements based on directly weighing samples (which suffer from problems similar to the pendulum experiments). However, the upper limit on η did not decrease much until the torsion balance experiments of Baron R. von Eötvös and his colleagues in the late nineteenth and early twentieth centuries [7]. Eötvös' apparatus was a torsion balance designed originally for gravity gradient measurements but modified by having masses of different materials at the two ends of the torsion arm. Suppose such a balance is placed in a field where a gravitational acceleration transverse to the torsion arm is balanced by a centrifugal acceleration. The gravitational acceleration acts on the passive gravitational masses of the two bodies and the centrifugal acceleration on their inertial masses. If the torsion head is rotated through 180° the two masses change places, and if the ratios of the gravitational to inertial masses are not identical for the two bodies, the result will be a slight difference in torque on the balance arm, which means that the balance arm rotates through an angle $(180° + \theta)$, where θ is a small difference angle proportional to η and inversely proportional to the torsion constant of the wire. Eötvös originally designed an experiment in which the driving acceleration was the component of the earth's gravity balanced against the centrifugal acceleration arising from the rotation of the earth. Eötvös eventually achieved a sensitivity of $\eta \sim 5 \times 10^{-8}$, and measured a variety of common and uncommon materials, including water, radium bromide, and snakewood.

The torsion balance technique was extended by Roll, Krotkov and Dicke in the early 1960's in the most careful measurement so far [8]. The experiment is subtly changed from that of Eötvös. Instead of using the earth's centrifugal acceleration, the Princeton experiment used the balance of gravitational attraction and centrifugal acceleration in the orbit around the sun. This results in a somewhat smaller signal size if there is a violation, because the centrifugal acceleration of the earth's orbit is about one-third that of the earth's rotation, but there are compensations. In particular, rotation of the apparatus was carried out very smoothly by the rotation of the earth, greatly reducing abuse of the torsion fiber and the resulting errors. The

Princeton measurements had a sensitivity of about $\eta = 3 \times 10^{-11}$. Since then, V. Braginsky and V. Panov have claimed an $\eta = 10^{-12}$ [9]; and in an experiment replacing the torsion fiber by a fluid support, Keiser and Faller have reported a measurement of 7×10^{-11} [10].

The present experiment grew from a suggestion for a thesis topic due to Francis Everitt. The questions were how good an equivalence principle experiment could be done in earth orbit and should one apply cryogenics to it? We found that ultimately it should be possible to do one million times better in orbit than has been done on earth, and that cryogenics, among other advanced technologies, is essential to the project.

2. THE ORBITAL EXPERIMENT

The motivation for doing an orbital experiment [11] comes largely from observing that the sensitivity is proportional to the accelerating field. Other things being equal, one should use the largest possible gravitational or centripetal field to increase the absolute size of any difference in acceleration of the test masses. Eötvös used the centripetal acceleration of the earth, about 1.7 cm/sec^2; the Princeton team used 0.6 cm/sec^2, the rate at which the earth falls toward the sun. In earth orbit, the acceleration is about 850 cm/sec^2, which could provide a 1000 fold improvement in sensitivity. Additionally, earth-based experiments are limited mostly by vibrational noise—microseisms—which, with care, can be eliminated in an earth satellite. One can gain another factor of 1000 in sensitivity from this reduction in disturbance.

In earth orbit the entire experiment may be regarded as falling continuously around the earth. One can then consider a revival of the Galileo 'free fall' experiment rather than a torsion balance experiment. Free fall experiments on the earth are greatly limited by the time of fall. In the 4 1/2 seconds it takes to fall 100 meters in the earth's gravity, the test masses would separate no more than 1 angstrom in a 10^{-12} experiment. The problem is not in accurately measuring this displacement, but in insuring that the masses are released simultaneously with no initial difference in velocity or position, and are not disturbed on the way down. In earth orbit, time of fall is no longer a limitation (the masses fall continuously), so that the setup is less critical at the same sensitivity. Even better, for the same effort in preparation we can have a much more sensitive experiment.

A torsion balance cannot be used for this experiment in earth orbit. The reason is the earth's gravity gradient. The earth's field decreases as the inverse square of the distance, so that the low end of a dumbbell is more strongly attracted than the high end. This produces a difference in acceleration (and corresponding torque on the dumbbell) of about

2×10^{-7} g per meter of length. To be able to compare small differences in the rate of fall of the ends, we must make this disturbance the same size as or smaller than the signal we want to measure, or it will swamp the measurement. At a sensitivity level of 10^{-12} the dumbbell needs to be less than 10^{-3} meters long! A more sophisticated version of this argument places similar, unreasonable limits on the shapes, sizes and matter distributions of all possible rigid rotors with dissimilar masses. The moments of inertia of the torsion balance have to be matched to an impossible degree.

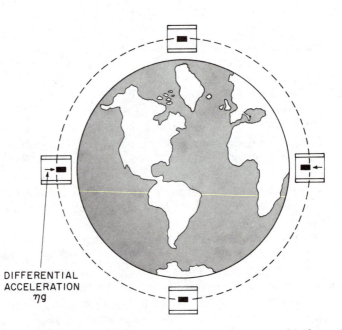

DIFFERENTIAL
ACCELERATION
ηg

FIGURE 1. Concept of the orbital experiment. Two test masses orbit the earth from the same initial conditions; if the equivalence principle is violated, they travel in slightly different orbits.

The proposed experiment uses two independent test masses, a short solid cylinder inside a short thick-walled tube. These orbit the earth, and if one falls a bit faster than the other it will tend to go in a slightly different orbit (figure 1). The net result is a periodic, once per orbit acceleration of one mass with respect to the other. In order not to be disturbed by the gravity gradient of the earth, it is necessary to center the test masses on each other *very* precisely. This can be done with movable masses but not with a dumbbell. The procedure is to carefully observe the relative acceleration of the masses. If they are not centered, they will have a relative acceleration proportional to the center of mass displacement, due to the gravity gradient of the earth. This acceleration can be measured and used as an error signal for a servo that controls the displacement. Better yet, it is possible to distinguish this gradient acceleration from any acceleration which arises from a violation of the equivalence principle. The gravity gradient acceleration due to a displacement occurs twice per orbit, whereas that from a violation occurs only once. If the masses can have their separation controlled to, say, 1 angstrom, the gravity gradient disturbance will be less than about 2×10^{-17} g, which would make a very good experiment. There are good possibilities for controlling the masses much more closely than this, so that other disturbances will limit the measurement.

The cylindrical shape was chosen to give access to the inner mass (which spherical shape does not) and also to provide some definite control over the orientation of the test masses. The test masses are constrained radially by superconducting magnetic bearings. This simplifies the control problem by restricting the motion to one dimension, but some provision must be made to ensure that the axes of the bearings are concentric. The mass positions are read out by a superconducting magnetometer which can detect displacements of 0.001 Å or less. As described above, the acceleration measurements will be used to calculate and correct the center of mass offset, probably to within something like the position sensitivity, if disturbances are small enough.

Some of the disturbances which arise are gas drag and solar wind perturbations on the orbit. These disturbances can be quite large compared to the forces in the experiment, and they can recur once per orbit. These can be greatly reduced by applying the concept of the drag-free satellite originally suggested by G. E. Pugh and developed in the Stanford Department of Aeronautics and Astronautics by B. O. Lange, D. B. DeBra and their colleagues. In this experiment a heavy mass floats within a hollow shell which is the body of the satellite, and which shields the mass from external disturbances. If the shell is moved off-center from the mass, the shell fires small jets to stay put, without ever disturbing the mass. Accordingly the mass and satellite follow a purely gravitational trajectory. Drag-free satellites have achieved average nongravitational accelerations of 10^{-11} g. Significant improvement should be possible over this by using linear thrusters. (At present, satellites use thrusters which are either off or on, with no in-between, and this limits their performance.) It is reasonable to expect no more than 10^{-5} of the residual acceleration to appear as a differential acceleration between the test masses, so if a drag-free satellite can achieve 10^{-12} g, it can carry an equivalence principle experiment sensitive to $\eta = 10^{-17}$ or better.

There are of course many other disturbances, but some of the most severe are internal and can be reduced only by careful design and execution of the experiment. Among the worst are the forces caused by residual gas in the apparatus. Residual gas can cause forces in several ways: by streaming effects, where an actual jet of gas impinges on a test mass; by emission from warmer surfaces more than from cold; and by coupling the motion of the satellite body to a test mass. It is here that the need for cryogenics really becomes felt. At the temperature of liquid helium, all gases but helium are frozen. By excluding helium, it becomes quite easy to get very good vacuums, and these forces are eliminated together with the gases.

Just what are the real needs for cryogenics and superconductivity in this experiment? After all, these are the trademarks of our group. We find five essential reasons for operating this experiment at low temperature.

(1) "Perfect" magnetic shielding is available by using superconductors. The earth's magnetic field would otherwise cause unacceptable disturbance of the test masses, even if iron shielding were used to reduce it.

(2) The dimensional and thermal stability of the apparatus is greatly increased. The thermal expansion of solids becomes very small at temperatures near absolute zero, and so does the time for a temperature gradient to relax. Together, these are the major cause of changes in shape at room temperature. Additionally, thermally activated creep is reduced.

(3) A very stable and highly sensitive position detector is available using SQUID technology. This is described below. Its stability depends on the great stability of supercurrents and the dimensional stability of the apparatus. Its sensitivity depends on the extent to which the flux quantum, already a small quantity, can be divided.

(4) Gas pressure and forces resulting from it can be greatly reduced.

(5) Radiation pressure disturbances are greatly reduced. At room temperature, the thermal radiation pressure resulting from a fraction of a millidegree temperature gradient around the test masses is enough to upset the measurement. The thermal radiation decreases as the fourth power of the temperature, and is negligible at four degrees.

When all of the relevant factors are considered in detail, it appears that an orbital equivalence principle test with a sensitivity of 10^{-17} or 10^{-18} should be possible. The main limitations will be noise caused by gas streaming and residual accelerations of the spacecraft. Currently I am engaged in a ground-based version of the experiment, which serves as a base for developing the technology necessary for the orbital experiment and should also add confidence to our predictions of the performance of the orbital experiment. The ground-based experiment should yield a determination of η good to between 10^{-12} and 10^{-13}.

3. THE GROUND-BASED EXPERIMENT

3.1 General Description

The difficulty in developing on earth technology intended for use in space is gravity. The experiment must be rescaled to take account of this force with the least possible loss in sensitivity. In particular the magnetic bearing design which is most suitable for the experiment has little lifting capacity, so that the test masses must be smaller by a factor of five or more than they can be in the orbital experiment. Gravity also gives the experiment an asymmetry in the vertical direction. The test masses can never be quite centered with respect to the bearings although they can still be centered on each other. This increases the coupling between the radial modes of the bearing and complicates the interaction with vibrational noise. Nonetheless, the projected sensitivity is better than $\eta = 10^{-12}$.

In the earth-based experiment in progress, two cylindrical test masses, one and five centimeters in diameter, are levitated in superconducting magnetic bearings, as illustrated in figure 2. The bearings are scaled for use in the earth's gravity field, and give extremely stiff support radially with maximum freedom of motion horizontally. The residual force in the sensitive horizontal direction is one of the limitations on the experiment. The position sensors are prototypes similar to those intended for the orbital

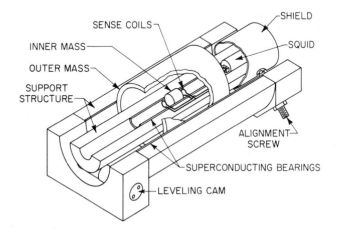

EQUIVALENCE PRINCIPLE ACCELEROMETER

FIGURE 2. Apparatus for the earth-based experiment. Two test masses are levitated in superconducting magnetic bearings. Their displacements are measured with SQUID magnetometers.

experiment, and the horizontal position is adjusted with control coils under the bearings, and also by restoring forces from the position detector coils. Superconducting shielding protects the apparatus from changes in external magnetic fields. The experiment is actively controlled by a smaller computer which monitors the test masses, provides the primary control signals for the masses, performs some of the setup procedure, and logs data.

As in Roll, Krotkov and Dicke's experiment, the sun is used as a source, but now we regard the test masses as falling toward the sun (together with the earth and everything on it). Because the apparatus is rotating once per day as it falls, the signal, if any, will change direction daily.

3.2   Period Matching and Position Subtraction

A serious problem with an earth-based experiment is mechanical vibration due to microseisms and in cities also due to human activity. It is essential that the masses respond identically to a motion of the apparatus; otherwise the slightest vibration could give a false signal. This is equally true in the orbital experiment. Regard the masses as coupled to horizontal ground motion by very soft springs. These "springs" may be the forces applied by the position detector coils, by imperfections in the bearings, or by the servo loops for keeping the masses centered. If the ratio of spring constant to inertia is the same for each test mass, the masses will respond identically if they start from the same initial conditions. Thus, the periods of the masses must be identical; when they are, the differential motion of the masses is not excited by seismic noise. If possible, this condition should be satisfied for all six degrees of freedom. As will be seen below, the design of the bearing does this in part for the radial oscillations.

There are several ways to adjust the longitudinal periods of the test masses to be equal. One is to design a servo loop, executed by the computer, which uses the control coils to keep the masses centered. The mass periods and dynamic characteristics are completely under the control of the experimenter and independent of the properties of the position detectors, which can have their gains adjusted separately for an accurate subtraction. This method has been tested and found to be feasible but is limited by the rate at which the computer can process data. A more promising method [12] is to use the restoring force from the position detectors to do the matching. In this case the mass periods are changed by adjusting the sensitivity of the position detectors and *vice versa*, but by repeating the setup procedure they may be adjusted independently over a wide range. The methods are ideally equivalent except in one regard. Adjustments to the position detector circuit alone cannot change the equilibrium position of the test mass without affecting the sensitivity and period. Therefore, the method for matching the periods and subtracting the mass positions must involve computer feedback to the control coils to adjust the mass positions. By using the position detectors for the critical matching and subtraction, the processing load on the computer is reduced to a manageable level.

3.3   Magnetic Bearings

The most difficult part of the earth-based experiment has been the development of the magnetic bearings. These bearings should be very stable, should have little cross coupling of the radial force into motion along the sensitive axial direction, should not interfere with the position detectors, and should have small dissipation so as not to add to thermal noise. The first and last requirements are satisfied by superconductors. The other requirements must be satisfied by good design and execution.

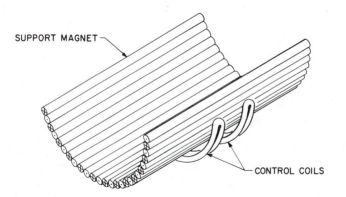

SUPPORT MAGNET

CONTROL COILS

FIGURE 3. Concept for the magnetic bearings. A persistent current flows in opposite directions in adjacent wires to provide the maximum stiffness radially with minimum force along the cylinder axis.

Figure 3 shows the concept of the magnetic bearings. The support wires are arranged parallel to the cylinder axis to eliminate bumps as the mass crosses the wires, and current flows in opposite directions in adjacent wires. The magnetic field of this configuration decreases very rapidly with distance, which stiffens the bearing and reduces interaction with the position detector. The large stiffness of the radial modes reduces cross coupling to the longitudinal mode by separating the resonant frequencies. A further benefit is that the vertical period of the mass in a bearing of this type depends almost entirely on the spacing of the wires and hardly at all on the mass or height of levitation; therefore the vertical periods of the inner and outer masses can be well matched simply by giving their respective bearings the same wire spacing.

If the bearings are slightly tapered or crooked, they will in general exert forces along the sensitive axis which can interfere with the measurement of their difference in acceleration. We can estimate how straight the bearings have to be. The test mass position measurement is uncertain by, say, 10^{-10} cm. The bearing will exert a force on the test mass which may vary along the axis although it is constant in time. Therefore, because of the uncertainty in position there is also an uncertainty in the acceleration measurement, proportional to the position uncertainty and the variation of force with distance. The variation of force is related to the straightness of the bearing, although other factors such as variation of the magnetic susceptibility can also contribute to it. The bearings must be straight to about 10^{-3} cm per meter of length for the earth-based experiment, which is a stringent but not impossible requirement. Because of the weightless environment, the bearings for the orbital experiment can be less precise.

The performance of the bearings is crucial to the experiment, and they have undergone substantial development. In addition to the obvious difficulty of manufacturing them to the required tolerance, there are more subtle problems of the distortion of the bearings as they are cooled to 4 K, of keeping them magnetically clean, and of charging them with up to 100 amperes of persistent current. For a long period I made use of bearings in which the superconducting wires were embedded in a thin layer of low thermal expansion fiberglass epoxy [13]. Despite extended developmental effort, these bearings proved ultimately unsatisfactory because of difficulties with differential contraction on cooling, and I have now developed a bearing of a different design in which the wires are tightly stretched like violin strings between two semicylindrical frets. With this design, thermal contraction is actually an aid because the wires can be arranged to tighten up on cooldown. Figure 4 shows measurements on a typical bearing of the newer design.

3.4 Position Detectors

The position detectors work by the change in inductance of field coils as a superconducting test mass moves near them. The basic circuit is shown in figure 5a. The test mass is between coils L_1 and L_3, which are connected in a superconducting loop. A third coil, L_2, is in parallel with L_1 and L_3. If the mass moves, it causes equal and opposite changes in L_1 and L_3. If a magnetic flux is trapped in the loop L_1–L_3, this change forces a current to flow in L_2 to keep the flux in the two superconducting loops constant. A SQUID (Superconducting Quantum Interference Device) magnetometer detects the current in L_2, which is proportional to the trapped flux and the displacement of the mass. Using the great sensitivity of the SQUID, this circuit can detect a change in position of the test mass of 0.001 Å or less. By counting flux quanta the dynamic range can be as great as 10^7.

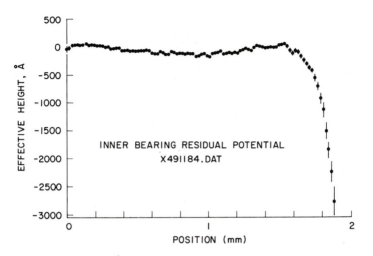

FIGURE 4. Smoothness of an inner bearing expressed as effective height in the earth's gravity field. The remaining curvature of about $2 \times 10^{-4}/$cm in the central region is believed due to forces from the position detector. These data were taken by averaging the acceleration of the test mass at each position in the bearing.

The position detector coils exert forces on the test mass which tend to move it to a position midway between them. The frequency of the harmonic oscillator thus formed depends directly on the sensitivity, that is, the smaller the minimum detectable displacement, the higher the frequency. The relationship is not one-to-one, but there is a range of sensitivities possible at a given frequency. One may, with some loss of strict accuracy, think of the sensitivity as being determined by the total flux in $L_1 \ldots L_3$, and the frequency by the way the flux is distributed on either side of L_2. This property is useful for matching the periods of the test masses while keeping the sensitivities unchanged. However, the equilibrium position of the mass is uniquely dependent on the frequency unless it can be changed by applying an extra force. Since it is important to keep the centers of mass of the test bodies at the same place, extra coils are provided beneath each mass to adjust the equilibrium positions. These provide a little extra lift under each end and can magnetically tilt the support given by the magnetic bearing. How much adjustment is needed is calculated by the computer controlling the experiment.

In addition to matching the periods of the test masses, the position detector circuit can perform the subtraction of the mass motion to very high accuracy. The basic circuit is modified by combining the circuits for each mass so that one inductance (L_2 in figure 5b) is in common. By putting opposite fluxes in the circuits on either side of the common inductance,

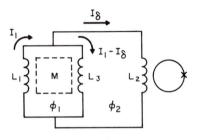

SIMPLIFIED POSITION DETECTOR CIRCUIT

FIGURE 5a. Basic circuit. A movement of the mass changes the inductances L_1 and L_3. Flux conservation forces a current proportional to the mass displacement to flow in L_2, and this is detected by a SQUID magnetometer.

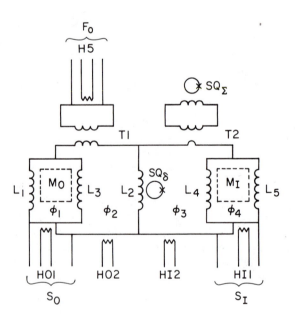

DETAILED POSITION DETECTOR CIRCUIT

FIGURE 5b. Subtraction and matching circuit. Two basic circuits are combined with an inductance in common. By adjusting the sensitivities of the two circuits to be equal and opposite, the current in L_2 is made to be an accurate difference of the mass motions. The mass periods may likewise be adjusted to be equal.

currents due to equal mass motions subtract. In principle the equality of the scale factors is limited only by the stability of the persistent currents, but in practice there are difficulties in matching the scale factors closer than the mass periods, currently about one part in 10^5 [14]. This limit is set by cross coupling between modes of the masses.

3.5 A 200 Meter Pendulum

A study of the literature reveals that with the probable exception of Eötvös' experiment, all of the modern experiments have been limited by seismic noise rather than any fundamental limit. This one will be no exception. The position subtraction circuit described above has a common-mode rejection ratio of 10^5, which means that one part in 10^5 of the common-mode amplitude appears as noise in the differential mode. The major noise in the common mode is natural seismic noise with frequency in the range of 1/2 to 1 cycle per day, due to the earth tides. Since the amplitude of this noise is about 10^{-7} cm/sec^2, the resulting noise in the equivalence principle measurement will be about 10^{-12} cm/sec^2. It is likely that most of the noise in this bandwidth will be due to frequency conversion from much higher frequency unless some isolation is provided.

Conventional antivibration devices are not much help. Most of them involve spring-mass systems with low resonant frequencies, and cannot isolate against noise below about 30 Hz. They have the additional problem of being quite compliant, so that as cryogenic fluids such as nitrogen and helium boil away the apparatus tends to move up and down and tilt a lot. The best isolation method for horizontal noise is a long simple pendulum. If the pendulum is made long enough, it becomes very easy to move the bob even if it is very heavy. Conversely, the bob becomes nearly decoupled from ground motion. We can say that for horizontal motions the bob of a long pendulum is effectively in zero gravity. The difficulty is that the pendulum must fit within an ordinary laboratory.

I had one side of a thick aluminum plate machined to a spherical surface with a radius of 50 meters, and made a closely matching support ring by pressing thick epoxy against it, with the appropriate channels for an air bearing. When pressurized with about 5 psi, the plate floats on a very thin film of air and is constrained to move on the surface of a sphere. Because of slight bending of the support framework, the bob acts as though it were really on a 200 meter long pendulum. This attenuates ground vibration by at least a factor of 30 at frequencies below about 60 Hz; at present, turbulence in the air bearing prevents the pendulum from providing better isolation than this. In principle it can isolate at zero frequency (because the bob always hangs straight down), but at frequencies less than 1/10 Hz it is limited by the connections to the bob.

4. STATUS AND CONCLUSION

The ground-based project is approaching the end of its development period, and, after completion of an improved set of bearings and position detectors, should begin to produce data. Figure 6 is an example of some preliminary data taken to estimate the sensitivity, although the period matching circuits were not operational. Curve 1 is the position of the outer test mass during a period of about 18 hours. The mass was disturbed by ground tilt as well as thermal distortion of the stand caused by an automatic liquid nitrogen transfer at 2 a.m. and the building air conditioning at 9 a.m. Curve 2 is the difference in position between the outer mass and the inner mass. The difference is constant to within 2×10^{-7} cm/sec^2 except for high frequency noise from the masses' natural motion. This noise will be reduced by the common-mode rejection ratio of the position subtraction and matching system, which is expected to be about 10^5. Some further improvement is likely because the data in figure 6 was taken without benefit of vibration isolation.

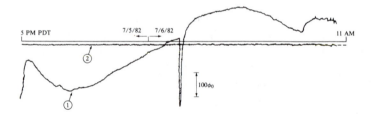

FIGURE 6. First data from the cryogenic equivalence principle apparatus. Curve 1 is the motion of an individual proof mass during a period of about 18 hours. Curve 2 is the difference measurement between the two masses in the same period, showing absence of any differential acceleration. The width of the noise is due to the mass motion with a common-mode rejection ratio of essentially unity, and has a component at 24 hour period of about 4×10^{-7} cm/sec^2.

The orbital experiment is in a study stage supported by NASA. The intent is to find if there is an inexpensive way to do the experiment on shuttle. A preliminary conclusion is that there is, and that it requires vibration isolation from the rather noisy orbiter by flying the orbiter in a drag-free mode about the experiment.

Acknowledgments

The work was supported partially by NSF grant PHY-23006 and partially by NASA contract NAS8-33796.

References

[1] L. I. Schiff, *Proc. Nat. Acad. Sci.* **45**, 69 (1959).

[2] M. Mathison, *Acta Phys. Polonica* **6**, 163 (1937).

[3] T. D. Lee and C. N. Yang, *Phys. Rev.* **98**, 1501 (1955).

[4] Galileo's "experiment" is recorded in his biography by Viviani. For a plausible interpretation of the event more as a qualitative demonstration to a class of students or a group of skeptics than as a quantitative measurement, see Stillman Drake, *Galileo at Work* (U. Chicago, Chicago, 1978), pp. 19, 414.

[5] Ioannes Philiponus, Commentary on Aristotle's Physics (5th century), quoted by M. Clagett, *Greek Science in Antiquity* (Abelard Schumann, New York, 1955), p. 153.

[6] G. Galileo, in *Two New Sciences*, translated by H. Crew and A. de Salvio (Dover, New York, 1954), p. 62.

[7] R. V. Eötvös, D. Pekar and E. Fekete, *Ann. Phys. (Leipzig)* **68**, 11 (1922).

[8] P. G. Roll, R. Krotkov and R. H. Dicke, *Ann. Phys.* **26**, 442 (1964).

[9] V. B. Braginsky and V. Panov, *Sov. Phys.–JETP* **34**, 463 (1972).

[10] G. M. Keiser and J. E. Faller, in *Proceedings of the Second Marcel Grossmann Meeting on General Relativity*, edited by R. Ruffini (North-Holland, Amsterdam, 1981), p. 969.

[11] The first serious proposal for an orbital equivalence principle experiment was by P. K. Chapman. See P. K. Chapman and A. J. Hanson, in *Proceedings of the Conference on Experimental Tests of Gravitation Theories*, edited by R. W. Davies (California Institute of Technology, Pasadena, 1970), p. 228. The first account of the present experiment was by P. W. Worden, Jr., and C. W. F. Everitt, in *Experimental Gravitation*, edited by B. Bertotti (Academic Press, New York, 1974), p. 381. See also P. W. Worden, Jr., *Acta Astronautica* **5**, 27 (1978).

[12] This method was originally described by E. Mapoles, H. J. Paik and K. Y. Wang, in *Future Trends in Superconducting Electronics (Charlottesville, 1978)*, edited by B. S. Deaver, Jr., C. M. Falco, J. H. Harris and S. A. Wolf (Am. Inst. of Physics, New York, 1978), p. 166.

[13] P. W. Worden, Jr., *Precision Engineering* **4**, 139 (1982).

[14] H. A. Chan, Ph.D. Thesis, University of Maryland, College Park, Maryland, 1982.

Measurements of
Small Gravity Changes
on the Surface of the Earth

J. M. Goodkind

1. INTRODUCTION

The Near Zero conference occurred just twenty years after the end of my close association with William Fairbank when I was a graduate student and a post doc. That is sufficient time to allow a judgment of his influence on the people who have worked with him. In my judgment, that influence is very strong and very positive. It results from his obvious delight in exploring the most fundamental questions and from his own very positive nature. That combination leads one to attack big questions and to assume that any experiment that looks interesting enough can be done. Few of us can succeed to the extent that Bill has with this approach, but physics is the more exciting for trying. The work which I will discuss below was begun shortly after I began work on my own at La Jolla. It certainly reflects the influence I have described, and after having been exposed to the Fairbank atmosphere at the Near Zero conference I came home excited that this work may yet succeed in answering some fundamental questions.

There is a natural extension of the notion of "near zero" which I would call "near constant." That is, if one can make a system in which some physical variable is almost constant, then by making difference measurements one can observe "near zero" fluctuations of that variable about some average value. In my case I wanted a constant force to balance the force of gravity so that I could measure small variations of gravity.

My motivation for doing so was originally a problem suggested by Professor Nina Beyers in a conversation with Bill Fairbank and me. In order to attack the problem I tried to think of a way in which I could measure the change in the gravitational force on a sphere when it was set into rotation. That effect, as predicted by general relativity, turned out to be too small to measure even for the most optimistic Fairbank-type thinking which I could summon. However, using a superconducting suspension it seemed that a new "near zero" regime could be achieved in which some new physics would emerge. Perhaps, for example, I thought one might be able to measure the time dependence of G predicted by Dirac. I therefore asked Walter Munk of the Institute for Geophysics and Planetary Physics at our campus if there would be any geophysical applications for such a device. He responded with great excitement over the possibility of measuring the long period tides of the solid earth, and that was sufficient motivation to convince me to begin. Using a niobium sphere which I borrowed from Bill Fairbank, and a small Sloan Foundation grant, I began measurements and calculations with an excellent graduate student, William Prothero. Ultimately this work has produced the most stable gravimeter in the world, and in the paragraphs which follow I will describe what we have done with it and what we are currently attempting to do with it.

2. THE SUPERCONDUCTING GRAVIMETER

A gravimeter is a device for measuring gravity changes over periods from a few seconds to years. The changes in gravity at a fixed location on the earth, with amplitude about 2×10^{-7} g, are dominated by the tides. However, in order to measure properties of the interior of the earth, these must be measured to a precision of 10^{-4}. Many other interesting phenomena could be observed if one could measure nonperiodic variations of 10^{-9} g per year. The only absolute gravimeter which has been built measures the acceleration of a freely falling mirror using laser interferometry. It has a resolution of a few parts in 10^8. All other gravimeters achieve their resolution by canceling the average force of gravity with a stable counterforce. In one form or another this is done by suspending a test mass on a spring. The spring must support the weight of the test mass but at the same time must be weak enough so that the smallest gravity changes to be measured will produce a measurable change in the length of the spring. The trouble with mechanical springs is that they are not as stable as one would like them to be. The superconducting gravimeter uses a magnetic "spring" consisting of a superconducting test mass suspended in the magnetic field of superconducting, persistent current magnets. The point is to make use of the inherent stability of supercurrents to produce a stable reference force.

The essential features of the instrument were described at early stages of its development by Prothero and Goodkind [1]. I will not discuss the instrument here but will use this space to describe what we have done with it. Historically our work has proceeded, naturally enough, from the largest or most obvious signals to increasingly subtle ones. The tides are the largest so I discuss them first. The next largest signal comes from the variation in atmospheric mass over the measuring site. Then there are a number of possible causes of "secular" changes in gravity which we have tried to observe with only limited success until recently. Our most recent work has compared three instruments at a time in the laboratory and two instruments in the laboratory with one located at a distance of 10 km and one located at a distance of 1000 km. Finally I will discuss the use of the instrument to measure forces which are produced deliberately in the laboratory.

3. TIDES

The top trace of figure 1 shows the raw signal from one of our instruments. In this figure we use the units of acceleration 1 gal = 1 cm/sec^2, or 1 μgal $\simeq 10^{-9}$ g. What you see is the tides of the solid earth. All but about 20% of this signal would appear on a perfectly rigid, oceanless earth. It results from the fact that the gravimeter is moving with the surface of the earth through the gradient of the gravitational fields of the moon and the sun. 16% of the signal results from the elastic yielding of the earth to the tidal forces and this means that the gravimeter moves up and down with respect to the center of the earth by about 30 cm. As much as 6% of the signal can result from the influence of the ocean tides on local gravity. The magnitude and phase of this component depends strongly on where the measurements are made, but it is a significant part of the signal everywhere on earth.

The upper trace of figure 2 shows a power spectrum of a year-long record of the tides. All studies of the tides involve comparing the amplitudes at the various discrete frequencies of the spectrum. The biggest problem in interpreting the tides is that the global ocean tides are not well determined so that their influence on the gravity tide cannot be calculated precisely. With our first measurements we attempted to see how well the best models of the ocean tides would predict the observations [2]. For some purposes the calculation can be useful, but in most cases it is necessary to work with terms which are sufficiently close in frequency so that the influence of the ocean can be assumed to be the same on all of them.

By examining the tidal spectrum with the new precision provided by our gravimeter, we searched for anomalous amplitudes or phases at specific

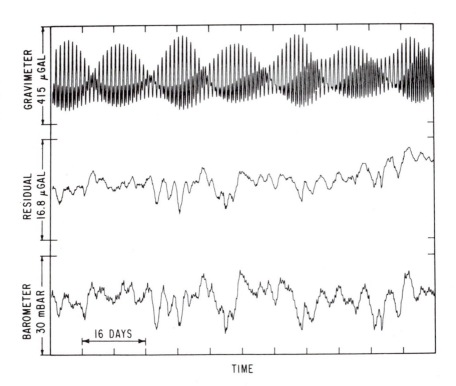

FIGURE 1. Unprocessed gravimeter signal and signal with tides subtracted. Peak amplitude is about 200 μgal (1 gal = 1 cm/sec^2). If the tides are subtracted, the residual signal fluctuates by about ± 5 μgal. This is due in part to the gravitational attraction of the atmosphere which results in a change in gravity of about 0.3 μgal/mbar change in pressure. This coefficient can vary as much as 20% and there can be a phase lag or lead between pressure and the related gravity change.

frequencies [3]. The one anomaly which was anticipated results from a resonant amplification of the tides near the frequency of the "nearly diurnal free wobble" [3,4,5]. This is a motion of the earth which is analogous to the Eulerian nutation of a rigid, rotating, spheroidal body. However, the earth is not a rigid body, but rather an elastic yielding mantle around a liquid core with an ellipsoidal interface. This condition alters the frequency of the wobble of the axis of rotation with respect to the body axis and of the associated nutation of the axis of rotation in free space. Also, it introduces an additional wobble and associated nutation which were not calculated by Euler. The original Euler mode is called the "Chandler wobble" for the earth and is measured by astronomical observations. The other mode is too small to be observed by astronomical means but can be detected as a resonant response of the earth to the tidal forcing function in the neighborhood of one cycle per sidereal day. Our measurements of this

GRAVITY POWER SPECTRUM

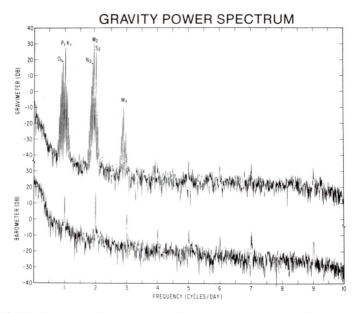

FIGURE 2. Spectrum of 1-year record from gravimeter showing the complexity of the tidal signal, the signal-to-noise ratio provided by our instrument, and the power spectrum of the atmospheric pressure for the same time period.

resonance were sufficiently precise to allow us to set an upper limit on the Q of the resonance and thereby determine that the energy dissipation in the mode is most likely in the oceans [6].

We also used this tide spectrum to search for evidence of a universal preferred reference frame [7]. Of the many theories in addition to general relativity which are not in conflict with experiment, some require the existence of such a frame. If these theories are reduced to their first-order corrections to Newtonian gravity in the form of the "parametrized post-Newtonian theory" (PPN), they will have a term in the force between two masses of the form [8,9]: $(\mathbf{r} \cdot \mathbf{v})^2$. $\mathbf{r}$ is the distance between the centers of mass and $\mathbf{v}$ is the velocity of the system with respect to the preferred reference frame. If the two masses are the earth and the test mass of a gravimeter, then $\mathbf{v}$ results from the earth orbiting the sun, the sun moving about the galactic center and the motion of the galaxy with respect to the preferred frame. The direction of $\mathbf{r}$ with respect to $\mathbf{v}$ varies periodically as the earth rotates about its axis. Therefore, this term leads to periodic variations of gravity at several frequencies close to one and two cycles per sidereal day. The frequencies and relative amplitudes of these variations were computed in reference [9], and they differ substantially from the relative amplitudes of the Newtonian tides at the same frequencies [10].

For this reason we searched for evidence of them by examining the relative amplitudes and phases of closely spaced tidal constituents near these frequencies. The results showed no evidence for a preferred frame and set upper limits on the associated post-Newtonian parameters of $\alpha_2 \lesssim 2 \times 10^{-3}$ and $\xi_W \lesssim 10^{-3}$.

4. GRAVITY OF THE ATMOSPHERE

The frequencies of the tidal constituents are known very precisely from the astronomical data on the sun and the moon. This means that they can be removed from the gravity signals by fitting and subtracting sinusoids with these frequencies. The middle curve in figure 1 shows the result of removing the tides from the curve shown at the top. The bottom curve in figure 1 shows the barometric pressure during this same time. The correlation is obvious and demonstrates that the gravitational attraction of the atmosphere, after the tides, is the next largest cause of gravity changes. The lower curve of figure 2 also demonstrates this by showing the spectrum of the pressure variations. The atmospheric tides at multiples of 1 cycle per solar day are clearly evident, and the increase in gravity noise at low frequencies can be seen to result mostly from the pressure variations. The pressure can be fitted and subtracted from the gravity data. The coefficient which one measures is always close to 0.3 μgal per millibar and this is what one calculates for a cylinder of uniform pressure with a radius large compared to the height of the atmosphere. Detailed examination of this phenomenon reveals additional gravity effects due to the deformation of the crust of the earth with the pressure, and the influence of the ocean response to the pressure changes [11,12]. Recently we have shown that the amplitude and phase relationships between gravity and pressure vary with the shape and the direction of motion of the pressure cells so that a simple fit of the pressure to the gravity is not an optimal removal of the pressure "noise." Better results will be achieved by more sophisticated techniques of analysis.

5. SECULAR VARIATIONS OF GRAVITY

The tides and the atmospheric gravity discussed above are of considerable interest, but it is only when they are subtracted from the raw gravity signal that we are "near zero." It is here that the superconducting gravimeter can measure phenomena which have never before been observable. Figure 3 is an example of a gravity event which became observable only after this subtraction. These data were obtained at the Big Geysers geothermal field near Santa Rosa, California, where we were attempting to measure gravity changes caused by the removal of water from the reservoir for power

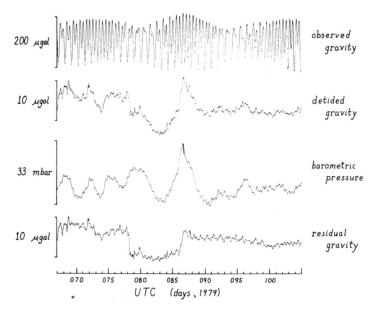

200 μgal — observed gravity

10 μgal — detided gravity

33 mbar — barometric pressure

10 μgal — residual gravity

070   075   080   085   090   095   100

UTC   (days , 1979)

FIGURE 3. The residual signal from the Big Geysers geothermal field in California, showing a gravity "event." This may have been related to a turnover of the condensate layer suspended on top of the less dense two-phase region in the reservoir.

generation. Comparison with other sites where no such effect is observed confirm our opinion that it is related to reservoir dynamics.

Measurement of very small changes of gravity over months or years would reveal vertical motion of the crust of the earth. This would be of special interest in regions of colliding plates where the motion would reveal something about plate tectonics. If a point on the surface of the earth moves vertically, gravity at that point will change since the distance to the center of mass of the earth has changed. If there is no accompanying redistribution of mass, gravity changes by 3 μgal per cm of elevation. It would also be of great interest in fault zones where slow changes in elevation could precede earthquakes. Several long records had been obtained prior to 1980 in an attempt to observe these "secular" variations of gravity. They all showed slow fluctuations of about ±5 μgal. However, in one case (figure 4) we observed a steady decrease of gravity over a year and a half. This was at a site in the region of the "Palmdale uplift" where a careful examination of old survey records had indicated vertical motion of as much as 30 cm between successive surveys one year apart [13].

Prior to 1980 we had only one instrument at each site so that we had no way to prove whether the drift and the 5 μgal fluctuations were real or instrumental. In 1980, we began to return all of the instruments to the

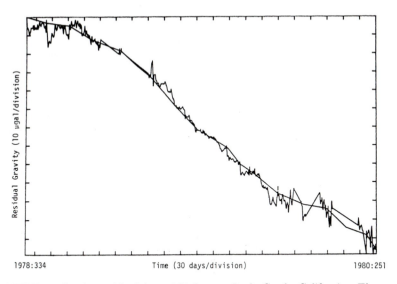

FIGURE 4. Gravity residual from 647 days at Lytle Creek, California. The curve consisting of straight line segments is drawn between the points when the instrument was aligned with the vertical so that errors from tilt are eliminated. It is not known if the drift was instrumental or real.

laboratory to study their behavior when several instruments were operated at the same time in the same place. Some sources of instrumental noise were identified and eliminated in this way; we can now obtain signals from two instruments which agree within about 0.1 μgal. We are continuing to explore the sources of instrument noise, and we are starting to make difference measurements with instruments separated by larger distances.

One instrument has been placed in the garage of the home of my co-worker, R. Warburton, in Del Mar, which is about 7 km from the lab. Since this site is at a different elevation and distance from the beach than the laboratory, the difference in gravity is dominated by the local ocean tides. This is not of much interest, but if we high pass the records with a cutoff frequency of 4 cycles per day, the tides are suppressed and the results begin to reveal new phenomena. Figure 5 shows the differences between two instruments in the lab, labeled SG7 and SG5, as well as the differences between each of them and the instrument in Del Mar. The differences between SG7 and SG5 are smaller, indicating that the differences between Del Mar and the lab are real gravity changes. Most of the differences in figure 5 do not correspond with pressure changes or with differences in pressure between the two locations.

There are several new phenomena which we are hoping to observe in this frequency range. The solid inner core of the earth should be able to oscillate

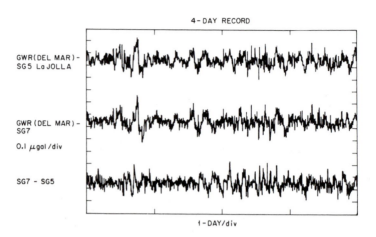

FIGURE 5. High passed records (4 cycles/day cutoff) from two instruments in the lab and one located 10 km to the north along the coast. Significant differences between the locations are observed at the level of 0.1 μgal.

about its equilibrium position at the center of the liquid core [14]. Too little is known of the properties of the liquid core to predict the frequency with precision, but all predictions are in the range of a few hours. Motion of the crust in the form of earthquakes is observed with seismographs, but there is reason to believe that there are "silent" earthquakes in which most of the motion takes place at frequencies below the sensitive band of seismographs. The normal modes of the earth have frequencies between about one and one hundred cycles per hour. They are excited by deep earthquakes, and the measurement of their frequencies and Q's have revealed a great deal about the deep interior of the earth [15]. Attempts, thus far unsuccessful, have been made to observe the excitation of the earth modes by gravity waves, thus using the earth as a very large gravity wave antenna. The earth could also serve as a nonresonant antenna at still lower frequencies where there could be radiation from rotating double stars or very large scale cosmic events.

6.   LABORATORY MEASUREMENTS

The work described thus far has all involved measurements of variations of local gravity due to geophysical phenomena of the tides. It is also possible to produce relatively large gravitational forces in the laboratory with large masses. We have filled a hollow steel sphere with mercury to produce a mass of 340 kg. At one-meter distance this results in an acceleration of 2.3 μgal. We are moving this mass between a position far from the gravimeter and positions directly underneath it once every 20 minutes. By averaging over two days of data at each position, our preliminary measurements indicated that we could test the inverse square law to within close to one part in 10^4. These measurements are almost completed, but the analysis is still in progress.

The motivation for testing the inverse square law is the observation by Long that the historical Cavendish-type experiments showed a trend toward smaller values of G at shorter distances [16]. He then performed an experiment which appeared to confirm the fact [17]. Other experiments since that time have produced contradictory results [18], and several other experiments at a range of distances are in the works. An interesting interpretation of a distance dependence of G offered by Long [19] is that it reflects the vacuum polarization of the low mass elementary particle which represents the quantized gravity field. We are also testing this notion directly by measuring the force from our mercury-filled sphere with and without a layer of lead bricks between it and the gravimeter.

These are just a few of the possible applications of a superconducting instrument to measure "near zero" variations of gravity.

References

[1]   W. A. Prothero, Jr., and J. M. Goodkind, *Rev. Sci. Inst.* **39**, 1257 (1968).

[2]   R. J. Warburton, C. J. Beaumont and J. M. Goodkind, *Geophys. J. R. Astr. Soc.* **43**, 707 (1975).

[3]   R. J. Warburton and J. M. Goodkind, *Geophys. J. R. Astr. Soc.* **52**, 117 (1978).

[4]   M. G. Rochester, O. G. Jensen and D. E. Smylie, *Geophys. J. R. Astr. Soc.* **38**, 349 (1974).

[5]   A. Toomre, *Geophys. J. R. Astr. Soc.* **38**, 335 (1974).

[6]   J. J. Olson and J. M. Goodkind, *Proceedings of the 9th International Symposium on Earth Tides* (E. Schweizerbart'sche Verlagsbuchhandlung, D7000 Stuttgart, 1983), p. 569.

[7]   R. J. Warburton and J. M. Goodkind, *Astrophys. J.* **208**, 881 (1976).

[8]   C. W. Misner, K. S. Thorne and J. A. Wheeler, *Gravitation* (W. H. Freeman and Company, San Francisco, 1973).

[9]   C. M. Will and K. Nordtvedt, Jr., *Astrophys. J.* **177**, 757 (1972).

[10]   D. E. Cartwright and R. J. Tayler, *Geophys. J. R. Astr. Soc.* **23**, 45 (1971); and D. E. Cartwright and A. C. Edden, *Geophys. J. R. Astr. Soc.* **33**, 253 (1973).

[11] R. J. Warburton and J. M. Goodkind, *Geophys. J. R. Astr. Soc.* **48**, 281 (1977).

[12] R. S. Spratt, *Geophys. J. R. Astr. Soc.* **71**, 173 (1982).

[13] R. O. Castle, *Earthquake Inform. Bull.* no. 10, 88 (U.S. Geological Survey, Reston, VA, 1978); W. E. Strange, *J. Geophys. Res.* **86**, 2809 (1981).

[14] L. B. Slichter, *Proc. Nat. Acad. Sci. USA* **47**, 186 (1961); I. S. Won and J. T. Kuo, *J. Geophys. Res.* **78**, 905 (1973).

[15] F. A. Dahlen, *Geophys. J. R. Astr. Soc.* **16**, 329 (1968); D. Agnew, J. Berger, R. Boland, W. Farrell and F. Gilbert, *EOS* **57**, 180 (1976).

[16] D. R. Long, *Phys. Rev. D* **9**, 850 (1974).

[17] D. R. Long, *Nature* **260**, 417 (1976).

[18] R. Spero, J. K. Hoskins, R. Newman, J. Pellam and J. Schultz, *Phys. Rev. Lett.* **44**, 1645 (1980).

[19] D. R. Long, *Il Nuovo Cimento* **55**, 252 (1980).

In Search of
Low Frequency Gravitational Radiation
Evolution of a Thought

V. S. Tuman

1. GENERAL DISCUSSION

The existence of gravitational radiation was predicted by Einstein early in this century. Presently all viable theories of gravity predict the existence of such radiation. Gravitational radiation interacts very weakly with ordinary matter and consequently its detection is extremely difficult. Joseph Weber predicted that some background gravitational radiation would exist in the millihertz region; for this reason he suggested using the earth as an antenna and utilizing the best available room temperature seismometers and gravity meters. A search was made by Forward and colleagues in 1961 to see if an anomalous signal was present among the even eigenvibrations of the earth, but no positive result was obtained [1]. In the meantime, with the cooperation of William Fairbank of Stanford University, I had embarked upon the development of a cryogenic gravity meter. The project was supported by the Air Force Cambridge Research Center. A progress report was presented at LT-12, Kyoto, 1970 [2].

After a long trial run with the apparatus, analog records were manually digitized at U. C. Berkeley, and a Fourier analysis was performed on the digital data. Some very interesting results were obtained. A large number of peaks corresponded to earth eigenvibrations. Furthermore, within a certain frequency band a parity structure was observed among these eigenmodes. Essentially, a number of even eigenmodes had more energy content than their neighboring odd eigenvibrations [3]. Theoretically one can design a source function to cause such an effect, but no realistic geophysical mechanism could be found which would give rise to such a specific anomalous result. It was at that time that I became aware of Weber's work and consequently speculated that a possible mechanism for the anomalous parity structure may be the excitation of earth eigenmodes by gravitational radiation [4–7]. This work has continued over the last twelve years, during which a cryogenic gravity meter has been developed and its performance has been improved. With the aid of the cryogenic gravity meter, ground motions are recorded. It is presumed that ground motions recorded by the instrument contain the following information:

(1) tidal signals

(2) signals due to earthquakes

(3) seismic noise

(4) electronic noise and instrument noise

(5) signals possibly due to gravitational radiation.

So far, attempts have been made to remove the tides from the time series and then, after complex demodulation (band-pass filtering) of the remaining time series, to examine the energy content of a single eigenmode. It is assumed that gravitational radiation, being of tensor form, would preferentially excite certain even eigenvibrations of the earth. Based on such an assumption, one can speculate that parity structures among the earth's eigenvibrations could also indicate the excitation of even eigenvibrations due to gravitational radiation.

We have been recording data on a daily basis, when numerous small earthquakes of magnitude 4 and 5 appear randomly in time and at different locations on the earth's surface. These earthquakes couple energy to different eigenmodes in an amount which is related to the seismic moment. The magnitude of displacement due to small earthquakes could not be measured with the present room temperature instruments. The smallest earthquake that has excited measurable eigenvibrations is magnitude $MS = 6.5$ [8]. With the cryogenic gravity meter it is possible to observe excited eigenmodes due to an earthquake of magnitude 5 or even less [9,10].

The problem of low frequency gravitational radiation detection, in which the entire earth is used as an antenna, has been studied extensively. A study summarizing the problems and some answers was presented to Jet Propulsion Laboratory in 1982 [11]. Within the total time series (obtained from the cryogenic gravity meter during an experiment) one can isolate the time series for an individual eigenvibration utilizing the technique of complex demodulation. In this manner, after removing the tide signal, the total time series of an individual eigenvibration at circular frequency ω_0 with damping factor Q is given by

$$Y(t) = X(t) + n_1(t) + n_2(t) + G(t) \quad , \tag{1}$$

where

$X(t)$ = signal due to earthquakes
$n_1(t)$ = seismic noise
$n_2(t)$ = instrument noise
$G(t)$ = signal due to gravity waves at frequencies ω_0.

Normally in the case of large earthquakes, when the signal to noise ratio is high, $X(t)$ can easily be determined. The signal due to earthquakes may be expressed as

$$X(t) = Z_0 \exp\left(\frac{-\omega_0 t}{2Q}\right) \cos\left(\omega_0 t + \phi_0\right) \quad , \tag{2}$$

where Z_0 is the initial amplitude. Using complex demodulation, ω_0, Q, Z_0 and ϕ_0 can be determined. Some 30 to 40 hours after a large earthquake, the amplitude of the oscillation decays so that the noise amplitudes $n_1(t)$ and $n_2(t)$, and possibly $G(t)$, become significant; thus beyond a certain length of time it is difficult to extract any useful information from the data using the power spectral density technique.

2. A SAMPLE OF EXPERIMENTAL RESULTS AND ADVANCED DATA ANALYSIS

On May 22, 1981, we started an experiment at the High Energy Physics Laboratory, Stanford University. We began collecting data at 16 h, 28 min, 13 sec, local California time. The instrument was tuned to have a resonance period of 0.81 seconds. Some 8 h, 29 min, 4 sec later, the Okinawa earthquake of MS = 5 at 28.6°N and 131.2°E at a depth of 35 km took place. The recording continued until May 24th, 18 h, 48 min. However, due to severe interference of the earthquake, the data analysis was performed some 16 hours after the quake. The data were filtered by a low pass 6-pole Butterworth filter. The filter had a gain of five and a corner frequency of 10 mHz. The digital unit of the instrument samples the output voltage up to ±10 volts, once every 9.5 seconds. A special digital bandpass filter is also used to remove the effect of glitches whenever they appear. After minor editing, the power spectral density was calculated using an FFT on 25 hours of data. Some of the results are given in figure 1. A total of

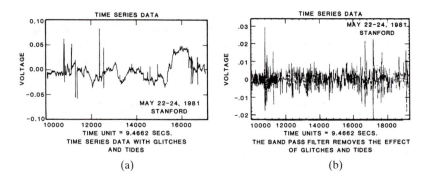

FIGURE 1. Time series data. (a) Glitches and tidal effects are included. (b) These effects have been removed by bandpass filtering.

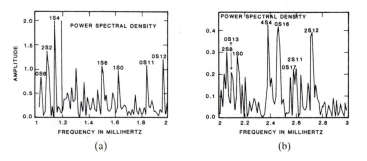

FIGURE 2. Power spectral density.

25 eigenmodes were identified in this experiment and assigned to 25 of the peaks on the record. (See figure.)

In order to evaluate the eigenmode excitation, geophysicists usually calculate the power spectral density for time series data from a gravity meter. When the peaks in these power spectral density data correspond to an eigenmode of the earth with significant signal-to-noise ratio, then it is claimed that the particular eigenmode has been excited. However, when signal-to-noise ratio is low, the power spectral density in these cases does not resolve the problem. The difficulty arises due to the fact that:

(1)   The power spectral density represents the average power at a particular frequency per unit time averaged over the entire length of the record. For an excited eigenmode of the earth, the power spectral density averaged over two or three days will not reflect the original energy content of the eigenmode.

(2)   The noncoherent noise has a finite power spectral density when the lag time is zero. (The noise is squared at any time and is summed and divided by the length of the records.)

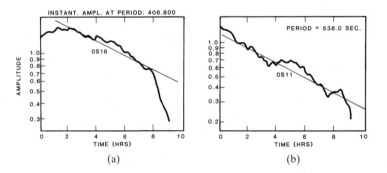

FIGURE 3. Amplitude *versus* time. (a) OS16 mode. (b) OS11 mode.

The power spectral density method allows us to have a total picture of earth excitation, examining a very wide frequency band, but denies us a detailed look at selective individual eigenmodes. Let us assume that gravitational radiation will excite only the even eigenvibrations with quadrupole and multipole moments. Then, if the energy coupled to these even modes is significant, the power spectral density may reflect an even-odd parity structure within the eigenmodes. However, if we are within the noise level, the skeptic may consider the whole domain as nothing but noise. On the other hand, if a selective bandpass filtering (also known as complex demodulation) is performed at the frequency of an eigenmode, and the results in the time domain are inspected, we can establish coherency of the oscillations, phase coherency, initial amplitude of the eigenmode, and its quality factor due to its amplitude decay.

As an example, see figure 3b, where the logarithm of the amplitude of the OS11 eigenmode is plotted against time, and figure 4b where we plot the phase coherency for the same eigenmode. It is evident that we observe vibrational coherency for some nine hours, representing a total of 60 oscillations. The amplitude is contaminated with noncoherent seismic and instrumental noise. Our theoretical modeling, and studies by B. Bolt [12] and R. Hansen [13], have found that the phase coherency drifts whenever the central frequency of the eigenmode is not precise. In fact, this has become a technique for determining more accurate central frequencies of the earth eigenmodes. In figures 3a and 4a we show amplitude and phase for the mode OS16. Note that the period of 406.8 seconds is a good choice for the mode OS16. We have a more stable phase coherency; it represents some 71 successsive oscillations. Note that the amplitude of OS16 is also contaminated with noise and possibly with a gravity wave signal. The autocorrelation of the eigenmode function in the time domain diminishes the noncoherent noise and enhances superimposed signals such as gravity waves which will be modulated due to the earth's rotation. (See figures 5a and 5b.)

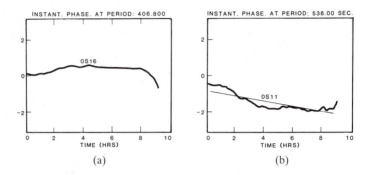

(a) (b)

FIGURE 4. Phase *versus* time. (a) OS16 mode. (b) OS11 mode.

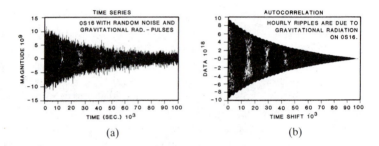

(a) (b)

FIGURE 5. (a) Time series data for the OS16 mode with random noise and simulated gravitational wave pulses. (b) Autocorrelation of time series data of figure 5a. The hourly ripples are due to the simulated gravitational wave pulses.

Furthermore, if an even eigenmode suddenly changed its amplitude and phase while the same change was not observed in the neighboring odd eigenmode, this would be considered as indirect evidence of excitation due to gravitational radiation. We have simulated the effects of a gravitational wave signal on a time series for OS16. This is shown in figure 5, where the initial amplitudes are $X(t) = 10^{10}$, $n_1(t) + n_2(t) = 10^9$. Gravitational wave pulses appear 180 seconds after $t = 0$ for the eigenmode and are repeated once every hour with $G(t) = 10^8$. From figure 5a it is evident that regular gravitational wave pulses would be completely buried within the noise. However, when an autocorrelation is performed on the time series, the simulated gravitational pulses are enhanced and noise is diminished. In this manner the gravitational radiation pulses stand out. (See figure 5b.) The larger beat frequencies (dark regions) are related to the frequency of OS16 and the mechanical frequency of the plotter.

Another interesting technique we are developing in cooperation with John D. Anderson [14] involves the cross correlation of our earth stationed gravity meter data with the satellite doppler shift data obtained by JPL.

With these new exciting advances in data analysis techniques, we have established that we can study the coherency of earth eigenvibrations which were stimulated some 16 hours before by a small earthquake of magnitude 5. For these eigenmodes, during a nine-hour observing period, a total of 60 to 70 successive oscillations were recorded. Additionally, we were able to differentiate between noncoherent noise and truly excited eigenmodes. We are also investigating techniques to differentiate small gravitational pulses interacting with an individual eigenmode. (See figures 5a and 5b.) These are important steps in the search for low frequency gravitational radiation.

Acknowledgments

I am very grateful to William Fairbank who has been a friend and constant source of help, guidance and encouragement for the past 20 years. I was introduced to low temperature physics and technology by William Fairbank and his associates, in particular his research associate Francis Everitt and his former graduate student Arthur Hebard. A portion of the cryogenic gravity meter was designed and put together in the low temperature lab of William Fairbank at Stanford. The Air Force Cambridge Research Center provided support in the early stages of research and development of the gravity meter. The Research Corporation provided funds for the Cottrell observatory and some instrumental developments.

I am grateful to the National Research Council for the opportunity to work at the Jet Propulsion Laboratory with John D. Anderson and his associates.

K. Seidman joined the project for one year, 1981–82, as a visiting research associate at JPL with a grant from the National Research Council. He and I were introduced to the technique of complex demodulation by Roger Hansen, a former graduate student of B. A. Bolt at University of California, Berkeley. The experimental results and the data analysis are due to the joint effort of Seidman and Tuman.

This paper has been edited by my wife, Turan Tuman, and my son, John Tuman. I am also grateful to Victoria Eden and to Florence Finney.

References

[1]   R. L. Forward, D. Zipoy, S. Weber, S. Smith and H. Benioff, *Nature* **189**, 473 (1961).

[2]   V. S. Tuman, in *Proceedings of the 12th International Conference on Low Temperature Physics*, edited by E. Kanda (Keigaku, Tokyo, 1970), p. 859.

[3] V. S. Tuman, *Nature* **229**, 618 (1971).

[4] V. S. Tuman, *Nature* **23**, 104 (1971).

[5] V. S. Tuman, "Observation of Earth Eigenvibrations Possibly Excited by Low Frequency Gravity Waves," at conference on experimental tests of gravitational theories, California Institute of Technology, Pasadena, California (1971).

[6] V. S. Tuman, *Gen. Rel. Grav.* **4**, 279 (1973).

[7] V. S. Tuman, in *Proceedings of the International School of Physics (Enrico Fermi), Course LVI, Experimental Gravitation*, edited by B. Bertotti (Academic Press, New York, 1974), p. 543.

[8] B. Block, J. Dartler and R. D. Moore, *Nature* **226**, 343 (1970).

[9] V. S. Tuman, "Detection of Earth Eigenvibrations from Small Earthquakes with a Cryogenic Gravity Meter," at American Geophysical Union Fall Meeting (1982).

[10] V. S. Tuman, *Acta Astronautica* **9**, 63 (1982).

[11] V. S. Tuman, *J.P.L. Report 413-259*, dedicated to Professor William Fairbank on the occasion of his 65th birthday, March, 1982.

[12] B. A. Bolt and D. R. Brillinger, *Geophy. J. R. Astron. Soc.* **59**, 593 (1979).

[13] R. A. Hansen, Ph.D. Thesis, University of California, Berkeley, 1981.

[14] J. D. Anderson, private communication.

A High Energy
Gamma Ray Astronomy Experiment

Robert Hofstadter

I am very happy to contribute to this "family-type" publication. I feel as if I am almost a part of the family. Felix Bloch and George Pake have both stated that they shared in bringing William Fairbank to Stanford, and I must lay a claim of my own. A long time ago, in approximately 1958, Bill Fairbank and I were on a national committee to select postdoctoral candidates. That was when I first met him and I was pretty impressed by Bill, particularly by the way he took me aside and talked about his ideas. I had to listen (as we heard at the conference from several speakers), but I thought Bill was a real find. I cannot remember the precise history, but I think I came back to Stanford and talked to Felix Bloch and Leonard Schiff about him. Of course, there were other candidates at the time—well, you know the consequences.

I am going to describe a rather big experiment that my colleagues and I are working on, involving NASA's Gamma Ray Observatory (GRO). The theme of this volume is "Near Zero" and that fits my topic, because we have no results. But GRO exemplifies the near zero principle in another way because it investigates new gamma ray phenomena by relying on the space program to take us into the region of zero interference above the earth's atmosphere. In its present form (figure 1), GRO has four experiments.

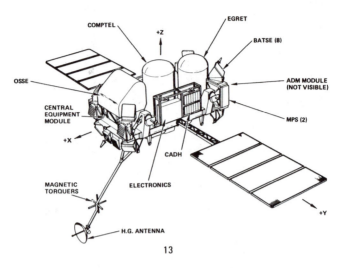

13

FIGURE 1. The Gamma Ray Observatory.

(1)   OSSE (Oriented Scintillation Spectrometer Experiment) by J. O. Kurfess and colleagues at the Naval Research Laboratory, using sodium iodide detectors in the range between 0.1 MeV and 10 MeV with energy resolutions from 8% to 3.2% in different ranges.

(2)   COMPTEL (Imaging Compton Telescope) by V. Schoenfelder and colleagues at the Max Planck Institute, which depends on the Compton effect employing an imaging technique to look in the energy range 1 to 30 MeV.

(3)   BATSE (Burst and Transient Source Experiment) by G. J. Fishman and colleagues at NASA Marshall Space Flight Center, looking in the range 50 keV to 600 keV.

(4)   EGRET (Energetic Gamma Ray Experiment Telescope), our own program, on which the principal investigators are Carl Fichtel of NASA Goddard Center, myself, and Klaus Pinkau of the Max Planck Institute, with various divided responsibilities. The EGRET instrument is essentially a spark chamber device and a sodium iodide energy detector, with a range of 20 MeV to perhaps 30,000 MeV with an energy resolution of 15% most of the way through the spectrum.

You may wonder how I became involved in GRO. It is a long story. In 1966 an old friend of mine, Urner Liddell, who was then working for NASA, said to me: "There is a lot of fundamental physics that can be done in space. How about proposing an experiment?" At first I thought about the infrared field but I could not think of any good ideas. Later, with two colleagues and former students of mine, Don Aitken and Hall Crannell,

I proposed to NASA a rocket experiment with millisecond time response to detect x-rays from objects such as the Crab nebula, but it was turned down. This was a great disappointment to us, especially when Herb Friedman detected x-rays from the Crab pulsar with millisecond timing after the pulsar was discovered. Later when HEAO (the High Energy Astronomical Observatory) was suggested NASA said, "Why don't you propose for it?" and we wrote a proposal to do what is essentially the experiment I am going to describe here. The proposal ultimately became one of the two winners of an open competition. The other winner was a proposal from NASA Goddard Center. NASA told us to get together. We did, and combined efforts, and have been together ever since with Carl Fichtel and his group at Goddard. I will not go into all the complicated history of the cutbacks in HEAO and the birth of GRO, but finally NASA has been able to support our experiment and we are starting to get some real activity going. Listed below[1] are the past and present members of the Stanford team, and our principal collaborators elsewhere. Among these people Barrie Hughes has been a leader; a great deal of the credit for what has been done belongs to him.

Why do we want to do this experiment? There are two main reasons, and these are so primitive that they are often overlooked. One is that gamma rays come directly from their source. This is not like the situation with cosmic rays. Cosmic ray charged particles are bent by magnetic fields and it is not possible to know where they came from. Gamma rays come along a straight line from the source and from way back in time. They penetrate through nebulae without interaction. They can be used as very good probes for what is out there in the sky. Second, we are talking about high energy gamma rays in the range of 20 MeV to 30,000 MeV. These gamma rays are characteristic of elementary particle physics. If you think about it, all of astronomy and astrophysics is based on classical physics. Very little is known about nuclear physics in stars or nebulae or quasars— practically nothing is known except for the little work that I shall describe below. It also appears that a large amount of the energy emitted from active galaxies, pulsars and so on, is in the form of high energy gamma rays. In fact, Ruffini and his associates predict that 50% of the emitted energy in many active systems should be in the gamma ray region at high energy. For these important reasons I think that gamma ray astronomy is going to be a really big new subject.

----

[1]Past and present members of the team include: *Stanford:* B. Beron, L. Campbell-Finman, R. Hofstadter, E. B. Hughes, J. Lepetich, A. Johansson, L. O'Neill, R. Parks, J. Rolfe, R. Schilling. *Goddard:* C. Fichtel, R. Hartman, D. Kniffen. *Max Planck Institute:* G. Kambach, H. Mayer-Hasselwander, K. Pinkau. *Grumman Aerospace Corp.:* A. Favale, E. Schneid.

You all know what x-ray astronomy has done for astrophysics, but x-rays, I remind you again, are still part of atomic physics. Gamma ray astronomy will take us even farther, because it will discover phenomena characteristic of particle physics. Very few of such phenomena have yet been seen in astronomical objects. We can make a list of some of the principal goals of the experiment as follows: (1) scan and search for localized ("point") γ-ray sources, (2) measure their intensity, energy spectrum, position and time variation, (3) determine more precisely locations *etc.* of known sources, (4) examine supernova remnants, (5) study diffuse high energy emission from our galaxy (spiral arms, clouds, galactic center, halo), (6) attempt to distinguish electron and nucleon effects from the shape of the spectra in this galaxy, (7) study other galaxies, normal and active, (8) examine the properties of diffuse celestial radiation, including whether it is isotropic. But of course the main point is that there is a lot out there that is unknown. Probably the most interesting things are going to be unexpected discoveries.

Before describing our instrument, let us consider some history. The first detection of gamma rays in space was made by Kraushaar in 1968. As a matter of fact, the Japanese physicist Hayakawa, in 1952, just after the discovery of the pion, predicted that there ought to be radiation from neutral pions coming from stars, and indeed there is. Kraushaar was the first person to see the relevant gamma rays, using a rather primitive and small instrument containing a Cerenkov counter and converter and an energy discriminator. The idea of the converter, of course, is to convert the gamma rays into pairs, and there is an anticoincidence shield on the outside so that charged particles do not register in the telescope. There is also a Cerenkov counter to tell whether the particle is going up or down. The sodium iodide detector's purpose is to detect the energy of the pairs made by the gamma rays and hence the energy of the gamma rays. This experiment established that there are gamma rays in the energy range above 50 MeV coming from the galactic disk—our own galaxy's disk—the Milky Way. Kraushaar detected 631 photons, and only a couple of hundred of those were above 100 MeV. Later there was a rather big step made by our collaborators, Carl Fichtel and his group, with an instrument (SAS-2) having a number of similarities to the earlier experiment of Kraushaar. Fichtel's device had an anticoincidence shield and spark chambers and a Cerenkov counter facing photomultipliers at the bottom. Subsequently I will show some of the results this instrument produced. The next launch was a European satellite experiment with a telescope which has proved to be a marvelous instrument. It was built more or less in the same way as the others, having an anticoincidence shield, spark chambers, a time-of-flight indicator and a cesium iodide detector at the bottom. The instrument produced data for 4 years and I think it is still producing very nice data.

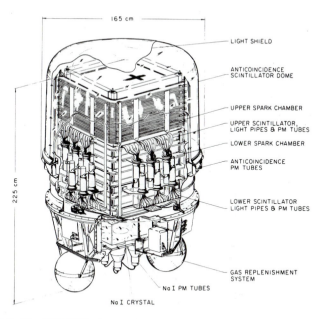

FIGURE 2. The EGRET instrument.

Figure 2 shows our instrument. It has an anticoincidence shield, upper spark chambers, and scintillators which allow time-of-flight measurement to be made and provide a trigger, and a set of lower spark chambers. The upper spark chambers are 28 in number and contain 27 interleaved tantalum plates, in which the conversion of the gamma rays occurs into pairs. The pairs then go through the rest of the scintillator tiles and through the lower spark chambers into a sodium iodide crystal at the bottom. The sodium iodide crystal assembly consists of 36 crystals weighing 900 lbs. We could have improved the sensitivity by adding more crystals but there was a weight limit.

Figure 3 shows what a gamma ray looks like when it converts. The figure shows an x view and a y view in the spark chamber. Every good triggered event will be examined. The spark chamber assembly consists of 1,000 wires in x and 1,000 wires in y; and it will obtain very good angular information. The anticoincidence shield is being built by Munich. This group is also building the calibration structure, a huge device which we hope to have here at Stanford in 1985. We will calibrate the EGRET instrument at SLAC.

In ten days of observation at various energies, the Crab nebula should yield the number of counts shown in table 1. This should give one a feeling for the kind of counting rates we expect to have. The Crab nebula is not the most powerful gamma ray emitter that has been seen. It is about

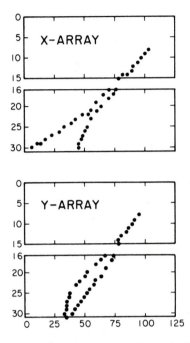

FIGURE 3. The appearance of a gamma ray pair production event taken from an observation of Fichtel and collaborators. The vertical axis is compressed by a factor of 2.7 relative to the horizontal axis.

one-third the intensity of the most powerful that has been seen, and there are probably a lot more powerful sources to be discovered. The sky really has not been completely scanned and most of the effort of observation has gone into the galactic plane, so that very little is known about what lies outside the Milky Way. In fact, so far as I know, only one quasar (3C273) which is a good gamma emitter has been discovered outside the galactic disk. Table 2 shows the kind of angular resolution we might expect to achieve in straightforward measurements. By splitting the peak we can do better. For strong point sources one might expect to be able to split the central region of a line to perhaps 10% of its half width, which would mean a resolution of 0.02° or about 1 arc-minute at 2000 MeV and 8 arc-minutes at 100 MeV. From intensity considerations it should be possible to recognize point sources more than two orders of magnitude weaker than the Crab nebula. The number of x-ray sources detectable should be over 100.

What kinds of physics can be done? Various processes that give rise to energetic gamma rays are sketched in figure 4. Curvature radiation has been described by our Stanford colleague Peter Sturrock. This process is

TABLE 1. SOME TYPICAL COUNTING RATES.

Number of Counts from Crab Pulsar in 10 Days of Observation

E(MeV)	$I (> E)$ photons $\cdot$ cm$^{-2}\cdot$s^{-1}	Number Observed*
20	1.6×10^{-5}	4.6×10^3
50	5.9×10^{-6}	4.0×10^3
100	2.7×10^{-6}	1.8×10^3
500	4.7×10^{-7}	3.4×10^2
2000	1.0×10^{-7}	0.7×10^2

Number of Galactic Plane γ-Rays with Energy Greater than 100 MeV That Could Be Observed in 2 Months

Direction	Intensity	Number Observed
Galactic plane (center)	1×10^{-4} cm$^{-2}\cdot$rad$^{-1}\cdot$s^{-1}	3×10^5
Galactic plane (away from center)	2×10^{-5} cm$^{-2}\cdot$rad$^{-1}\cdot$s^{-1}	6×10^4

Expected Number of "Isotropic" γ-Rays for a 6-Month Viewing Period

E(MeV)	$J(> E)$ cm$^{-2}\cdot$ster$^{-1}\cdot$s^{-1}	Number Observed $(> E)$*
20	1.8×10^{-4}	6.5×10^5
50	3.3×10^{-5}	1.9×10^5
100	9.0×10^{-6}	6.6×10^4
500	4.6×10^{-7}	3.4×10^3

* Effective collection time $\sim$ 45%

TABLE 2. EXPECTED ANGULAR RESOLUTION FROM EGRET.

100 MeV	1.6°
500 MeV	0.6°
2000 MeV	0.2°

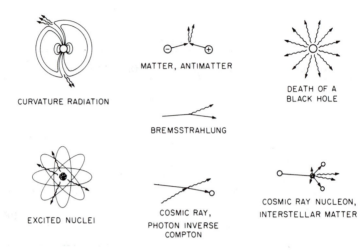

FIGURE 4. Possible source processes for astrophysical gamma rays.

very important for the emission of gamma rays. Electrons and positrons can also annihilate and produce gamma rays. There is the additional possibility of Hawking radiation, involving the death of a black hole. The inverse Compton effect, nuclear gamma rays, bremsstrahlung, and the source of gamma rays originally postulated by Hayakawa, neutral pions—all of these processes can occur. I would like to call attention to an additional possibility. Our Crystal Ball experiment at SLAC also uses sodium iodide as a gamma detector, and has produced the spectrum of gamma radiation from the decay of charmonium shown in figure 5. Charmonium is a meson just like the pion, except it is more massive and is made of a charmed quark and an antiquark. The energy states in charmonium, as well as the gamma rays, are shown in the figure. The latter range from 100 MeV up to 700 MeV; there is one at 640 MeV. These gamma rays must exist in the sky, too. In fact, I think the best accelerators in the world are up there in the sky, and they must be making charmonium and everything else—electron/positron collisions and annihilations at all energies; and so we ought to eventually see such radiations. The gamma rays may be broadened, of course, by the Doppler effect, but we ought to see them at some future time, and I am predicting that we will. I hope Remo Ruffini or other theoreticians will calculate the appropriate rates.

Figure 6 shows the observations from SAS-2 along the galactic plane. The Vela pulsar is the strongest emitter, as far as I know; the Crab is also shown. The new γ195+5 was discovered; no one knew this existed previously. Also note Cygnus X-3. A little later in time, along came COS-B, the European satellite. Figure 7 shows what it produced. As you can see,

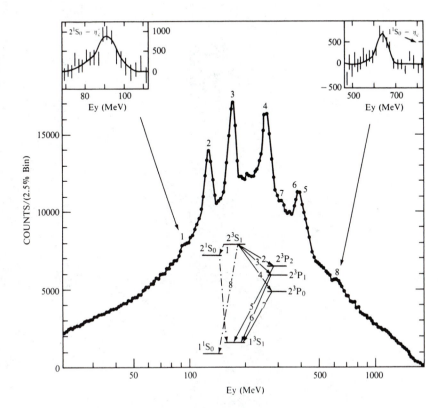

FIGURE 5. The gamma ray spectrum of the first excited state of charmonium.

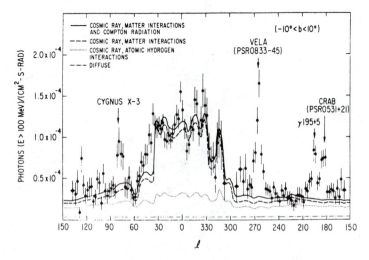

FIGURE 6. Gamma ray data in the galactic plane obtained by SAS-2.

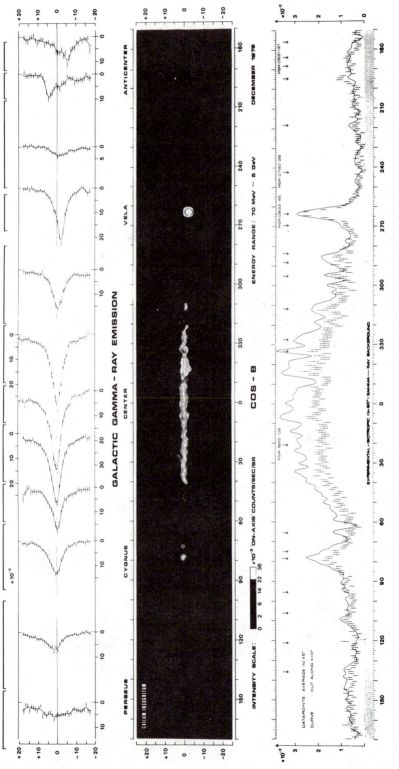

FIGURE 7. Galactic gamma ray emission data obtained by COS-B.

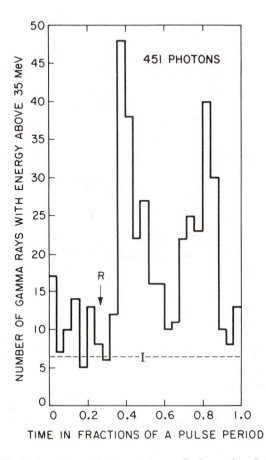

FIGURE 8a. The Vela pulsar and its emissions. *R* shows the phase of the radio emissions.

there are a lot of sources which fit very nicely with previous information on SAS-2, and also new sources. More work has been done on COS-B, and there are probably about 25 or more sources now known. It is not clear whether they are "point" sources or whether they are just localized sources because the angular resolution has not yet been quite good enough to tell. Since our instrument will have better angular resolution than any previous one, we may be able to answer this question.

Figure 8a shows the Vela pulsar and presents the pulse period. There are two gamma ray emissions in time, and note where the radio pulse appears. How do you explain that? Obviously, we have to have a model of the pulsar to explain it. In figure 8b we see the way the energy is distributed in time in various time intervals; all of that has to be explained. Figure 9a shows the Vela pulsar again. Notice that what is shown is the

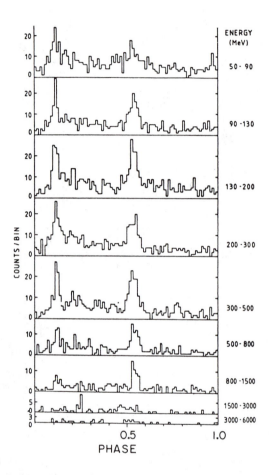

FIGURE 8b. The distribution of energy in various time intervals of the period of the Vela pulsar.

differential gamma ray spectrum of the *pulsed* emission. Figure 9b shows the *total* emission, and it turns out that the two curves are exactly the same for Vela. The emission is all pulsed. But not so for the Crab pulsar. The SAS-2 data for the Crab are shown in figure 10. There is something different going on with the Crab. You can compare the Vela pulsar with the Crab pulsar. In the Crab, everything is repeated in phase: in the optical spectrum, in the radio spectrum, and in the x-ray spectrum. In the Vela, there is some similarity between the optical and the x-ray results, but the radio pulse occurs at a different time. So there is real physics involved.

Peter Sturrock and Peter Goldreich are two of the individuals who tried to explain how a pulsar works, and further work has been done by

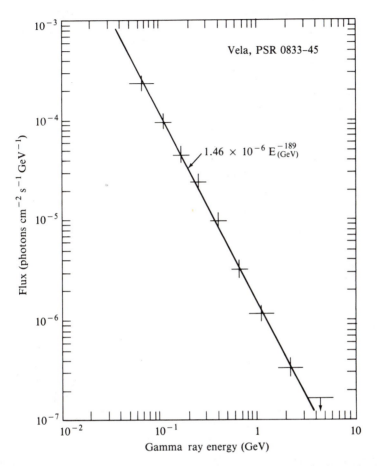

FIGURE 9a. The COS-B differential gamma ray spectrum of the *pulsed* emission from the Vela pulsar.

Russian authors. Figure 11 shows a model made by the Russian authors, L. M. Ozernoy and V. V. Usov, of a pulsar which is rotating around an axis perpendicular to the plane of the figure with angular velocity Ω. The pulsar radius is R_* and the dipole magnetic field around the pulsar that has been postulated is represented in the figure. In the neighborhood of the pulsar/neutron star the magnetic field is approximately 10^{12} gauss, and it has a dipole field strength of that order near the surface. At the velocity of light radius c/Ω there are currents streaming out. For example, an electron current will stream out to the right and a proton current will stream out to the left. The electron current in the magnetosphere in the plasma generates high energy gamma rays, which come out in a beam labeled 1, and the

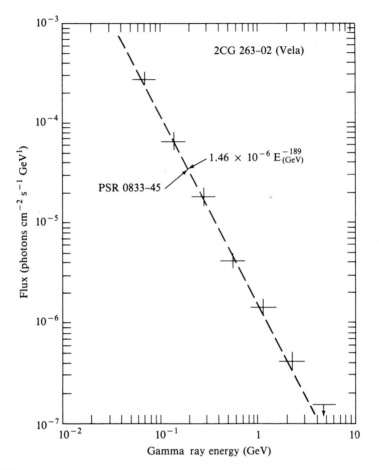

FIGURE 9b. The COS-B differential gamma ray spectrum of the *total* emission from the Vela pulsar.

radio waves come out in a beam labeled 2. There is a great deal of classical physics as well as particle physics involved in the model. I am not sure, of course, that this model represents the correct explanation. Positrons play a large role in the production of radio waves through the showering mechanism, and in the magnetosphere it seems possible to produce lots of particles and waves.

Figure 12a shows the spectrum of gamma ray emission in a quasar (3C273). Figure 12b shows a picture of the diffuse radiation coming from outside the galaxy. All these observations need to be explained. Finally I come to the burst recorded by our own Vela satellites that keep looking for nuclear explosions everywhere. The same event was seen by a Russian satellite, and was an enormous burst that was observed on March 5, 1979.

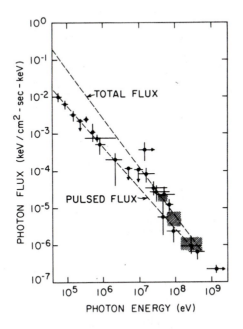

FIGURE 10. The SAS-2 gamma ray flux data from the Crab nebula.

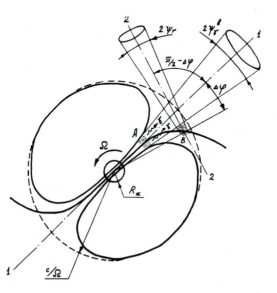

FIGURE 11. A model of a pulsar made by Russian authors, L. M. Ozernoy and V. V. Usov.

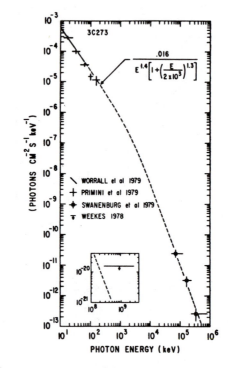

FIGURE 12a. The energy spectrum of the quasar 3C273.

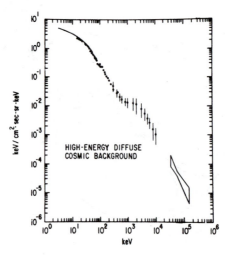

FIGURE 12b. The diffuse gamma radiation spectrum.

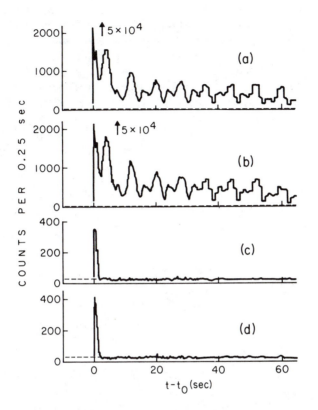

FIGURE 13. The Russian observations, by means of the Venera spacecraft, of the huge burst on March 5, 1979. The data show the time structure of the bursts with resolution 1/4 and 1 sec in the energy range 50–150 keV. (a) 5 March, Venera 12. (b) 5 March, Venera 11. (c) 6 March, Venera 12: $t_0 = 6$ h 17 min 30 sec, 455 UT (Universal Time). (d) 6 March, Venera 11: $t_0 = 6$ h 17 min 25 sec, 200 UT. Dashed line indicates background count rate. Points before t_0 show previous history of the bursts.

The ringing that occurred after the burst is shown in figure 13, and this represents the Russian observation of the event. Furthermore, there is an energy spectrum associated with those bursts.

What I want to stress is that the GRO, if it flies, will get a great deal of new information of the kind that I have described, and it is wonderful to think of all the spectacular information that will be made available. We should get data in 1990, almost a quarter of a century after we began thinking of the experiment.

I hope I make it, and I hope I can come back to Bill Fairbank's 75th birthday party and tell him about the results. I hope that his and Francis Everitt's relativity gyroscope experiment will also be producing results by that time.

Let me make one more remark. Many times I hear scientists say that astrophysics is a difficult subject because one cannot do controlled experiments. This may be true, but I can think of astrophysics in a different way. All the experiments have already been done—out there—and we only have to observe them and analyze the data. What a challenging future!

I thank Dr. Carl Fichtel for providing me with many of the figures and data used in this paper.

Low Temperature Optics for the Space Infrared Telescope Facility

Fred C. Witteborn, Jacob H. Miller and Michael W. Werner

1. INTRODUCTION

Low temperature techniques find numerous applications in instrumentation for infrared astronomy. Sensitive infrared detectors require liquid helium cooling and cryogenic electronics. Even ^{3}He refrigerators are used at the focal planes of astronomical telescopes to permit detector cooling to 0.3 K. A fundamental limitation on the sensitivity of infrared instruments stems from statistical fluctuations in the thermal background radiation. This background consists of radiation emitted along the entire path between the astronomers' cold detectors and the distant sources they wish to study. The unwanted radiation comes from the earth's atmosphere, dust in the solar system (zodiacal dust) and the telescope used to collect the desired radiation. Airborne and balloon-borne infrared telescopes, which have already reduced the spectral absorption of the atmosphere by flying above most of the water vapor, are nevertheless severely hampered by thermal emission from the atmosphere and the telescope optics.

The advent of space flight offers the possibility of greatly increased sensitivity for infrared measurements through the use of cryogenic techniques.

As seen from figure 1, the atmospheric background is reduced many orders of magnitude by observing from space, where the principal natural limitation is emission from zodiacal dust. To take full advantage of the low infrared background in space, however, it is necessary to cool the telescope to reduce its thermal emission below the level of the natural background. Following preliminary surveys by rocket-borne cooled telescopes [1], a complete survey of the sky in four passbands spanning the range 8–120 μm has been performed by the IRAS, the superfluid helium cooled Infrared Astronomy Satellite [2] launched early in 1983. Detailed studies of individual objects with better spectroscopic and spatial resolution are now essential. They require a moderate aperture telescope (80 cm or more) operating in a near zero thermal background with good image quality and precise pointing capability. Infrared astronomers, scientists and engineers at the NASA Ames Research Center and commercial aerospace and optics firms, have spent several years studying the scientific requirements [3,4,5] and design feasibility [6,7] of an observatory-class telescope with such capabilities for use on the space shuttle. The resulting system, the Space Infrared Telescope Facility (SIRTF), will achieve 100 to 1000 times the sensitivity of any existing telescope in the 2 to 200 μm spectral range by exploiting the low background of the space environment [8].

SIRTF will be designed for on the order of 2 years of unattended operation in earth orbit. It will be delivered to orbit by the shuttle which

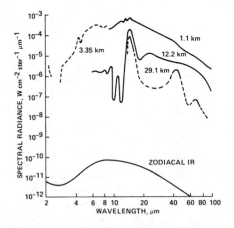

FIGURE 1. Infrared background radiations at various altitudes. Atmospheric radiance is plotted *versus* wavelength for various altitudes based on data from balloon, airborne and ground-based observations. The zodiacal IR curve, representative of the background radiation in earth orbit, corresponds to radiation from the ecliptic plane, 90° from the sun [16,17,18].

can also be used to service or retrieve it. A combination of gyro-controlled stabilization and star tracking systems will point the telescope to a fraction of an arc-second. Routine operation of the telescope will be controlled by an on-board computer with preprogrammed instructions for most observations. Updates to these instructions can be entered, with suitable lead time, by an investigator on the ground. Data will be broadcast to the ground during portions of each orbit *via* the Tracking and Data Relay Satellites.

2. THE SIRTF ENVIRONMENT

Before examining the design of the telescope, we first review the shuttle environment which sets constraints on the achievable infrared sensitivity. SIRTF will be operated at an altitude of at least 400 km and no higher than 900 km. This altitude is low enough to avoid the most intense part of the inner Van Allen belt, whose high energy particles cause serious noise in infrared detectors, yet high enough to make atmospheric infrared background radiation negligible throughout almost all of the infrared. The only atmospheric radiation likely to be seen above SIRTF is from the 63.1 μm and 147 μm lines of atomic oxygen [9]. In addition, there will be radiation arising from contaminant gases and dust particles evolving from the telescope and its platform. The radiation from contaminant molecules is predicted to be comparable to or less than the zodiacal infrared background throughout the infrared (figure 2; also [10]) for shuttle-borne operation, and much less for unattended operation. Preliminary analysis of the IRAS data shows no degradation of sensitivity by dust particles or molecular deposition [11]. In the 900 km IRAS orbit high energy particle bombardment is sufficiently high near the South Atlantic Anomaly to prevent useful data-taking over about 10% of each orbit.

The background environment in which SIRTF operates is the sum of the sources listed above plus the radiation emitted and scattered from the telescope optics and those portions of the baffles and telescope structure that lie in the fields of view of the focal plane instruments. The noise arises from fluctuations in the total background radiation plus intrinsic noise in the detectors and their associated electronics. Photons obey Bose-Einstein statistics, so that the noise $\overline{\Delta N_{rms}}$ in a measurement of N photons of wavelength λ from a background at temperature T is found from [12]:

$$\overline{\Delta N_{rms}^2} = \overline{N} \left[\frac{\exp(hc/\lambda kT) - 1 + \eta\theta\epsilon}{\exp(hc/\lambda kT) - 1} \right] \quad , \tag{1}$$

where η is detector quantum efficiency, ϵ is emissivity of the photon source, and θ is transmission of the optics. For most practical cases in the infrared this is very closely approximated by $\overline{\Delta N_{rms}} = N^{1/2}$.

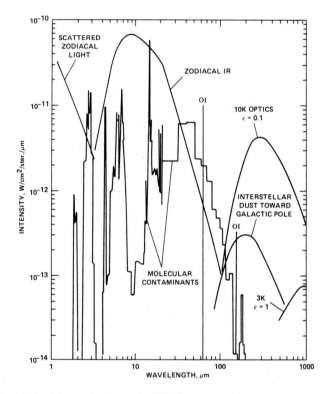

FIGURE 2. Infrared emission from the SIRTF environment. The zodiacal infrared in the ecliptic plane, 90° from the sun, is approximated by a surface at $T = 300$ K with emissivity of 6×10^{-8}. The intensity at the ecliptic poles is thought to be at least 3 times lower. The estimates of molecular contaminant emission [9 and 10] are based on an H_2O column density of 1×10^{12} cm^{-2} and CO_2 column density of 2×10^{11} cm^{-2}.

Figure 3 is a plot of the background noise as a function of telescope temperature, seen by a detector having unit quantum efficiency (*i.e.*, it detects every photon in its spectral bandwidth), a diffraction limited field of view in object space (the sky) and a 50% spectral bandwidth. Detectors with noise-equivalent-power less than 10^{-17} W/Hz$^{1/2}$ have already been reported [13], and continued improvements can be anticipated. Some of the most exciting investigations proposed for SIRTF require this high sensitivity, which is permitted by the expected minima in background noise (figure 2) near 3 μm and 100 μm.

3. THE DESIGN OF SIRTF

Reaching the sensitivity limit set by the naturally occurring zodiacal infrared background emission requires a stable cryogenic environment below

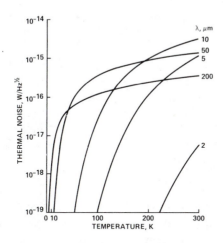

FIGURE 3. Photon fluctuation noise *versus* temperature of telescope optics at several wavelengths λ. The radiation is assumed to come from optics and structure with combined emissivity 0.2, in a diffraction limited field of view and in a bandwidth of 0.5 λ centered on λ.

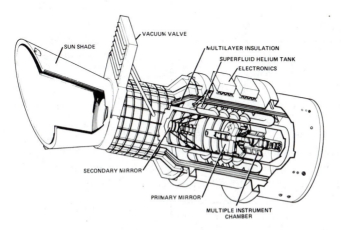

FIGURE 4. SIRTF telescope concept. The primary mirror diameter is 90 cm. The overall length is 7.4 meters.

10 K, good rejection of off-axis infrared radiation and good image quality (dependent on both the optics and the pointing stability). The general layout in shown in figure 4. The telescope is essentially a large cryostat surrounding an optical system that focuses visible and infrared energy from distant sources onto a variety of instruments. Superfluid helium will be used to cool the instrument chamber and optics. Provision for cooling below 2 K will be provided internally by those instruments which require it. The telescope will be installed in the space shuttle and cooled down several weeks prior to launch. The large, vacuum-tight aperture cover (essentially a gate valve) will be closed prior to vacuum pump-out and cooldown of the telescope interior. The valve will not be opened until at least 24 hours after reaching orbit, so that most contaminants will have had time to disperse before the optics and baffles are exposed. The entrance to the telescope is protected further by an aperture shade which reduces the heat load on the cold telescope baffle by reflecting radiation from the earth and sun back out of the entrance. (The telescope line of sight never comes closer than 60° to any part of the sun, the earth's troposphere, or the shuttle.)

Consider now the light path through the telescope after its protective cover has been removed. Photons from an object of interest pass through the entrance aperture and then strike the 90 cm diameter concave primary which reflects them to the secondary mirror. The present concept incorporates Ritchey-Chrétien (an aplanatic Cassegrain) optics. In this as in other Cassegrain variations, the secondary is convex and the photons are reflected to a focus well past a beam splitter in the multiple instrument chamber. The beam splitter allows 50% or more of the visible light to pass through to a focus on a two-dimensional array of detectors used as a fine guidance sensor. Image motion detected by this array can be used to provide error signals to the gyro stabilization system. The beam splitter reflects infrared photons into any one of several instruments in the multiple instrument chamber. These instruments will include photometers, detector array cameras, polarimeters and spectrometers operating in various bands throughout the 1 to 1000 μm range. Signals from the instruments will be amplified, digitized and multiplexed for broadcast to earth *via* the Tracking and Data Relay Satellites.

The signal from the source can be extracted from the average background power by oscillating the secondary mirror at frequencies in the range 2 Hz to 20 Hz and measuring only the phase and frequency correlated signals from the detectors. This is done by pointing the telescope so that the desired object is seen by the detector at one position of the oscillating mirror and an equal angular area of "empty sky" is seen at the other position. The effects of possible gradients in background intensity are removed by periodically moving the telescope through an angle equal to the amplitude of secondary oscillation, so that the object is seen with the mirror in the opposite position. These techniques of "chopping" and "beam switching" are used commonly in earth-bound infrared observing, where signals as small as 1 part in 10^6 of the average background are routinely detected. To achieve optimum discrimination against background in a cryogenic space telescope requires special attention to problems of thermal control. This problem is particularly acute for the secondary mirror. Since the detection system measures as signal the difference in energy received in two successive time intervals, fluctuations in system temperature in the bandwidth of detection will appear as noise.

One important source of background at long wavelengths is emission from the secondary mirror itself. If the secondary mirror temperature $T = 7$ K, it can be shown for $\lambda = 200$ μm that to keep the power fluctuations below 10^{-17} W/Hz$^{1/2}$ the temperatures of the mirror and nearby baffle must change at rates no higher than a few mK per 0.25 second. Since P_λ is very wavelength dependent, the constraint on temperature variations becomes much more stringent at the longer wavelengths. The heat load on the secondary mirror is from the motor which oscillates it and off-axis radiation that strikes the baffle assembly surrounding the secondary mirror. In the current design, cooling is provided by helium gas flow from the superfluid helium tanks. An additional device available to the designer to reduce the temperature fluctuations is to increase the heat capacity of the secondary-mirror/baffle assembly. A 500 cm^3 sealed container filled with 18 grams of He would have a heat capacity of more than 50 J/K. If this were properly heat sunk to intercept heat flow from other sources, the secondary mirror temperature variations could be kept below 50 μK per 0.25 second for heat exchange imbalances (heating minus cooling) as high as 10 mW. The primary mirror has a larger thermal mass and relatively smaller variations in heat flow at the oscillation frequency. Its thermal problems are different from those of the secondary and are treated in the next section.

In discussing the path of a photon through the telescope we considered above only photons from sources in the field of view. For the telescope to achieve its desired sensitivity, photons that do not originate in the field of view must not be permitted to reach the detectors. The main problem is from large (in angular size) off-axis sources such as the sun and the earth. The aperture shade provides the first stage of off-axis rejection, and the cold baffles the second stage. The length of the cold baffle section relative to the secondary mirror to primary mirror distance was selected to minimize stray radiation. Edges, corners and rough surfaces, when illuminated by off-axis radiation sources, themselves become sources of off-axis radiation. Estimates of the off-axis energy reaching the detector show that natural background limited performance is achieved by the present design for off-axis sources at angles greater than 60°.

4. THE USE OF GLASSY MATERIALS
FOR THE SIRTF PRIMARY

4.1   Thermal Considerations

The foregoing discussion leads us to the following requirements on our primary mirror material. (a) It must be capable of achieving a good optical figure (diffraction-limited performance at 2 μm wavelength to exploit the background minimum at 3 μm) at a temperature of less than 10 K. (b) It must not exhibit thermal variations large enough to produce excessive noise at the focal plane. (c) It must have a smooth surface to minimize scattering of off-axis infrared radiation.

Early studies of SIRTF mirror materials favored the use of beryllium because of its high thermal conductivity and high strength-to-weight ratio, but in spite of years of efforts to adapt metal mirrors to all the needs of space optics, no manufacturers can promise large metal mirrors with adequate figure quality for SIRTF.

The use of glassy materials, particularly fused silica and quartz, in other space optics applications led us to consider fused quartz for the SIRTF primary. The density of fused quartz, 2.2 g/cm^3, is only a little higher than that of beryllium, but its thermal conductivity of 6×10^{-3} W/cm/K near 10 K compares poorly with values expected for a pure metal. On the other hand, fused quartz can be figured to much closer tolerances than beryllium. In addition, the high homogeneity of commercially available glasses, such as Hereaus-Amersil T08E, offers the promise of low distortion after cooling to low temperature. Furthermore, glassy materials can be polished to much lower surface roughness than can bare beryllium and thus scatter less off-axis energy. To establish the usefulness of fused quartz in the SIRTF application we first show that its thermal conductivity is adequate for the application and then review results of a measurement of the optical figure of a 50 cm diameter fused quartz mirror cooled to 10.5 K.

In the SIRTF application we are concerned about the time required for a mirror to reach a new equilibrium temperature distribution after the heat load to its front surface changes. This change will be on the order of a few milliwatts. It stems mainly from sunlight scattered off the inside of the aperture shade, which is subsequently absorbed by the cooled baffles in the interior of the telescope and reradiated. This extra heat load disappears on the night side of the orbit. Since the orbital period is 90 minutes, the loss of 2 or 3 minutes per half orbit to allow for thermal relaxation would not impact the telescope usage severely, but ten times greater loss would be unacceptable. If we simplify the geometry by regarding the mirror as an infinite slab of thickness $L = 15$ cm, heat sunk on one side and heated on the other side uniformly, then the thermal relaxation time is on the order of $\frac{4L^2\rho c}{\pi^2 k} = 150$ sec, where the density $\rho = 2.2$ g/cm^3, the heat capacity $c = 4.5 \times 10^{-3}$ J/g/K, and the thermal conductivity $k = 6 \times 10^{-3}$ J/cm/K/sec. This is a sufficiently small fraction of an orbit to be acceptable.

One must also be concerned with temperature gradients along the surface of the mirror. These will occur when the power absorbed by the mirror is nonuniformly distributed. A detector in the focal plane does not see all of the primary mirror at one time. As the secondary mirror oscillates, it reflects light to the detector from slightly different parts of the primary. The fraction of the primary not common to the two extreme positions can be as large as 9%. This places a constraint on the maximum allowable temperature variations across the primary mirror, a constraint that becomes more severe at longer wavelengths. Opposite edges of the mirror must have a temperature difference less than 2 mK if the error in signal is to be less than 10^{-17} W at 200 μm. This error tends to be canceled by the beam switching, but systematic variations on time scales of the order of the beam switching time must be avoided.

The main heat load on the primary is radiation from the baffles which operate at 10 K to 20 K most of the time. Less than 0.42 mW is absorbed in the mirror. Under this heat load a mirror with the thickness and conductivity used above would have an average equilibrium front-to-back temperature difference of less than 0.2 mK. It should not be difficult, therefore, to keep opposite edge temperature differences below 2 mK if a well-designed heat sink is used to cool the back side of the primary mirror.

4.2   Optical Testing

Optical testing of a sample fused quartz (Hereaus-Amersil T08E) mirror was performed in the liquid helium cooled vacuum test chamber shown in figure 5. The 50 cm diameter fused quartz spherical f/2 mirror was ground, polished and figured by the Optical Science Center, University of Arizona. Thermal contact to the mirror was provided by 48 copper straps arranged so that each would cool an equal volume of the mirror. Each strap was 3 cm long and consisted of 16 strands of 0.005 inch diameter copper. Each strap was attached to the glass with a 3 mm diameter drop of epoxy (Torr Seal). The other end of each strap was soldered to large copper buss bars which were bolted to the aluminum bottom of the LHe reservoir. The entire inside of the helium cooled radiation shield surrounding the mirror was painted black to reduce the thermal load on the mirror caused by scattered radiation from the warmer sections of the dewar. Cooling from room temperature to 100 K using LN$_2$ required 96 hours. After LHe was added, cooling from 100 K to 10.5 K took 36 hours. The results of the optical interference tests [14] are summarized in figure 6. The figure changed by only about 1/30 of a 2 μm wave in cooling from room temperature to 10.5 K. Including the

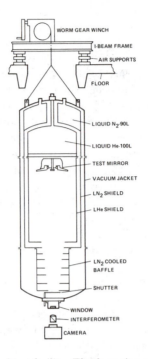

FIGURE 5. Cryogenic optical test facility. The dewar is suspended from a vibrationally isolated platform. Mirrors with diameters up to 60 cm and focal lengths up to 120 cm may be tested at temperatures down to about 10 K. Mirror surface contours can be measured to about 0.1 visible light wave.

test at 10.5 K, the mirror has been temperature cycled six times from room temperature to less than 100 K. No degradation of figure quality resulted from these cycles.

To date, the mirror has been tested on a 3-point support which would not be adequate for the shuttle launch environment. Later tests will utilize a mirror cell capable of holding the mirror in place against oscillating forces of up to 3 g's in any direction. From results already obtained, however, we can conclude that the use of a glass mirror will permit the achievement of the SIRTF goal of diffraction limited performance at 2 μm and will also have acceptable thermal and surface quality characteristics.

5. SIRTF AND INFRARED ASTRONOMY

5.1 The Status of Infrared Astronomy

The principal goals of astronomical research include the study of the formation and evolution of galaxies, stars and planetary systems. Observations

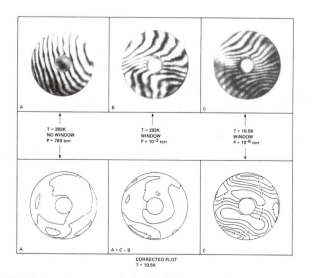

FIGURE 6. (*Top*) Interferograms of 50 cm diameter fused quartz mirror using cryogenic optical test facility. (*Bottom*) Contour plots of mirror surface calculated from interferograms. The contours at intervals of 0.2 (6328 Å) wave represent deviations from a perfectly spherical shape. Contour plot A + C − B represents the mirror condition at 10.5 K after subtracting dewar window distortions.

in the infrared (1–1000 μm) band of the electromagnetic spectrum make unique contributions to the study of these problems for the following general reasons.

(1)   The expansion of the universe implies that progressively more of the optical/ultraviolet radiation from more distant, younger objects is Doppler shifted into the infrared. Thus infrared observations provide a crucial link to the early stages of galactic evolution and the evolution of the universe.

(2)   Cool objects with temperatures between 3 and 3000 K radiate primarily in the infrared. These include planets, both very young and very old stars, and much of the interstellar medium between the stars.

(3)   Many objects are hidden from view at optical and ultraviolet wavelengths by obscuring clouds of interstellar dust grains. The dust clouds are more transparent at longer wavelengths, and infrared observations are a primary means of probing heavily attenuated regions, such as the central portions of our Milky Way galaxy.

(4)   The fundamental vibration-rotation transitions and most rotational transitions of all molecules lie in the infrared, as do the fine-structure transitions of many important atoms and ions. Infrared spectroscopy permits study of a variety of astronomical environments, including planetary

atmospheres, circumstellar gas shells and clouds of neutral and ionized gas in the interstellar medium.

5.2   The Contribution of SIRTF

5.2.1   Anticipated Gains of SIRTF

The SIRTF observatory will provide a 100 to 1000 fold gain in sensitivity over existing capabilities (including IRAS) at wavelengths between 2 and 200 μm. SIRTF will provide the same gain in infrared capability that other advanced facilities, such as the Hubble Space Telescope, the Gamma Ray Observatory, and the Very Large Array, are providing or will provide in other spectral bands; SIRTF thus ensures our essential capability to study astrophysical phenomena with high sensitivity across the entire electromagnetic spectrum. The potential scientific impact of SIRTF can best be illustrated by discussion of specific examples from several areas of astronomy.

5.2.2   The Energy Distribution of Quasars

Quasars are the most distant, most luminous and most mysterious objects in the universe [19]. Light from the most distant quasars comes to us from a time when the universe was much less than one-half of its present age and size. The luminosity of a quasar often exceeds 10^{13} times the solar luminosity, or 100 to 1000 times the luminosity of an entire galaxy of stars, even though quasars are very compact, often showing rapid brightness variations suggestive of sizes of a few light days. Neither the processes which generate such enormous amounts of power within such a small volume, the exact details of the radiation mechanisms, nor the relationship between quasars and galaxies is well understood. Recent data [20] suggest that at least some quasars are located at the centers of galaxies, and it is possible that the nucleus of every galaxy, including our own, contains an extinct or dormant quasar.

A major gap in our empirical understanding of quasars exists at infrared wavelengths between 120 μm and 1000 μm; until the IRAS mission only one or two of the brightest quasars had been detected in the 20 to 120 μm portion of the spectrum and relatively few had been detected at any infrared wavelength [21]. As illustrated in figure 7, the infrared region is of particular interest for the study of quasars for several reasons. (a) Some quasars emit the bulk of their luminosity in the infrared. Figure 7 shows that SIRTF can determine this contribution for quasars 100 to 1000 times fainter than the bright objects currently detectable in the infrared. (b) Observations in the infrared are essential for understanding the distinction between "radio-quiet" and "radio-loud" quasars. As is also shown in figure 7, radio-quiet and radio-loud quasars are indistinguishable on the basis of ultraviolet, optical and near-infrared observations. IRAS was able to provide 4-band photometry from 8 to 120 μm for several quasars [24]. Excess 100 μm

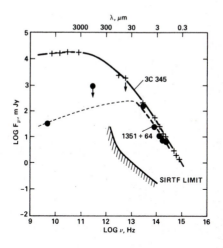

FIGURE 7. The optical-to-radio energy distributions of a very bright radio-loud quasar, 3C345, and a very bright radio-quiet quasar, 1351+64, are compared with the limiting flux detectable from SIRTF (10 σ in $\sim$ 1000 sec) from 3 to 300 μm. The fluxes are in units of mJy (1 mJy = 10^{-29} W m^{-2} Hz^{-1}). The SIRTF performance assumes a detector sensitivity of 10^{-17} W/$\sqrt{\text{Hz}}$ for $\lambda \leq 100$ μm and 10^{-16} W/$\sqrt{\text{Hz}}$ at 300 μm. The data for 3C345 are adapted from [22], and those for 1351+64 are from [23].

emission from the radio-quiet quasars suggests that they may have a cool, dusty component. Somewhere beyond 120 μm, however, the emission from the radio-quiet quasars drops dramatically. SIRTF will have the sensitivity required to define the position and nature of this spectral break, which may be related to the orientation and properties of the jet of relativistic particles which is thought to be responsible for the radio emission from the radio-loud quasars [25]. Observations from SIRTF may thus tell us whether radio-loud and radio-quiet quasars are totally unrelated phenomena, or whether they merely represent different views of the same underlying object.

5.2.3   Brown Dwarfs and the Missing Mass

A pervasive problem in astrophysics is the discrepancy between the mass of an astronomical system determined from dynamical considerations and that determined from a census of the objects it is observed to contain. In many cases, the former is considerably greater than the latter, giving rise to the problem of the "missing mass," matter which makes its presence felt gravitationally but is not in a readily observable form. The astrophysical systems which exhibit the missing mass problem include individual galaxies, clusters of galaxies and perhaps the universe itself. At each level a significant amount of the mass of the system appears to be concealed in an as yet unobservable form. The suggestions for the state of this unobservable matter have run the gamut from primordial black holes to massive neutrinos. Perhaps a more conservative and readily testable hypothesis is that the missing mass exists in the form of low mass stars, "brown dwarfs" with masses below 0.085 solar masses. Theoretical considerations indicate that stars of this mass or less will not generate enough internal heat and pressure to initiate the nuclear burning of hydrogen which powers stars through most of their lifetimes. Such an object would, however, radiate the energy liberated in its gravitational collapse and would have a substantial internal energy source, just as the planets Saturn and Jupiter do, appearing as a faint, cool object detectable only at infrared wavelengths. SIRTF, with its high infrared sensitivity, offers a major improvement in our ability to detect brown dwarfs with temperatures in the range 250 to 1500 K and thereby to either define or markedly constrain the properties of the missing mass. Such objects could be 10 or more times more numerous than ordinary stars in the solar neighborhood and still remain undetected by existing infrared techniques.

Figure 8 shows the maximum distance to which SIRTF could detect brown dwarfs as a function of the temperature of the brown dwarf. The relation between temperature and luminosity required to generate these curves was taken from the theoretical models of Stevenson [26]. Also shown are the distances to which the IRAS survey and an ambient temperature 3 m ground or space telescope operating in the infrared could detect the same stars. The difference between the curves is a measure of the gain in sensitivity to be realized from SIRTF. The data shown in the figure can be used to design an observing program for SIRTF which would either detect an astrophysically interesting number of brown dwarfs or set useful limits on their properties. Although the time required for this very important experiment may be too long for a short shuttle flight, it is very plausible for a long duration SIRTF mission. It is interesting to note also that this investigation benefits from the excellent short wavelength performance of SIRTF which seems attainable on the basis of the mirror test results cited above.

5.2.4   Planet Formation

So far we know of only one star that has planets. It would be extremely difficult to detect planets near another star, but it may be possible to detect evidence of planet formation. Some theories predict that planets are formed by accretion of cold dust and gas over a period of 10^8 years after the birth of the parent star. If such theories are correct some of the stars in young galactic clusters should be surrounded by tenuous debris clouds made up of ejecta from collisions between lumps of material not yet accreted into planetary bodies. Such impacts are thought to be responsible for the craters observed on nearly all solar system bodies with solid surfaces. The debris cloud of ejected and unaccreted matter would have a much higher surface to mass ratio than the planets and consequently be

MAXIMUM DETECTABLE DISTANCE OF
BROWN DWARFS

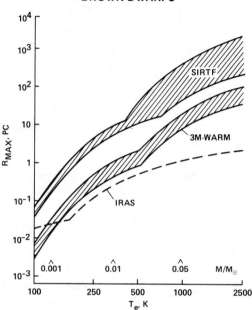

FIGURE 8. The maximum distance R_{max} to which SIRTF could detect brown dwarfs (with S/N 10:1 in $\sim$ 1000 sec) of a given temperature T_e is compared with the performance of a 3 meter ground or space telescope at 273 K and with that of the IRAS all sky survey. The upper envelope of the SIRTF area corresponds to background-limited performance, while the lower assumes 10^{-17} W/$\sqrt{\text{Hz}}$ detector sensitivity. The "3 m warm" band reflects an analogous range in assumptions regarding instrument capability for an uncooled 3 meter telescope. The numbers above the x-axis give the brown dwarf mass, in solar masses, corresponding to a given temperature, assuming a cooling time of 5×10^9 yr [26]. The vertical scale is in parsecs (1 pc = 3.3 light years = 3×10^{18} cm). We thank Dr. R. Probst for assistance in preparing this figure.

much more easily detected in the infrared. If theories of cold accretion are correct, planetary systems less than 10^8 years old would exhibit infrared luminosities sufficient to add 20% or more to the 10 μm luminosities of their parent stars [15]. IRAS has detected infrared excesses around α Lyrae and possibly a few other bright, main-sequence stars [27]. This tantalizing discovery shows that small, cold particles do surround some stars long after their formation. Unfortunately IRAS was not able to study solar-type stars of known age that are younger than 10^8 years old such as those in the α Persei cluster. SIRTF could search the known young solar-type stars for luminosity excesses at wavelengths of 10 μm and longer. It could then determine the composition of the excesses spectroscopically to verify that they were characteristic of planetary debris. This search could provide us with our first information about how common planet formation really is, a fundamental factor in assessing the likelihood of other life in the universe.

6. CONCLUDING REMARKS

The technical reasons for building a cryogenically cooled infrared telescope for use in space are manifest. A moderate sized, fused quartz mirror has satisfactory thermal and optical qualities for operation near 10 K in such a telescope. The few examples we have given of the astronomical problems that SIRTF could address with its 100 to 1000 fold sensitivity advantage over existing infrared telescopes are clearly of great importance, but if the recent history of astrophysical exploration is a guide, the most exciting discoveries to be made with such a telescope are totally unanticipated.

References

[1]   S. D. Price and R. G. Walker, *The AFGL Four Color Infrared Sky Survey: Catalog of Observations at 4.2, 11.0, 19.8 and 27.4 μm* (Air Force Geophysics Laboratory, Hanscom AFB, Massachusetts, 1976), AFGL-TR-76-0208.

[2]   H. H. Aumann and R. G. Walker, *Optical Engineering* **16**, 537 (1977).

[3]   F. C. Witteborn and L. S. Young, *J. Spacecraft and Rockets* **13**, 667 (1976).

[4]   M. W. Werner and K. R. Lorell, *Proc. (SPIE)* **265**, 344 (1981).

[5]   *Focal Plane Instruments and Requirements Science Team* (SIRTF Shuttle Infrared Telescope Facility, 1979). (Unpublished copies available from L. Young, Ames Research Center, CA.)

[6]   *Shuttle Infrared Telescope Facility (SIRTF) Preliminary Design Study Final Report* (Hughes Aircraft Company, Culver City, CA, August, 1976), Contract NAS 2-8494.

[7]   *SIRTF Design Optimization Study Final Technical Report* (Perkin-Elmer, Beechcraft, SAI, September, 1979), ER-408, Contract NAS 2-10066.

[8]   G. B. Field, *Physics Today* **35**, 47 (1982).

[9]   J. P. Simpson, *Infrared Emission From the Atmosphere Above 200 km* (NASA, 1976), TN D-8138.

[10]   J. P. Simpson and F. C. Witteborn, *Applied Optics* **16**, 2051 (1977).

[11]   H. Aumann, B. Brown, F. Gillett, W. Irace, D. Langford, P. Mason and R. Salazar, *Infrared Astronomical Satellite In-Orbit Performance Assessment* (Jet Propulsion Laboratory, Pasadena, CA, 1983), JPL D-871.

[12]   M. G. Hauser, *Calculation of Expected Infrared Signals and Background Induced Noise Limitations* (NASA Goddard Space Flight Center, Greenbelt, MD 20771, 1981), X-693-80-23; K. M. van Vliet, *Applied Optics* **6**, 1145 (1967).

[13]   E. T. Young and F. J. Low, *Applied Optics* **23**, 1308 (1984).

[14]   J. H. Miller, F. C. Witteborn and H. J. Garland, *Proc. SPIE* **332**, 413 (1982).

[15]   F. C. Witteborn, J. D. Bregman, D. F. Lester and D. M. Rank, *Icarus* **50**, 63 (1982).

[16]   D. A. Briotta, Jr., *Rocket Infrared Spectroscopy of the Zodiacal Dust Cloud*, Thesis, Cornell U., Ithaca, N. Y., 1976, CRSR 635.

[17]   A. F. M. Moorwood, in *Infrared Detection Techniques for Space Research*, edited by V. Manno and J. Ring (D. Reidel Publishing Co., Dordrecht, Holland, 1972).

[18]   E. E. Bell, L. Eisner, J. Young and R. A. Oetijen, *J.O.S.A.* **50**, 1313 (1960).

[19]   For recent reviews of quasar properties, see articles by R. J. Weymann, R. F. Carswell and N. G. Smith, and by K. I. Kellerman and I. I. K. Pauliny-Toth in *Ann. Rev. Astr. Ap.* **19** (1981).

[20]   J. B. Hutchings, D. Crampton, B. Campbell and C. Pritchet, *Astrophys. J.* **247**, 743 (1981); S. Wyckoff, P. A. Wehinger and T. Gehren, *Astrophys. J.* **247**, 750 (1981).

[21]   G. H. Rieke and M. J. Lebofsky, *Ann. Rev. Astr. Ap.* **17**, 477 (1979); B. T. Soifer and G. Neugebauer, in *Infrared Astronomy*, edited by C. G. Wynn-Williams and D. P. Cruikshank (D. Reidel Publishing Co., Dordrecht, Holland, 1981), IAN Symposium No. 96.

[22]   P. M. Harvey, B. A. Wilking and M. Joy, *Astrophys. J.* **254**, L29 (1982).

[23]   D. J. Ennis, G. Neugebauer and M. W. Werner, *Astrophys. J.* **262**, 460 (1982).

[24]   G. Neugebauer, B. T. Soifer, G. Miley, E. Young, C. A. Beichman, P. E. Clegg, H. Habing, S. Harris, F. Low and M. Rowan-Robinson, *Astrophys. J. Lett.* **278**, L83 (1984).

[25]   R. D. Blandford, M. C. Begelman and M. J. Rees, *Sci. Am.* **246**, No. 5, 124 (1982).

[26]   D. J. Stevenson, *Proc. Astr. Soc. Australia* **3**, 227 (1978).

[27]   H. H. Aumann, F. C. Gillett, C. A. Beichman, T. de Jong, J. R. Houck, F. Low, G. Neugebauer, R. G. Walker and P. Wesselius, *Astrophys. J. Lett.* **278**, L23 (1984).

CHAPTER VII

Surface Shielding

(*Top*) John Henderson, James Lockhart and Fred Witteborn with the electron-positron free fall apparatus. (*Bottom*) William Fairbank, Wayne Rigby, Henry Fairbank and John Bardeen discussing the electron surface state on copper during the Near Zero conference at Stanford in 1982.

Surface Shielding: Introduction

F. C. Witteborn

The lure of exploration of the gravitational properties of elementary particles led indirectly to the discovery of the unusual shielding properties of low temperature copper surfaces. Publications by Morrison and Gold [1,2] suggesting the possibility of antigravitational properties for antimatter, and comments by Brice DeWitt on the need for more experiments on gravity inspired William Fairbank to conceive the idea of comparing the gravitational force on freely falling electrons with that on positrons. The charged particles were to be constrained to move along the axis of a long vertical metal tube by a constant magnetic field maintained by a superconducting solenoid. The interaction between the magnetic moments resulting from spin and orbital motion with magnetic gradients was to be minimized by using only those electrons whose spin and orbital moments cancelled. What appeared to be lacking in this approach was a method for reducing electrostatic fields resulting from the patch fields [3] of individual crystal faces along the surface of the metal tube. For his Ph.D. thesis project, Fred Witteborn, under Fairbank's direction, constructed a small pilot model for determining the feasibility of studying the free fall of charged particles with energies on the order of 10^{-10} eV. The variation in patch potential in the small vertical tube, expected to be on the order of 10^{-3} eV, was found to be absent in experiments conducted at 4.2 K. The full scale free fall apparatus [4,5] exhibited potential variations less than 10^{-10} eV. James Lockhart modified the apparatus to study the temperature dependence of the shielding effect. The paper by Lockhart and Witteborn reviews the free fall studies and the temperature dependence measurements.

The successful measurements of a temperature transition for the surface shielding effect prompted Fairbank to consider totally different methods of studying the surface charge state that could be responsible for the shielding. Chris Waters, at that time one of Fairbank's graduate students, had built a microwave-cavity-stabilized oscillator system which could be used to measure with high precision the cavity characteristics at resonance. These characteristics (resonant frequency and Q) would be sensitive to changes in the surface properties of the cavity. Application of his system to oxidized copper cavities became part of Waters' thesis, and was followed by further work with different cavity surface preparations and altered experimental conditions by Henry Fairbank, James Lockhart and K. W. Rigby. The studies of the microwave surface impedance of copper by all four investigators is the subject of the next paper in this chapter. It presents not only further evidence that oxidized copper surfaces have unusual properties at low temperatures, but also the dependence of these properties on temperature and magnetic field strength.

Theoretical discussions of the surface shielding effect have been advanced by Hanni and Madey [6], by Hutson [7] and by John Bardeen in the final paper in this chapter. Bardeen discusses a model in which a conducting layer is formed above the oxide layer of the copper tube, providing a reduction of patch potential differences. Such a conducting (but not superconducting) layer would still permit the production of uniform axial electric fields in the drift tube when axial currents are applied, and is thus consistent with the behavior seen in the free fall experiments. A complete theory, explaining quantitatively all observed properties, has not yet been written. Further studies, both experimental and theoretical, are clearly needed in this exciting and difficult field.

References

[1]   P. Morrison and T. Gold, *Essays on Gravity* (Gravity Research Foundation, New Boston, N. H., 1957), p. 45.

[2]   P. Morrison, *Am. J. Phys.* **26**, 358 (1958).

[3]   C. Herring and M. H. Nichols, *Rev. Mod. Phys.* **21**, 185 (1949).

[4]   F. C. Witteborn and W. M. Fairbank, *Phys. Rev. Lett.* **19**, 1949 (1967).

[5]   F. C. Witteborn and W. M. Fairbank, *Nature* **220**, 436 (1968).

[6]   R. S. Hanni and J. M. J. Madey, *Phys. Rev. B* **17**, 1976 (1978).

[7]   A. R. Hutson, *Phys. Rev. B* **17**, 1934 (1978).

Near Zero Electric Field: Free Fall of Electrons and Positrons and the Surface Shielding Effect

James M. Lockhart and F. C. Witteborn

The use of low temperature techniques to make possible a direct measurement of the force of gravity on elementary particles and antiparticles shows clearly the efficacy of William Fairbank's approach to studying the enormous range of physical systems which his interests span. His method of exploiting the advantages of low temperatures to explore previously untouched areas of experimental physics has certainly been a powerful one. We feel privileged to have been a part of this foray into new frontiers of physics.

In 1957 Fairbank attended a conference on gravity in which Brice DeWitt pointed out that there was very little in the way of experimental data on gravity; in fact, there was no direct experimental verification of the assumption that antimatter has normal gravitational properties. Not too long after that, Fairbank conceived the basic outlines of an experiment in which the force of gravity on positrons would be determined by measuring their flight time along a vertical path.

Such measurements would be of interest for several reasons. It is true that currently accepted theory predicts ordinary gravitational properties for all particles and antiparticles, charged or uncharged. For example, the equivalence principle of general relativity states that the inertial and gravitational masses are equal. However, general relativity has not been tested in a system containing comparable amounts of matter and antimatter. Some anomaly in the gravitational properties of antimatter might provide a way of explaining the enormous discrepancy in the abundance of matter relative to antimatter in the nearby universe [1]. There is strong indirect evidence from experiments on unstable particles such as the K° and the anti-K° mesons [2], but the fact remains that no direct gravitational measurements have been made on stable antimatter. As it has turned out, some other very interesting results have developed on the way to that goal.

Shortly after Bill Fairbank came to Stanford, he had one of the authors, Fred C. Witteborn, as a student in a graduate class. When Bill mentioned he had a challenging project that was probably too hard for a graduate student, this particular student soon set to work, along with Larry Knight, on the design of a pilot model apparatus for doing time of flight measurements on electrons in order to look for solutions to some of the anticipated problems. The basic difficulty in performing the experiment can be appreciated by realizing that the gravitational force exerted by the earth on an electron is equaled by the electrical force exerted on it by just one other electron five meters away. That says something about the kind of near zero electric field that one needs. Since mg/e is 5.6×10^{-11} volts per meter (where mg is the force of gravity on the electron and e its charge), one would like to have an electric field region of less than 10^{-12} volts per meter to do a significant measurement. Clearly, the advantages of cryogenic techniques (thermal and dimensional stability, superconducting magnets and leads, cryopumping, and low Johnson noise) were called for.

Obviously some sort of enclosure for the experimental region is needed; very high vacuum is required in order not only to avoid collisions, but even to avoid induced dipole moment interactions with gas present near the beam path. The basic idea of this experiment was to look at the time of flight of either electrons or positrons on a vertical flight path, and to deduce the force of gravity from that measurement. An insulating container would be a poor choice, since it would need dimensions of hundreds of meters in order to keep the charge on the surface far enough away from the test electron. Also, no shielding of external fields from the environment would be provided. A solution that came after some thought was to use a long, narrow metal tube. If one goes well inside the tube along the axis, the image charge exerts no net force on the moving electron. That seemed to be a reasonable solution, but there are problems. A rather special tube is required. It needs to be a single crystal of metal, preferably a meter or two long, and it needs an infinite Young's modulus and massless conduction electrons. To the extent that these conditions do not obtain, some of the interesting physics of the problem emerges.

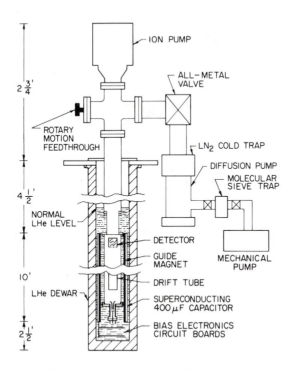

FIGURE 1. An overall view of the electron free fall apparatus.

The metal shielding tube, or drift tube, was the basis of the final design of the experiment. The important features of the apparatus can be seen in figure 1, which shows the final version of the apparatus designed by one of the authors, F. C. Witteborn, and later updated by the other, James M. Lockhart, [3–5]. The experiment is performed under a 10^{-10} torr or better vacuum in the 4 K environment of a 14 foot long dewar. Figure 2 shows the electron flight path. Electrons are emitted in a short burst from a pulsed tunnel cathode with about a 1 eV spread of energies, and are guided through the drift tube region by the axial magnetic field of a persistent superconducting magnet. (The use of a superconducting magnet allows a very stable guide field to be obtained.) The electrons continue on to a windowless photomultiplier detector. Voltage pulses from the photomultiplier corresponding to electron arrivals are amplified and stored in a multi-channel scaler according to the elapsed time since the source was pulsed. This process is repeated over and over again to accumulate a distribution of arrival times whose parameters are determined by the energy spread of the source and the effect of gravity.

In order to avoid Coulomb repulsion effects, the cathode emission and the electrical bias of the drift tube relative to the cathode are adjusted so as to

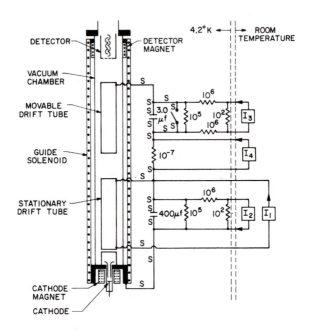

FIGURE 2. A schematic diagram of the apparatus. Wires and capacitors labeled S are superconducting.

have only one slow electron in the tube per cathode pulse, on the average. The data taking system rejects data not meeting this requirement. Thermoelectric emf's are minimized by carefully controlling thermal gradients along the tube.

The problem of unacceptably large forces exerted on the test electrons by unavoidable inhomogeneities in the guide magnetic field was dealt with by using magnetic ground state electrons. Such electrons, which have cyclotron and spin magnetic moments which are opposite and equal in magnitude except for an amount corresponding to the deviation of the electron g-factor from 2, interact so weakly with field gradients that residual inhomogeneities are no longer a problem. Ground state electrons can be state-selected by setting up a large magnetic field gradient between the cathode and the drift tube. This gradient will accelerate all non-ground state electrons so that they rapidly pass through the tube, leaving only ground state particles moving slowly.

Let us begin looking at expected flight time distributions by considering electrons which enter the drift tube with kinetic energies of say 10^{-2} to 10^{-6} eV. For these electrons gravity has a negligible effect on the transit time through the tube and thus the distribution of arrival times should reflect just the energy distribution of the electrons emitted from the cathode.

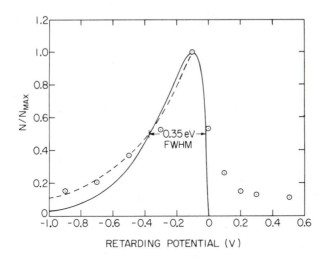

RETARDING POTENTIAL (V)

FIGURE 3. The energy distribution of electrons emitted from the tunnel cathode as obtained from retarding potential measurements.

A typical source distribution is shown in figure 3. We adjust the electrical bias of the cathode so that the electrons in the peak of the distribution move slowly through the tube. The time scale is set so that the "fast" particles (those with initial kinetic energies of more than 10^{-2} eV or so) all arrive in the first time interval on the multi-channel scaler. The narrow slice of the source distribution which represents all of the slow electrons which have enough energy to get through the tube is essentially flat and is quite accurately described by

$$dN(E_0 < E < E_0 + dE) = C(E_0)dE   ,   \tag{1}$$

where E_0 is the energy an electron must have in order to enter the tube, $dN(E_0 < E < E_0 + dE)$ is the number of electrons which enter the tube with energies between E_0 and $E_0 + dE$, and $C(E_0)$ is a constant determined by cathode emission and bias parameters. Since the kinetic energy of an electron in the tube is the amount by which its total energy exceeds E_0, integration of both sides of the above equation gives the number of electrons which have a kinetic energy in the tube less than E_k for small values of E_k:

$$N(E < E_k) = C(E_0)E_k   .   \tag{2}$$

These electrons will all arrive after a time t determined from

$$E_k = (1/2)m(s/t)^2   ,   \tag{3}$$

where s is the drift tube length. Therefore, the number of electrons arriving at time t or later is given by

$$N(t) = (1/2)C(E_0)ms^2t^{-2} = C_1(E_0)t^{-2} \quad ; \tag{4}$$

so that if we plot $N(t)$ *versus* t, we expect a t^{-2} curve. Figure 4 shows some early data from the pilot model; the solid curve is a t^{-2} fit.

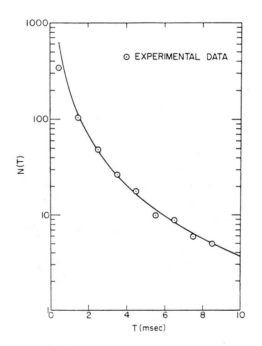

FIGURE 4. Early data from the pilot model free fall apparatus.

When first obtained, these data were startling. In fact, the multi-channel analyzer was not yet available and the only data output was detector pulses on an oscilloscope. Fortunately, W. O. Hamilton was around to suggest that a scope picture be taken immediately. It showed some electrons taking 10 ms or more to get through the tube, a result that seems impossible when one considers the true nature of the shielding tube (*i.e.*, the ways in which it fails to meet the list of conditions mentioned earlier). A drift tube structure would be expected to have ambient electric fields in it that would prevent the passage of such low energy electrons. First, the tube is polycrystalline; it is long enough that single crystals become hard to obtain. Since it is polycrystalline, there are work function variations from crystallite to crystallite (different faces are exposed, there are varying strains within the system, and so forth). There are thus contact potential differences along the surface of the tube and, therefore, patch effect electric fields. Even though the tube was electroformed in a way that produced very small crystals in order to get the maximum spatial averaging of these random fields, calculations indicated that the patch fields ought to have rms values of 10^{-5} or 10^{-6} volts per meter at the very least, maybe more. The existence of such patch effect fields should eliminate electrons with flight times in excess of one microsecond. In fact, consequences of the patch effect field were expected to be so serious that, to minimize its effect, provisions had been made in the Varian physics building to allow if necessary the installation of a large diameter drift tube and dewar extending from the basement through four floors to the roof. Astonishingly, this did not turn out to be necessary.

The effect of gravity on the metal of the tube must also be taken into account. First, the conduction electron gas in the tube tends to undergo a sedimentation effect; that is, in order for the conduction of electrons to be stable in the gravitational field, the downward force of gravity must be balanced by an electric field which exerts a compensating upward force. This field, of magnitude mg/e (5.6×10^{-11} V/m), is produced by a slight density gradient in the vertical distribution of conduction electrons. This effect was first predicted by Schiff and Barnhill [6].

A second gravitational effect involving the tube is the differential compression of the ion lattice which comes about because the lower portion of the lattice must support the weight of the upper portion. This effect was predicted by Dessler, Michel, Rorschach and Trammell [7]. The electric field resulting from this strain gradient (of magnitude about 10^{-5} to 10^{-6} V/m in copper) is compensated within the metal by the shielding action of the conduction electrons. The original calculations by Schiff and Barnhill indicated that there would thus be no field on the tube axis from this source. However, the shielding is incomplete at the tube surface partially because of the limited compressibility of the Fermi gas (as also pointed out by Herring [8]), and a field of about 10% of the unshielded value should exist along the tube axis. Thus the total ambient field in the tube should be dominated by the lattice distortion field of about 10^{-6} V/m and the maximum observable flight time should be limited to about one millisecond. In view of these facts, the initial data were certainly stimulating. On the strength of those data, the improved apparatus was built and more data taken.

Let us now look in more detail at the data analysis. The principal effect of gravity on the time of flight distribution is to define a maximum flight time, or *cutoff* time, the transit time of an electron which is just barely able to climb the gravitational hill:

$$t_{\max} = \sqrt{2ms/Fg} \quad . \tag{5}$$

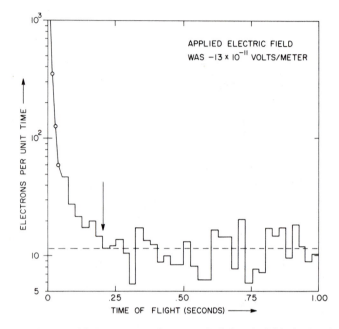

FIGURE 5a. A time of flight spectrum for an applied electric field of 1.3×10^{-10} V/m.

We can test this by running a current through the tube in order to produce a uniform electric field along the flight path and then looking for a cutoff in the time of flight distribution. Figure 5a shows the distribution obtained with an applied field of 1.3×10^{-10} V/m. Figure 5b shows arrival time distributions for the same applied field and a smaller and a larger applied field, all plotted together. Thus it is possible to resolve 10^{-10}–10^{-11} V/m values of applied field. One can plot the observed force on the electron (as inferred from the cutoff times) as a function of applied electric field to obtain figure 6. There are two very interesting points about this set of results. The slope of the line should give the charge/mass ratio of the electron, which it does to quite good agreement. Also, and even more interestingly, the results show that the total force on the electron in the tube is just that due to the applied field; the curve extrapolates to zero observed force at zero applied field. This means that an electron on the tube axis moves inertially; it feels only applied forces. The gravitational field in the tube must be balanced by an ambient electric field in the tube of magnitude mg/e, just the Schiff-Barnhill electron redistribution field. Neither the lattice distortion field nor the patch effect fields are observed, a decidedly surprising result. An additional argument for the proper functioning of the apparatus was provided by Larry Knight, who was able to use a time of flight apparatus with a magnetic gradient coil to measure $g - 2$

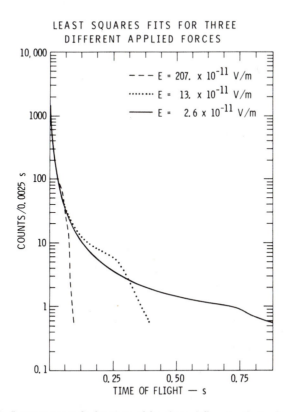

FIGURE 5b. Least squares fit for time of for three different values of applied electric field. Constant background has been subtracted out.

of electrons to within 10%, where in this case g refers to the gyromagnetic ratio [9]. Experiments with the moveable drift tube [10] also confirmed the single drift tube results, as did measurements of the force of gravity on helium ions [2].

After the electron results were published, other experimenters performed a variety of room temperature experiments on metals subjected to large strain gradients. They observed fields essentially consistent with the Dessler, Michel, Rorschach, Trammell prediction. It seemed that an explanation of the electron results would require the existence of a shielding layer on the inside surface of the tube, a situation which might occur only at low temperatures. Such a possibility was suggested by John Bardeen [11]. The next step seemed to be to try to operate the free fall apparatus at room temperature with enough sensitivity (after the loss of cryogenic advantages) to see 10^{-6} V/m fields. One of the authors, Lockhart, joined the project at about that time. The results of the room temperature work are summarized in

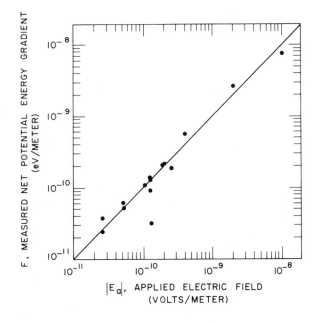

FIGURE 6. A plot of observed force *versus* applied electric field at 4.2 K.

figure 7, which shows the fractional effect on the time of flight distribution of various applied fields in the tube. The room temperature data suggest that there are 10^{-6} V/m ambient fields in the tube, in agreement with predictions and other room temperature experiments. The apparatus was then set up to operate at low temperature while allowing independent control of the drift tube temperature from 4 K up to 30–40 K. We had already done some measurements at 77 K which indicated that large ambient fields were present at that temperature, so we designed the system to operate between there and liquid helium temperature. The heating arrangement was chosen so as to have very small thermal gradients along the drift tube.

The variable temperature work consisted of setting up a stable drift tube temperature and then accumulating enough data to allow a plot of observed force inferred from cutoff times *versus* applied electric field. This work proved to be difficult because of the problem of electrons released from electrostatic traps along the flight path; the signals from the arrival of these electrons smeared out the cutoffs on the time of flight distributions. This difficulty, which is always present, seemed to be more troublesome at the elevated temperatures. Quite a lot of data had to be accumulated at each applied force value and at each temperature. Representative data for 4.2 K and 6.3 K drift tube temperatures are shown in figures 8a and 8b,

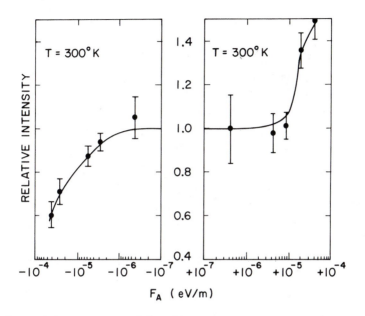

FIGURE 7. Relative intensity of slow electrons with various applied forces at room temperature.

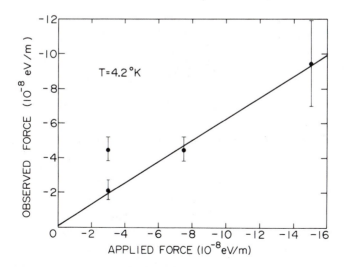

FIGURE 8a. An observed force *versus* applied force plot for 4.2 K from the heated drift tube work.

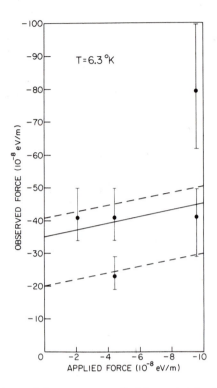

FIGURE 8b. An observed force *versus* applied force plot for 6.3 K from the heated drift tube work.

respectively. By fitting these data, we were able to obtain values for the ambient electric field in the tube at each temperature [12,4]. The results of that study are shown in figure 9a. There is a dramatic change, or transition, in the ambient field level at about 4.5 K. Figure 9b shows a magnified view of that temperature region. The error bars on these plots are often large because of the cutoff smearing effects of the electrons released from traps and the limited amount of data, but the nature of the change in ambient field can be clearly discerned. This plot, which was essentially retraced in several independent sets of runs, is rather amazing. The thermometry was not precisely calibrated, but the transition is certainly in the 4–5 K region.

This work, and Witteborn's original electron free fall results at 4.2 K and the room temperature work, clearly indicate that a temperature-dependent surface shielding mechanism is present in the tube. In order to be consistent with the observations, the shielding layer would have to lie at the very surface of the copper drift tube, would have to be decoupled from the strain gradients and the band structure of the metal, and would have to consist of

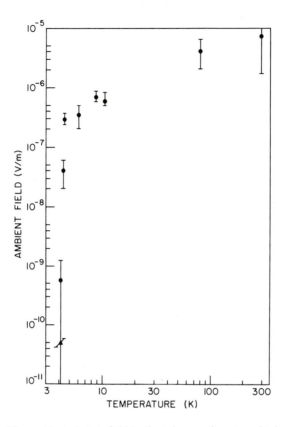

FIGURE 9a. The ambient electric field in the tube as a function of tube temperature. The solid circles show the heated drift tube results and the triangle shows the absolute value of Witteborn's earlier result.

particles with the electron's charge/mass ratio (in order to show the Schiff-Barnhill field). One might consider Tamm states (electron states existing at the surface of the metal because of the abrupt lattice discontinuity), but such states do not give the necessary decoupling. The critical factor (see John Bardeen's paper entitled "Comments on Shielding by Surface States" elsewhere in this volume [13]) appears to be the oxide layer on the inside surface of the tube. The 20–40 angstrom layer of copper oxide present there can provide the decoupling mechanism if electron states are present on the oxide surface. One would expect to see the Schiff-Barnhill sedimentation field in that case, since the electrons on the oxide surface feel gravity and redistribute appropriately.

Some theoretical work on the characteristics required of a shielding layer was done by Richard Hanni and John Madey [14] at Stanford; in addition

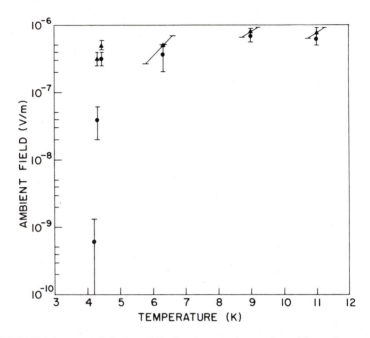

FIGURE 9b. An expanded view of the low temperature region of figure 9a.

to their models, other models were suggested by Andy Hutson [15] of Bell Labs, Thomas Collins [16] of the Air Force Office of Scientific Research, and more recently by John Bardeen in this volume [13].

Some additional experiments, described in the paper entitled "Low Temperature Transition in the Microwave Surface Impedance of Copper" by Waters, Fairbank, Lockhart and Rigby in this book, imply that the shielding mechanism may have some features in common with conventional superconductivity.

During all of this, the aim of doing experiments with positrons had not been forgotten. In fact, interest in the positron experiment was heightened since it would provide an excellent verification of the shielding effect results. John Madey began working on the positron source design around 1965 and developed most of the operating principles [17]. The problem of creating a source of low energy positrons is a severe one. A one-curie ^{58}Co $\beta+$ emitter, for example, would have a probability of less than 10^{-13}/sec of emitting into a 10^{-7} eV range. Conservative forces such as those provided by electric fields can shift the energy of the beam, but by Liouville's theorem cannot reduce its energy-time phase-space density. Dissipative techniques which do not result in excessive annihilation must be used. Madey [18] developed a channeled polyethylene absorber which, when bombarded with positrons

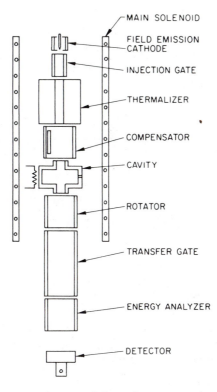

MAIN SOLENOID

FIELD EMISSION
CATHODE

INJECTION GATE

THERMALIZER

COMPENSATOR

CAVITY

ROTATOR

TRANSFER GATE

ENERGY ANALYZER

DETECTOR

FIGURE 10a. A schematic diagram of the positron source electrodes.

from a radioactive source, emits a usable fraction of them with an energy spread of a few eV. This is the first stage of the source.

We call the next stage of the source the thermalizer since its purpose is to thermalize the positron (or electron) beam to a low temperature without contact (*i.e.*, beam cooling). The thermalizer and the remaining stages of the source have been brought to reality by Glen Westenskow as a part of his Ph.D. work [19]. The apparatus is shown schematically in figure 10a. As in the main free fall apparatus, low temperatures are employed and a magnetic field is used to guide the particles along the beam path. Rotation and contraction of the beam are employed to stabilize the beam against stray electric fields.

The linear energy of the particles is reduced by passing them through a resistive-walled tube and making use of eddy-current damping. Westenskow has been able to trap particles in the resistive tube region in order to increase the time for damping. A clever technique is used here to get around the fact that there are no convenient materials with the proper

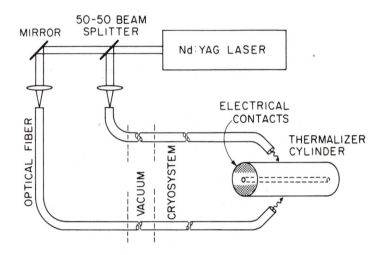

FIGURE 10b. A diagram of the thermalizer optics system.

resistivity at low temperature to provide a good "impedance match" to the beam. The resistive sleeve is made of very pure silicon (an almost perfect insulator at 4 K) and the resistivity is adjusted to the proper value by varying the intensity of laser light which is used to create charged carriers in the silicon. The laser light is brought in *via* fiber optics. Refer to figure 10b.

The next stage of the thermalizer is a microwave cavity tuned to the cyclotron resonance frequency of electrons in the magnetic field of the apparatus (usually 1–2 tesla). The particles are trapped in the cavity so that their cyclotron energy can be radiated off to the 4 K walls. The last stage of the thermalizer is an adiabatic expansion in which trapped particles escape over a slowly lowered potential barrier. Glen Westenskow and John Henderson now have this source about ready for use with positrons, and Henderson is preparing to interface the source to the free fall apparatus in order to achieve the original goal of the program and also to get a very interesting check on the surface shielding effect. If positrons do indeed have normal gravitational properties, then the Schiff-Barnhill field would produce a force on the positrons in the same direction as gravity and we should measure an acceleration of $-2\ g$ for positrons in the tube.

We would like to return to a discussion of some of the other evidence for unusual surface effects in copper. One such relevant experiment was the set of precision capacitance measurements made by Van Degrift [20] who found indications of unusual surface charge states on copper at low temperatures under certain conditions of surface preparation. At Stanford, Chris Waters turned his precision microwave cavity stabilized oscillator system, capable of determining cavity resonant frequencies to better than a part in

10^{10}, to the task of examining the microwave surface impedance of copper as a function of temperature in the 1.5–10 K range [21]. Others, including H. A. Fairbank, James M. Lockhart and K. W. Rigby, have since worked on the experiment, which is described elsewhere in this book. (See paper entitled "Low Temperature Transition in the Microwave Surface Impedance of Copper" by Waters, Fairbank, Lockhart and Rigby.) Results so far include the observation of an anomalous shift in the microwave surface impedance of copper at temperatures of about 3.5 K and 7 K, and a fascinating magnetic hysteresis effect suggestive of the presence of some sort of persistent current.

Finally, remanent magnetization measurements made by Blas Cabrera at liquid helium temperatures indicated the existence of currents with extremely long decay times in a copper sample [22]. The decay times were far in excess of what could be obtained *via* ordinary conduction processes in copper of the highest conductivity known.

All of the work discussed here indicates that some sort of new physical phenomenon is at work on copper surfaces at low temperatures. We think the achievement of a complete understanding of this effect will be extremely exciting. The theoretical work of John Bardeen described in his paper in this book is a major step in that direction. With the results that we have seen so far and the prospect of a theoretical model for this unusual effect in the offing, the reader will agree that this new frontier of physics near zero electric field is a most exciting one.

Acknowledgments

This work has been supported by the Air Force Office of Scientific Research.

References

[1]   P. Morrison and T. Gold, *Essays on Gravity* (Gravity Research Foundation, New Boston, N. H., 1957), p. 45; P. Morrison, *Am. J. Phys.* **26**, 358 (1958).

[2]   M. L. Good, *Phys. Rev.* **121**, 311 (1961).

[3]   F. C. Witteborn, Ph.D. Thesis, Stanford University (1965).

[4]   F. C. Witteborn and W. M. Fairbank, *Rev. Sci. Instr.* **48**, 1 (1977).

[5]   J. M. Lockhart, Ph.D. Thesis, Stanford University (1976).

[6]   L. I. Schiff and M. V. Barnhill, *Phys. Rev.* **151**, 1067 (1966).

[7]   A. J. Dessler, F. C. Michel, H. E. Rorschach and G. T. Trammell, *Phys. Rev.* **168**, 737 (1968).

[8]   C. Herring *Phys. Rev.* **171**, 1361 (1968).

[9] L. V. Knight, Ph.D. Thesis, Stanford University (1965).

[10] F. C. Witteborn and W. M. Fairbank, *Nature* **220**, 436 (1968).

[11] J. Bardeen, comment during invited session of Washington meeting of American Physical Society, 1968.

[12] J. M. Lockhart, F. C. Witteborn and W. M. Fairbank, *Phys. Rev. Lett.* **38**, 1220 (1977).

[13] J. Bardeen, "Comments on Shielding by Surface States," in this publication.

[14] R. S. Hanni and J. M. J. Madey, *Phys. Rev. B* **17**, 1976 (1978).

[15] A. R. Hutson, *Phys. Rev. B* **17**, 1934 (1978).

[16] T. C. Collins, private communication.

[17] J. M. J. Madey, Ph.D. Thesis, Stanford University (1970).

[18] J. M. J. Madey, *Phys. Rev. Lett.* **22**, 784 (1969).

[19] G. A. Westenskow, Ph.D. Thesis, Stanford University (1981).

[20] C. T. Van Degrift, Ph.D. Thesis, University of California, Irvine (1975).

[21] C. A. Waters, Ph.D. Thesis, Stanford University (1979).

[22] B. Cabrera (private communication).

Low Temperature Transition
in the Microwave Surface Impedance
of Copper

C. A. Waters, H. A. Fairbank,
J. M. Lockhart and K. W. Rigby

1. INTRODUCTION

In another paper in this volume the unusual temperature dependent surface shielding effect observed in a copper tube at low temperatures in the
electron/positron free fall experiment is described [1]. There has been an
effort for the past two or three years to find other manifestations of the
surface charge state which is responsible for this shielding. One of the authors, C. A. Waters, had developed for another experimental program a
high precision microwave cavity stabilized oscillator system which allows
determination of the resonant frequency of a microwave cavity to better
than one part in 10^{10} and determination of the Q to about five parts in
10^4 [2]. This stabilized oscillator was patterned after the one developed by
Stein and Turneaure and described elsewhere in this book [3]. It seemed
that interesting new information about the nature of the charge state on
copper surfaces could be obtained by making resonant frequency and Q
measurements on copper cavities as a function of temperature with this
system. Cavities made of other materials could also be measured and
then compared with copper. Since the resonant frequency and the Q are

related to the surface impedance of the cavity, such measurements should be sensitive to the sort of varying surface charge state seen in the free fall experiment. In fact, since for good conductors the microwaves are confined to an extremely thin surface layer, the cavity measurements should be a particularly good way of searching for unusual surface electronic states with a technique that allows data to be obtained reasonably rapidly.

Previous experiments involving copper cavities (e.g., those of Smith [4]) did not detect unusual effects. However, unusual surface electronic properties in copper samples have been found by Van Degrift [5] and others.

It is interesting to calculate the expected change in cavity resonant frequency and Q produced by a surface charge layer of density $10^{17}/m^2$ (as assumed by Hanni and Madey [6] in their theoretical study of the surface shielding effect seen in the electron free fall experiment) if the surface layer were to make a transition to a frictionless state.

For this calculation, we begin with the asymptotic value of the surface impedance in the extreme anomalous skin effect region. We then place a relatively small number of frictionless electrons into a two-dimensional layer just outside the surface. The motion of these electrons in the background field is used to calculate a perturbation to the magnetic field at the cavity surface, and from this the shift in surface impedance is derived. Since the number of electrons within one skin depth in the copper is $2.1 \times 10^{22}/m^2$, it is clear that even as many as $10^{17}/m^2$ can be treated as a small effect.

Since the electrons are constrained to move parallel to the surface, the force equation can be written in terms of the electron velocity v as

$$mdv/dt = qE_t \quad , \tag{1}$$

where q and m are the charge and mass of the electron, respectively, and E_t is the transverse electric field. The small perpendicular magnetic field may be ignored. If we let the transverse magnetic field $H_t = H_0 e^{j\omega t}$ and use $E_t = Z_s H_t$, where Z_s is the surface impedance, we may then integrate to solve for v:

$$v_0 = qZ_s H_0/j\omega m \quad , \tag{2}$$

where $v = v_0 e^{j\omega t}$.

Designating n' as the surface density of the electrons, the perturbation to the magnetic field, H', becomes

$$H' = n'qv_0 = -jn'q^2 H_0 Z_s/m\omega \quad . \tag{3}$$

The shift in Z_s is [2]

$$\Delta Z_s = \frac{H'}{H_0} Z_s = \frac{-2n'q^2 R_\infty^2(\sqrt{3}+j)}{m\omega} \quad , \tag{4}$$

where we take $Z_s = R_\infty(1 + j\sqrt{3})$; R_∞ is the proportionality constant corresponding to the extreme anomalous skin effect region (see [2]). Substitution of numbers gives

$$\Delta Z_s = -(\sqrt{3} + j) \times 2.36 \times 10^{-6} \, \text{ohms} \quad , \tag{5}$$

and the corresponding shifts in frequency and Q of

$$\Delta f = 16 \, \text{Hz} \quad \text{and} \quad \Delta Q/Q_0 = 8.3 \times 10^{-4} \quad . \tag{6}$$

The shift in Q would be near the noise level for these experiments, but the change in frequency could easily be detected.

2. APPARATUS AND METHOD

2.1 Cavities and Cryogenic System

The cavities investigated in these experiments were right circular cylinders with nominal length and radius of 1 inch and were excited in the TE-011 and TM-111 modes by a frequency near 9 GHz. The cavities were coupled to a rectangular X-band waveguide by an iris located in the center of the cavity side wall and normal to the guide axis. These two modes are degenerate in an ideal cylinder, and a mode-suppression groove in one end plate was used to create a frequency separation of about 7 MHz between the modes.

Two constructions were used for the cavities. Some cavities were of the two-piece construction, consisting of a body section and a removable endplate. The others had both endplates removable from the body.

The cavity under test was mounted by its coupling waveguide inside an evacuated cylindrical copper can which was suspended by two waveguides and three support rods from the probe top-plate. The probe could be cooled to about 1.5 K in an eight-inch diameter liquid helium dewar shown in figure 1.

2.2 System for Measurement of Cavity Resonant Frequency

The system for measuring changes in resonant frequency with temperature, shown in figure 2, consists of three sections. A servo-control loop is used to lock a microwave oscillator to the cavity resonance, thus forming a cavity-stabilized oscillator (CSO). A high-stability quartz crystal oscillator operating at 5 MHz drives a frequency multiplication chain to produce a reference signal of 9120 MHz. The microwave oscillator signal is double heterodyned against the reference frequency and then against the signal from a frequency synthesizer to produce a difference frequency in the audio range, which is then demodulated with a phase-locked loop.

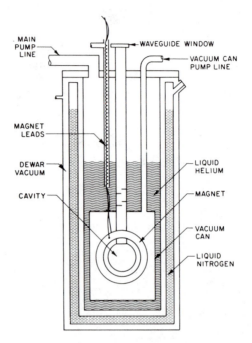

FIGURE 1. Schematic view of cryostat for surface impedance measurements.

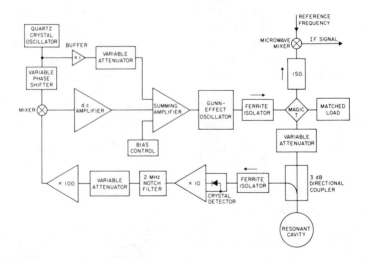

FIGURE 2. Detailed block diagram of the cavity stabilized oscillator.

Most of the frequency measurements were made at an incident power level of 70 μW, with about half the power at carrier frequency, and 70-80% of this power absorbed by the cavity. Other power levels were used to determine whether the effects observed were dependent on microwave field strength.

2.3   System for Measurement of Cavity Q

Measurements of absolute Q were made by plotting the reflected power against frequency. A crystal detector, driving the Y channel of the recorder, provided the reflected power signal P, while the adjustable portion of the V_{bias} of the Gunn effect oscillator provided the frequency (X) axis signal. These measurements were of low accuracy, but provided values of Q_0, the natural Q of the cavity, and β, from which to determine the size of Q variations with temperature.

For these measurements the Gunn effect oscillator was locked onto the sum of the synthesizer and the 9120 MHz reference signal, which was tuned to the cavity resonance. The crystal signal, representing the reflected power, was plotted against temperature.

Since $\beta = Q_0/Q_E$, and Q_E, the external Q (see [5]), is a function of coupling iris geometry, we can measure Q_0 through β:

$$\frac{\Delta P}{P} = \frac{\beta}{1 - \beta^2} \frac{\Delta \beta}{\beta} = \frac{\beta}{1 - \beta^2} \frac{\Delta Q_0}{Q_0} \quad . \tag{7}$$

The fractional precision of the $\Delta Q_0/Q_0$ measurement was 2×10^{-4}.

2.4   Cavity Temperature Measurement and Heating

The temperature of the experimental cavity was measured by using a carbon resistance thermometer with an ac lock-in comparison bridge. The output of the bridge was used to drive the X-axis of the X-Y recorder.

The cavity temperature was varied over the experimental range of 1.8 K to 8.0 K by balancing the heat delivered by a carbon resistance heater and that lost to the helium bath which was held at about 1.5 K. In order to traverse the experimental temperature range and to repeatably set power levels to the heater, a precision voltage source with programmable sweep rate was used. A six-decade voltage divider was used to set the heater voltage. When the setting was changed, the output voltage ramped to the new setting at a rate adjustable from 2 to 20 mV sec^{-1}. The rate corresponding to 10 mV sec^{-1} was slow enough to eliminate hysteresis effects due to thermal nonequilibrium.

2.5   Cavity Magnet

The effect of magnetic field on the surface impedance of the cavity was observed by placing a superconducting Helmholtz coil pair around the cavity and coaxial with it. The radius and separation of the coils were 5.72 cm. The magnet was operated up to an induction of 0.045 tesla.

3.   DATA ANALYSIS AND RESULTS

In the first measurement of the first all-copper cavity, unexpected slope changes were found in the curve of resonant frequency as a function of temperature, $f(T)$ (figure 3). The succeeding experiments were then performed to discover the cause of the phenomenon, and to establish its relationship, if any, to the surface shielding effect.

Analysis of the frequency data requires consideration of the dimensional change due to thermal expansion, and the impedance variation due to the anomalous skin effect. The work of Reuter and Sondheimer [8], Dingle [9], and Chambers [10] was adapted to calculate the expected frequency shift due to anomalous skin depth variation with temperature. In the range of 0 to 7 K this shift is less than 1 Hz, and increases to 15 Hz only at 12 K. The thermal expansion effect is much greater. Data of Kroeger and Swenson [11] for high purity copper and aluminum samples below 15 K were used to produce an expected frequency variation curve, TX, which was used for comparison with our data. In the temperature range above 8 K, all $f(T)$ data compared well with the appropriate TX curves. By matching the data in this range, the slope discontinuities could be easily seen. The frequency axes in the data figures have been referred to the limiting value of TX at 0 kelvin; that is, we plot $f(t) - f_{TX}(0)$.

3.1   Early Data

The behavior of the first all-copper cavity is illustrated in figures 3 and 4. In early runs (1, 3 and 5), a slope change was seen at about 3.48 K, below which temperature the $f(T)$ curve departed from TX and varied greatly between runs. Four months later in time (run 12), this same cavity showed an abrupt slope change at 7.12 K as well as at 3.48 K. The temperatures at which the anomalous slope changes were seen have been designated T_1 and T_2 as in figure 4.

Several tests were made on this cavity to determine what factors might alter the shape of $f(T)$ or influence T_1 and T_2. In figure 5 the effect of four incident microwave power levels is shown. The effect of reducing the power below 70 μwatt is small, except to increase the noise level, and this power was used for most of the experiments. The peak rf magnetic induction at the cavity surfaces is about 1.2×10^{-6} tesla.

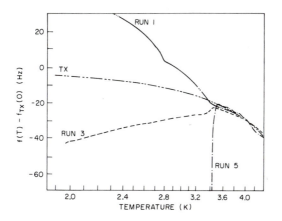

FIGURE 3. Change in resonant frequency *versus* temperature for several runs on a copper microwave cavity. The TX curve is the theoretical curve due to changes in dimensions from thermal expansion.

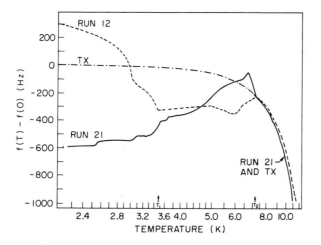

FIGURE 4. Further runs on the same cavity as in figure 3. The upper transition at 7.3 K is clearly discernible. (The TX notation on $f(0)$ is omitted on figures 4 and 5.)

The effect of adsorbed helium gas was determined by incrementally adding controlled amounts of He to the vacuum can during the run. Gas equivalent to an estimated 13 atomic layers on the exposed surfaces was admitted in 11 increments. No change in $f(T)$ was observed other than that expected from the increased dielectric constant within the cavity.

A striking transition in the cavity Q as a function of temperature has also been observed. Figure 6 shows this transition and also indicates the

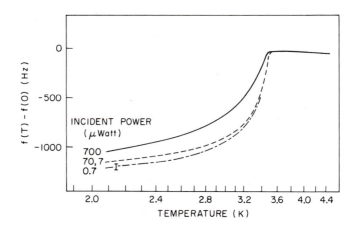

FIGURE 5. Change in resonant frequency *versus* temperature for the copper cavity as a function of incident power.

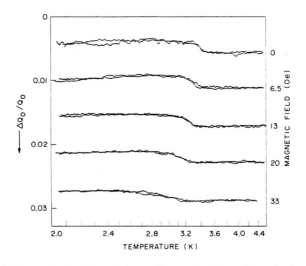

FIGURE 6. Change in Q of the copper cavity as a function of temperature for various applied magnetic fields.

response of the break-point temperature and shape of the transition to an applied magnetic field perpendicular to the cavity endplates. Note that increasing values of magnetic field tend to shift the Q transition to lower temperatures and to broaden it. Similar magnetic field effects are seen in the resonant frequency response, as will be discussed later.

3.2   Recent Data

Recent work, especially a series of studies carried out by one of the authors, Henry Fairbank, has emphasized the role of surface preparation of the cavities in determining the nature of the temperature dependence of the cavity resonant frequency.

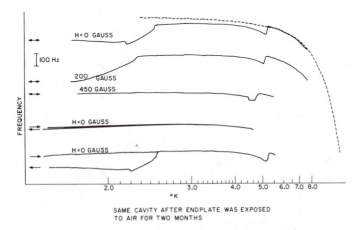

SAME CAVITY AFTER ENDPLATE WAS EXPOSED
TO AIR FOR TWO MONTHS

FIGURE 7. Change in resonant frequency *versus* temperature and magnetic field for another copper cavity with a copper endplate. The dotted curve is for a freshly prepared endplate. The solid curve was taken after the endplate had been exposed to air for two months. After six more months, the data again fit the dotted curve.

Work by another of the authors, James M. Lockhart, presented in figure 7, shows clearly the relevance of surface condition and possibly of oxide thickness in determining the existence and form of the anomalous transition. The difference between a freshly prepared endplate, shown as a dashed curve in figure 7, and the same endplate after exposure to air for two months, shown as a solid curve, is quite striking. In a series of runs over two more months, the transition shown by the solid curves remained. After four additional months of exposure to air at room temperature, the transition disappeared and the frequency *versus* temperature curve returned to the dashed curve shown in figure 7 for the freshly prepared sample. Measurements with an ellipsometer failed to yield meaningful results on oxide thickness. Auger spectroscopy will be used to determine oxide thickness and surface composition. In this context we note that such measurements determined the oxide thickness to be 30 Å for a sample of the copper drift tube used in the free fall experiment, which showed a transition to a shielded surface state at 4.5 K [1].

Figure 7 also shows a magnetic hysteresis effect which is observed in some runs. Starting at the top of the figure and moving down, we see that the application of a magnetic field (again perpendicular to the end-plates) lowers the break-point temperatures of both the upper and lower transitions as well as broadening them. When the applied field reached 450 gauss, the lower transition was no longer visible in the available temperature range. This is similar to the magnetic field dependence of the cavity Q discussed earlier. The particularly interesting effect shows up in the fourth curve from the top, where we see that removal of the applied magnetic field does not bring about a return of the lower transition. Note that in this curve we do not reach the upper transition temperature, and that the results of ramping the temperature both up and down are shown. Finally, in the lowermost figure we see that the upward temperature ramp does not yield a lower transition, but that the upper transition occurs at its expected temperature. As we then ramp down in the temperature, *having exceeded the upper transition temperature*, we then regain the lower transition. It seems that the system retains some memory of the magnetic field after the field has been turned off. This memory remains until the upper transition temperature has been exceeded.

4.   DISCUSSION AND CONCLUSIONS

We assume that the transitions seen in the copper cavities are closely related to the surface shielding transition seen in the free fall experiment described by Lockhart and Witteborn elsewhere in this volume [1]. The transition in the free fall experiment always occurred above 4.2 K, the temperature of the helium bath, which is appreciably higher than the lower transition temperature in the copper cavities. We assume, therefore, that the upper transition in the copper cavities is the one which is related to the transition in the free fall experiment. It seems reasonable to assume, therefore, that the upper transition in the copper cavities represents the same kind of sudden alteration in the charge on the copper surface apparently required for the shielding transition seen in the free fall experiment. Possible explanations for the sort of surface charge state alteration implied by the results of the electron free fall experiment have been surveyed by Hanni and Madey [6]. Another possible explanation for the anomalous surface shielding effect has been advanced by Bardeen elsewhere in this volume [12]. He discusses the possibility of a metal-insulator transition taking place at the surface of the copper oxide layer present in the drift tubes of the free fall experiment. We are now attempting to analyze the frequency and Q shifts which would be expected if the cavity underwent such a metal-insulator transition at its surface.

The lower transition poses an additional puzzle. As we have mentioned, one sample has shown magnetic memory (see figure 7). One example of a magnetic memory is trapped flux in a superconductor. It occurred to us that if the upper transition actually represents a transition to a superconducting-like state at the surface of the metal, the lower transition might then be explainable in terms of a Kosterlitz-Thouless [13] fluxon-antifluxon pair-breaking transition. A conceivable mechanism for the observed magnetic hysteresis effect would then be as follows. When the cavity temperature is brought below the upper transition, its surface goes into a superconducting-like state. (This would explain also the magnetic field dependence of the upper transition.) At a lower temperature (with no applied magnetic field) there is a K-T transition. Application of an external magnetic field lowers the K-T transition temperature, just as it does in the case of granular thin film superconductors [14]. Since we have a superconducting-like situation, magnetic flux is trapped by the cavity surface and persists when the external field is removed. This trapped flux continues to depress the K-T transition temperature until it is allowed to decay by causing the cavity to be warmed above the upper transition temperature. This hypothetical mechanism seems consistent with the observed phenomena. However, if Bardeen's metal-insulator transition theory correctly explains the upper transition, the idea of a Kosterlitz-Thouless transition would no longer be applicable. In that event, one might speculate whether the situation could be a metal-insulator transition at the upper temperature and a superconducting-like transition at the lower temperature. In this case, it is not obvious how to explain the magnetic hysteresis effect. Perhaps the lower transition just represents some sort of additional ordering in the metallic phase.

Experimental work is continuing with a rebuilt vacuum chamber and pumping system designed to greatly reduce the likelihood of contamination. Work is also continuing on the fitting of various models to the observed behavior of the two transitions, with particular interest in trying to fit the Bardeen model to the lower transition.

Addendum

The data presented in this paper have been taken on one copper cavity (figures 3–6) and one copper endplate (figure 7). At the time of the Near Zero conference, measurements on other copper cavities and endplates and on aluminum cavities and endplates showed no evidence of a transition.

However, since this paper was first submitted, one of the authors, K. W. Rigby, has completed a series of experiments with a new apparatus [15]. An improved analysis allows anomalies as small as 1 Hz to be detected unambiguously. Sharp anomalies have been found in two aluminum samples near 3.8 K and in three copper samples near 3.6 K and 7.2 K. These anomalies have been reproducible over several runs and many months. The frequency shifts have been positive in all of the samples. The anomalies below 4 K are eliminated in fields of about 300 G. Some hysteresis has been observed, but the "memory" effect shown in figure 7 has not been seen. The surfaces of some of these samples have been examined using the back-scattering mode of an electron microscope [16] to detect elements of high atomic number. The elements are then identified by their x-ray fluorescence spectra. Using this method, particles of tin have been found on the same two aluminum samples and one of the copper endplates referred to in this paragraph. We have estimated [16] that 0.01 cm^2 of tin would give the upward frequency shift of 100 Hz seen on one of these aluminum endplates. After examining 9% of the total surface of this aluminum endplate, 0.001 cm^2 of tin was found. Approximately 5% of the surface of the copper endplate was examined; the amount of tin found was a factor of two lower than that estimated to give the frequency shift observed on this sample. No lead contamination was observed on any of these surfaces. However, it appears quite likely that the upward frequency shifts seen in this last series of experiments discussed in this paragraph are due to superconducting contamination of the samples.

Cyclotron resonance of surface electrons was looked for in these experiments. A calculation [15] shows that surface densities of 10^{12}–10^{14} electrons/cm^2 would produce observable frequency shifts at resonance. No surface cyclotron resonance was observed, but resonance of electrons in the center of the cavity was found when the peak microwave electric field in the cavity was above a few volts/cm. This has been shown to be due to secondary electron emission.

An experiment has recently been developed by Mark Rzchowski [16,17] to search directly for a copper shielding effect. Using a variation of the Kelvin method, this experiment monitors spatial variations in the surface potential as a function of temperature by recording the charge flow between a copper sample and a small electrode rotating above it. The lower temperature range is 2.9 K. A transition to a shielded state has not yet been observed.

Acknowledgments

This work was supported in part by the Air Force Office of Scientific Research.

References

[1] J. M. Lockhart and F. C. Witteborn, "Near Zero Electric Field: Free Fall of Electrons and Positrons and the Surface Shielding Effect," in this

volume. See also J. M. Lockhart, F. C. Witteborn and W. M. Fairbank, *Phys. Rev. Lett.* **38**, 1220 (1977).

[2] C. A. Waters, Ph.D. Thesis, Stanford University (1979).

[3] J. P. Turneaure and S. R. Stein, "Development of the Superconducting Cavity Oscillator," in this volume.

[4] T. I. Smith, Stanford University, private communication.

[5] C. T. Van Degrift, Ph.D. Thesis, University of California, Irvine (1974).

[6] R. S. Hanni and J. M. J. Madey, *Phys. Rev. B* **17**, 1976 (1978).

[7] R. A. Baugh, in *Proceedings of the 26th Annual Symposium on Frequency Control* (Elec. Ind. Assoc., Washington, D.C., 1972), p. 50.

[8] G. E. H. Reuter and E. H. Sondheimer, *Proc. Roy. Soc. A* **195**, 336 (1948).

[9] R. B. Dingle, *Physica* **19**, 311 (1953).

[10] R. G. Chambers, *Proc. Roy. Soc. A* **215**, 481 (1952).

[11] F. R. Kroeger and C. A. Swenson, *J. Appl. Phys.* **48**, 853 (1977).

[12] J. Bardeen, "Comments on Shielding by Surface States," in this volume.

[13] J. M. Kosterlitz and D. J. Thouless, *J. Phys. C* **6**, 1181 (1973).

[14] A. F. Hebard and A. T. Fiory, in *Ordering in Two Dimensions*, edited by S. K. Sinha (North-Holland, New York, 1980), p. 181.

[15] K. W. Rigby and W. M. Fairbank, in *Proceedings of the 17th International Conference on Low Temperature Physics*, edited by U. Eckern, A. Schmid, W. Weberand and H. Wühl (North-Holland, Amsterdam, 1984), p. 1363; K. W. Rigby, Ph.D. Thesis, Stanford University (1986); and papers in preparation.

[16] M. S. Rzchowski, Ph.D. Thesis, Stanford University (1988).

[17] M. S. Rzchowski, K. W. Rigby and W. M. Fairbank, *Jap. J. Appl. Phys.* **26**, 651 (1987), supplement 26-3.

Comments on Shielding by Surface States

John Bardeen

The remarkable shielding of patch fields observed in free fall experiments at low temperatures by William Fairbank and his students, Fred Witteborn and Jim Lockhart [1–3], has long been a puzzle. When I first heard about the experiments, I suggested that the shielding is due to electrons in states on the outer surface of the oxide layer, but did not try to work out a detailed theory. It would have been difficult to do so without the later experiments on the temperature dependence of the shielding, described by Lockhart and Witteborn in the paper entitled "Near Zero Electric Field: Free Fall of Electrons and Positrons and the Surface Shielding Effect" elsewhere in this book.

I am grateful to the organizers of the Near Zero conference honoring Bill's 65th birthday for inviting me to participate. When I was asked to give a talk, I submitted a safe title, "Comments on Shielding by Surface States," because I did not know whether or not I would have anything new to say. With the incentive of the talk, I put in some hard thinking and have come up with a model that appears to account for most of the experiments. While the model may not be right, it does show that the remarkable shielding properties can be accounted for by a reasonable model with reasonable values for the parameters involved.

Before describing the model, I will summarize the facts that have to be explained. It is known that there are variations in work function of copper of the order of 0.5 V for different crystal faces. These variations are reduced to the order of 0.1 V just outside the surface of copper with an oxide layer. The oxide on the copper used in the drift tube is 20–30 Å thick. The fields on the axis (estimated from those of a single patch) are of the order of 10^{-5} V/m at room temperature, 300 K, but decrease with temperature to about 3×10^{-7} V/m at 4.2 K where there is a sudden drop of two or three orders of magnitude to values of the order of 10^{-10} V/m or smaller. Further, the field on the axis reflects a voltage gradient from a current flowing along the tube. If the tube is used as part of a microwave cavity, a small change in Q occurs at 4.2 K, probably from the same cause that gives a dramatic drop in the field on the axis in the free fall experiments. Magnetic hysteresis effects are observed with the changes in Q.

I first became involved with surface states [4] when I suggested that they could shield the interior of a semiconductor from externally applied fields and thus account for the failure of early attempts to make a field effect transistor. I suggested that the shielding is due to a distribution of surface states in energy on the surface of the semiconductor. The effectiveness of the shielding then depends on the energy density of the surface states, $dN_s/d\mu$. In an excellent analysis of shielding in the free fall experiments, Hanni and Madey [5] have extended this model to try to account for shielding of patch fields on the copper tube. The values of $dN_s/d\mu$ required are so large that they suggested there might be some sort of Bose transition occurring with electrons in surface states at 4.2 K.

I would like to propose a somewhat different model in which shielding is due to states in normal sites on the outermost layer of the oxide, described roughly as $O^{--} \rightarrow O^-$ or $Cu^+ \rightarrow Cu^{++}$. When occupied they have the normal charge for a neutral surface, but become positively charged (a hole state) when unoccupied. The atoms involved (of the order of $10^{15}/cm^2$) are presumed to have similar local surroundings. Above 4.2 K, the hole states are discrete and are individually occupied, but at 4.2 K there is a phase transition to a metallic state. A narrow band opens up which is partially occupied by holes. The holes give a positive surface charge and a 2-D conducting layer on the surface of the oxide. Electron tunneling through the oxide is presumed to be sufficient to maintain the Fermi level of the surface layer the same as that of electrons in the interior of the copper. At the phase change, the surface atoms all come to the same electrostatic potential, the value of which is determined by the Fermi level in the copper. Fields outside the metallic surface layer are very small.

The model of the copper surface with the oxide layer is illustrated in figure 1. It is assumed that if there were no voltage drop across the oxide, the hole state would be occupied (positively charged) and that it would have an energy eV_0 above the Fermi level, E_F, of the copper. Since only a small fraction is charged, a voltage drop must occur across the oxide such

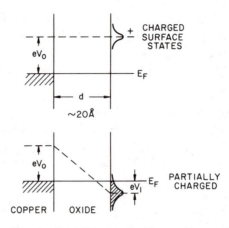

FIGURE 1. Schematic energy level diagram of oxide on copper surface. Upper: Surface states positively charged when there is no voltage drop across oxide. Lower: Equilibrium with states partially occupied as voltage drop across oxide brings states below Fermi level.

as to bring the level to eV_1 below E_F, giving a total voltage drop across the oxide of $V_0 + V_1$.

The value of V_1 depends on the level width and is such that the surface charge of holes, σ, gives the required voltage drop:

$$V_0 + V_1 = (4\pi\sigma/\kappa)d \quad . \tag{1}$$

Here κ is the dielectric constant of the oxide and d is the thickness of the layer. The surface charge density is

$$\sigma = N_s f e \quad , \tag{2}$$

where $N_s \sim 10^{15}/\mathrm{cm}^2$ is the density of surface atoms and $f \sim 10^{-2}$ is the fraction that are charged.

The value of V_1 depends on the line shape. If the energy, ϵ, is measured from the center of the line, we assume that the weight in $d\epsilon$ is

$$df = \frac{(\omega/\pi)d\epsilon}{\omega^2 + \epsilon^2} \quad , \tag{3}$$

where ω, the line width, is determined by thermal broadening and is approximately $k_B T$. In terms of the dimensionless variable, $x = \epsilon/\omega$, with $x_1 = eV_1/\omega$,

$$f = \int_{x_1}^{\infty} \frac{(1/\pi)dx}{1 + x^2} \simeq \frac{1}{\pi x_1} \quad (x_1 \gg 1) \quad . \tag{4}$$

We shall take a specific example to show that the numbers are reasonable. We assume $d = 20$ Å, $N_s = 10^{15}/\mathrm{cm}^2$, $\kappa = 4$ and $V_0 = 2$ V. If V_1 is neglected, equation (1) requires that $N_s f = 2 \times 10^{13}/\text{cm}^2$ and $x_1 = 16$. The latter requires that $V_1 = 0.4$ V at 300 K.

Patch effects are assumed to give variations of order $\delta V = 0.5$ V in V_0. These give variations of order $5 \times 10^{12}/\text{cm}^2$ in $N_s f$ and of about 4 in x_1 ($12 < x_1 < 20$). At room temperature, this would give variations in contact potential of the order of $4k_B T/e$, or of ± 0.1 V at 300 K, just outside the oxide surface, as observed.

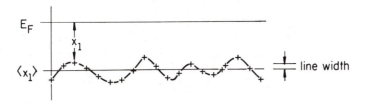

FIGURE 2. Profile of variations of electrostatic potential of surface states, $V_1 \sim (k_B T/e)x_1$, from patch effects. The values of x_1 are temperature independent for a fixed voltage drop across the oxide layer.

Shown schematically in figure 2 are the variations in x_1 that result from patch effects. The variations are shown in a one-dimensional profile, although they actually occur in two dimensions. The important point is that because of patch effect variations the individual levels are separated by more than the line width, so there is little interaction between them. This justifies the assumption that they are occupied independently.

Note that for the same surface charge, x_1 is independent of temperature. This implies that the same diagram applies at all temperatures. Since $eV_1 = k_B T x_1$, potential changes from variations in V_1 should be proportional to T. Within rather large error bars, the data of Lockhart *et al.* show that this is indeed the case. In figure 3, their data for the temperature variation of the field on the axis are replotted on a linear scale for $T < 80$ K. In this range, V_1 can be neglected in comparison with V_0. Also shown is the point for 300 K with both coordinates divided by a factor of 10. The axis field is about 2/3 that corresponding to the straight line. This departure can be accounted for if V_1 is not neglected at this temperature.

At sufficiently low temperatures it becomes favorable for all the surface atoms to come to the same potential so that states in different atoms can interact with one another and form a 2-D band. This presumably accounts for the phase transition at 4.2 K. If W is the width of the hole-band, this should occur when

$$W/2 > eV_1 = k_B T x_1 \quad . \tag{5}$$

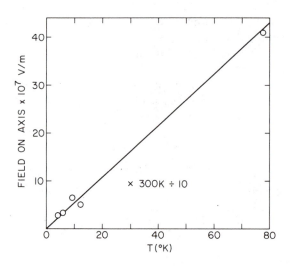

FIGURE 3. A replot on a linear scale of the data of Lockhart *et al.* [3] of the temperature variation of the field on the axis of the copper tube. Error bars are not shown.

For $T = 4.2$ K and $x_1 = 16$, this corresponds to $W = 0.015$ eV. Only the lowest states of the hole-band are occupied by holes, since of $\sim 10^{15}/\mathrm{cm}^2$ total states, only $\sim 2 \times 10^{13}/\mathrm{cm}^2$ are occupied. The Fermi level is near the band edge.

One way of estimating shielding by the metallic surface layer is to use the method of Hanni and Madey [5] that involves the density of states in energy. The density in a 2-D band is independent of energy, giving for our case $10^{15}/W$ or $\sim 7 \times 10^{16}$ states/eV. A change in hole concentration of $\sim 5 \times 10^{12}/\mathrm{cm}^2$, corresponding to a change δV in V_0 of 0.5 V, would thus result in a change in surface potential relative to the Fermi energy of

$$\delta V_1 = \frac{5 \times 10^{12}}{7 \times 10^{16}} = 7 \times 10^{-5} \,\mathrm{V} \quad . \tag{6}$$

The value of δV_1 just above the transition is $\sim 4k_b T/e$ or about 1.5×10^{-3} V. This estimate would give an additional screening factor of about 5×10^{-2} from the metal-insulator transition and an overall decrease from 300 K by a factor of 7×10^{-4}.

This screening, while considerable, is still one or two orders of magnitude less than observed. Lockhart *et al.* found that the field on the axis drops by a factor of about 2×10^{-3} at 4.2 K and that the overall drop from 300 K is by a factor 10^{-4} or less. Even smaller fields by an order of magnitude were found by Witteborn and Fairbank in their early experiments.

There are a couple of possible explanations for this discrepancy. One is that the long wavelength Fourier components that give the field on the axis are shielded more than those of shorter wavelength. If the surface layer is conducting, it will shield patch effects by charge transfer in neighboring regions. The patch fields will then drop more rapidly with distance from the oxide surface [6].

Another possibility is that the conducting layer will effectively screen fields from the interior of the oxide. It is known that the charges that give the image field from a metallic surface and effectively screen the field of an external charge are on the outermost regions of the electron charge cloud of the metal. These charges arise from a modification by the external field of the tail of the wave functions of the electrons extending into the vacuum, not from a change in population of the levels. A similar thing could be occurring in the 2-D metallic layer on the oxide. The potentials and wave functions extending into the vacuum from the oxide could be unaffected and the charge density at the surface could come from modification of the wave functions of the surface states extending into the oxide.

A small change in Q of the microwave cavity would be expected at the metal-insulator transition. The effects of magnetic fields on Q are still to be explained.

The field-free drift experiments of Bill Fairbank and coworkers are an excellent example of plunging into the unknown in a difficult experiment and finding remarkable new effects. Although from the beginning they were well aware of patch effects, this did not deter them from going ahead and hopefully finding some way out of the difficulties. In this case nature was kind and the first experiments were successful, but the reason for the remarkable screening they found has long remained an outstanding puzzle. It is hoped that the suggestions I have made will help resolve the problem and lead to a greater understanding of surface states and surface conduction.

Addendum

Since the above was presented, an article by A. R. Hutson [7] proposing a different mechanism for the screening has come to my attention. He suggests that the screening is due to extra electrons produced by the cathode which are drawn by residual fields that remain near the surface of the tube and help to neutralize them. The neutralizing charges would reside in metastable states at the oxide surface. Presumably those drawn to the surface would have very low velocities because the residual surface fields after screening are extremely small. If at room temperature the fields come from patches with voltage fluctuations of the order of 0.5 V, and the difference in screening between room temperature and 4.2 K is a factor of 10^4, the voltage fluctuations at the lower temperature would only be the order of 0.5×10^{-4} V, corresponding to electron energies of less than 1 K. In our model there is additional screening from the insulator-metal transition of the surface layer, a collective effect that screens long range fluctuations more than short range. No model may be regarded as more than suggestive at the present time and it is hoped that further experiments will help elucidate the remarkable shielding observed.

References

[1] F. C. Witteborn and W. M. Fairbank, *Nature* **220**, 436 (1968).

[2] F. C. Witteborn and W. M. Fairbank, *Rev. Sci. Inst.* **48**, 1 (1977).

[3] J. M. Lockhart, F. C. Witteborn and W. M. Fairbank, *Phys. Rev. Lett.* **38**, 1220 (1977).

[4] J. Bardeen, *Phys. Rev.* **71**, 717 (1947).

[5] R. S. Hanni and J. M. J. Madey, *Phys. Rev.* **17B**, 1976 (1978).

[6] The potential of a patch of radius a drops with distance z from the surface as a/z if the outer surface is nonconducting, and as $(a/z)^3$ if it is conducting.

[7] A. R. Hutson, *Phys. Rev.* **B17**, 1934 (1978).

CHAPTER VIII

Some Thoughts on
Future Frontiers of Physics

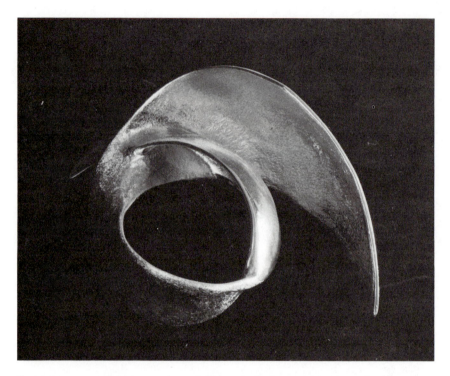

Sculpture representing the envelope of orbits of a body precessing around a rotating Kerr-Newman black hole, as calculated by D. C. Wilkins. (See paper VI.4, figure 2.) The sculpture, by Attilio Pierelli, was presented to William Fairbank at the fourth Marcel Grossmann meeting on general relativity in Rome, Italy, in 1985.

Some Thoughts on Future Frontiers of Physics

William M. Fairbank

1. INTRODUCTION

We live in one of the most exciting times in the history of physics. Because of the development of new technologies, the space program, high energy accelerators, lasers, computers, electronics, and ever improving near zero techniques, we are able to explore new regions of physics never before possible and old regions to a depth only dreamed of in the past. With this new exploration come new concepts which affect the very foundations of physics and give a glimpse of where we came from, back to the beginning of the universe.

That such a renaissance in physics should come with the improved ability to look where we have never looked before is not surprising. Kapitza, who was one of the founders of our low temperature field, wrote in an article in 1961 entitled "The Future of Science" [1]: "Have all the important unexpected discoveries in physics been made? Have the secrets of nature been exhausted?" He said that in the past 150 years seven new phenomena had been discovered in physics which could "neither be foreseen nor explained on the basis of existing theoretical concepts." He cited, for example, the discovery of magnetic fields around a current by Oersted, the discovery of the photoelectric effect by Hertz, and the discovery of radioactivity by Becquerel. These experiments have changed the course of physics. Certainly one would also have to include in the unexpected the results of cooling to very low temperatures, specifically, the discovery of superconductivity and superfluidity.

Kapitza did not believe that this process of discovery had been exhausted. The nature of these discoveries is that they cannot be predicted and they occur in regions of physics where no one has looked before. Hence, Kapitza would say that physics is not dead because we are continuing to look in new regions and will continue to make new unexpected discoveries. Certainly the discoveries that have been made in physics since Kapitza wrote his article in 1961 have amply borne out his contention.

What are these new frontiers in physics? One frontier involves elementary particles at very high energies. One is astronomical exploration in space extending over the complete electromagnetic spectrum. Other frontiers involve new experiments with nuclei, atoms, molecules, condensed matter, or living systems, made possible by modern technology. A particular frontier which allows one to explore new areas of physics involves, as we have seen in this book, reducing one or more of the variables of physics very close to zero. Of course, the most obvious variable in cryogenics is temperature. Other variables include magnetic field, electric field, electric charge, effective force of gravity, effective noise level and specific heat. The advantage of reducing one or more of these variables nearly to zero is that it allows one to look for effects otherwise camouflaged.

It is instructive to recall a comment by Lord Kelvin in a lecture entitled "Nineteenth Century Clouds over the Dynamical Theory of Heat and Light" [2], delivered at the Royal Institution of Great Britain on April 27, 1900: "The beauty and clearness of the dynamical theory, which asserts heat and light to be modes of motion, is at present obscured by two clouds. I. The first came into existence with the undulatory theory of light, and was dealt with by Fresnel and Dr. Thomas Young; it involved the question, How could the earth move through an elastic solid, such as essentially is the luminiferous ether? II. The second is the Maxwell-Boltzmann doctrine regarding the (equal) partition of energy." Lord Kelvin concludes after discussing aberration of starlight and the Michelson-Morley experiment that "I am afraid we must still regard cloud No. I. as very dense." After considering several theoretical models and the ratios of specific heats of gases, he concludes with respect to cloud II, "What would appear to be wanted is some escape from the destructive simplicity of the general conclusion. The simplest way of arriving at this desired result is to deny the conclusion; and so in the beginning of the twentieth century, to lose sight of a cloud which has obscured the brilliance of the molecular theory of heat and light during the last quarter of the nineteenth century." These two little clouds produced special relativity and quantum mechanics.

There are some very interesting clouds forming right now, as well as some interesting regions of physics where no one has looked before which are now becoming accessible to experiment. I would like to consider some of these challenging problems, especially those which can be addressed by near zero techniques.

What are some of the major frontiers in physics that make this such an exciting time? Certainly one is the discovery through experiments with high energy accelerators that there is a new world of quarks and gluons inside the proton and neutron, a new world governed by a new theory, quantum chromodynamics. Theory has predicted the unification of the weak and electromagnetic forces at energies of 100 GeV and the existence of the Z and W particles through which the unification is possible [3]. These particles have recently been discovered in the proton-antiproton collider at CERN [4]. It is predicted that the electroweak and strong forces will be unified at very much higher energies (approximately 10^{14} GeV), the unification involving a new particle of 10^{14} GeV mass (10^{-8} grams) at a distance scale of 10^{-29} cm [5]. Finally the ultimate dream of physics may be realized at 10^{19} GeV and distances of 10^{-33} cm with the unification of the strong, the weak and the electromagnetic forces with gravity. It has been predicted that at this early time, the first 10^{-43} seconds after the big bang, the time of the Planck mass, the universe may have involved quantized gravity, perhaps in the form of quantized strings in a supersymmetrical ten-dimensional space [6]. Other theories have been proposed, including eleven-dimensional space [7], 26 dimensional space [8], other supersymmetrical theories [9], technicolor [10], and SO-18 [11]. It has been proposed that the early universe underwent a very rapid expansion called inflation [12].

How can these theories be tested experimentally, and how did the universe evolve into its present state with its galaxies, stars, planets and life? I believe that, in the continuing unfolding of this drama, experiments near zero will play a crucial role.

2.   HIGH ENERGY PHYSICS; GRAND UNIFICATION

To extend the known region of high energies to the order of 20 TeV, a 90 km superconducting supercollider (SSC) is being planned [13]. This accelerator will make crucial use of the zero-resistive properties of superconducting magnets. The design calls for ninety kilometers of 65 kG superconducting magnets. This certainly is the largest low temperature project ever seriously contemplated. Although the particle energy expected from this accelerator is still far below the particle energies of the early universe, 10^{19} GeV, this accelerator will extend the known region of high energy physics into a regime where new particles may appear which shed light on the grand unified theories.

However, even this heroic accelerator is still 15 orders of magnitude lower in energy than the predicted energies at the time of the big bang. An accelerator with a gradient of the Stanford two-mile linear accelerator would need to be a hundred thousand light years long to accelerate particles to this energy. Certainly we cannot explore this very high energy region with man-made accelerators alone. We must depend on observations of either events predicted by the theory of the early universe and observable at lower energies or relics left over from this earliest beginning. One prediction that is being tested vigorously is proton decay. The standard SU-5 unification theory predicts a proton lifetime of $4.5 \times 10^{29 \pm 1.7} (< 2.3 \times 10^{31})$ years [14]. To date, this prediction of proton decay has been tested with negative results down to a lifetime of 2×10^{32} years for a proton decay into e^+ and π^0 [14]. These results on proton decay seem to favor other theories that predict longer proton lifetimes [6–11].

3. COSMOLOGY; RELICS FROM THE BIG BANG

Relics left over from the big bang that have already been observed include black body radiation, matter observable by its electromagnetic radiation, and dark matter not observable by its electromagnetic radiation. Important information about these relics includes temperature and distribution of the black body radiation, the photon-to-baryon ratio, the relative abundance of the elements, the all-prevasive presence of real matter with no evidence of primordial antimatter, the quantity, distribution, velocity, spectrum and other characteristics of visible matter, and the distribution, quantity and other characteristics of dark matter. Together these relics give us the best evidence for our present picture of the early universe and will continue to play a very important role. Still this evidence is very incomplete. For example, black body radiation, which gives us the most definitive evidence for the big bang theory [15], condensed out relatively late in the history of the universe (10^5 years after the big bang). We cannot see directly events that took place before the black body radiation condensed out.

What other relics might be left over, and perhaps detectable? Included in this volume are descriptions of near zero techniques to detect four possible relics: fractional electric charge [16,17], magnetic monopoles [18], heavy atoms [17] and stochastic gravitational radiation [19]. (The sensitivity of gravitational radiation experiments to stochastic radiation will be discussed below.) Mentioned later in this paper are near zero techniques to detect massive neutrinos, other weakly interacting massive particles (WIMPS), and axions. The search for relics, such as those mentioned in this paragraph, is, in my opinion, one of the most important problems in physics. Discovery of any one of these relics would give definitive experimental information obtainable in no other way.

3.1 Fractional Electric Charge

Fractional electric charge is an important relic not only because its discovery in the free state would prove the existence of free fractionally charged particles which could be identified and measured, but also because the discovery or lack of discovery of a free fractionally charged quark would give evidence obtainable in no other way about the prediction of quantum chromodynamics (QCD) that quarks are completely confined and cannot be observed in the free fractionally charged state [20]. It is believed that the universe started with a sea of quarks. Unless quarks are completely confined, some quarks must have escaped the big bang [21]. As long as charge is conserved, a fractionally charged quark cannot lose its fractional charge unless it finds appropriate fractionally charged partner(s) with which it can combine to form a zero or integer charge. Thus a search for fractional electric charge is a very important experiment, whether it yields positive or negative results. A test of matter, as reported in this volume [16], can detect a single fractional charge on as many as 10^{20} nucleons.

A fractionally charged particle left over from the early universe does not have to be a quark. Some theories, for example SU-9, predict fractionally charged leptons [22]. Some supersymmetric ten-dimensional string models predict fractionally charged particles when the ten dimensions curl up into four dimensions [24]. Such particles would have a mass of the order of the Planck mass, or at least the grand unification mass.

The experiments reported in one of the papers [16] in this volume seem to demonstrate the existence of fractional charges on niobium spheres. Whether this is really true or a subtle artifact of the experiment is crucially important for physics. A subtle magnetic effect has recently been discovered when the apparatus is not exactly vertical, which could give spurious results [16,25,26]. At the present time this effect is being studied, and techniques for eliminating it have been developed [26].

If the new experiments continue to show fractional electric charges of $1/3$ e, then crucial questions for the future are: What atoms are these charges on? What is their e/m? What are the properties of these fractionally charged particles? To answer these questions, the techniques to detect single atoms and investigate their spectrum, described by William M. Fairbank, Jr., elsewhere in this volume [17], would be an important tool.

If the new experiments on fractional charges on niobium spheres should demonstrate that the observed fractional charges were merely a subtle magnetic artifact, then it is important that these or other experiments be continued until either fractional charges are discovered or it is proven to the best possible level that free fractional charges do not exist.

To date the total amount of mass which has been definitively searched for fractional charges using Millikan oil drop techniques, taking into account all the worldwide experiments, is 0.01 grams, consisting mainly of niobium, iron, mercury and water [16,27]. These experiments measure directly the total charge. Other experiments involving enrichment procedures and measurement of properties such as e, e/m or the atomic spectrum are very important and have looked at larger quantities of matter [28]. However, these experiments involve assumptions about the properties of the fractionally charged particles and do not cover all possibilities, for example, a very large mass. In his review article on quark searches [28], L. W. Jones states, "Had quarks been discovered as a result of enrichment schemes, these would have proven of obvious validity. However the negative results might mean either that no quarks exist or that they do not behave in stable matter in the manner assumed by the experimenters."

Since the distribution of fractional charges is not expected to be uniform and since fractionally charged atoms might be very rare if they have a large mass [29], it is very important that experiments be continued until the largest possible mass with the maximum variety of samples has been tested.

3.2   Magnetic Monopoles

Also in this volume Cabrera, Taber and Felch [18] discuss the present and projected experiments to look for another relic, magnetic monopoles. Magnetic monopoles represent an absolutely unique particle in nature and are present in all grand unified theories. The unique signature of a magnetic monopole passing through a superconducting loop independent of the monopole's velocity makes heroic searches with very large area superconducting detectors very important.

If a monopole has actually been observed, as discussed in [18], then measuring the properties of the monopole becomes extremely important. The magnitude of the magnetic charge comes automatically from the size of the signal in the superconducting loop, and the velocity could be measured with time of flight experiments with more than one detector.

The mass would be more difficult to measure; determining it would involve measuring the deflection in a magnetic field. Blas Cabrera has described the method using the trapped magnetic flux in the superconducting shields to detect the passage of monopoles through the shields. The trapped magnetic flux would identify the exact location where a monopole passed through each shield. Two shields could accurately determine the direction in which the particles passed through the shields. Knowledge of the direction before and after passage through a large magnetic field would in principle make possible the determination of the mass. The biggest problems with such a detector are the small solid angle available for the passage of a monopole through two detectors separated by a magnetic field, and the very small deflection of the particles in this magnetic field when the monopole mass is as large as predicted by the grand unified theories (GUT). A monopole with mass predicted by GUT of 10^{16} GeV traveling with the velocity of particles in our galaxy, 10^{-3} c, where c is the velocity of light, would be deflected by 2×10^{-7} radians in a magnetic field of 20 kilogauss 100 meters long perpendicular to the direction of flight. A monopole with the Planck mass, 10^{19} GeV, would be deflected by only 2×10^{-10} radians. However, if monopoles are definitively observed, a great effort to determine the mass would certainly be worthwhile, and could involve very large scale near zero techniques.

In a talk given at the "Inner Space/Outer Space" conference at Fermi Lab in 1983, Bjorken pointed out that monopoles and antimonopoles of the SU-5 (GUT) mass could provide a way to look at energetic events at the GUT scale [30]. In looking toward the future he said: "If one had one SU(5) GUT monopole and one antimonopole and could bring them together to annihilate (nontrivial but thinkable), then the single annihilation could produce several SU(5) X and Y bosons, which could decay into observable, indeed lethal, hadron jets and leptons of energy 10^{14}–10^{15} GeV. A single event would be a radiation hazard. While far out, maybe this at least indicates that high energy experimental physics is a very long way from becoming a sterile discipline." Although such an experiment seems completely impossible, it illustrates the tremendous physics that might be learned if monopoles could be found and studied.

3.3 Coexistence of Fractional Charges and Dirac Monopoles

If both free fractional electric charges and Dirac monopoles exist, (for example, if the results of both the experiments with niobium spheres and Blas Cabrera's monopole experiment are confirmed), we have a dilemma. Dirac predicted a monopole to explain the quantization of electric charge [31]. He reasoned that the string which connects a north and a south pole monopole must be invisible to all experiments if a true monopole exists. This requires that the phase of a charged particle's wave function change by an integral number of wavelengths if the particle is carried around the monopole string. Otherwise, two charged particles passing on the two sides of the string would exhibit interference and the string would be observable. The invisibility of the string requires that $ge^* = n\hbar c/2$, where g is the monopole strength, e^* is the electric charge, and $\hbar$ is Planck's constant. In order for this to be true at all distances from the string, the field from the charged particle must fall off as $1/r^2$. If the niobium sphere experiment and the monopole experiment are both correct, then $ge^* = \hbar c/6$ for $e^* = 1/3$ e or $ge^* = \hbar c/3$ for $e^* = 2/3$ e. This is obviously a violation of the Dirac quantum condition.

However, several physicists have pointed out that the existence of a fractional charge and a Dirac monopole does not have to violate the Dirac hypothesis because the monopoles in SU-5 theory also contain color fields [22,23]. The color monopole interacting with the color charge associated with the fractional charge produces an extra term such that $ge^* + g_c e_c = \hbar c/2$, where g_c is the color monopole strength and e_c is the color charge. However, quarks are confined and colored monopoles are screened in the SU-5 theory. If they were not confined and the Dirac hypothesis were to hold for free quarks and Dirac monopoles, then the color field would have to fall as $1/r^2$ at large distances. Inside the nucleus it is known that this force is independent of distance. Preskill has argued that if the color force falls off as $1/r^2$ to infinite distance then the resulting massless hadrons should be observable in accelerator experiments [22].

What then is the solution? Is either the Stanford quark experiment or the monopole experiment certain to be wrong because together they disagree with a fundamental tenet of quantum mechanics? Is it impossible to have both a free fractional charge and a Dirac monopole? Not necessarily, as several have said [22,23]. If SU-5 is not the correct grand unification theory (and proton decay does not seem to agree with the simple SU-5 theory), then they suggest that SU-9 might be the correct grand unified theory. In this theory there is a V monopole associated with the Dirac monopole and a V charge associated with a fractionally charged lepton. This V force between fractionally charged leptons falls off as $1/r^2$. The V force and the V monopole exactly compensate for the Dirac monopole and fractional electric charge such that the Dirac quantum condition is satisfied for free fractionally charged leptons and free Dirac monopoles, that is, $q_e g + q_v g_v = \hbar c/2$, where q_e is the electric charge of the fractionally charged leptons, g is the Dirac monopole charge, q_v is the V charge on the fractionally charged lepton, and g_v is the V monopole charge.

The existence of a V force between fractionally charged particles would make it much easier for a free fractional charge to find a partner and form a neutral molecule. Fractionally charged particles can never be neutralized by integer charges, so they are readily dissolvable in the ocean. The properties of the salt water screen the electric charge from other fractionally charged atoms. The V force would not be screened, and fractionally charged particles with opposite V charges would attract each other. The resulting molecules would be neutral and quite stable [32]. Thus most of the fractionally charged nuclei might exist in quite stable neutral molecules and be quite unobservable, except in experiments sensitive to mass [17].

How could this V force be observed? If there is an imbalance in the $\pm V$ forces in the sun or on the earth, then an equivalence principle experiment at very high accuracy might show up the new force if materials with different V charges were used to perform the experiment. The possibility of finding a new force illustrates the importance of a space experiment to measure the equivalence principle to 1 part in 10^{17}, as discussed in [33].

3.4   Fractional Charge; A Practical Application

George Zweig has suggested that if fractional charges exist, they can be used to catalyze fusion at room temperature [34]. If negative fractionally charged particles were mixed with deuterium, then when the fractionally charged particles were absorbed on a deuterium nucleus until it was negative instead of positive, it would react with another deuterium nucleus without repulsion. The resulting fusion process would give up the fractionally charged particles and produce energy as deuterium is changed to helium. If the quark remains attached to the final helium nucleus, it could be knocked off by thermal neutrons.

Zweig estimates that 10 moles of quarks would be required to supply all the energy needs of the U.S. If fractionally charged quarks or leptons were to solve our energy problems, that would be a striking example of very esoteric fundamental research into the foundations of physics resulting in the unanticipated solution of a very practical problem of society. Of course, all of these experiments with fractionally charged atoms are meaningless unless the fractionally charged particles exist outside the nucleus, as reported in [16].

3.5   Stochastic Gravitational Radiation

A third relic from the big bang might be stochastic gravitational radiation from (a) quantum effects before the big bang [35], (b) first order cosmological phase transitions in the early universe [36], or (c) decaying cosmic strings [37]. Salam stated in an invited talk opening the Fourth Marcel Grossmann meeting held in Rome in July 1985, "... it is well to remember that the gravity waves—so hard to detect today—could become routine tools of the 21st century high energy physics—providing new windows on Nature." [38] The detection of stochastic gravitational radiation from the initial phase transition in the early universe would be a fulfillment of this prediction.

Cosmic strings are linear topological defects which might be formed in a phase transition in the early universe. It has been predicted that the decaying strings might produce at the surface of the earth stochastic gravitational radiation with a spectral energy density Ω_g equal to 10^{-7} of closure density [37]. This would yield, over a bandwidth equal to the frequency f of the gravitational wave radiation, a strain $h \sim 3 \times 10^{-22}$ 1 Hz$/f$ [39]. A stochastic background of comparable magnitude could be left over as a relic from the quantum effects before the big bang [35]. Such gravitational radiation, if it could be observed and identified, might give the only direct evidence for the nature of the universe at the time of the big bang (10^{-43} seconds).

To detect stochastic gravitational radiation one needs to measure correlations between the noise in two gravitational wave antennas. If the second antenna is slowly rotated, the correlated stochastic gravitational wave signals will be modulated, making it possible to identify a stochastic gravitational wave background in the presence of spurious correlated background noise [39]. Even at the quantum limit of sensitivity (a strain spectral density $h/\text{Hz}^{1/2} = 7 \times 10^{-23}$), present bar gravitational wave detectors operating at a frequency of 840 Hz and a bandwidth of $\Delta f = f$, require 10^{10} seconds (~ 300 years) of integration time to detect the above predicted gravitational wave stochastic background of 3×10^{-22} 1 Hz/f. This assumes a signal-to-noise ratio of unity.

However, the situation improves dramatically as one reduces the frequency. Peter Michelson has described a low frequency gravitational wave detector composed of two masses separated by a high-Q microwave cavity, where the two masses act as free masses [40]. This detector would make use of the very near zero loss ($Q = 10^{11}$) observed in superconducting niobium cavities by John Turneaure [41]. Assuming each gravitational wave antenna to be made up of two aluminum masses, each of length 20 feet and diameter 34 inches, the sensitivity at the quantum limit at 100 Hz is a strain spectral density $h/\text{Hz}^{1/2} = 7 \times 10^{-23}$. Such systems, operating over a bandwidth $\Delta f = f$, could detect the above mentioned stochastic gravitational radiation in 3×10^5 seconds with a signal-to-noise ratio of 1 [39]. The same systems operating at the quantum limit at 10 Hz could detect the above mentioned stochastic gravitational radiation with a unity signal-to-noise ratio in 3×10^4 seconds. Even better sensitivity could be obtained with the long baseline laser interferometer gravitational wave detectors presently being planned [42] if they could be operated at the theoretical noise level, assuming 100 watts of input power.

The major problem to be solved for operation at low frequencies is vibration isolation. Solving this problem is one of the major challenges for the future gravitational wave detectors. Ideally very low frequency gravitational wave detectors might be operated in space without the prohibitive problem of vibration isolation, but with, of course, many new problems of space. Low frequency bar detectors might conceivably be operated in conjunction with the space station. Peter Bender [43] has proposed the ultimate low frequency experiment of two interferometers made of free flying satellites in a stable orbit around the sun.

Bond and Carr [44] have considered several different possibilities by which collapse of massive population III stars or coalescence of black hole binaries could lead to a stochastic background of frequency ≤ 100 Hz if black holes of mass greater than 100 solar masses make up a major part of the dark matter. Such a background would be observable with two bar detectors operating at the quantum limit or with long baseline interferometer gravitational wave detectors.

Certainly the detection of stochastic background radiation poses one of the major challenges and opportunities for physics in the next few decades.

3.6 Massive Particles

A fourth relic which might be left from the big bang is a heavy charged or neutral particle, for example, a heavy color quark or "techniquark" of mass ~ 1 TeV [17]. The search for single atoms with a spectrum indicating a heavy mass, as described by William M. Fairbank, Jr., in this volume [17], is designed to detect such particles if they exist. To date, the most thorough search for such heavy particles has been made by Smith *et al.* looking for heavy hydrogen [45].

The observation of any of the relics mentioned, fractionally charged particles, magnetic monopoles, stochastic gravitational radiation from the early universe, or heavy particles, would give perhaps the first experimental evidence of the real nature of the early universe and shed new light on the foundations of physics.

4. NATURE OF GRAVITY

The true nature of gravity is certainly one of the very interesting and important questions for future physics experiments. Newton's universal law of gravitation is one of the most significant discoveries in the history of mankind, and Einstein's theory of general relativity in which gravity is explained in terms of the geometry of space time is one of the most beautiful theories ever proposed. So what is there to test? The validity of a physics theory does not rest on beauty alone, and even general relativity must be checked in all possible ways by experiment. Central to the understanding of the early universe and the possible unification of the four forces of nature, including gravity, is a reexamination of the nature of gravity, including new experiments. The necessity for such checks is made more urgent as one tries to unify gravity with the other forces of nature and to quantize gravity at the time of the big bang.

4.1 Rotation: Gyro Relativity Experiment

As described in this volume [46], the Stanford gyro relativity experiment, known as Gravity Probe B or the Schiff gyro experiment, will make a completely new kind of check on general relativity. It will check for the first time a rotational term, the dragging of inertial frames by the rotating earth. As the earth makes one rotation, a gyroscope sitting above the North Pole is dragged with the rotating earth by 3×10^{-9} radians per revolution of the earth. If the gyroscope were in a polar orbit around the earth and its spin axis oriented perpendicular to the axis of the earth, then the gyroscope, even if it were perfect, would find its spin axis dragged in the direction of the rotation of the earth by 0.05 seconds of arc per year.

If there is any discrepancy between the answer obtained by the gyro experiment and Einstein's prediction for the dragging of inertial frames, it will most likely not refute Einstein's theory but, rather, show how rotational terms like spin and torsion can be introduced into gravitational theory [47,48]. It is, therefore, very important to push the accuracy of the experiment as far as possible. By using the most sensitive SQUID's, coupled with a lower value of acceleration in the zero-g part of the gyro experiment ($\sim 10^{-11}$ g), it may be possible to improve the projected gyro sensitivity by an order of magnitude to 10^{-4} seconds of arc, a very significant improvement if it could be made.

4.2   Strong Fields, Gravitational Radiation and High Time Resolution X-Ray Astronomy

A very important question about gravity involves the exact behavior of gravity in very strong fields near the surface of a neutron star or black hole. A suggestion by Peter Michelson and a calculation by Robert Wagoner [49] predict that an accreting neutron star with a small enough magnetic field ($< 10^8$ gauss) could be spun up by the accreting mass until it became relativistically unstable. At this point a nonaxisymmetric distortion would circulate around the star. This would modulate the x-rays produced by the accreting mass and cause an emission of gravitational radiation.

Peter Michelson, Kent Wood, Herbert Freedman, Herbert Gursky, Mason Yearian and Paul Boynton have proposed to fly on the space station a large x-ray antenna of 200 square meters of proportional counters with an effective area for x-rays of 100 square meters (XLA) [50]. This would be able to detect a sufficient number of photons/second to observe a modulation as small as 1 part in 10^4 of the x-rays from some of the sources in our galaxy. If the modulation were observed, then the gravitational wave antenna could phase-sensitively look for the emission of gravity waves synchronized with the x-ray output. With such synchronization the sensitivity of a quantum limited bar could be pushed to $h = 10^{-26}$ with integration time of several weeks. Such a large x-ray detector could also look at x-rays from other objects where the gravitational fields are very high, for example, near the surface of a black hole like Cygnus X-1. It might be able to measure the characteristics of the innermost orbit of the accreting disk and prove that Cygnus X-1 is a black hole. By using the moon as a knife edge, the angular diameter of an x-ray star that was occulted by the moon could be measured with 10^{-3} seconds of arc resolution. Many other experiments probing the physics of very strong gravitational fields seem possible with such an x-ray antenna.

Gravitational radiation from coalescing binaries, including black holes, and from collapsing stars offers a unique way to probe gravitational fields in the very strong field limit near the surface of a neutron star or black hole.

4.3 Inverse Square Law

Another very important aspect of gravity involves the inverse square law at laboratory distances. The inverse square law has been very accurately verified at astronomical distances by the elliptical form of the orbits of planets around the sun and of the moon around the earth. It is much more difficult to verify at laboratory distances. If there is another particle with finite mass which mediates an interaction between matter then there would be an extra exponential term added to gravity which would not obey the inverse square law. Long has pointed out that the existing measurements of G made before 1974 give larger values as the distance is increased from 5 to 30 centimeters [51]. If this were really true it would seem to point to the existence of a new force in nature coupled to a new particle.

Long has made measurements that confirm his earlier prediction that the inverse square law is violated at laboratory distances [52]. Several other experimenters have reported confirmation of the inverse square law in the range 5–100 cm ([53] and references therein). In particular, Hoskins, Newman, Spero and Schultz and, independently, Chen, Cook and Metherell have reported verification of the inverse square law over this distance range to an accuracy of better than one part in 10^4 as compared with Long's observed discrepancy of 2 parts in 10^3. Stacey, in Australia, has made measurements of g in deep mines and under water, and has come to the conlusion that the inverse square law is violated at distances of the order of 200 meters [54].

Obviously, the inverse square law remains a very exciting subject for future investigations. As reported in this volume, Ho Jung Paik has developed a three-dimensional gradiometer to measure $\Delta^2 \phi$ which must equal zero for an inverse square law force of gravity [55]. This is a very useful type of measurement for distances longer than a few meters.

4.4   Equivalence Principle

Fischbach and his coauthors [56,57] have recently pointed out that Stacey's data show a violation of the inverse square law that could be fit by a fifth force between baryons if that force had an exponentially decaying range of $\lambda = 200$ meters and a strength $\alpha = 6 \times 10^{-3}$ compared to gravity. He calls that force hypercharge. Fischbach *et al.* also pointed out that such a hypercharge force should cause an apparent violation of the equivalence principle when measurements are made with respect to mass within a few hundred meters of the experiment. Thus the experiment of Eötvös [58], in which measurements were made with respect to the acceleration of the earth, but not the experiments of Roll, Krotkov and Dicke [59] or Braginsky and Panov [60], in which measurements were made with respect to the sun, should show a violation.

Fischbach and his coauthors [56,57] have plotted the Eötvös results as a function of mass/baryon of each sample. They find a straight line with nonzero slope as would be expected from a fifth hypercharge force between baryons. However, interpretation of the results is very sensitive to the geometry of the nearby masses and cannot even qualitatively be compared with Stacey's results without an exact knowledge of the Eötvös experimental conditions (*i.e.*, a nearby wall, a basement below the experiment, *etc.*).

The work cited in the above paragraph is stimulating many experiments (listed in [57]), including torsion pendulum experiments near a cliff and a repeat of the Galileo free fall experiment using modern optical techniques with lasers and corner reflectors. Such experiments are very important and could prove the existence and nature of a new force of laboratory range.

Further experiments on the inverse square law of gravity and improvements on the equivalence experiments, including the Galileo experiment comparing two masses in free fall toward the earth, are very important for the future and could give us our first proof of the nature of the new unification theories being pursued so vigorously theoretically. The equivalence experiment of Worden and Everitt [33], when performed in space, could not only improve the Eötvös experiment by several orders of magnitude but also, as Worden has pointed out [61], be used to check a force of a range of a few hundred meters. Such a check could be made if the equivalence principle experiment were flown on a satellite near the shuttle and the shuttle used as the mass to generate the short range force. As the theorists try harder and harder to understand gravity and unify it with the other forces in nature, it becomes increasingly important to make new kinds of meaurements on gravity with very high precision.

4.5 Antimatter and Gravity

One question which arose in the 1950's was, "What is the force of gravity on antimatter?" The paper by Lockhart and Witteborn [62] in this volume describes a very sensitive experiment to measure the gravitational force on the electron and positron. In 1957 Morrison and Gold stated that a necessary and sufficient condition to separate antimatter from real matter in the early universe was the existence of a gravitational force of repulsion between them [63]. This was the original incentive for the electron-positron free fall experiment. Now proton decay and the resulting baryon nonconservation have shown another possible way to make the universe predominantly real matter [14] without the aid of a repulsive force between antimatter and real matter.

Goldman, Hughes and Nieto, however, have pointed out [64] that spin-0 particles and spin-1 particles predicted by quantum gravity could contribute to the gravitational force between matter and matter two additional forces, an attractive force associated with the spin-0 particle and a repulsive force associated with the spin-1 particle, which nearly cancel. These terms would add, however, for antimatter, increasing the gravitational force between antimatter and real matter, perhaps by a factor of three. These forces could be of finite range if the spin-0 and spin-1 particles acquired mass through a phase transition in the early universe. If they do not exactly cancel, they might account for the fifth force suggested by Fischbach and discussed above.

A large group of scientists has been formed to try to measure the force of gravity on the antiproton at CERN using thermalized antiprotons [65] in an apparatus like the one developed to measure the force of gravity on the electron and positron [62,66]. Because of the lack of a slow positron source, the electron-positron free fall experiment has been used only on electrons and has not yet been used to measure the force of gravity on positrons. However, such measurements now appear feasible since a pulsed source of slow positrons is now possible using an electron accelerator to produce the positrons, and the thermalizing techniques that have been developed for them [67]. Gravitational experiments on both positrons and antiprotons will add to our fundamental understanding of gravity and may represent a unique way of looking at the spin-1 and spin-0 particles of quantum gravity, if they exist.

4.6 Experiments with Electrons; Surface Shielding Effect

Already in the experiments on electrons [62] a new electron surface state on copper oxide has apparently been discovered, which undergoes a transition at about 4.5 K to a state in which the electrons shield the patch effect field on the surface of the copper oxide drift tube. This not only makes possible the electron-positron free fall experiment, but points the way to a new effect in solid state physics. John Bardeen [68] points out in this volume that the surface states on copper oxide could become metallic, which may be the first step in a theory. Madey and Hanni [69] have suggested a Bose-Einstein condensation of the surface electrons.

The unraveling of the details of this apparent shielding effect on copper oxide is one of the very interesting problems in solid state physics and requires the very best of the near zero techniques. Recent microwave experiments with copper and aluminum cavities [70,71] indicate that a transition in the surface impedance of a microwave cavity takes place on copper and also on aluminum. The transition may be related to the transition seen in the copper drift tube experiments. An experiment to measure directly the patch effect field of copper and aluminum at liquid helium temperatures is being performed [72].

5. DARK MATTER

An important problem in physics today is the nature of the dark matter in the universe. In the last ten years astronomers have discovered, by careful observation of the motion of stars, that around our galaxy there exists a spherical halo composed of more than ten times as much matter as makes up the visible part of our elliptical, pancake-shaped galaxy [73,74]. Furthermore, between the galaxies there seems to be extra dark matter. In fact, there may even be enough dark matter to close the universe, a situation required by some of the theories involving an inflationary universe.

What is the dark matter and why can't we see it? Several possibilities have been considered. One suggests that it may consist of stars too small to produce detectable electromagnetic radiation, for example, stars the size of Jupiter. It is difficult to see how enough stars of this size could have been formed [75]. Similarly it is difficult to square with experimental evidence a model of the dark matter composed of dust, or gas, or balls of hydrogen [75]. Another possibility is that the dark matter consists of collapsed stars in the form of neutron stars or black holes. If the dark matter in the halo were made of collapsed stars of 10 solar masses or less, more metallic elements would exist throughout the galaxy than are observed. (These heavy elements would have been distributed around the galaxy during collapse of the stars.) Olive and Hegyi [75] have come to the conclusion that not more than 35% of the galactic halo mass can be hidden in the form of baryonic mass even if all of the above possibilities are included.

If this is true, then to explain the galactic halo one is left with either massive black holes ($>$ 100 solar masses, which accrete their ejected mass and avoid the problem of too much distributed heavy elements [44]) or nonbaryonic mass, such as massive neutrinos or new exotic particles. Bond and Carr have suggested that the dark matter might in fact be made up of such massive black holes (in excess of 100 solar masses) formed before the galaxies [44].

Concerning dark matter in the universe as a whole, many theorists reason from the relative abundance of primordial hydrogen, deuterium, ^{3}He, ^{4}He and ^{7}Li that the total mass of normal matter (baryons), at the time of element formation and therefore also now, cannot be greater than 2/10's of the closure mass [76]. Thus, if there is enough dark matter to close the universe, it cannot all be in the form of baryons, but must be made, at least in part, of either massive neutrinos or some new kind of particle.

One suggestion is that these particles are axions, proposed by Quinn and Peccei to account for parity conservation in baryons [77]. The mass of the axions would be very small, of the order of 10^{-11} of the electron mass. Other candidates for the dark matter are the new particles suggested by the supersymmetry theories of gravity [78]; in these theories there is for every fermion a symmetrical boson, and *vice versa*. One or more of the supersymmetrical particles, spin-0 quarks, sleptons, photinos, w-inos, z-inos, gluinos, Higgsinos or gravitinos, could form the dark matter. One such particle, with mass $>$ 1 GeV and weakly interacting, has been called a WIMP (weakly interacting massive particle) [79,80].

Certainly a major question is the nature of the dark matter; for its resolution, near zero techniques are not only important but in many cases essential.

5.1   Black Holes; Detection by Gravitational Radiation

In suggesting that large black holes of 10^2 to 10^5 solar masses might make up the dark matter of the universe, Bond and Carr have concluded that 50% of the black holes could be in the form of binaries [44]. They predict that the typical separation of these binaries is such that a pair of binary black holes will radiate away sufficient energy in the form of gravitational waves to cause collapse into each other within the age of the universe. Such a corotating pair of black holes would give a gravitational wave signal with increasing frequency as the orbit shrinks, until at coalescence of the two black holes the gravitational wave signal reaches the maximum in frequency and amplitude.

Bond and Carr have calculated the gravitational wave energy that would be radiated, for example, on the coalescence of two 1000 solar mass black holes. At a distance of 30 kiloparsecs in the halo around the galaxy, this would give a gravitational wave signal of frequency 10 hertz with a strain at the earth of $h = 7 \times 10^{-16}$. Such black holes coalescing anywhere in the visible universe could be detected, with a signal to noise ratio greater than two, by a 20,000 kilogram 12 meter split bar detector [40] operating at 10 hertz at the quantum limit of $h = 2 \times 10^{-21}$. The above estimate neglects change of frequency due to red shift.

The possible observation of such an event makes it worthwhile to put a major effort into building gravitational wave observatories covering the range from 10 to 1000 hertz, operating with a wide bandwidth as close as possible to the quantum limit. The gravitational wave programs using bar detectors (see, for example, [19]), as well as the laser projects with potentially even greater sensitivity [42], are advancing in this direction.

Many other astronomical events, such as supernovae forming neutron stars, the collapse of a star to a black hole, and the spin up of a very rapidly rotating neutron star by accretion, are potential candidates for gravitational wave observation. This is certainly one of the pioneering fields for the future of physics. To observe a gravitational wave at 10 hertz at the quantum limit, one must be able to detect in a tenth of a second a single photon with frequency 10 hertz and energy $h\nu = 6 \times 10^{-33}$ joules, equivalent to the thermal energy of vibration of a single atom at a temperature of 5×10^{-10} K. The unique property of gravitational waves which makes it possible to attempt these heroic efforts of observation is the capacity of a gravitational wave to pass unmolested through any conceivable earth-bound shield, even a shield that successfully forms an isolated region with almost no interference from the outside world and near zero in temperature and thermal noise.

5.2   Massive Neutrinos, WIMPS and Axions

Another candidate for the dark matter is a massive neutrino. This has been discussed by Remo Ruffini elsewhere in this volume [74]. Many experiments on earth are underway to measure the mass of different varieties of neutrinos [81]. All such experiments except the radioactive decay experiments by the ITEP group in the Soviet Union [82] have yielded negative results for the mass. The ITEP group reported the mass of the electron neutrino to be > 20 electron volts.

Theory predicts that when stars collapse, large quantities of neutrinos as well as gravitational waves are emitted at the instant of collapse, the massless neutrinos and the gravitational waves both traveling outward with the velocity of light. As predicted by the special theory of relativity, *neutrinos with a rest mass* travel with less than the velocity of light. The comparison of the arrival time of the gravitational wave and the neutrinos, when properly accounting for delayed emission of neutrinos after collapse, would give a measure of the rest mass of the neutrino. For example, a neutrino of 10 electron volts rest mass and one million electron volts energy, emitted from the center of our galaxy, would arrive at the earth 45 seconds after the simultaneously emitted gravitational wave, both having traveled for 30,000 years. Different energy neutrinos would arrive at different times. A neutrino with the same 10 electron volt mass but with 10 kilovolts of energy would arrive 5.2 days later than the gravitational wave, and the same neutrino with 100 MeV energy would arrive 4.5×10^{-3} seconds after the gravitational wave.

Cabrera, Martoff and Neuhauser are developing a phonon detector using very pure silicon and superconducting tunnel junctions maintained below 100 millikelvins to detect such neutrinos and their energy by measuring the phonons created when the neutrinos are elastically scattered from electrons or nuclei [83]. A similar detector is being developed using tiny superconducting spheres, which undergo a transition from superconducting to normal state when the energy of recoil from a scattered neutrino is deposited in a sphere [84]. Such neutrino detectors could also look at neutrinos from the sun to check the results of Davis [85], who finds too few neutrinos emitted from the sun. It has been suggested that such detectors could also observe axions coming from the sun with the thermal energy of the temperature of the sun, or WIMP's as components of dark matter [86,87]. The axions from the dark matter could in principle be detected by means of the microwave photons predicted to be given off when an axion passes through a very high magnetic field produced by a superconducting magnet in the presence of a microwave cavity [88,89]. Such experiments are underway [90] and constitute a beautiful example of the use of near zero techniques.

5.3 Supersymmetrical Particles

How might one be able to observe the supersymmetrical particles, if they do exist but cannot be observed in accelerator experiments or in experiments (such as those described above) devised to detect such particles coming in from outer space? Other methods do exist and offer exciting possibilities. It has been predicted that there might be a charged stable particle of mass in excess of 200 times the proton mass. The atomic spectrum of an atom containing such a particle would be changed very slightly. Single atom detection experiments such as those by William M. Fairbank, Jr., described elsewhere in this book [17], represent a possible method of detecting such particles. By this method one could also detect a massive stable neutral particle such as a spin-0 quark if it were bound to the nucleus with strong forces.

6. CP & T VIOLATION:
THE ELECTRIC DIPOLE MOMENT OF ^{3}HE

A very important question in physics today is the cause of CP & T violation in the K$^\circ$-K̄$^\circ$ experiments. This is crucial to baryon nonconservation and the explanation of matter-antimatter asymmetry in the universe. The best experimental hope for more information on the subject would come from the discovery of an electric dipole moment on the nucleus. It was this motivation that led us to initiate an experiment to measure the electric dipole moment on the ^{3}He nucleus many years ago [91]. N. F. Ramsey and his colleagues and I. S. Altarev et al. have made measurements on the electric dipole moment of the neutron and placed a limit of $p(n) < 6 \times 10^{-25}$ e cm at 90% confidence level for its value [92]. This is a continuing experiment.

The neutron has the advantage over ^{3}He that an electric field can be applied directly to the neutral particle. In the case of ^{3}He, since the nucleus is charged it moves off center in an applied electric field and cancels the electric field to one part in 10^7 [91]. However, the longer intrinsic observing time (about 150 hours) and the larger number of polarized ^{3}He nuclei available make possible in principle measurements on ^{3}He of comparable accuracy to the measurements on the neutron, given the near zero magnetic field environment ($< 10^{-8}$ gauss) obtained by Blas Cabrera [93].

Recently, Avishiai and Fabre de la Ripella [94] have calculated that the electric dipole moment of ^{3}He due to the possible currents of P and T nonconserving nucleon-nucleon interactions, as in the Kabayashi-Maskawa model [95], should be 60 times larger than the electric dipole moment on the neutron from this model. This points out the importance of measurements of the electric dipole moment on other nuclei than the neutron, and is an added reason to complete the ^{3}He electric dipole experiment. It is a very difficult experiment and the theoretical estimate of $p = 1.68 \times 10^{-30}$ e cm [94] for the electric dipole moment of ^{3}He may never be achievable, but it is an important region of physics to continue to pursue. It will check the measurements made on the neutron, and even a negative result in an experiment done well enough will set new limits on the electric dipole moment.

Timothy Chupp is contemplating an experiment to compare the electric dipole moment of ^{3}He and a heavier inert gas by comparing their nuclear magnetic resonant frequencies in the same magnetic field in the presence and absence of an electric field [96]. This appears to be a very accurate way to check the difference in the electric dipole moments of light and heavy nuclei.

7. CONDENSED MATTER PHYSICS

Several developments in condensed matter physics promise to open up new vistas in the future. Not only does the field offer solutions to very fundamental many-body problems, but it continues to make practical contributions to all of physics including the near zero techniques. Obviously important for the experiments discussed here is the continued investigation of macroscopic quantization and in particular the noise temperature of the SQUID. Approaching the quantum limit of a low frequency amplifier and reducing the $1/f$ noise is crucial to experiments such as the gyro relativity experiment and gravitational radiation experiments. One of the most interesting practical developments is the tunneling electron microscope [97]. The beautiful pictures of individual atoms and charge density waves are indications that many new effects with surfaces may be observed with this technique.

One of the most amazing phenomena in nature is the sharpness of the lambda point of liquid helium. A change of only 10^{-8} K makes a qualitative difference in the behavior of the liquid [98]. As discussed by Buckingham [99] and Lipa [98] in this volume, the asymptotic behavior near zero temperature difference has been explained theoretically by the renormalization group theory for which Kenneth Wilson received the Nobel prize in 1982. John Lipa points out that to fully test this theory requires experiments within 10^{-10} K of the lambda point in an environment where gravitational gradients do not smear out the results of the measurements. The experiment planned in the shuttle [98] will accurately check this very fundamental prediction.

The use of satellites in space to make fundamental experiments on physical and chemical processes in the absence of gravity may become an important near zero field of research [100]. However, it will, require the careful theoretical considerations and experimental development which has made possible the beautiful experiments on the lambda point [98].

A very interesting development in near zero physics is illustrated by the recent experiments which have shown that it is now possible to cool and trap neutral atoms with laser light [101]. Sodium atoms have been cooled and confined to a temperature of $T = 2.4 \times 10^{-4}$ K in a viscous molasses of photons tuned to the yellow "D-line" resonance of sodium. This technique opens up the possibility of doing experiments with dilute gases at very low temperature, for example, experiments to look for Bose-Einstein condensation in such gases. It has been estimated [101] that atoms could be cooled to temperatures of 10^{-6}–10^{-10} K.

8.   PRACTICAL APPLICATIONS

It is interesting to consider some of the practical consequences of near zero technology. The superconducting accelerator [102] made possible the development of the free electron laser by John Madey [103] and promises to play an important part in its further development. Of particular interest is the recirculation idea of Todd Smith and Alan Schwettman. After an electron beam has been accelerated by superconducting cavities and passed through a series of magnets to produce laser light, the electron beam is brought back through the cavities and gives up its energy to the cavities before the electron beam is thrown away. Such a system could produce a relatively simple tunable free electron laser with 50% efficiency, and has already been demonstrated using a Van de Graaff generator [104]. These developments are of very great practical importance for the future. Superconducting cavities are now planned for the intermediate energy accelerator to be built in Virginia [105].

Medical applications of near zero techniques appear to be very promising. The paper on π mesons in this volume [106] describes the cancer treatment project started at Stanford and now being carried on in Switzerland at SIN using a superconducting magnet to direct a beam of pions into a cancer to destroy the cancer without killing the patient. For cancers which grow large before detection, this appears to be a very important development for the future. The free electron laser [103] gives the possibility of another medical treatment technique, based on evaporating material from blood vessels by tuning the lasers to just the right frequency for absorption by a particular molecule and adjusting the pulse length and amplitude of the laser beam to optimize the selective evaporation.

Magnetic fields have historically played a very important part in the history of physics, but a relatively minor part in human biology. This is partly because of the small size of the magnetic signal from currents in the body. The heart generates a magnetic field of approximately 10^{-6} gauss. The brain generates a magnetic field at the surface of the brain of 10^{-8} gauss. The small size of such fields has until recently made them impractical for diagnostic purposes in medicine. Now the availability of very sensitive SQUID magnetometers has made possible measurements of both currents in the heart [107] and currents in the brain [108]. John Wikswo has succeeded in detecting currents from a single axon [107]. John Philo and Tad Day have made use of the SQUID to make very sensitive high time resolution susceptibility measurements with the SQUID and superconducting magnets [109].

Finally, I would like to mention a development of tremendous potential in medicine which represents the largest practical use of low temperature physics outside of high energy physics: nuclear magnetic resonance imaging (MRI), and nuclear magnetic resonance chemical shifts (magnetic resonance spectroscopy, MRS). In 1946 Felix Bloch and Edward Purcell announced with their colleagues the discovery of nuclear magnetic resonance. Over the years this technique has become an indispensable tool in chemistry, and provided the means for Walters and myself to discover the phase separation in ^{3}He and ^{4}He solutions and the Fermi degeneracy temperature [110]. In the phase separation experiment we divided the ^{3}He-^{4}He sample chamber into three cavities, one above the other, separated by capillary tubing. When a vertical magnetic field gradient was applied to the sample, three separate ^{3}He resonance peaks were observed in the three cavities. The size of each of the peaks enabled us to discover and observe quantitatively the phase separation curve as the ^{3}He isotope separated from the ^{4}He and floated to the top container. This seems to have been the first case of NMR imaging with a magnetic field gradient.

Since 1973 the technique has been used very successfully on the human body to image the protons, differentiated not only by number but also by the relaxation times T1 and T2 and the flow velocities [111]. Observation of chemical shifts of phosphorous and now also of protons has enabled the determination *in vivo* of such important processes as phosphorous metabolism in the presence of oxygen [111]. Superconducting magnets are playing an increasingly important part as whole body magnets. Fields as high as 4 tesla are contemplated for magnetic resonance spectroscopy. I think this field, especially magnetic resonance spectroscopy with high field magnets, will continue to grow in importance.

A practical problem with the very high fields is the fringe field from the superconducting full body magnet. If this can be successfully eliminated by self-shielding the magnet, it will open the door for general use in hospitals without the necessity of a separate building to isolate the fringe fields of the magnet. One possible solution to this problem is a double magnet designed so that the fields fall off as $1/r^9$ instead of $1/r^3$ [112].

9. SUMMARY

In speculating about the future we have discussed research that represents a marriage of astrophysics, high energy physics and low temperature physics. This interdisciplinary field obviously contains some important unsolved problems. Why do low temperatures, and near zero techniques in particular, play an important role in the solution of these problems? The key to the answer is the very interesting phenomenon of macroscopic quantization [113]. Fluxoid quantization leads to persistent current magnets, the London moment used for very accurate gyro readouts, superconducting shields and, in particular, the Josephson effect which makes possible SQUID magnetometers with noise temperatures approaching the quantum limit of one photon $(10^{-7}$ K$)$ for frequencies of 1000 hertz. These in turn make possible searches for the very rare relics left over from the big bang, the very small effects involved in making fundamental measurements of gravity, and new forces that may exist. Semiconductor amplifiers are limited to a noise temperature of the order of 1 K by the zero-point motion of the individual electrons. SQUID's, on the other hand, involve transitions between macroscopic quantized states involving very large numbers of electrons, making possible a noise temperature 10^6 lower than for a semiconductor.

In summary, what is in store for physics using near zero techniques? Certainly a host of very fundamental problems. Are there new particles left behind as relics from the big bang (quarks, fractionally charged leptons, axions, heavy stable particles, monopoles, WIMPS)? Do they constitute the dark matter of the universe, or is it perhaps massive black holes? Can these be detected by low temperature gravitational wave observations? What is the nature of black holes like Cygnus X-1? Can important aspects of its nature be learned from very careful observation of the high frequency modulation of the x-rays given off by the black hole? Can the modulation in the x-rays at a few hundred hertz enable one to detect a rapidly rotating neutron star and perhaps eventually the simultaneous gravitational radiation? Are two-dimensional electron states on the surface of copper oxide and possibly other materials undergoing a new kind of transition? Can we detect the neutrinos from the sun and collapsing stars by low temperature techniques? Do new particles exist which would prove the validity of the supersymmetry theory of gravity? Can they be detected in the energy range of the proposed 20 TeV colliding beam proton accelerator, or will their detection come only from observations of the relics left over from the big bang?

It is an exciting time in the history of physics, and questions not even dreamed of regarding the nature of the universe may be answered in the next few decades, some of them by near zero techniques.

One of the most challenging problems in science involves evolution: first, evolution of the universe; then, ultimately, evolution of life. Stimulated by the interaction of high energy physics and astrophysics, great progress is being made in the understanding of evolution on a cosmic scale.

Tremendous progress has also been made in understanding life and its evolution. Selection through the survival of the fittest obviously plays a vital role. Still, the puzzle remains: How do the processes of evolution of life work so perfectly and so fast? The restrictions of quantum mechanics explain why, in the evolution of the inanimate universe, particular elements are formed. If they were to be formed over again in a new universe with the same physics, the same elements would be formed; other elements are forbidden by the exclusion principle of quantum mechanics. Is it possible that some exclusion principle works on the evolution of life such that the infinite number of choices is limited and life proceeds to develop along a much more restricted course? Does W. A. Little's search for a room temperature superconductor [114] have any bearing on this problem, as once speculated by F. London [115]? Certainly evolution from the big bang to the human being continues to pose exciting problems for the future.

So I close this article with the strong belief that the universe remains an exciting and challenging place to investigate, and that it is not possible for us to so completely learn her secrets that there is nothing fundamentally new to discover. Making some of the variables of physics very close to zero is a tool which has unlocked doors and will shed new light on the future. I have been able to participate in this research experience with some success and a great deal of pleasure and satisfaction because of the many superb students and excellent colleagues who have worked with me and shown me the way. I also owe a major part of my success to my wife, Jane Fairbank, who has been a partner in my life and in my work. She has played a major role in the completion of this book. My career in physics has been and remains a very exciting experience. I know that the future holds important opportunities for those with the talent, desire and capability of sustained effort necessary to succeed. I consider myself fortunate to have had the opportunities I have had, both in the problems that have been available to solve and the colleagues who have been there to solve them with me. I look forward with anticipation to the future.

Addendum

The information from the supernova seen on February 23, 1987 (SN 1987A) has a direct bearing on sections 4.2, 5.1 and 5.2 of this paper. The reader is referred to the addendum to paper VI.5 in this volume for a short discussion of this supernova and its effect on gravitational wave astronomy.

References

[1]   P. L. Kapitza, *Bull. Atom. Sci.* **18**, 3 (1962).

[2]   Lord Kelvin, *Phil. Mag.* S. 6 V. 2 No. 7, 1 (1901).

[3]   S. Weinberg, *Phys. Rev. Lett.* **19**, 1267 (1967); A. Salam, in *Elementary Particle Physics*, edited by N. Svartholm (Almquist and Wiksells, Stockholm, 1969), p. 367; S. L. Glashow, *Nucl. Phys.* **22**, 579 (1961); S. L. Glashow, J. Iliopoulos and L. Maiani, *Phys. Rev. D* **2**, 1285 (1970).

[4]   P. Bagnaia *et al.*, *Phys. Lett.* **129B**, 130 (1983); G. Arnison *et al.*, *ibid.*, 273.

[5]   H. Georgi and S. L. Glashow, *Phys. Rev. Lett.* **32**, 438 (1974).

[6]   See, *e.g.*, E. Witten, in *Fourth Workshop on Grand Unification*, edited by H. A. Weldon, P. Langacker and P. J. Steinhardt (Birkhauser, Boston, 1983), p. 395; J. Schwarz, *Phys. Rep.* **89**, 223 (1982); M. B. Green, *Surveys in High Energy Physics* **3**, 127 (1983).

[7]   See, *e.g.*, M. J. Duff, B. E. W. Nilsson and C. N. Pope, in *Fourth Workshop on Grand Unification*, edited by H. A. Weldon *et al.* (Birkhauser, Boston, 1983), p. 341.

[8]   A. M. Polyakov, *Phys. Lett.* **103B**, 207 (1981).

[9]   See, *e.g.*, J. Polchinski, in *Inner Space/Outer Space*, edited by E. W. Kolb *et al.* (University of Chicago Press, Chicago, 1986), p. 447.

[10]   For review, see E. Farhi and L. Susskind, *Phys. Rep.* **74**, 277 (1981).

[11]   J. Bagger and S. Dimopoulos, *Nucl. Phys.* **B244**, 247 (1984).

[12]   A. H. Guth, *Phys. Rev. D* **23**, 347 (1981); A. H. Guth, in *Inner Space/Outer Space*, edited by E. W. Kolb *et al.* (University of Chicago Press, Chicago, 1986), p. 287.

[13]   J. D. Jackson, M. Tigner and S. Wojcicki, *Sci. Am.* **254**, 66 (March, 1986).

[14]   For a summary, see P. Langacker, in *Inner Space/Outer Space*, edited by E. W. Kolb *et al.* (University of Chicago Press, Chicago, 1986), p. 3.

[15]   R. H. Dicke, P. J. E. Peebles, P. G. Roll and D. T. Wilkinson, *Ap. J.* **142**, 414 (1965).

[16]   A. F. Hebard, G. S. LaRue, J. D. Phillips and C. R. Fisel, paper V.3 in this volume.

[17]   W. M. Fairbank, Jr., paper V.4 in this volume, and references therein.

[18]   B. Cabrera, M. Taber and S. B. Felch, paper V.6 in this volume.

[19]   P. F. Michelson, M. Bassan, S. Boughn, R. P. Giffard, J. N. Hollenhorst, E. R. Mapoles, M. S. McAshan, B. E. Moskowitz, H. J. Paik and R. C. Taber, paper VI.5 in this volume; W. O. Hamilton, paper VI.6 in this volume; E. Amaldi, paper VI.7 in this volume.

[20]   For a review, see M. Bander, *Phys. Rep.* **75**, 205 (1981).

[21]   R. V. Wagoner, paper V.5 in this volume.

[22]   J. Preskill, in *Magnetic Monopoles*, edited by R. A. Carrington, Jr., and W. P. Trower (Plenum Press, New York, 1982), p. 111; J. Pantaleone, Cornell University preprint (1982).

[23]   A. E. Strominger, Institute for Advanced Study preprint (1982); S. M. Barr, D. B. Reiss and A. Zee, University of Washington preprint 40048-31-P2 (1982); S. Aoyama, Y. Fujimoto and Zhao Zhiyong, International Center for Theoretical Physics, Trieste, preprint (1982).

[24] X. G. Wen and E. Witten, Princeton preprint (1985), submitted to *Nucl. Phys. B.*; M. B. Green, J. H. Schwarz and E. Witten, *Superstring Theory* (Cambridge University Press, Cambridge, 1987), Vol. 2, p. 506.

[25] J. D. Phillips, Ph.D. Thesis, Stanford University, 1983, unpublished.

[26] W. M. Fairbank and J. D. Phillips, in *Inner Space/Outer Space*, edited by E. W. Kolb *et al.* (University of Chicago Press, Chicago, 1986), p. 563.

[27] M. Marinelli and G. Morpurgo, *Phys. Rep.* **85**, 163 (1982); D. Liebowitz, M. Binder and K. O. H. Ziock, *Phys. Rev. Lett.* **50**, 1640 (1984); P. F. Smith, G. J. Homer, J. D. Lewin, H. E. Walford and W. G. Jones, *Phys. Lett.* **153B**, 188 (1985); C. L. Hodges *et al.*, *Phys. Rev. Lett.* **51**, 731 (1983); M. A. Lindgren *et al.*, *Phys. Rev. Lett.* **51**, 1621 (1983).

[28] For a review of quark searches, see L. W. Jones, *Rev. Mod. Phys.* **49**, 717 (1977), and references in S. J. Freedman and J. Napolitano, paper V.2 in this volume. Also, see paper V.2 for limits on accelerator and cosmic ray searches.

[29] R. V. Wagoner and G. Steigman, *Phys. Rev. D* **20**, 825 (1979).

[30] J. D. Bjorken, in *Inner Space/Outer Space*, edited by E. W. Kolb *et al.* (University of Chicago Press, Chicago, 1986), p. 621.

[31] P. A. M. Dirac, *Proc. Roy. Soc. (London) A* **133**, 60 (1931); *Phys. Rev.* **74**, 817 (1948).

[32] L. J. Schaad, B. A. Hess, Jr., J. P. Wikswo, Jr., and W. M. Fairbank, *Phys. Rev. A* **23**, 1600 (1981).

[33] P. W. Worden, Jr., paper VI.9 in this volume.

[34] G. Zweig, *Science* **201**, 973 (1978).

[35] A. A. Starobinsky, *JETP Lett.* **30**, 682 (1979).

[36] E. Witten, *Phys. Rev. D* **30**, 272 (1984); C. J. Hogan, *Mon. Not. Roy. Astron. Soc.* **218**, 629 (1986).

[37] See, *e.g.*, A. Vilenkin, *Phys. Rep.* **121**, 265 (1985); C. J. Hogan and M. J. Rees, *Nature* **311**, 109 (1984).

[38] A. Salam, in *Proceedings of the IV Marcel Grossmann Meeting*, edited by R. Ruffini (North-Holland, Amsterdam, 1986), p 5.

[39] P. F. Michelson, submitted to *Mon. Not. Roy. Astron. Soc.*, 1987.

[40] P. F. Michelson, *Phys. Rev. D* **34**, 2966 (1986).

[41] J. Turneaure and S. R. Stein, paper IV.5 in this volume.

[42]   See National Research Council Physics Survey Committee report, *Gravitation, Cosmology and Cosmic Ray Physics*, of the series *Physics Through the 1990's* (National Academy Press, Washington, D.C., 1986), p. 49.

[43]   *Ibid.*, p. 56.

[44]   J. R. Bond and B. J. Carr, *Mon. Not. Roy. Astron. Soc.* **207**, 585 (1984); B. J. Carr, *Astron. Astrophys.* **89**, 6 (1980).

[45]   P. F. Smith, J. R. J. Bennett, G. J. Homer, J. D. Lewis, H. E. Walford and W. A. Smith, *Nucl. Phys.* **B206**, 333 (1982).

[46]   C. W. F. Everitt, paper VI.3(A); J. A. Lipa and G. M. Keiser, paper VI.3(B); J. T. Anderson, paper VI.3(C); J. P. Turneaure, E. A. Cornell, P. D. Levine and J. A. Lipa, paper VI.3(D); R. A. Van Patten, paper VI.3(E); J. V. Breakwell, paper VI.3(F); and D. B. DeBra, paper VI.3(G), in this volume.

[47]   See reference to statement by C. N. Yang in section 2.3 of paper VI.3(A) in this volume.

[48]   L. Halpern, *Intl. J. Theor. Phys.* **23**, 843 (1984).

[49]   R. V. Wagoner, *Ap. J.* **278**, 345 (1984).

[50]   See preliminary report of the Space Science Board Task Group on Fundamental Physics and Chemistry of the National Academy of Sciences, p. 47. That report is being prepared for *Major Directions for Space Sciences 1995-2015*.

[51]   D. R. Long, *Phys. Rev. D* **9**, 850 (1974).

[52]   D. R. Long, *Nature* **260**, 417 (1976).

[53]   J. K. Hoskins, R. D. Newman, R. Spero and J. Schultz, *Phys. Rev. D* **32**, 3084 (1985); Y. T. Chen, A. H. Cook and A. J. F. Metherell, *Proc. Roy. Soc. (London) A* **394**, 47 (1984).

[54]   F. D. Stacey and G. J. Tuck, *Nature* **292**, 230 (1981); S. C. Holding and G. J. Tuck, *Nature* **307**, 714 (1984); S. C. Holding, F. D. Stacey and G. J. Tuck, *Phys. Rev. D* **33**, 3487 (1986).

[55]   H. J. Paik, paper VI.8 in this volume.

[56]   E. Fischbach, D. Sudarsky, A. Szafer, C. Talmadge and S. H. Aronson, *Phys. Rev. Lett.* **56**, 3 (1986).

[57]   E. Fischbach, D. Sudarsky, A. Szafer, C. Talmadge and S. H. Aronson, in *Proceedings, 2nd Conference on Interactions between Particle and Nuclear Physics*, Lake Louise, Canada, May, 1986.

[58]   R. V. Eötvös, D. Pekar and E. Fekete, *Ann. Physik* **678**, 11 (1922).

[59] P. G. Roll, R. Krotkov and R. H. Dicke, *Ann. Phys.* **26**, 442 (1967).

[60] V. B. Braginsky and V. I. Panov, *Sov. Phys.–JETP* **34**, 463 (1972).

[61] P. W. Worden, Jr., presentation to PACE Committee of NASA, 1986.

[62] J. M. Lockhart and F. C. Witteborn, paper VII.2 in this volume.

[63] P. Morrison and T. Gold, in *Essays on Gravity* (Gravity Research Foundation, New Boston, N.H., 1957), p. 45; P. Morrison, *Am. J. Phys.* **26**, 358 (1958).

[64] T. Goldman, R. J. Hughes and M. M. Nieto, *Phys. Lett.* B **171**, 217 (1986).

[65] J. H. Billen, K. R. Crandall and T. P. Wangler, in *Physics with Antiprotons at LEAR in the ACOL Eri*, edited by U. Gastaldi, R. Klepisch, J. M. Richard and J. Tran Thank Van (Editions Frontiers, Gif sur Yvette, France, 1985), p. 107; T. Goldman and M. M. Nieto, *ibid.*, p. 639; M. V. Hynes, *ibid.*, p. 657.

[66] J. R. Henderson, Ph.D. Thesis, Stanford University, 1986, unpublished; J. R. Henderson and W. M. Fairbank, *Bull. Am. Phys. Soc.* **31**, 372 (1986).

[67] A. P. Mills, Jr., *Science* **218**, 335 (1982), and references therein; B. L. Brown, W. S. Crane and A. P. Mills, Jr., AT&T Bell Labs reprint (1985); R. H. Howell, R. A. Alvarez and M. Stanek, *Appl. Phys. Lett.* **40**, 751 (1982).

[68] J. Bardeen, paper VII.4 in this volume.

[69] R. S. Hanni and J. M. J. Madey, *Phys. Rev.* B **17**, 1976 (1978).

[70] C. A. Waters, H. A. Fairbank, J. M. Lockhart and K. W. Rigby, paper VII.3 in this volume.

[71] K. W. Rigby, Ph.D. Thesis, Stanford University, 1986, unpublished; K. W. Rigby and W. M. Fairbank, *Bull. Am. Phys. Soc.* **31**, 342 (1986).

[72] M. S. Rzchowski and W. M. Fairbank, *ibid.*, 371.

[73] See, *e.g.*, S. M. Faber and J. S. Gallagher, *Ann. Rev. of Astron. and Astrophys.* **17**, 135 (1979); D. Burstein and V. C. Rubin, *Ap. J.* **297**, 423 (1985).

[74] R. Ruffini, paper VI.4 in this volume.

[75] K. A. Olive and D. J. Hegyi, in *Inner Space/Outer Space*, edited by E. W. Kolb *et al.* (University of Chicago Press, Chicago, 1986), p. 112.

[76] See, *e.g.*, G. Steigman, in *ibid.*, p. 25, and references therein.

[77] R. D. Peccei and H. R. Quinn, *Phys. Rev. Lett.* **38**, 1440 (1977); *Phys. Rev.* D **16**, 1791 (1977); S. Weinberg, *Phys. Rev. Lett.* **40**, 223 (1978); F. Wilczek, *ibid.*, p. 279.

[78]   J. Polchinski, in *Inner Space/Outer Space*, edited by E. W. Kolb *et al.* (University of Chicago Press, Chicago, 1986), p. 447, and references therein.

[79]   B. W. Lee and S. Weinberg, *Phys. Rev. Lett.* **39**, 165 (1977); M. W. Goodman and E. Witten, *Phys. Rev. D* **31**, 3059 (1985).

[80]   R. L. Gilliland, J. Faulkner, W. H. Press and D. N. Spergel, *Ap. J.* **306**, 703 (1986).

[81]   See review by F. Sciulli in *Inner Space/Outer Space*, edited by E. W. Kolb *et al.* (University of Chicago Press, Chicago, 1986), p. 495.

[82]   V. A. Lubimou *et al.*, *Phys. Lett.* **94B**, 266 (1980); S. Boris *et al.*, *Proceedings of HEP '83*, edited by J. Guy and C. Costain, Rutherford Appleton Lab.

[83]   B. Cabrera, L. M. Krauss and F. Wilczek, *Phys. Rev. Lett.* **55**, 25 (1985); B. Cabrera, C. J. Martoff and B. J. Neuhauser, submitted to *Nucl. Inst. and Methods*.

[84]   A. Drukier and L. Stodolsky, *Phys. Rev. D* **30**, 2295 (1984).

[85]   R. Davis, in *Proceedings of the Telemark Mass Miniconference*, edited by V. Barger and D. Cline (University of Wisconsin Press, Madison, Wisconsin, 1980).

[86]   S. Dimopoulos, G. D. Starkman and B. W. Lynn, *Phys. Lett.* **168B**, 145 (1986).

[87]   I. Wasserman, Cornell University Center for Radiophysics and Space Research reprint CPSR 834 (1985).

[88]   P. Sikivie, *Phys. Rev. Lett.* **51**, 1415 (1983).

[89]   L. M. Krauss, J. Moody, F. Wilczek and D. Morris, *Phys. Rev. Lett.* **55**, 1797 (1985).

[90]   A. Melissinos, Brookhaven National Laboratory, private communication.

[91]   M. Taber and B. Cabrera, paper V.7 in this volume.

[92]   N. F. Ramsey, *Phys. Rep.* **43**, 409 (1978); I. S. Altarev *et al.*, *Phys. Lett.* **102B**, 13 (1981); J. M. Pendlebury *et al.*, *Phys. Lett.* **136B**, 327 (1984).

[93]   B. Cabrera, paper III.5 in this volume.

[94]   Y. Avishiai and M. Fabre de la Rippela, *Phys. Rev. Lett.* **56**, 2121 (1986).

[95]   M. Kabayashi and T. Maskawa, *Prog. Theor. Phys.* **49**, 652 (1973).

[96]   T. E. Chupp, Harvard University, private communication.

[97] G. Binning and H. Rohrer, *Sci. Am.* **253**, 50 (August, 1985).

[98] J. A. Lipa, paper II.9 in this volume.

[99] M. J. Buckingham, paper II.8 in this volume.

[100] See section on "Science in Microgravity Environment" in the preliminary report of the Space Science Board Task Group on Fundamental Physics and Chemistry of the National Academy of Sciences. That report is being prepared for *Major Directions for Space Sciences 1995–2015.*

[101] S. Chu, L. Hollberg, J. E. Bjorkholm, A. Cable and A. Ashkin, *Phys. Rev. Lett.* **55**, 48 (1985); S. Chu, J. E. Bjorkholm, A. Ashkin, L. Hollberg and A. Cable, in *Methods of Laser Spectroscopy*, to be published by Plenum Press.

[102] H. A. Schwettman, paper IV.2 in this volume; T. I. Smith, C. M. Lyneis, M. S. McAshan, R. E. Rand, H. A. Schwettman and J. P. Turneaure, paper IV.3 in this volume.

[103] J. M. J. Madey, paper IV.6 in this volume.

[104] L. P. Elias and G. J. Ramian, *Intl. Soc. for Optical Engr.* **453** (1984); see also "Search and Discovery" section of *Physics Today* **37**, 21 (Nov. 1984).

[105] Conceptual Design Report on the Continuous Electron Beam Accelerator Facility, Project 87-R-203, Newport News, Virginia, February, 1986.

[106] P. Fessenden, M. A. Bagshaw, D. P. Boyd, D. A. Pistenmaa, H. A. Schwettman and C. F. von Essen, paper IV.4 in this volume.

[107] J. P. Wikswo, Jr., and M. C. Leifer, paper IV.8 in this volume.

[108] See chapter 12 of *Biomagnetism: An Interdisciplinary Approach*, edited by S. J. Williamson, G.-L. Romani, L. Kaufman and I. Modena (Plenum Press, New York, 1983).

[109] E. P. Day and J. S. Philo, paper IV.7 in this volume.

[110] G. K. Walters, paper II.2 in this volume.

[111] See, *e.g.*, P. Mansfield and P. G. Morris, *NMR Imaging in Biomedicine* (Academic Press, New York, 1982).

[112] T. I. Smith, M. S. McAshan and W. M. Fairbank, U.S. Patent #4,595,899, issued June 17, 1986.

[113] B. S. Deaver, Jr., papers III.1 and III.4 in this volume.

[114] W. A. Little, paper III.9 in this volume.

[115] F. London, *Superfluids* (Dover Publications, Inc., New York, 1961), p. 8.

PHOTO GALLERY

Left: William Fairbank, quarter miler, 1937; *right*: Jane Davenport and William M. Fairbank, 1937.

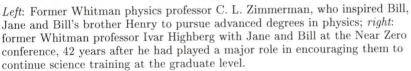

Left: Former Whitman physics professor C. L. Zimmerman, who inspired Bill, Jane and Bill's brother Henry to pursue advanced degrees in physics; *right*: former Whitman professor Ivar Highberg with Jane and Bill at the Near Zero conference, 42 years after he had played a major role in encouraging them to continue science training at the graduate level.

Top left: Mr. and Mrs. William M. Fairbank, August 16, 1941, Seattle, Washington; *top right*: Amherst days: Bill, Jane and Bill, Jr., 1947; *right and bottom*: the Fairbank family in full force at the Near Zero conference, even third generation William Henry with his father; in the photo below, *left to right*: Bob, Karen, Bill, Jr., Donna, Bill, Jane, Rich and Chris.

ASSOCIATES AND MENTORS
AT THE RADIATION LABORATORY, YALE AND DUKE

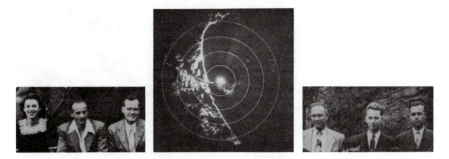

Left: J. D. Fairbank, R. G. Herb and R. E. Meagher, who were in the Radiation Laboratory Ground and Ship division headed by Herb, E. C. Pollard and J. C. Street; *middle*: 1944 radar view of Massachusetts coastline near eastern entrance to Cape Cod Canal, with breakwater; *right*: N. E. Edlefsen, J. McGregor and W. M. Fairbank, who worked in ship applications groups led by R. M. Emberson, Meagher, J. S. Hall and E. L. Hudspeth.

Recalling Yale days: Professor C. T. Lane (*third from left*) and, *left to right*, three of his former students: Joseph M. Reynolds, Ted G. Berlincourt and William M. Fairbank, 1973.

Left: Yale theoretical chemist Lars Onsager; *right*: Duke theoretical physicist Fritz London.

SCENES AT THE STANFORD PHYSICS LABORATORIES

Stanford's first low temperature gravitational wave detector, 1975. In front of the apparatus, *left to right*: B. O. Reese, M. S. McAshan, S. P. Boughn, R. Sears, R. C. Taber, H. D. Butler, R. P. Giffard and W. M. Fairbank; perched atop the equipment, *left*, H. J. Paik and *right*, J. N. Hollenhorst.

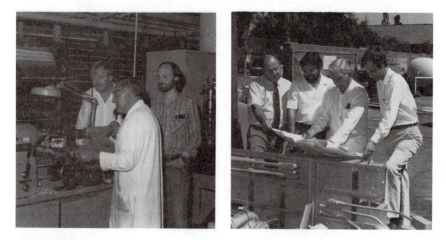

Left: M. Taber, at right, consulting with W. Jung and J. Ascorra in the physics machine shop; *right*: M. D. O'Neill, D. Bravo, S. Bravo and P. F. Michelson reviewing design plans.

Right: William Little modeling, with computer graphics, the molecular interactions of a high temperature superconductor, 1972; *below*: Paul W. Worden, Jr., working on the cryogenic equivalence principle experiment.

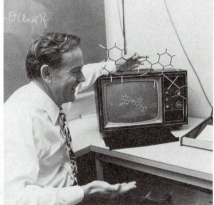

William Fairbank and George LaRue adjusting the quark apparatus, 1977.

Top left: John Turneaure and G. M. Keiser with new gimbaled gyro test facility; *above right*: Richard Potter and Brad Parkinson examining a quartz housing to be used in the Stanford relativity gyroscope experiment.

Jack Gilderoy, John Lipa, John Anderson and Blas Cabrera standing in front of the low field gyro test facility.

Top left: Luis W. Alvarez discussing the Stanford monopole experiment with Blas Cabrera and Susan B. Felch; *top right*: Luis Elias and John Madey at the Near Zero conference; *above*: Kip Thorne presenting a paper at a conference session chaired by John Goodkind.

Top: W. O. Hamilton presenting a paper at the Near Zero conference; *middle*: Mark Leifer, William Fairbank and John Wikswo at the Near Zero conference; *bottom*: C. N. Yang, Felix Bloch and Edoardo Amaldi discussing a point at the Near Zero conference.

Top: The Near Zero conference banquet: Francis Everitt reading a message from John Wheeler. Also shown are R. Ruffini (*left*), and R. Decher, N. Hofstadter, R. Hofstadter and A. T. Schawlow at the speakers' table; *middle*: Arthur Schawlow, Arthur Hebard and William Fairbank at the Near Zero banquet; *bottom*: Henry Fairbank, William Fairbank, Robert Hofstadter and William Fairbank, Jr., at the Near Zero conference.

Top: A run for the future: William Fairbank and his track associates competing in the 50 meter dash for men 65 years and older at the Cow Palace in San Francisco in 1983; *bottom*: local organizing committee for Near Zero conference, *left to right*: B. Cabrera, J. D. Fairbank, B. S. Deaver, Jr., D. Polyhronakis, C. W. F. Everitt, L. Henneberg, P. F. Michelson, P. Crandall, A. L. Schawlow, J. C. Tice, M. B. Willers, J. M. Lockhart and J. M. J. Madey; shown separately: N. Shick, *upper photo*, and F. Alkemade.

NAME INDEX

SUBJECT INDEX